MICHELIN

road atlas

USA | CANADA | MEXICO

ROAD MAPS are organized geographically. *United States, Canada, and Mexico road maps are organized in a grid layout, starting in the northwest of each country. To find your way, use either the **Key to Map Pages** inside the front cover, the **Listing of State and City Maps** on page 3, or the **index** in the back of the atlas.*

COUNTRY COLORS
Colors represent countries throughout the atlas.
Red → Canada
Green → Mexico
Blue → United States
Purple → United States (Northeast Corridor)

MAP SCALES
Scale bars are shown at a constant length throughout the atlas for quick and easy scale comparison between regions.

DRIVING DISTANCES
Use this chart to check driving distances between major cities within each map. Refer to distance and driving time information at the back of the atlas for travel over greater distances.

LOCATOR MAPS
A quick glance at this miniature map lets you check which states and/or provinces are shown on each page.

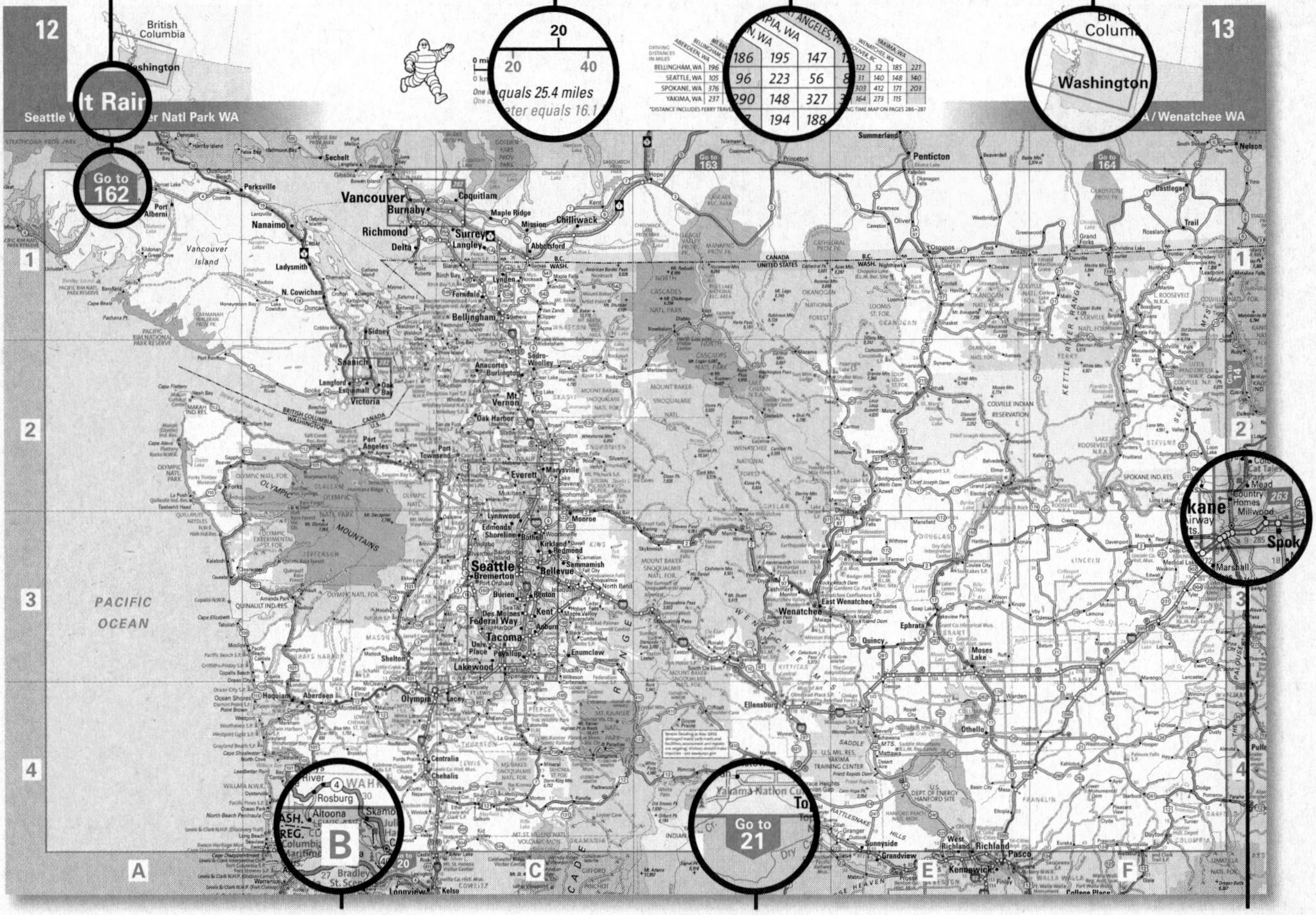

GRID REFERENCES
Use grid references to locate places listed in the index. For instance, Rosburg WA is listed in the index with "12" and "B4", indicating that the town may be found on page 12 in grid square B4.

"GO TO" POINTERS
Handy page tabs point the way to the next map, making navigation a breeze.

INSET MAP BOXES
These color-coded boxes outline areas that are featured in greater detail in the index section. The tab with "263" (above) indicates that a detailed map of Spokane may be found on page 263 (below).

HOW THE INDEX WORKS

Cities and towns are listed alphabetically, with separate indexes for the United States, Canada, and Mexico. Figures after entries indicate population, page number, and grid reference. Entries in bold color indicate cities with detailed inset maps. The U.S. index also includes counties and parishes, which are shown in bold black type.

INSET MAP INDEXES
Many inset maps have their own indexes. Metro area inset map indexes list cities and towns; downtown inset map indexes list points of interest.

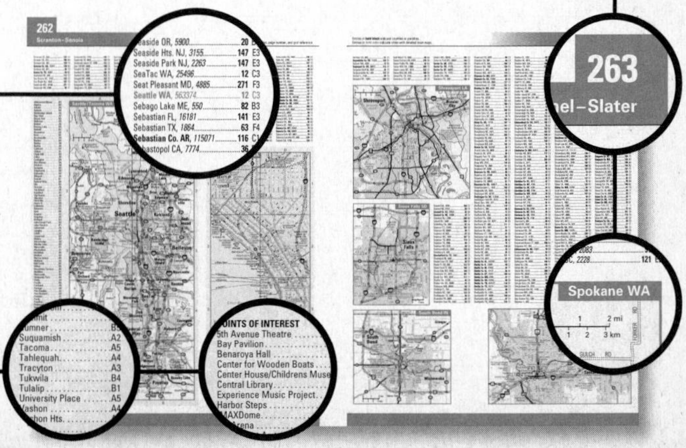

United States Interstate Map

0 mi 125 250 375
0 km 125 250 375 500

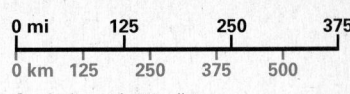

One inch equals 217 miles
One centimeter equals 138 kilometers

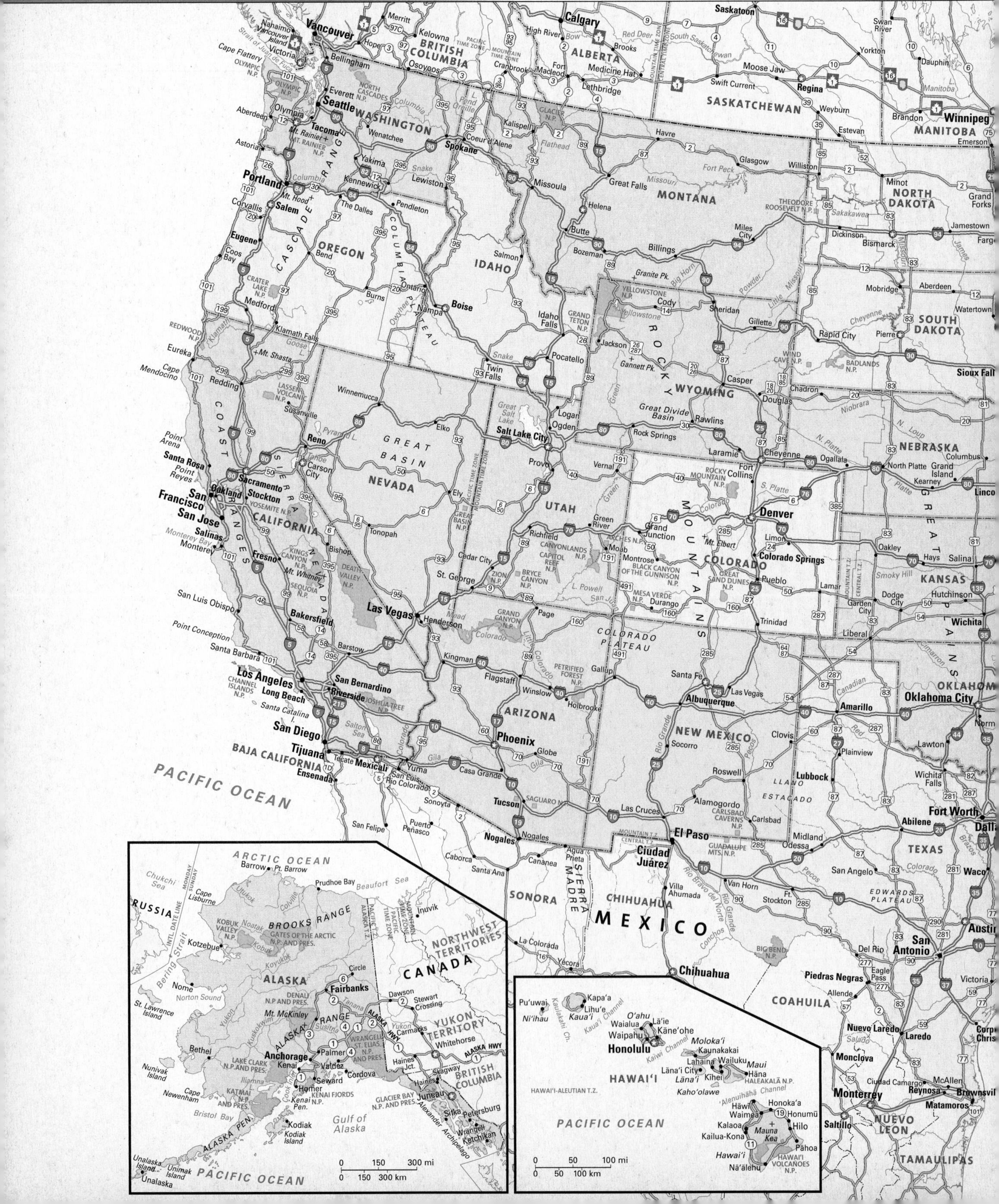

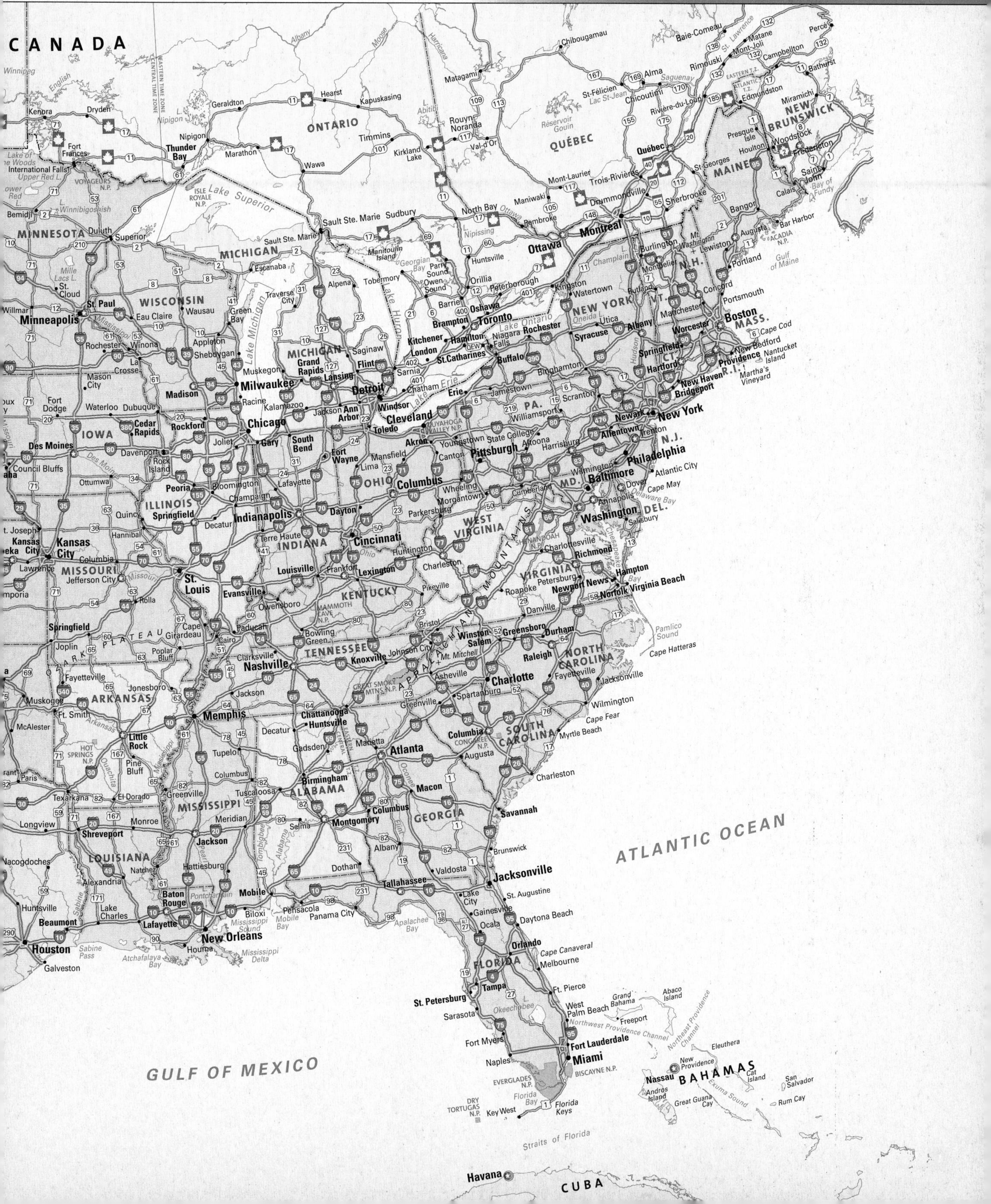
CANADA
ONTARIO
QUÉBEC
NEW BRUNSWICK
MAINE
N.H.
VT.
MASS.
R.I.
CT.
NEW YORK
PA.
N.J.
DEL.
MD.
WEST VIRGINIA
VIRGINIA
NORTH CAROLINA
SOUTH CAROLINA
GEORGIA
FLORIDA
ALABAMA
MISSISSIPPI
LOUISIANA
ARKANSAS
TENNESSEE
KENTUCKY
OHIO
INDIANA
ILLINOIS
MICHIGAN
WISCONSIN
MINNESOTA
IOWA
MISSOURI
BAHAMAS
CUBA
ATLANTIC OCEAN
GULF OF MEXICO
Lake Superior
Lake Michigan
Lake Huron
Lake Erie
Lake Ontario
APPALACHIAN MOUNTAINS
Straits of Florida
Winnipeg
Kenora
Dryden
Thunder Bay
Geraldton
Hearst
Kapuskasing
Timmins
Wawa
Sault Ste. Marie
Sudbury
North Bay
Ottawa
Montreal
Québec
Chibougamau
Baie-Comeau
Rimouski
Alma
Chicoutimi
Trois-Rivières
Sherbrooke
Fredericton
Saint John
Bangor
Portland
Augusta
Bar Harbor
Toronto
Hamilton
London
Windsor
Detroit
Buffalo
Rochester
Syracuse
Albany
Burlington
Boston
Providence
Hartford
New Haven
Bridgeport
New York
Newark
Philadelphia
Allentown
Scranton
Harrisburg
Pittsburgh
Baltimore
Washington
Richmond
Norfolk
Virginia Beach
Newport News
Raleigh
Durham
Greensboro
Winston-Salem
Charlotte
Columbia
Charleston
Savannah
Atlanta
Macon
Jacksonville
Tallahassee
Orlando
Tampa
St. Petersburg
Fort Myers
Fort Lauderdale
Miami
Key West
Havana
Nassau
Birmingham
Montgomery
Mobile
Huntsville
Chattanooga
Knoxville
Nashville
Memphis
Jackson
New Orleans
Baton Rouge
Shreveport
Little Rock
Houston
Galveston
Beaumont
Louisville
Lexington
Cincinnati
Columbus
Cleveland
Toledo
Akron
Dayton
Indianapolis
Chicago
Milwaukee
Madison
Minneapolis
St. Paul
Duluth
Des Moines
Cedar Rapids
Kansas City
St. Louis
Springfield
Peoria
Evansville
Charleston
Lansing
Grand Rapids
Flint
Saginaw

0 mi 200 400
0 km 200 400 600
One inch equals 250 miles/Un pouce équivaut à 250 milles
One cm equals 159 km/Un cm équivaut à 159 km

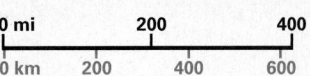

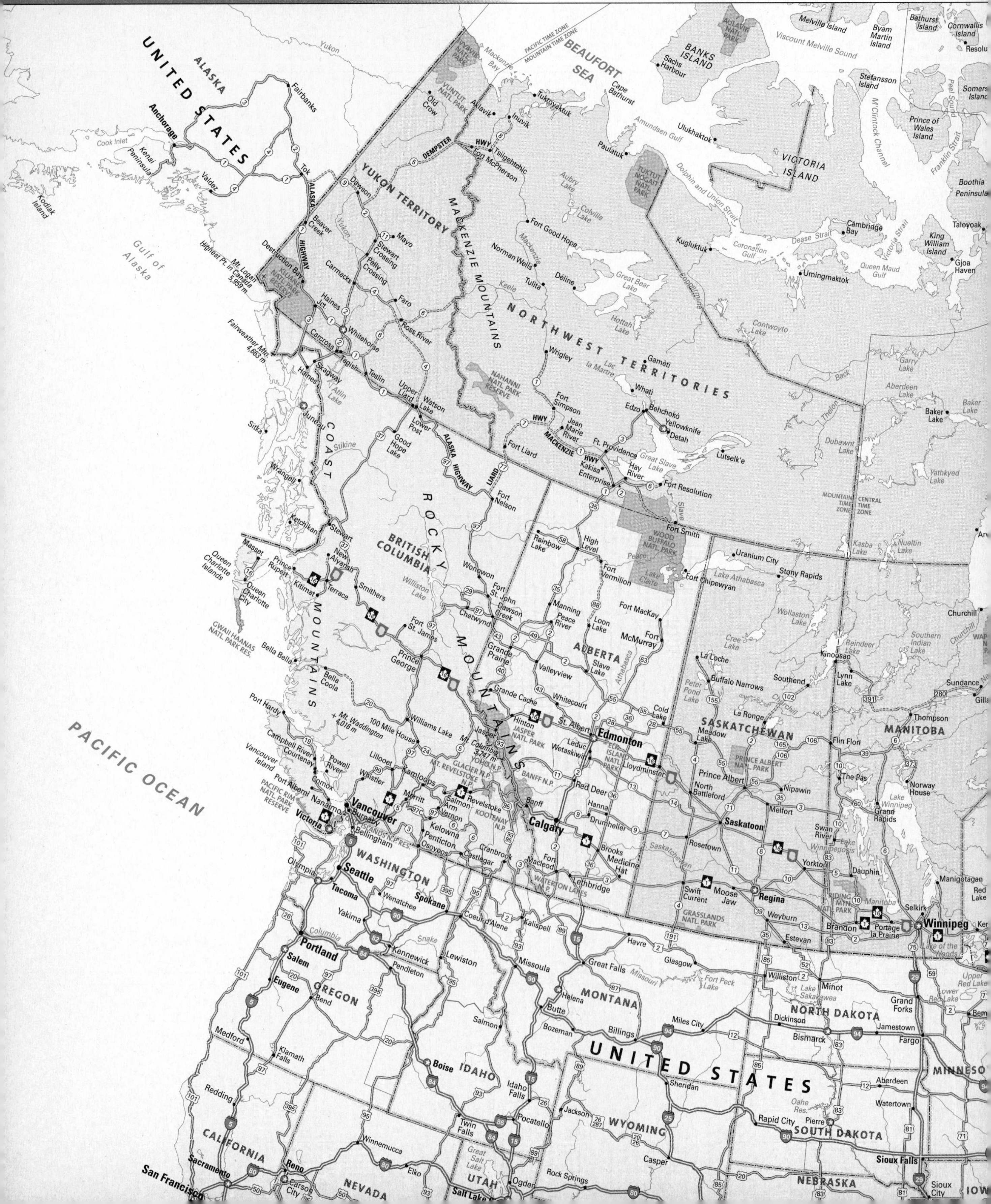

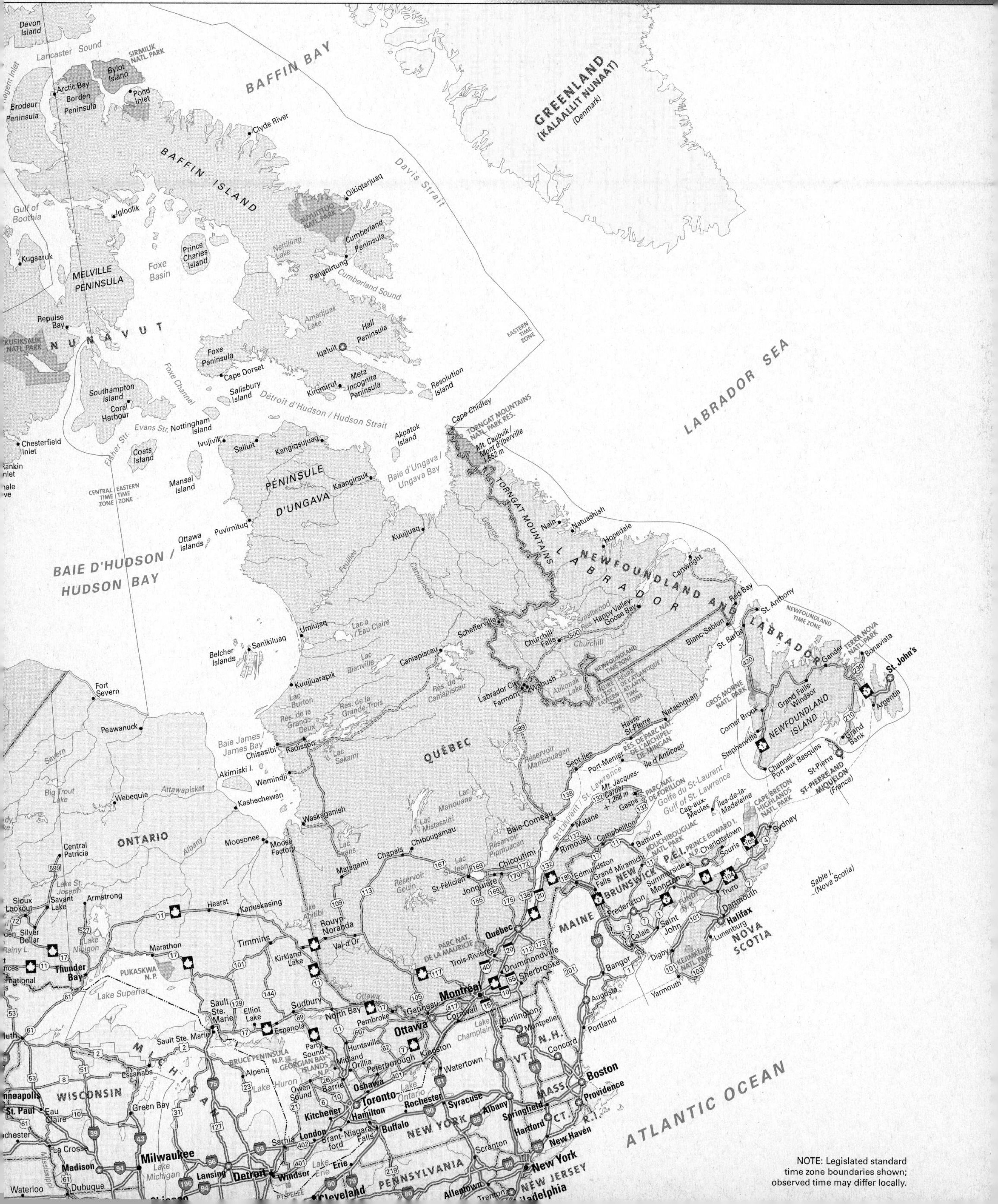

NOTE: Legislated standard time zone boundaries shown; observed time may differ locally.

Experience the thrill of the open roads of North America with these great Scenic Drives from Michelin. The famous star ratings highlight natural and cultural attractions along the way.

★★★ **Highly recommended**
★★ **Recommended**
★ **Interesting**

Michelin Scenic Drives are indicated by a green and yellow dashed line (▬ ▬ ▬) on corresponding atlas maps for easy reference. The following 17 drives are also plotted for your use.

North Coast/Mendocino★★ .... Maps 259, 36
Napa Valley/Sonoma★★ .... Map 36
Willamette Valley★ .... Maps 251, 20, 257, 212
Florida Panhandle★ .... Maps 265, 137, 136, 247
Coastal Carolinas★★ .... Maps 115, 123, 275, 237, 131, 200, 261
San Juan Country★★ .... Map 40
Snake River Valley★ .... Maps 221, 23, 22, 196, 30
Hill Country★★ .... Maps 192, 61, 60, 257
Southern New Mexico★ .... Maps 211, 56, 226, 48, 49, 57
New Jersey Pine Barrens★ .... Maps 191, 147, (105, 104)
Litchfield Hills★★ .... Maps 148, 94
Hudson River Valley★★★ .... Maps 240, 148, 94
Thompson Canyon★ .... Maps 282, 163
Alberta Badlands★★★ .... Maps 277, 164, 165
Saint John River Valley★★ .... Maps 277, 180, 178
Annapolis Valley/Atlantic Shore★★ .... Maps 180, 181, 277
Blue Heron Drive★★ .... Maps 277, 181

A Calendar of Events on page 290 shows popular festivals for several cities in the scenic drives.

ABBREVIATIONS			
N	North	**NHS**	National Historic Site
E	East	**NL**	National Lakeshore
S	South	**NM**	National Memorial/ National Monument
W	West	**NMP**	National Military Park
NE	Northeast	**NP**	National Park
NW	Northwest	**NPR**	National Park Reserve
SE	Southeast	**NRA**	National Recreation Area
SW	Southwest	**NWR**	National Wildlife Refuge
Hwy.	Highway	**PP**	Provincial Park
Pkwy.	Parkway	**SHP**	State Historical Park
Rte.	Route	**SHS**	State Historic Site
Mi	Miles	**SP**	State Park
Km	Kilometers	**SR**	State Reserve
Sq Ft	Square Feet	**VC**	Visitor Center

For detailed coverage of the attractions, and for suggestions of places to dine and stay overnight, see Michelin's North America **Regional Atlas Series**, designed for the way you drive, and Michelin's **Green Guide Collection**, the ultimate guidebooks for the independent traveler.

NORTHWEST

Anchorage/Fairbanks/Denali★★★

892 miles/1,436 kilometers Maps 189, 154, 155

From **Anchorage★**, Alaska's largest city, take Rte. 1 (Glenn Hwy. and Tok Cutoff) N and then E through the broad Matanuska Valley to the small town of **Tok**. The route passes agricultural communities, the **Matanuska Glacier** and the Wrangell Mountains before heading up the Copper River Basin. From Tok, take the Alaska Hwy. (Rte. 2) NW to **Fairbanks★**, a friendly town with a frontier feel. The road passes the **Trans-Alaska Pipeline** and **Big Delta SHP** then parallels the Tanana River. From Fairbanks, opt for Rte. 3 W that crosses the river at Nenana, then veers S to **Denali NP★★★**, home of spruce forests, grassy tundra, grizzlies, moose and North America's highest peak, **Mount McKinley** (20,320ft). Return S to Anchorage via Rtes. 3 and 1.

Badlands NP

Badlands★★

164 miles/264 kilometers Maps 253, 26

From **Rapid City★**, South Dakota, drive SE on Rte. 44 through Farmingdale and Scenic, then east to Interior to enter **Badlands NP★★**. Take Rte. 377 NE 2mi to Cedar Pass and stop at the park's **Ben Reifel VC**. From there, **Cliff Shelf Nature Trail★★** (.5mi) is popular for its shady junper trees and **Castle Trail★★★** (4.5mi) is spectacular in early morning when the moonscape valley and pointed spires get first light. Turn left onto Rte. 240, **Badlands Loop Road★★★**, along the northern rim, where prairie grasslands give way to buttes and hoodoos. **Pinnacles Overlook★★** is a sweeping viewpoint to the south. Drive N to I-90, and cross the Interstate N to Wall. On Main St. visit **Wall Drug★**, a "drug store" with over 20 shops filled with historical photos, 6,000 pairs of cowboy boots, wildlife exhibits and Western art displayed in five dining rooms. In the backyard a roaring, 80ft **Tyrannosaurus** sends toddlers running. Leave Wall on I-90, driving W. Take Exit 66 to Ellsworth Air Force Base, where the **South Dakota Air and Space Museum** displays stealth bombers and other aircraft. Continue W on I-90 back to Rapid City to conclude the tour.

Black Hills★★

244 miles/393 kilometers Maps 253, 26, 25

From **Rapid City★**, drive S on US-16 then US-16A S past Keystone. Take Rte. 244 W to **Mount Rushmore NM★★★**. Continue W on Rte. 244 to the junction of US-16/385. En route S to Custer, **Crazy Horse Memorial★** honors the famous Sioux chief. From Custer, head S on US-385 through Pringle to the junction of Rte. 87. Take Rte. 87 N through **Wind Cave NP★★** and into **Custer SP★★**. Follow **Wildlife Loop Road★★** (access S of Blue Bell, across from Rte. 342 junction) E and N to US-16A. Then travel W to join scenic **Needles Highway★★** (Rte. 87) NW to US-16/385 N. Where US-16 separates, continue N on US-385 to **Deadwood★★**, a former gold camp. Turn left onto US-14A, driving SW through **Lead★**, site of the former **Homestake Gold Mine★★**, to Cheyenne Crossing. Drive N on US-14A to I-90, turning SE back to Rapid City.

Columbia River Gorge★★

83 miles/134 kilometers Maps 251, 20, 21

From **Portland★★**, Oregon's largest city, take I-84 E to Exit 17 in Troutdale. There, head E on the winding **Historic Columbia River Highway★★** (US-30), which skirts the steep cliffs above the river. For great **views★★**, stop at **Vista House at Crown Point**. You'll pass the 620ft **Multnomah Falls★★** and moss-draped **Oneonta Gorge**. At Ainsworth State Park (Exit 35), rejoin I-84 and travel E to Mosier (Exit 69), where US-30, with its hairpin turns, begins again. Continue E on US-30, stopping at **Rowena Crest Viewpoint★★** for grand vistas—and wildflowers. Just past the Western-style town called **The Dalles**, take US-197 N to conclude the tour at **The Dalles Lock and Dam VC★★**.

Grand Tetons/Yellowstone★★★

224 miles/361 kilometers Map 24

Note: parts of this tour are closed in winter.

From **Jackson★★**, drive N on US-26 to Moose. Turn left onto Teton Park Rd. to access **Grand Teton NP★★★** and **Jenny Lake Scenic Drive★★★**. From Teton Park Rd., drive N to the junction of US-89/191/287 (**John D. Rockefeller Jr. Memorial Pkwy.**) and follow the parkway into **Yellowstone NP★★★** to **West Thumb**. Take Grand Loop Rd. W to **Old Faithful★★★**, the world's most famous geyser. Continue N on the Grand Loop Rd., passing **Norris Geyser Basin★★** en route to **Mammoth Hot Springs★★★**. Turn E on Grand Loop Rd. to Tower Junction, then S into **Grand Canyon of the Yellowstone★★★**. Continue S from Canyon Village through **Hayden Valley★★** to Lake Junction. Head SW, back to West Thumb, to conclude to tour.

Pacific Coast/Olympic Peninsula★★★

419 miles/675 kilometers Maps 245, 12

From the state capital of **Olympia**, drive N on US-101 to Discovery Bay. Detour on Rte. 20 NE to **Port Townsend★★**, a well-preserved Victorian seaport. From Discovery Bay, head W on US-101 through **Port Angeles** to the **Heart O' the Hills** park entrance for **Olympic NP★★★** to see **Hurricane Ridge★★★**. Back on US-101, head E then S to the park entrance that leads to **Hoh Rain Forest★★★**. Follow US-101 S, then E after Queets to **Lake Quinault★**, home to bald eagles, trumpeter swans and loons. Continue S on US-101 to Aberdeen, taking Rte. 105 to the coast. At Raymond, return to US-101 heading S to **Long Beach**. Follow Rte. 103 N past the former cannery town of **Oysterville** to **Leadbetter Point★** on Willapa Bay, where oysters are still harvested. Return S to **Ilwaco** and drive E and S on US-101 to Astoria, Oregon, to end the tour.

The Oregon Coast★★

368 miles/592 kilometers Maps 20, 28

Leave **Astoria★**, Oregon's first settlement, via US-101, heading SW. **Fort Clatsop National Memorial★★** recalls Lewis and Clark's historic stay. **Cannon Beach★** boasts a sandy **beach★** and tall coastal rock. At the farming community of **Tillamook★**, go west on 3rd St. to **Cape Meares** to begin **Three Capes Scenic Drive★★**. Continue S, rejoining US-101 just beyond Pacific City. Drive S on US-101 through **Newport★**, then **Yachats★**, which neighbors **Cape Perpetua Scenic Area★★**. From **Florence** to **Coos Bay★** stretches **Oregon Dunes National Recreation Area★★**. At Coos Bay, take Cape Arago Hwy. W to tour the gardens of **Shore Acres State Park★**. Drive S on the highway to rejoin US-101. Pass **Bandon★**, known for its cheese factory, and **Port Orford**, with its fishing fleet. Farther S, **Boardman State Park★** shelters Sitka spruce, Douglas fir and **Natural Bridge Cove**. End the tour at **Brookings**.

Oregon Coast at Bandon

SOUTHWEST

Big Bend Area★★

581 miles/935 kilometers Maps 211, 56, 57, 62, 60

Head S from **El Paso★** via I-10, then E to Kent. Take Rte. 118 S to Alpine, passing **McDonald Observatory★** (telescope tours) and **Fort Davis NHS★★**. Continue S to Study Butte to enter **Big Bend NP★★★**, edged by the Rio Grande River and spanning 1,252sq mi of spectacular canyons, lush bottomlands, sprawling desert and mountain woodlands. The park has more species of migratory and resident birds than any other national park. Travel E to the main VC at **Panther Junction** in the heart of the park (US-385 and Rio Grande Village Dr.). Then take US-385 N to Marathon. Turn E on US-90 to Langtry, site of **Judge Roy Bean VC★**. Continue E to **Seminole Canyon SP★★**, with its 4,000-year-old pictographs. Farther E, **Amistad NRA★** is popular for water sports. Continue on US-90 to conclude the tour in Del Rio.

Canyonlands of Utah★★★

481 miles/774 kilometers Maps 39, 40

From **St. George★**, drive NE on I-15 to Exit 16. Take Rte. 9 E to Springdale, gateway to **Zion NP★★★**, with its sandstone canyon, waterfalls and hanging gardens. Continue E on Rte. 9 to Mt. Carmel Junction, turn left onto US-89 and head N to the junction with Rte. 12. Take Rte. 12 SE to **Bryce Canyon NP★★★**, with its colored rock formations. Continue SE on Rte. 12 to Cannonville, then S to **Kodachrome Basin SP★★**, where sandstone chimneys rise from the desert floor. Return to Cannonville, and drive NE on Rte. 12 through Boulder to Torrey. Take Rte. 24 E through **Capitol Reef NP★★**—with its unpaved driving roads and trails—then N to I-70. Travel E on I-70 to Exit 180, then S on US-191 to Rte. 313 into **Canyonlands NP★★★** to **Grand View Point Overlook**. Return to US-191, turning S to access **Arches NP★★★**—the greatest concentration of natural stone arches in the country. Continue S on US-191 to **Moab★** to end the tour.

Arches NP

Central Coast/Big Sur★★★

118 miles/190 kilometers Maps 236, 44

From **Cannery Row★** in **Monterey★★**, take Prescott Ave. to Rte. 68. Turn right and continue to Pacific Grove Gate (on your left) to begin scenic **17-Mile Drive★★**, a private toll road. Exit at Carmel Gate to reach the upscale artists' colony of **Carmel★★**, site of Carmel **mission★★★**. The town's **Scenic Road** winds S along the beachfront. Leave Carmel by Hwy. 1 S. Short, easy trails at **Point Lobos SR★★** line the shore. Enjoy the wild beauty of the **Big Sur★★★** coastline en route to San Simeon, where **Hearst Castle★★★**, the magnificent estate of a former newspaper magnate, overlooks the Pacific Ocean. Continue S on Hwy. 1 to **Morro Bay**, where the tour ends.

Colorado Rockies★★★

499 miles/803 kilometers Maps 209, 41, 33, 40

Note: Rte. 82 S of Leadville to Aspen is closed mid-Oct to Memorial Day due to snow.

From **Golden★★**, W of **Denver★★★**, drive W on US-6 along Clear Creek to Rte. 119, heading N on the **Peak to Peak Highway★★** to **Nederland★**. Continue N on Rte. 72, then follow Rte. 7 N to the town of **Estes Park★★**. Take US-36 to enter **Rocky Mountain NP★★★**. Drive **Trail Ridge Road★★★** (US-34) S to the town of **Grand Lake★**. Continue S to Granby, turn left on US-40 to I-70 at Empire. Head W on I-70 past **Georgetown★** and through **Eisenhower Tunnel**. You'll pass ski areas **Arapahoe Basin**, **Keystone Resort★** and **Breckenridge★★**. At Exit 195 for **Copper Mountain Resort★**, take Rte. 91 S to **Leadville★★**, Colorado's former silver capital. Then travel S on US-24 to Rte. 82 W over **Independence Pass★★** to **Aspen★★★**. Head NW to I-70, passing **Glenwood Springs★★** with its **Hot Springs Pool★★**. Drive E on I-70 along **Glenwood Canyon★★** and the Colorado River to **Vail★★**. Continue E on I-70 to the old mining town of **Idaho Springs** to return to Golden via Rte. 119.

Lake Tahoe

Lake Tahoe Loop★★

71 miles/114 kilometers Map 37

Begin in **Tahoe City** at the intersection of Rtes. 89 and 28. Drive S on Rte. 89. **Ed Z'berg-Sugar Pine Point State Park★** encompasses a promontory topped by **Ehrman Mansion★** and other historic buildings. Farther S, **Emerald Bay State Park★★** surrounds beautiful **Emerald Bay★★**. At the bay's tip stands **Vikingsholm★★**, a mansion that resembles an ancient Nordic castle. At **Tallac Historic Site★★**, preserved summer estates recall Tahoe's turn-of-the-19C opulence. From Tahoe Valley, take Rte. 50 NE. **South Lake Tahoe**, the lake's largest town, offers lodging, dining and shopping. High-rise hotel-casinos characterize neighboring **Stateline** in Nevada. Continue N to Spooner Junction. Then follow Nevada Rte. 28 N to **Sand Harbor** (7mi), where picnic tables and a sandy beach fringe a sheltered cove. Continue through Kings Beach to end the tour at Tahoe City.

Maui's Hana Highway★★

62 miles/100 kilometers Map 153

Leave **Kahului** on Rte. 36 E toward **Paia**, an old sugar-plantation town. Continue E on Rte. 36, which becomes Rte. 360, the **Hana Highway★★**. The road passes **Ho'okipa Beach Park**, famous for windsurfing, and **Puohokamoa Falls**, a good picnic stop, before arriving in **Hana**, a little village on an attractive bay. If adventurous, continue S on the Pulaui Highway to **Ohe'o Gulch★★** in **Haleakala NP★★★**, where small waterfalls tumble from the SE flank of the dormant volcano Haleakala. Past the gulch the grave of aviator **Charles Lindbergh** can be found in the churchyard at Palapala Hoomau Hawaiian Church. End the tour at Kipahulu.

Redwood Empire★★

182 miles/293 kilometers Maps 36, 28

In **Leggett**, S of the junction of Hwy. 1 and US-101, go N on US-101 to pass through a massive redwood trunk at **Chandelier Drive-Thru Tree Park**. To the N, see breathtaking groves along 31mi **Avenue of the Giants★★★**. **Humboldt Redwoods SP★★** contains **Rockefeller Forest★★**, the world's largest virgin redwood forest. From US-101, detour 4mi to **Ferndale★**, a quaint Victorian village. N. along US-101, **Eureka★** preserves a logging camp cookhouse and other historic sites. The sleepy fishing town of **Trinidad★** is home to a marine research lab. **Patrick's Point SP★★** offers dense forests, agate-strewn beaches and clifftop **views★★**. At **Orick**, enter the **Redwood National and State Parks★★**, which protect a 379ft-high, 750-year-old **tree★**. The tour ends in **Crescent City**.

Santa Fe Area★★★

267 miles/430 kilometers Maps 189, 48, 260, 49

From **Albuquerque★**, drive E on I-40 to Exit 175 and take Rte. 14, the **Turquoise Trail★★**, N to **Santa Fe★★★**. This 52mi back road runs along the scenic Sandia Mountains and passes dry washes, arroyos and a series of revived "ghost towns." Continue N on US-84/285, turning NE onto Rte. 76, the **High Road to Taos★★**. East of Vadito, take Rte. 518 N to Rte. 68 N into the rustic Spanish colonial town of **Taos★★**, a center for the arts. Head N on US-64 to the junction of Rte. 522. Continue W on US-64 for an 18mi round-trip detour to see the 1,200ft-long, three-span **Rio Grande Gorge Bridge** over the river. Return to Rte. 522 and take this route, part of the **Enchanted Circle★★** Scenic Byway, N to **Questa**, starting point for white-water trips on the Rio Grande. Turn onto Rte. 38, heading E to the old mining town of **Eagle Nest**. There, detour 23mi E on US-64 to **Cimarron**, a Wild West haunt. Back at Eagle Nest, travel SW on US-64, detouring on Rte. 434 S to tiny **Angel Fire**. Return to Taos on US-64 W to end the tour.

Sedona/Grand Canyon NP★★★

482 miles/776 kilometers Maps 249, 54, 47, 213

Drive N from **Phoenix★** on I-17 to Exit 298 and take Rte. 179 N toward **Sedona★★** in the heart of **Red Rock Country★★★**. The red-rock formations are best accessed by four-wheel-drive vehicle via 12mi **Schnebly Hill Road★** (off Rte. 179, across Oak Creek bridge from US-89A "Y" junction), which offers splendid **views★★★**. Then head N on Rte. 89A through Sedona to begin 14mi drive of **Oak Creek Canyon★★**. Continue N on Rte. 89A and I-17 to **Flagstaff★**, commercial hub for the region. Take US-180 NW to Rte. 64, which leads N to the **South Rim★★★** of **Grand Canyon NP★★★**. Take the shuttle (or drive, if permitted) along **West Rim Drive★★** to **Hermits Rest★**. Then travel **East Rim Drive★★★** (Rte. 64 E) to **Desert View Watchtower★** for **views★★★** of the canyon. Continue to the junction with US-89 at Cameron. Return S to Flagstaff, then S to Phoenix via I-17.

Grand Canyon NP

NORTHEAST

The Berkshires★★★

57 miles/92 kilometers Map 94

From **Great Barrington**, take US-23 E to Monterey, turning left onto Tyringham Rd., which becomes Monterey Rd., to experience scenic **Tyringham Valley★**. Continue N on Main Rd. to Tyringham Rd., which leads to **Lee**, famous for its marble. Then go NW on US-20 to **Lenox★**, with its inviting inns and restaurants. Detour on Rte. 183 W to **Tanglewood★**, site of a popular summer music festival. Return to Lenox and drive N on US-7 to **Pittsfield**, the commercial capital of the region. Head W on US-20 to enjoy **Hancock Shaker Village★★★**, a museum village that relates the history of a Shaker community established here in 1790. Rte. 41 S passes West Stockbridge, then opt for Rte. 102 SE to **Stockbridge★★** and its picturesque **Main Street★**. Follow US-7 S to the junction with Rte. 23, passing **Monument Mountain★** en route. Return to Great Barrington.

Cape Cod★★★

164 miles/264 kilometers Maps 151, 95

At US-6 and Rte. 3, cross **Cape Cod Canal** via Sagamore Bridge and turn onto Rte. 6A to tour the Cape's **North Shore★★**. Bear right onto Rte. 130 to reach **Sandwich★**, famous for glass manufacture. Continue on Rte. 6A E to **Orleans**. Take US-6 N along **Cape Cod National Seashore★★★**, with its wooded and marshland trails, to reach **Provincetown★★**, a resort town offering **dune tours★★** and summer theater. Return to Orleans and take Rte. 28 S through **Chatham★**, then W to **Hyannis**, where ferries depart for **Nantucket★★★**. Continue to quaint **Falmouth★**. Take Surf Rd., which becomes Oyster Pond Rd. to nearby **Woods Hole**, a world center for marine research and departure point for ferries to **Martha's Vineyard★★**. Take Woods Hole Rd. N to Rte. 28. Cross the canal via Bourne Bridge and head E on US-6 to end the tour at Rte. 3.

Maine Coast★★

238 miles/383 kilometers Maps 82, 251, 83

From **Kittery**, drive N on US-1 to **York★**, then along US-1A to see the 18C buildings of **Colonial York★★**. Continue N on coastal US-1A to **Ogunquit★**. Rejoin US-1 and head N to Rte. 9, turn right, and drive to **Kennebunkport**, with its colorful shops. Take Rte. 9A/35 to **Kennebunk**. Then travel N on US-1 to **Portland★★**, Maine's largest city, where the **Old Port★★** brims with galleries and boutiques. Take US-1 N through the outlet town of **Freeport**, then on to **Brunswick**, home of **Bowdoin College**. Turn NE through **Bath★**, **Wiscasset**, **Rockland**, **Camden★★**, **Searsport** and **Bucksport**. At Ellsworth, take Rt. 3 S to enter **Acadia NP★★★** on **Mount Desert Island★★★**, where **Park Loop Road★★★** *(closed in winter)* parallels open coast. From the top of **Cadillac Mountain★★★**, the **views★★★** are breathtaking. The tour ends at **Bar Harbor★**, a popular resort village.

Bar Harbor from Cadillac Mountain

Mohawk Valley★

114 miles/184 kilometers Maps 188, 94, 80

From the state capital of **Albany★**, take I-90 NW to Exit 25 for I-890 into **Schenectady**, founded by Dutch settlers in 1661. Then follow Rte. 5 W along the Mohawk River. In Fort Hunter, **Schoharie Crossing SHS★** stretches along a canal towpath. Near Little Falls, **Herkimer Home SHS** (Rte. 169 at Thruway Exit 29A) interprets colonial farm life. Rte. 5 continues W along the Erie Canal to **Utica**. From Utica, drive W on Rte. 49 to **Rome**, where the river turns N and peters out. The tour ends in Rome, site of **Fort Stanwix NM★**.

South Shore Lake Superior★

530 miles/853 kilometers Maps 211, 64, 65, 69

From **Duluth★**, drive SE on I-535/US-53 to the junction of Rte. 13 at Parkland. Follow Rte. 13 E to quaint **Bayfield**, gateway to **Apostle Islands NL★★**, accessible by boat. Head S to the junction of US-2, and E through Ashland, Ironwood and Wakefield. There, turn left onto Rte. 28, heading NE to Bergland, and turning left onto Rte. 64. Drive N to Silver City and take Rte. M-107 W into **Porcupine Mountains Wilderness SP★**. Return to Rte. 64 and go E to Ontonagon. Take Rte. 38 SE to Greenland, then follow Rte. 26 NE to Houghton. Cross to Hancock on US-41 and continue NE to Phoenix. Turn left onto Rte. 26 to Eagle River and on to Copper Harbor via **Brockway Mountain Drive★★**. Return S to Houghton via US-41, then travel S and E past Marquette, turning left onto Rte. 28. Head E to Munising, then take County Road H-58 E and N through **Pictured Rocks NL★**. End the tour at Grand Marais.

Villages of Southern Vermont★★

118 miles/190 kilometers Map 81

Head N from the resort town of **Manchester★** by Rte. 7A. At Manchester Center, take Rte. 11 E past **Bromley Mountain**, a popular ski area, to Peru. Turn left on the back road to **Weston★**, a favorite tourist stop along Rte. 100. Continue to **Chester**, turning right onto Rte. 35 S to reach **Grafton★**, with its **Old Tavern**. Farther S, Rte. 30 S from Townshend leads to **Newfane** and its lovely village **green★**. Return to Townshend, then travel W, following Rte. 30 through West Townshend, passing **Stratton Mountain** en route to Manchester. S of Manchester by Rte. 7A, the crest of Mt. Equinox is accessible via **Equinox Skyline Drive** (fee). Then continue S on Rte. 7A to end the tour at **Arlington**, known for its trout fishing.

The White Mountains★★★

127 miles/204 kilometers Map 81

From the all-season resort of **Conway**, drive N on Rte. 16 to **North Conway★**, abundant with tourist facilities. Continue N on US-302/Rte. 16 through **Glen**, passing **Glen Ellis Falls★** and **Pinkham Notch★★** en route to Glen House. There, drive the **Auto Road** to the top of **Mount Washington★★★** (or take guided van tour). Head N on Rte. 16 to Gorham, near the Androscoggin River, then W on US-2 to Jefferson Highlands. Travel SW on Rte. 115 to Carroll, then S on US-3 to Twin Mountain. Go SW on US-3 to join I-93. Head S on I-93/Rte.3, passing scenic **Franconia Notch★★★** and **Profile Lake★★**. Bear E on Rte. 3 where it separates from the interstate to visit **The Flume★★**, a natural gorge 90ft deep. Rejoin I-93S to the intersection with Rte. 112. Head E on Rte. 112 through **Lincoln** on the **Kancamagus Highway★★★** until it joins Rte. 16 back to Conway.

SOUTHEAST

Blue Ridge Parkway★★

574 miles/924 kilometers Maps 102, 112, 111, 190, 121

From **Front Royal**, take US-340 S to begin **Skyline Drive★★**, the best-known feature of **Shenandoah NP★★**. The drive follows former Indian trails along the **Blue Ridge Parkway★★**. **Marys Rock Tunnel to Rockfish Entrance Station★★** passes the oldest rock in the park and **Big Meadows★**. The Drive ends at **Rockfish Gap** at I-64, but continue S on the Parkway. From Terrapin Hill Overlook, detour 16mi W on Rte. 130 to see **Natural Bridge★★**. Enter NC at **Cumberland Knob**, then pass **Blowing Rock★**, **Grandfather Mountain★★** and **Linville Falls★★**. Detour 4.8mi to **Mount Mitchell SP★** to drive to the top of the tallest mountain (6,684ft) E of the Mississippi. At mile 382, the **Folk Art Center** stocks high-quality regional crafts. Popular **Biltmore Estate★★** in **Asheville★** (North Exit of US-25, then 4mi N) includes formal **gardens★★**. The rugged stretch from **French Broad River to Cherokee** courses 17 tunnels within two national forests. **Looking Glass Rock★★** is breathtaking. The Parkway ends at **Cherokee**, gateway to **Great Smoky Mountains NP★★★** and home of Cherokee tribe members.

Central Kentucky★★

379 miles/610 kilometers Maps 230, 100, 214, 227, 110

From **Louisville★★**, home of the **Kentucky Derby★★★**, take I-64 E to **Frankfort**, the state capital. Continue E to **Lexington★★**, heart of **Bluegrass Country★★** with its rolling meadows and white-fenced horse farms. Stop at the **Kentucky Horse Park★★★** (4089 Iron Works Pkwy.) for the daily **Parade of Breeds**. Then head S on I-75 through Richmond to the craft center/college town of **Berea★**. Return to Lexington and follow the Blue Grass Parkway SW to Exit 25. There, US-150 W leads to Bardstown, site of **My Old Kentucky Home SP★**, immortalized by **Stephen Foster** in what is now the state song. Drive S from Bardstown on US-31E past **Abraham Lincoln Birthplace NHS★**. Turn right onto Rte. 70 to Cave City, then take US-31W to Park City, gateway to **Mammoth Cave NP★★★**, which features the world's longest cave system. Return to Louisville via I-65 to end the tour.

Florida's Northeast Coast★★

174 miles/280 kilometers Maps 222, 139, 141, 232

From **Jacksonville★**, drive E on Rte. 10 to **Atlantic Beach**, the most affluent of Jacksonville's beach towns. Head S on Rte. A1A through residential **Neptune Beach**, blue-collar **Jacksonville Beach** and upscale **Ponte Vedra Beach** to reach **St. Augustine★★★**, the oldest city in the US and former capital of Spanish Florida. Farther S, car-racing mecca **Daytona Beach** is known for its **international speedway**. Take US-92 across the Intracoastal Waterway to US-1, heading S to **Titusville**. Take Rte. 402 across the Indian River to **Merritt Island NWR★★** to begin **Black Point Wildlife Drive★**. Return to Titusville and follow Rte. 405 to **Kennedy Space Center★★★**, one of Florida's top attractions, to end the tour.

Florida Keys★★

168 miles/270 kilometers Maps 143, 142

*Note: Green **mile-marker** (MM) posts, sometimes difficult to see, line US-1 (Overseas Hwy.), showing distances from Key West (MM 0). Much of the route is two-lane, and traffic can be heavy in December to April and on weekends. Allow 3hrs for the drive. Crossing 43 bridges and causeways (only one over land), the highway offers fine views of the Atlantic Ocean (E) and Florida Bay (W).*

Drive S from **Miami★★★** on US-1. Near **Key Largo★**, **John Pennekamp Coral Reef SP★★** habors tropical fish, coral and fine snorkeling waters. To the SW, **Islamorada** is known for **charter fishing**. At **Marathon** (MM 50), **Sombrero Beach** is a good swimming spot, but **Bahia Honda SP★★** (MM 36.8) is considered the best **beach★★** in the Keys. Pass **National Key Deer Refuge★** (MM 30.5), haven to the 2ft-tall deer unique to the lower Keys. End at **Key West★★★**, joining others at **Mallory Square Dock** to view the **sunset★★**.

Florida Keys

The Ozarks★

343 miles/552 kilometers Maps 227, 117, 219, 107, 106

From the state capital of **Little Rock**, take I-30 SW to Exit 111, then US-70 W to Hot Springs. Drive N on Rte. 7/Central Ave. to **Hot Springs NP★★** to enjoy the therapeutic waters. Travel N on Rte. 7 across the Arkansas River to Russellville. Continue on **Scenic Highway 7★** N through **Ozark National Forest** and across the **Buffalo National River** to Harrison. Take US-62/65 NW to Bear Creek Springs, continuing W on US-62 through **Eureka Springs★**, with its **historic district**, to **Pea Ridge NMP★**, a Civil War site. Return E on US-62 to the junction of Rte. 21 at Berryville. Travel N on Rte. 21 to Blue Eye, taking Rte. 86 E to US-65, which leads N to the entertainment hub of **Branson**, Missouri, to end the tour.

River Road Plantations★★

200 miles/323 kilometers Maps 239, 134, 194

From **New Orleans★★★**, take US-90 W to Rte. 48 along the Mississippi River to Destrehan. At no. 13034, **Destrehan★★** is considered the oldest plantation house in the Mississippi Valley. Continue NW on Rte. 48 to US-61 to Laplace to connect to Rte. 44. Head N past **San Francisco Plantation★**, built in 1856. At Burnside, take Rte. 75 N to St. Gabriel. En route, watch for **Houmas House★** (40136 Hwy. 942). Take Rte. 30 to **Baton Rouge★**, the state capital. Then drive S along the **West Bank★★** on Rte. 1 to White Castle, site of **Nottoway★**, the largest plantation home in the South. Continue to Donaldsonville, then turn onto Rte. 18. Travel E to Gretna, passing **Oak Alley★** (no. 3645) and **Laura Plantation★★** (no. 2247) along the way. From Gretna, take US-90 to New Orleans, where the tour ends.

CANADA

Gaspésie, Québec★★★

933 kilometers/578 miles Maps 178, 179

Leave **Sainte-Flavie** via Rte. 132 NE, stopping to visit **Reford Gardens★★** en route to **Matane**. After **Cap-Chat**, take Rte. 299 S to **Gaspésie Park★** for expansive **views★★**. Back on Rte. 132, follow the **Scenic Route from La Martre to Rivière-au-Renard★★**. Continue to **Cap-des-Rosiers**, entrance to majestic **Forillon NP★★**. Follow Rte. 132 along the coast through **Gaspé★**, the administrative center of the peninsula, to **Percé★★★**, a coastal village known for **Percé Rock★★**, a mammoth offshore rock wall. Drive SW on Rt. 132 through **Paspébiac** to **Carleton**, which offers a **panorama★★** from the summit of **Mont Saint-Joseph**. Farther SW, detour 6km/4mi S to see an array of fossils at **Parc de Miguasha★**. Back on Rte. 132, travel W to Matapédia, then follow Rte. 132 N, passing **Causapscal**—a departure point for salmon fishing expeditions—to end the tour at Sainte-Flavie.

North Shore Lake Superior★★

275 kilometers/171 miles Map 169

From the port city of **Thunder Bay★★**—and nearby **Old Fort William★★**—drive the Trans-Canada Hwy. (Rte. 11/17) E to Rte. 587. Detour to **Sleeping Giant PP★**, which offers fine **views★** of the lake. Back along the Trans-Canada Hwy., **Amethyst Mine** (take E. Loon Rd.) is a rock hound's delight (fee). Farther NE, located 12km/8mi off the highway, **Ouimet Canyon★★** is a startling environment for the area. Just after the highway's Red Rock turnoff, watch for **Red Rock Cuesta**, a natural formation 210m/690ft high. Cross the Nipigon River and continue along **Nipigon Bay★★**, enjoying **views★★** of the rocky, conifer-covered islands. The **view★★** of Kama Bay through Kama Rock Cut is striking. Continue to Schreiber to end the tour.

Nova Scotia's Cabot Trail★★

338 kilometers/210 miles Map 181

From **Baddeck★**, follow Hwy. 105 S to the junction with the road to **North East Margaree** in salmon-fishing country. Take this road NW to Margaree Harbour, then N to **Chéticamp**, an enclave of Acadian culture. Heading inland, the route enters **Cape Breton Highlands NP★★**, combining seashore and mountains. At Cape North, detour N around Aspy Bay to **Bay St. Lawrence★★**. Then head W to tiny **Capstick** for shoreline **views★**. Return S to Cape North, then drive E to South Harbour. Take the coast road, traveling S through the fishing villages of **New Haven** and **Neils Harbour★**. Rejoin Cabot Trail S, passing the resort area of the **Ingonishs**. Take the right fork after Indian Brook to reach St. Ann's, home of **Gaelic College★**, specializing in bagpipe and Highland dance classes. Rejoin Hwy. 105 to return to Baddeck.

Capstick

Lake Louise

Canadian Rockies★★★

467 kilometers/290 miles Map 164

Leave **Banff★★** by Hwy. 1, traveling W. After 5.5km/3.5mi, take **Bow Valley Parkway★** (Hwy. 1A) NW within **Banff NP★★★**. At **Lake Louise Village**, detour W to find **Lake Louise★★★**. Back on Hwy. 1, head N to the junction of Hwy. 93, turn W and follow Hwy. 1 past **Kicking Horse Pass** into **Yoho NP★★**. Continue through **Field**, and turn right onto the road N to **Emerald Lake★★★**. Return to the junction of Rte. 93 and Hwy. 1, heading N on Rte. 93 along the **Icefields Parkway★★★**. Pass **Crowfoot Glacier★★** and **Bow Lake★★** on the left. **Peyto Lake★★★** is reached by spur road. After **Parker Ridge★★**, massive **Athabasca Glacier★★★** looms on the left. Continue to **Jasper★** and **Jasper NP★★★**. From Jasper, turn left onto Hwy. 16 and head into **Mount Robson PP★★**, home to **Mount Robson★★★** (3,954m/12,972ft.). End the tour at Tête Jaune Cache.

Vancouver Island★★★

337 kilometers/209 miles Maps 282, 163, 162

To enjoy a scenic drive that begins 11mi N of **Victoria★★★**, take Douglas St. N from Victoria to the Trans-Canada Highway (Hwy. 1) and follow **Malahat Drive★** (between Goldstream PP and Mill Bay Rd.) for 12mi. Continue N on Hwy. 1 past Duncan, **Chemainus★**—known for its murals—and Nanaimo to Parksville. Take winding Rte. 4 W (Pacific Rim Hwy.) passing **Englishman River Falls PP★** and **Cameron Lake**. Just beyond the lake, **Cathedral Grove★★** holds 800-year-old Douglas firs. The road descends to **Port Alberni**, departure point for cruises on Barkley Sound, and follows Sproat Lake before climbing Klitsa Mountain. The route leads to the Pacific along the Kennedy River. At the coast, turn left and drive SE to Ucluelet. Then head N to enter **Pacific Rim NPR★★★**. Continue to road's end at **Tofino★** to end the tour.

Yukon Circuit★★

1,485 kilometers/921 miles Map 155

From **Whitehorse★**, capital of Yukon Territory, drive N on the **Klondike Hwy.** (Rte. 2), crossing the Yukon River at **Carmacks**. After 196km/122mi, small islands divide the river into fast-flowing channels at **Five Finger Rapids★**. From Stewart Crossing, continue NW on Rte. 2 to **Dawson★★**, a historic frontier town. Ferry across the river and drive the **Top of the World Hwy.★★** (Rte. 9), with its **views★★★**, to the Alaska border. Rte. 9 joins Rte. 5, passing tiny **Chicken**, Alaska. At Tetlin Junction, head SE on Rte. 2, paralleling **Tetlin NWR**. Enter Canada and follow the **Alaska Highway★★** (Rte. 1) SE along **Kluane Lake★★** to **Haines Junction**, gateway to **Kluane NPR★★**, home of **Mount Logan**, Canada's highest peak (5,959m/19,550ft). Continue E to Rte. 2 to return to Whitehorse.

0 mi 20 40
0 km 20 40 60
One inch equals 25.4 miles
One centimeter equals 16.1 kilometers

Seattle WA / Mt Rainier Natl Park WA

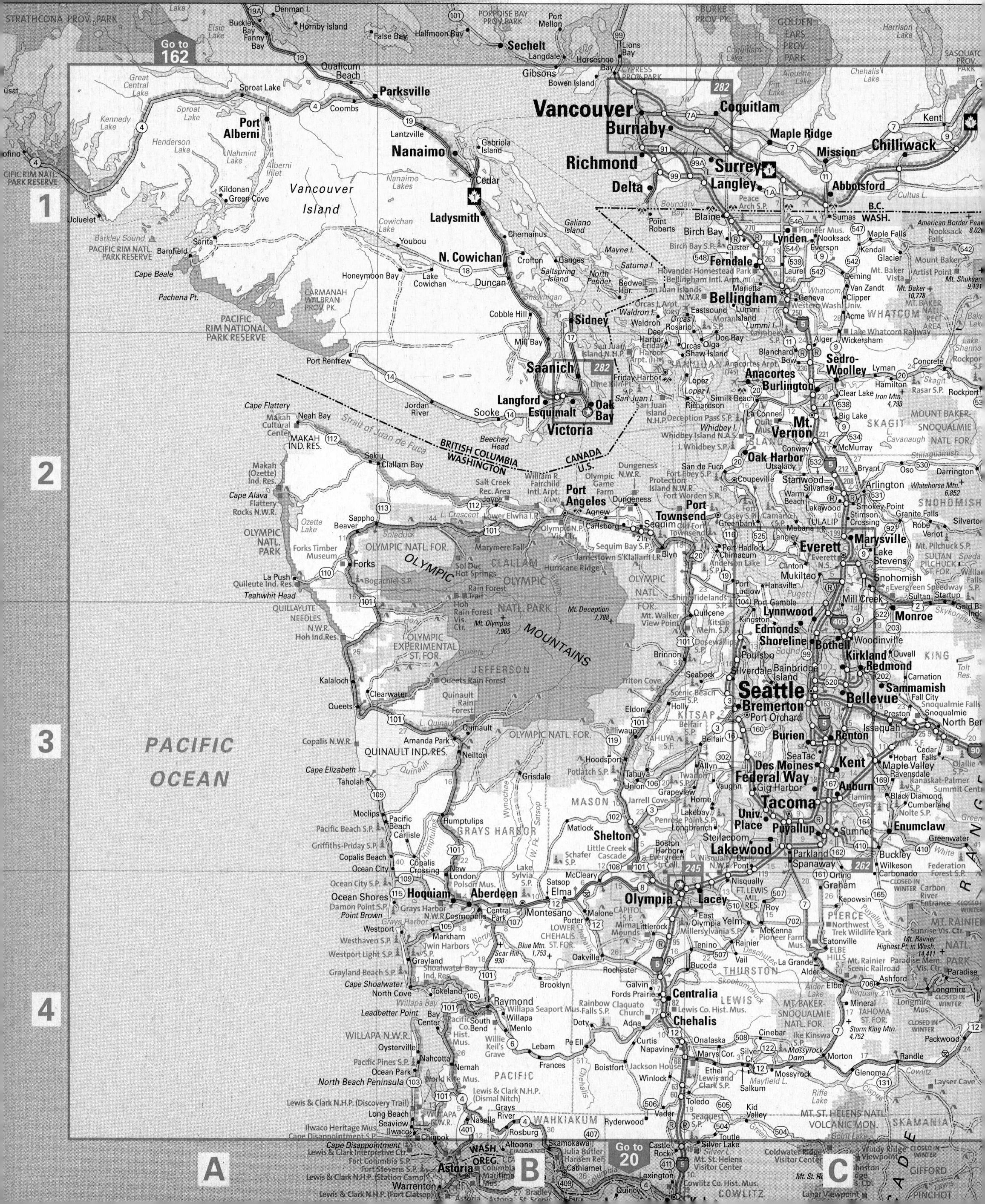

DRIVING DISTANCES IN MILES

	ABERDEEN, WA	BELLINGHAM, WA	MT. RAINIER NP, WA	OKANOGAN, WA	OLYMPIA, WA	PORT ANGELES, WA	SEATTLE, WA	SPOKANE, WA	TACOMA, WA	VANCOUVER, BC	WENATCHEE, WA	YAKIMA, WA
BELLINGHAM, WA	196		186	195	147	127*	88	360	122	52	185	221
SEATTLE, WA	105	88	96	223	56	83*		278	31	140	148	140
SPOKANE, WA	376	360	290	148	327	362*	278		303	412	171	203
YAKIMA, WA	237	221	87	194	188	223*	140	203	164	273	115	

*DISTANCE INCLUDES FERRY TRAVEL

SEE ALSO DISTANCE AND DRIVING TIME MAP ON PAGES 286–287

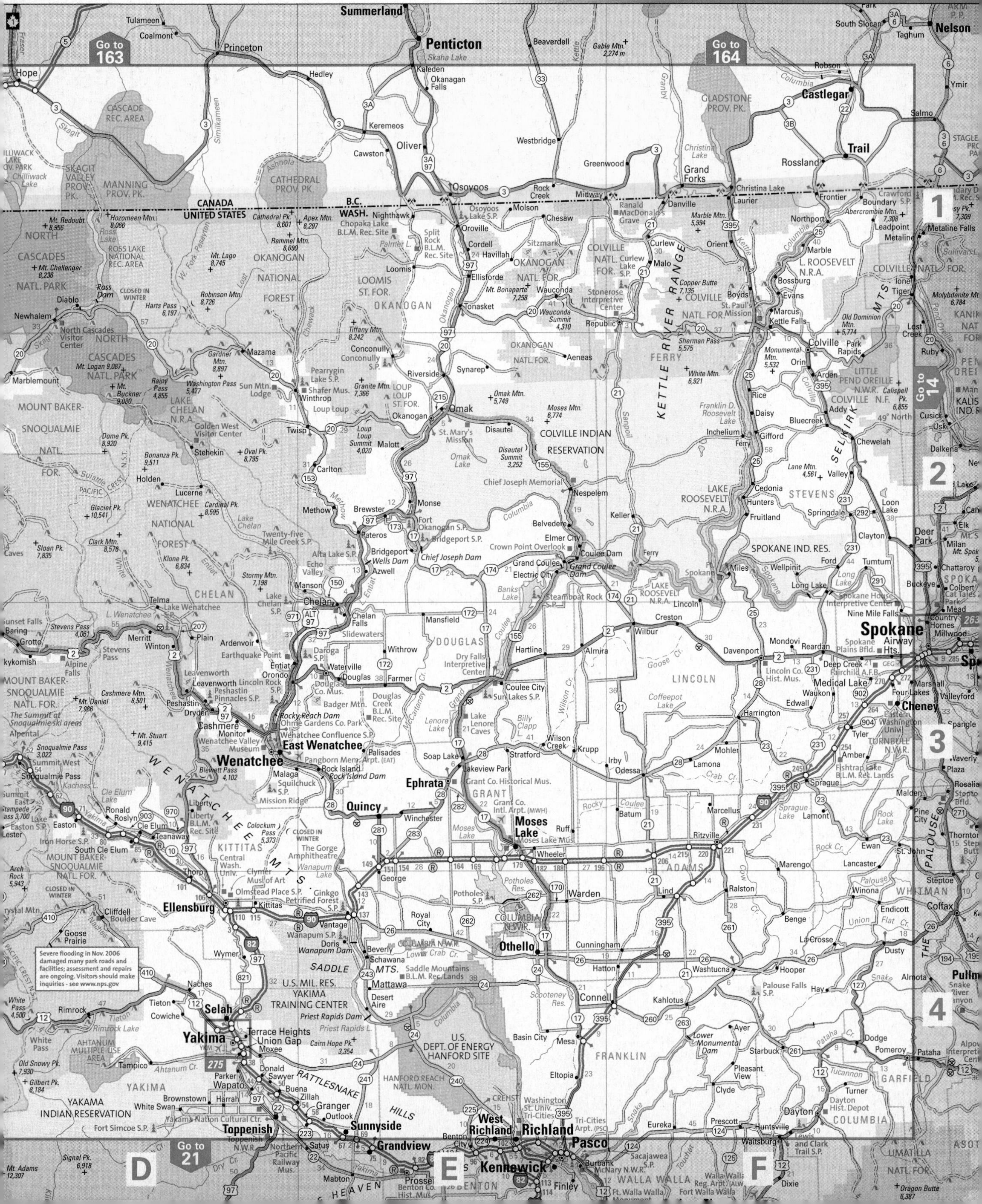

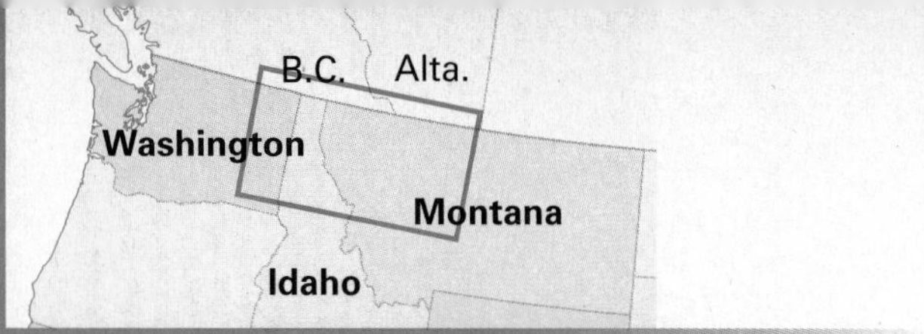

0 mi 20 40
0 km 20 40 60

One inch equals 25.4 miles
One centimeter equals 16.1 kilometers

Spokane WA / Coeur d'Alene ID

DRIVING DISTANCES IN MILES

	BONNERS FERRY, ID	BROWNING, MT	COEUR D'ALENE, ID	COLVILLE, WA	GREAT FALLS, MT	HELENA, MT	KALISPELL, MT	LEWISTON, ID	MISSOULA, MT	SHELBY, MT	SPOKANE, WA	WEST GLACIER, MT
GREAT FALLS, MT	369	124	364	471		85	222	420	199	82	398	192
LEWISTON, ID	196	413	118	176	420	334	315		221	448	103	348
MISSOULA, MT	244	201	167	274	199	114	116	221		227	201	136
SPOKANE, WA	110	336	34	73	398	313	238	103	201	426		271

SEE ALSO DISTANCE AND DRIVING TIME MAP ON PAGES 286–287

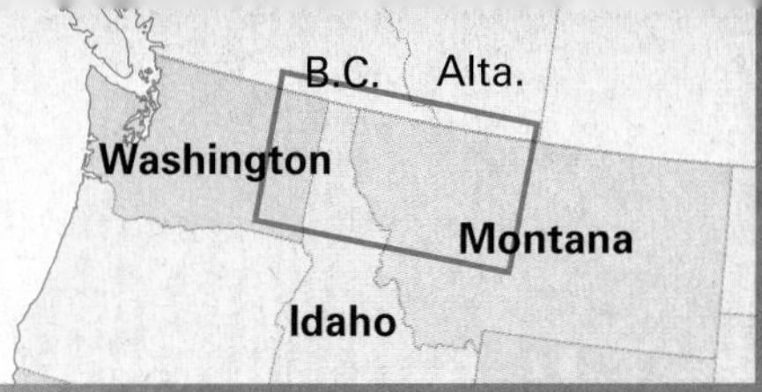

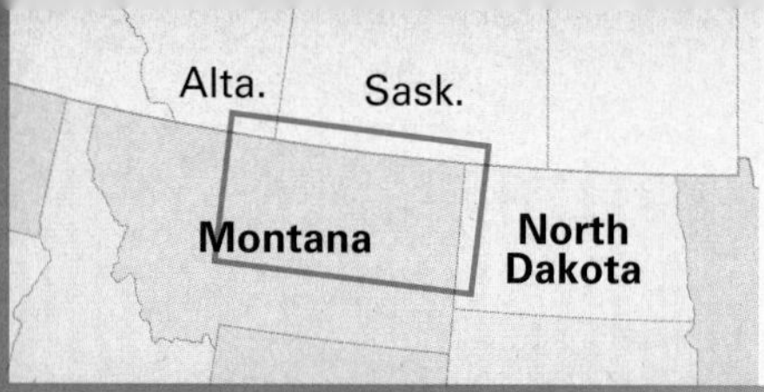

0 mi 20 40
0 km 20 40 60
One inch equals 25.4 miles
One centimeter equals 16.1 kilometers

Great Falls MT / Havre MT

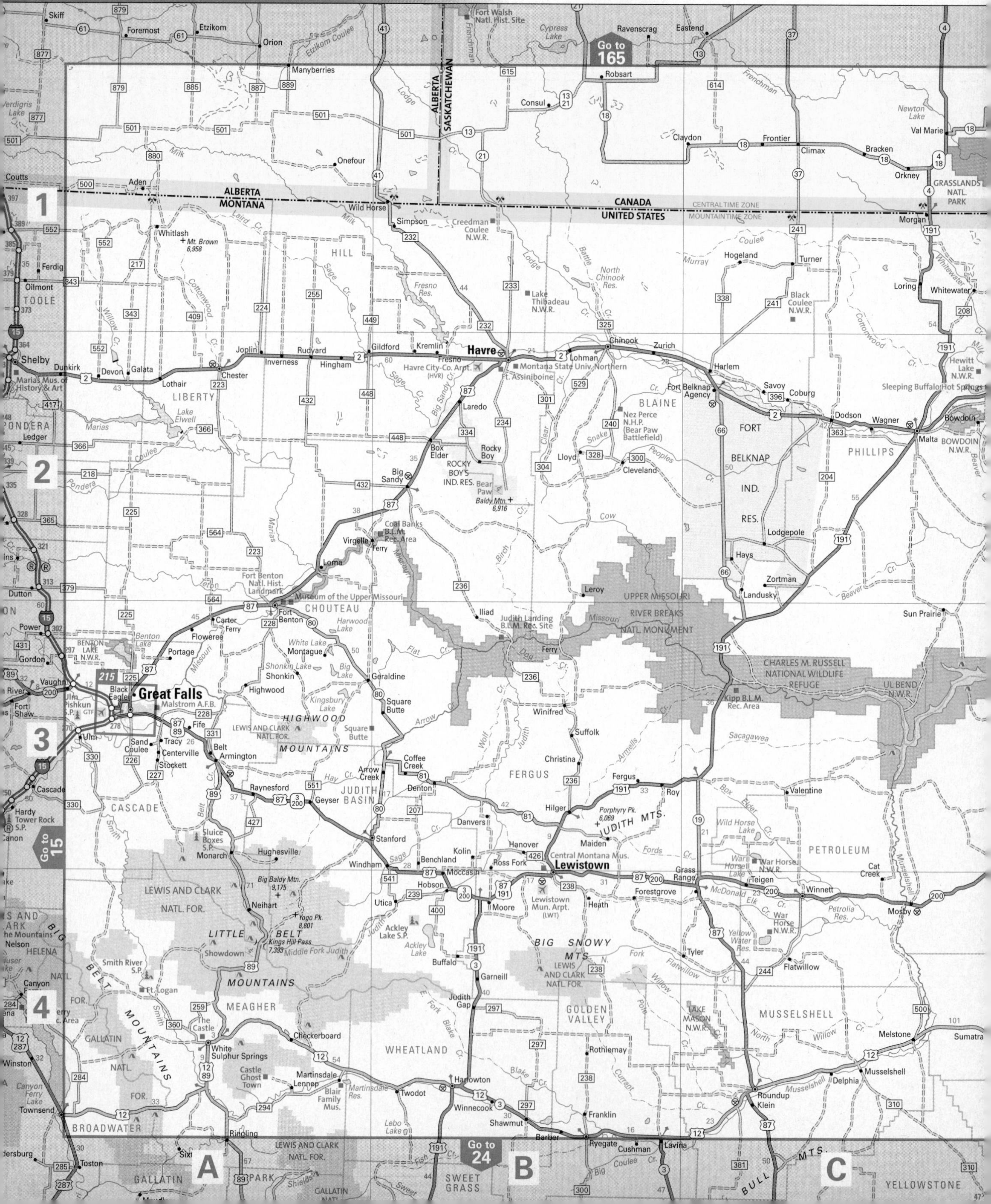

DRIVING DISTANCES IN MILES

	GLASGOW, MT	GLENDIVE, MT	GREAT FALLS, MT	HARLOWTON, MT	HAVRE, MT	LEWISTOWN, MT	MALTA, MT	MILES CITY, MT	ROUNDUP, MT	SHELBY, MT	WILLISTON, ND	WOLF POINT, MT
GLENDIVE, MT	147		351	309	306	242	217	74	219	408	106	98
GREAT FALLS, MT	277	351		133	118	109	207	329	183	82	422	326
HAVRE, MT	159	306	118	210		175	89	345	198	102	304	208
WILLISTON, ND	145	106	422	415	304	324	215	180	325	406		96

SEE ALSO DISTANCE AND DRIVING TIME MAP ON PAGES 286–287

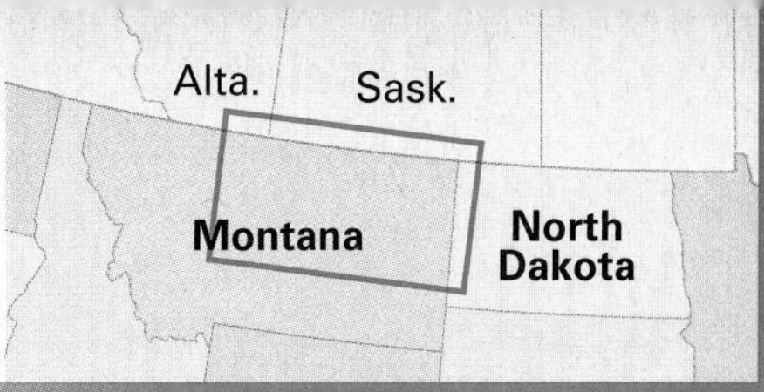

0 mi 20 40
0 km 20 40 60
One inch equals 25.4 miles
One centimeter equals 16.1 kilometers

Bismarck ND / Minot ND

DRIVING DISTANCES IN MILES

	BISMARCK, ND	BOTTINEAU, ND	DETROIT LAKES, MN	DICKINSON, ND	FARGO, ND	GRAND FORKS, ND	JAMESTOWN, ND	MINOT, ND	PEMBINA, ND	RUGBY, ND	THIEF RIVER FALLS, MN	WILLISTON, ND
BISMARCK, ND		189	244	97	199	274	105	116	347	153	319	229
FARGO, ND	199	271	45	291		79	97	268	152	221	113	424
GRAND FORKS, ND	274	198	125	367	79		173	212	77	148	61	340
MINOT, ND	116	76	313	178	268	212	171		238	64	276	128

SEE ALSO DISTANCE AND DRIVING TIME MAP ON PAGES 286–287

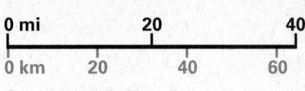

0 mi 20 40
0 km 20 40 60
One inch equals 25.4 miles
One centimeter equals 16.1 kilometers

Portland OR / Eugene OR

DRIVING DISTANCES IN MILES	ASTORIA, OR	BEND, OR	BURNS, OR	COOS BAY, OR	EUGENE, OR	KENNEWICK, WA	LA GRANDE, OR	NEWPORT, OR	PORTLAND, OR	SALEM, OR	THE DALLES, OR	WALLA WALLA, WA
BEND, OR	252		142	227	115	245	295	183	158	134	137	276
EUGENE, OR	216	115	257	105		328	377	101	112	65	198	359
KENNEWICK, WA	306	245	256	440	328		111	328	212	264	131	49
PORTLAND, OR	97	158	299	224	112	212	261	116		48	82	243

SEE ALSO DISTANCE AND DRIVING TIME MAP ON PAGES 286–287

Boise ID / La Grande OR

0 mi 20 40
0 km 20 40 60
One inch equals 25.4 miles
One centimeter equals 16.1 kilometers

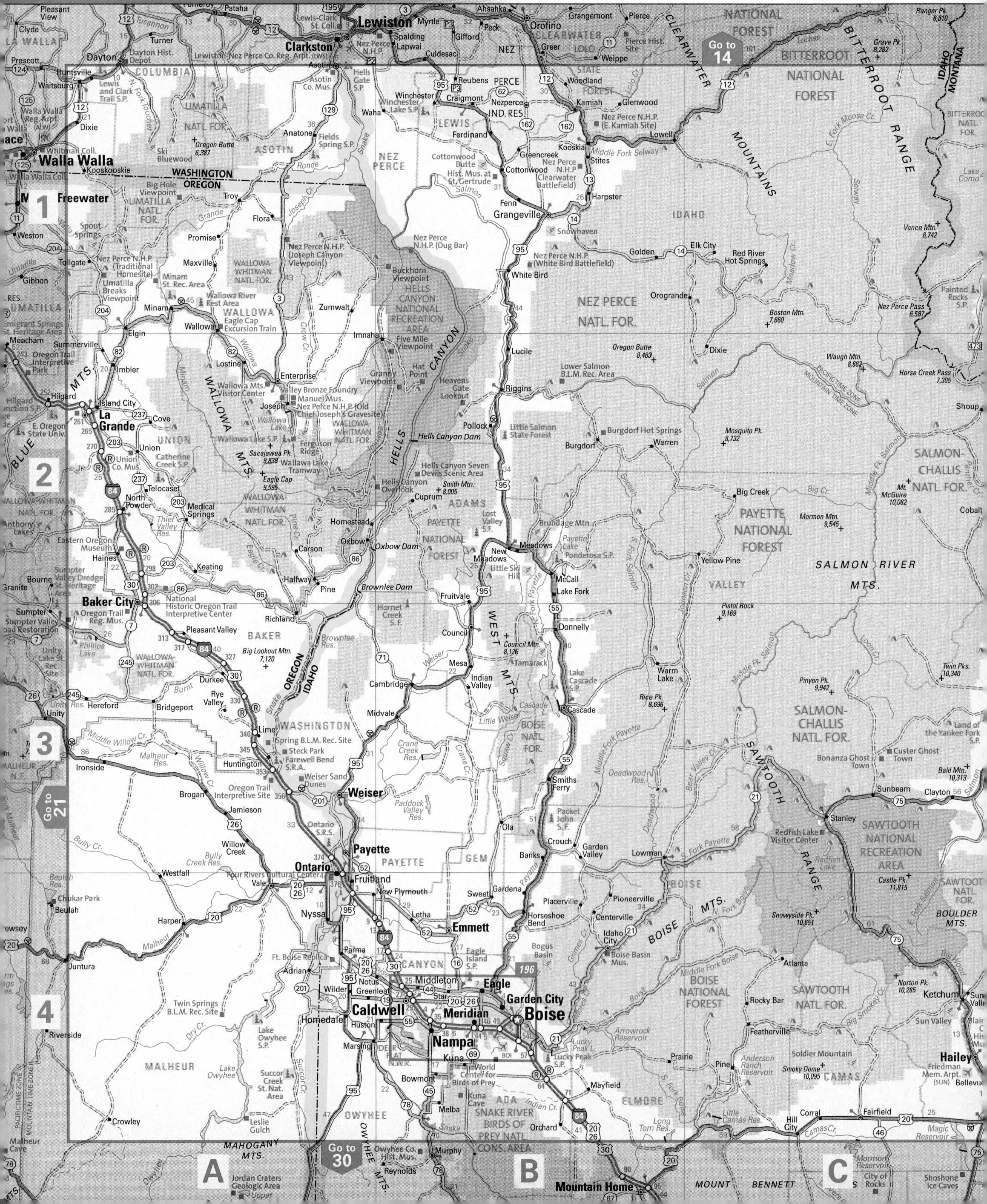

DRIVING DISTANCES IN MILES

	BOISE, ID	BOZEMAN, MT	BUTTE, MT	GRANGEVILLE, ID	HAMILTON, MT	IDAHO FALLS, ID	JACKSON, WY	LA GRANDE, OR	ONTARIO, OR	SALMON, ID	SUN VALLEY, ID	W. YELLOWSTONE, MT
BOISE, ID		485	486	202	339	288	378	170	58	247	163	395
BUTTE, MT	486	81		290	103	203	275	566	541	150	312	162
IDAHO FALLS, ID	288	199	203	483	272		92	455	342	168	153	109
W. YELLOWSTONE, MT	395	90	162	451	264	109	128	562	449	244	252	

SEE ALSO DISTANCE AND DRIVING TIME MAP ON PAGES 286–287

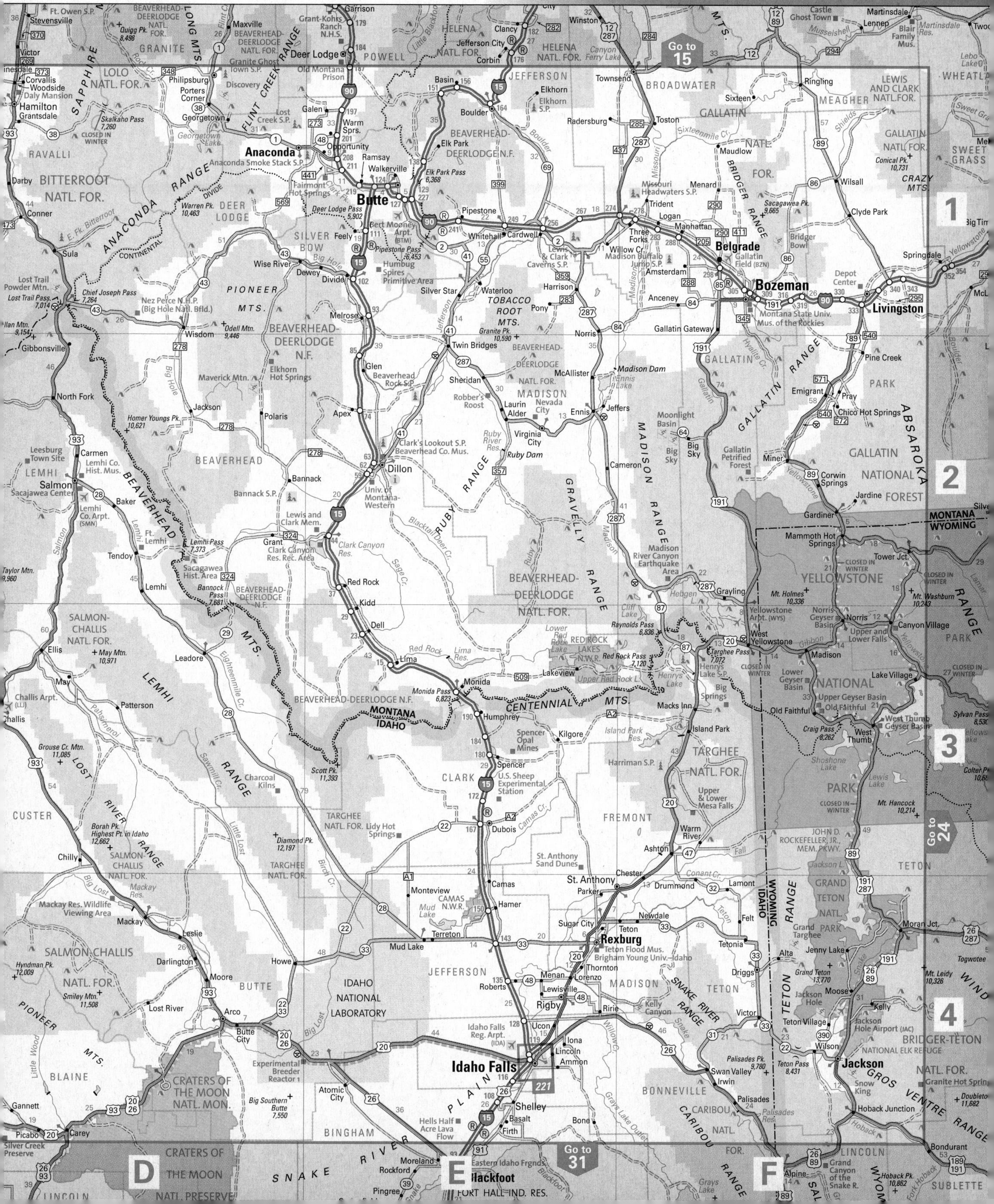

Montana — North Dakota
Idaho — South Dakota
Wyoming

0 mi 20 40
0 km 20 40 60
One inch equals 25.4 miles
One centimeter equals 16.1 kilometers

BROADWATER — MEAGHER — WHEATLAND — MUSSELSHELL — TREASURE — GOLDEN VALLEY — SWEET GRASS — STILLWATER — CARBON — YELLOWSTONE — BIG HORN — CROW INDIAN RESERVATION — CUSTER NATL. FOR. — BIGHORN CANYON NATL. REC. AREA — BIGHORN NATL. FOR.

GALLATIN NATL. FOR. — LEWIS AND CLARK NATL. FOR. — CRAZY MTS. — BRIDGER RANGE — ABSAROKA RANGE — BEARTOOTH RANGE — GALLATIN RANGE — YELLOWSTONE NATL. PARK — PARK — SHOSHONE NATIONAL FOREST — BIGHORN BASIN — GRAND TETON NATL. PARK — TETON RANGE — WIND RIVER RANGE — GROS VENTRE RANGE — BRIDGER-TETON NATL. FOR. — ROCKY MTS. — WIND RIVER IND. RES. — HOT SPRINGS — WASHAKIE — COPPER MTN. — BOYSEN S.P. — SUBLETTE — LINCOLN — FREMONT — BIG HORN

Townsend, Toston, Maudlow, Sixteen, Ringling, Wilsall, Clyde Park, Springdale, Big Timber, Greycliff, Reed Point, Columbus, Park City, Laurel, Billings, Shepherd, Huntley, Ballantine, Worden, Pompeys Pillar, Custer, Hardin, Toluca, Crow Agency, Garryowen, Lodge Grass, Wyola, Parkman, Dayton, Ranchester

Belgrade, Bozeman, Livingston, Gallatin Gateway, Big Sky, Emigrant, Pray, Chico Hot Springs, Corwin Springs, Jardine, Gardiner, Mammoth Hot Springs, Cooke City, Silver Gate, Tower Jct., Canyon Village, Norris, Madison, Lake Village, West Thumb, Old Faithful, West Yellowstone

Martinsdale, Lennep, Harlowton, Twodot, Winnecook, Shawmut, Franklin, Barber, Ryegate, Lavina, Cushman, Broadview, Hailstone N.W.R., Comanche, Rapalje, Halfbreed Lake N.W.R., Acton, Molt, Laurel, Columbus, Silesia, Joliet, Edgar, Boyd, Fromberg, Bridger, Belfry, Warren, Frannie, Deaver, Cowley, Lovell, Kane, Byron, Garland, Powell, Ralston, Willwood, Emblem, Burlington, Otto, Greybull, Basin, Shell, Manderson, Hyattville, Worland, Ten Sleep, Winchester, Kirby, Gebo, Lucerne, Thermopolis, East Thermopolis, Hamilton Dome, Grass Creek, Meeteetse, Pitchfork, Sunshine, Cody, Wapiti, Valley

Roundup, Klein, Delphia, Musselshell, Lavina, Billings, Laurel, Park City, Silesia, Red Lodge, Bearcreek, Washoe, Roscoe, Luther, Roberts, Fishtail, Nye, Dean, Alpine, Limestone, Absarokee, Columbus

Livingston, Pine Creek, Emigrant, Pray, Chico Hot Springs, Miner, Jardine, Gardiner

Grayling, West Yellowstone, Island Park, Ashton, Tetonia, Driggs, Felt, Alta, Victor, Teton Village, Wilson, Jackson, Hoback Junction, Alpine, Moran Jct., Moose, Kelly, Jenny Lake, Dubois, Crowheart, Burris, Pavillion, Kinnear, Riverton, Shoshoni, Lysite, Lost Cabin, Moneta, Bonneville, Midvale, Morton

Sacagawea Pk. 9,665; Conical Pk. 10,731; Mt. Holmes 10,336; Mt. Washburn 10,243; Saddle Mtn. 10,670; Electric Pk. 10,992; Pilot Pk. 11,708; Beartooth Pass 10,947; Granite Pk. Highest Pt. in Montana 12,799; Red Lodge Mountain; Colter Pass 8,000; Cooke City; Dead Indian Summit 8,060; Heart Mtn. 8,123; Trout Pk. 12,244; Fortress Mtn. 12,085; Colter Pk. 10,683; Carter Mtn.; Francs Pk. 13,153; Younts Pk. 12,156; Washakie Needles 12,518; Mt. Leidy 10,326; Togwotee Pass 9,658; Union Pass 9,210; Grand Teton 13,770; Mt. Hancock 10,214; Mt. Moran; Doubletop Pk. 11,682; Gannett Pk. Highest Point in Wyoming 13,804; Guffy Pk. 8,046; Teton Pass 8,431; Snow King; Caribou Mtn. 9,805; Palisades Pk. 9,780; Targhee Pass 7,072; Craig Pass 8,262

Grand Teton 13,770; Mt. Leidy 10,326

Go to 16
Go to 23
Go to 32

MONTANA / WYOMING

CONTINENTAL DIVIDE

A — B — C
1 — 2 — 3 — 4

Billings MT / Yellowstone Natl Park WY

DRIVING DISTANCES IN MILES	BILLINGS, MT	BOZEMAN, MT	BUFFALO, WY	CODY, WY	GILLETTE, WY	JACKSON, WY	MILES CITY, MT	RAPID CITY, SD	SHERIDAN, WY	SPEARFISH, SD	W. YELLOWSTONE, MT	WORLAND, WY
BILLINGS, MT		141	165	111	233	287	144	379	131	333	232	161
BUFFALO, WY	165	306		180	70	342	237	216	34	170	396	91
SPEARFISH, SD	333	474	170	350	100	512	186	53	202		564	261
W. YELLOWSTONE, MT	232	90	396	147	464	128	376	610	363	564		236

SEE ALSO DISTANCE AND DRIVING TIME MAP ON PAGES 286–287

0 mi 20 40
0 km 20 40 60
One inch equals 25.4 miles
One centimeter equals 16.1 kilometers

DRIVING DISTANCES IN MILES

	ABERDEEN, SD	BROOKINGS, SD	HOT SPRINGS, SD	HURON, SD	MITCHELL, SD	MOBRIDGE, SD	PIERRE, SD	RAPID CITY, SD	SIOUX FALLS, SD	WAHPETON, ND	WALL, SD	WATERTOWN, SD
ABERDEEN, SD		150	412	90	146	99	160	357	204	154	303	98
PIERRE, SD	160	188	247	115	155	107		193	226	301	138	189
RAPID CITY, SD	357	390	56	313	275	243	193		346	543	55	436
SIOUX FALLS, SD	204	57	401	127	73	303	226	346		210	292	103

SEE ALSO DISTANCE AND DRIVING TIME MAP ON PAGES 286–287

0 mi 20 40
0 km 20 40 60
One inch equals 25.4 miles
One centimeter equals 16.1 kilometers

Medford OR / Redding CA

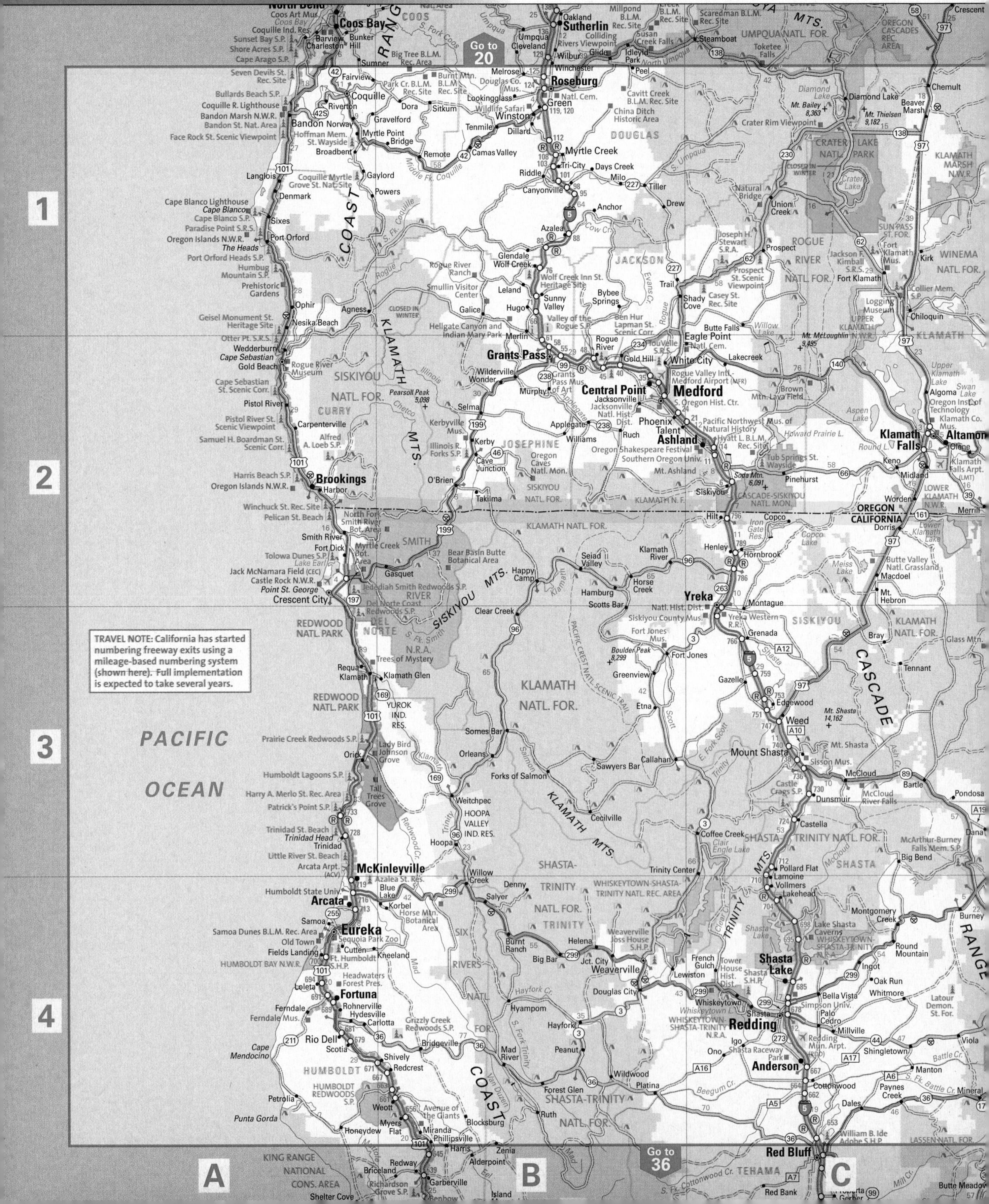

DRIVING DISTANCES IN MILES	ALTURAS, CA	CRATER LAKE NP, OR	CRESCENT CITY, CA	EUREKA, CA	KLAMATH FALLS, OR	LAKEVIEW, OR	LASSEN VOLCANIC NP, CA	MEDFORD, OR	REDDING, CA	ROSEBURG, OR	SUSANVILLE, CA	WINNEMUCCA, NV
LAKEVIEW, OR	56	153	282	332	98		192	171	199	265	161	212
MEDFORD, OR	176	80	111	192	76	171	208		148	94	221	383
REDDING, CA	143	198	189	133	141	199	63	148		242	114	364
SUSANVILLE, CA	105	226	303	247	170	161	74	221	114	315		250

SEE ALSO DISTANCE AND DRIVING TIME MAP ON PAGES 286–287

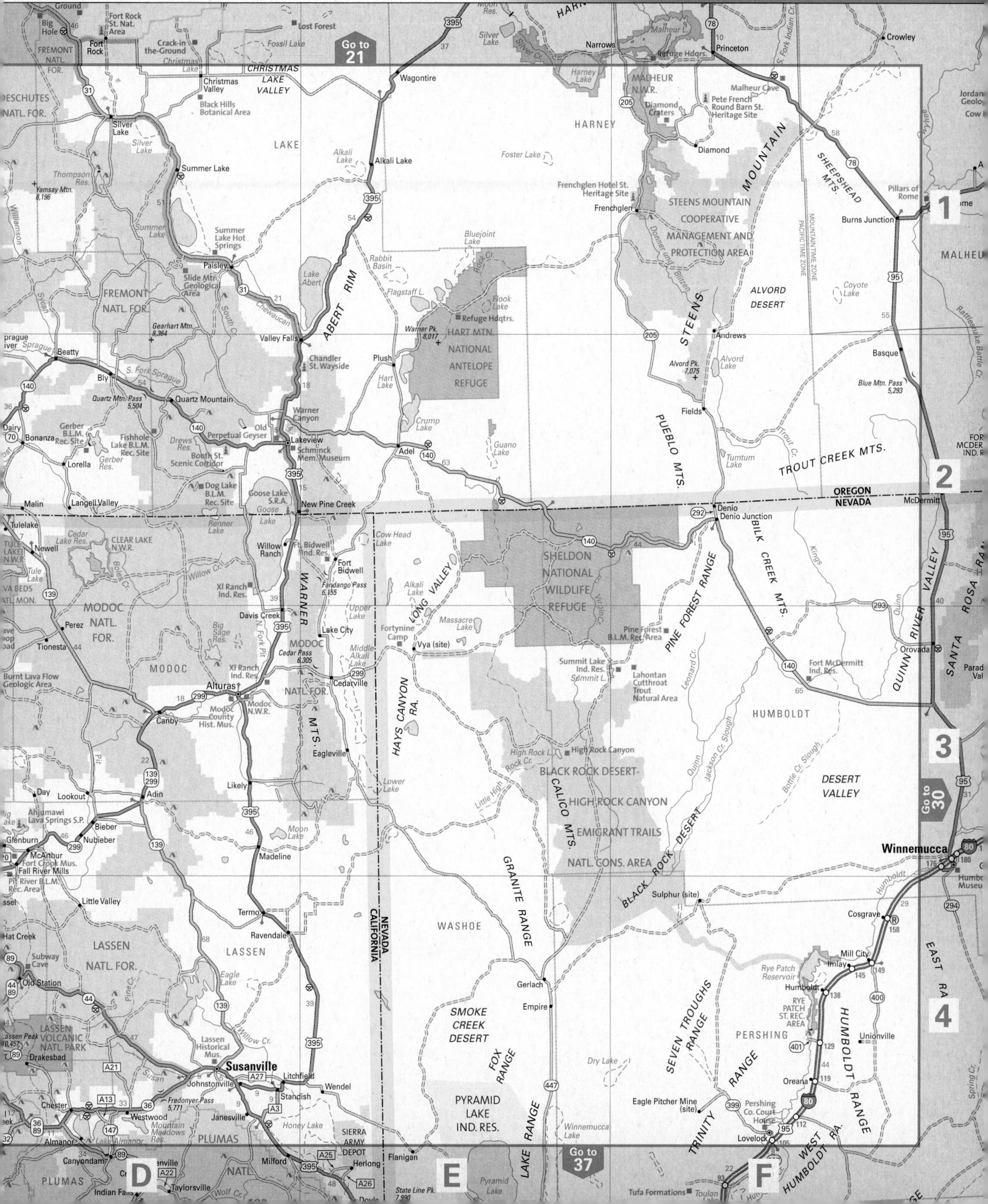

0 mi 20 40
0 km 20 40 60
One inch equals 25.4 miles
One centimeter equals 16.1 kilometers

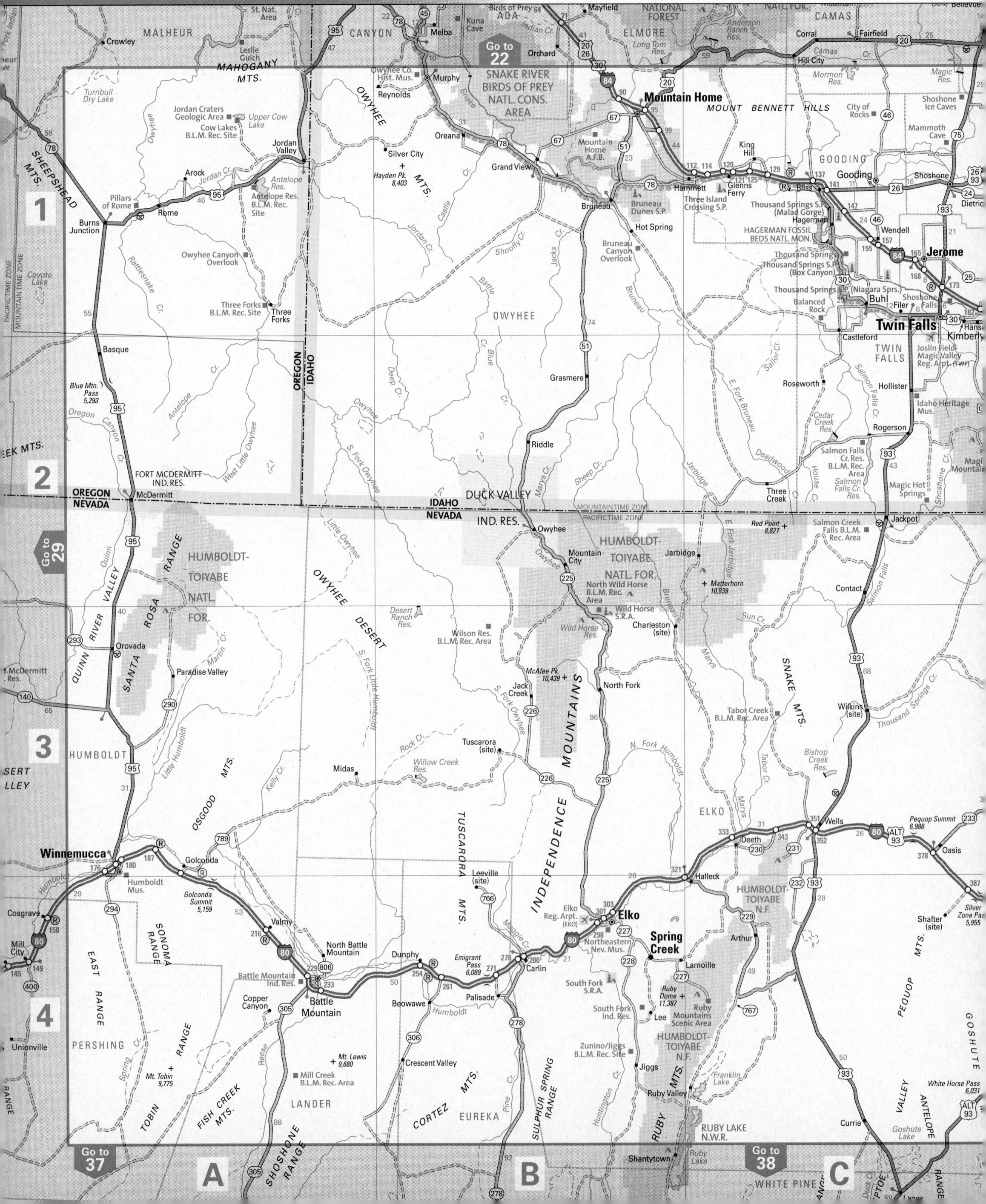

DRIVING DISTANCES IN MILES

	BRIGHAM CITY, UT	ELKO, NV	EVANSTON, WY	MONTPELIER, ID	MOUNTAIN HOME, ID	OGDEN, UT	POCATELLO, ID	PROVO, UT	SALT LAKE CITY, UT	TWIN FALLS, ID	WELLS, NV	WINNEMUCCA, NV
ELKO, NV	286		314	375	194	267	283	279	232	167	50	127
POCATELLO, ID	107	283	200	87	193	127		205	159	116	233	410
SALT LAKE CITY, UT	56	232	82	145	295	37	159	47		217	182	359
TWIN FALLS, ID	165	167	259	204	86	185	116	264	217		117	294

SEE ALSO DISTANCE AND DRIVING TIME MAP ON PAGES 286–287

Salt Lake City UT / Pocatello ID

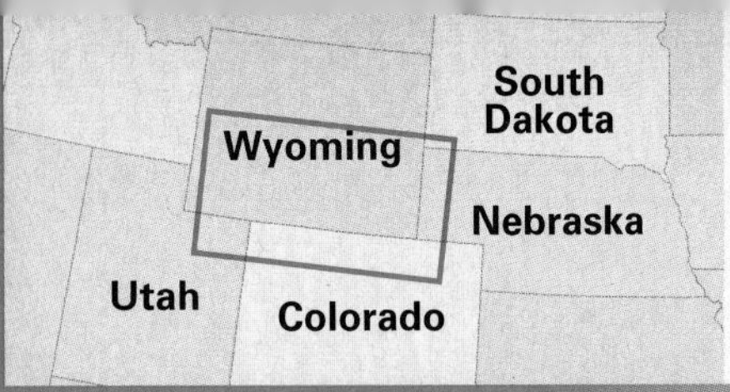

0 mi 20 40
0 km 20 40 60
One inch equals 25.4 miles
One centimeter equals 16.1 kilometers

Rock Springs WY / Craig CO

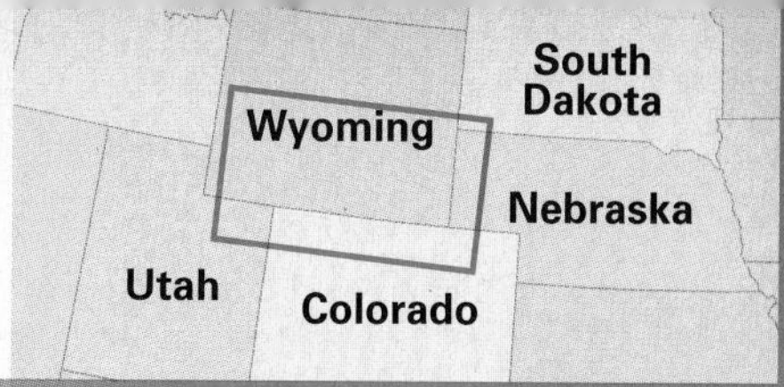

DRIVING DISTANCES IN MILES

	CASPER, WY	CHEYENNE, WY	CRAIG, CO	FORT COLLINS, CO	KEMMERER, WY	LANDER, WY	LARAMIE, WY	PINEDALE, WY	RAWLINS, WY	ROCK SPRINGS, WY	SCOTTSBLUFF, NE	VERNAL, UT
CASPER, WY		175	234	217	297	144	148	271	117	214	173	322
CHEYENNE, WY	175		221	44	342	276	52	355	151	260	111	367
CRAIG, CO	234	221		194	257	221	171	269	117	149	331	123
ROCK SPRINGS, WY	214	260	149	273	86	118	210	98	110		370	111

SEE ALSO DISTANCE AND DRIVING TIME MAP ON PAGES 286–287

0 mi 20 40
0 km 20 40 60
One inch equals 25.4 miles
One centimeter equals 16.1 kilometers

North Platte NE / Chadron NE

DRIVING DISTANCES IN MILES

	CHADRON, NE	GRAND ISLAND, NE	LINCOLN, NE	MCCOOK, NE	NORFOLK, NE	NORTH PLATTE, NE	OGALLALA, NE	OMAHA, NE	SCOTTSBLUFF, NE	SIOUX CITY, IA	STERLING, CO	YANKTON, SD
GRAND ISLAND, NE	373		95	147	105	143	196	150	318	180	281	167
LINCOLN, NE	453	95		226	119	223	275	58	397	153	361	218
NORTH PLATTE, NE	230	143	223	67	248		53	278	175	373	138	310
OMAHA, NE	508	150	58	281	115	278	330		452	99	416	163

SEE ALSO DISTANCE AND DRIVING TIME MAP ON PAGES 286–287

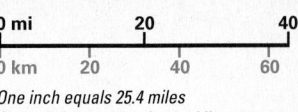

0 mi 20 40
0 km 20 40 60
One inch equals 25.4 miles
One centimeter equals 16.1 kilometers

San Francisco CA / Sacramento CA

Reno NV / Yosemite Natl Park CA

DRIVING DISTANCES IN MILES	AUSTIN, NV	CHICO, CA	MERCED, CA	RENO, NV	SACRAMENTO, CA	SAN FRANCISCO, CA	SAN JOSE, CA	S. LAKE TAHOE, CA	STOCKTON, CA	TONOPAH, NV	UKIAH, CA	YOSEMITE VIL., CA
RENO, NV	171	164	243		132	217	245	59	177	237	261	199
SACRAMENTO, CA	302	88	118	132		87	115	100	48	329	153	170
SAN FRANCISCO, CA	387	182	131	217	87		43	185	82	352	116	183
YOSEMITE VIL., CA	280	257	79	199	170	183	168	180	123	199	289	

SEE ALSO DISTANCE AND DRIVING TIME MAP ON PAGES 286–287

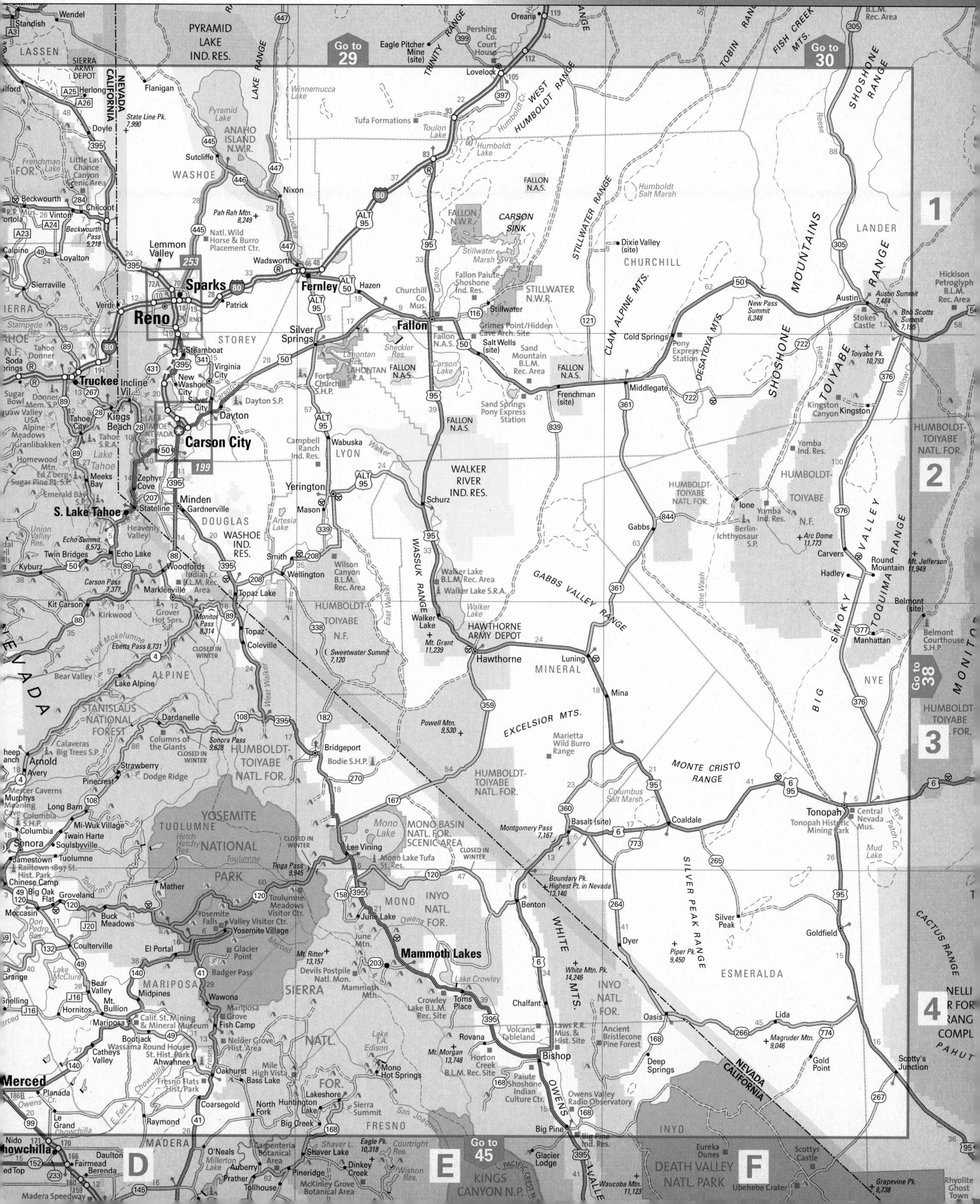

0 mi 20 40
0 km 20 40 60
One inch equals 25.4 miles
One centimeter equals 16.1 kilometers

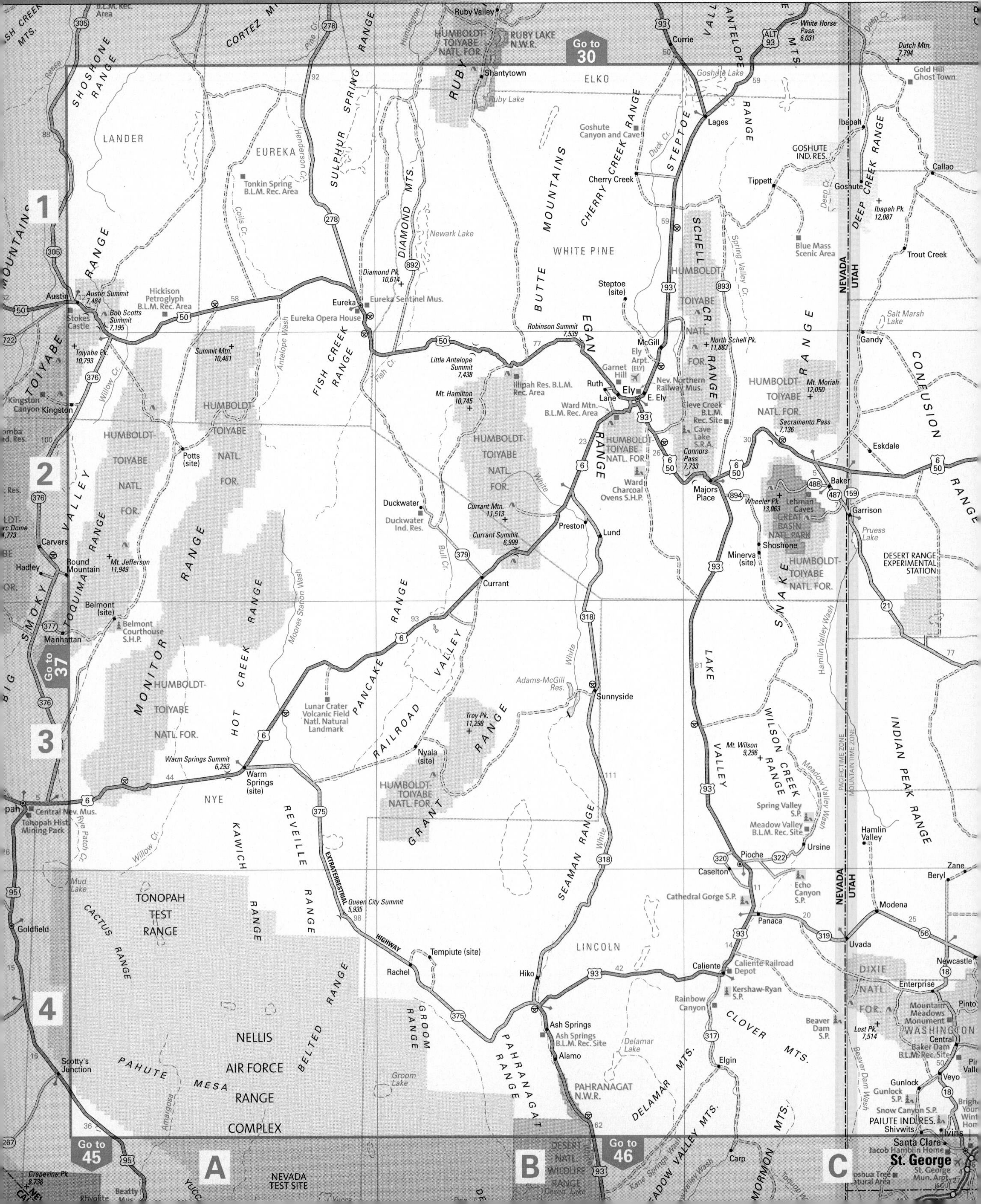

DRIVING DISTANCES IN MILES

	AUSTIN, NV	BAKER, NV	CEDAR CITY, UT	DELTA, UT	ELY, NV	GREEN RIVER, UT	PROVO, UT	ST. GEORGE, UT	SALINA, UT	SPRINGDALE, UT	TONOPAH, NV	TORREY, UT
ELY, NV	147	68	198	156		332	243	216	224	261	167	307
PROVO, UT	426	193	204	88	243	137		256	94	266	410	172
SALINA, UT	371	187	128	68	224	108	94	180		190	411	78
SPRINGDALE (ZION), UT	408	193	64	205	261	297	266	45	190		339	191

SEE ALSO DISTANCE AND DRIVING TIME MAP ON PAGES 286–287

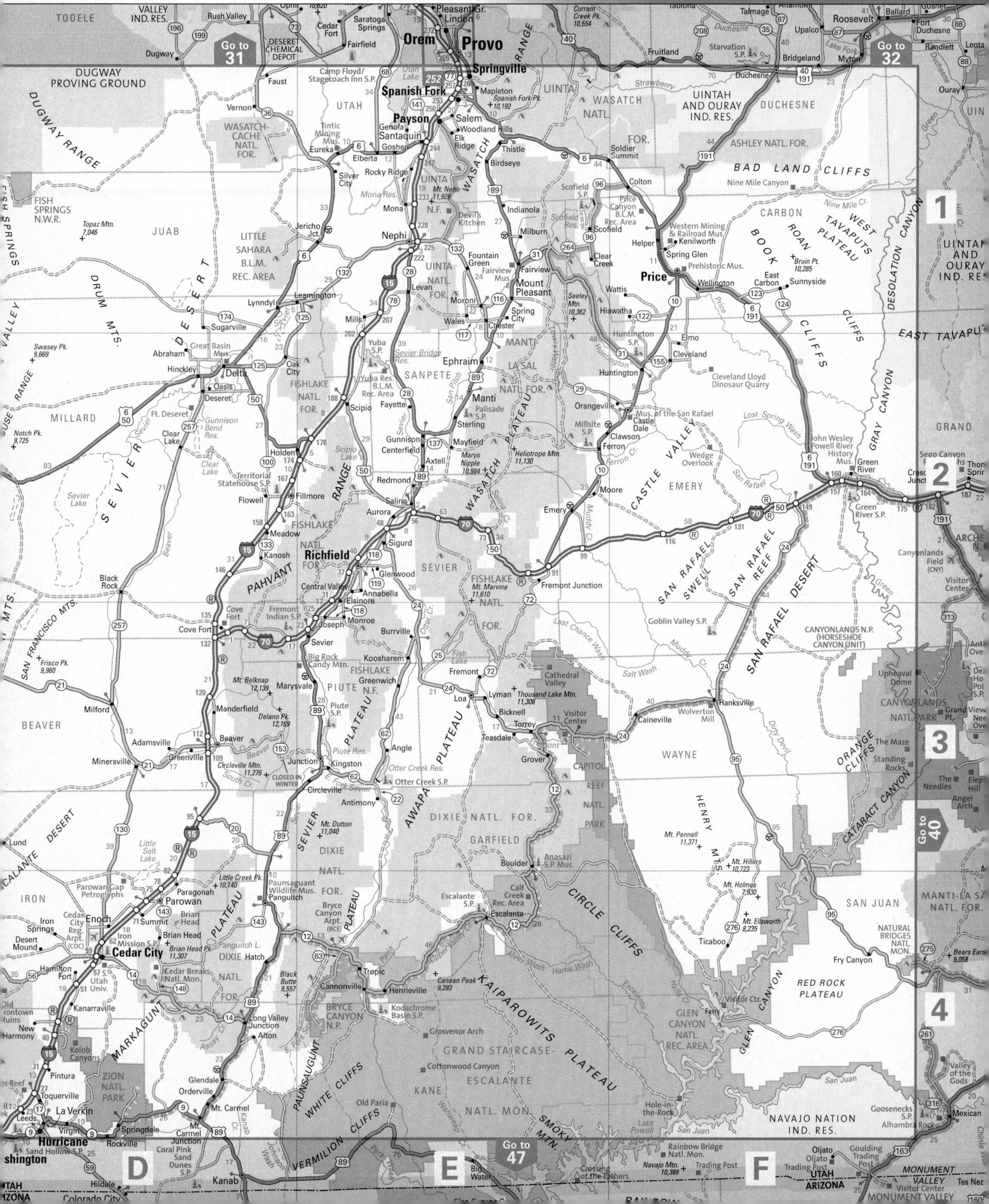

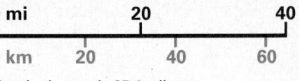

0 mi 20 40
0 km 20 40 60

One inch equals 25.4 miles
One centimeter equals 16.1 kilometers

Grand Junction CO / Durango CO

DRIVING DISTANCES IN MILES	ALAMOSA, CO	ASPEN, CO	COLORADO SPRS., CO	CORTEZ, CO	DENVER, CO	DURANGO, CO	GRAND JUNCTION, CO	GREEN RIVER, UT	MOAB, UT	MONTROSE, CO	PUEBLO, CO	TRINIDAD, CO
COLORADO SPRS., CO	162	157		359	70	314	318	418	404	236	43	127
DENVER, CO	230	164	70	452		337	250	350	337	277	111	196
DURANGO, CO	152	244	314	45	337		169	214	160	107	271	260
GRAND JUNCTION, CO	261	135	318	203	250	169		102	88	62	360	444

SEE ALSO DISTANCE AND DRIVING TIME MAP ON PAGES 286–287

0 mi 20 40
0 km 20 40 60
One inch equals 25.4 miles
One centimeter equals 16.1 kilometers

Garden City KS / Dodge City KS

DRIVING DISTANCES IN MILES

	BURLINGTON, CO	DODGE CITY, KS	EMPORIA, KS	GARDEN CITY, KS	HAYS, KS	LAMAR, CO	MANHATTAN, KS	McCOOK, NE	OAKLEY, KS	SALINA, KS	TOPEKA, KS	WICHITA, KS
GARDEN CITY, KS	167	52	290		139	98	272	167	79	204	311	205
OAKLEY, KS	88	136	293	79	87	156	247	88		179	286	268
SALINA, KS	266	164	118	204	93	335	72	240	179		111	92
WICHITA, KS	354	153	85	205	181	303	131	329	268	92	137	

SEE ALSO DISTANCE AND DRIVING TIME MAP ON PAGES 286–287

0 mi 20 40
0 km 20 40 60
One inch equals 25.4 miles
One centimeter equals 16.1 kilometers

TRAVEL NOTE: California has started numbering freeway exits using a mileage-based numbering system (shown here). Full implementation is expected to take several years.

Nevada
California

DRIVING DISTANCES IN MILES	BAKERSFIELD, CA	BISHOP, CA	DEATH VALLEY, CA	FRESNO, CA	RIDGECREST, CA	SALINAS, CA	SAN FRANCISCO, CA	SAN JOSE, CA	SAN LUIS OBISPO, CA	STOCKTON, CA	TONOPAH, NV	YOSEMITE VIL, CA
BAKERSFIELD, CA		215	236	111	99	209	287	245	119	243	318	200
BISHOP, CA	215		169	219	141	302	283	269	333	223	119	130
FRESNO, CA	111	219	333		196	145	190	153	134	130	288	90
SAN JOSE, CA	245	269	437	153	344	61	43		191	68	338	168

SEE ALSO DISTANCE AND DRIVING TIME MAP ON PAGES 286–287

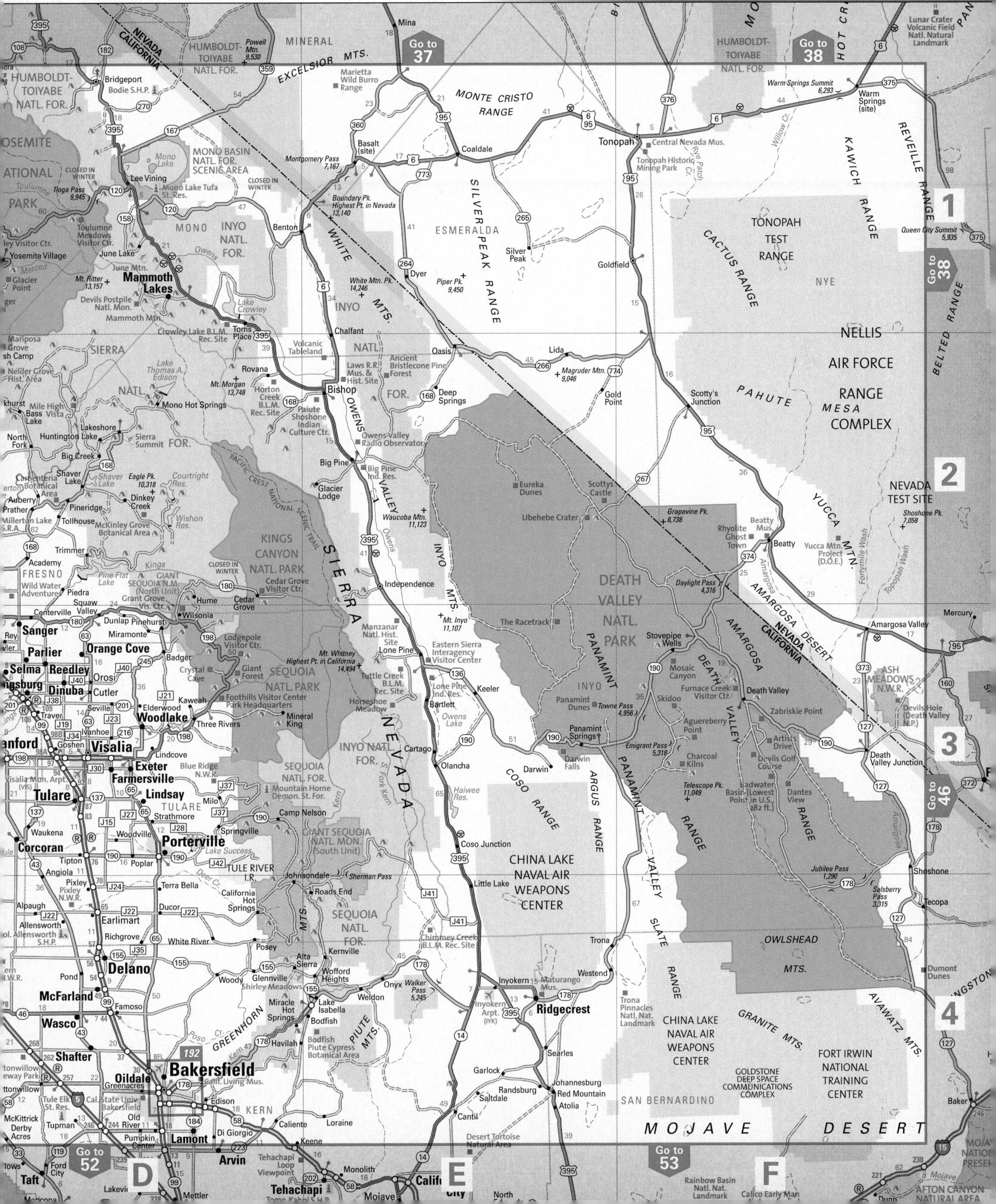

0 mi 20 40
0 km 20 40 60
One inch equals 25.4 miles
One centimeter equals 16.1 kilometers

Las Vegas NV / St George UT

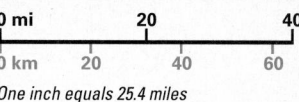

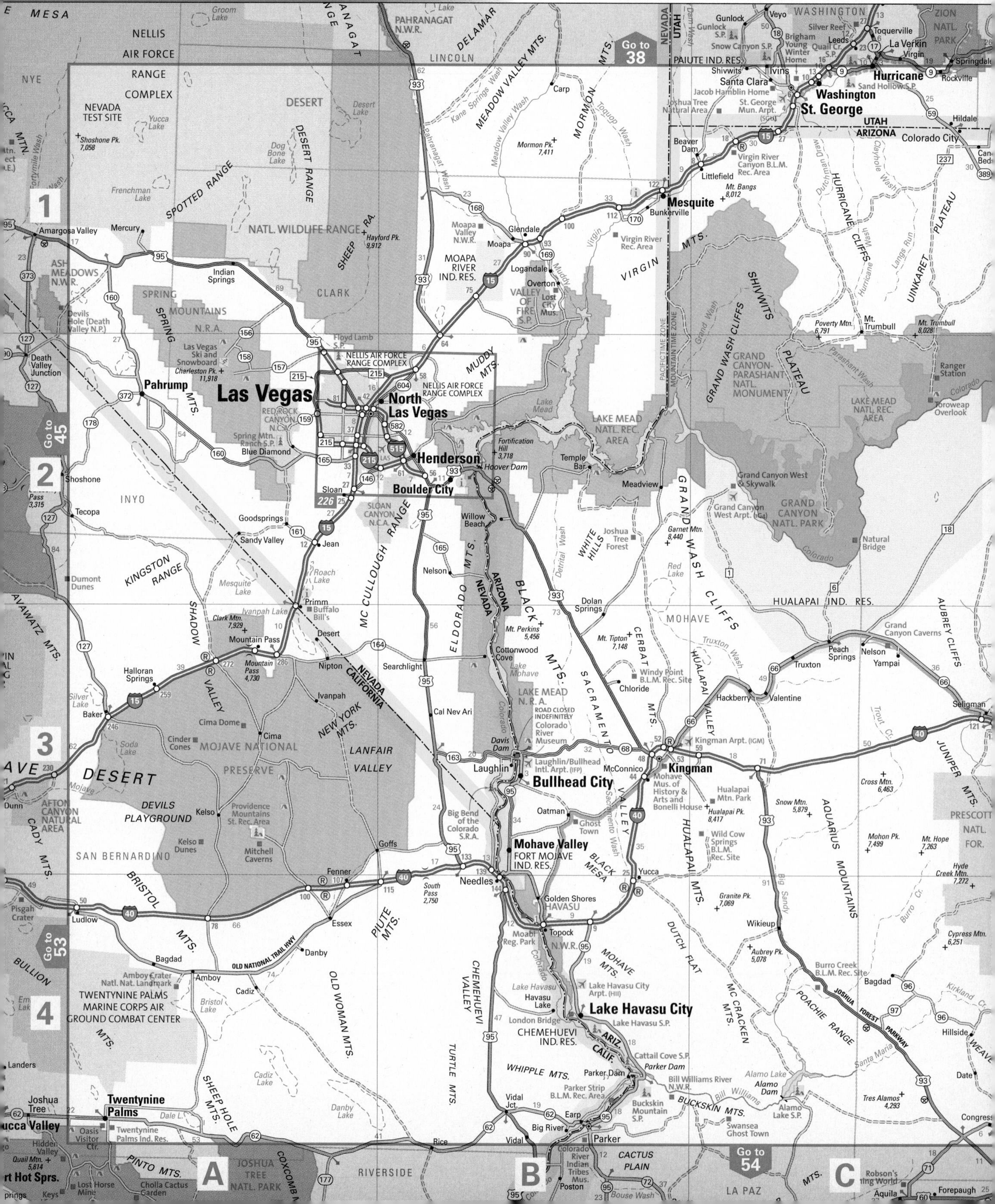

DRIVING DISTANCES IN MILES

	CHINLE, AZ	FLAGSTAFF, AZ	GRAND CANYON, AZ	HOLBROOK, AZ	KAYENTA, AZ	KINGMAN, AZ	LAKE HAVASU CITY, AZ	LAS VEGAS, NV	LAUGHLIN, NV	PAGE, AZ	PRESCOTT, AZ	ST. GEORGE, UT
FLAGSTAFF, AZ	216		89	93	152	148	209	249	182	135	89	271
GRAND CANYON, AZ	232	89		182	153	175	236	276	209	136	131	272
LAS VEGAS, NV	465	249	276	341	374	103	154		94	277	251	118
ST. GEORGE, UT	358	271	272	353	255	221	272	118	212	159	369	

SEE ALSO DISTANCE AND DRIVING TIME MAP ON PAGES 286–287

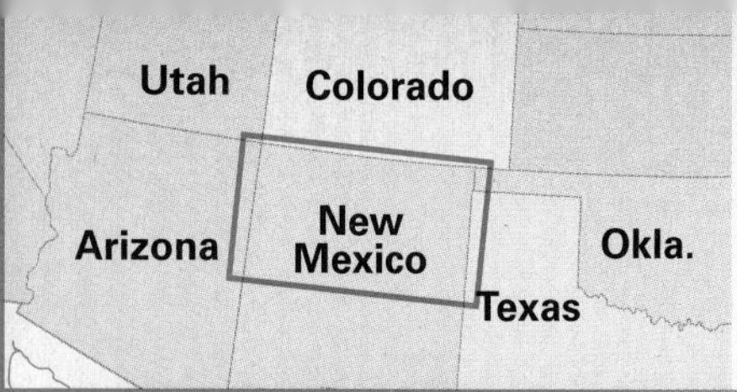

0 mi 20 40
0 km 20 40 60
One inch equals 25.4 miles
One centimeter equals 16.1 kilometers

Albuquerque NM / Farmington NM

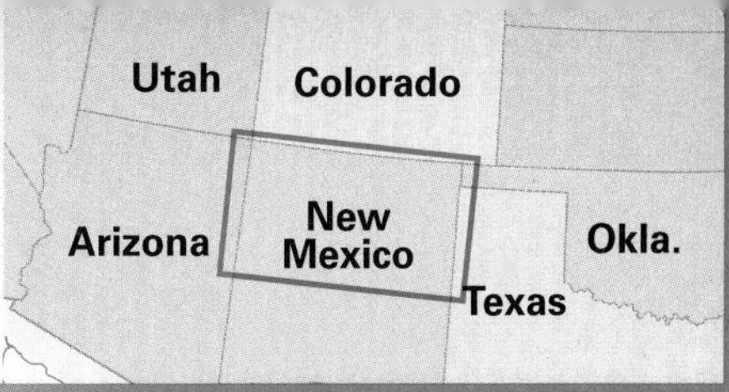

DRIVING DISTANCES IN MILES	ALBUQUERQUE, NM	CLAYTON, NM	CLOVIS, NM	DURANGO, CO	FARMINGTON, NM	GALLUP, NM	SANTA FE, NM	SOCORRO, NM	TAOS, NM	TRINIDAD, CO	TUCUMCARI, NM	VAUGHN, NM
ALBUQUERQUE, NM		266	220	212	181	141	55	77	123	242	174	104
FARMINGTON, NM	181	368	401	50		120	205	263	211	300	355	284
SANTA FE, NM	55	216	213	207	205	197		132	68	192	167	96
TUCUMCARI, NM	174	111	82	386	355	316	167	251	195	198		98

SEE ALSO DISTANCE AND DRIVING TIME MAP ON PAGES 286–287

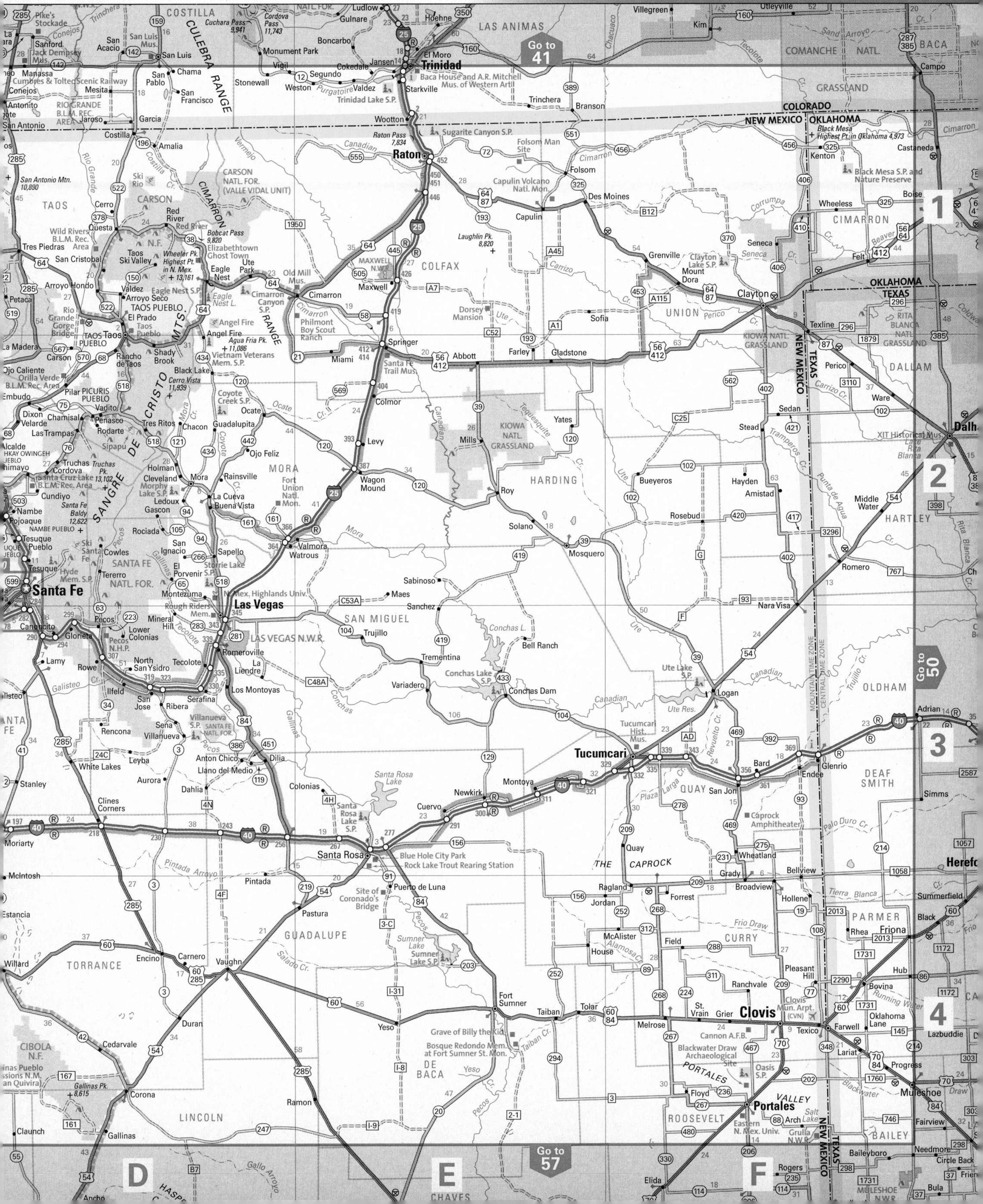

Amarillo TX / Liberal KS

0 mi 20 40
0 km 20 40 60
One inch equals 25.4 miles
One centimeter equals 16.1 kilometers

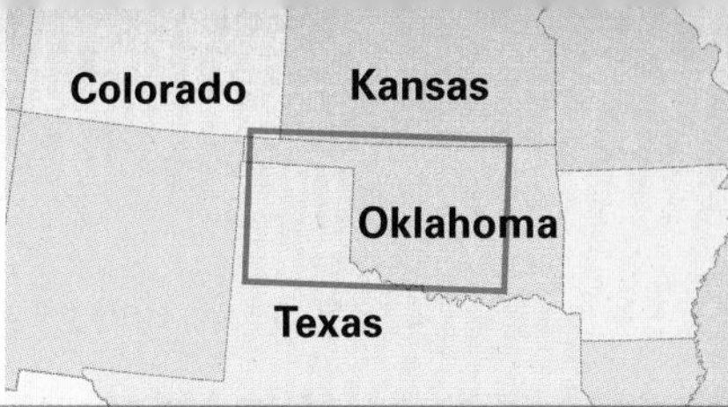

DRIVING DISTANCES IN MILES

	AMARILLO, TX	ARDMORE, OK	BARTLESVILLE, OK	CHILDRESS, TX	CLINTON, OK	ENID, OK	LAWTON, OK	LIBERAL, KS	OKLAHOMA CITY, OK	STILLWATER, OK	TULSA, OK	WOODWARD, OK
AMARILLO, TX		361	419	118	177	298	240	165	262	329	371	177
LAWTON, OK	240	103	243	124	98	142		287	85	152	194	175
OKLAHOMA CITY, OK	262	99	157	225	85	84	85	259		67	109	143
TULSA, OK	371	206	48	334	194	117	194	321	109	71		205

SEE ALSO DISTANCE AND DRIVING TIME MAP ON PAGES 286–287

0 mi 20 40
0 km 20 40 60
One inch equals 25.4 miles
One centimeter equals 16.1 kilometers

Los Angeles CA / Santa Barbara CA

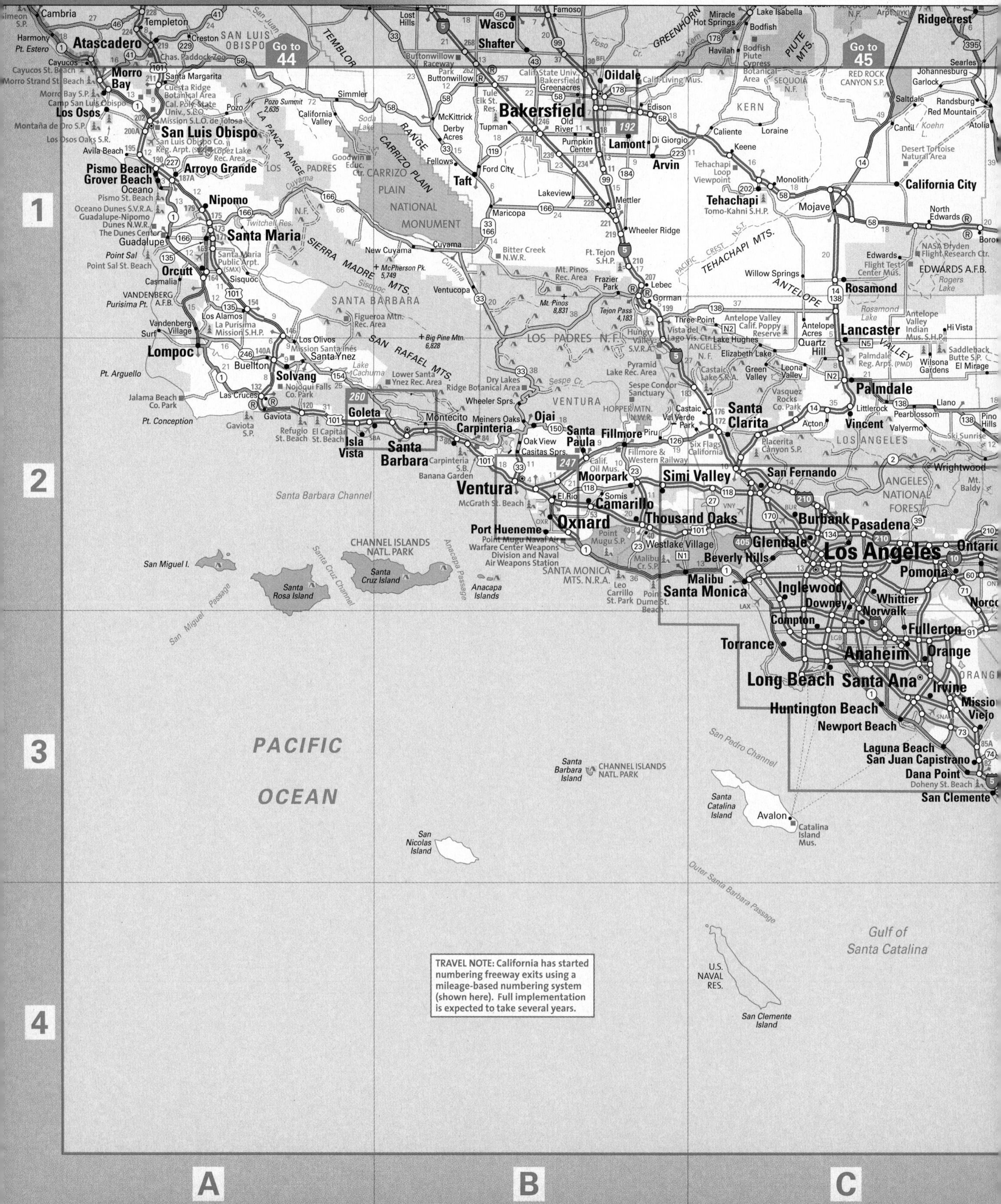

DRIVING DISTANCES IN MILES	BAKERSFIELD, CA	BARSTOW, CA	BLYTHE, CA	EL CENTRO, CA	LOS ANGELES, CA	NEEDLES, CA	PALM SPRINGS, CA	SAN BERNARDINO, CA	SAN DIEGO, CA	SAN LUIS OBISPO, CA	SANTA BARBARA, CA	YUMA, AZ
LOS ANGELES, CA	111	118	230	234		263	110	62	124	190	97	294
SAN DIEGO, CA	234	181	211	117	124	326	143	111		314	221	177
SANTA BARBARA, CA	150	213	325	330	97	358	205	157	221	93		391
YUMA, AZ	403	294	103	65	294	187	171	225	177	483	391	

SEE ALSO DISTANCE AND DRIVING TIME MAP ON PAGES 286–287

California
Arizona
New Mexico
Mexico

0 mi 20 40
0 km 20 40 60
One inch equals 25.4 miles
One centimeter equals 16.1 kilometers

Phoenix AZ / Yuma AZ

Go to 46

Vidal Jct.
Parker Strip B.L.M. Rec. Area
River N.W.R.
Alamo Dam
Tres Alamos 4,293
Peeples Valley
Towers Mtn. 7,628
AGUA FRIA NATL. MON.
MAZATZAL MOUNTAINS
SAN BERNARDINO
Earp
Buckskin Mountain S.P.
BUCKSKIN MTS.
Bill Williams
Alamo Lake S.P.
Yarnell
Wagoner
Crown King
Bumble Bee
North Peak 7,449
Rice
Vidal
Big River
Parker
Swansea Ghost Town
Congress
Stanton Ghost Town
Octave
Black Canyon City
TONTO NATL. FOR.
Mazatzal Peak 7,903
Colorado River Indian Tribes Mus.
CACTUS PLAIN
PACIFIC TIME ZONE
MOUNTAIN TIME ZONE
Poston
Robson's Mining World
Desert Caballeros Western Mus.
YAVAPAI
Constellation
Rock Springs
Castle Hot Springs
Horseshoe Res.
Horseshoe Dam
Bouse Wash
Bouse
Forepaugh
Aquila
Wickenburg
Hassayampa River Preserve
HARCUVAR MTS.
New River
RIVERSIDE
LA PAZ
Gladden
COLORADO RIVER IND. RES.
Midland
MC COY
Blythe Intaglios
Wenden
Vulture Mine
Morristown
Lake Pleasant
Camp Creek
Bartlett Res.
Cave Creek
HARQUAHALA MTS.
Harquahala Mtn. 5,681
Salome
Harcuvar
Historic Harquahala Observatory
Hi Jolly Monument
Vicksburg
Hope
Quartzsite
Brenda
Wittmann
Pioneer Arizona Living Hist. Museum
Carefree
VULTURE MTS.
Sugarloaf Mtn. 3,418
Rio Verde
Ft. McDowell
FORT MCDOWELL IND. RES.
Fountain Hills
Saguaro Lake
Blythe
Ehrenberg
DOME ROCK MTS.
La Posa Long-Term Visitor Area
Black Mesa 3,639
HARQUAHALA PLAIN
BIG HORN MTS.
Surprise
Sun City
Luke A.F.B.
Peoria
Tortilla Flat
Go to 53
Ripley
Colorado
Tonopah
MARICOPA
Glendale
Scottsdale
Lost Dutchman S.P.
Palo Verde
Palm Canyon
KOFA N.W.R.
EAGLETAIL MTS.
Centennial Wash
Wintersburg
Goodyear
Avondale
Phoenix
Mesa
Apache Junction
Cibola
YUMA PROVING GROUND
Polaris Mtn. 3,624
Buckeye
Liberty
Tempe
Gilbert
Chandler
CIBOLA N.W.R.
Red Raven Wash
Arlington
Palo Verde
Komatke
Sun Lakes
Queen Creek
CALIFORNIA
ARIZONA
TRIGO MTS.
Nortbusch Wash
GILA BEND MTS.
TANK MTS.
Rainbow Valley
249
Chandler Heights
IMPERIAL N.W.R.
Castle Dome Pk. 3,788
CASTLE DOME MTS.
Hoodoo Wash
YUMA
Painted Rocks Petroglyph Site
MARICOPA MTS.
GILA RIVER IND. RES.
Bapchule
Olberg
IMPERIAL
SONORAN
Casa Blanca
Maricopa
Sacaton
Florence
Picacho S.R.A.
TOHONO O'ODHAM (GILA BEND) IND. RES.
Mobile
Martinez Lake
Hyder
CITRUS VALLEY
Valley Farms
Agua Caliente
Ak-Chin
MARICOPA (AK-CHIN) IND. RES.
Imperial Dam Long Term Visitor Area
YUMA PROVING GROUND
HYDER VALLEY
Horn
SENTINEL PLAIN
Gila Bend
Paloma
Bosque
DESERT
Casa Grande
Casa Grande Ruins N.M.
Coolidge
Randolph
La Palma
Ogilby
FORT YUMA IND. RES.
Winterhaven
Bard
Betty's Kitchen Interpretive Site
Dome
Gila
Dateland
Aztec
Sentinel
Stanfield
Arizona
Toltec
Picacho
Eloy
Andrade
Yuma
Blaisdell
Ligurta
Dome Valley Mus.
Wellton
Roll
Tacna
Mohawk
NATL. MON.
SAND TANK MTS.
Table Top Mtn. 4,373
Chuichu
Cucklebur
Arizona City
Friendly Corners
Picacho
Cocopah Mus.
276
The Peanut Patch
BARRY M. GOLDWATER AIR FORCE RANGE
SAUCEDA MTS.
Somerton
Cocopah Ind. Res.
Tenmile Wash
Jackrabbit
Kohatk
Picacho Peak S.P.
San Luis
YUMA DESERT
GILA MTS.
LECHUGUILLA DESERT
Coyote Pk. 2,808
MOHAWK MTS.
MOHAWK VALLEY
San Cristobal Wash
GRANITE MTS.
Growler Wash
GROWLER MTS.
Kaka
Ventana
SANTA ROSA VALLEY
North Komelik
IRONWOOD FOREST NATL. MON.
San Luis Rio Colorado
Lagunitas
CABEZA PRIETA MTS.
TULE DESERT
Childs
Ajo
Hickiwan
Vaya Chin
Anegam
Santa Rosa
AGUIRRE VALLEY
CABEZA PRIETA NATL. WILDLIFE REFUGE
Why
Charco
TOHONO O'ODHAM INDIAN RESERVATION
Ak Chin
Queens Well
ARIZONA
SONORA
Gunsight
Sil Nakya
ORGAN PIPE CACTUS NATL. MON.
Quijotoa
Covered Wells
Schuchk
San Luis
San Pedro
DESIERTO DE ALTAR
AGUA DULCE MTS.
AJO RANGE
Gu Vo
Kaihon Kug
Pisinimo
Comobabi
Haivana Nakya
Pan Tak
Los Vidrios
Pia Oik
Vamori Wash
PIMA
Kitt Peak Natl. Observatory
Gu Oidak
Sells
Little Tucson
Sonoyta
Lukeville
Ali Chuk
Cowlic
Topawa
Fresnal Canyon
Baboquivari Pk. 7,736
BABOQUIVARI MTS.
Golfo de Santa Clara
Río Sonoyta
UNITED STATES
MEXICO
Vamori
South Komelik
Choulic
San Miguel
BUENOS AIRES N.W.R.
SIERRA LA ESPUMA
Sasabe
Sásabe
Puerto Peñasco
Gulf of California
Bahía San Jorge
MOUNTAIN TIME ZONE
PACIFIC TIME ZONE
El Plomo
SIERRA EL HUMO
Río El Sásabe
Los Molinos
San Felipe
El Arenoso
Campodóni
Las Enchilayas
SIERRA LA C...

1 2 3 4
A B C

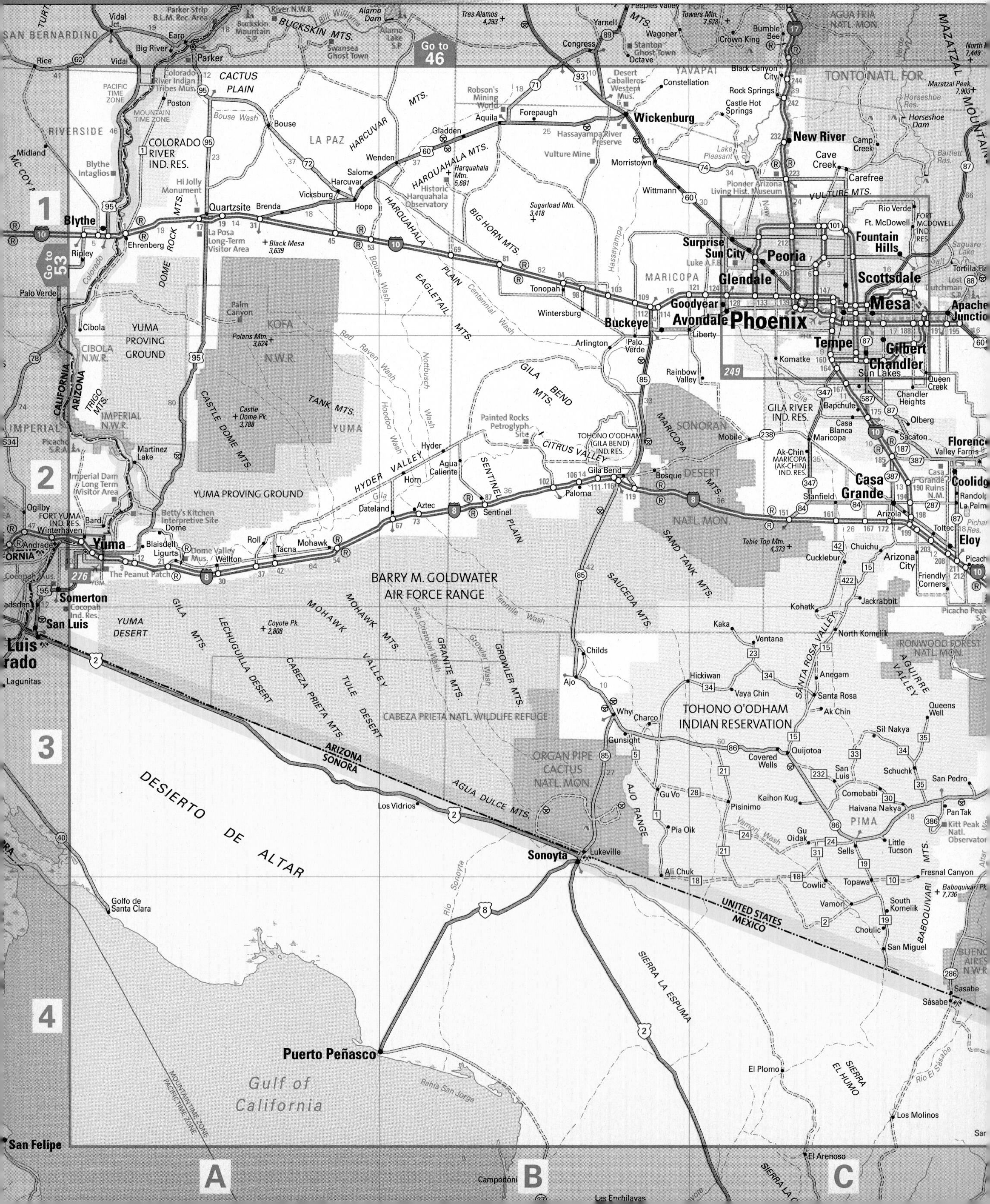

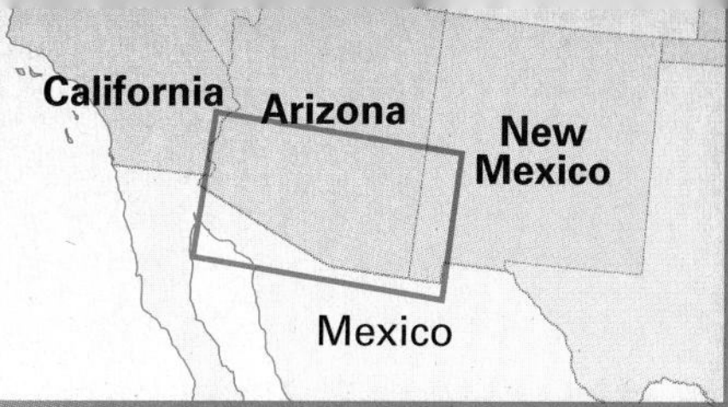

DRIVING DISTANCES IN MILES

	BLYTHE, CA	CASA GRANDE, AZ	DOUGLAS, AZ	EAGAR, AZ	GLOBE, AZ	LORDSBURG, NM	NOGALES, AZ	PHOENIX, AZ	SAFFORD, AZ	SILVER CITY, NM	TUCSON, AZ	YUMA, AZ
LORDSBURG, NM	417	228	101	184	155		185	278	77	45	161	401
PHOENIX, AZ	140	50	237	227	92	278	181		169	322	118	183
TUCSON, AZ	258	68	120	242	106	161	65	118	128	205		241
YUMA, AZ	103	179	360	401	265	401	304	183	368	446	241	

SEE ALSO DISTANCE AND DRIVING TIME MAP ON PAGES 286–287

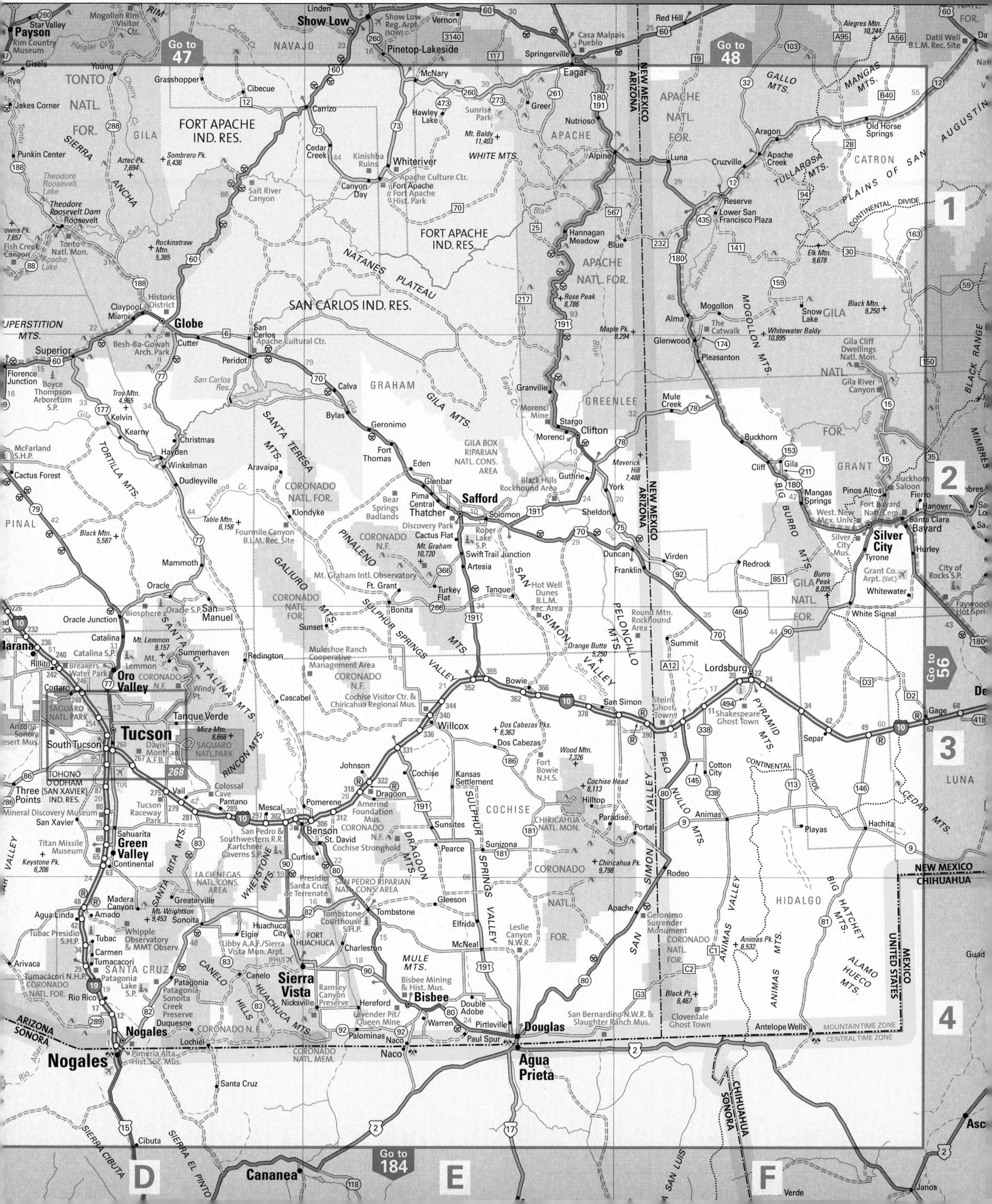

El Paso TX / Las Cruces NM

0 mi 20 40
0 km 20 40 60
One inch equals 25.4 miles
One centimeter equals 16.1 kilometers

DRIVING DISTANCES IN MILES	ALAMOGORDO, NM	CARLSBAD, NM	EL PASO, TX	HOBBS, NM	LAS CRUCES, NM	LORDSBURG, NM	ODESSA, TX	PECOS, TX	PORTALES, NM	ROSWELL, NM	SILVER CITY, NM	SOCORRO, NM
CARLSBAD, NM	144		162	70	203	321	137	87	168	76	311	241
EL PASO, TX	86	162		232	42	160	285	209	295	203	150	190
LAS CRUCES, NM	65	203	42	250		122	325	250	274	182	111	146
ROSWELL, NM	117	76	203	117	182	304	201	163	92		293	164

SEE ALSO DISTANCE AND DRIVING TIME MAP ON PAGES 286–287

0 mi 20 40
0 km 20 40 60
One inch equals 25.4 miles
One centimeter equals 16.1 kilometers

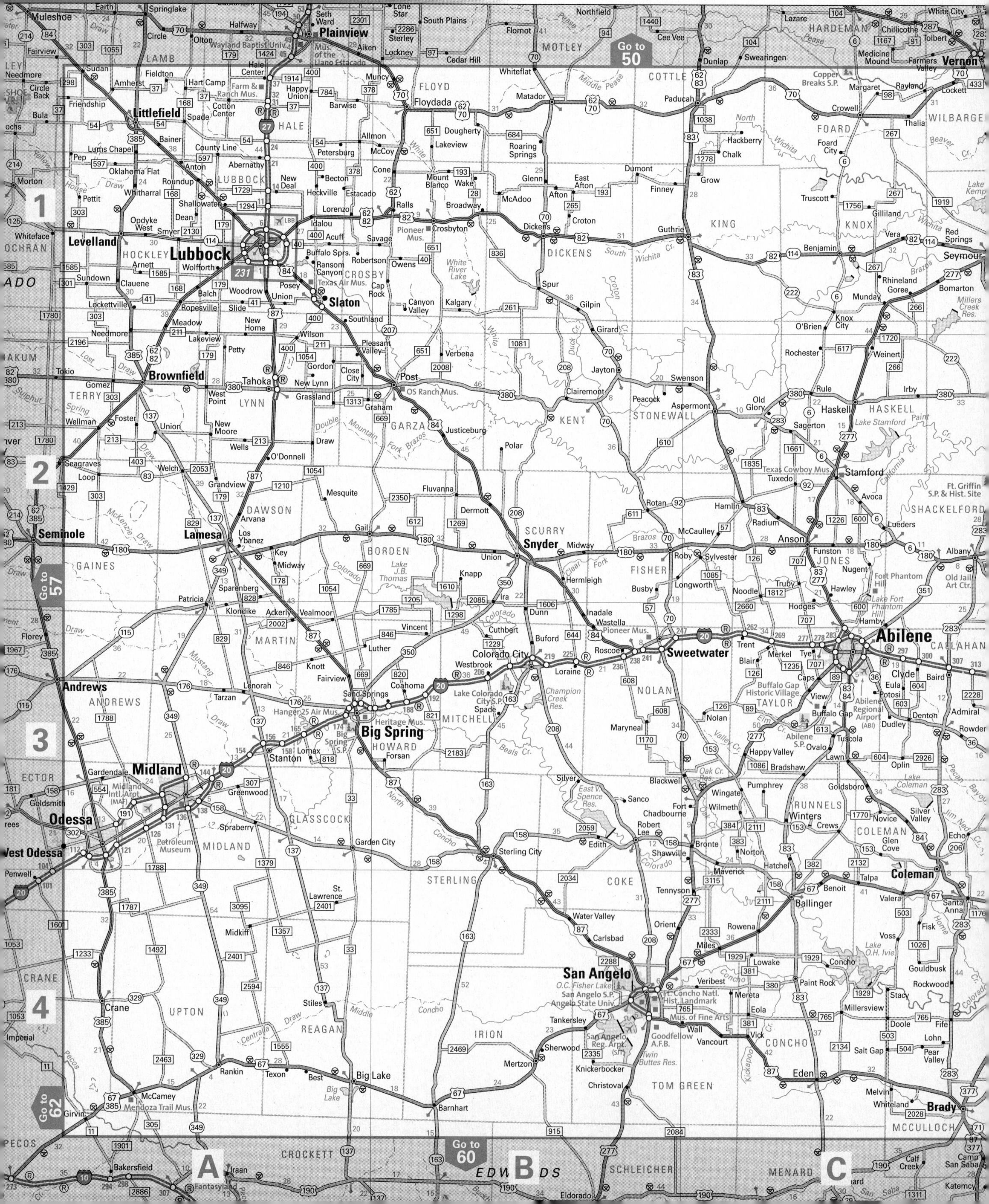

DRIVING DISTANCES IN MILES	ABILENE, TX	BIG SPRING, TX	BROWNWOOD, TX	DALLAS, TX	FORT WORTH, TX	LUBBOCK, TX	ODESSA, TX	SAN ANGELO, TX	SHERMAN, TX	TEMPLE, TX	WACO, TX	WICHITA FALLS, TX
ABILENE, TX		110	78	191	153	166	176	91	249	194	235	144
DALLAS, TX	191	298	190		32	354	364	265	64	130	94	141
LUBBOCK, TX	166	106	247	354	317		142	185	322	358	399	207
WACO, TX	235	343	124	94	87	399	409	219	159	40		201

SEE ALSO DISTANCE AND DRIVING TIME MAP ON PAGES 286–287

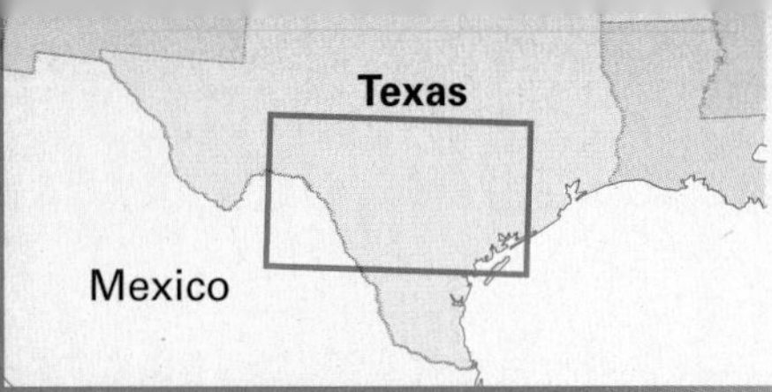

0 mi 20 40
0 km 20 40 60
One inch equals 25.4 miles
One centimeter equals 16.1 kilometers

Del Rio TX / Eagle Pass TX

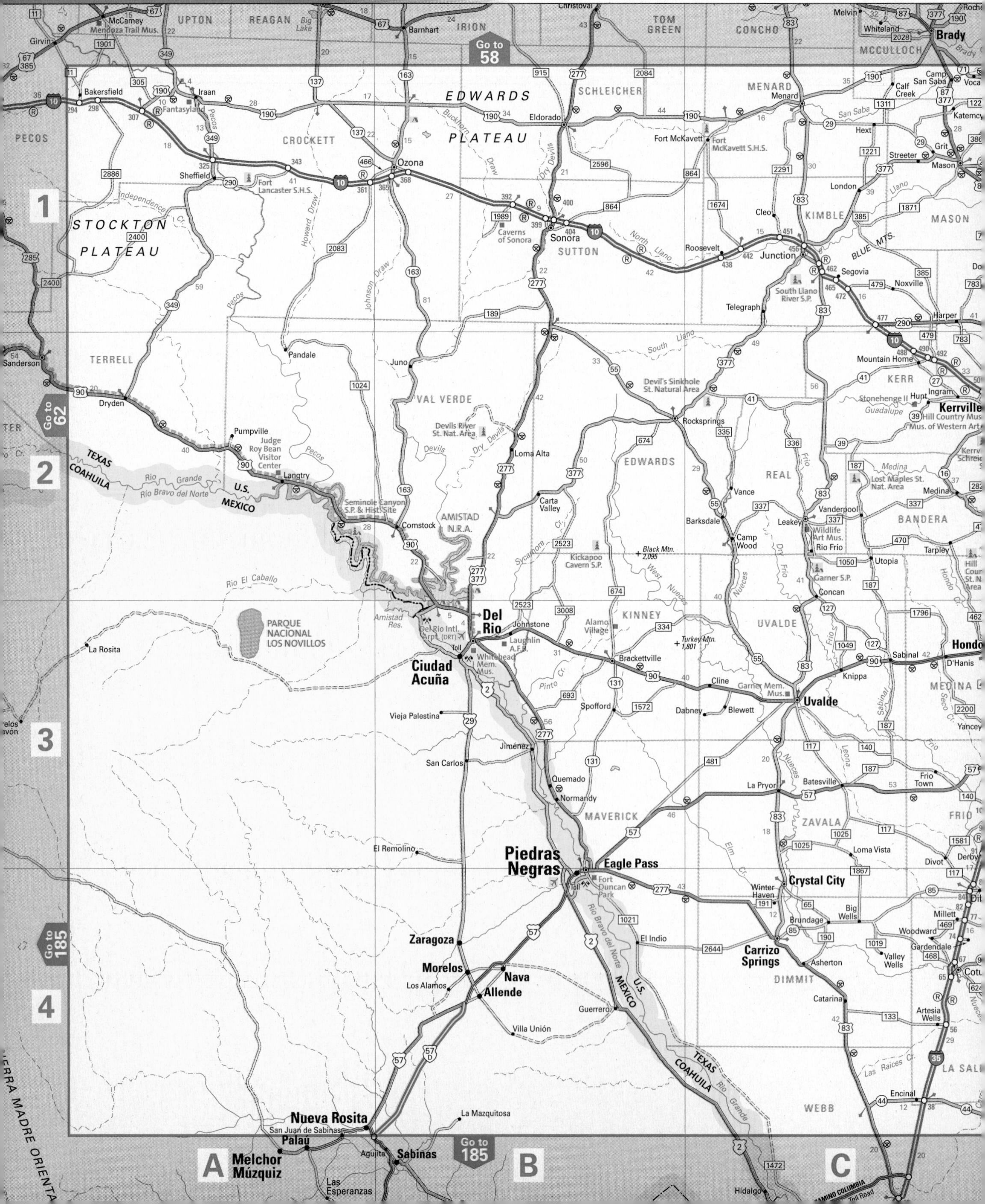

DRIVING DISTANCES IN MILES

	AUSTIN, TX	BEEVILLE, TX	COLLEGE STATION, TX	COLUMBUS, TX	DEL RIO, TX	EAGLE PASS, TX	FREDERICKSBURG, TX	SAN ANTONIO, TX	SONORA, TX	TEMPLE, TX	UVALDE, TX	VICTORIA, TX
AUSTIN, TX		136	108	92	229	226	78	78	244	67	159	123
DEL RIO, TX	229	235	322	277		55	178	152	89	295	70	268
SAN ANTONIO, TX	78	110	171	128	152	145	67		172	144	82	118
VICTORIA, TX	123	56	160	87	268	254	186	118	292	187	198	

SEE ALSO DISTANCE AND DRIVING TIME MAP ON PAGES 286–287

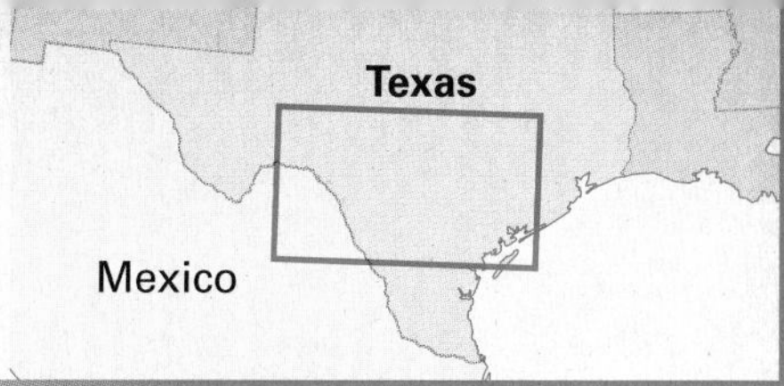

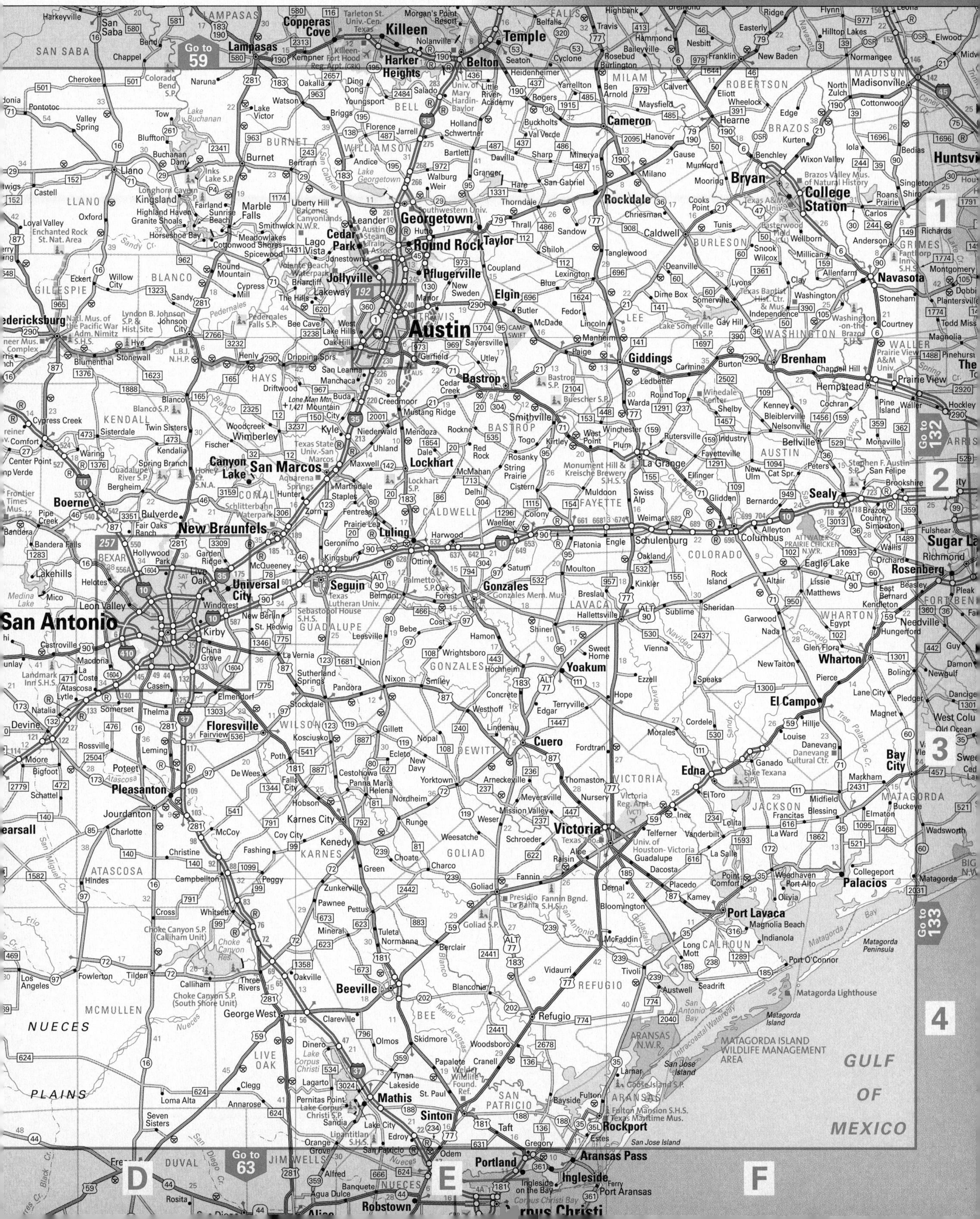

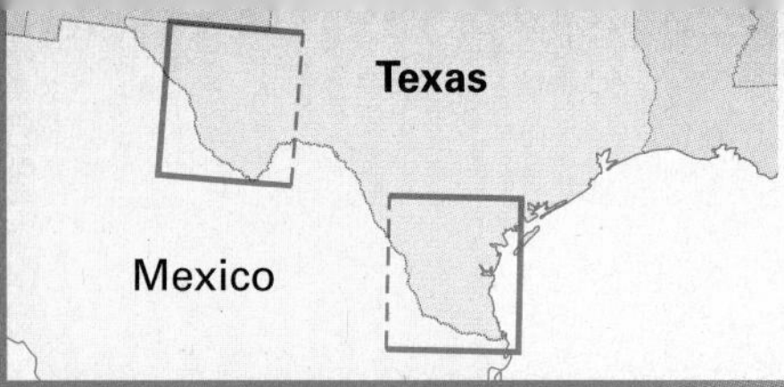

DRIVING DISTANCES IN MILES

	ALPINE, TX	BIG BEND NP, TX	FORT STOCKTON, TX	ODESSA, TX	PECOS, TX	VAN HORN, TX
ALPINE, TX		97	65	151	96	110
FORT STOCKTON, TX	65	123		86	58	119
ODESSA, TX	151	209	86		76	163
VAN HORN, TX	110	207	119	163	87	

SEE ALSO DISTANCE AND DRIVING TIME MAP ON PAGES 286–287

0 mi 10 20 30
0 km 20 40

One inch equals 25.4 miles
One centimeter equals 16.1 kilometers

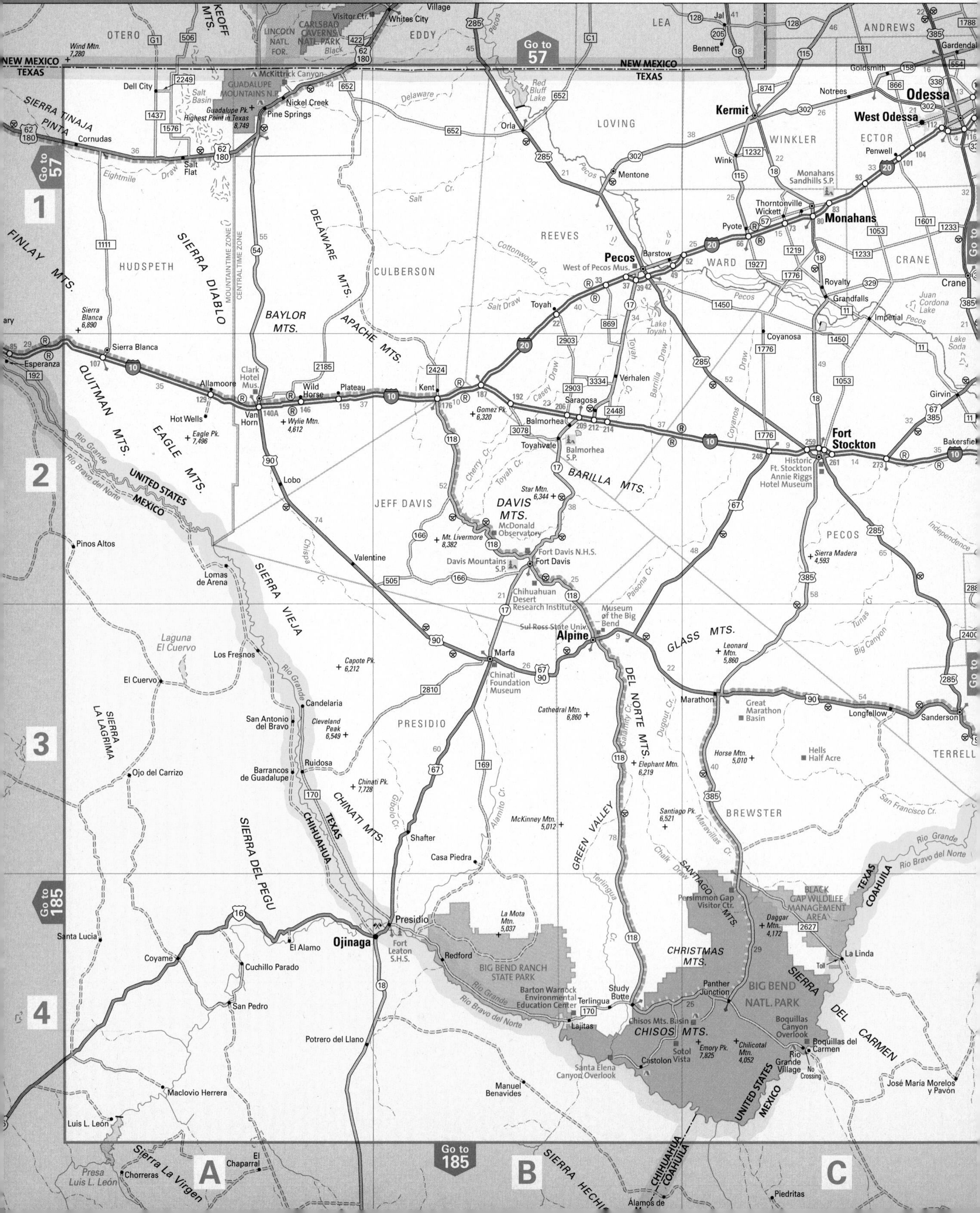

DRIVING DISTANCES IN MILES	BEEVILLE, TX	BROWNSVILLE, TX	CARRIZO SPRS., TX	CORPUS CHRISTI, TX	HARLINGEN, TX	KINGSVILLE, TX	LAREDO, TX	MCALLEN, TX	VICTORIA, TX
BROWNSVILLE, TX	192		282	157	27	119	202	61	226
CORPUS CHRISTI, TX	59	157	199		131	38	141	152	94
LAREDO, TX	130	202	79	141	176	124		144	186
MCALLEN, TX	168	61	223	152	35	114	144		221

SEE ALSO DISTANCE AND DRIVING TIME MAP ON PAGES 286–287

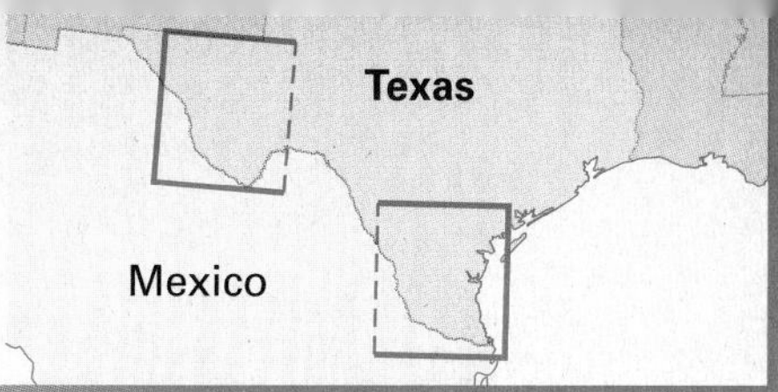

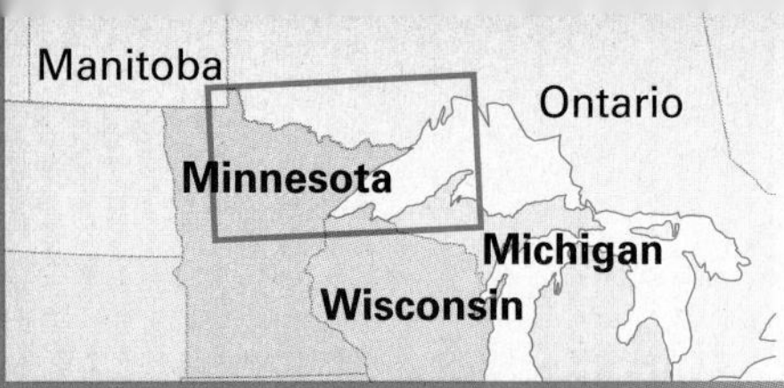

0 mi 20 40
0 km 20 40 60
One inch equals 25.4 miles
One centimeter equals 16.1 kilometers

DRIVING DISTANCES IN MILES	ASHLAND, WI	BEMIDJI, MN	BRAINERD, MN	DETROIT LAKES, MN	DULUTH, MN	GRAND PORTAGE, MN	HOUGHTON, MI	INTERNAT'L FALLS, MN	IRONWOOD, MI	ISHPEMING, MI	THUNDER BAY, ON	VIRGINIA, MN
BEMIDJI, MN	239		96	91	153	295	362	109	254	384	314	124
DULUTH, MN	92	153	116	202		143	215	157	107	238	183	61
HOUGHTON, MI	132	362	325	412	215	358		370	108	87	654	274
INTERNAT'L FALLS, MN	247	109	190	200	157	245	370		262	393	205	97

SEE ALSO DISTANCE AND DRIVING TIME MAP ON PAGES 286–287

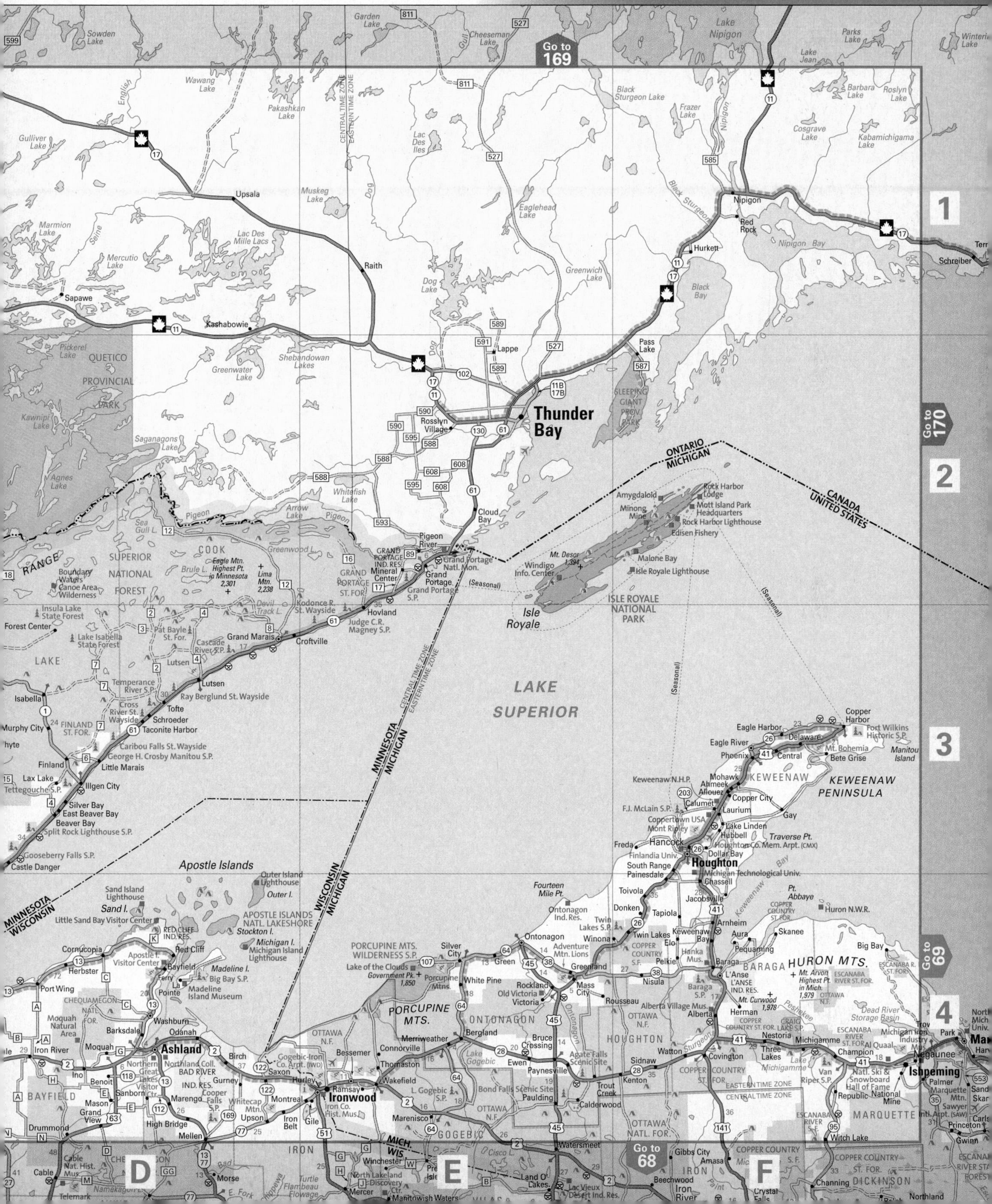

0 mi 10 20 30 40
0 km 10 20 30 40 50 60

One inch equals 18.4 miles
One centimeter equals 11.7 kilometers

Go to 19 Go to 64

Detroit Lakes · Frazee · Evergreen · Menahga · Hubbard · Huntersville · Backus · Outing · LAND O' LAKES ST. FOR.

Cormorant · Vergas · Perham · New York Mills · WADENA · Sebeka · Nimrod · FOOTHILLS ST. FOR. · Pine River · Chickamaw Beach · Manhattan Beach · Fifty Lakes · Emily

Pelican Rapids · Dent · Richville · Bluffton · Wadena · Leader · Nisswa · Crosslake · Pequot Lakes · Breezy Point · CROW WING ST. FOR.

OTTER TAIL · Ottertail · Deer Creek · Verndale · Staples · E. Gull Lake · Baxter · Brainerd · Merrifield · Deerwood · Cuyuna · Crosby

Fergus Falls · Henning · Hewitt · Motley · Aldrich · Philbrook · CAMP RIPLEY MIL. RES. · Fort Ripley · CROW WING

Dalton · Vining · Almora · Bertha · Clarissa · Lincoln · Cushing · Randall · Pine Center · Vineland · MILLE LACS LAKE

Ashby · Millerville · Leaf Valley · Rose City · Miltona · Clotho · Browerville · Long Prairie · Flensburg · Little Falls · Genola · MORRISON · Harding

TRAVERSE · Evansville · Brandon · Garfield · DOUGLAS · Carlos · Gutches Grove · Round Prairie · Sobieski · Buckman · RUM RIVER ST. FOR.

GRANT · Hoffman · Alexandria · Nelson · Osakis · Grey Eagle · Swanville · Elmdale · Royalton · Little Rock · Morrill · Ramey · Milaca

Herman · Kensington · Holmes City · Forada · West Union · Burtrum · Upsala · Bowlus · Gilman · Foreston · BENTON

Donnelly · Farwell · Lowry · Villard · Westport · Sauk Centre · St. Rosa · Holdingford · Rice · Ronneby · Oak Park · Pease

Morris · Cyrus · Starbuck · Glenwood · Long Beach · Padua · Melrose · Freeport · Albany · Avon · Collegeville · St. Stephen · Watab · Foley · Estes Brook

STEVENS · Hancock · POPE · Grove Lake · Sedan · Greenwald · New Munich · St. Wendel · Sartell · Sauk Rapids · St. Cloud · Duelm · Princeton

Alberta · Chokio · Terrace · Brooten · Spring Hill · St. Martin · Farming · Richmond · Cold Spring · St. Joseph · Waite Park · St. Augusta · Clearwater · Becker · SHERBURNE N.W.R.

BIG STONE · Clontarf · Benson · Belgrade · Georgeville · Roscoe · Paynesville · Eden Valley · Marty · Rockville · Clear Lake · Santiago · Zimmerman

Danvers · De Graff · Sunburg · Swift Falls · Regal · Hawick · Watkins · Kimball · South Haven · Silver Creek · Big Lake · Monticello

Appleton · Murdock · Kerkhoven · New London · Spicer · Mannannah · Annandale · Maple Lake · Albertville · St. Michael · Otsego

Correll · Louisburg · Hagen · Big Bend City · Pennock · Kandiyohi · Atwater · Grove City · WRIGHT · Kingston · Buffalo · Hanover · Rogers

Milan · Watson · Willmar · Litchfield · Darwin · Dassel · Cokato · Montrose · Delano · Maple Plain · HENNEPIN

Montevideo · CHIPPEWA · Raymond · Roseland · Svea · Lake Lillian · Cosmos · Cedar Mills · Stockholm · Howard Lake · Waverly · Independence · Loretto

Maynard · Prinsburg · Blomkest · Corvuso · Silver Lake · New Germany · Mayer · Watertown · Mound · Orono

Wegdahl · Renville · Danube · Olivia · Bird Island · Hector · Brownton · New Auburn · Glencoe · Norwood Young America · Waconia · Chanhassen · Shakopee · Chaska

LAC QUI PARLE · Granite Falls · Sacred Heart · RENVILLE · Buffalo Lake · Stewart · Plato · Cologne · CARVER · MINN. VALLEY N.W.R.

Clarkfield · Hazel Run · Wood Lake · Echo · Belview · Delhi · Franklin · Fairfax · Gibbon · Winthrop · Arlington · Green Isle · Belle Plaine · Jordan

YELLOW MEDICINE · Cottonwood · Redwood Falls · Morton · LOWER SIOUX IND. RES. · SCOTT

Marshall · Milroy · Lucan · Wabasso · Clements · Morgan · Sleepy Eye · New Ulm · St. George · Le Sueur · NICOLLET

Go to 19 · Go to 27 · Go to 72

SOUTH DAKOTA / MINNESOTA

1 · 2 · 3 · 4 · A · B · C

DRIVING DISTANCES IN MILES	ASHLAND, WI	BRAINERD, MN	DULUTH, MN	EAU CLAIRE, WI	FERGUS FALLS, MN	MARSHALL, MN	MINNEAPOLIS, MN	MORRIS, MN	RICE LAKE, WI	ST. CLOUD, MN	ST. PAUL, MN	WILLMAR, MN
EAU CLAIRE, WI	167	220	155		267	236	93	247	57	156	83	193
MINNEAPOLIS, MN	196	129	158	93	176	148		156	103	64	10	92
ST. CLOUD, MN	205	62	149	156	117	131	64	98	155		73	63
WILLMAR, MN	263	112	206	193	113	68	92	57	196	63	102	

SEE ALSO DISTANCE AND DRIVING TIME MAP ON PAGES 286–287

Wisconsin

Michigan

Green Bay WI / Wausau WI

0 mi 10 20 30 40

0 km 10 20 30 40 50 60

One inch equals 18.4 miles

One centimeter equals 11.7 kilometers

Go to 65

Go to 67

Go to 74

LAKE SUPERIOR

PORCUPINE MTS.

PORCUPINE MTS. WILDERNESS S.P.

Lake of the Clouds

Government Pk. + 1,850

ONTONAGON

HOUGHTON

BARAGA

HURON MTS.

Mt. Arvon Highest Pt. in Mich. 1,979 +

+ Mt. Curwood 1,978

COPPER COUNTRY S.F.

OTTAWA N.F.

OTTAWA NATL. FOR.

GOGEBIC

IRON

MICHIGAN WISCONSIN

EASTERN TIME ZONE

CENTRAL TIME ZONE

Ashland

Ironwood

Iron Co. Hist. Mus.

Hurley

Bessemer

Wakefield

Bergland

Merriweather

Ewen

Bruce Crossing

Paynesville

Watersmeet

Land O' Lakes

Lac Vieux Desert Ind. Res.

Iron River

Crystal Falls

Iron Mountain

Kingsford

Norway

Amasa

Covington

L'Anse

Baraga

Alberta

Herman

Nestoria

Three Lakes

Michigamme

Republic

BAD RIVER IND. RES.

Mellen

Morse

Glidden

Clam Lake

ASHLAND

Butternut

Park Falls

Fifield

CHEQUAMEGON NATL. FOR.

Phillips

PRICE

Prentice

Brantwood

Tripoli

Minocqua

Woodruff

Arbor Vitae

St. Germain

Eagle River

Conover

Phelps

Land O' Lakes

Star Lake

Sayner

VILAS

Lac du Flambeau

LAC DU FLAMBEAU IND. RES.

NORTHERN HIGHLAND AMERICAN LEGION ST. FOR.

Manitowish Waters

Mercer

Presque Isle

Winchester

Rhinelander

ONEIDA

Three Lakes

Tomahawk

Crandon

FOREST

NICOLET NATL. FOR.

FLORENCE

Florence

Laona

FOREST CO. POTAWATOMI IND. COMM.

Wabeno

MARINETTE

Athelstane

Wausaukee

Timms Hill Highest Pt. in Wisconsin 1,951

Ogema

Spirit

LINCOLN

Merrill

Antigo

LANGLADE

Elton

Mountain

Medford

TAYLOR

Gilman

Rib Lake

Athens

Abbotsford

Colby

Wausau

Schofield

Weston

Rothschild

Rib Mtn. S.P.

Rib Mtn. 1,924

Granite Peak at Rib Mountain

MARATHON

Mosinee

Marshfield

Spencer

Stratford

Neillsville

CLARK

Greenwood

Loyal

Wittenberg

Birnamwood

MENOMINEE

MENOMINEE IND. RES.

STOCKBRIDGE-MUNSEE IND. RES.

Keshena

Shawano

SHAWANO

Gillett

Oconto Falls

OCONTO

Pulaski

Suamico

Green Bay

Howard

Ashwaubenon

De Pere

ONEIDA IND. RES.

Seymour

Clintonville

Marion

WAUPACA

Iola

Stevens Point

Plover

Amherst

Wisconsin Rapids

PORTAGE

WOOD

JACKSON

Ho-Chunk Ind. Res.

Waupaca

New London

Shiocton

OUTAGAMIE

Manawa

Scandinavia

Central Wisconsin Arpt. (CWA)

Univ. of Wis.-Stevens Point

1 2 3 4

A B C

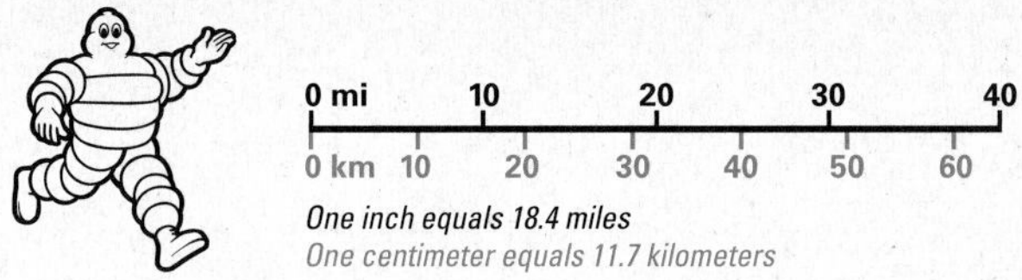

DRIVING DISTANCES IN MILES

	ESCANABA, MI	GREEN BAY, WI	IRON MOUNTAIN, MI	IRONWOOD, MI	L'ANSE, MI	MANISTIQUE, MI	MARINETTE, WI	MARQUETTE, MI	RHINELANDER, WI	STEVENS POINT, WI	TRAVERSE CITY, MI	WAUSAU, WI
ESCANABA, MI		111	52	178	134	54	57	65	132	185	252	171
GREEN BAY, WI	111		96	202	178	165	54	175	124	87	363	93
MARQUETTE, MI	65	175	79	145	70	86	122		147	238	269	204
WAUSAU, WI	171	93	133	121	176	225	112	204	58	35	423	

SEE ALSO DISTANCE AND DRIVING TIME MAP ON PAGES 286–287

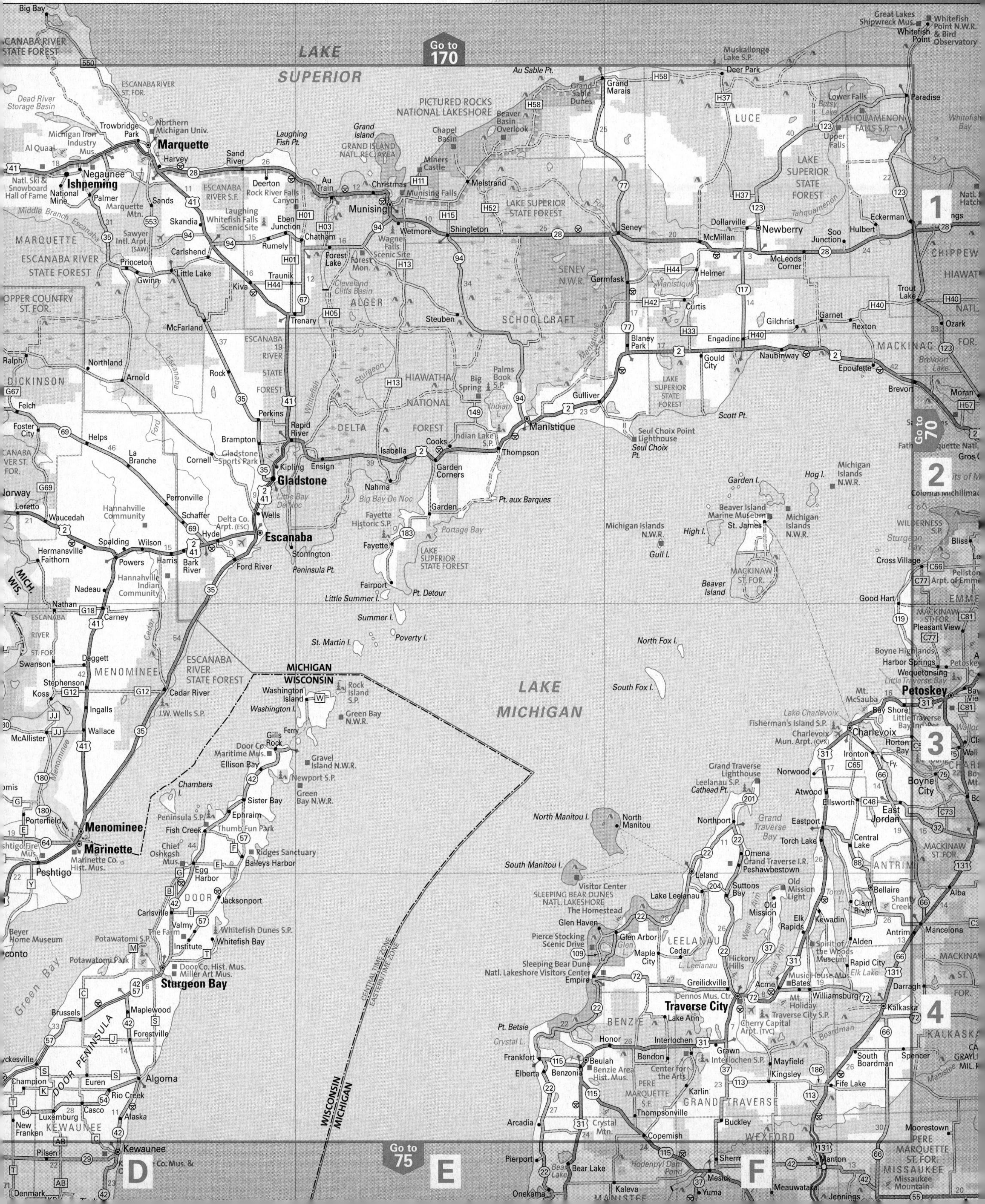

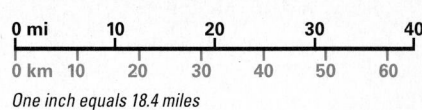

0 mi 10 20 30 40
0 km 10 20 30 40 50 60

One inch equals 18.4 miles
One centimeter equals 11.7 kilometers

Sault Ste Marie MI / Traverse City MI

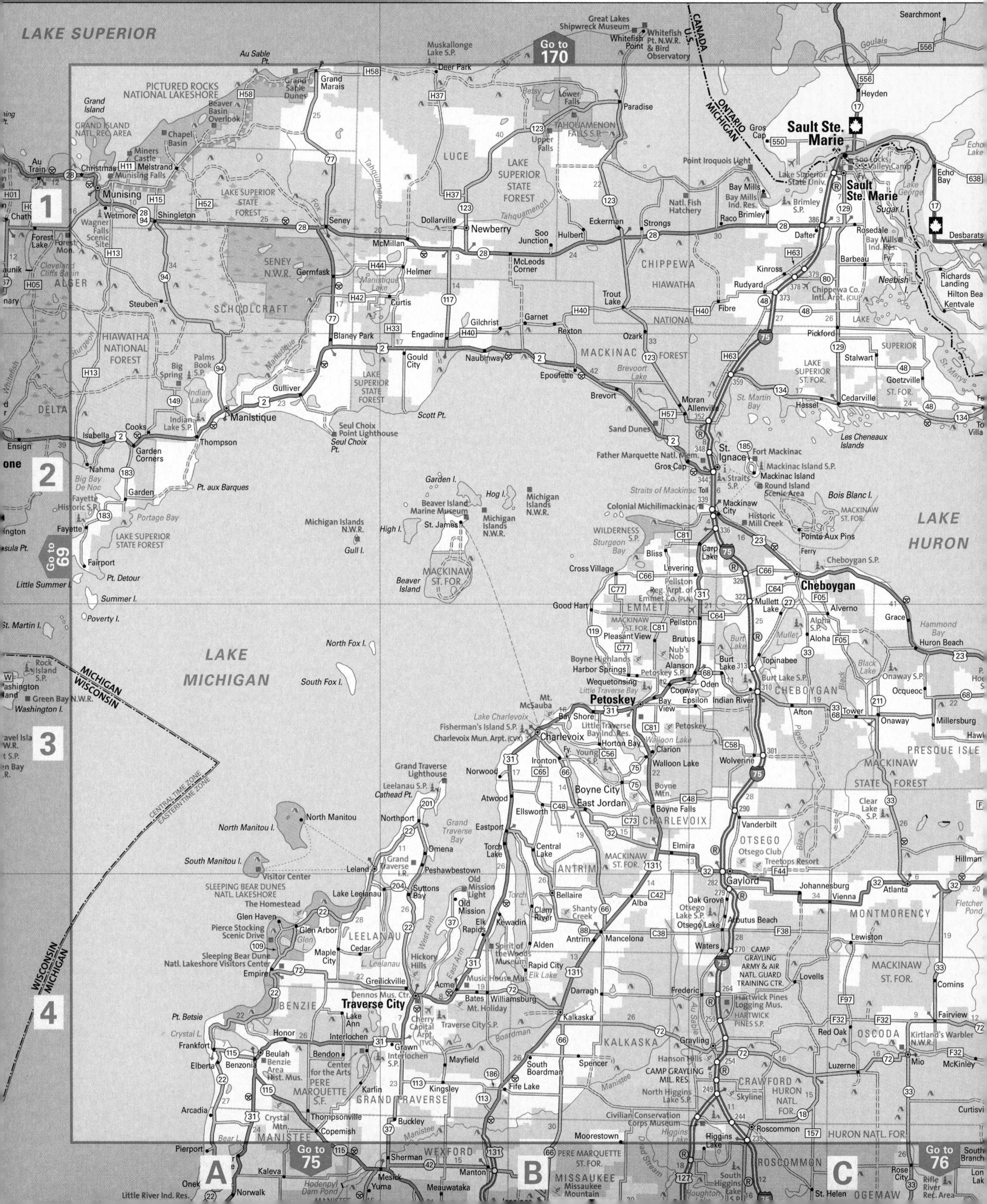

DRIVING DISTANCES IN MILES

	ALPENA, MI	CHEBOYGAN, MI	GAYLORD, MI	GRAYLING, MI	MACKINAW CITY, MI	MANISTIQUE, MI	MUNISING, MI	PETOSKEY, MI	ROGERS CITY, MI	SAULT STE. MARIE, MI	SUDBURY, ON	TRAVERSE CITY, MI
ALPENA, MI		78	76	95	94	187	215	101	38	148	334	141
MACKINAW CITY, MI	94	16	60	87		95	123	38	58	57	242	106
SAULT STE. MARIE, MI	148	71	114	142	57	120	120	93	112		186	160
TRAVERSE CITY, MI	141	115	65	52	106	198	226	67	135	160	346	

SEE ALSO DISTANCE AND DRIVING TIME MAP ON PAGES 286–287

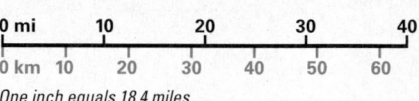

0 mi 10 20 30 40
0 km 10 20 30 40 50 60
One inch equals 18.4 miles
One centimeter equals 11.7 kilometers

Mankato MN / Fort Dodge IA

DRIVING DISTANCES IN MILES	ALBERT LEA, MN	DECORAH, IA	DUBUQUE, IA	FORT DODGE, IA	LA CROSSE, WI	MANKATO, MN	MASON CITY, IA	ROCHESTER, MN	SPENCER, IA	WATERLOO, IA	WINONA, MN	WORTHINGTON, MN
FORT DODGE, IA	124	186	200		245	138	97	183	95	108	225	148
MANKATO, MN	56	151	253	138	149		100	80	123	186	128	108
ROCHESTER, MN	62	68	170	183	71	80	103		189	116	51	174
WATERLOO, IA	130	79	93	108	138	186	79	116	189		144	244

SEE ALSO DISTANCE AND DRIVING TIME MAP ON PAGES 286–287

0 mi 10 20 30 40
0 km 10 20 30 40 50 60
One inch equals 18.4 miles
One centimeter equals 11.7 kilometers

Milwaukee WI / Madison WI

DRIVING DISTANCES IN MILES

	CADILLAC, MI	DUBUQUE, IA	GRAND RAPIDS, MI	GREEN BAY, WI	KALAMAZOO, MI	MADISON, WI	MILWAUKEE, WI	MUSKEGON, MI	OSHKOSH, WI	ROCKFORD, IL	SHEBOYGAN, WI	TOMAH, WI
GRAND RAPIDS, MI	99	364		393	53	335	277	40	363	271	332	424
GREEN BAY, WI	492	229	393		362	135	115	400	50	211	61	162
MADISON, WI	434	93	335	135	304		78	341	86	78	132	98
MILWAUKEE, WI	377	167	277	115	247	78		285	87	95	54	168

SEE ALSO DISTANCE AND DRIVING TIME MAP ON PAGES 286–287

0 mi 10 20 30 40
0 km 10 20 30 40 50 60
One inch equals 18.4 miles
One centimeter equals 11.7 kilometers

Detroit MI / Lansing MI

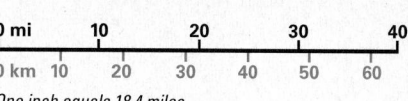

DRIVING DISTANCES IN MILES	ANN ARBOR, MI	BAD AXE, MI	BATTLE CREEK, MI	CADILLAC, MI	DETROIT, MI	FLINT, MI	HAMILTON, ON	LANSING, MI	LONDON, ON	MT. PLEASANT, MI	PORT HURON, MI	SAGINAW, MI
DETROIT, MI	42	107	116	209		62	203	86	128	149	58	97
LANSING, MI	63	140	56	131	86	53	270		191	67	117	86
PORT HURON, MI	101	81	175	211	58	64	154	117	75	155		100
SAGINAW, MI	87	64	142	116	97	36	253	86	174	60	100	

SEE ALSO DISTANCE AND DRIVING TIME MAP ON PAGES 286–287

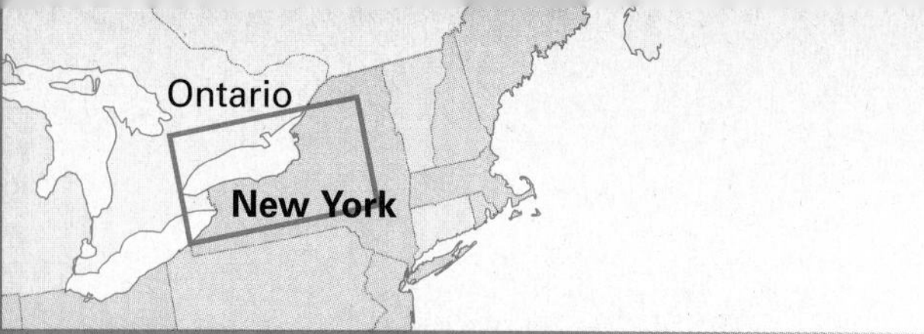

0 mi 10 20 30 40
0 km 10 20 30 40 50 60

One inch equals 18.4 miles
One centimeter equals 11.7 kilometers

Buffalo NY / Rochester NY

DRIVING DISTANCES IN MILES

	BATH, NY	BUFFALO, NY	ITHACA, NY	NIAGARA FALLS, NY	ONEONTA, NY	OSWEGO, NY	ROCHESTER, NY	SYRACUSE, NY	TORONTO, ON	TUPPER LAKE, NY	UTICA, NY	WATERTOWN, NY
BUFFALO, NY	113		153	20	263	158	74	152	106	321	199	210
ROCHESTER, NY	78	74	89	88	200	73		88	181	257	135	146
SYRACUSE, NY	105	152	59	166	118	38	88		260	176	53	65
UTICA, NY	152	199	108	213	65	81	135	53	307	131		86

SEE ALSO DISTANCE AND DRIVING TIME MAP ON PAGES 286–287

0 mi 10 20 30 40
0 km 10 20 30 40 50 60
One inch equals 18.4 miles
One centimeter equals 11.7 kilometers

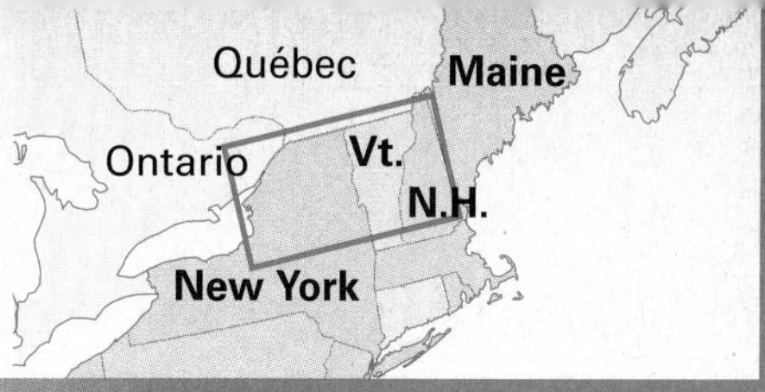

Burlington VT / Concord NH

DRIVING DISTANCES IN MILES

	BURLINGTON, VT	CONCORD, NH	LAKE PLACID, NY	OGDENSBURG, NY	PLATTSBURGH, NY	RUTLAND, VT	ST. JOHNSBURY, VT	SARATOGA SPRS., NY	SYRACUSE, NY	UTICA, NY	WATERTOWN, NY	WHITE RIVER JCT., VT
BURLINGTON, VT		150	68	208	51	69	76	115	230	183	195	91
CONCORD, NH	150		215	357	198	104	104	173	280	228	312	59
LAKE PLACID, NY	68	215		96	49	133	141	106	192	148	126	156
WATERTOWN, NY	195	312	126	68	167	244	319	179	65	86		289

SEE ALSO DISTANCE AND DRIVING TIME MAP ON PAGES 286–287

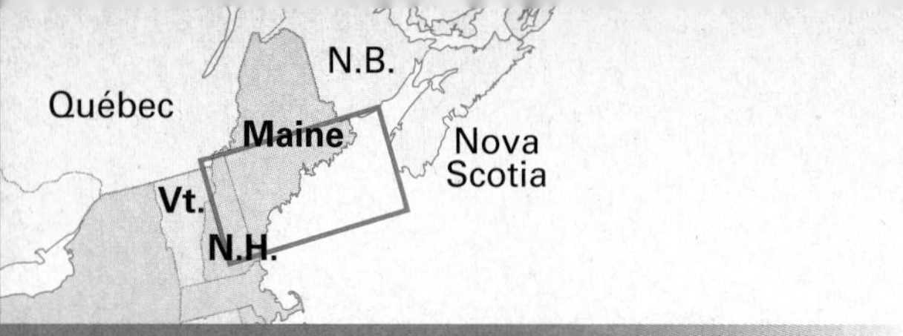

0 mi 10 20 30 40
0 km 10 20 30 40 50 60
One inch equals 18.4 miles
One centimeter equals 11.7 kilometers

Portland ME / Augusta ME

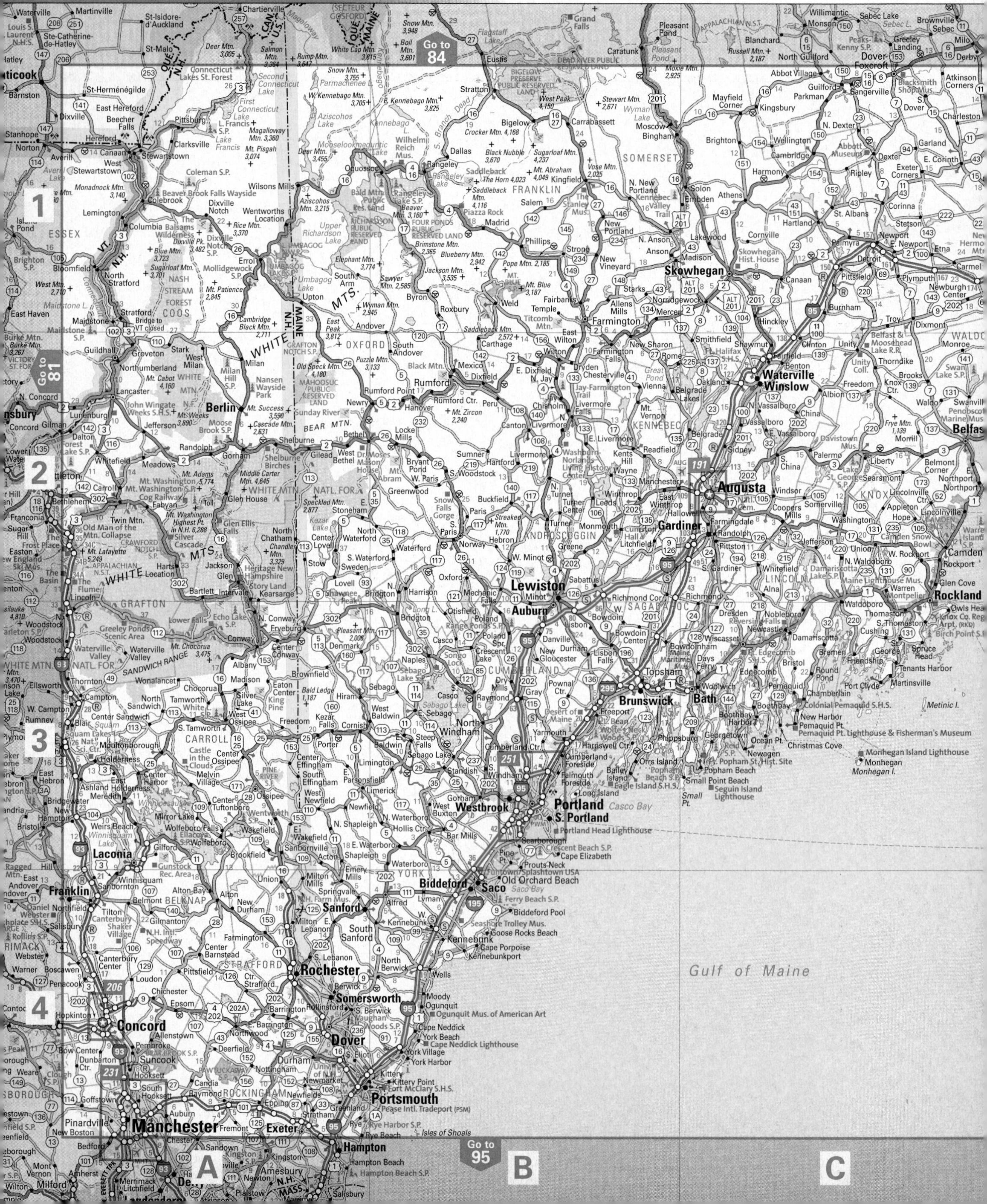

DRIVING DISTANCES IN MILES

	AUGUSTA, ME	BANGOR, ME	BAR HARBOR, ME	BERLIN, NH	CALAIS, ME	CONCORD, NH	CONWAY, NH	LEWISTON, ME	MACHIAS, ME	PORTLAND, ME	PORTSMOUTH, NH	WATERVILLE, ME
AUGUSTA, ME		77	120	110	173	141	97	35	158	58	110	20
BANGOR, ME	77		45	160	97	214	170	108	83	131	184	56
BAR HARBOR, ME	120	45		204	112	257	214	151	71	175	227	100
PORTLAND, ME	58	131	175	93	228	83	62	36	213		53	84

SEE ALSO DISTANCE AND DRIVING TIME MAP ON PAGES 286–287

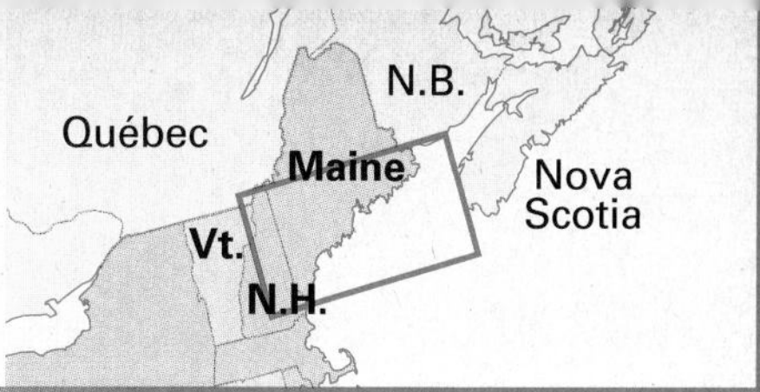

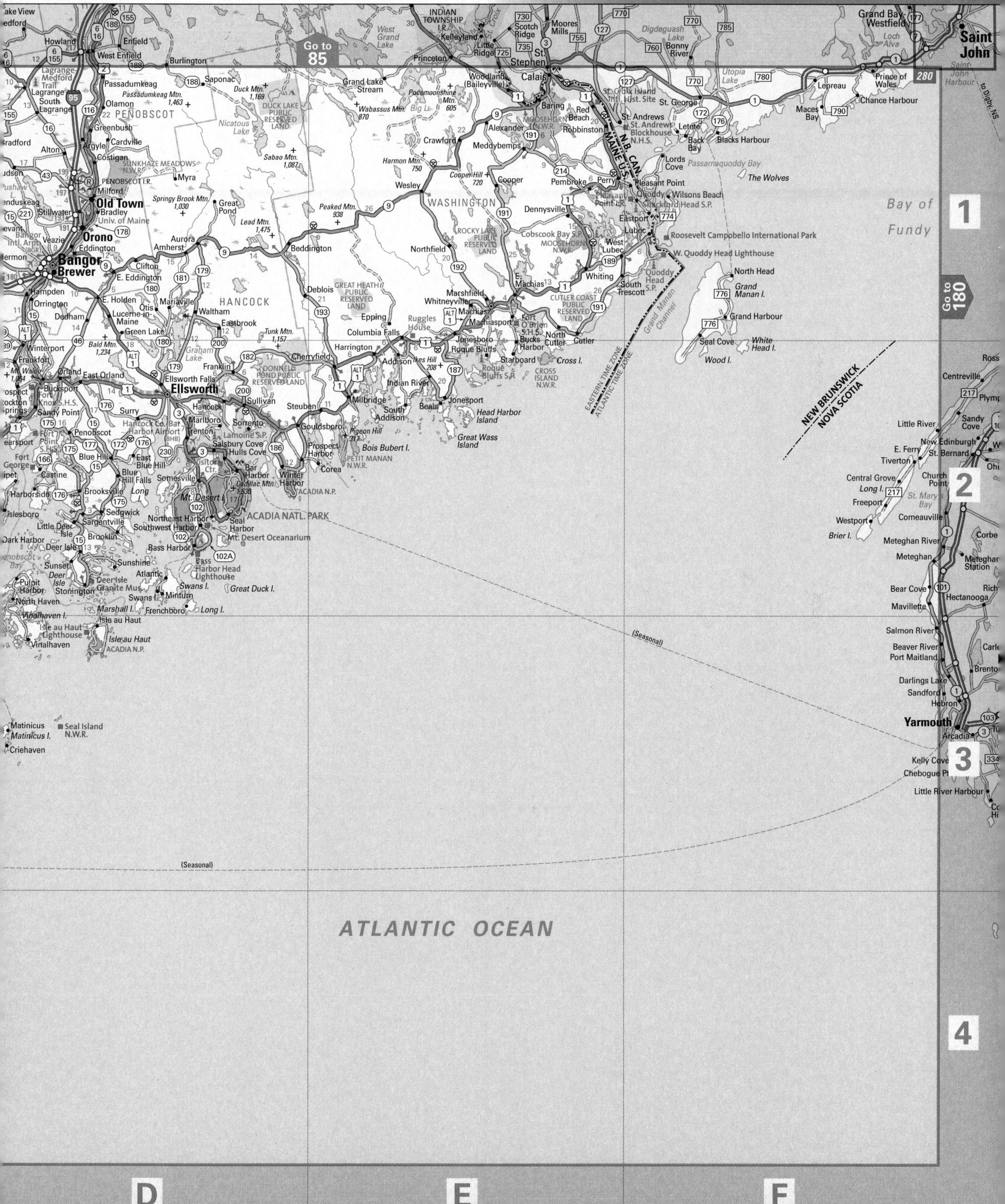

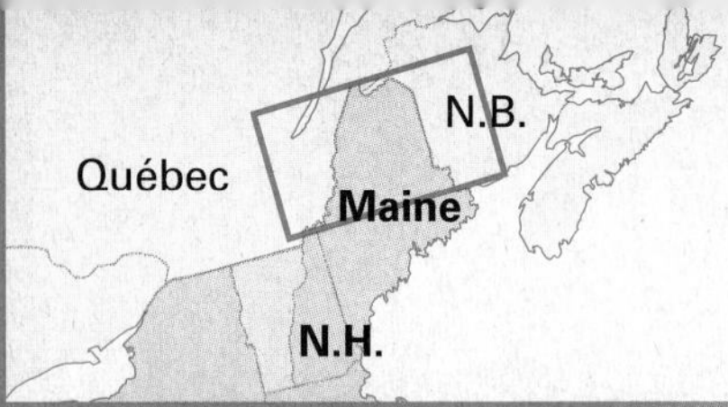

0 mi 10 20 30 40
0 km 10 20 30 40 50 60
One inch equals 18.4 miles
One centimeter equals 11.7 kilometers

DRIVING DISTANCES IN MILES	BANGOR, ME	CALAIS, ME	CARIBOU, ME	FREDERICTON, NB	GREENVILLE, ME	HOULTON, ME	JACKMAN, ME	LINCOLN, ME	MADAWASKA, ME	MILLINOCKET, ME	PRESQUE ISLE, ME	QUÉBEC, QC
HOULTON, ME	122	91	55	73	155		204	83	102	73	42	286
LINCOLN, ME	51	77	135	114	83	83	132		174	35	122	231
MADAWASKA, ME	214	207	50	167	212	102	269	174		164	62	182
PRESQUE ISLE, ME	162	133	13	113	166	42	215	122	62	113		246

SEE ALSO DISTANCE AND DRIVING TIME MAP ON PAGES 286–287

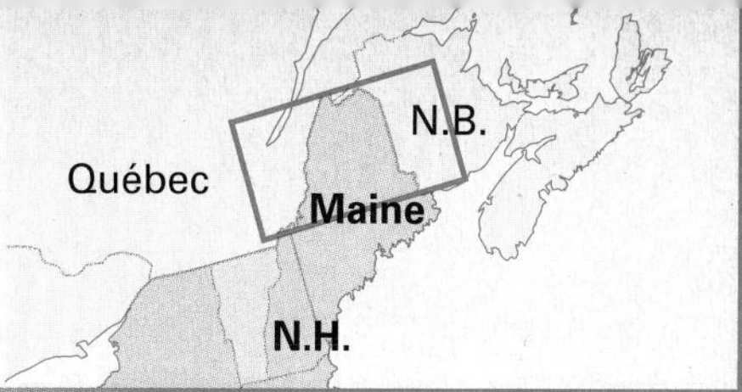

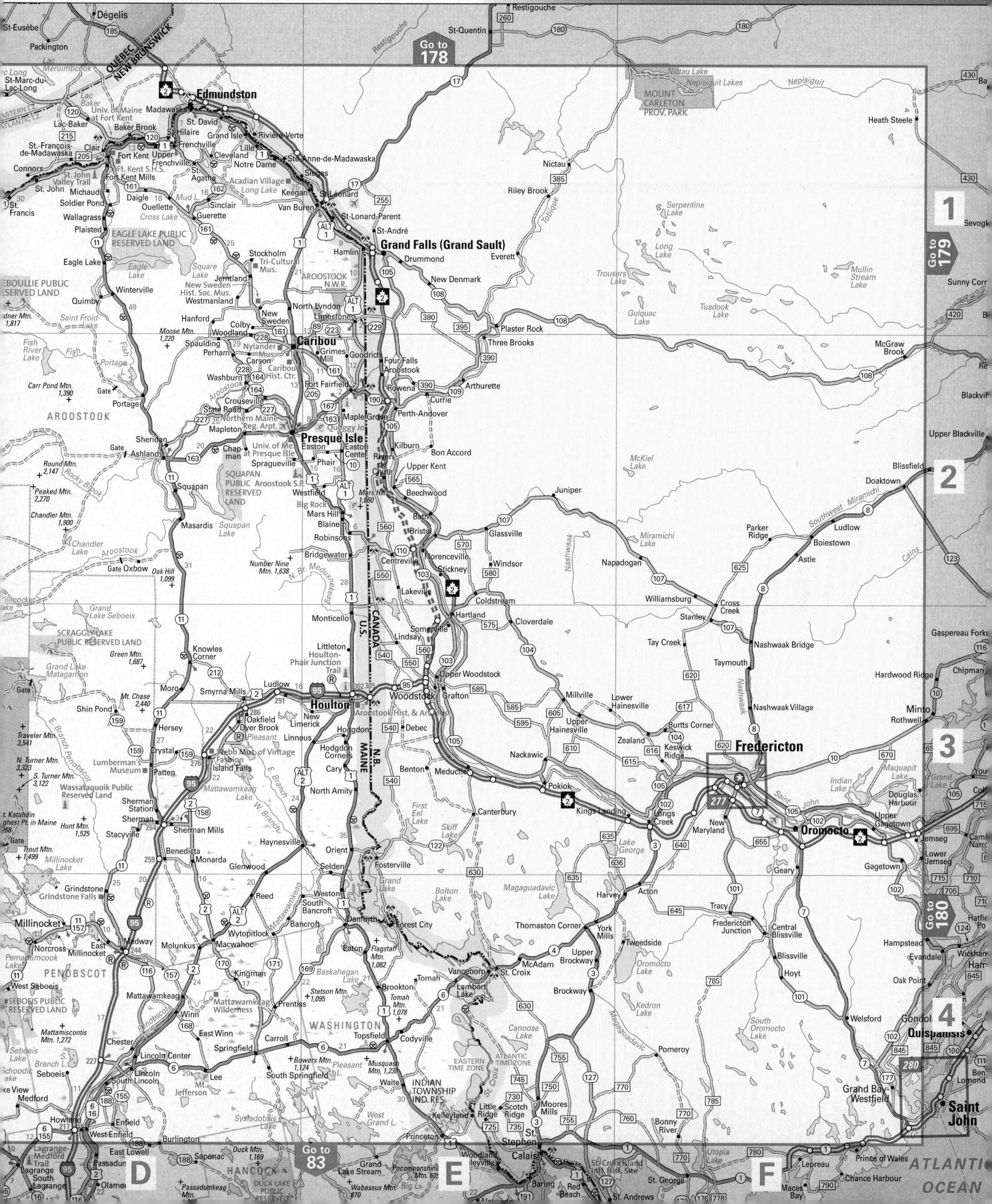

0 mi 10 20 30 40

0 km 10 20 30 40 50 60

One inch equals 18.4 miles

One centimeter equals 11.7 kilometers

Des Moines IA / Omaha NE

DRIVING DISTANCES IN MILES	AMES, IA	BURLINGTON, IA	CARROLL, IA	CEDAR RAPIDS, IA	CRESTON, IA	DAVENPORT, IA	DES MOINES, IA	IOWA CITY, IA	KIRKSVILLE, MO	MARYVILLE, MO	OMAHA, NE	OTTUMWA, IA
CEDAR RAPIDS, IA	108	106	173		211	87	129	28	170	276	266	111
DES MOINES, IA	34	157	90	129	81	171		113	145	146	136	86
IOWA CITY, IA	136	82	195	28	195	59	113		143	260	250	83
OMAHA, NE	171	328	97	266	98	308	136	250	275	112		221

SEE ALSO DISTANCE AND DRIVING TIME MAP ON PAGES 286–287

Davenport IA / Peoria IL

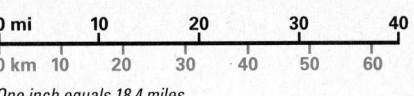

0 mi 10 20 30 40
0 km 10 20 30 40 50 60
One inch equals 18.4 miles
One centimeter equals 11.7 kilometers

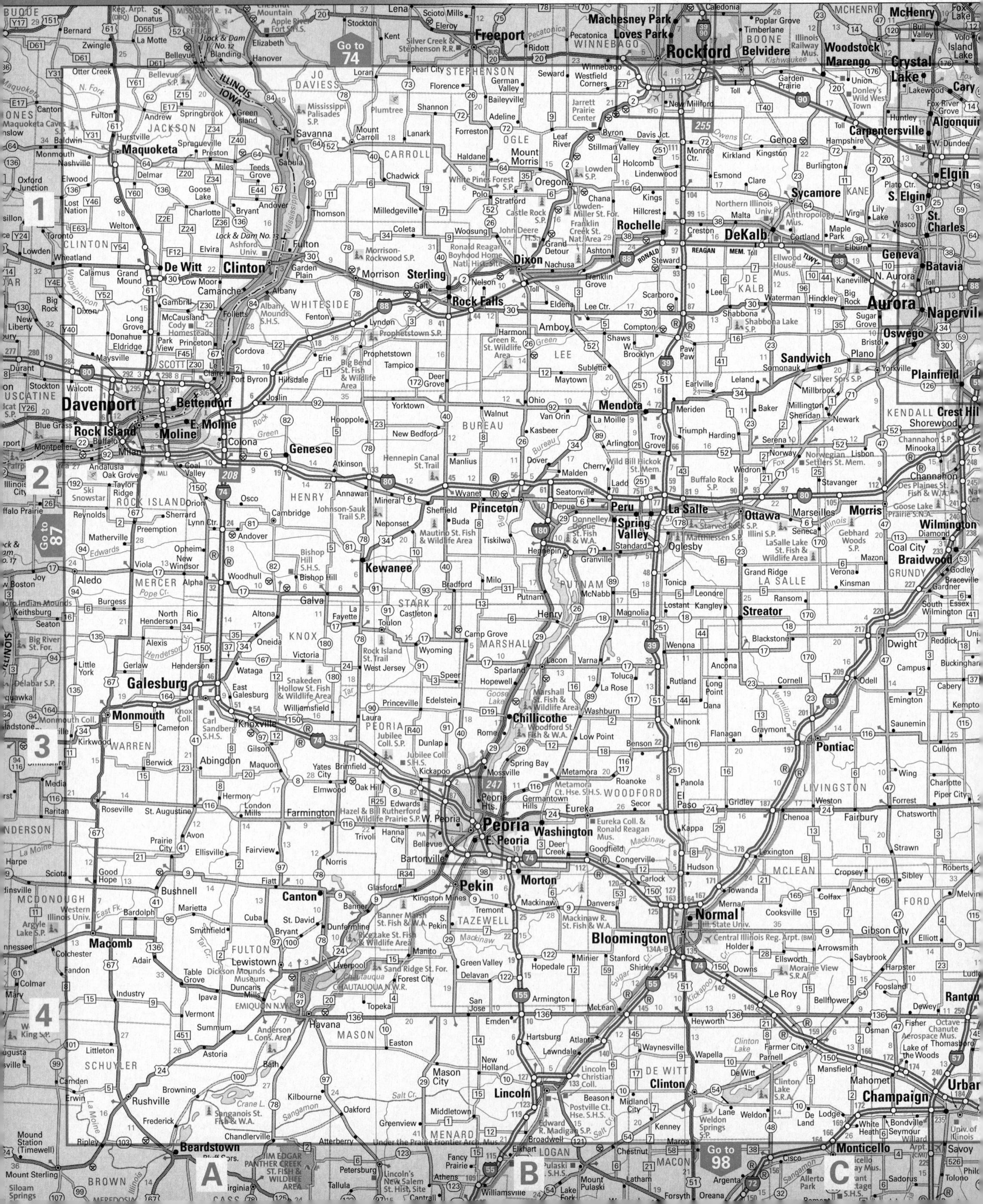

DRIVING DISTANCES IN MILES

	BLOOMINGTON, IL	CHAMPAIGN, IL	CHICAGO, IL	DAVENPORT, IA	JOLIET, IL	KALAMAZOO, MI	KOKOMO, IN	LAFAYETTE, IN	LA SALLE, IL	PEORIA, IL	ROCKFORD, IL	SOUTH BEND, IN
CHAMPAIGN, IL	54		141	192	115	255	145	94	117	94	189	198
CHICAGO, IL	135	141		170	40	150	158	121	98	168	86	93
PEORIA, IL	41	94	168	99	132	291	235	184	63		135	234
SOUTH BEND, IN	201	198	93	248	105	76	86	104	164	234	183	

SEE ALSO DISTANCE AND DRIVING TIME MAP ON PAGES 286–287

0 mi 10 20 30 40
0 km 10 20 30 40 50 60
One inch equals 18.4 miles
One centimeter equals 11.7 kilometers

Fort Wayne IN / Toledo OH

DRIVING DISTANCES IN MILES

	AKRON, OH	CLEVELAND, OH	COLUMBUS, OH	DETROIT, MI	ERIE, PA	FORT WAYNE, IN	LIMA, OH	MANSFIELD, OH	MUNCIE, IN	TOLEDO, OH	WHEELING, WV	YOUNGSTOWN, OH
CLEVELAND, OH	38		144	171	106	214	163	81	287	119	16	275
FORT WAYNE, IN	237	214	186	170	322		66	151	75	109	290	274
MANSFIELD, OH	66	81	67	156	179	151	93		209	105	141	112
TOLEDO, OH	142	119	148	60	227	109	83	105	180		261	179

SEE ALSO DISTANCE AND DRIVING TIME MAP ON PAGES 286–287

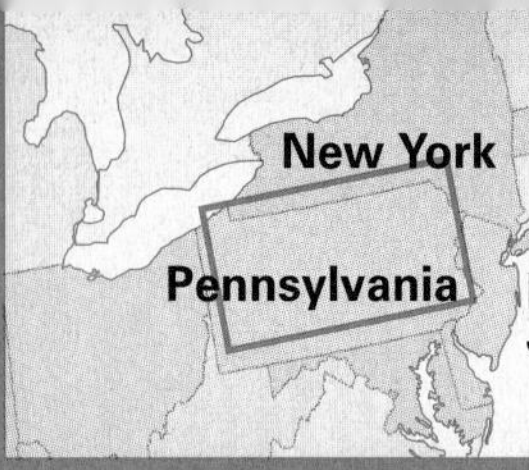

Pittsburgh PA / Erie PA

0 mi 10 20 30 40
0 km 10 20 30 40 50 60
One inch equals 18.4 miles
One centimeter equals 11.7 kilometers

DRIVING DISTANCES IN MILES	ALLENTOWN, PA	ALTOONA, PA	BINGHAMTON, NY	ELMIRA, NY	ERIE, PA	HARRISBURG, PA	JOHNSTOWN, PA	PITTSBURGH, PA	READING, PA	SCRANTON, PA	STATE COLLEGE, PA	WILLIAMSPORT, PA
ALLENTOWN, PA		218	132	188	361	82	217	284	37	76	165	116
HARRISBURG, PA	82	140	181	157	298		138	205	65	119	88	83
PITTSBURGH, PA	284	99	363	284	126	205	73		262	301	139	215
SCRANTON, PA	76	185	61	117	317	119	233	301	103		149	83

SEE ALSO DISTANCE AND DRIVING TIME MAP ON PAGES 286–287

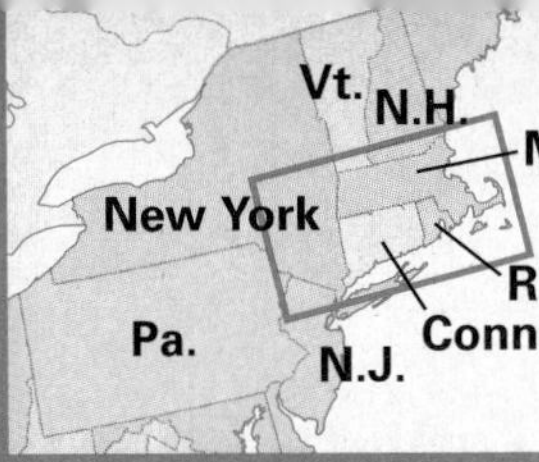

0 mi 10 20 30 40
0 km 10 20 30 40 50 60
One inch equals 18.4 miles
One centimeter equals 11.7 kilometers

Hartford CT / Albany NY

DRIVING DISTANCES IN MILES

	ALBANY, NY	BOSTON, MA	HARTFORD, CT	MANCHESTER, NH	NEWBURGH, NY	NEW HAVEN, CT	NEW YORK, NY	ONEONTA, NY	PROVIDENCE, RI	PROVINCETOWN, MA	SPRINGFIELD, MA	WORCESTER, MA
ALBANY, NY		172	111	145	89	150	151	81	170	271	86	133
BOSTON, MA	172		102	54	201	139	215	251	52	117	95	46
HARTFORD, CT	111	102		131	99	39	115	190	73	200	25	62
NEW YORK, NY	151	215	115	245	56	78		193	177	292	141	176

SEE ALSO DISTANCE AND DRIVING TIME MAP ON PAGES 286–287

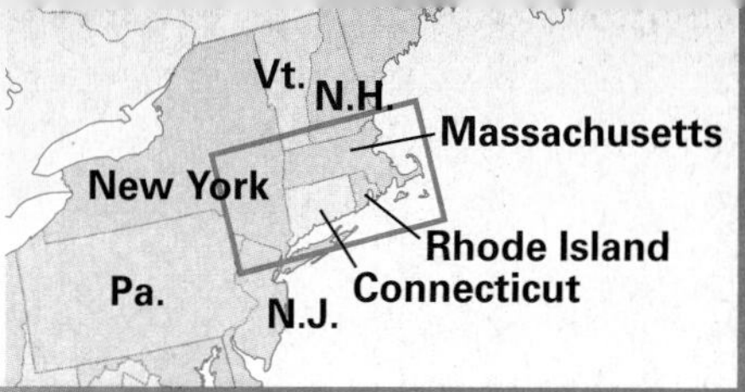

0 mi 10 20 30 40
0 km 10 20 30 40 50 60
One inch equals 18.4 miles
One centimeter equals 11.7 kilometers

Kansas City MO / Topeka KS

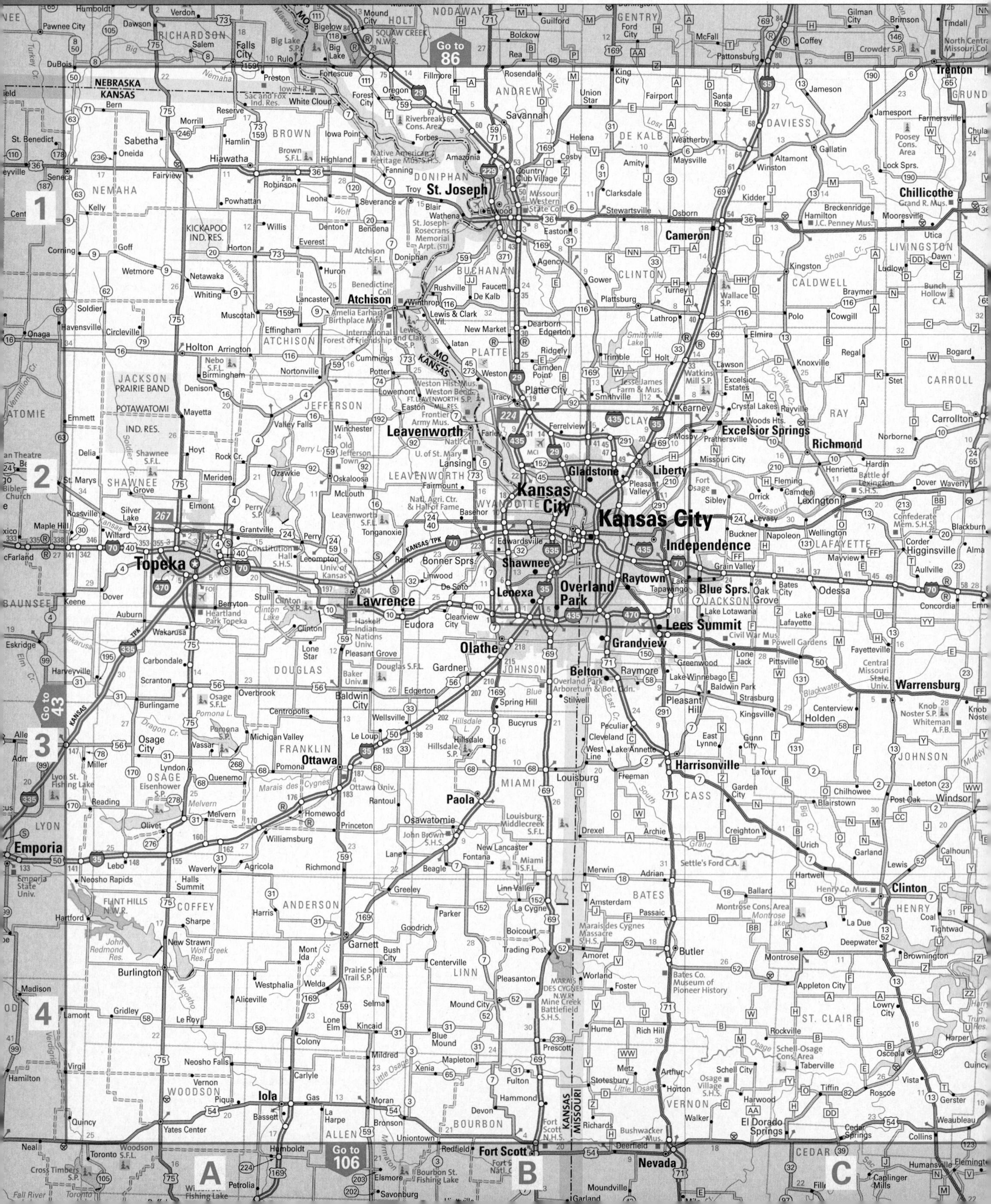

DRIVING DISTANCES IN MILES	COLUMBIA, MO	IOLA, KS	JEFFERSON CITY, MO	KANSAS CITY, MO	LAWRENCE, KS	MACON, MO	OSAGE BEACH, MO	QUINCY, IL	ROLLA, MO	ST. JOSEPH, MO	SEDALIA, MO	TOPEKA, KS
JEFFERSON CITY, MO	32	263		161	198	88	44	131	65	217	64	225
KANSAS CITY, MO	129	106	161		37	148	173	251	226	56	97	63
ST. JOSEPH, MO	185	154	217	56	76	131	229	210	282		153	71
TOPEKA, KS	193	100	225	63	26	209	236	314	289	71	161	

SEE ALSO DISTANCE AND DRIVING TIME MAP ON PAGES 286–287

0 mi 10 20 30 40
0 km 10 20 30 40 50 60
One inch equals 18.4 miles
One centimeter equals 11.7 kilometers

St Louis MO / Springfield IL

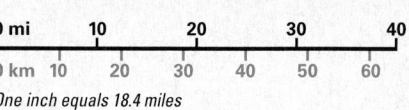

DRIVING DISTANCES IN MILES	BLOOMINGTON, IN	CHAMPAIGN, IL	DECATUR, IL	EFFINGHAM, IL	EVANSVILLE, IN	INDIANAPOLIS, IN	LOUISVILLE, KY	MT. VERNON, IL	ST. LOUIS, MO	SPRINGFIELD, IL	TERRE HAUTE, IN	VINCENNES, IN
EVANSVILLE, IN	117	192	184	117		166	114	90	170	247	107	51
INDIANAPOLIS, IN	47	123	177	137	166		112	205	239	212	77	123
ST. LOUIS, MO	223	179	116	103	170	239	264	81		97	169	185
SPRINGFIELD, IL	209	87	40	89	247	212	326	158	97		155	169

SEE ALSO DISTANCE AND DRIVING TIME MAP ON PAGES 286–287

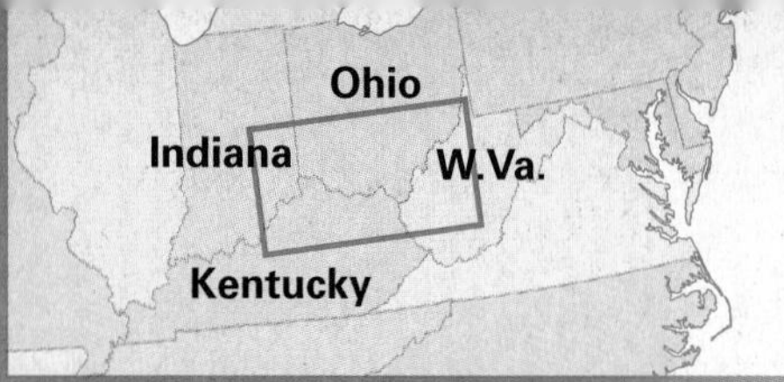

0 mi 10 20 30 40
0 km 10 20 30 40 50 60
One inch equals 18.4 miles
One centimeter equals 11.7 kilometers

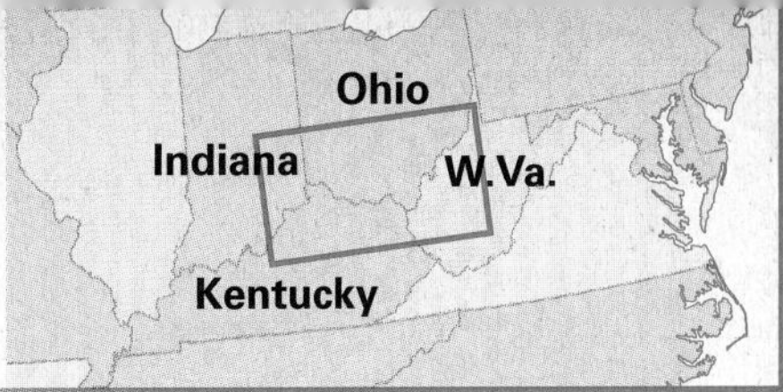

DRIVING DISTANCES IN MILES	CHARLESTON, WV	CHILLICOTHE, OH	CINCINNATI, OH	COLUMBUS, OH	DAYTON, OH	HUNTINGTON, WV	LEXINGTON, KY	LOUISVILLE, KY	MAYSVILLE, KY	PARKERSBURG, WV	WHEELING, WV	ZANESVILLE, OH
CHARLESTON, WV		121	202	168	198	52	176	251	155	73	176	155
CINCINNATI, OH	202	108		109	52	150	85	100	63	191	235	164
COLUMBUS, OH	168	47	109		70	135	193	207	114	108	130	58
LEXINGTON, KY	176	191	85	193	135	126		80	67	249	319	247

SEE ALSO DISTANCE AND DRIVING TIME MAP ON PAGES 286–287

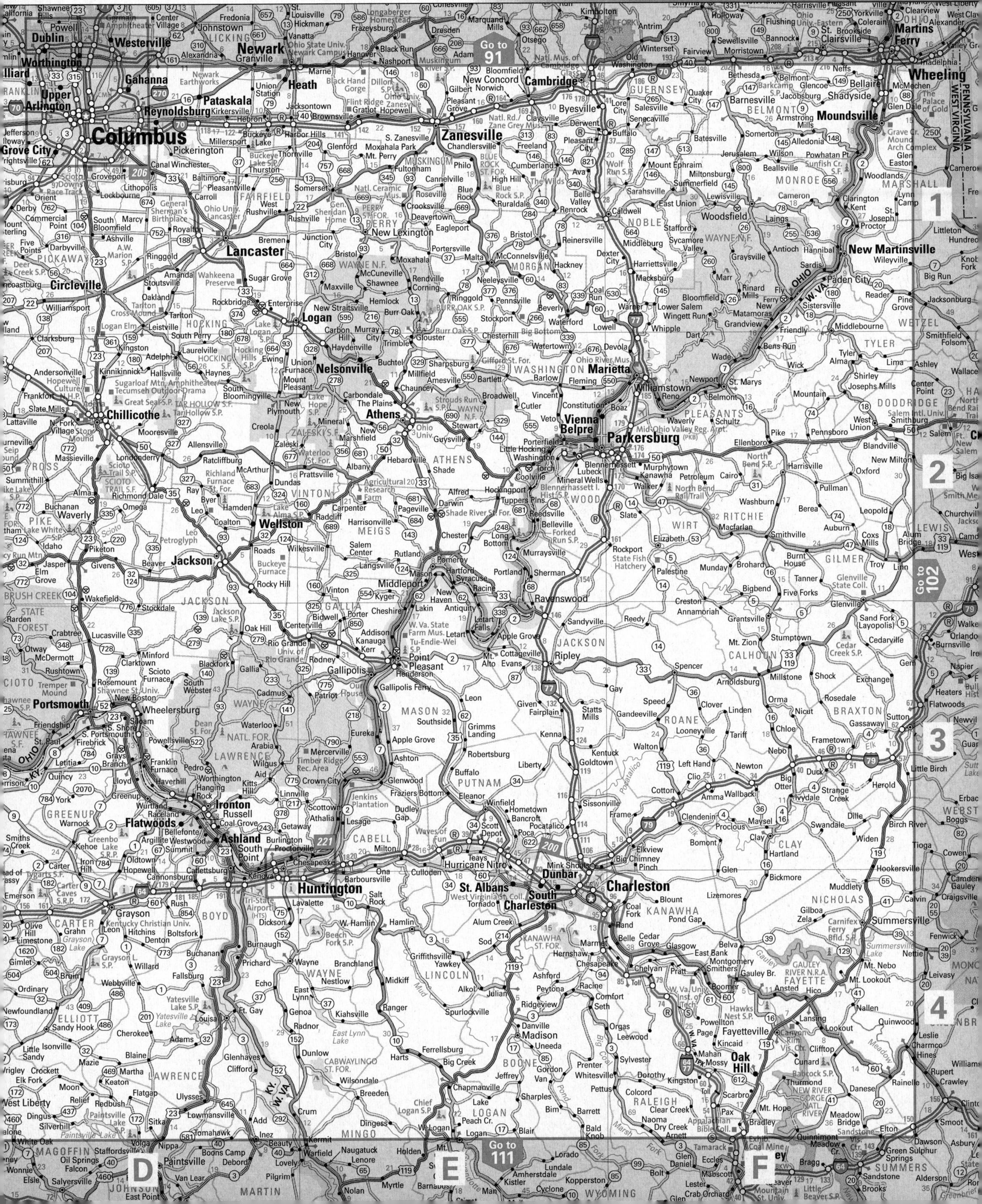

0 mi 10 20 30 40
0 km 10 20 30 40 50 60
One inch equals 18.4 miles
One centimeter equals 11.7 kilometers

Charlottesville VA / Morgantown WV

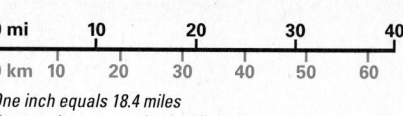

DRIVING DISTANCES IN MILES	BALTIMORE, MD	CHARLOTTESVILLE, VA	CUMBERLAND, MD	ELKINS, WV	FREDERICKSBURG, VA	FRONT ROYAL, VA	GETTYSBURG, PA	HAGERSTOWN, MD	MORGANTOWN, WV	SALISBURY, MD	WASHINGTON, DC	WHEELING, WV
BALTIMORE, MD		161	140	229	98	110	62	76	211	106	38	290
CHARLOTTESVILLE, VA	161		163	142	70	74	190	141	204	235	118	279
MORGANTOWN, WV	211	204	71	62	252	161	181	138		317	205	76
WASHINGTON, DC	38	118	134	192	54	73	80	70	205	115		284

SEE ALSO DISTANCE AND DRIVING TIME MAP ON PAGES 286–287

Washington DC / Baltimore MD

0 mi 10 20 30 40
0 km 10 20 30 40 50 60
One inch equals 18.4 miles
One centimeter equals 11.7 kilometers

Philadelphia PA / Harrisburg PA

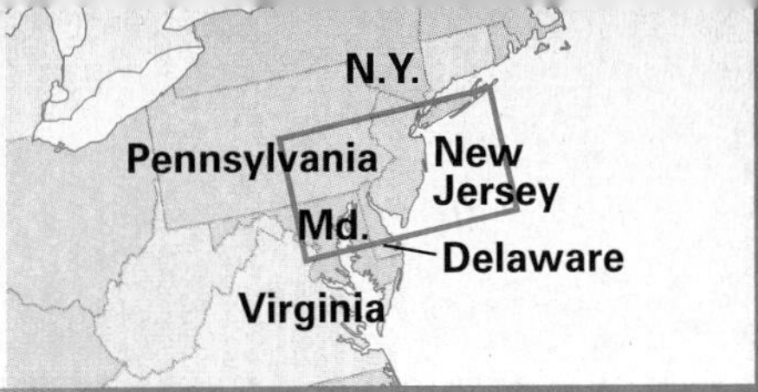

DRIVING DISTANCES IN MILES

	ALLENTOWN, PA	ATLANTIC CITY, NJ	BALTIMORE, MD	DOVER, DE	HARRISBURG, PA	LANCASTER, PA	NEWARK, NJ	NEW YORK, NY	PHILADELPHIA, PA	TRENTON, NJ	WASHINGTON, DC	WILMINGTON, DE
HARRISBURG, PA	82	171	83	126		44	154	165	109	135	123	102
NEW YORK, NY	84	125	192	160	165	165	11		91	55	228	120
PHILADELPHIA, PA	63	62	104	74	109	79	80	91		34	140	30
WASHINGTON, DC	188	186	38	94	123	123	218	228	140	179		110

SEE ALSO DISTANCE AND DRIVING TIME MAP ON PAGES 286–287

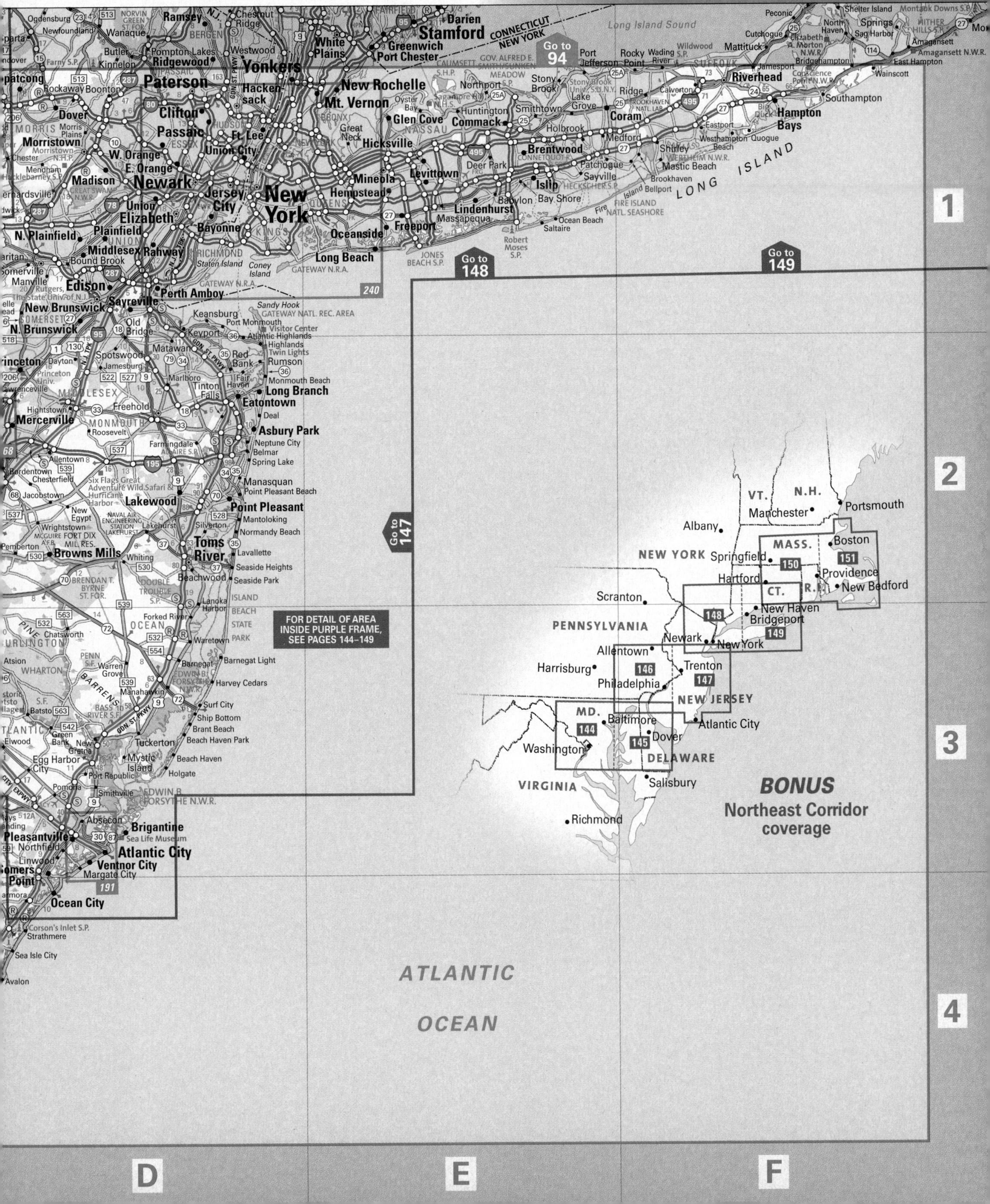

0 mi 10 20 30 40
0 km 10 20 30 40 50 60
One inch equals 18.4 miles
One centimeter equals 11.7 kilometers

Tulsa OK / Fayetteville AR

DRIVING DISTANCES IN MILES

	BARTLESVILLE, OK	BRANSON, MO	FAYETTEVILLE, AR	INDEPENDENCE, KS	JOPLIN, MO	MOUNTAIN HOME, AR	MUSKOGEE, OK	NEWPORT, AR	ROLLA, MO	SPRINGFIELD, MO	TULSA, OK	WEST PLAINS, MO
BRANSON, MO	213		95	188	111	84	181	178	147	41	225	109
FAYETTEVILLE, AR	154	95		165	88	127	86	241	227	121	113	182
SPRINGFIELD, MO	177	41	121	153	70	112	193	219	110		189	109
TULSA, OK	48	225	113	86	116	237	52	344	295	189		293

SEE ALSO DISTANCE AND DRIVING TIME MAP ON PAGES 286–287

0 mi 10 20 30 40
0 km 10 20 30 40 50 60

One inch equals 18.4 miles
One centimeter equals 11.7 kilometers

Jonesboro AR / Cape Girardeau MO

DRIVING DISTANCES IN MILES

	BOWLING GREEN, KY	CAPE GIRARDEAU, MO	CARBONDALE, IL	CLARKSVILLE, TN	DYERSBURG, TN	HOPKINSVILLE, KY	JACKSON, TN	JONESBORO, AR	NASHVILLE, TN	OWENSBORO, KY	PADUCAH, KY	POPLAR BLUFF, MO
BOWLING GREEN, KY		199	206	63	217	63	196	349	68	76	135	239
CAPE GIRARDEAU, MO	199		46	155	112	136	161	155	197	168	67	75
JONESBORO, AR	349	155	199	268	101	249	160		285	304	178	81
NASHVILLE, TN	68	197	204	46	178	68	132	285		141	133	237

SEE ALSO DISTANCE AND DRIVING TIME MAP ON PAGES 286–287

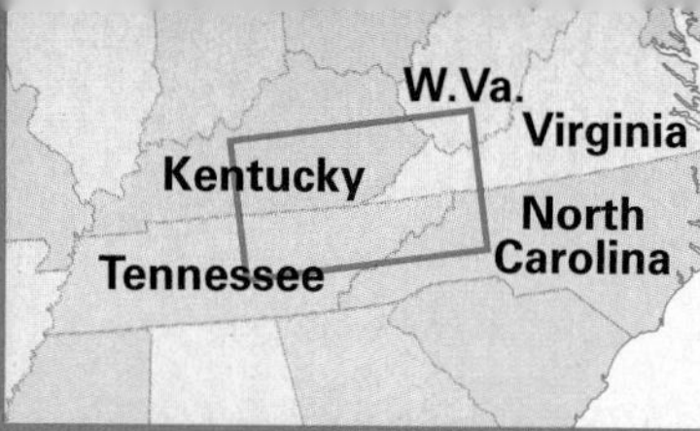

0 mi 10 20 30 40
0 km 10 20 30 40 50 60
One inch equals 18.4 miles
One centimeter equals 11.7 kilometers

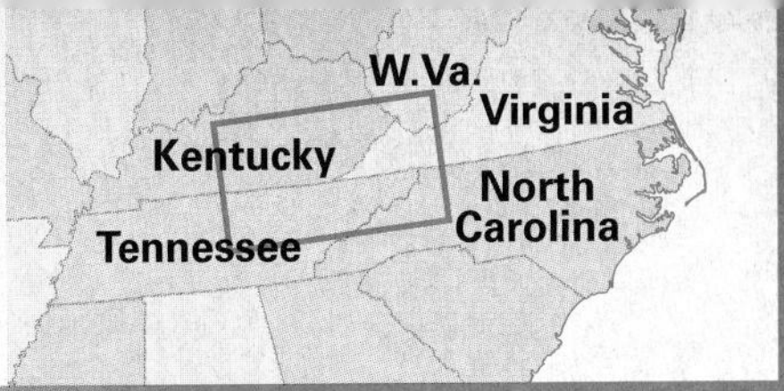

DRIVING DISTANCES IN MILES

	ASHEVILLE, NC	BECKLEY, WV	BRISTOL, TN/VA	COOKEVILLE, TN	GATLINBURG, TN	HICKORY, NC	JOHNSON CITY, TN	KNOXVILLE, TN	LONDON, KY	MAMMOTH CAVE NP, KY	PIKEVILLE, KY	RICHMOND, KY
BRISTOL, TN/VA	83	140		224	118	98	24	117	213	348	116	265
HICKORY, NC	78	196	98	291	147		98	185	280	415	214	332
KNOXVILLE, TN	109	256	117	107	40	185	107		100	234	202	151
LONDON, KY	205	287	213	129	136	280	203	100		136	121	53

SEE ALSO DISTANCE AND DRIVING TIME MAP ON PAGES 286–287

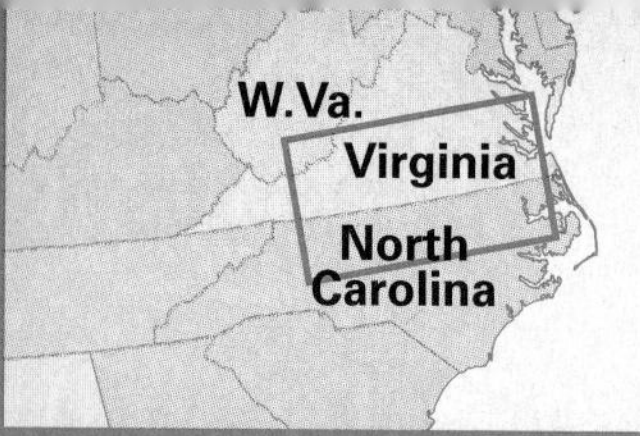

0 mi 10 20 30 40

0 km 10 20 30 40 50 60

One inch equals 18.4 miles

One centimeter equals 11.7 kilometers

DRIVING DISTANCES IN MILES

	DANVILLE, VA	GREENSBORO, NC	LYNCHBURG, VA	NORFOLK, VA	RALEIGH, NC	RICHMOND, VA	ROANOKE, VA	ROANOKE RAPIDS, NC	ROCKY MOUNT, NC	WILLIAMSBURG, VA	WINSTON-SALEM, NC	WYTHEVILLE, VA
GREENSBORO, NC	46		106	230	69	200	101	132	124	237	30	120
RALEIGH, NC	89	69	140	179		157	156	84	54	204	96	186
RICHMOND, VA	160	200	114	91	157		192	91	127	49	228	256
ROANOKE, VA	83	101	55	285	156	192		190	211	243	107	78

SEE ALSO DISTANCE AND DRIVING TIME MAP ON PAGES 286–287

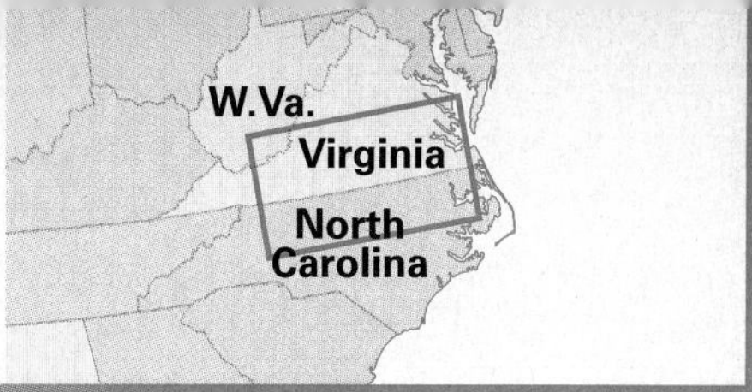

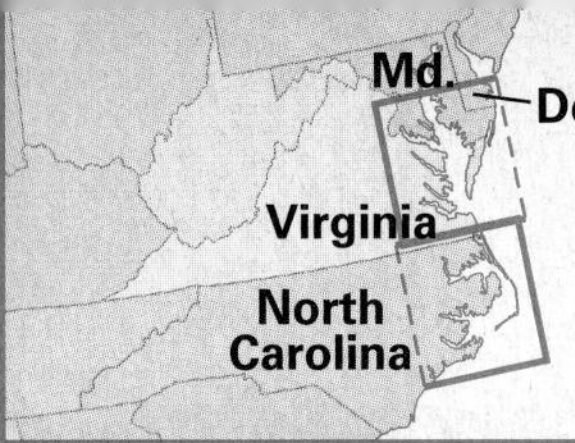

0 mi 10 20 30 40
0 km 10 20 30 40 50 60

One inch equals 18.4 miles

One centimeter equals 11.7 kilometers

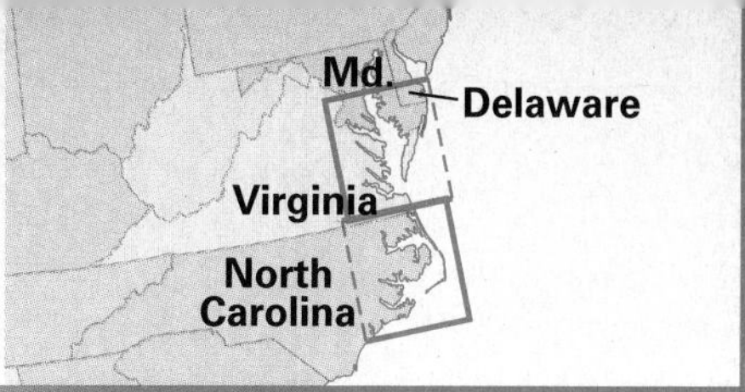

DRIVING DISTANCES IN MILES

	ELIZABETH CITY, NC	GREENVILLE, NC	MOREHEAD CITY, NC	NAGS HEAD, NC	NEW BERN, NC	NORFOLK, VA	OCEAN CITY, MD	RICHMOND, VA	SALISBURY, MD	VIRGINIA BEACH, VA	WASHINGTON, DC	WILLIAMSBURG, VA
MOREHEAD CITY, NC	150	82		184	35	185	326	241	321	206	352	221
NAGS HEAD, NC	59	135	184		149	82	214	179	209	94	284	131
NORFOLK, VA	50	130	185	82	151		138	91	133	18	196	43
WASHINGTON, DC	243	270	352	284	317	196	139	108	115	212		153

SEE ALSO DISTANCE AND DRIVING TIME MAP ON PAGES 286–287

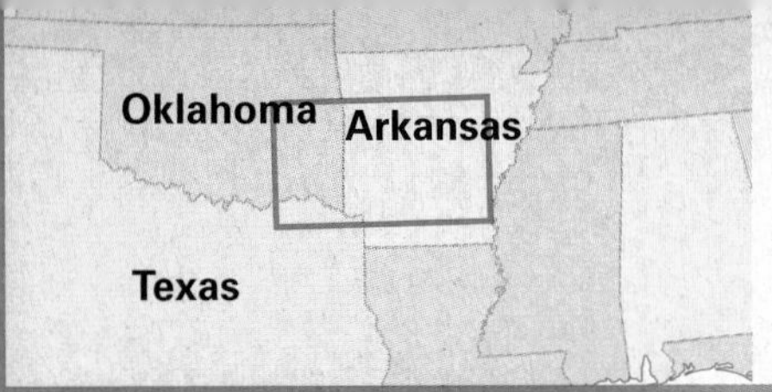

0 mi 10 20 30 40
0 km 10 20 30 40 50 60

One inch equals 18.4 miles
One centimeter equals 11.7 kilometers

Fort Smith AR / Texarkana AR–TX

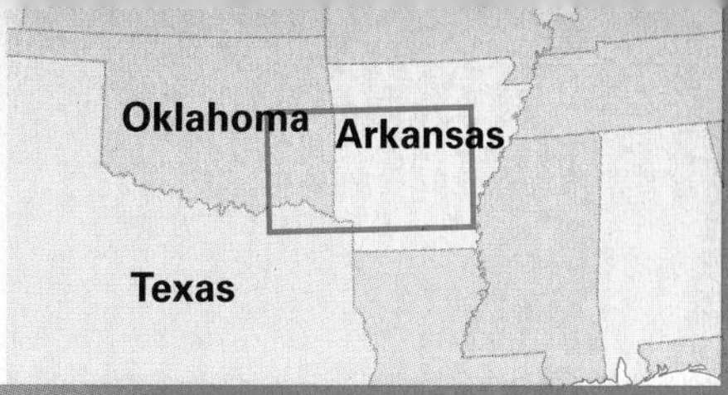

DRIVING DISTANCES IN MILES	ARKADELPHIA, AR	FORT SMITH, AR	HENRYETTA, OK	HOT SPRINGS, AR	LITTLE ROCK, AR	MCALESTER, OK	MENA, AR	NEWPORT, AR	PARIS, TX	PINE BLUFF, AR	RUSSELLVILLE, AR	TEXARKANA, AR/TX
FORT SMITH, AR	152		100	126	165	114	81	220	214	210	87	180
HOT SPRINGS, AR	37	126	224		65	193	75	154	207	76	67	117
LITTLE ROCK, AR	72	165	263	65		278	141	89	242	45	81	153
TEXARKANA, AR/TX	83	180	227	117	153	188	99	241	92	163	180	

SEE ALSO DISTANCE AND DRIVING TIME MAP ON PAGES 286–287

0 mi 10 20 30 40
0 km 10 20 30 40 50 60

One inch equals 18.4 miles
One centimeter equals 11.7 kilometers

Memphis TN / Oxford MS

DRIVING DISTANCES IN MILES

	BIRMINGHAM, AL	CLARKSDALE, MS	COLUMBIA, TN	COLUMBUS, MS	DECATUR, AL	FLORENCE, AL	GREENVILLE, MS	HUNTSVILLE, AL	JACKSON, TN	MEMPHIS, TN	OXFORD, MS	TUPELO, MS
BIRMINGHAM, AL		248	161	122	83	121	286	101	223	241	185	136
HUNTSVILLE, AL	101	260	79	163	25	65	318		205	216	196	148
MEMPHIS, TN	241	76	210	175	191	156	148	216	91		85	109
TUPELO, MS	136	113	159	66	123	92	172	148	107	109	50	

SEE ALSO DISTANCE AND DRIVING TIME MAP ON PAGES 286–287

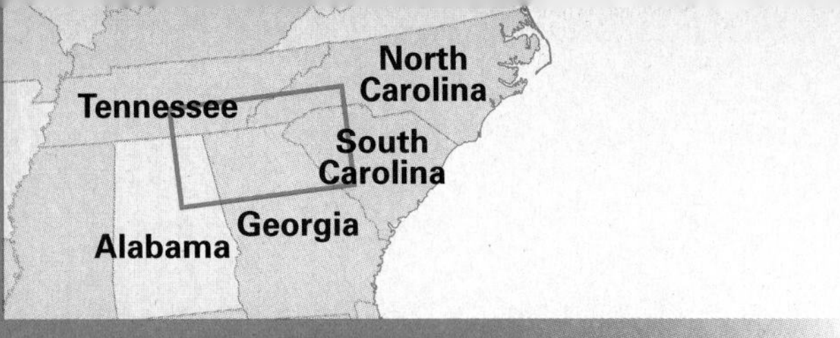

Atlanta GA / Chattanooga TN

0 mi 10 20 30 40
0 km 10 20 30 40 50 60

One inch equals 18.4 miles
One centimeter equals 11.7 kilometers

DRIVING DISTANCES IN MILES	ANNISTON, AL	ASHEVILLE, NC	ATHENS, GA	ATLANTA, GA	AUGUSTA, GA	CHATTANOOGA, TN	GADSDEN, AL	GATLINBURG, TN	GREENVILLE, SC	HUNTSVILLE, AL	MANCHESTER, TN	SPARTANBURG, SC
ATLANTA, GA	91	207	70		149	113	117	187	146	191	180	173
AUGUSTA, GA	240	179	97	149		266	266	240	110	334	333	118
CHATTANOOGA, TN	120	225	170	113	266		94	156	245	109	69	272
GREENVILLE, SC	238	64	104	146	110	245	264	125		313	311	30

SEE ALSO DISTANCE AND DRIVING TIME MAP ON PAGES 286–287

0 mi 10 20 30 40
0 km 10 20 30 40 50 60
One inch equals 18.4 miles
One centimeter equals 11.7 kilometers

Charlotte NC / Columbia SC

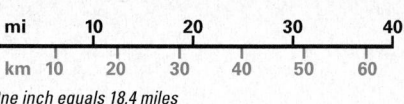

DRIVING DISTANCES IN MILES	CHARLOTTE, NC	COLUMBIA, SC	FAYETTEVILLE, NC	FLORENCE, SC	GOLDSBORO, NC	HICKORY, NC	LUMBERTON, NC	MOREHEAD CITY, NC	MYRTLE BEACH, SC	ROCK HILL, SC	SUMTER, SC	WILMINGTON, NC
CHARLOTTE, NC		91	139	107	208	47	128	298	173	26	115	205
COLUMBIA, SC	91		170	80	240	139	139	289	146	70	45	199
MYRTLE BEACH, SC	173	146	116	66	170	220	83	165		181	93	71
WILMINGTON, NC	205	199	92	120	100	292	77	95	71	220	158	

SEE ALSO DISTANCE AND DRIVING TIME MAP ON PAGES 286–287

0 mi 10 20 30 40
0 km 10 20 30 40 50 60

One inch equals 18.4 miles
One centimeter equals 11.7 kilometers

Shreveport LA / Tyler TX

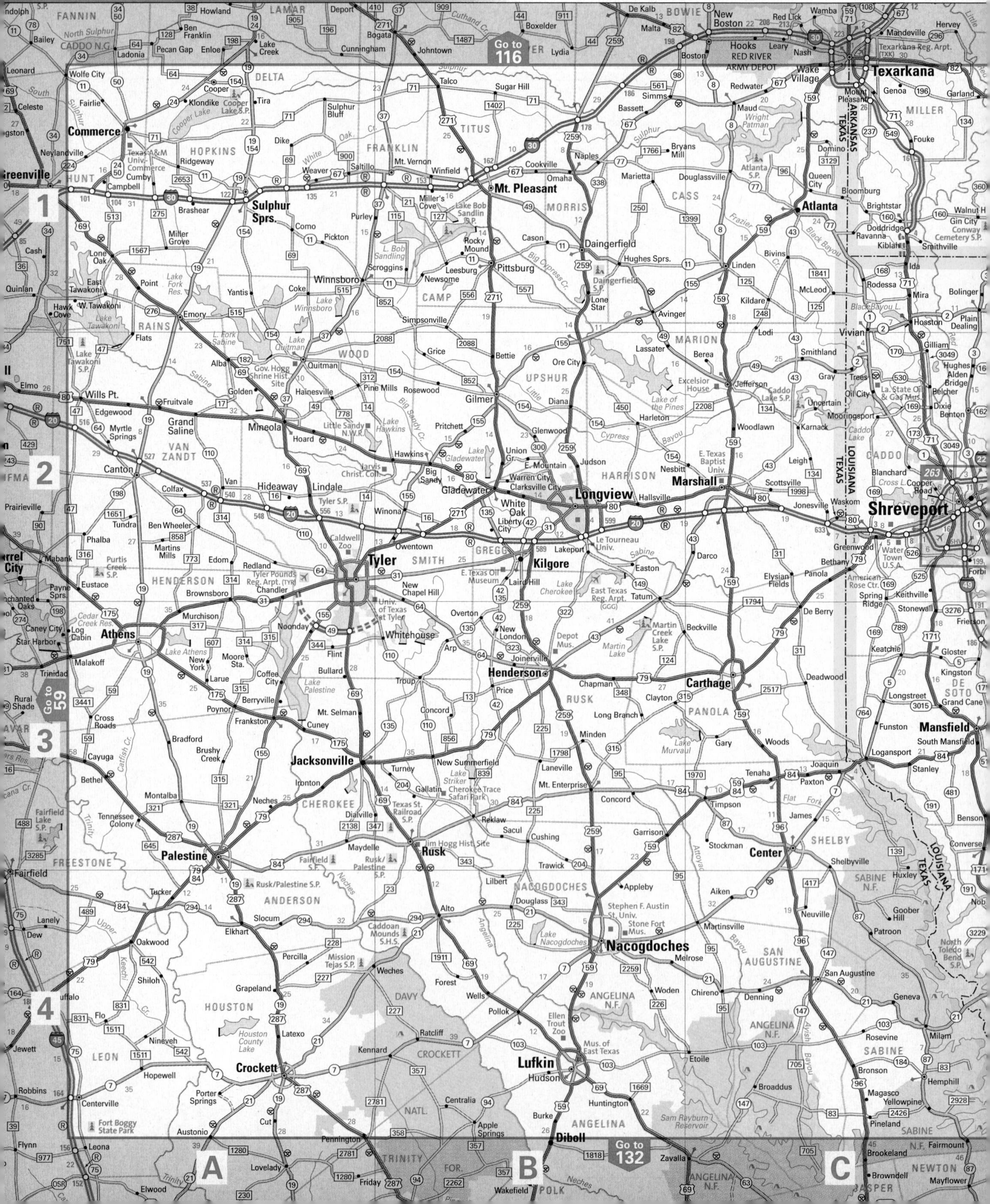

DRIVING DISTANCES IN MILES

	ALEXANDRIA, LA	EL DORADO, AR	GREENVILLE, TX	LONGVIEW, TX	LUFKIN, TX	MONROE, LA	NACOGDOCHES, TX	NATCHEZ, MS	NATCHITOCHES, LA	SHREVEPORT, LA	TEXARKANA, AR/TX	TYLER, TX
ALEXANDRIA, LA		147	276	179	160	96	167	76	55	121	190	213
MONROE, LA	96	86	267	170	223		203	95	100	103	172	204
SHREVEPORT, LA	121	96	165	68	121	103	101	198	73		69	102
TYLER, TX	213	196	77	42	82	204	76	288	164	102	118	

SEE ALSO DISTANCE AND DRIVING TIME MAP ON PAGES 286–287

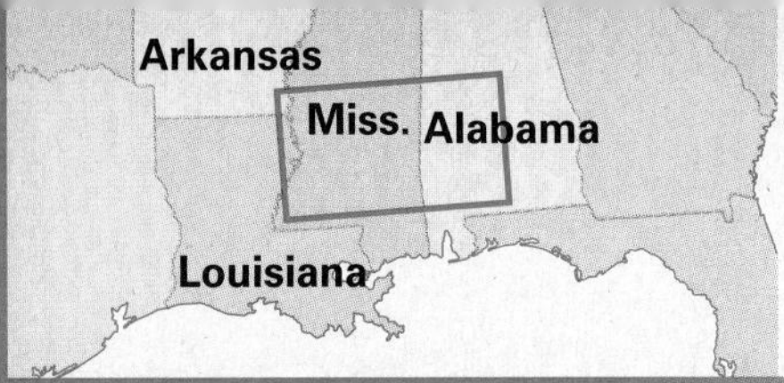

0 mi 10 20 30 40
0 km 10 20 30 40 50 60
One inch equals 18.4 miles
One centimeter equals 11.7 kilometers

Jackson MS / Hattiesburg MS

DRIVING DISTANCES IN MILES

	BIRMINGHAM, AL	EVERGREEN, AL	GREENVILLE, MS	HATTIESBURG, MS	JACKSON, MS	MCCOMB, MS	MERIDIAN, MS	NATCHEZ, MS	SELMA, AL	TUSCALOOSA, AL	VICKSBURG, MS	WINONA, MS
HATTIESBURG, MS	239	184	215		90	75	89	142	193	183	132	180
JACKSON, MS	241	243	125	90		76	91	102	195	185	42	94
MERIDIAN, MS	149	152	216	89	91	167		194	104	94	133	113
TUSCALOOSA, AL	61	211	225	183	185	261	94	287	82		227	144

SEE ALSO DISTANCE AND DRIVING TIME MAP ON PAGES 286–287

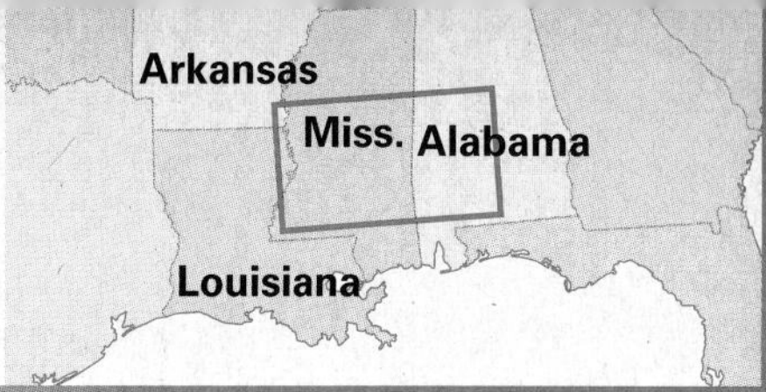

0 mi 10 20 30 40
0 km 10 20 30 40 50 60
One inch equals 18.4 miles
One centimeter equals 11.7 kilometers

Montgomery AL / Columbus GA

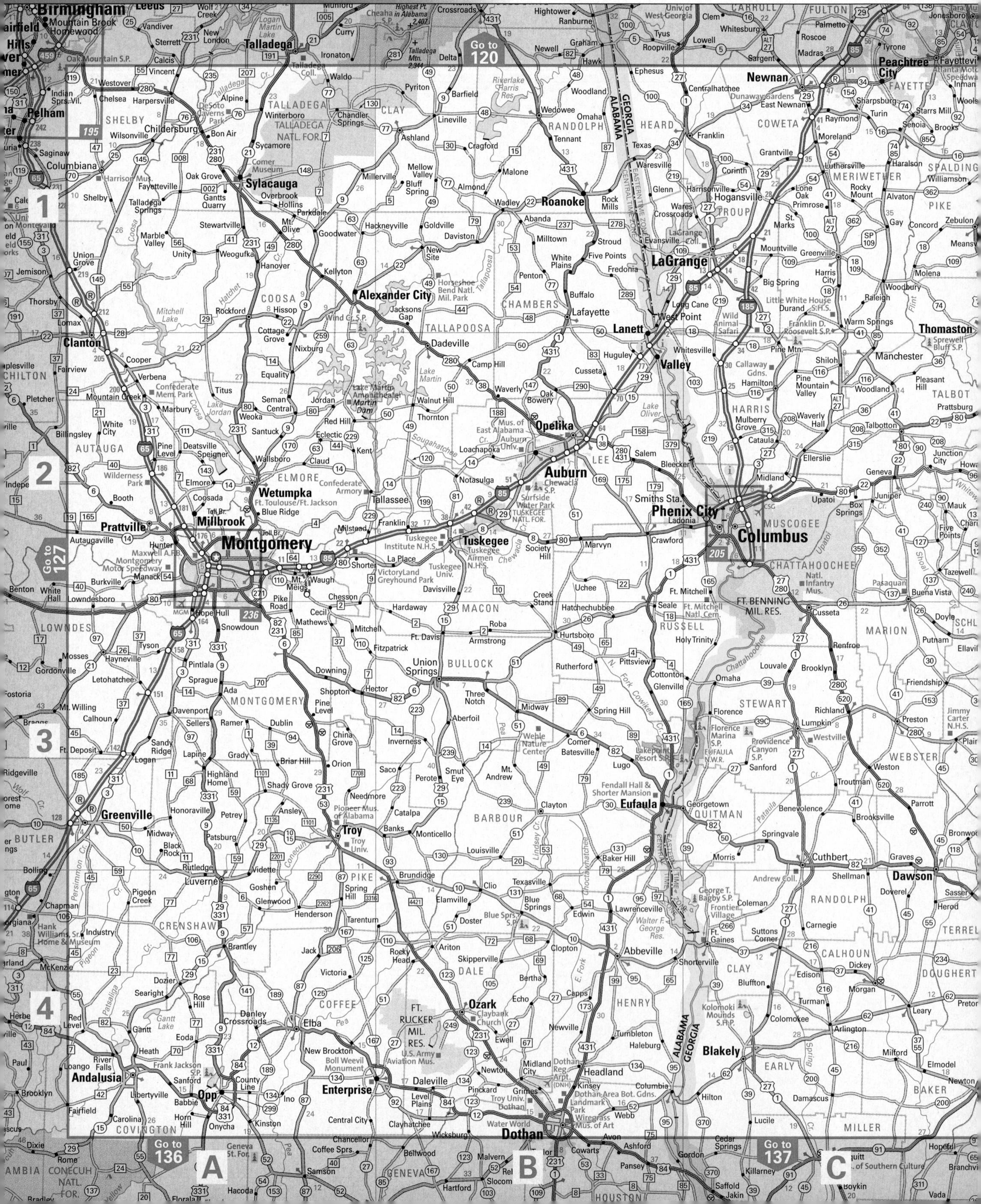

DRIVING DISTANCES IN MILES

	ALBANY, GA	ATLANTA, GA	AUBURN, AL	AUGUSTA, GA	BIRMINGHAM, AL	COLUMBUS, GA	DOTHAN, AL	LA GRANGE, GA	MACON, GA	MONTGOMERY, AL	TIFTON, GA	WAYCROSS, GA
ALBANY, GA		180	121	226	253	86	83	129	102	165	43	116
COLUMBUS, GA	86	106	34	249	167		97	46	95	79	135	208
MACON, GA	102	84	151	123	234	95	186	114		203	102	159
MONTGOMERY, AL	165	158	54	301	88	79	103	95	203		214	287

SEE ALSO DISTANCE AND DRIVING TIME MAP ON PAGES 286–287

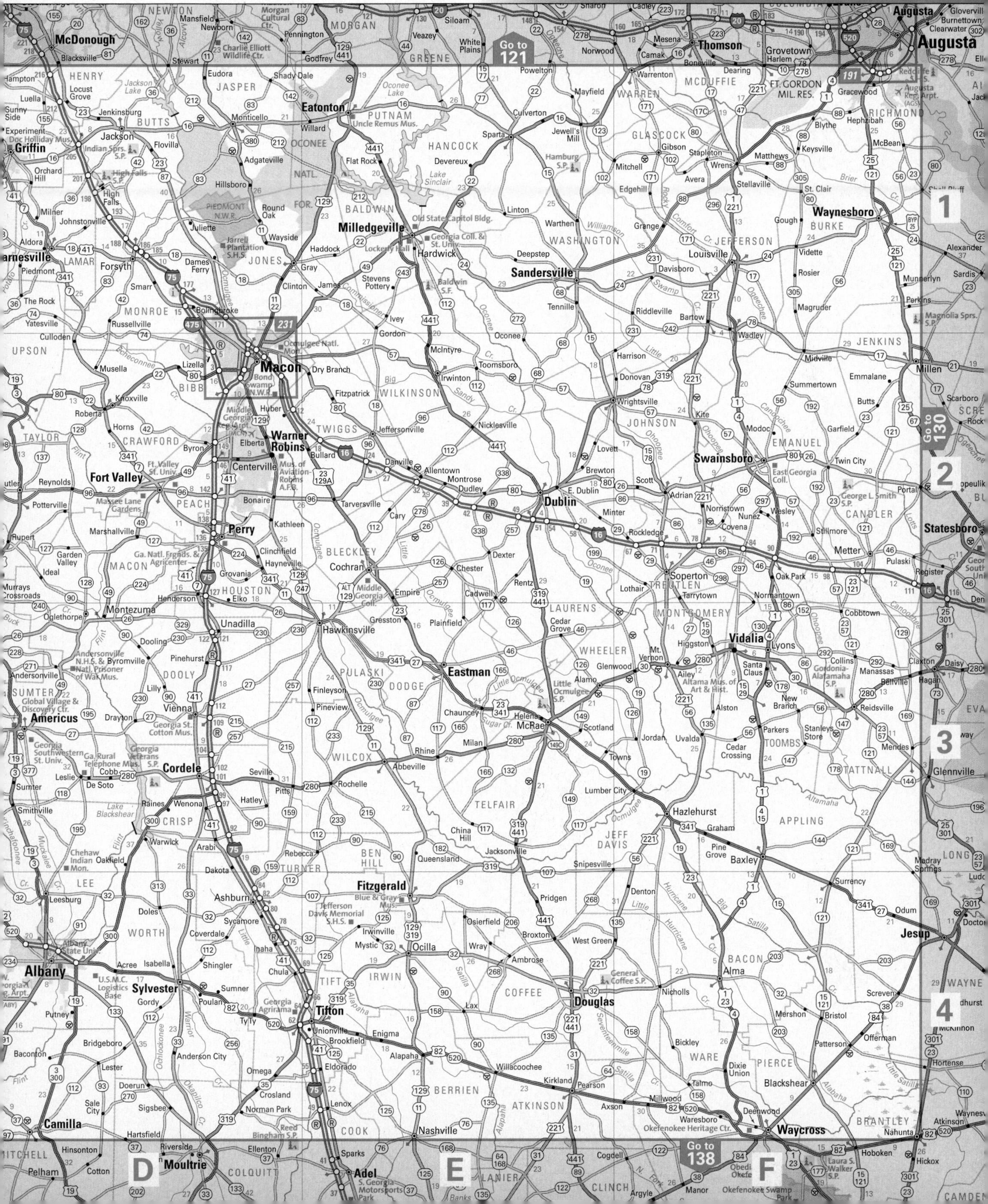

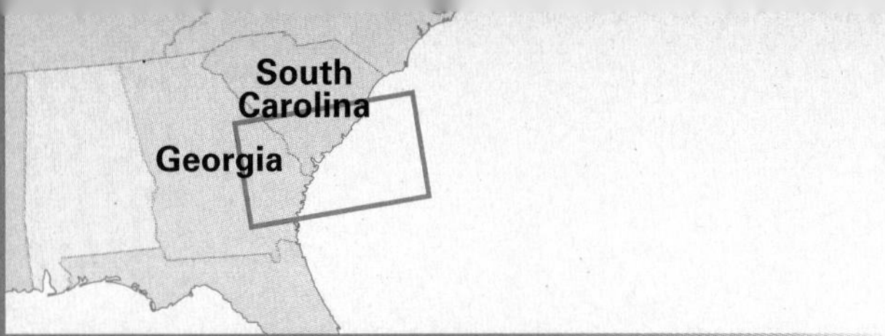

0 mi 10 20 30 40
0 km 10 20 30 40 50 60
One inch equals 18.4 miles
One centimeter equals 11.7 kilometers

Savannah GA / Hilton Head Island SC

DRIVING DISTANCES IN MILES

	AUGUSTA, GA	BEAUFORT, SC	BRUNSWICK, GA	CHARLESTON, SC	GEORGETOWN, SC	HILTON HEAD I., SC	HINESVILLE, GA	ORANGEBURG, SC	SAVANNAH, GA	STATESBORO, GA	WALTERBORO, SC	WAYCROSS, GA
AUGUSTA, GA		126	194	142	181	127	157	74	135	81	111	184
CHARLESTON, SC	142	66	175		58	95	138	73	107	150	51	203
HILTON HEAD I., SC	127	32	113	95	157		75	116	35	88	64	141
SAVANNAH, GA	135	42	78	107	163	35	41	123		53	71	106

SEE ALSO DISTANCE AND DRIVING TIME MAP ON PAGES 286–287

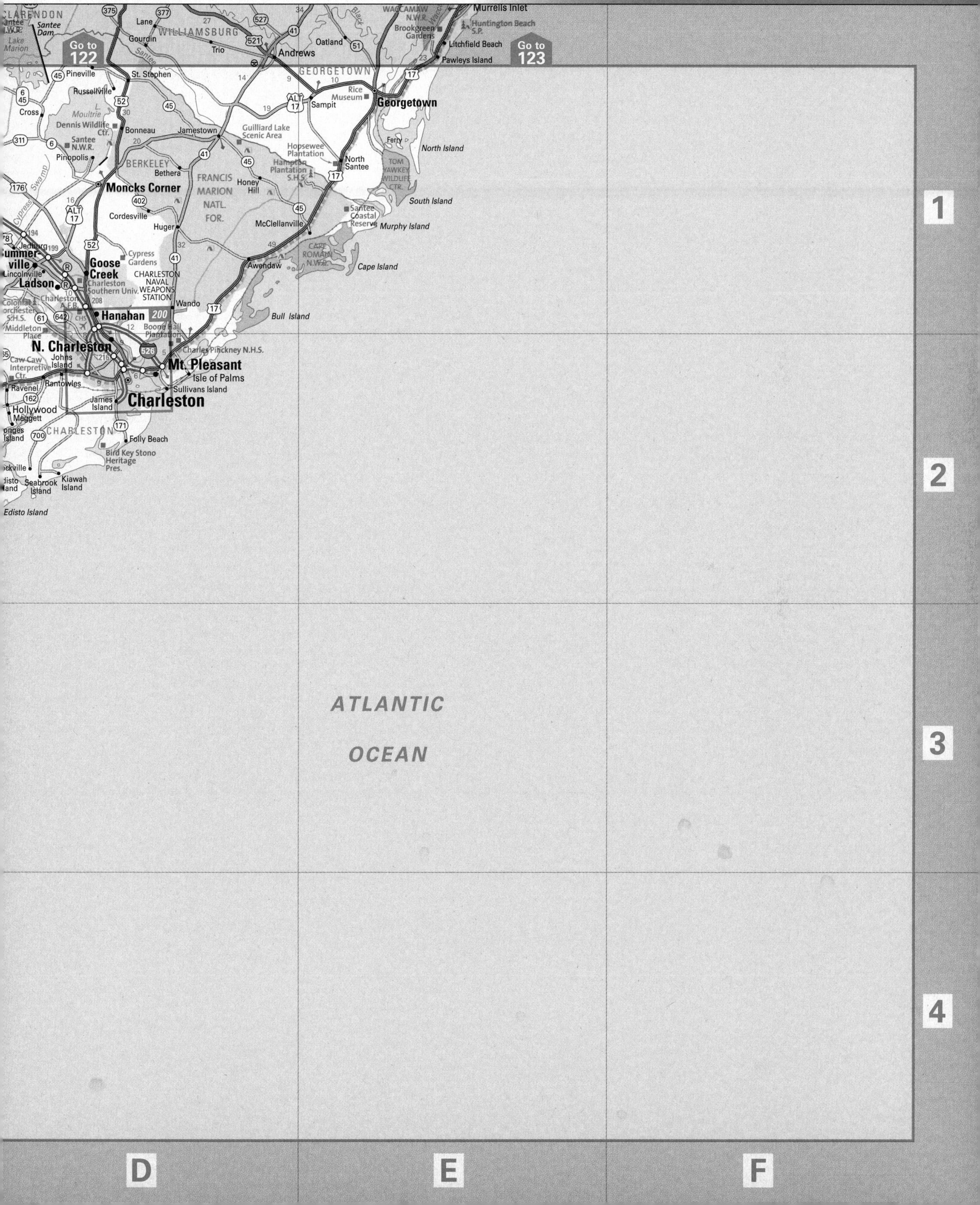

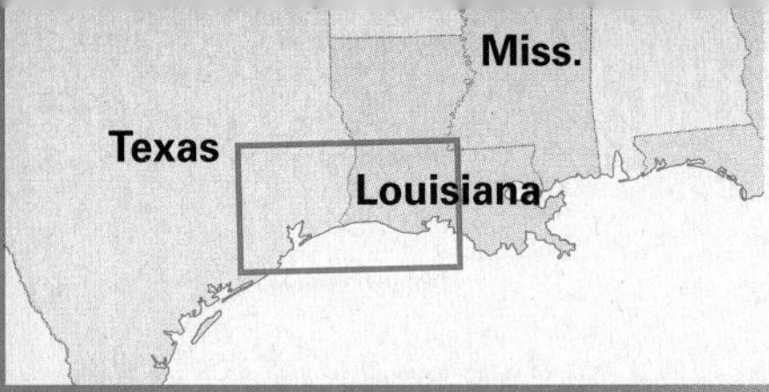

0 mi 10 20 30 40
0 km 10 20 30 40 50 60

One inch equals 18.4 miles
One centimeter equals 11.7 kilometers

DRIVING DISTANCES IN MILES

	ALEXANDRIA, LA	BEAUMONT, TX	DE RIDDER, LA	FREEPORT, TX	GALVESTON, TX	HOUSTON, TX	HUNTSVILLE, TX	LAFAYETTE, LA	LAKE CHARLES, LA	LUFKIN, TX	OPELOUSAS, LA	PORT ARTHUR, TX
BEAUMONT, TX	157		82	143	75	84	157	133	57	112	144	18
HOUSTON, TX	241	84	166	61	53		75	217	141	121	228	93
LAFAYETTE, LA	87	133	119	276	208	217	290		76	216	27	130
LAKE CHARLES, LA	100	57	49	200	132	141	214	76		140	87	54

SEE ALSO DISTANCE AND DRIVING TIME MAP ON PAGES 286–287

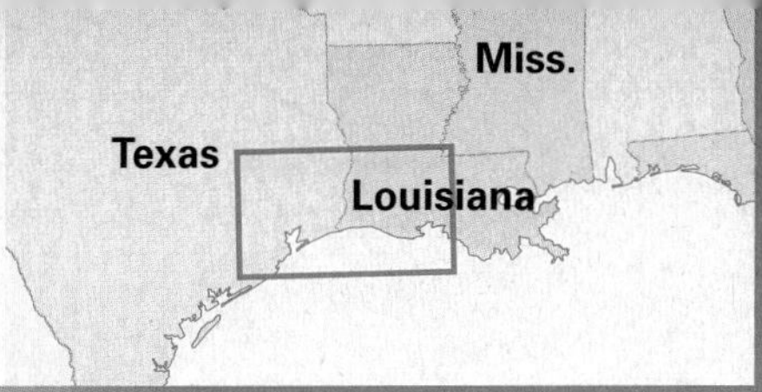

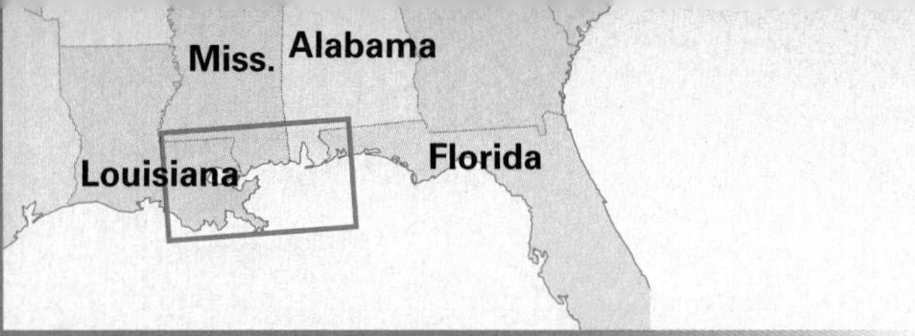

0 mi 10 20 30 40
0 km 10 20 30 40 50 60

One inch equals 18.4 miles
One centimeter equals 11.7 kilometers

New Orleans LA / Baton Rouge LA

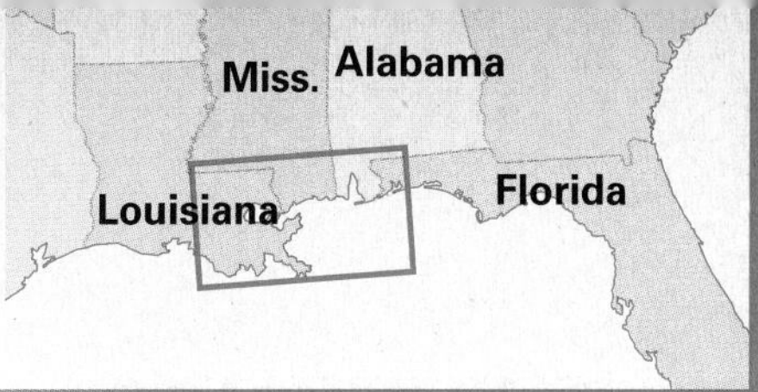

DRIVING DISTANCES IN MILES

	BATON ROUGE, LA	BILOXI, MS	GULFPORT, MS	GULF SHORES, AL	HAMMOND, LA	HATTIESBURG, MS	HOUMA, LA	MCCOMB, MS	MOBILE, AL	NEW ORLEANS, LA	PASCAGOULA, MS	PENSACOLA, FL
BATON ROUGE, LA		151	140	254	51	174	101	102	205	91	170	264
BILOXI, MS	151		12	110	106	82	148	161	61	93	20	120
MOBILE, AL	205	61	75	48	159	97	201	215		146	41	58
NEW ORLEANS, LA	91	93	81	195	57	115	57	111	146		112	205

SEE ALSO DISTANCE AND DRIVING TIME MAP ON PAGES 286–287

0 mi 10 20 30 40
0 km 10 20 30 40 50 60

One inch equals 18.4 miles
One centimeter equals 11.7 kilometers

Pensacola FL / Panama City FL

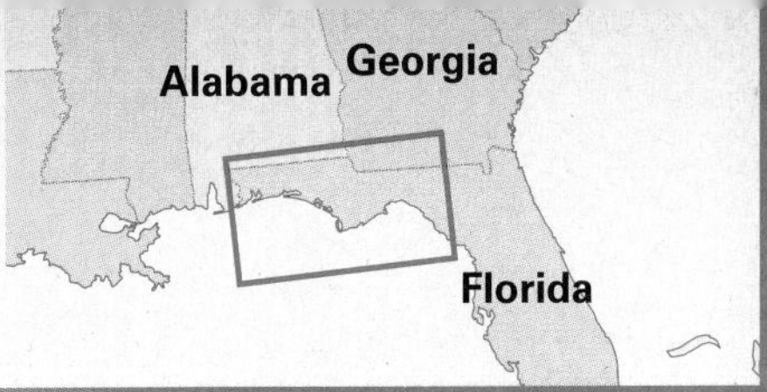

DRIVING DISTANCES IN MILES

	BREWTON, AL	DE FUNIAK SPRS., FL	DOTHAN, AL	FT. WALTON BEACH, FL	MARIANNA, FL	MOBILE, AL	PANAMA CITY, FL	PENSACOLA, FL	PERRY, FL	TALLAHASSEE, FL	THOMASVILLE, GA	VALDOSTA, GA
PANAMA CITY, FL	143	65	82	64	61	160		102	160	104	134	186
PENSACOLA, FL	57	82	152	39	138	58	102		256	200	230	282
TALLAHASSEE, FL	201	123	110	166	68	247	104	200	52		35	85
VALDOSTA, GA	283	204	133	247	149	329	186	282	66	85	42	

SEE ALSO DISTANCE AND DRIVING TIME MAP ON PAGES 286–287

Georgia

Florida

0 mi 10 20 30 40

0 km 10 20 30 40 50 60

One inch equals 18.4 miles

One centimeter equals 11.7 kilometers

Gainesville FL / Ocala FL

Go to 129

Go to 137

Go to 140

GEORGIA

FLORIDA

GULF OF MEXICO

Tallahassee

Valdosta

Moultrie

Thomasville

Cairo

Camilla

Adel

Nashville

Waycross

Homerville

Lakeland

Quitman

Monticello

Madison

Perry

Live Oak

Jasper

Lake City

Starke

Gainesville

Ocala

Hernando

OKEFENOKEE N.W.R.

OSCEOLA N.F.

LOWER SUWANNEE NATIONAL WILDLIFE REFUGE

ST. MARKS N.W.R.

Apalachee Bay

1 2 3 4

A B C

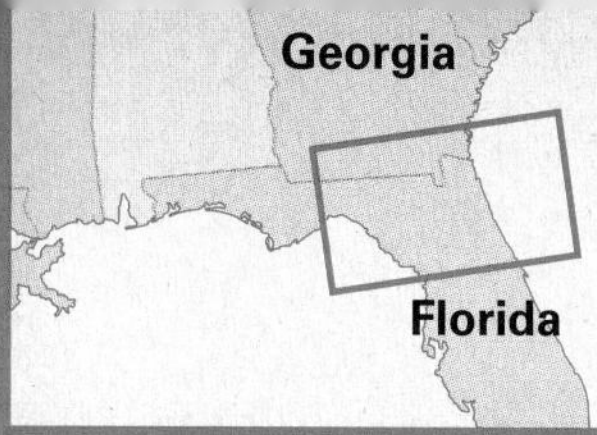

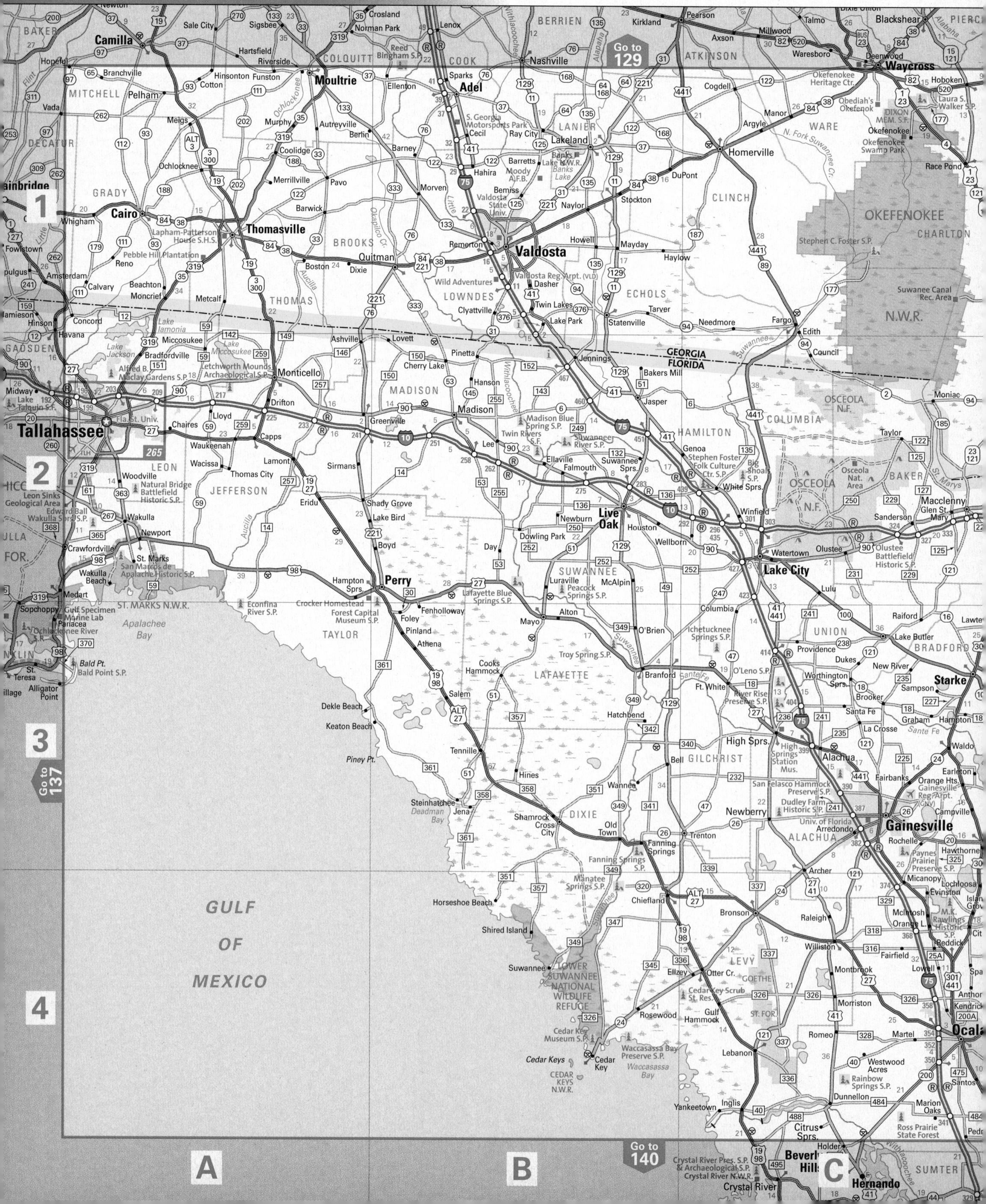

DRIVING DISTANCES IN MILES

	BRUNSWICK, GA	DAYTONA BEACH, FL	GAINESVILLE, FL	JACKSONVILLE, FL	LAKE CITY, FL	OCALA, FL	PERRY, FL	ST. AUGUSTINE, FL	STARKE, FL	TALLAHASSEE, FL	VALDOSTA, GA	WAYCROSS, GA
DAYTONA BEACH, FL	160		99	91	154	77	225	53	92	258	209	173
JACKSONVILLE, FL	69	91	70		62	101	133	41	45	166	117	78
OCALA, FL	171	77	40	101	80		120	81	57	186	137	170
TALLAHASSEE, FL	235	258	152	166	109	186	52	207	145		85	146

SEE ALSO DISTANCE AND DRIVING TIME MAP ON PAGES 286–287

0 mi 10 20 30 40
0 km 10 20 30 40 50 60
One inch equals 18.4 miles
One centimeter equals 11.7 kilometers

Tampa FL / Sarasota FL

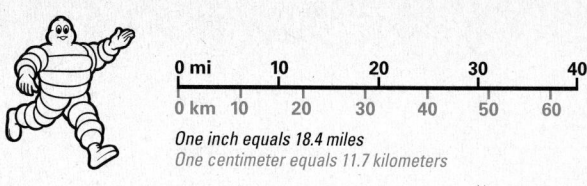

DRIVING DISTANCES IN MILES	FORT MYERS, FL	FORT PIERCE, FL	LAKELAND, FL	MELBOURNE, FL	OKEECHOBEE, FL	ORLANDO, FL	PUNTA GORDA, FL	ST. PETERSBURG, FL	SARASOTA, FL	TAMPA, FL	TITUSVILLE, FL	W. PALM BEACH, FL
FORT PIERCE, FL	126		122	57	36	120	127	197	150	172	95	57
ORLANDO, FL	155	120	56	72	108		131	107	130	82	40	169
SARASOTA, FL	74	150	85	190	114	130	50	35		60	170	184
TAMPA, FL	123	172	37	142	162	82	99	25	60		121	223

SEE ALSO DISTANCE AND DRIVING TIME MAP ON PAGES 286–287

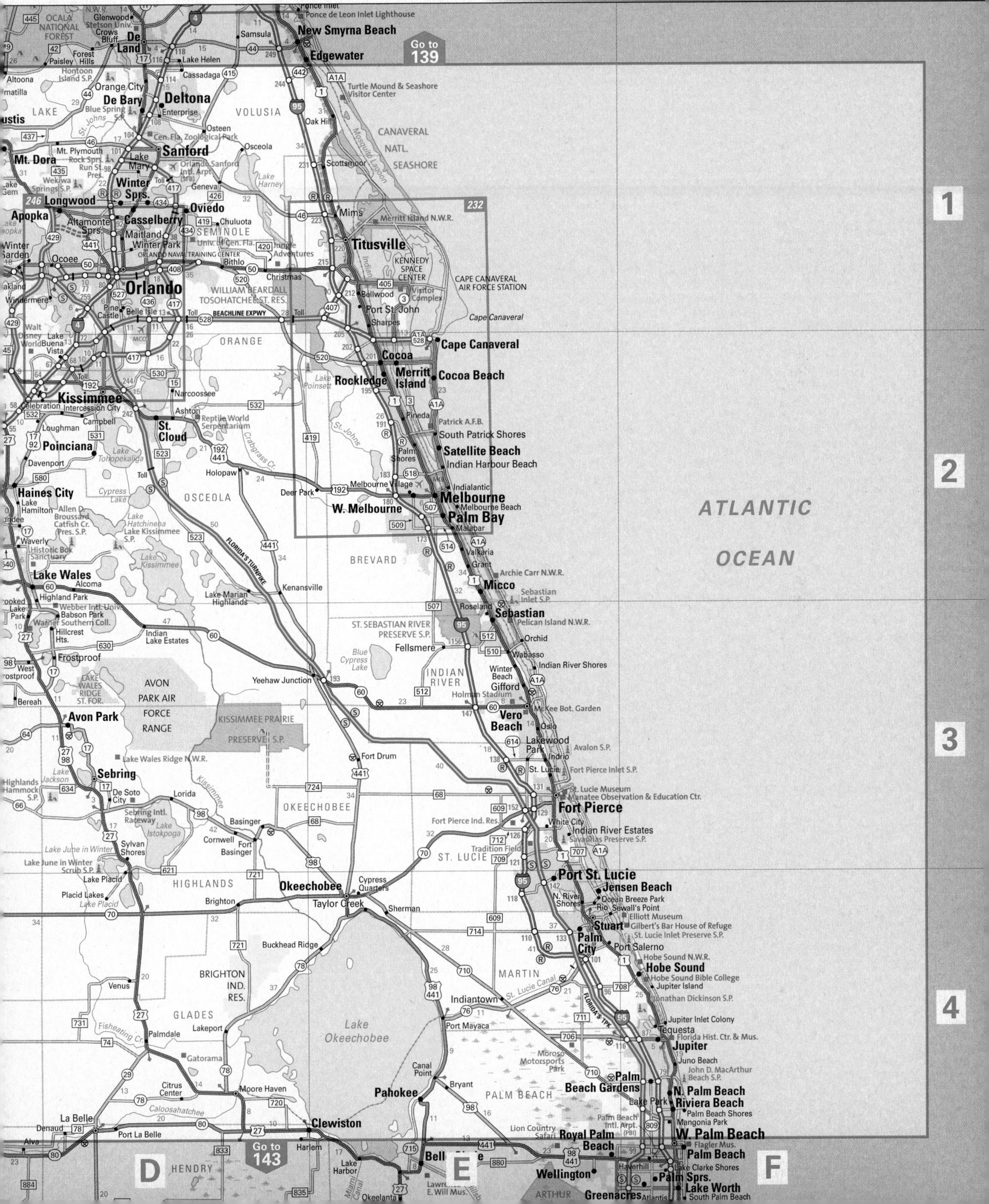

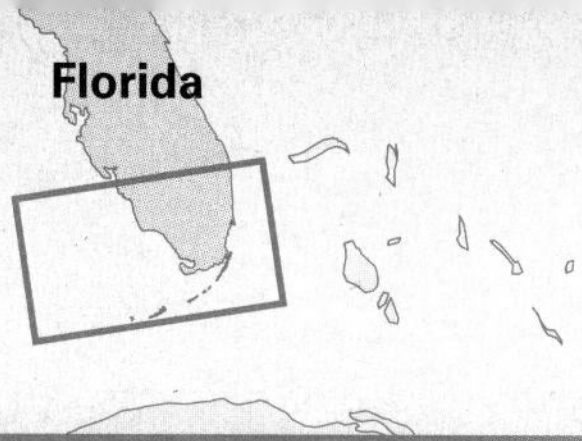

0 mi 10 20 30 40
0 km 10 20 30 40 50 60
One inch equals 18.4 miles
One centimeter equals 11.7 kilometers

Fort Myers FL / Key West FL

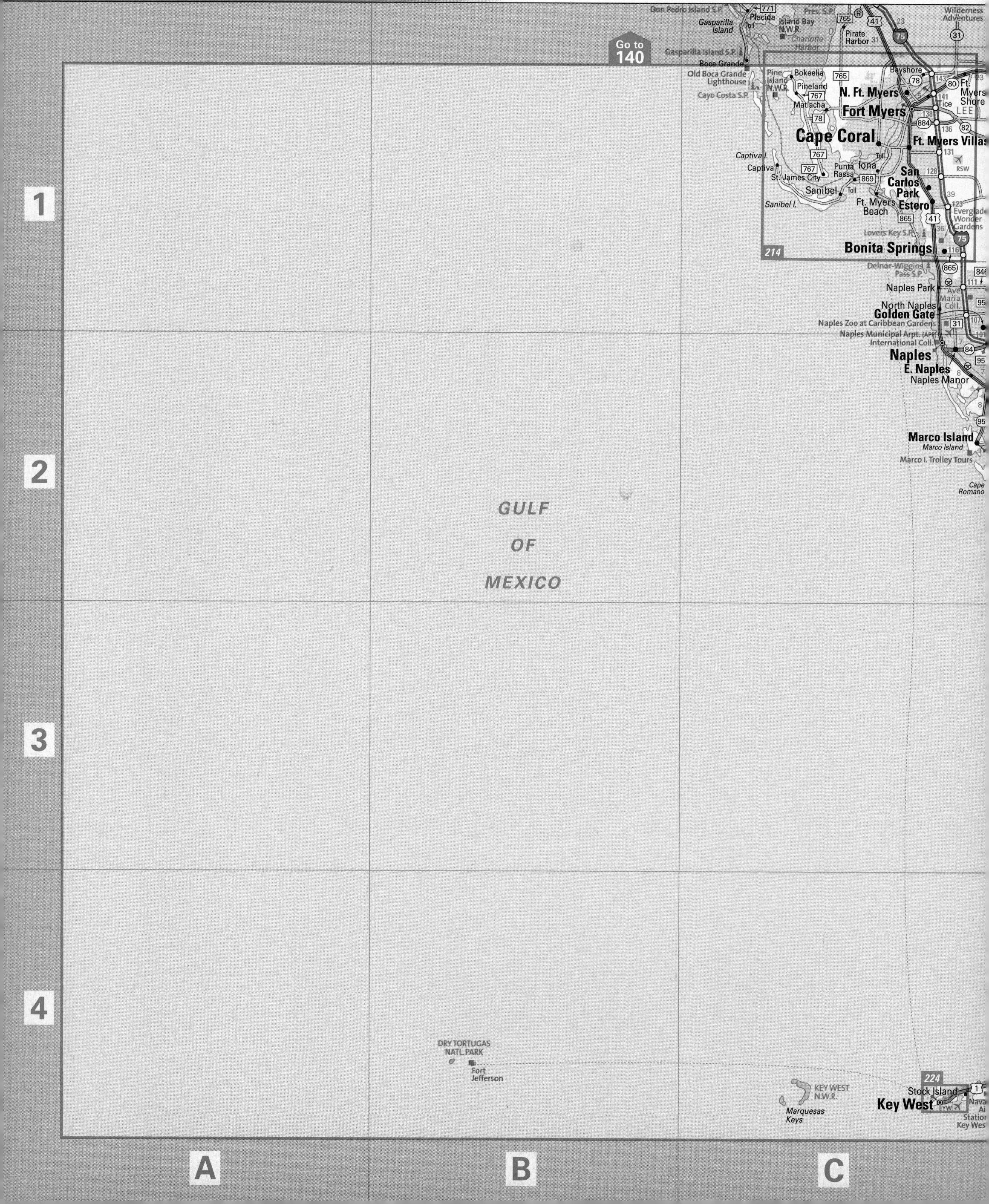

DRIVING DISTANCES IN MILES	BELLE GLADE, FL	BOCA RATON, FL	FLAMINGO, FL	FORT LAUDERDALE, FL	FORT MYERS, FL	HOMESTEAD, FL	KEY LARGO, FL	KEY WEST, FL	MARATHON, FL	MIAMI, FL	NAPLES, FL	W. PALM BEACH, FL
FORT MYERS, FL	84	155	227	139		174	195	308	260	155	36	125
KEY WEST, FL	235	211	181	190	308	133	113		48	168	273	234
MIAMI, FL	83	44	87	23	155	34	55	168	120		121	67
W. PALM BEACH, FL	41	28	153	48	125	100	121	234	186	67	144	

SEE ALSO DISTANCE AND DRIVING TIME MAP ON PAGES 286–287

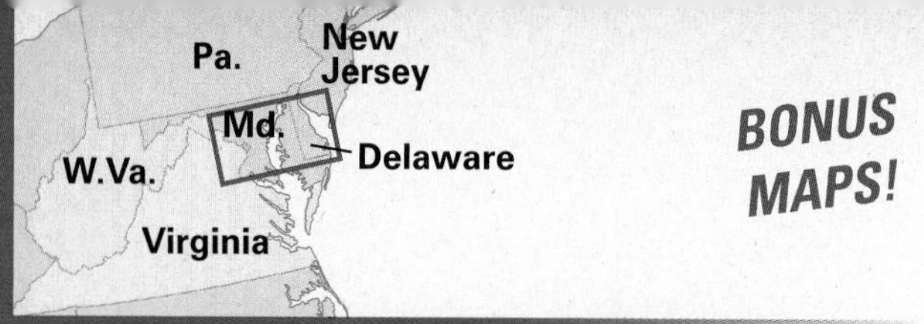

BONUS MAPS!

0 mi 5 10 15 20
0 km 5 10 15 20 25 30
One inch equals 9.85 miles
One centimeter equals 6.25 kilometers

Northeast Corridor / Washington DC

DRIVING DISTANCES IN MILES

	ANNAPOLIS, MD	BALTIMORE, MD	CAMBRIDGE, MD	DOVER, DE	ELKTON, MD	FREDERICK, MD	HAGERSTOWN, MD	LEESBURG, VA	MANASSAS, VA	REHOBOTH BEACH, DE	VINELAND, NJ	WASHINGTON, DC
BALTIMORE, MD	25		78	98	58	51	76	71	67	111	109	38
DOVER, DE	62	98	64		40	135	160	135	131	43	77	94
FREDERICK, MD	73	51	128	135	106		28	25	61	161	158	44
WASHINGTON, DC	31	38	87	94	94	44	70	38	31	120	145	

SEE ALSO DISTANCE AND DRIVING TIME MAP ON PAGES 286–287

BONUS MAPS!

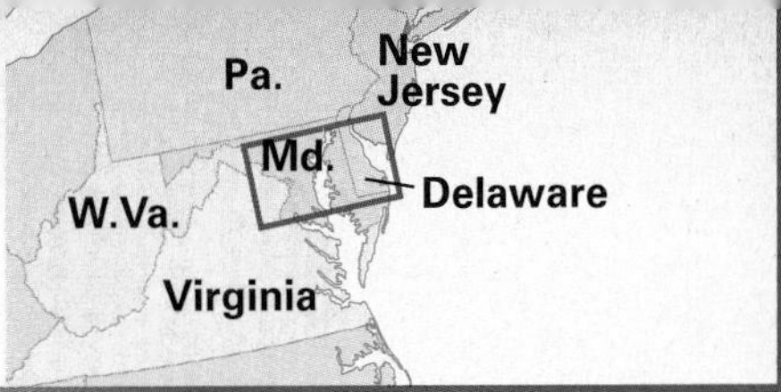

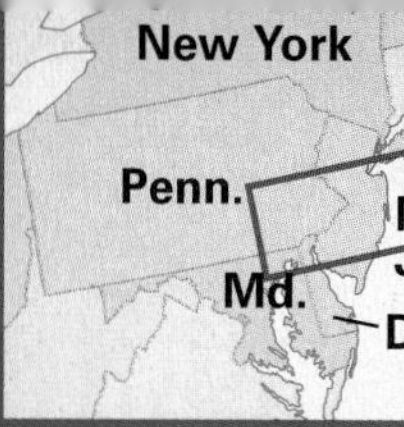

BONUS MAPS!

0 mi 5 10 15 20
0 km 5 10 15 20 25 30
One inch equals 9.85 miles
One centimeter equals 6.25 kilometers

Northeast Corridor / Philadelphia PA

BONUS MAPS!

DRIVING DISTANCES IN MILES

	ALLENTOWN, PA	ATLANTIC CITY, NJ	ELKTON, MD	LANCASTER, PA	LONG BRANCH, NJ	NEW BRUNSWICK, NJ	NEW YORK, NY	PHILADELPHIA, PA	READING, PA	TOMS RIVER, NJ	TRENTON, NJ	WILMINGTON, DE
NEW YORK, NY	84	125	137	165	55	34		91	118	75	55	120
PHILADELPHIA, PA	63	62	50	79	77	55	91		63	58	34	30
TRENTON, NJ	66	77	88	105	53	22	55	34	89	48		68
WILMINGTON, DE	77	86	20	53	106	90	120	30	56	85	68	

SEE ALSO DISTANCE AND DRIVING TIME MAP ON PAGES 286–287

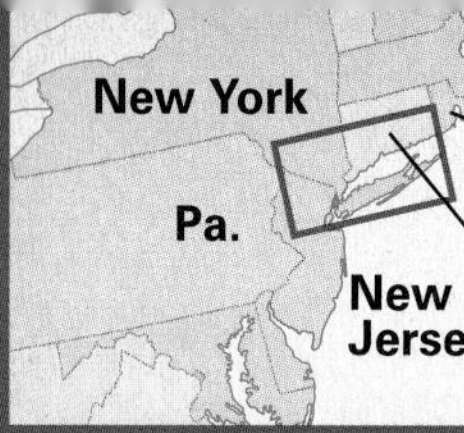

BONUS MAPS!

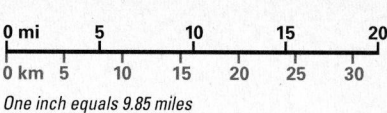

0 mi 5 10 15 20
0 km 5 10 15 20 25 30
One inch equals 9.85 miles
One centimeter equals 6.25 kilometers

DRIVING DISTANCES IN MILES	BRIDGEPORT, CT	DANBURY, CT	HARTFORD, CT	NEWARK, NJ	NEWBURGH, NY	NEW HAVEN, CT	NEW LONDON, CT	NEW YORK, NY	PATERSON, NJ	RIVERHEAD, NY	STAMFORD, CT	WATERBURY, CT
BRIDGEPORT, CT		31	56	69	73	19	64	60	71	115	21	33
NEWARK, NJ	69	79	125		66	88	134	11	18	88	48	108
NEW HAVEN, CT	19	35	39	88	78		46	78	89	133	40	30
NEW YORK, NY	60	69	115	11	56	78	124		16	78	38	99

SEE ALSO DISTANCE AND DRIVING TIME MAP ON PAGES 286–287

BONUS MAPS!

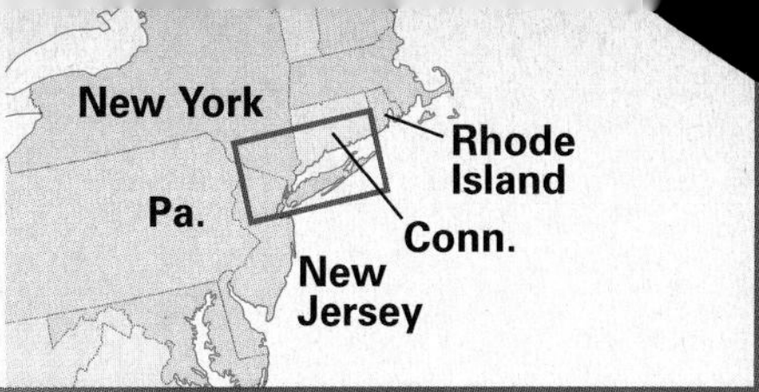

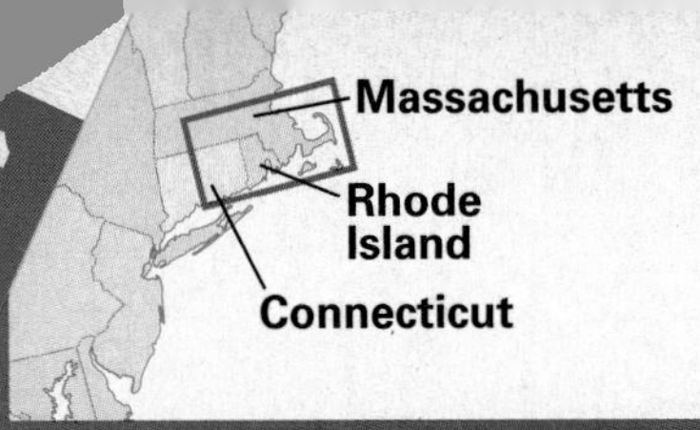

BONUS MAPS!

0 mi 5 10 15 20
0 km 5 10 15 20 25 30
One inch equals 9.85 miles
One centimeter equals 6.25 kilometers

Northeast Corridor / Hartford CT

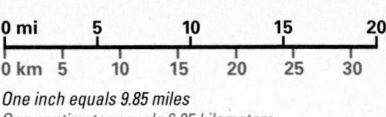

BONUS MAPS!

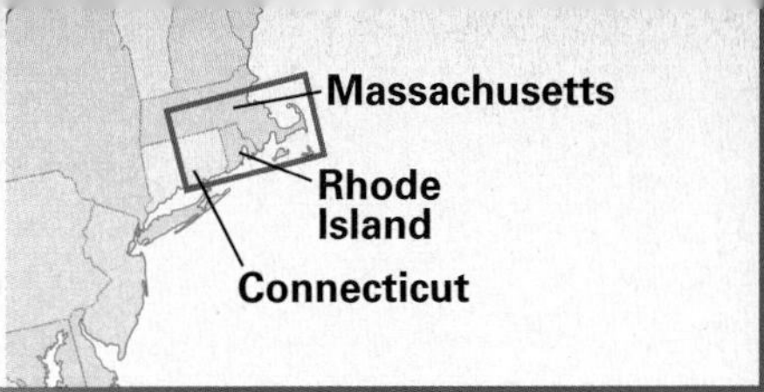

DRIVING DISTANCES IN MILES	BOSTON, MA	GLOUCESTER, MA	HARTFORD, CT	HYANNIS, MA	NEW BEDFORD, MA	NEW LONDON, CT	NEWPORT, RI	PLYMOUTH, MA	PROVIDENCE, RI	PROVINCETOWN, MA	SPRINGFIELD, MA	WORCESTER, MA
BOSTON, MA		35	102	72	60	109	73	41	52	117	95	46
HARTFORD, CT	102	136		155	104	46	85	127	73	200	25	62
PROVIDENCE, RI	52	92	73	71	33	58	33	41		117	75	43
SPRINGFIELD, MA	95	129	25	148	127	71	111	120	75	193		55

SEE ALSO DISTANCE AND DRIVING TIME MAP ON PAGES 286–287

CANADA
UNITED STATES
MEXICO
HAWAII

Honolulu HI / Lihue HI

0 mi 10 20 30 40
0 km 10 20 30 40 50 60
One inch equals 18.4 miles
One centimeter equals 11.7 kilometers

1

PACIFIC OCEAN

Kaua'i
Princeville
Hanalei N.W.R.
Kilauea Pt. N.W.R.
Hā'ena S.P.
Hā'ena
Kalihiwai
Kilauea
Hanalei
NĀPALI COAST
HONO'O NĀPALI NAT. AREA RES.
MOLOAA FOR. RES.
Anahola
NĀPALI COAST S.P.
KU'IA NAT. AREA RES.
Pu'uokila Lookout
PU'UKAPELE FOR. RES.
KŌKE'E S.P.
Wai'ale'ale (World's Rainiest Spot) 5,148
MAKALEHA MTS.
KEALIA FOR. RES.
Polihale S.P.
NĀPALI-KONA FOR. RES.
Kealia
Nohili Pt.
BARKING SANDS PACIFIC MISSILE RANGE FACILITY
WAIMEA CANYON S.P.
Waimea Canyon Lookout
Kawaikini 5,243
NONOU FOR. RES.
Kapa'a
Mānā
Wailua River S.P.
Wailua
Olu Pua Botanical Gardens & Plantation
Hanamā'ulu
Ahukini St. Rec. Pier
Kekaha
Waimea
Līhu'e
Līhu'e Arpt. (LIH)
Pākālā Village
Kalāheo
Puhi
Kaumakani
'Ōma'o
Hanapēpē
Kōloa
Hule'ia N.W.R.
'Ele'ele
Numila
Lāwa'i
Kukui'ula
Po'ipū
Spouting Horn
KAUA'I COUNTY

Lehua
Kīkepa Pt.
Keawanui Bay
Pāni'au 1,281
Pu'uwai
Pueo Pt.
Ni'ihau (RESTRICTED PUBLIC ACCESS)
Kawaihoa
Kaulakahi Channel
Kauai Channel

FOR CONTINUATION SEE MAP BELOW

FOR CONTINUATION SEE MAP ABOVE

2

3

4

PACIFIC OCEAN

Kauai Channel
O'ahu
Kahuku Pt.
James C. Campbell N.W.R.
Kawela Bay
Waiale'e
Kahuku
Sunset Beach
Waimea
Mālaekahana S.R.A.
Lā'ie
Pūpūkea
Polynesian Cultural Center
Kawailoa Beach
Waimea Falls
Hau'ula
Kawailoa
Pu'u Ka'inapua'a 2,360
SACRED FALLS S.P.
Ka'ena Pt. S.P.
Mokulē'ia
Punalu'u
Ka'ena Pt.
Hale'iwa
Kahana
Ka'a'awa
Waialua
Kamo'oloa
Ka'a'awa Beach Park
MĀKUA MIL. RES.
Whitmore Village
KAHANA VALLEY S.P.
KUALOA REG. PARK
Wahiawā
Waikāne
Mililani Town
Kahalu'u
CLOSED TO PUBLIC
'Āhuimanu
Mākaha
Waipi'o Acres
O'AHU FOREST N.W.R.
HAWAII MARINE CORPS BASE
Wai'anae
LUALUALEI NAVAL RES.
Pearl City
He'eia
Mā'ili
Kāne'ohe
Waipahu
Kailua
Nānākuli
Makakilo City
Maunawili
Honokai Hale
Waimānalo Bay Park
Kapolei
'Ewa Villages
Waimānalo
Waimānalo Beach
U.S. NAVAL RES.
Sea Life Park
'Ewa Beach
Makapu'u Pt.
HNL
Blow Hole
Waimalu
219
Hanauma Bay St. Underwater Park
HONOLULU COUNTY
Honolulu

Kaiwi Channel
'Īlio Pt.
Kalaupapa Airport (LUP)
Kahi'u Pt.
KALAWAO COUNTY
Pālā'au S.P.
KALAUPAPA NATL. HIST. PARK
Hipuapua Falls
Hālawa Bay
Pāpōhaku Beach
Moloka'i Airport (MKK)
Ho'olehua
Kalaupapa
Hālawa
Cape Hālawa
Maunaloa
Pu'u Nānā
Kualapu'u
MOLOKA'I FOR. RES.
Mo'a'ula Falls
Moloka'i Ranch Headquarters
Pu'u Ali'i Nat. Area Res.
Kamakou 4,970
'Ili'ili'ōpae Heiau
Oloku'i Nat. Area Res.
Lā'au Pt.
Waialua
Moloka'i
Kaunakakai
Kakahai'a N.W.R.
Kamalō
Pūko'o
'Ualapu'e
Pailolo Channel
Kalohi Channel
MAUI COUNTY
Honolua Bay
Nākālele Pt.
Honokahua
Honokōhau
Kahana
Kahakuloa
Haleki'i-Pihana Heiau St. Mon.
Shipwreck Beach
Honokōwai
WEST MAUI NAT. AREA RES.
Kahului Bay
Kā'anapali
Waihe'e
Lahaina
Waiehu
Kahului
Keanapapa Pt.
Garden of the Gods
Lāna'i
Kā'anapali & Pacific R.R.
Pu'u Kukui 5,788
Auau Channel
Lahaina
Wailuku
Pa'ia
Waikapū
Pu'unene Sugar Mus.
Lahaina Hist. Dist.
Lāna'i City
Keōmuku Village
WEST MAUI FOR. RES.
Olowalu
Mā'alaea
Kealia Pond N.W.R.
Kaumalapau
Lāna'i Airport (LNY)
Lāna'ihale 3,370
Maui
Papawai Pt.
Kīhei
Maui Ocean Ctr.
Mā'alaea Bay
Kaunolū Village
Pu'u Pehe
Pālaoa Pt.
Hulopo'e Beach Park
Kama'ole
Keōkea
'Ulupalakua
Mākena
Kealaikahiki Channel
Mākena Beach N.W.R.
Mākena S.P.
Molokini
Lao o Kukui
'ĀHIHI-KĪNA'U NAT. AREA RES.
Pu'u Moa'ulanui 1,483
Alalākeiki Channel
Kaho'olawe
Lae o Kealaikahiki
Lae o Kākā

FOR CONTINUATION SEE MAP AT RIGHT

A B C

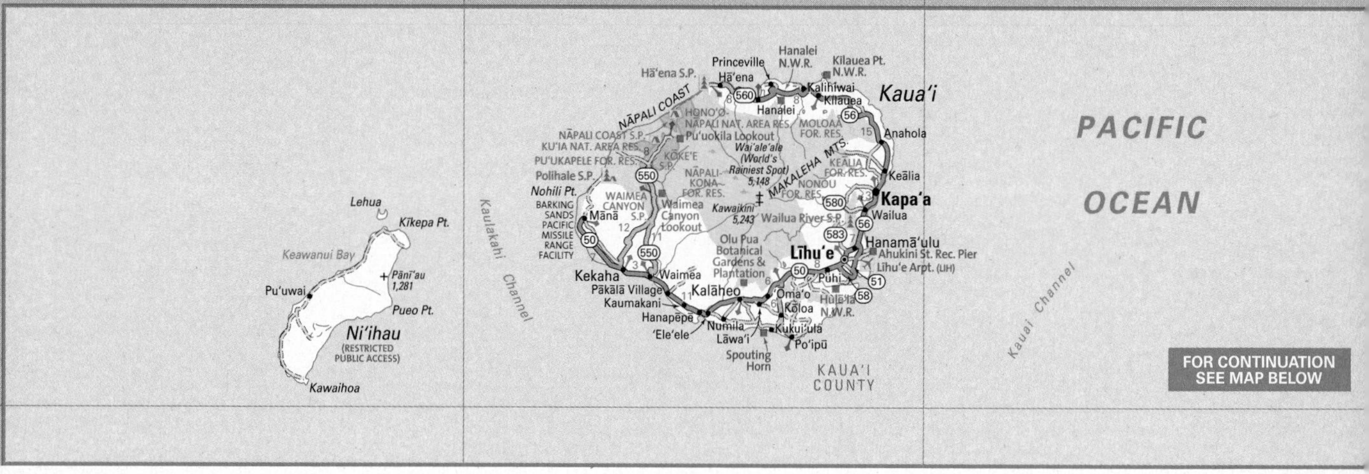

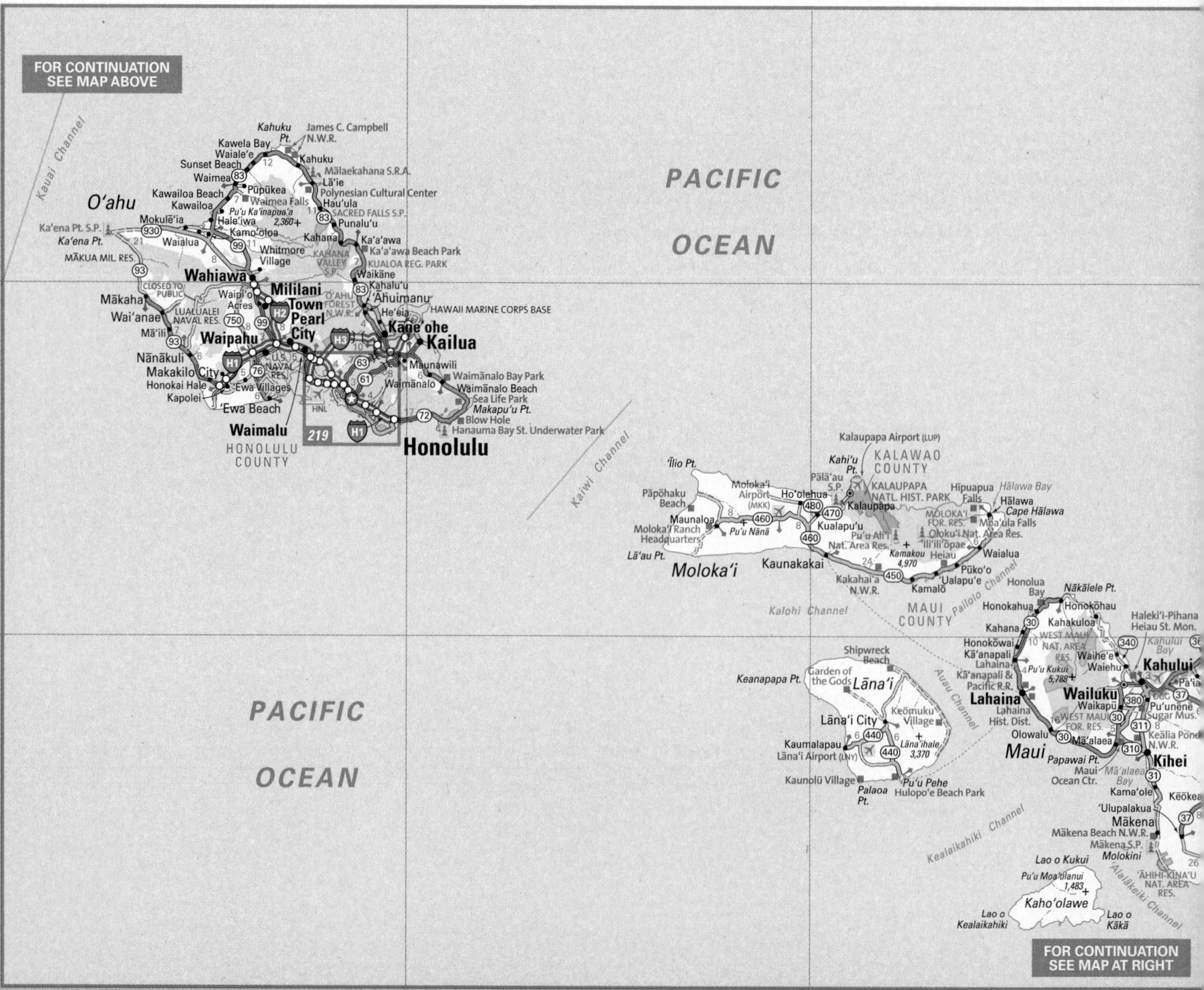

DRIVING DISTANCES IN MILES

	HĀNA	HILO	HONOLULU	HO'OLEHUA	KAHULUI	KAILUA	KAILUA-KONA	LAHAINA	LANAI CITY	LIHUE	WAHIAWĀ	WAIMEA
HILO	149*		217*	169*	121*	235*	88	142*	155*	319*	234*	54
HONOLULU	129*	217*		54*	101*	14	185*	92*	74*	102*	23	172*
KAHULUI	42	121*	101*	76*		119*	109*	23	57*	202*	118*	79*
LIHUE	230*	319*	102*	156*	202*	120*	285*	225*	176*		119*	174*

*DISTANCE INCLUDES AIR TRAVEL

SEE ALSO DISTANCE AND DRIVING TIME MAP ON PAGES 286–287

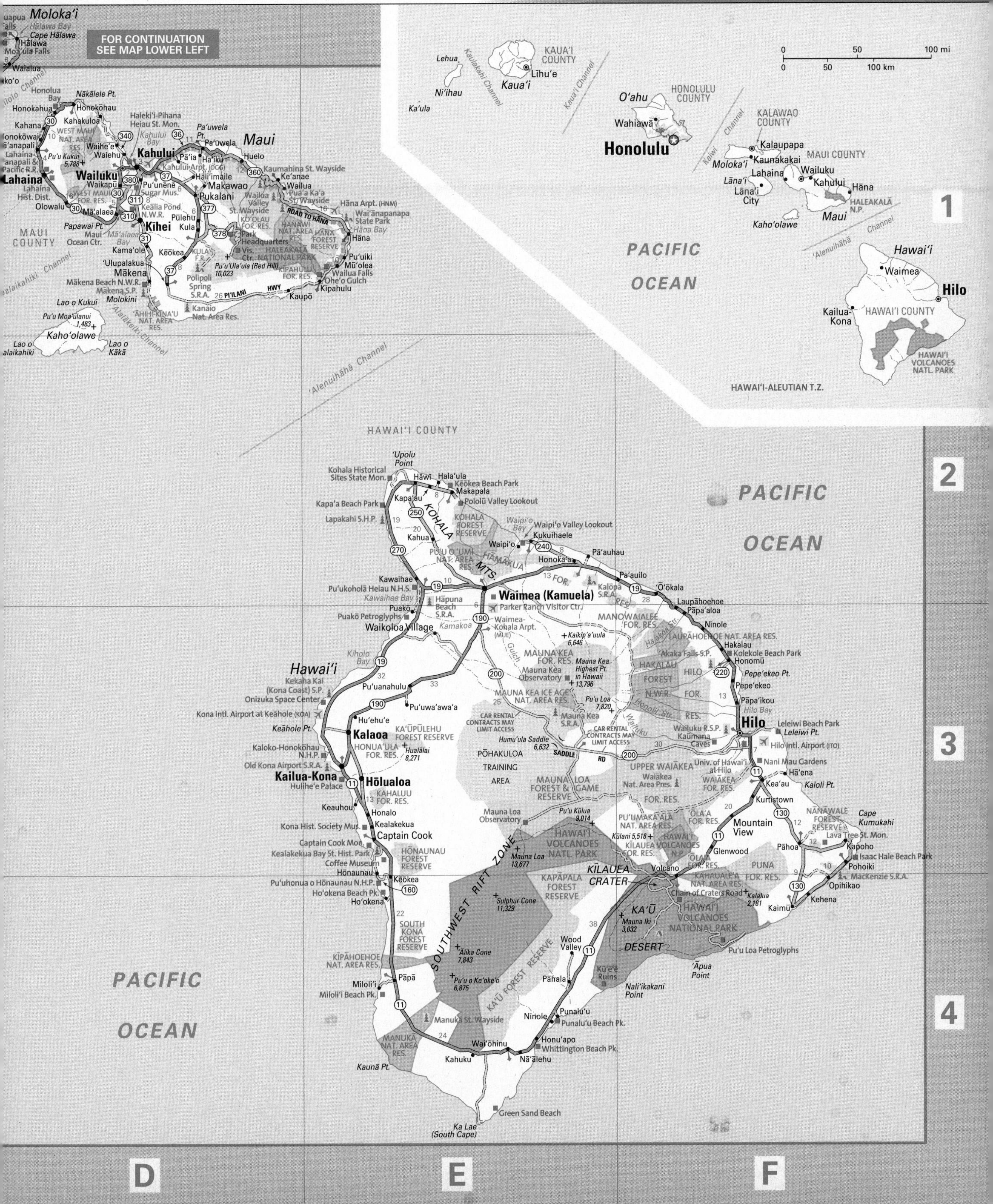

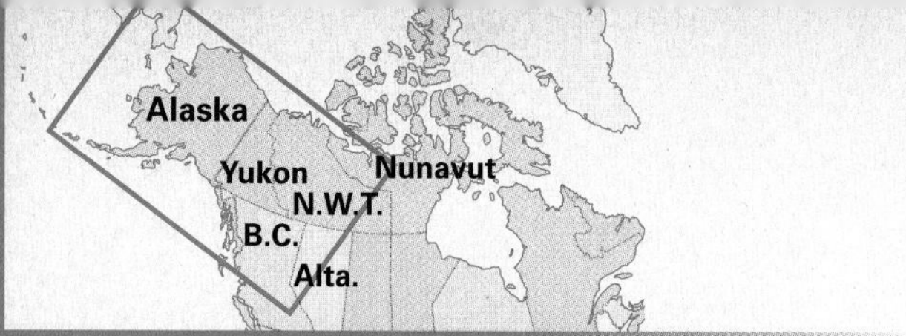

0 mi 100 200
0 km 100 200 300
One inch equals 142 miles
One centimeter equals 90 kilometers

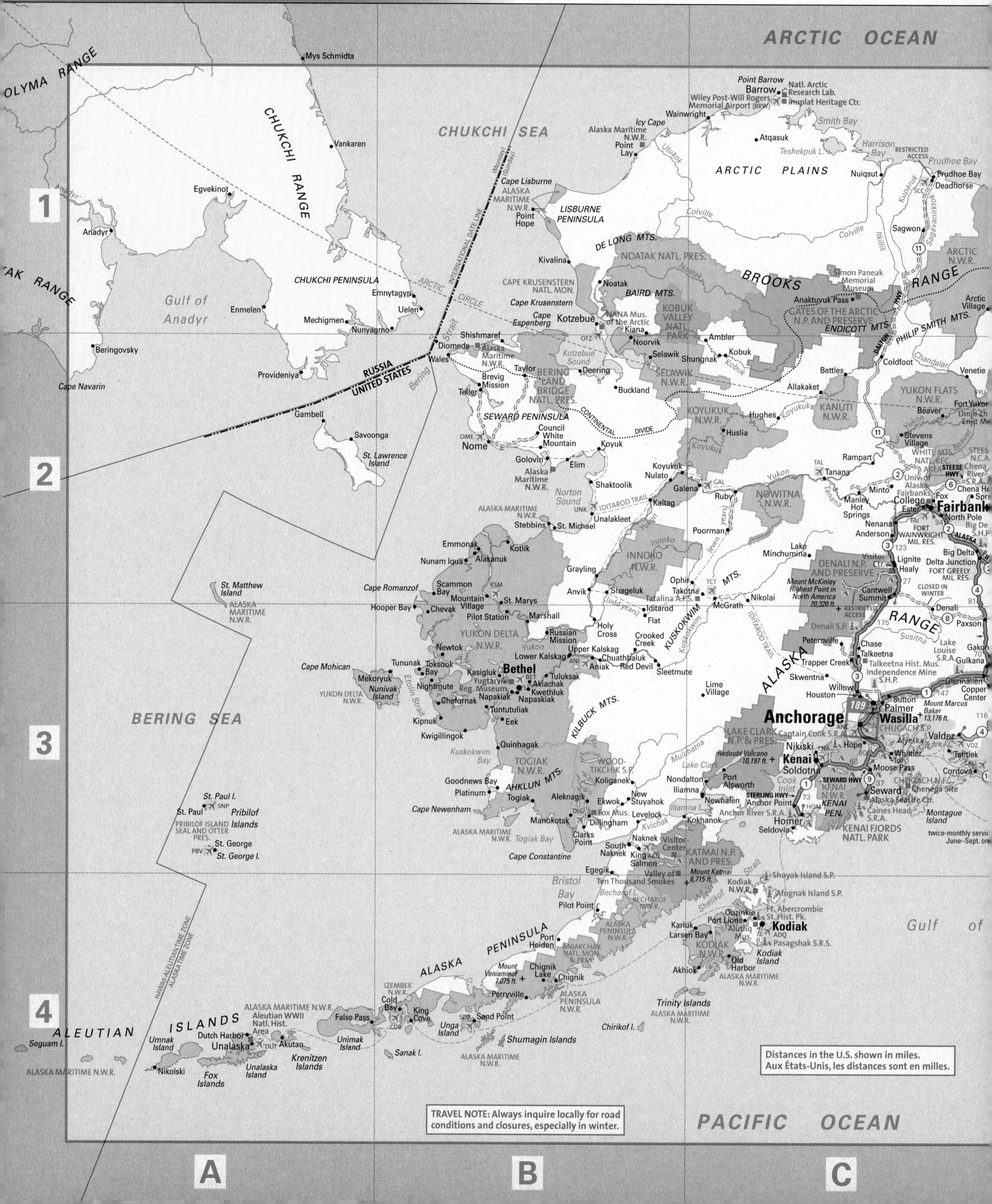

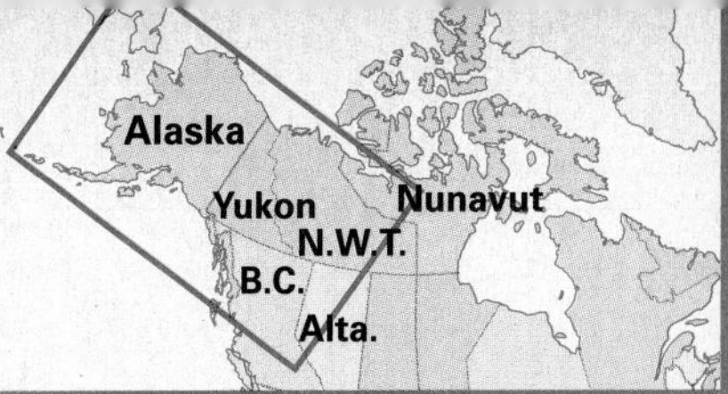

DRIVING DISTANCES IN MILES

	ANCHORAGE, AK	DAWSON CREEK, BC	DENALI NP, AK	FAIRBANKS, AK	HOMER, AK	JUNEAU, AK	PRINCE GEORGE, BC	PRINCE RUPERT, BC	SKAGWAY, AK	TOK, AK	WHITEHORSE, YT	YELLOWKNIFE, NT
ANCHORAGE, AK		1516	275	378	225	841*	1679	1514	807	323	697	1844
DAWSON CREEK, BC	1516		1503	1400	1740	963*	224	625	862	1193	819	741
FAIRBANKS, AK	378	1400	103		603	726*	1564	1398	691	207	581	1729
WHITEHORSE, YT	697	819	684	581	921	211*	982	817	110	374		1147

*DISTANCE INCLUDES FERRY TRAVEL

SEE ALSO DISTANCE AND DRIVING TIME MAP ON PAGES 286–287

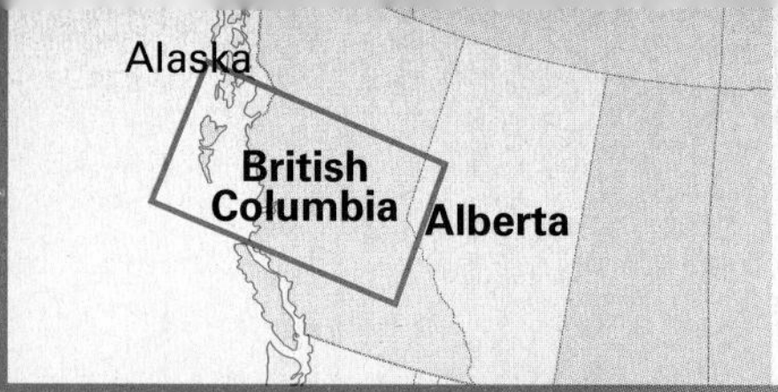
Alaska
British Columbia
Alberta

0 mi 20 40 60
0 km 20 40 60 80
One inch equals 40.3 miles/Un pouce équivaut à 40.3 milles
One centimeter equals 25.4 km/Un cm équivaut à 25.4 km

Prince Rupert BC / Terrace BC

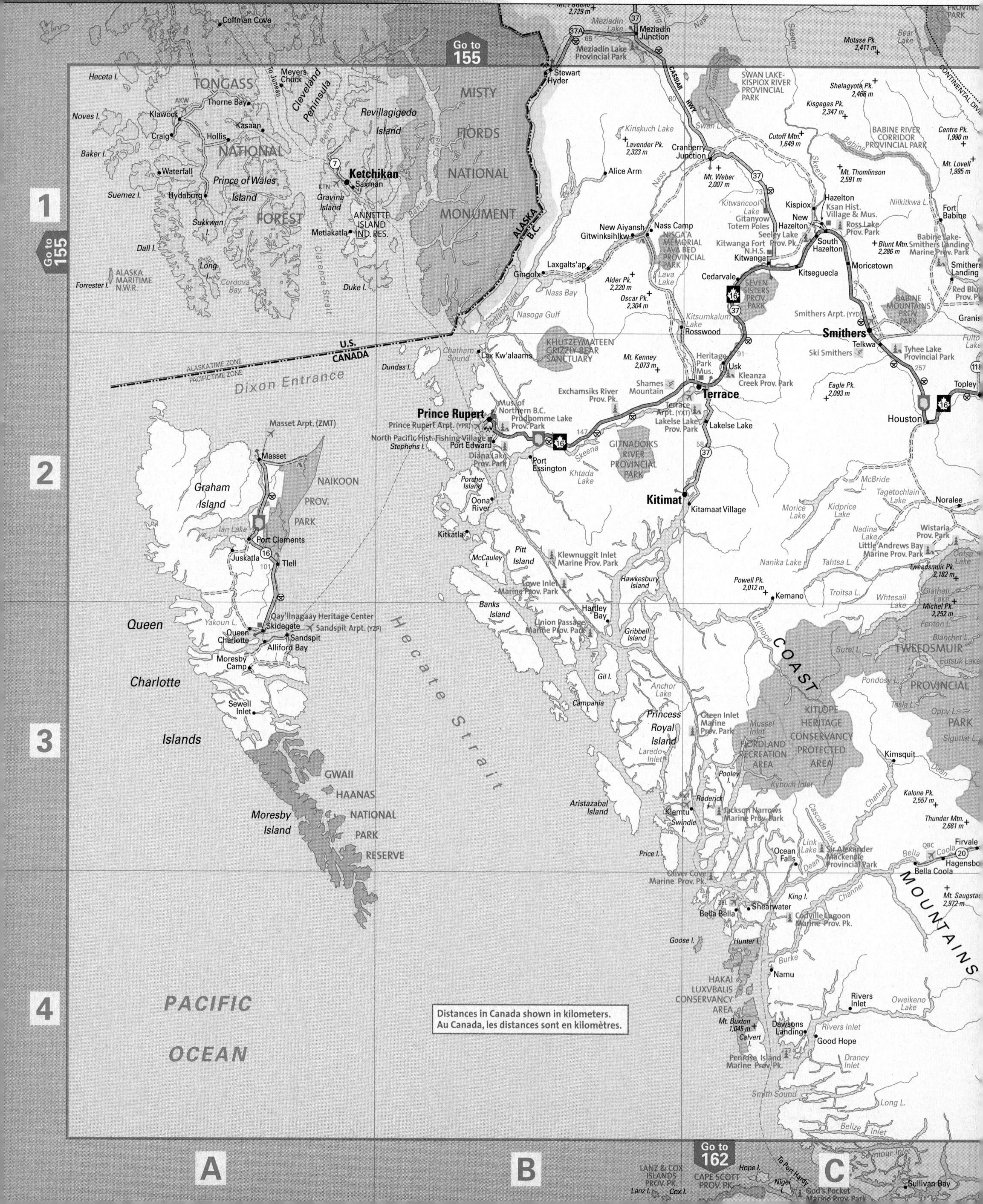
Go to 155
Go to 162
A
B
C
1
2
3
4
TONGASS
NATIONAL
FOREST
MISTY
FIORDS
NATIONAL
MONUMENT
Ketchikan
Saxman
Metlakatla
ANNETTE ISLAND IND. RES.
Revillagigedo Island
Cleveland Peninsula
Prince of Wales Island
Klawock
Craig
Hollis
Kasaan
Thorne Bay
Hydaburg
Waterfall
Coffman Cove
Meyers Chuck
Heceta I.
Noyes I.
Baker I.
Suemez I.
Dall I.
Long I.
Forrester I.
ALASKA MARITIME N.W.R.
Duke I.
Clarence Strait
Gravina Island
Stewart
Hyder
Meziadin Junction
Meziadin Lake Provincial Park
ALASKA
B.C.
U.S.
CANADA
ALASKA TIME ZONE
PACIFIC TIME ZONE
Dixon Entrance
CASSIAR HWY
SWAN LAKE-KISPIOX RIVER PROVINCIAL PARK
Cranberry Junction
Alice Arm
Kinskuch Lake
Lavender Pk. 2,323 m
Mt. Weber 2,007 m
Cutoff Mtn. 1,649 m
Kitwancool Lake
Gitanyow Totem Poles
Kispiox
Hazelton
New Hazelton
South Hazelton
Ksan Hist. Village & Mus.
Ross Lake Prov. Park
Seeley Lake Prov. Pk.
Kitwanga Fort N.H.S.
Kitwanga
Kitseguecla
Moricetown
Cedarvale
SEVEN SISTERS PROV. PARK
New Aiyansh
Gitwinksihlkw
Nass Camp
NISGA'A MEMORIAL LAVA BED PROVINCIAL PARK
Laxgalts'ap
Gingolx
Nass Bay
Nasoga Gulf
Alder Pk. 2,220 m
Oscar Pk. 2,304 m
KHUTZEYMATEEN GRIZZLY BEAR SANCTUARY
Lax Kw'alaams
Chatham Sound
Dundas I.
Kitsumkalum Lake
Rosswood
Mt. Kenney 2,073 m
Heritage Park Mus.
Usk
Kleanza Creek Prov. Park
Shames Mountain
Exchamsiks River Prov. Pk.
Terrace
Terrace Arpt. (YXT)
Lakelse Lake Prov. Park
Lakelse Lake
Prince Rupert
Mus. of Northern B.C.
Prudhomme Lake Prov. Park
Prince Rupert Arpt. (YPR)
North Pacific Hist. Fishing Village
Port Edward
Stephens I.
Diana Lake Prov. Park
Port Essington
Skeena
GITNADOIKS RIVER PROVINCIAL PARK
Khtada Lake
Kitimat
Kitamaat Village
Smithers
Smithers Arpt. (YYD)
Telkwa
Tyhee Lake Provincial Park
Ski Smithers
BABINE MOUNTAINS PROV. PARK
Houston
Topley
Eagle Pk. 2,093 m
BABINE RIVER CORRIDOR PROVINCIAL PARK
Mt. Thomlinson 2,591 m
Blunt Mtn. 2,286 m
Babine Lake-Smithers Landing Marine Prov. Park
Smithers Landing
Fort Babine
Nilkitkwa L.
Shelagyote Pk. 2,466 m
Kisgegas Pk. 2,347 m
Motase Pk. 2,411 m
Centre Pk. 1,990 m
Mt. Lovell 1,995 m
Bear Lake
Masset Arpt. (ZMT)
Masset
Graham Island
NAIKOON PROV. PARK
Ian Lake
Port Clements
Juskatla
Tlell
Queen Charlotte Islands
Yakoun L.
Qay'llnagaay Heritage Center
Skidegate
Sandspit Arpt. (YZP)
Queen Charlotte
Sandspit
Alliford Bay
Moresby Camp
Sewell Inlet
GWAII HAANAS NATIONAL PARK RESERVE
Moresby Island
Hecate Strait
PACIFIC OCEAN
Porcher Island
Oona River
Kitkatla
McCauley I.
Pitt Island
Klewnuggit Inlet Marine Prov. Park
Lowe Inlet Marine Prov. Park
Banks Island
Union Passage Marine Prov. Park
Hartley Bay
Hawkesbury Island
Gribbell Island
Gil I.
Campania
Anchor Lake
Princess Royal Island
Laredo Inlet
Aristazabal Island
Green Inlet Marine Prov. Park
Mussel Inlet
FIORDLAND RECREATION AREA
KITLOPE HERITAGE CONSERVANCY PROTECTED AREA
Kemano
Powell Pk. 2,012 m
Kitlope
COAST MOUNTAINS
McBride Lake
Tagetochlain Lake
Morice Lake
Kidprice Lake
Noralee
Nadina Lake
Wistaria Prov. Park
Little Andrews Bay Marine Prov. Park
Nanika Lake
Tahtsa L.
Troitsa L.
Whitesail Lake
Tweedsmuir Pk. 2,182 m
Michel Pk. 2,252 m
TWEEDSMUIR PROVINCIAL PARK
Kimsquit
Pooley I.
Roderick I.
Jackson Narrows Marine Prov. Park
Klemtu
Swindle I.
Kynoch Inlet
Price I.
Ocean Falls
Link Lake
Sir Alexander Mackenzie Provincial Park
Oliver Cove Marine Prov. Pk.
Kalone Pk. 2,557 m
Thunder Mtn. 2,681 m
Firvale
Hagensborg
Bella Coola
Mt. Saugstad 2,972 m
King I.
Bella Bella
Shearwater
Codville Lagoon Marine Prov. Pk.
Goose I.
Hunter I.
Burke
Namu
HAKAI LUXVBALIS CONSERVANCY AREA
Mt. Buxton 1,045 m
Calvert I.
Dawsons Landing
Rivers Inlet
Good Hope
Oweikeno Lake
Penrose Island Marine Prov. Pk.
Draney Inlet
Smith Sound
Long L.
Belize Inlet
Seymour Inlet
Sullivan Bay
LANZ & COX ISLANDS PROV. PK.
CAPE SCOTT PROV. PK.
Hope I.
Nigei
God's Pocket Marine Prov. Park
To Port Hardy
Distances in Canada shown in kilometers.
Au Canada, les distances sont en kilomètres.

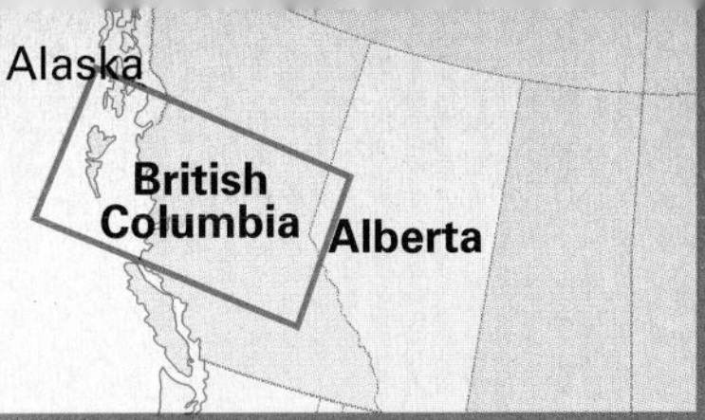

DRIVING DISTANCES IN KM / DISTANCES ROUTIÈRES EN KM

	DAWSON CREEK, BC	GRANDE PRAIRIE, AB	KAMLOOPS, BC	KITIMAT, BC	100 MILE HOUSE, BC	PRINCE GEORGE, BC	PRINCE RUPERT, BC	SMITHERS, BC	STEWART, BC	TERRACE, BC	VALEMOUNT, BC	WILLIAMS LAKE, BC
DAWSON CREEK, BC		124	931	1041	734	406	1130	777	1109	983	642	644
PRINCE GEORGE, BC	406	530	525	635	328		724	371	703	577	295	238
PRINCE RUPERT, BC	1130	1254	1249	205	1052	724		353	463	147	1019	962
WILLIAMS LAKE, BC	644	768	287	873	90	238	962	609	941	815	332	

SEE ALSO DISTANCE AND DRIVING TIME MAP ON PAGES 286–287 / VOIR AUSSI CARTE DES DISTANCES ET DES TEMPS DE PARCOURS PAGES 286–287

0 mi 20 40 60
0 km 20 40 60 80
One inch equals 40.3 miles/Un pouce équivaut à 40.3 milles
One centimeter equals 25.4 km/Un cm équivaut à 25.4 km

Grande Prairie AB / Jasper AB

DRIVING DISTANCES IN KM / DISTANCES ROUTIÈRES EN KM

	DAWSON CREEK, BC	EDMONTON, AB	FORT McMURRAY, AB	GRANDE PRAIRIE, AB	JASPER, AB	LLOYDMINSTER, AB/SK	MEADOW LAKE, SK	N. BATTLEFORD, SK	PEACE RIVER, AB	SLAVE LAKE, AB	VALEMOUNT, BC	WHITECOURT, AB
EDMONTON, AB	597		439	462	367	238	415	375	484	251	488	177
GRANDE PRAIRIE, AB	124	462	756		397	700	824	837	197	318	518	279
JASPER, AB	521	367	796	397		605	782	742	578	464	121	271
N. BATTLEFORD, SK	972	375	814	837	742	137	158		866	633	863	559

SEE ALSO DISTANCE AND DRIVING TIME MAP ON PAGES 286–287 / VOIR AUSSI CARTE DES DISTANCES ET DES TEMPS DE PARCOURS PAGES 286–287

Edmonton AB / Fort McMurray AB

0 mi 20 40 60
0 km 20 40 60 80
One inch equals 40.3 miles/Un pouce équivaut à 40.3 milles
One centimeter equals 25.4 km/Un cm équivaut à 25.4 km

Prince Albert SK / La Ronge SK

DRIVING DISTANCES IN KM / DISTANCES ROUTIÈRES EN KM

	FLIN FLON, MB	GILLAM, MB	GRAND RAPIDS, MB	LA LOCHE, SK	LA RONGE, SK	LYNN LAKE, MB	MEADOW LAKE, SK	NIPAWIN, SK	N. BATTLEFORD, SK	PRINCE ALBERT, SK	THE PAS, MB	THOMPSON, MB
FLIN FLON, MB		676	402	889	613	703	633	388	571	375	141	380
MEADOW LAKE, SK	633	1309	867	305	496	1336		399	158	258	569	1013
PRINCE ALBERT, SK	375	1051	609	514	238	1078	258	141	196		311	781
THOMPSON, MB	380	296	328	1269	697	323	1013	640	977	781	470	

SEE ALSO DISTANCE AND DRIVING TIME MAP ON PAGES 286–287 / VOIR AUSSI CARTE DES DISTANCES ET DES TEMPS DE PARCOURS PAGES 286–287

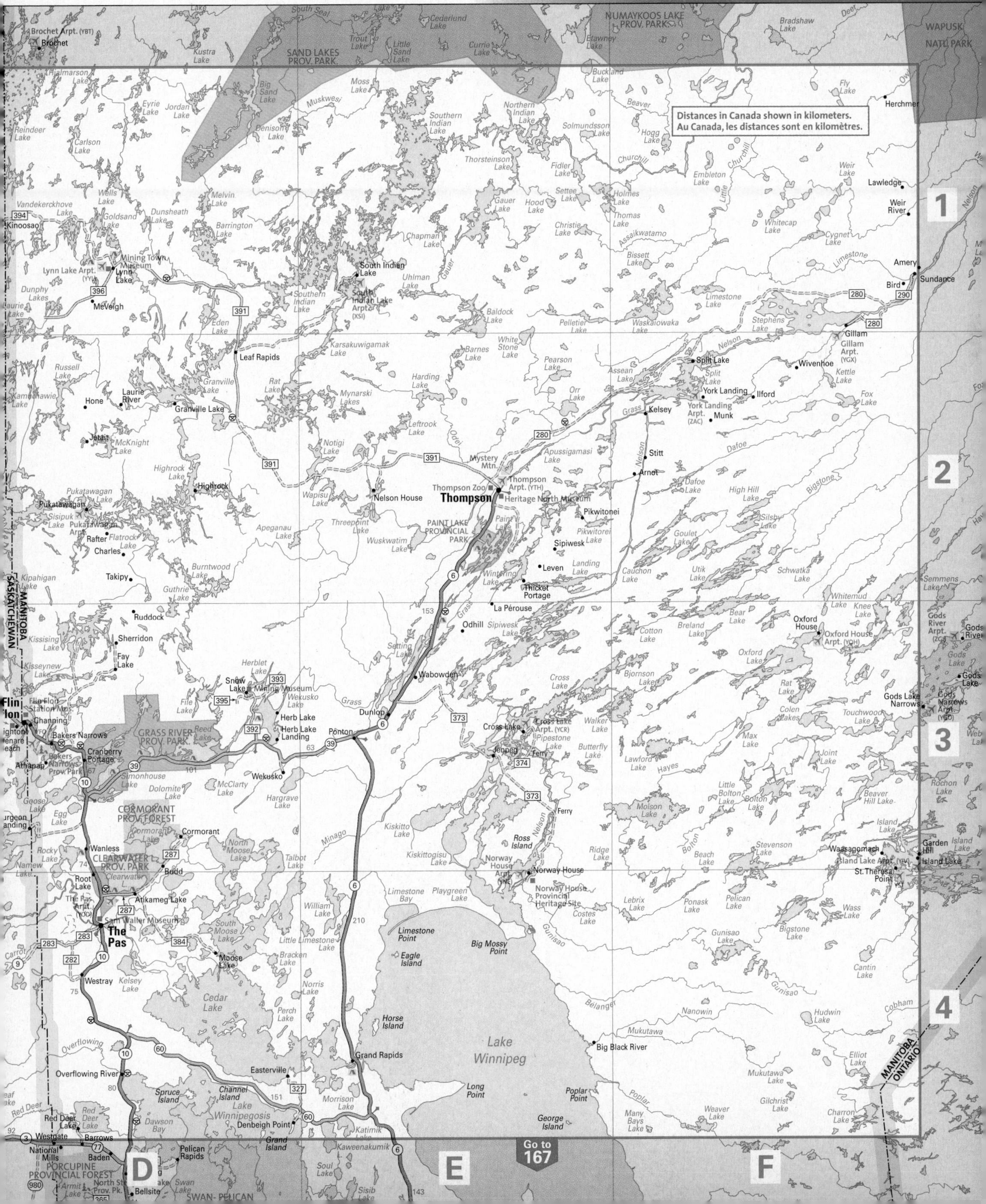

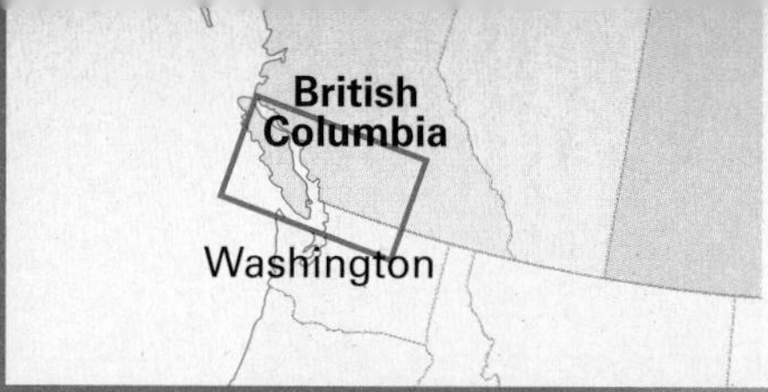

0 mi 20 40
0 km 20 40 60
One inch equals 25.4 miles/Un pouce équivaut à 25.4 milles
One cm equals 16.1 km/Un cm équivaut à 16.1 km

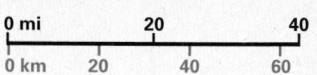

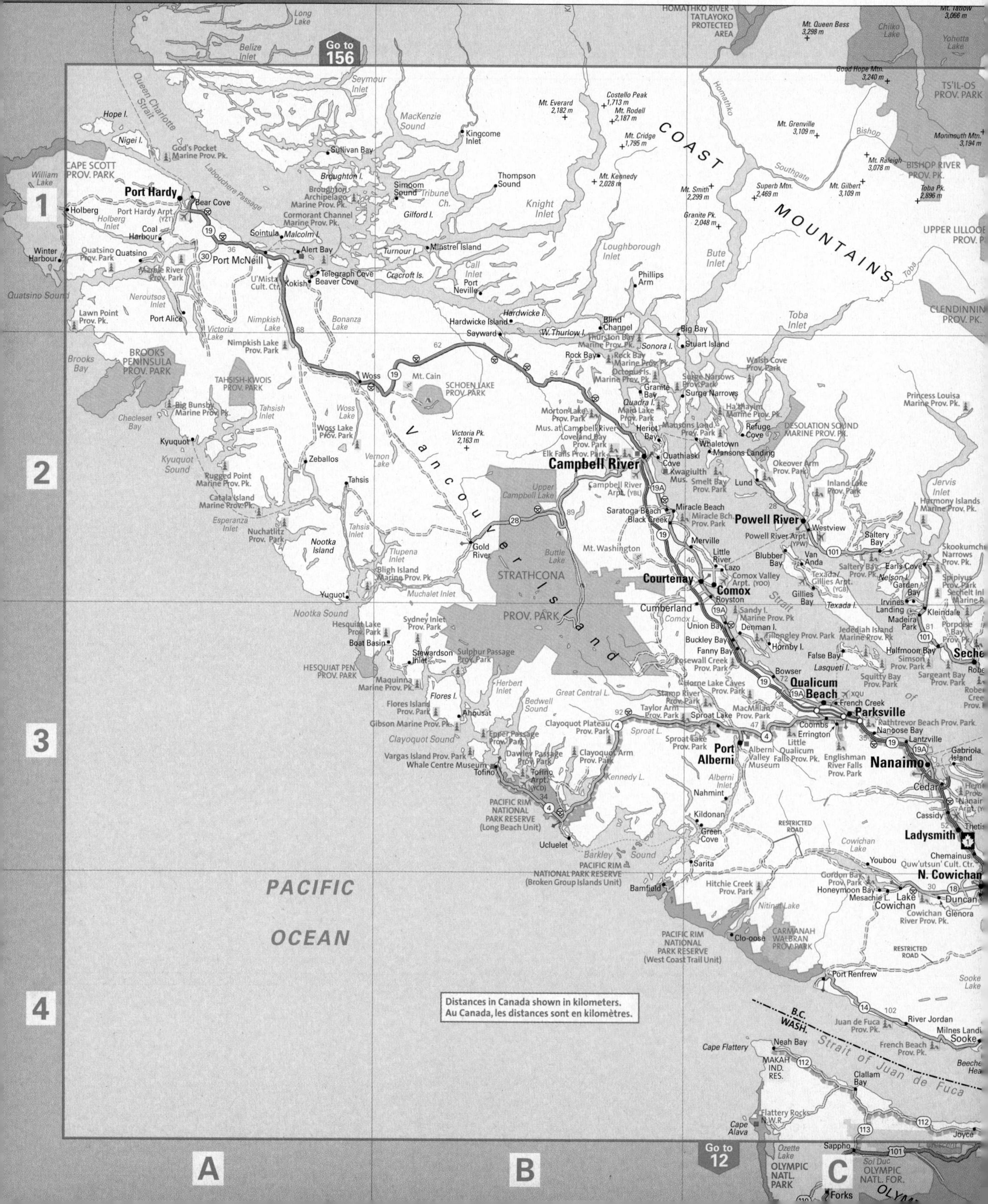

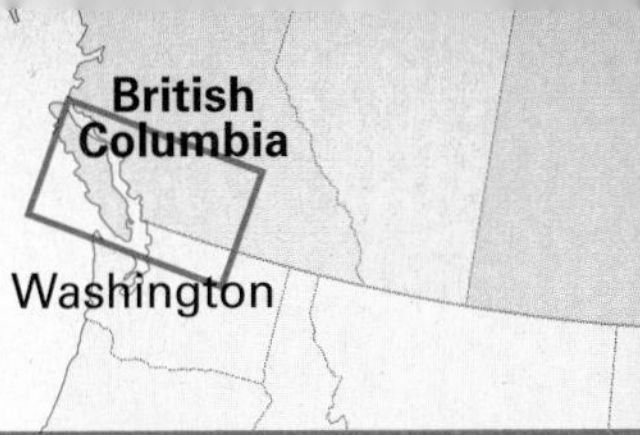

DRIVING DISTANCES IN KM / DISTANCES ROUTIÈRES EN KM

	CAMPBELL RIVER, BC	KAMLOOPS, BC	KELOWNA, BC	MERRITT, BC	NANAIMO, BC	OSOYOOS, BC	PORT ALBERNI, BC	PORT HARDY, BC	SALMON ARM, BC	VANCOUVER, BC	VICTORIA, BC	WHISTLER, BC
KAMLOOPS, BC	512		163	87	363	231	441	750	108	355	393	475
NANAIMO, BC	153	363	403	279		404	82	391	471	23	113	104
VANCOUVER, BC	172	355	395	271	23	396	101	410	463		69	123
VICTORIA, BC	266	393	433	309	113	434	195	504	501	69		192

SEE ALSO DISTANCE AND DRIVING TIME MAP ON PAGES 286–287 / VOIR AUSSI CARTE DES DISTANCES ET DES TEMPS DE PARCOURS PAGES 286–287

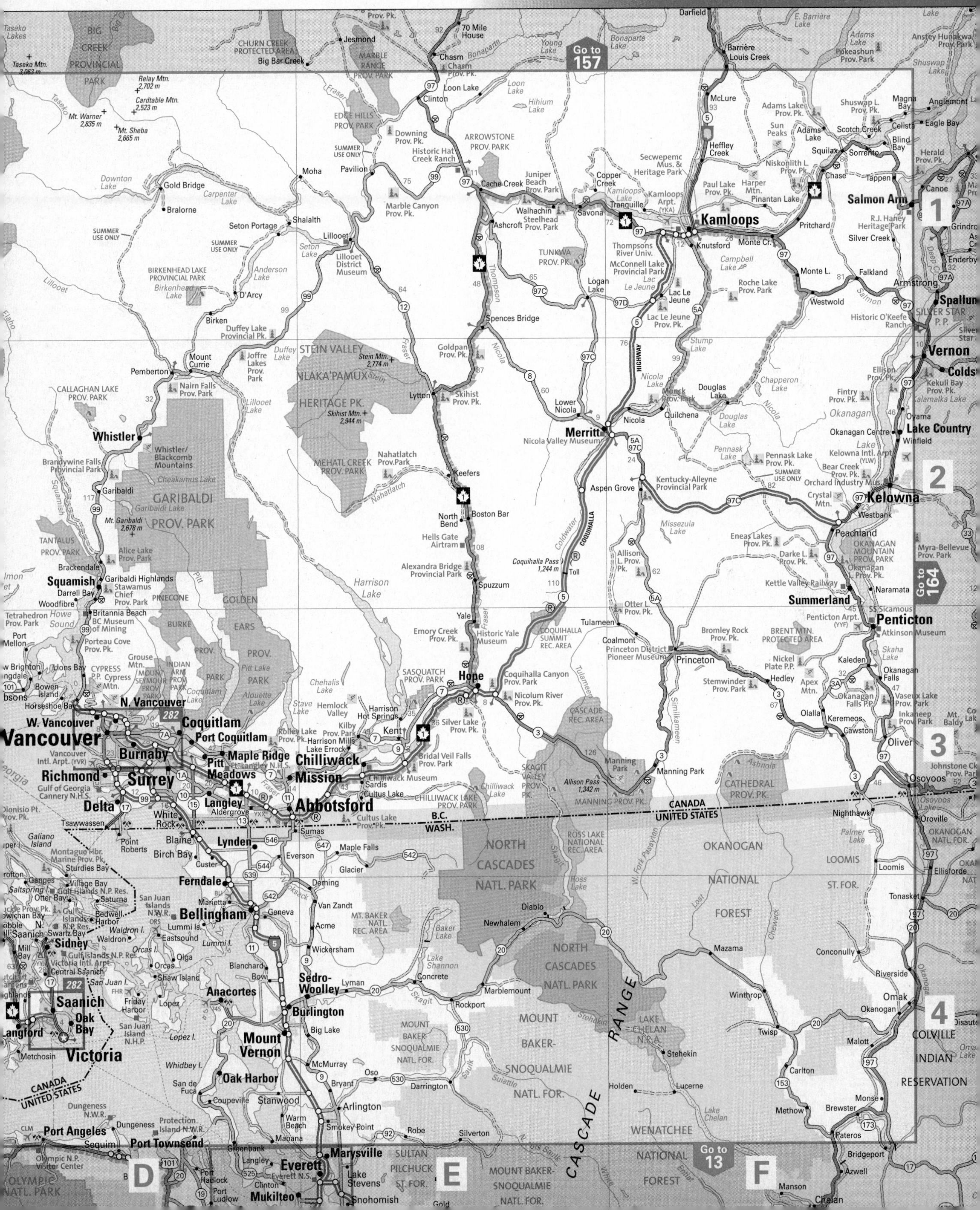

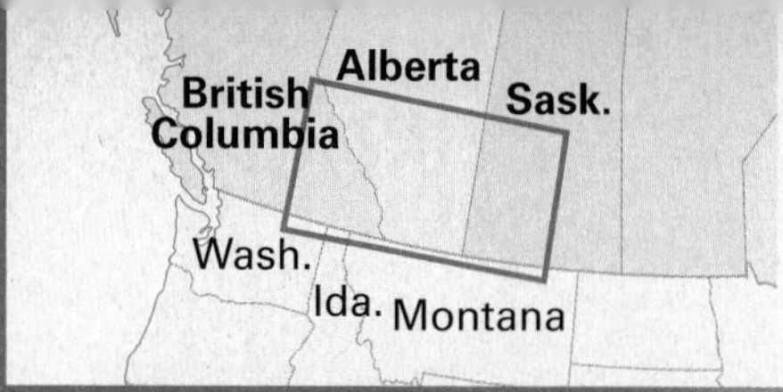

0 mi 20 40 60
0 km 20 40 60 80

One inch equals 40.3 miles/Un pouce équivaut à 40.3 milles
One centimeter equals 25.4 km/Un cm équivaut à 25.4 km

DRIVING DISTANCES IN KM / DISTANCES ROUTIÈRES EN KM

	BANFF, AB	CALGARY, AB	CRANBROOK, BC	EDMONTON, AB	JASPER, AB	KELOWNA, BC	LETHBRIDGE, AB	LLOYDMINSTER AB/SK	MEDICINE HAT, AB	RED DEER, AB	SASKATOON, SK	SWIFT CURRENT, SK
CALGARY, AB	128		383	296	396	638	216	534	285	145	620	503
EDMONTON, AB	412	296	679		367	934	512	238	579	150	513	676
LETHBRIDGE, AB	344	216	306	512	612	809		605	164	360	650	382
SASKATOON, SK	748	620	969	513	880	1255	650	275	486	639		267

SEE ALSO DISTANCE AND DRIVING TIME MAP ON PAGES 286–287 / VOIR AUSSI CARTE DES DISTANCES ET DES TEMPS DE PARCOURS PAGES 286–287

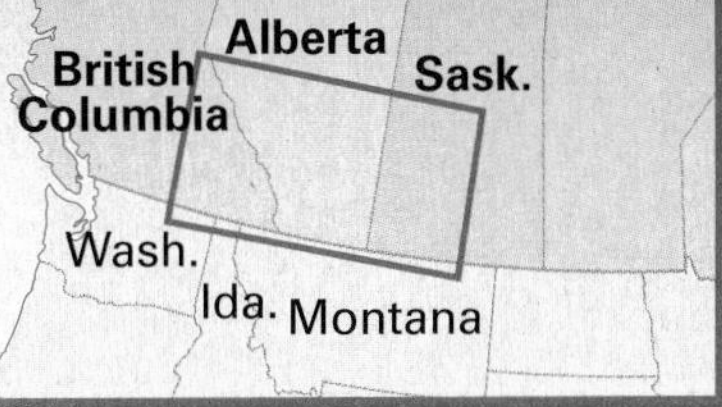

Lethbridge AB / Lloydminster AB–SK

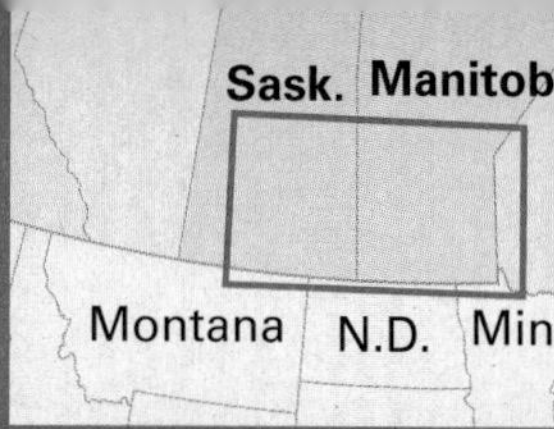

0 mi 20 40 60
0 km 20 40 60 80
One inch equals 40.3 miles/Un pouce équivaut à 40.3 milles
One centimeter equals 25.4 km/Un cm équivaut à 25.4 km

DRIVING DISTANCES IN KM / DISTANCES ROUTIÈRES EN KM

	BRANDON, MB	DAUPHIN, MB	GRAND RAPIDS, MB	MOOSE JAW, SK	PORTAGE LA PRAIRIE, MB	PRINCE ALBERT, SK	REGINA, SK	SASKATOON, SK	SWIFT CURRENT, SK	THE PAS, MB	WINNIPEG, MB	YORKTON, SK
BRANDON, MB		166	525	448	134	745	377	639	618	570	216	270
REGINA, SK	377	366	787	68	511	368		261	241	557	593	195
SASKATOON, SK	639	502	689	224	691	141	261		267	578	773	331
WINNIPEG, MB	216	322	430	664	82	819	593	773	834	611		442

SEE ALSO DISTANCE AND DRIVING TIME MAP ON PAGES 286–287 / VOIR AUSSI CARTE DES DISTANCES ET DES TEMPS DE PARCOURS PAGES 286–287

0 mi 20 40 60
0 km 20 40 60 80

One inch equals 40.3 miles/Un pouce équivaut à 40.3 milles
One centimeter equals 25.4 km/Un cm équivaut à 25.4 km

Kenora ON / Fort Frances ON

DRIVING DISTANCES IN KM / DISTANCES ROUTIÈRES EN KM

	DRYDEN, ON	FORT FRANCES, ON	GERALDTON, ON	GRAND FORKS, ND	HEARST, ON	KENORA, ON	MARATHON, ON	NIPIGON, ON	STEINBACH, MB	THUNDER BAY, ON	WAWA, ON	WINNIPEG, MB
FORT FRANCES, ON	190		627	315	845	215	641	445	310	335	805	420
KENORA, ON	140	215	772	429	990		786	585	184	480	950	205
THUNDER BAY, ON	340	335	292	650	510	480	306	110	664		470	685
WINNIPEG, MB	345	420	977	228	1195	205	991	790	55	685	1155	

SEE ALSO DISTANCE AND DRIVING TIME MAP ON PAGES 286–287 / VOIR AUSSI CARTE DES DISTANCES ET DES TEMPS DE PARCOURS PAGES 286–287

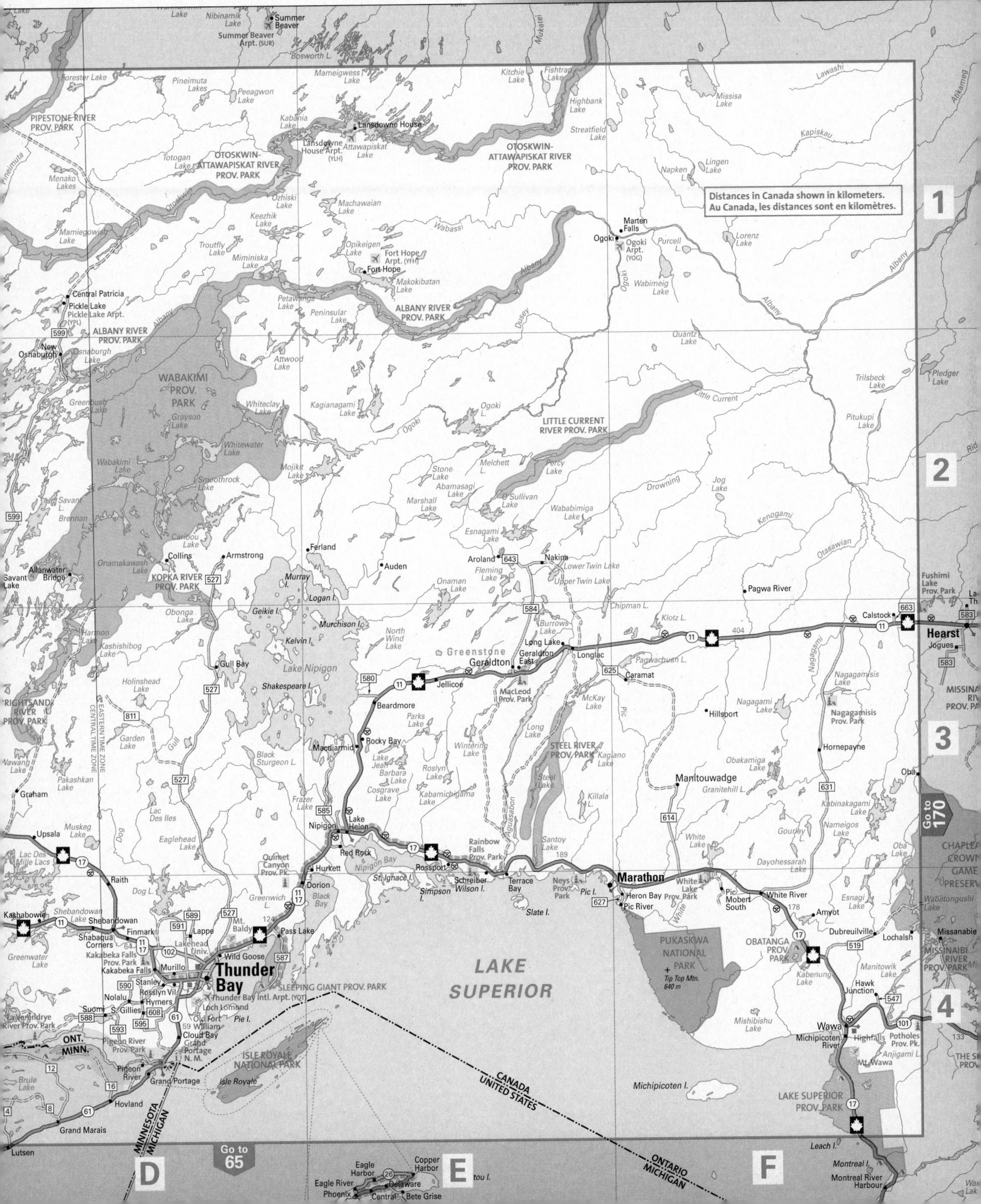

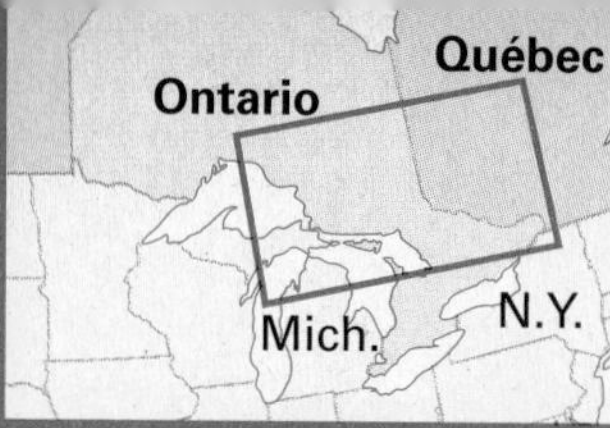

0 mi 20 40 60
0 km 20 40 60 80
One inch equals 40.3 miles/Un pouce équivaut à 40.3 milles
One centimeter equals 25.4 km/Un cm équivaut à 25.4 km

Sudbury ON / Sault Ste Marie ON

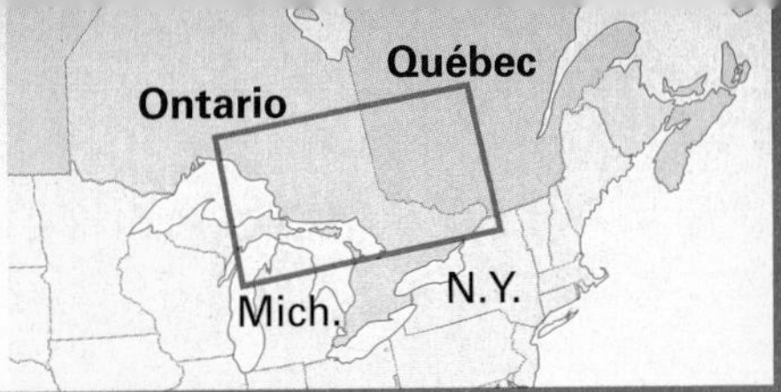

DRIVING DISTANCES IN KM / DISTANCES ROUTIÈRES EN KM	HEARST, ON	HUNTSVILLE, ON	KIRKLAND LAKE, ON	MONT-LAURIER, QC	NORTH BAY, ON	ORILLIA, ON	OTTAWA, ON	ROUYN-NORANDA, QC	SAULT STE. MARIE, ON	SUDBURY, ON	TIMMINS, ON	WAWA, ON
KIRKLAND LAKE, ON	370	370		505	250	578	610	154	580	315	140	475
OTTAWA, ON	955	350	610	209	364	415		456	787	488	730	1015
SAULT STE. MARIE, ON	545	560	580	1004	430	562	787	734		305	440	225
SUDBURY, ON	550	250	315	699	124	263	488	469	305		290	530

SEE ALSO DISTANCE AND DRIVING TIME MAP ON PAGES 286–287 / VOIR AUSSI CARTE DES DISTANCES ET DES TEMPS DE PARCOURS PAGES 286–287

North Bay ON / Rouyn-Noranda QC

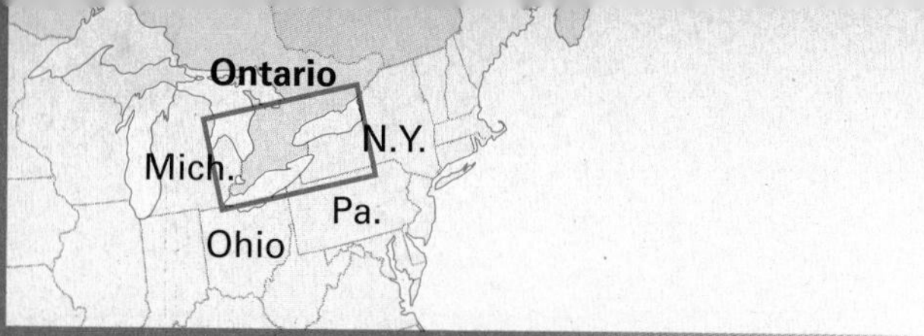

London ON / Windsor ON

0 mi 20 40
0 km 20 40 60

One inch equals 25.4 miles/Un pouce équivaut à 25.4 milles
One cm equals 16.1 km/Un cm équivaut à 16.1 km

RIVING
ISTANCES IN KM /
ISTANCES ROUTIÈRES EN KM

	BARRIE, ON	HAMILTON, ON	KINGSTON, ON	KITCHENER, ON	LONDON, ON	NIAGARA FALLS, ON	ORILLIA, ON	OWEN SOUND, ON	PETERBOROUGH, ON	SARNIA, ON	TORONTO, ON	WINDSOR, ON
KINGSTON, ON	350	330		430	430	390	317	430	180	530	260	620
NIAGARA FALLS, ON	200	68	390	130	190		237	260	260	290	130	380
TORONTO, ON	90	70	260	105	185	130	127	190	135	280		370
WINDSOR, ON	430	310	620	285	190	380	467	390	490	160	370	

EE ALSO DISTANCE AND DRIVING TIME MAP ON PAGES 286–287 / VOIR AUSSI CARTE DES DISTANCES ET DES TEMPS DE PARCOURS PAGES 286–287

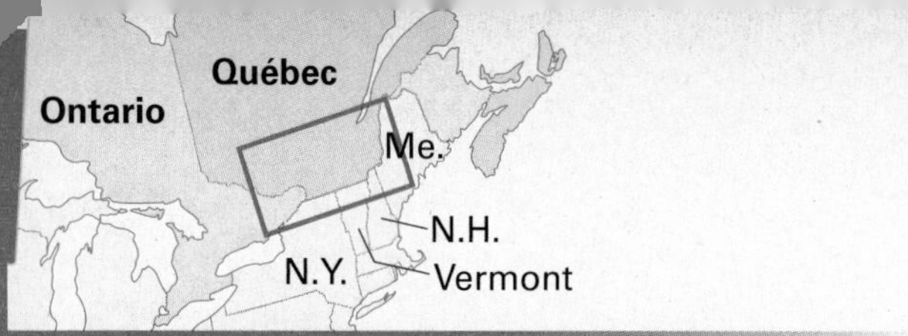

0 mi 20 40
0 km 20 40 60
One inch equals 25.4 miles/Un pouce équivaut à 25.4 milles
One cm equals 16.1 km/Un cm équivaut à 16.1 km

Ottawa ON / Montréal QC

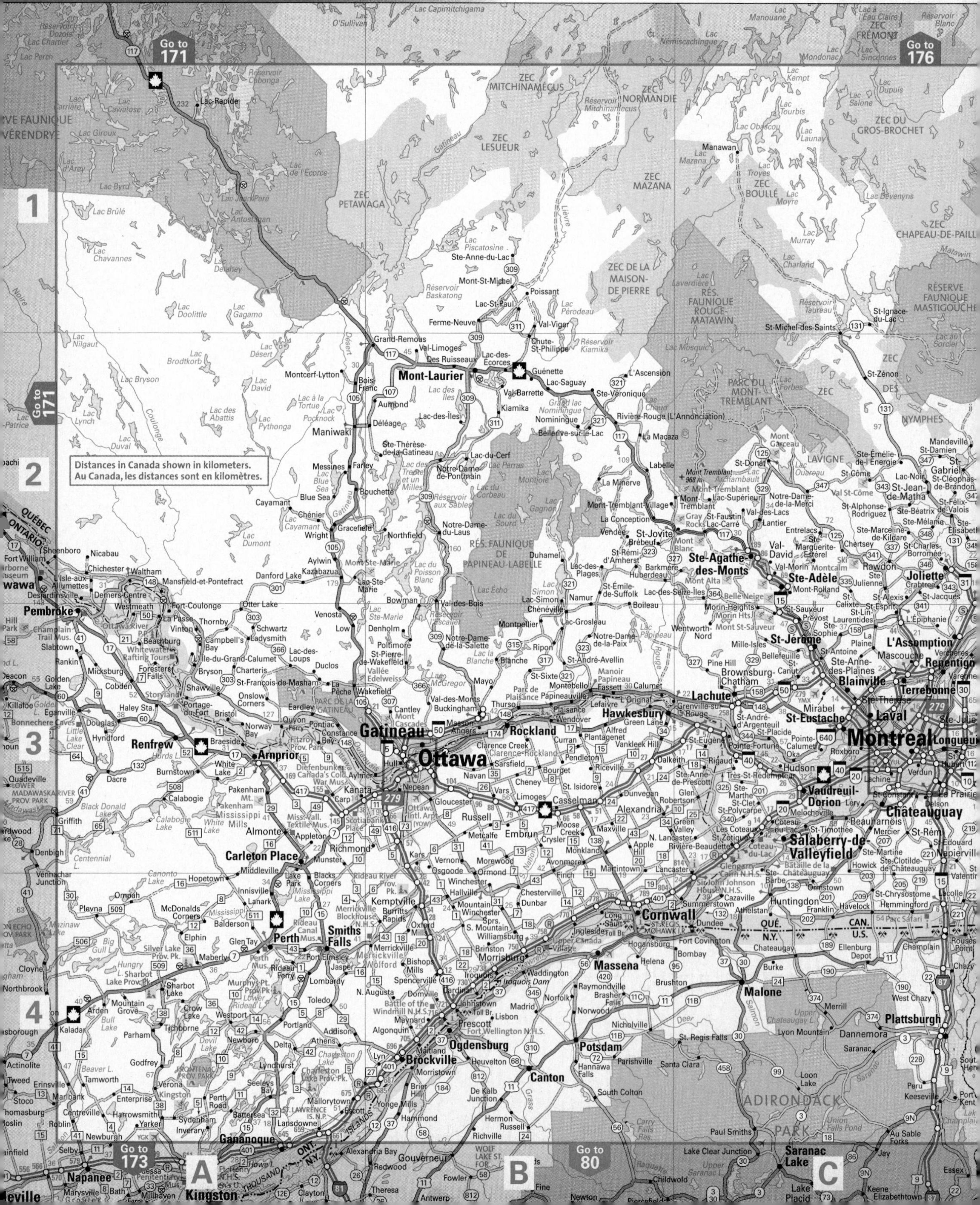

DRIVING DISTANCES IN KM / DISTANCES ROUTIÈRES EN KM

	BURLINGTON, VT	CORNWALL, ON	DRUMMONDVILLE, QC	KINGSTON, ON	MONT-LAURIER, QC	MONTRÉAL, QC	MONT-TREMBLANT, QC	OTTAWA, ON	QUÉBEC, QC	ST-GEORGES, QC	SHERBROOKE, QC	TROIS-RIVIÈRES, QC
MONTRÉAL, QC	153	103	116	283	230		126	194	250	325	143	146
OTTAWA, ON	360	97	310	175	209	194	208		444	485	337	340
QUÉBEC, QC	394	353	151	533	445	250	298	444		102	233	135
SHERBROOKE, QC	174	246	82	426	402	143	269	337	233	148		158

SEE ALSO DISTANCE AND DRIVING TIME MAP ON PAGES 286–287 / VOIR AUSSI CARTE DES DISTANCES ET DES TEMPS DE PARCOURS PAGES 286–287

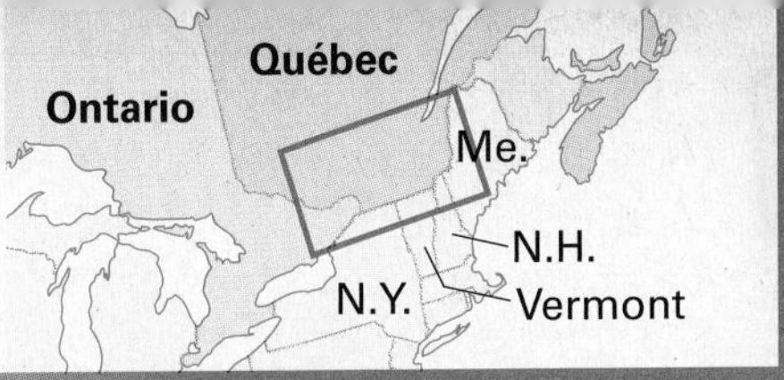

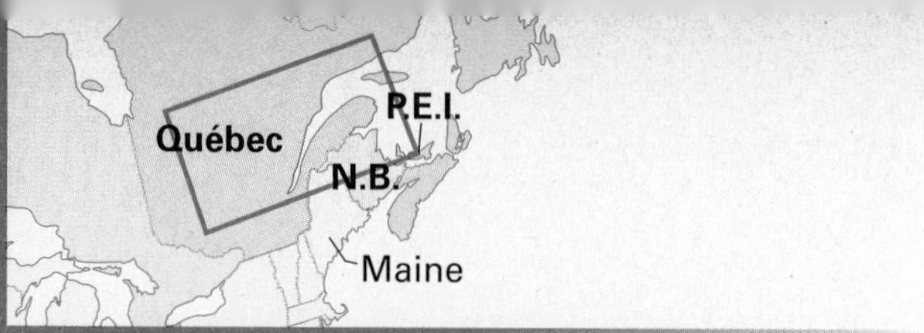

0 mi 20 40 60
0 km 20 40 60 80

One inch equals 40.3 miles/Un pouce équivaut à 40.3 milles
One centimeter equals 25.4 km/Un cm équivaut à 25.4 km

DRIVING DISTANCES IN KM / DISTANCES ROUTIÈRES EN KM

	BAIE-COMEAU, QC	CAMPBELLTON, NB	CHIBOUGAMAU, QC	CHICOUTIMI, QC	EDMUNDSTON, NB	GASPÉ, QC	HAVRE-ST-PIERRE, QC	MATANE, QC	MIRAMICHI, NB	QUÉBEC, QC	RIMOUSKI, QC	SEPT-ÎLES, QC
CHICOUTIMI, QC	435	444	359		269	771	884	348	622	211	253	667
EDMUNDSTON, NB	368	188	628	269		534	817	249	268	317	180	600
GASPÉ, QC	287	340	1130	771	534		743	294	518	706	389	526
QUÉBEC, QC	408	508	570	211	317	706	857	412	582		507	640

SEE ALSO DISTANCE AND DRIVING TIME MAP ON PAGES 286–287 / VOIR AUSSI CARTE DES DISTANCES ET DES TEMPS DE PARCOURS PAGES 286–287

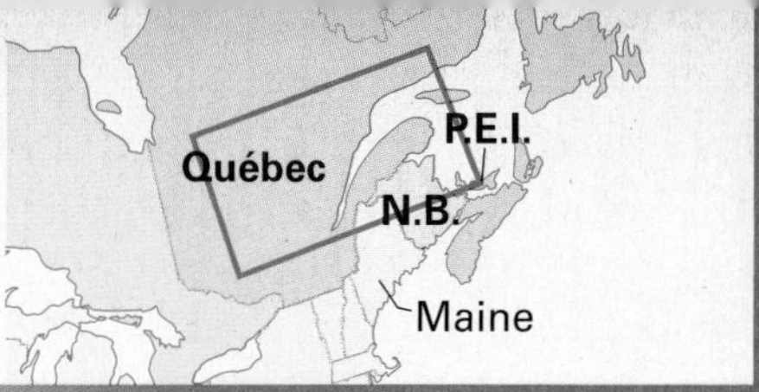

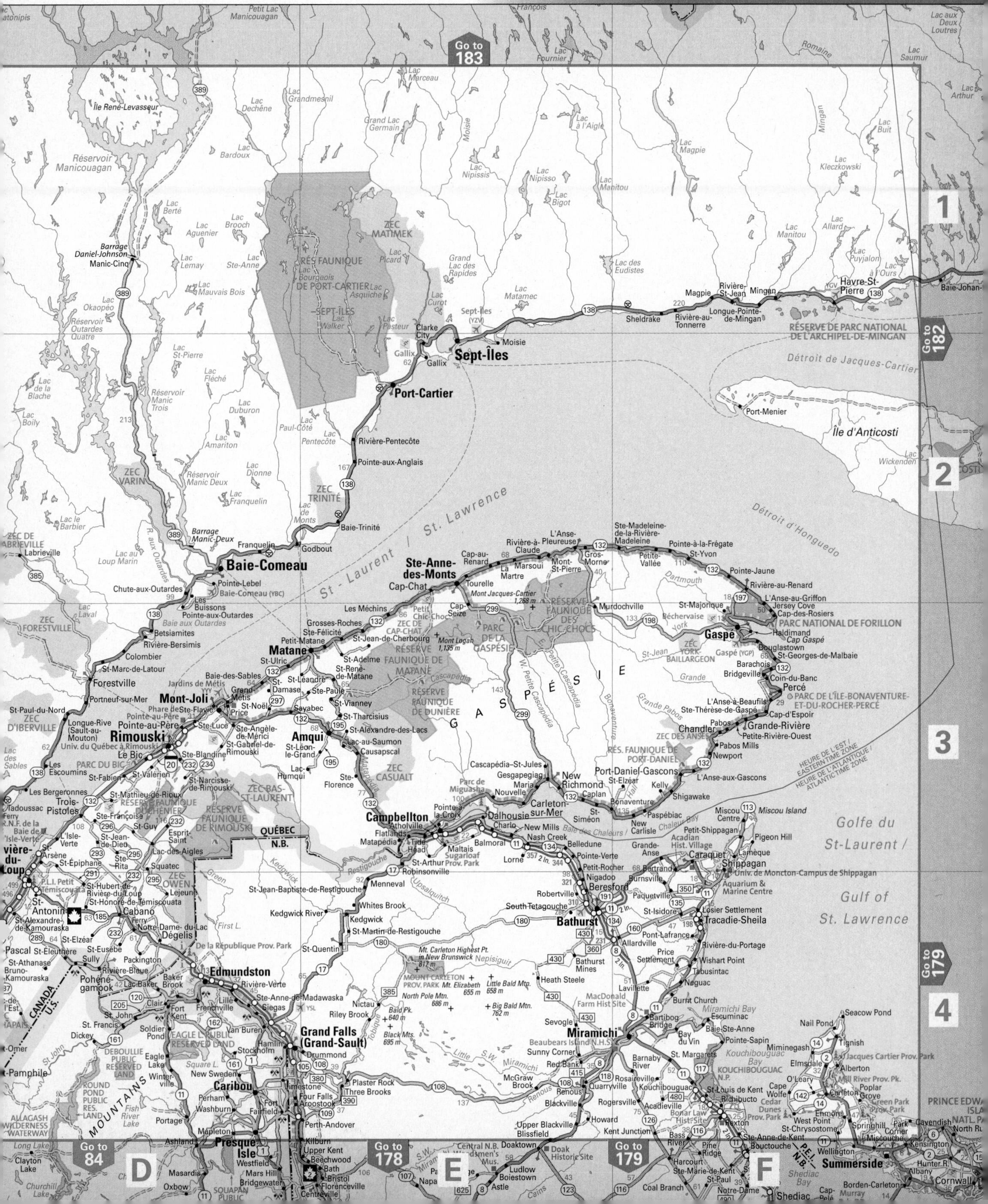

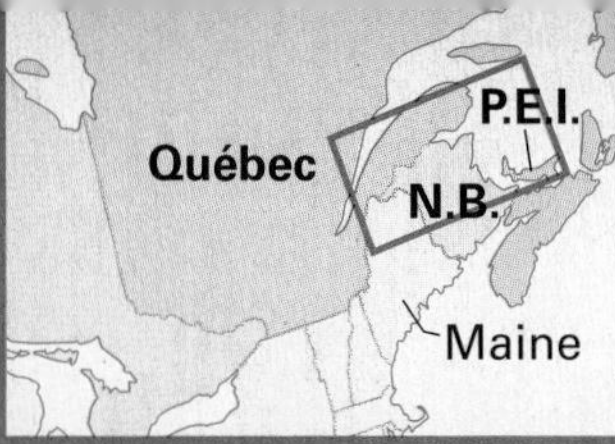

0 mi 20 40
0 km 20 40 60

One inch equals 25.4 miles/Un pouce équivaut à 25.4 milles
One cm equals 16.1 km/Un cm équivaut à 16.1 km

Rimouski QC / Edmundston NB

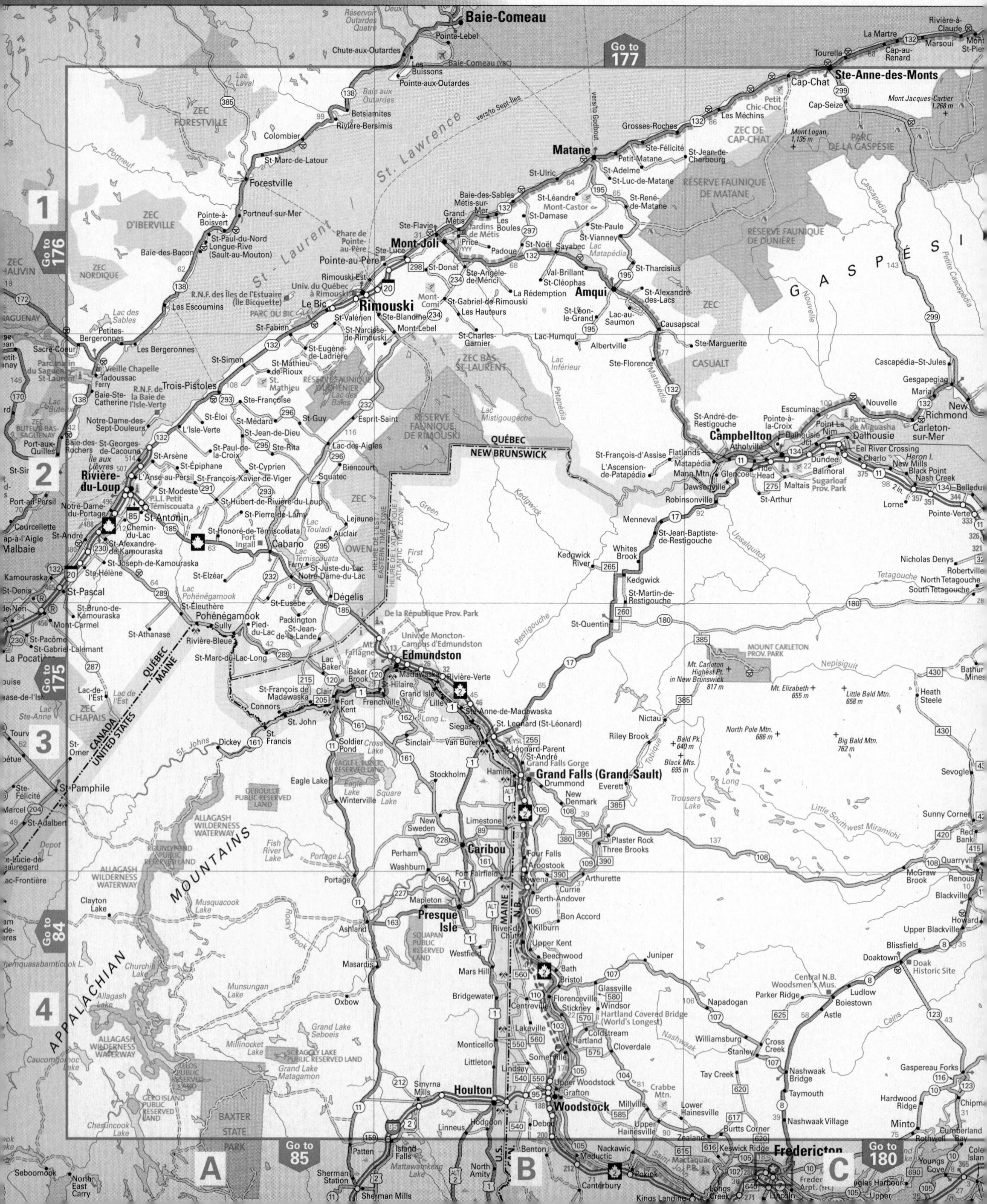

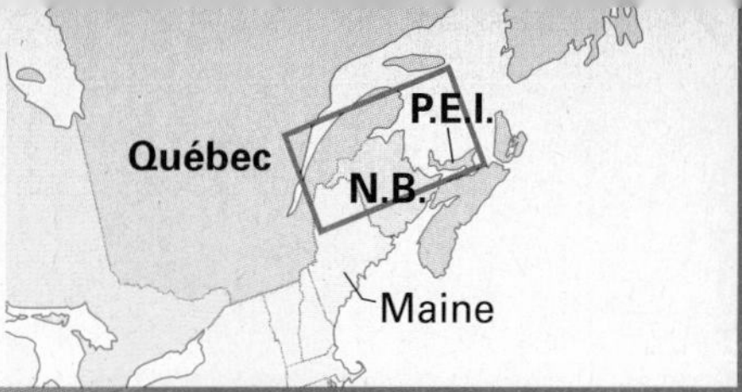

DRIVING DISTANCES IN KM / DISTANCES ROUTIÈRES EN KM

	BATHURST, NB	BORDEN-CARLETON, PE	CAMPBELLTON, NB	CHARLOTTETOWN, PE	EDMUNDSTON, NB	FREDERICTON, NB	GASPÉ, QC	GRAND FALLS, NB	MATANE, QC	MIRAMICHI, NB	MONCTON, NB	RIMOUSKI, QC
CHARLOTTETOWN, PE	338	56	438		629	362	791	581	562	273	164	596
EDMUNDSTON, NB	189	428	188	638		279	534	57	249	268	447	180
MATANE, QC	262	506	168	562	249	553	294	331		346	487	95
MONCTON, NB	206	108	306	164	447	170	659	390	487	141		502

SEE ALSO DISTANCE AND DRIVING TIME MAP ON PAGES 286–287 / VOIR AUSSI CARTE DES DISTANCES ET DES TEMPS DE PARCOURS PAGES 286–287

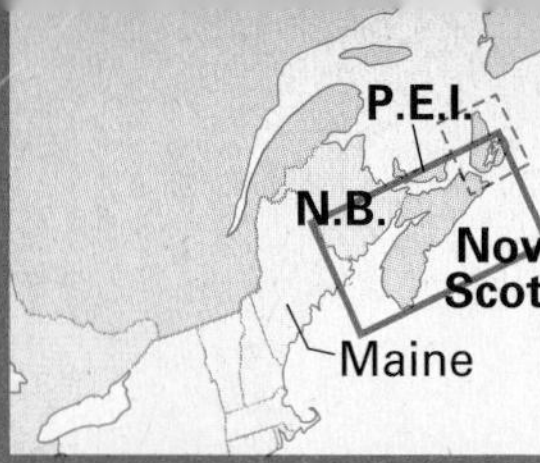

0 mi 20 40
0 km 20 40 60

One inch equals 25.4 miles/Un pouce équivaut à 25.4 milles
One cm equals 16.1 km/Un cm équivaut à 16.1 km

Saint John NB / Yarmouth NS

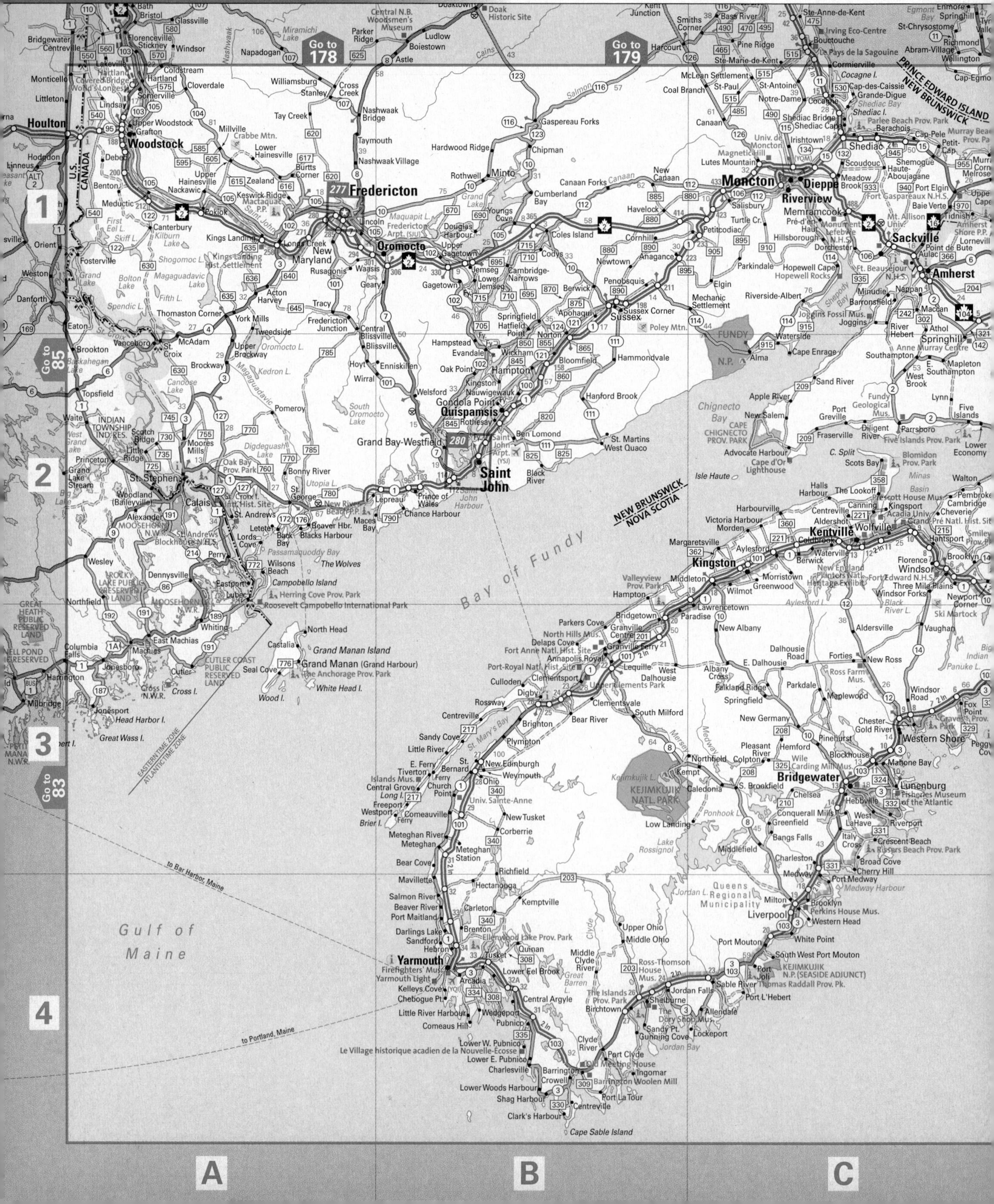

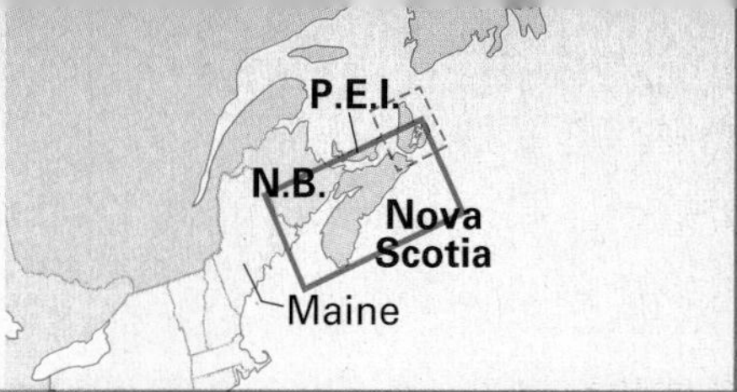

DRIVING DISTANCES IN KM / DISTANCES ROUTIÈRES EN KM

	CHARLOTTETOWN, PE	CHÉTICAMP, NS	DIGBY, NS	FREDERICTON, NB	HALIFAX, NS	MONCTON, NB	PORT HAWKESBURY, NS	SAINT JOHN, NB	ST. STEPHEN, NB	SYDNEY, NS	TRURO, NS	YARMOUTH, NS
HALIFAX, NS	322	425	235	462		260	265	410	515	415	89	339
MONCTON, NB	164	481	231	170	260		374	150	278	497	182	599
SAINT JOHN, NB	350	640	72	114	410	150	497		119	647	321	176
SYDNEY, NS	374	173	623	689	415	497	123	647	766		326	727

SEE ALSO DISTANCE AND DRIVING TIME MAP ON PAGES 286–287 / VOIR AUSSI CARTE DES DISTANCES ET DES TEMPS DE PARCOURS PAGES 286–287

Distances in Canada shown in kilometers.
Au Canada, les distances sont en kilomètres.

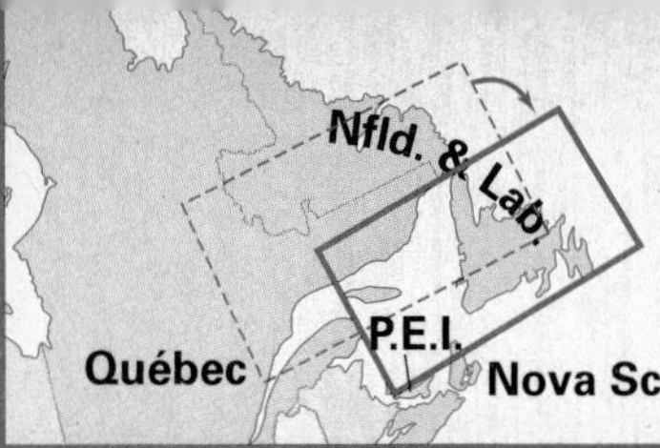

Corner Brook NL / Channel-Port aux Basques NL

0 mi 20 40 60
0 km 20 40 60 80
One inch equals 40.3 miles/Un pouce équivaut à 40.3 milles
One centimeter equals 25.4 km/Un cm équivaut à 25.4 km

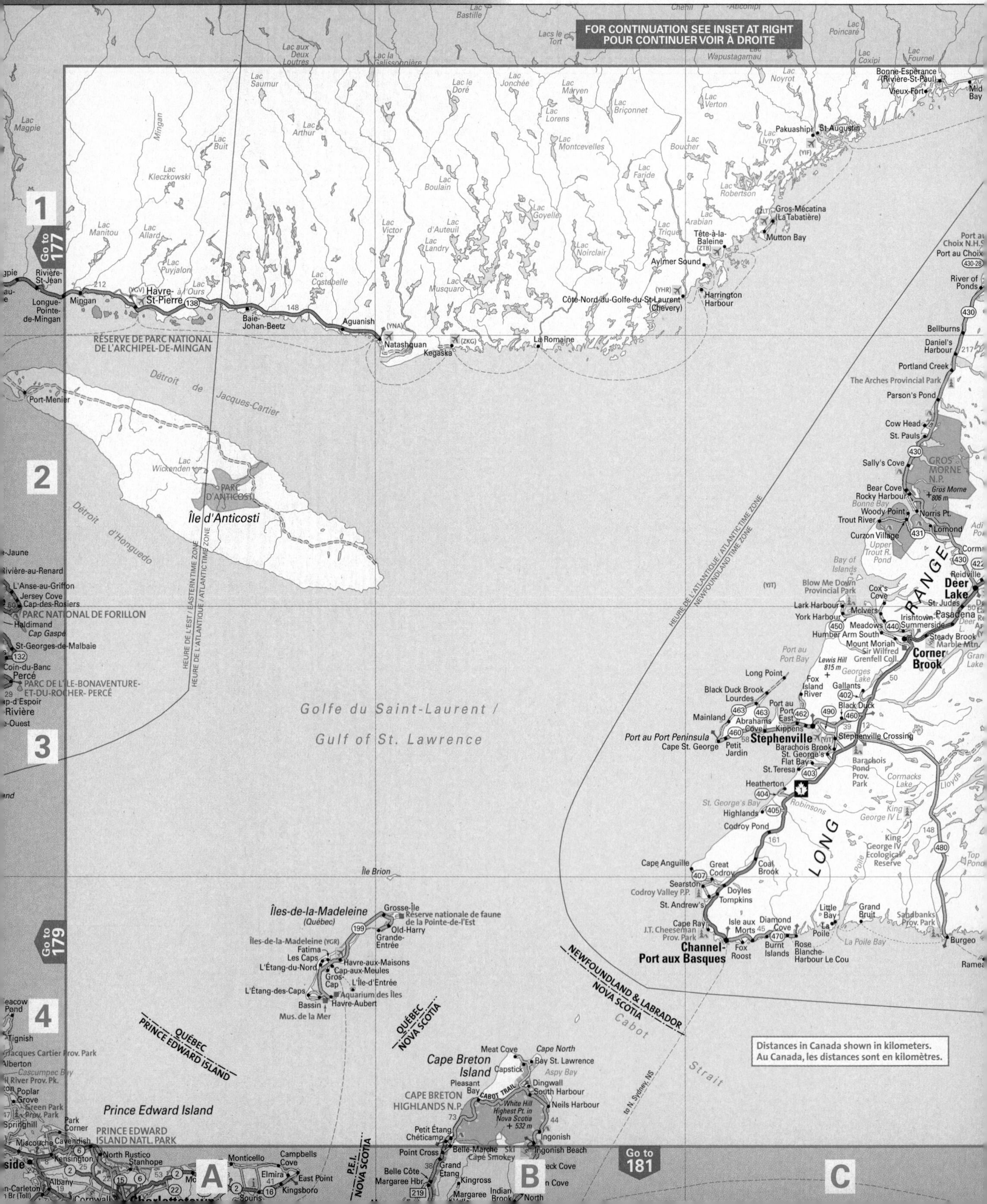

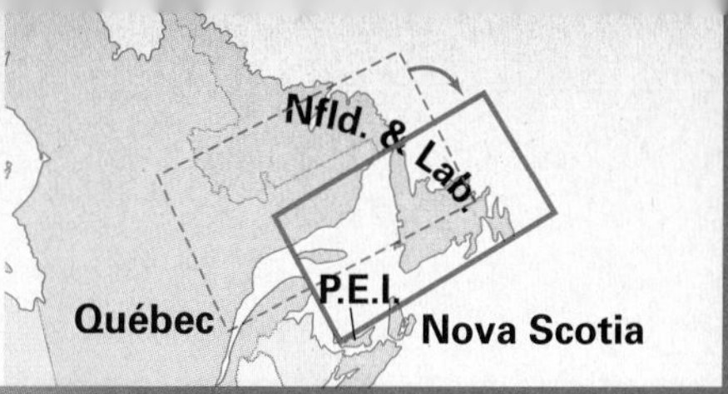

DRIVING DISTANCES IN KM / DISTANCES ROUTIÈRES EN KM

	ARGENTIA, NL	BISHOP'S FALLS, NL	BONAVISTA, NL	CHAN.-PT. AUX BASQUES, NL	CORNER BROOK, NL	DEER LAKE, NL	GANDER, NL	GRAND FALLS-WINDSOR, NL	MARYSTOWN, NL	ST. ANTHONY, NL	ST. JOHN'S, NL	STEPHENVILLE, NL
BISHOP'S FALLS, NL	363		307	482	280	225	72	18	384	628	393	339
CHAN.-PT. AUX BASQUES, NL	845	482	789		202	257	554	464	866	660	875	151
CORNER BROOK, NL	643	280	587	202		55	352	262	664	458	673	59
ST. JOHN'S, NL	134	393	296	875	673	618	321	411	293	1021		732

SEE ALSO DISTANCE AND DRIVING TIME MAP ON PAGES 286–287 / VOIR AUSSI CARTE DES DISTANCES ET DES TEMPS DE PARCOURS PAGES 286–287

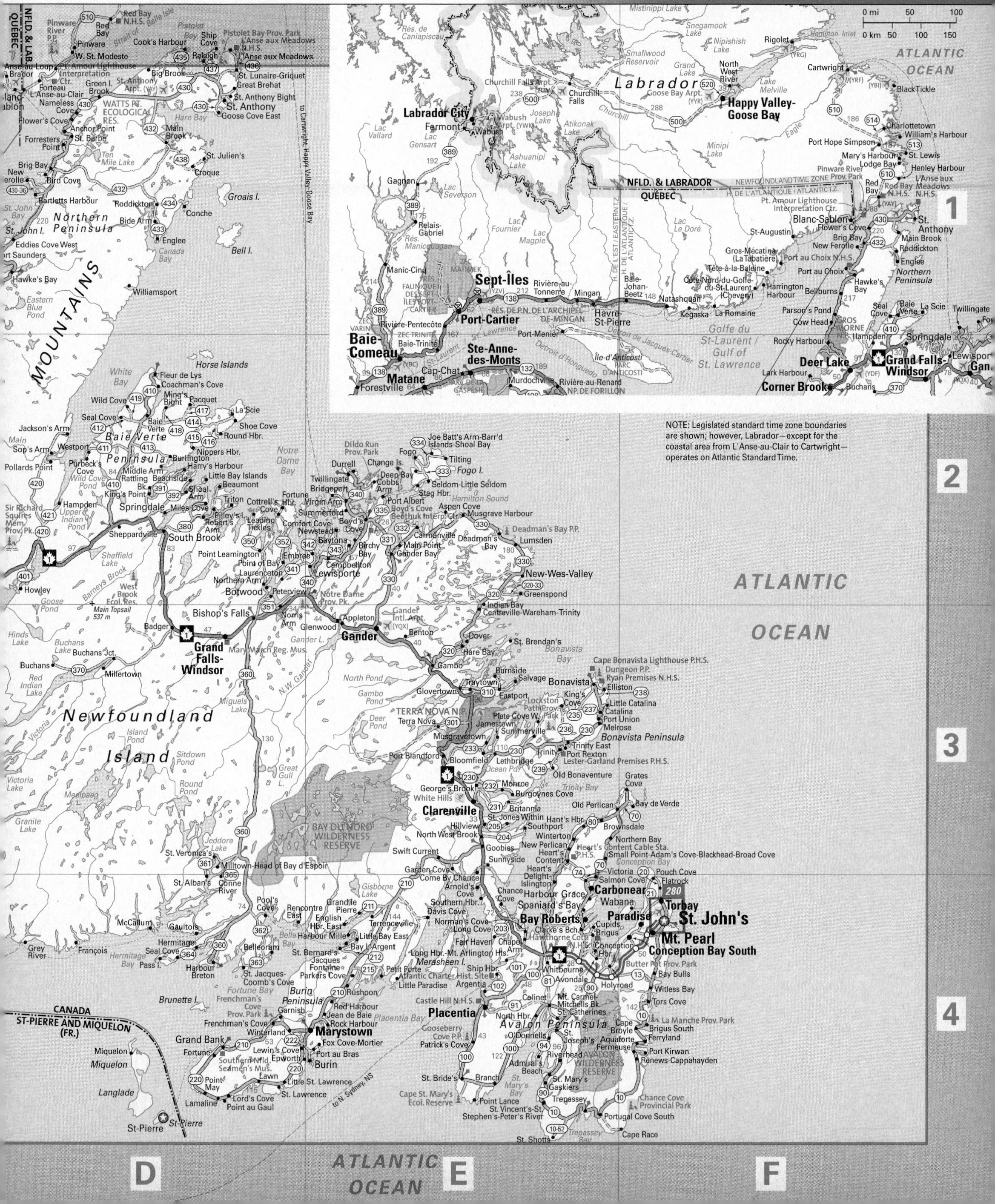

0 mi 50 100 150
0 km 50 100 150 200

One inch equals 83.75 miles/Una pulgada igual a 83.75 millas
One centimeter equals 53 km/Un centímetro igual a 53 km

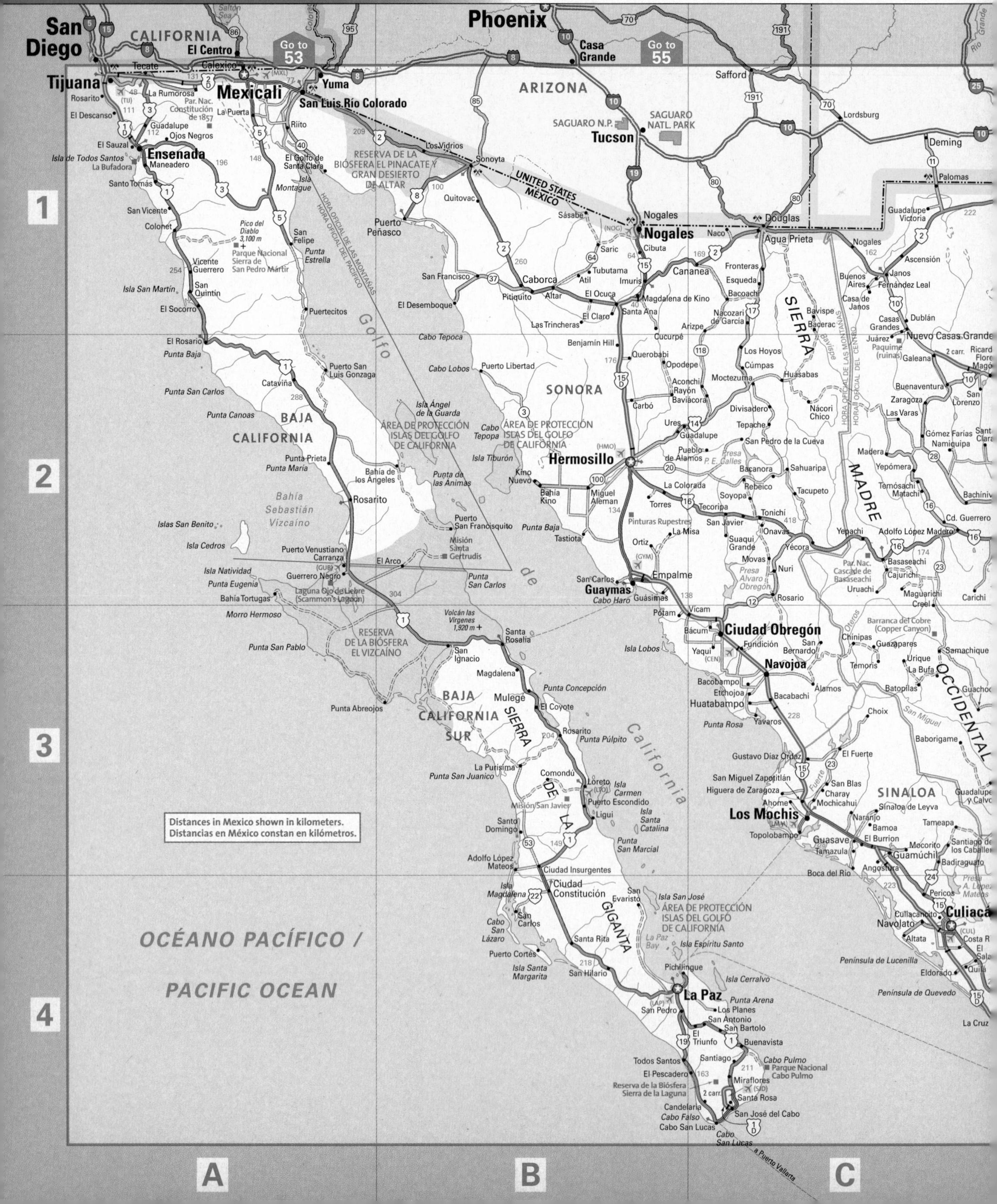

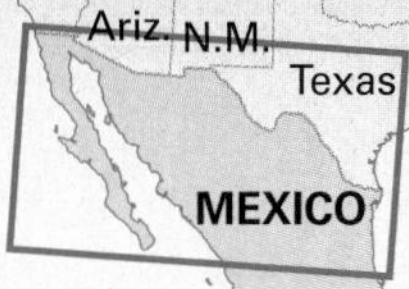

DRIVING DISTANCES IN KM / DISTANCIAS DE MANEJO EN KM	CHIHUAHUA	CIUDAD JUÁREZ	CIUDAD VICTORIA	CULIACÁN	DURANGO	HERMOSILLO	MAZATLÁN	MÉXICO	MONTERREY	SAN LUIS POTOSÍ	TIJUANA	TORREÓN
CHIHUAHUA		385	1086	919	686	579	1209	1538	808	1155	1456	449
HERMOSILLO	579	795	1666	706	941		729	1810	1387	1416	884	1028
MONTERREY	808	1236	288	924	689	1387	901	892		509	2362	359
TORREÓN	449	834	637	914	266	1028	892	1089	359	706	1905	

SEE ALSO DISTANCE AND DRIVING TIME MAP ON PAGES 286–287 / CONSULTE, PARA DISTANCIAS Y TIEMPO DE MANEJO, EN LAS PÁGINAS 286-287

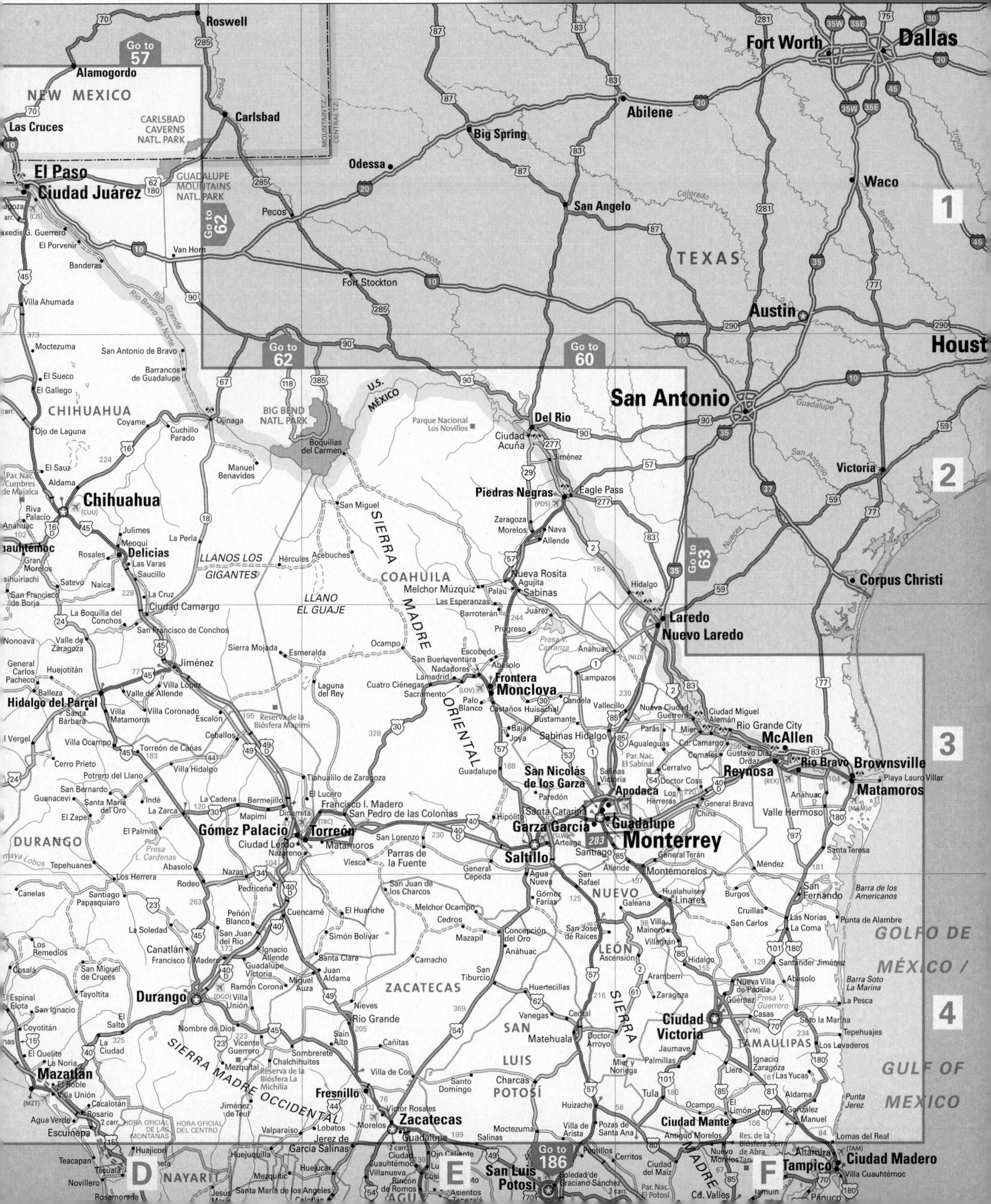

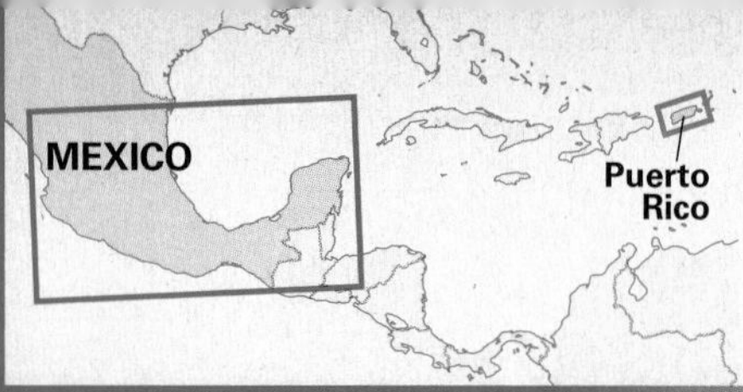

0 mi 50 100 150
0 km 50 100 150 200

One inch equals 83.75 miles/Una pulgada igual a 83.75 millas
One centimeter equals 53 km/Un centímetro igual a 53 km

México MEX / Guadalajara MEX

DRIVING DISTANCES IN KM / DISTANCIAS DE MANEJO EN KM

	ACAPULCO	CANCÚN	CIUDAD VICTORIA	DURANGO	GUADALAJARA	MAZATLÁN	MÉRIDA	MÉXICO	PUEBLA	SAN LUIS POTOSÍ	TUXTLA GUTIÉRREZ	VERACRUZ
GUADALAJARA	897	2275	774	599		523	1904	578	691	336	1510	943
MÉRIDA	1777	321	1725	2182	1904	2408		1326	1282	1707	786	995
MÉXICO	422	1736	682	856	578	1081	1326		133	381	932	365
SAN LUIS POTOSÍ	834	2161	438	475	336	687	1707	381	496		1313	747

SEE ALSO DISTANCE AND DRIVING TIME MAP ON PAGES 286–287 / CONSULTE, PARA DISTANCIAS Y TIEMPO DE MANEJO, EN LAS PÁGINAS 286-287

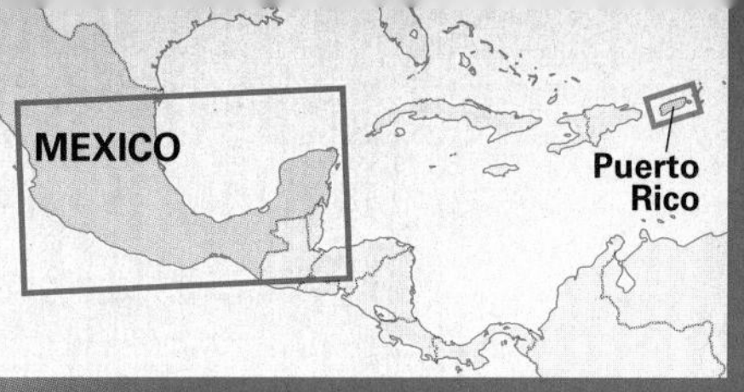

PUERTO RICO

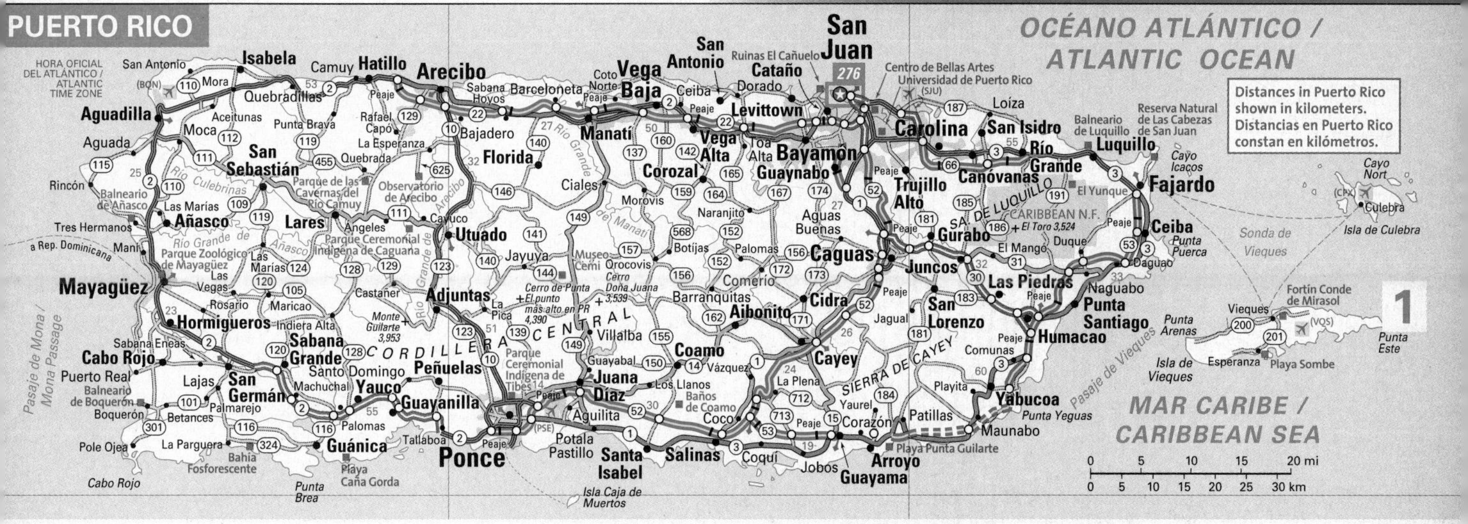

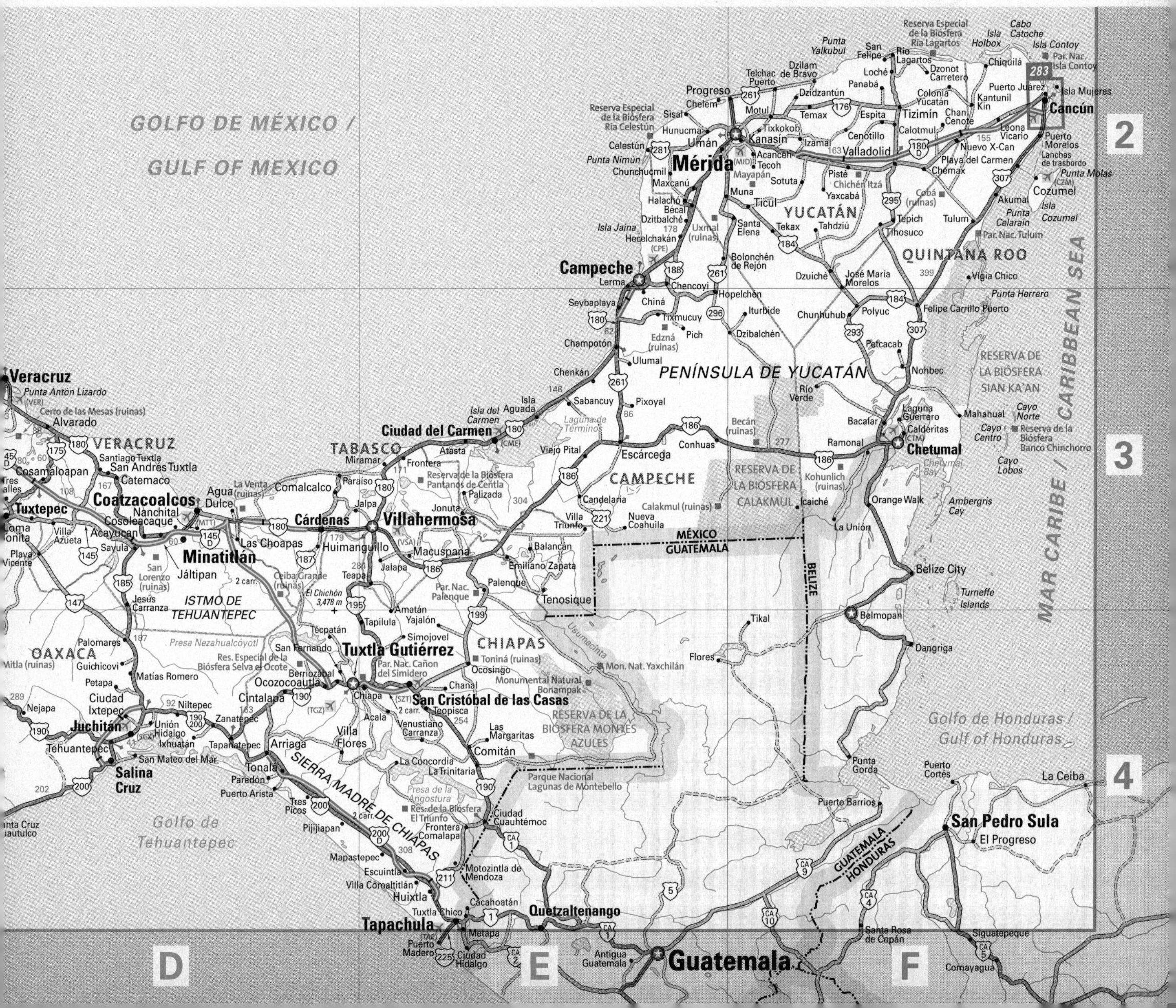

Figures after entries indicate population, page number, and grid reference.

UNITED STATES

A

Abbeville AL, 2987 ... 128 B4
Abbeville GA, 2298 ... 129 E3
Abbeville LA, 11887 ... 133 F3
Abbeville MS, 423 ... 118 C3
Abbeville SC, 5840 ... 121 E3
Abbeville Co. SC, 26167 ... 121 E3
Abbotsford WI, 1956 ... 68 A4
Abbottstown PA, 905 ... 103 E1
Abercrombie ND, 296 ... 19 F4
Aberdeen ID, 1840 ... 31 E1
Aberdeen MD, 13842 ... 145 D1
Aberdeen MS, 6415 ... 119 D4
Aberdeen NC, 3400 ... 122 C1
Aberdeen OH, 1603 ... 100 C3
Aberdeen SD, 24658 ... 27 E2
Aberdeen WA, 16461 ... 12 B4
Abernathy TX, 2839 ... 58 A1
Abilene KS, 6543 ... 43 E2
Abilene TX, 115930 ... 58 C3
Abingdon IL, 3612 ... 88 A3
Abingdon MD, 950 ... 145 D1
Abingdon VA, 7780 ... 111 E3
Abington MA, 14605 ... 151 D2
Abita Sprs. LA, 1957 ... 134 B2
Absarokee MT, 1234 ... 24 B2
Absecon NJ, 7638 ... 147 F4
Acadia Par. LA, 58861 ... 133 E2
Accokeek MD, 7349 ... 144 B4
Accokeek Acres MD, 1500 ... 144 B4
Accomac VA, 547 ... 114 C3
Accomack Co. VA, 38305 ... 114 C3
Accord MA, 2300 ... 151 D2
Accord NY, 622 ... 94 A3
Achille OK, 506 ... 59 F1
Achilles VA, 650 ... 113 F2
Ackerman MS, 1696 ... 126 C1
Ackley IA, 1809 ... 73 D4
Acme MI, 650 ... 69 F4
Acomita NM, 288 ... 48 B3
Acton CA, 2390 ... 52 C2
Acton MA, 2700 ... 150 C1
Acushnet MA, 3171 ... 151 D3
Acworth GA, 13422 ... 120 C3
Ada MN, 1657 ... 19 F3
Ada OH, 5582 ... 90 B3
Ada OK, 15691 ... 51 F4
Ada Co. ID, 300904 ... 22 B4
Adair IA, 839 ... 86 B2
Adair OK, 704 ... 106 A3
Adair Co. IA, 8243 ... 86 B2
Adair Co. KY, 17244 ... 110 B2
Adair Co. MO, 24977 ... 87 D4
Adair Co. OK, 21038 ... 106 B4
Adairsville GA, 2542 ... 120 B3
Adair Vil. OR, 536 ... 20 B3
Adairville KY, 920 ... 109 F3
Adams MA, 5784 ... 94 C1
Adams MN, 800 ... 73 D2
Adams NE, 489 ... 35 F4
Adams NY, 1624 ... 79 E2
Adams OR, 297 ... 21 F1
Adams TN, 566 ... 109 E3
Adams WI, 1914 ... 74 A1
Adams Ctr. NY, 1500 ... 79 E2
Adams Co. CO, 348618 ... 41 F1
Adams Co. ID, 3476 ... 22 B2
Adams Co. IL, 68277 ... 87 F4
Adams Co. IN, 33625 ... 90 A3
Adams Co. IA, 4482 ... 86 B3
Adams Co. MS, 34340 ... 126 A4
Adams Co. NE, 31151 ... 35 D4
Adams Co. ND, 2593 ... 26 A1
Adams Co. OH, 27330 ... 100 C3
Adams Co. PA, 91292 ... 103 E1
Adams Co. WA, 16428 ... 13 F4
Adams Co. WI, 18643 ... 74 A2
Adamston NJ, 4900 ... 147 E3
Adamstown MD, 650 ... 144 A2
Adamstown PA, 1203 ... 146 A2
Adamsville AL, 4965 ... 119 F4
Adamsville RI, 550 ... 151 D4
Adamsville TN, 1983 ... 119 D1
Addis LA, 2238 ... 134 A2
Addison AL, 723 ... 119 E3
Addison IL, 35914 ... 203 C4
Addison ME, 300 ... 83 E2
Addison MI, 627 ... 90 B1
Addison NY, 1797 ... 93 D1
Addison TX, 14166 ... 207 D1
Addison Co. VT, 35974 ... 81 D3
Adel GA, 5307 ... 137 F1
Adel IA, 3435 ... 86 C2
Adelanto CA, 18130 ... 53 D2
Adelphi MD, 14998 ... 270 E1
Adelphia NJ, 700 ... 147 E2
Adena OH, 815 ... 91 F4
Adrian GA, 579 ... 129 F2
Adrian MI, 21574 ... 90 B1
Adrian MN, 1234 ... 72 A2
Adrian MO, 1780 ... 96 B4
Advance IN, 562 ... 99 E1
Advance MO, 1244 ... 108 B2
Adwolf VA, 1457 ... 111 F2
Affton MO, 20535 ... 256 B3
Afton IA, 917 ... 86 C3
Afton NY, 836 ... 93 F1
Afton OK, 1118 ... 106 B3
Afton WY, 1818 ... 31 F1
Agawam MA, 28144 ... 150 A2
Agency IA, 622 ... 87 E3
Agency MO, 599 ... 96 B1
Agoura Hills CA, 20537 ... 228 A2
Agua Dulce TX, 737 ... 63 E2
Agua Fria NM, 2051 ... 49 D2
Aguilar CO, 593 ... 41 E4
Ahoskie NC, 4523 ... 113 F3
Ahsahka ID, 600 ... 14 B4
Ahuimanu HI, 8506 ... 152 A3
Aiken SC, 25337 ... 121 F4
Aiken Co. SC, 142552 ... 122 A4
Ainsworth IA, 524 ... 87 F2
Ainsworth NE, 1862 ... 34 C1
Airmont NY, 7799 ... 148 B3
Airport Drive MO, 622 ... 106 B2
Airway Hts. WA, 4500 ... 13 F3
Aitkin MN, 1984 ... 64 B4
Aitkin Co. MN, 15301 ... 64 B4
Ajo AZ, 3705 ... 54 B3
Ak-Chin AZ, 669 ... 54 C2
Akiachak AK, 585 ... 154 B3
Akins OK, 449 ... 116 B1
Akron AL, 521 ... 127 E1
Akron CO, 1711 ... 41 F1
Akron IN, 1076 ... 89 F3
Akron IA, 1489 ... 35 F1
Akron MI, 461 ... 76 B2
Akron NY, 3085 ... 78 B3
Akron OH, 217074 ... 91 E3
Akron PA, 4046 ... 146 A2
Akutan AK, 713 ... 154 A4
Alabaster AL, 22619 ... 127 F1
Alachua FL, 6098 ... 138 C3
Alachua Co. FL, 217955 ... 138 C3
Alakanuk AK, 652 ... 154 B2
Alamance Co. NC, 130800 ... 112 C4
Alameda CA, 72259 ... 259 C3
Alameda NM, 4200 ... 48 C3
Alameda Co. CA, 1443741 ... 36 B4
Alamo CA, 15626 ... 259 D2
Alamo GA, 1943 ... 129 E3
Alamo NM, 1183 ... 48 B4
Alamo TN, 2392 ... 108 C4
Alamo TX, 14760 ... 63 E4
Alamogordo NM, 35582 ... 56 C2
Alamo Hts. TX, 7319 ... 257 E2
Alamosa CO, 7960 ... 41 D4
Alamosa Co. CO, 14966 ... 41 D4
Alanson MI, 785 ... 70 C3
Alapaha GA, 682 ... 129 E4
Alba MO, 588 ... 106 B2
Albany CA, 16444 ... 259 C2
Albany GA, 76939 ... 129 D4
Albany IL, 895 ... 88 A1
Albany IN, 2368 ... 90 A4
Albany KY, 2220 ... 110 B3
Albany LA, 865 ... 134 B2
Albany MN, 1796 ... 66 B2
Albany MO, 1937 ... 86 B4
Albany NY, 95658 ... 94 B1
Albany OH, 808 ... 101 E2
Albany OR, 40852 ... 20 B3
Albany TX, 1921 ... 58 C2
Albany WI, 1191 ... 74 B4
Albany Co. NY, 294565 ... 94 B1
Albany Co. WY, 32014 ... 33 E2
Albemarle NC, 15680 ... 122 B1
Albemarle Co. VA, 79236 ... 102 C4
Albers IL, 878 ... 98 B3
Albert City IA, 709 ... 72 B4
Albert Lea MN, 18356 ... 72 C2
Alberton MT, 374 ... 15 D4
Albertville AL, 17247 ... 120 A3
Albertville MN, 3621 ... 66 C3
Albia IA, 3706 ... 87 D3
Albin VA, 700 ... 102 C2
Albion ID, 262 ... 31 D2
Albion IL, 1933 ... 99 D4
Albion IN, 2284 ... 90 A2
Albion IA, 592 ... 87 D1
Albion MI, 9144 ... 76 A4
Albion NE, 1797 ... 35 E3
Albion NY, 7438 ... 78 B3
Albion PA, 1607 ... 91 F1
Albion WA, 616 ... 14 A4
Albuquerque NM, 448607 ... 48 C3
Alburg VT, 488 ... 81 D1
Alburnett IA, 559 ... 87 E1
Alburtis PA, 2117 ... 146 B1
Alcalde NM, 377 ... 49 D2
Alcester SD, 880 ... 35 F1
Alcoa TN, 7734 ... 110 C4
Alcona Co. MI, 11719 ... 71 D4
Alcorn MS, 1200 ... 126 A3
Alcorn Co. MS, 34558 ... 119 D2
Alda NE, 652 ... 35 D4
Aldan PA, 4313 ... 248 B4
Alden IA, 904 ... 72 C4
Alden MN, 652 ... 72 C2
Alden NY, 2666 ... 78 B3
Alderson WV, 1091 ... 112 A1
Alderwood Manor WA, 15329 ... 262 B2
Aldine TX, 13979 ... 220 C1
Aledo IL, 3613 ... 87 F2
Aledo TX, 1726 ... 59 E2
Alex OK, 635 ... 51 E3
Alexander AR, 614 ... 117 E2
Alexander ND, 217 ... 17 F2
Alexander City AL, 15008 ... 128 A1
Alexander Co. IL, 9590 ... 108 C2
Alexander Co. NC, 33603 ... 112 A4
Alexandria AL, 3692 ... 120 A4
Alexandria IN, 6260 ... 89 F4
Alexandria KY, 8286 ... 100 B3
Alexandria LA, 46342 ... 125 E4
Alexandria MN, 8820 ... 66 B2
Alexandria SD, 563 ... 27 E4
Alexandria TN, 814 ... 110 A4
Alexandria VA, 128283 ... 144 B3
Alexandria Bay NY, 1088 ... 79 E1
Alexis IL, 863 ... 88 A3
Alfalfa Co. OK, 6105 ... 51 D1
Alford FL, 466 ... 136 C1
Alfred ME, 700 ... 82 B4
Alfred NY, 3954 ... 92 C1
Alger OH, 888 ... 90 B3
Alger Co. MI, 9862 ... 69 E1
Algoa TX, 900 ... 132 B4
Algodones NM, 688 ... 48 C3
Algoma MS, 508 ... 118 C3
Algoma WI, 3357 ... 69 D4
Algona IA, 5741 ... 72 B3
Algona WA, 2460 ... 262 B5
Algonac MI, 4613 ... 76 C4
Algonquin IL, 23276 ... 88 C1
Algood TN, 2942 ... 110 A3
Alhambra CA, 85804 ... 228 D2
Alhambra IL, 630 ... 98 B3
Alice TX, 19010 ... 63 E2
Aliceville AL, 2567 ... 127 E1
Ali Chuk AZ, 450 ... 54 B3
Aliquippa PA, 11734 ... 91 F3
Aliso Viejo CA, 40596 ... 229 G6
Allamakee Co. IA, 14675 ... 73 F3
Allamuchy NJ, 3125 ... 94 A4
Allardt TN, 642 ... 110 B3
Allegan MI, 4838 ... 75 F4
Allegan Co. MI, 105665 ... 75 F4
Allegany NY, 1883 ... 92 C1
Allegany Co. MD, 74930 ... 102 C1
Allegany Co. NY, 49927 ... 78 C4
Alleghany Co. NC, 10677 ... 111 F3
Alleghany Co. VA, 17215 ... 102 A4
Allegheny Co. PA, 1281666 ... 92 A4
Allen NE, 411 ... 35 F2
Allen OK, 951 ... 51 F3
Allen SD, 419 ... 26 B4
Allen TX, 43554 ... 59 F2
Allen Co. IN, 331849 ... 90 A3
Allen Co. KS, 14385 ... 96 A4
Allen Co. KY, 17800 ... 109 F2
Allen Co. OH, 108473 ... 90 B3
Allendale MI, 11555 ... 75 F3
Allendale NJ, 6699 ... 148 B3
Allendale SC, 4052 ... 130 B1
Allendale Co. SC, 11211 ... 130 B1
Allenhurst GA, 788 ... 130 B3
Allenhurst NJ, 718 ... 147 F2
Allen Par. LA, 25440 ... 133 E1
Allen Park MI, 29376 ... 210 B4
Allenspark CO, 496 ... 41 D1
Allenton RI, 1400 ... 150 C4
Allenton WI, 850 ... 74 C2
Allentown NJ, 1882 ... 147 D2
Allentown PA, 106632 ... 146 B1
Allenwood NJ, 935 ... 147 E2

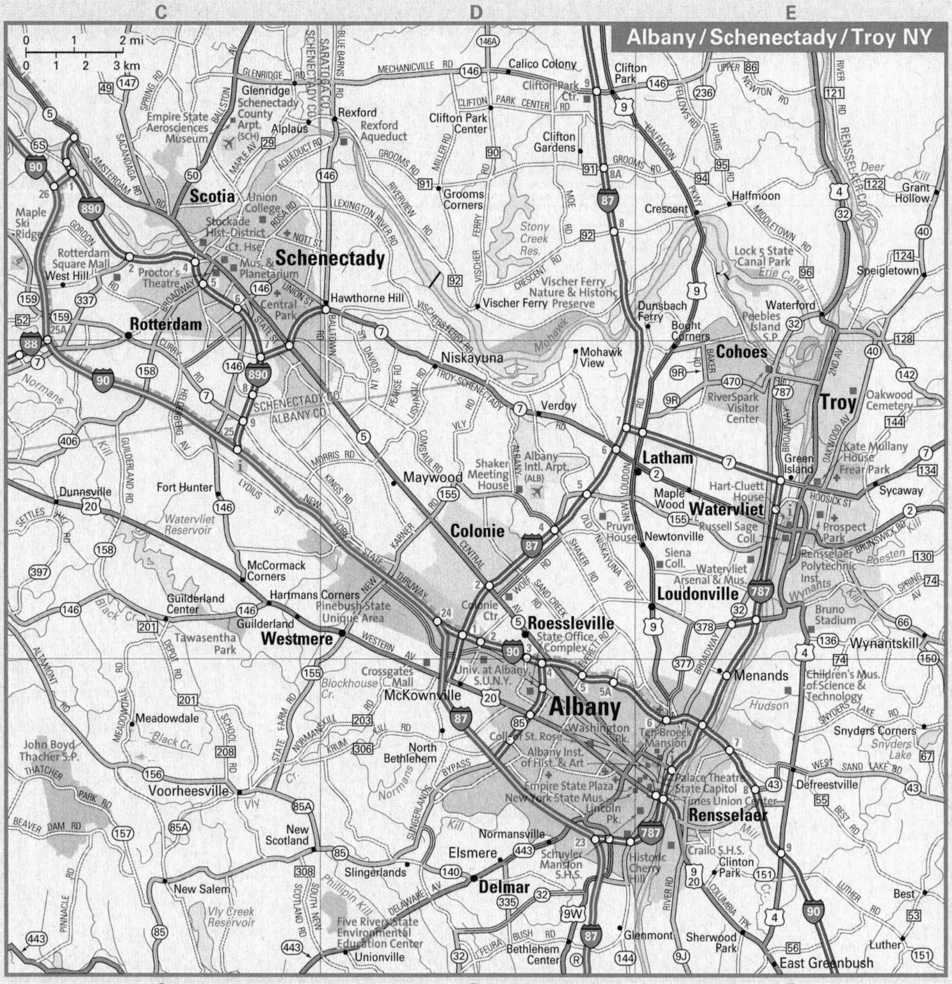

Albany ... D3
Alplaus ... C1
Best ... E3
Bethlehem Ctr. ... D3
Boght Corners ... E1
Calico Colony ... D1
Clifton Gardens ... D1
Clifton Park ... D1
Clifton Park Ctr. ... D1
Clinton Park ... E3
Cohoes ... E2
Colonie ... D2
Crescent ... E1
Defreestville ... E3
Delmar ... D3
Dunnsville ... C2
Dunsbach Ferry ... E1
E. Greenbush ... E3
Elsmere ... D3
Ft. Hunter ... C2
Glenmont ... D3
Glenridge ... C1
Grant Hollow ... E1
Green Island ... E2
Grooms Corners ... D1
Guilderland ... C2
Guilderland Ctr. ... C2
Halfmoon ... E1
Hartmans Corners ... C2
Hawthorne Hill ... D1
Latham ... E2
Loudonville ... E2
Luther ... E3
Maple Wood ... E2
Maywood ... D2
McCormack Corners ... C2
McKownville ... D3
Meadowdale ... C3
Menands ... E3
Mohawk View ... D2
New Salem ... C3
New Scotland ... C3
Newtonville ... E2
Niskayuna ... D2
Normanville ... D3
N. Bethlehem ... D3
Rensselaer ... E3
Rexford ... D1
Roessleville ... D2
Rotterdam ... C1
Schenectady ... C1
Scotia ... C1
Sherwood Park ... E3
Slingerlands ... D3
Snyders Corners ... E3
Speigletown ... E1
Sycaway ... E2
Troy ... E2
Unionville ... D3
Verdoy ... D2
Vischer Ferry ... D1
Voorheesville ... C3
Waterford ... E1
Watervliet ... E2
W. Hill ... C1
Westmere ... D2
Wynantskill ... E2

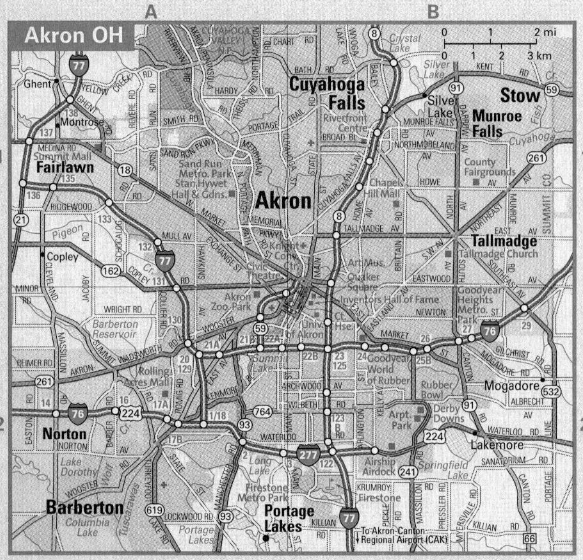

Akron ... A1
Barberton ... A2
Copley ... A1
Cuyahoga Falls ... B1
Fairlawn ... A1
Ghent ... A1
Lakemore ... B2
Mogadore ... B2
Montrose ... A1
Munroe Falls ... B1
Norton ... A2
Portage Lakes ... A2
Silver Lake ... B1
Stow ... B1
Tallmadge ... B1

Entries in **bold black** indicate counties or parishes.
Entries in **bold color** indicate cities with detailed inset maps.

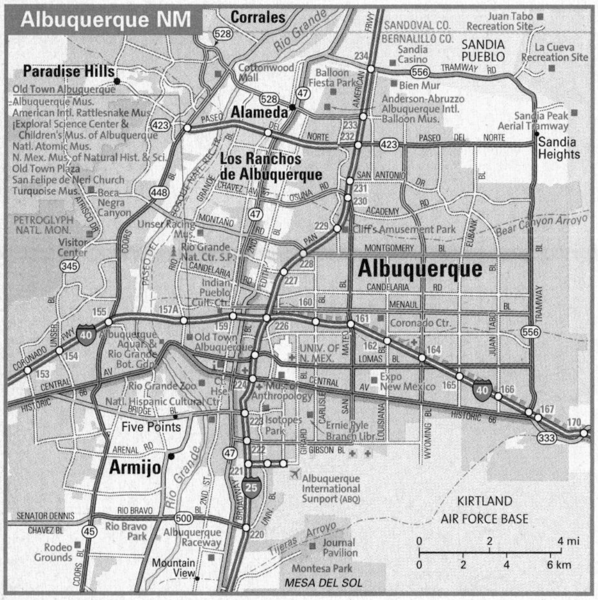

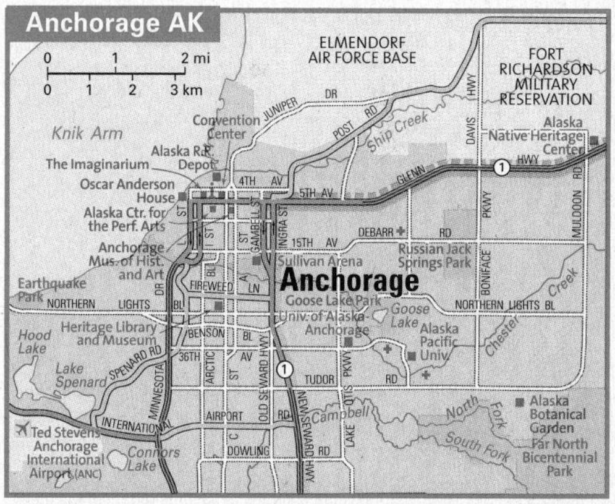

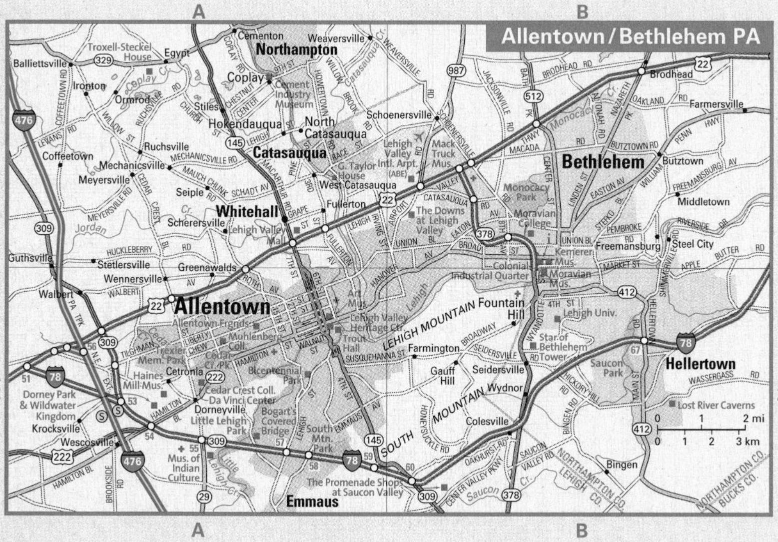

Allentown ... A2	Colesville ... B2	Gauff Hill ... B2	Northampton ... A1	Stiles ... A1
Balliettsville ... A1	Coplay ... A1	Greenawalds ... A2	N. Catasauqua ... A1	Walbert ... A2
Bethlehem ... B1	Dorneyville ... A2	Guthsville ... A1	Ormrod ... A1	Weaversville ... A1
Bingen ... B2	Egypt ... A1	Hellertown ... B2	Ruchsville ... A1	Wennersville ... A2
Brodhead ... B1	Emmaus ... A2	Hokendauqua ... A1	Scherersville ... A1	Wescosville ... A2
Butztown ... B1	Farmersville ... B1	Ironton ... A1	Schoenersville ... B1	W. Catasauqua ... A1
Catasauqua ... A1	Farmington ... B2	Krocksville ... A2	Seidersville ... B2	Whitehall ... A1
Cementon ... A1	Fountain Hill ... B2	Mechanicsville ... A1	Seiple ... A1	Wydnor ... B2
Cetronia ... A2	Freemansburg ... B1	Meyersville ... A1	Steel City ... B1	
Coffeetown ... A1	Fullerton ... A1	Middletown ... B1	Stetlersville ... A1	

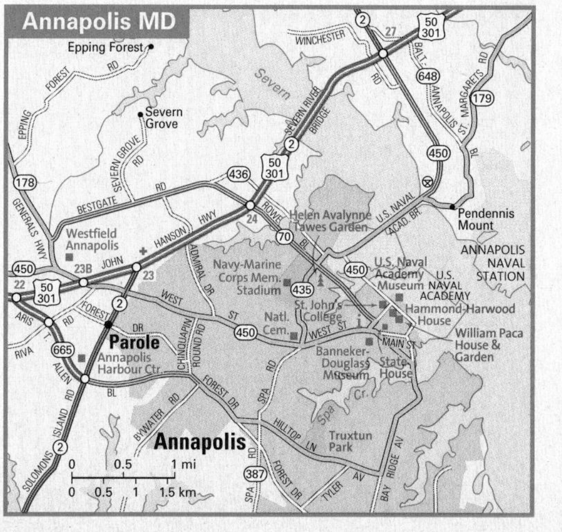

Figures after entries indicate population, page number, and grid reference.

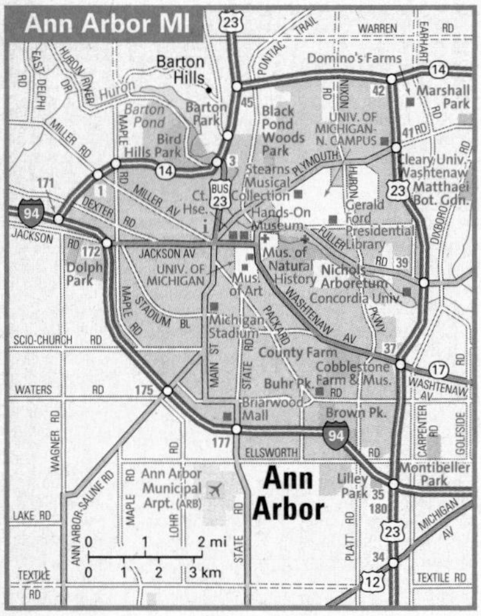

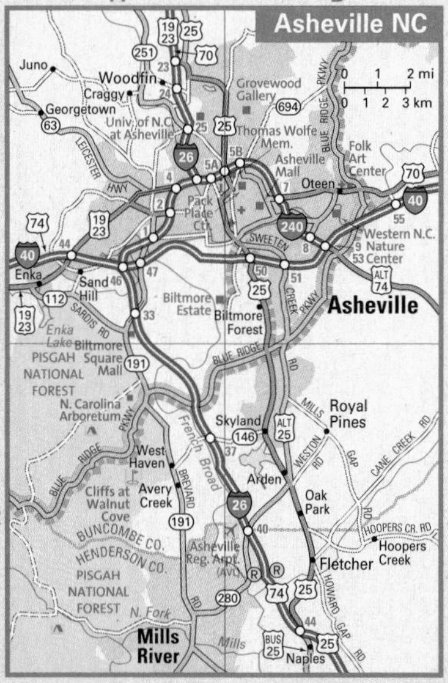

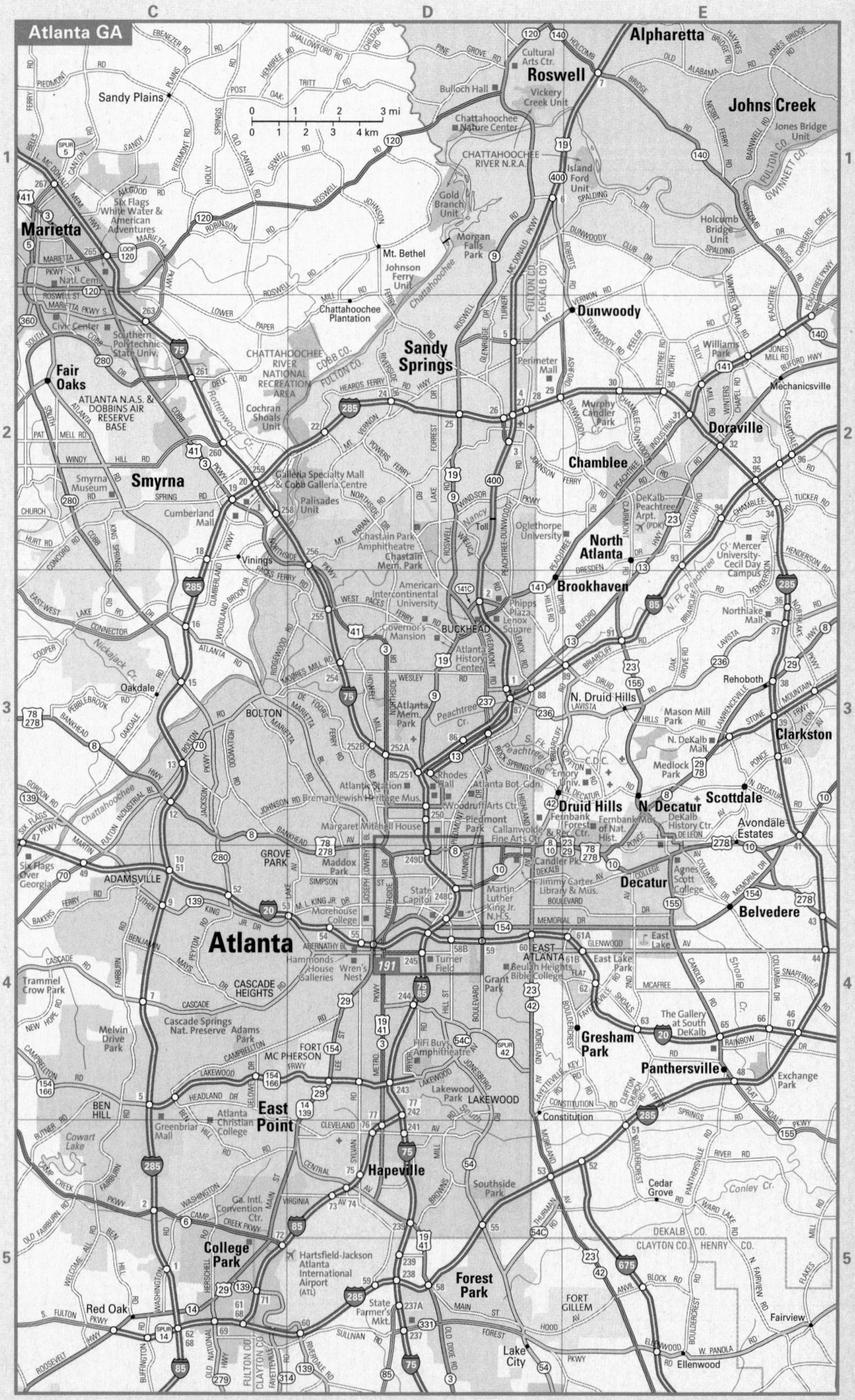

Entries in **bold black** indicate counties or parishes.
Entries in **bold color** indicate cities with detailed inset maps.

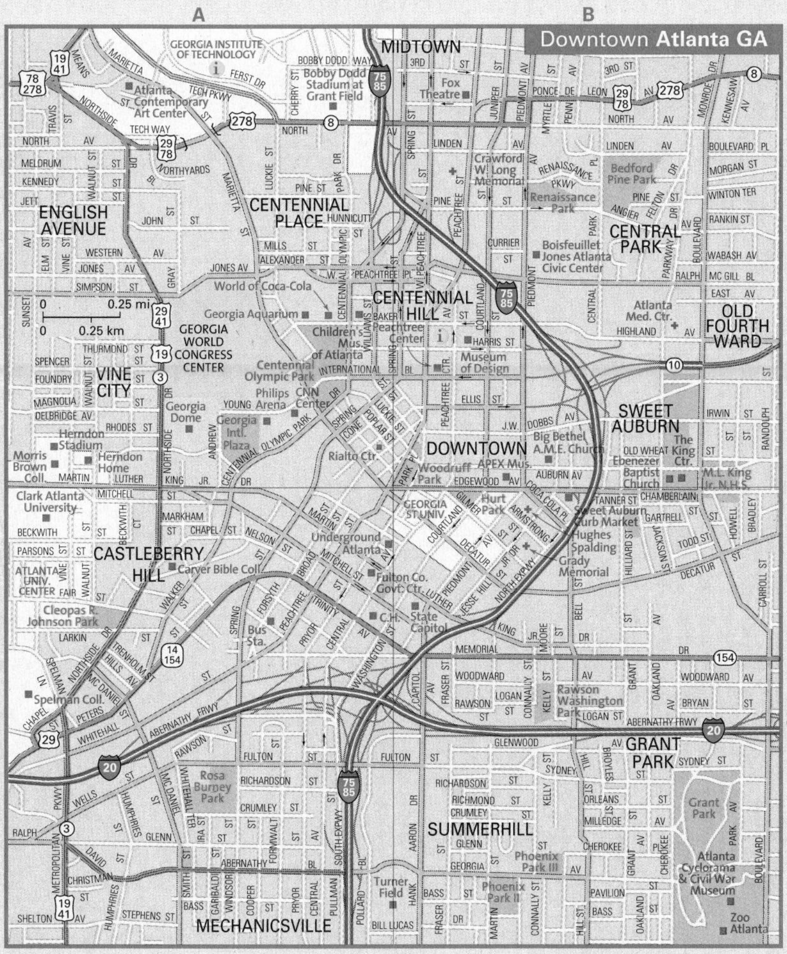

POINTS OF INTEREST

APEX Museum B1
Atlanta Contemporary Art Center A1
Atlanta Cyclorama & Civil War Museum B2
Atlanta University Center A2
Big Bethel African Meth. Episcopal Church B1
Bobby Dodd Stadium at Grant Field A1
Boisfeuillet Jones Atlanta Civic Center B1
Carver Bible College A2
Children's Museum of Atlanta A1
City Hall A2
Clark Atlanta University A2
CNN Center A1
Ebenezer Baptist Church B1
Fox Theatre B1
Georgia Aquarium A1
Georgia Dome A1
Georgia Institute of Technology A1
Georgia State University B2
Georgia World Congress Center A1
Herndon Home A1
Herndon Stadium A1
The King Center B1
Martin Luther King, Jr. Natl. Hist. Site B1
Morris Brown College A1
Museum of Design B1
Peachtree Center B1
Philips Arena A1
Rialto Center A1
Spelman College A2
State Capitol B2
Sweet Auburn Curb Market B2
Turner Field A2
Underground Atlanta A2
World of Coca-Cola A1
Zoo Atlanta B2

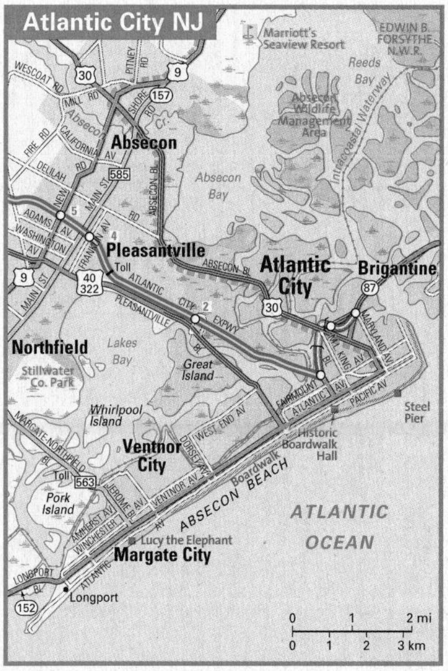

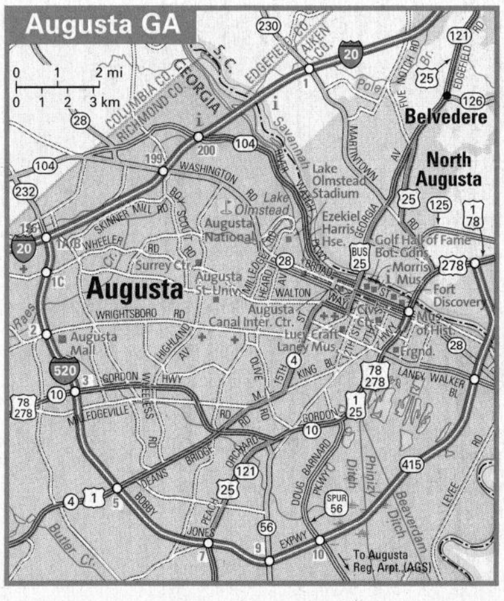

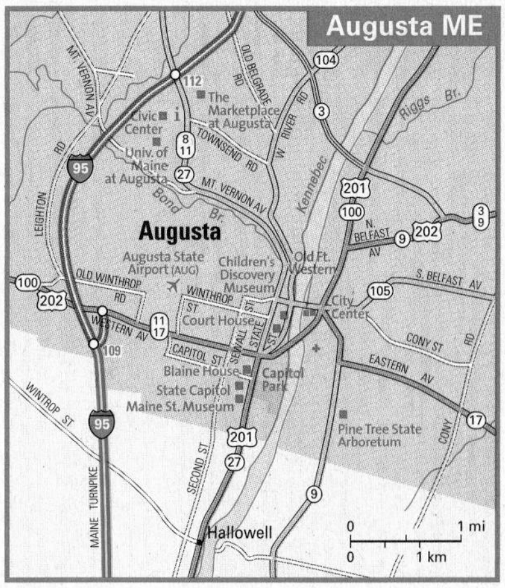

Figures after entries indicate population, page number, and grid reference.

Augusta Co. VA, *65615* 102 B4
Aulander NC, *888* 113 E3
Ault CO, *1432* 33 E4
Aumsville OR, *3003* 20 B2
Aurelia IA, *1062* 72 A4
Aurora CO, *276393* 41 E1
Aurora IL, *142990* 88 C1
Aurora IN, *3965* 100 B2
Aurora MN, *1850* 64 C3
Aurora MO, *7014* 106 C2
Aurora NE, *4225* 35 E4
Aurora NY, *720* 79 D4
Aurora NC, *583* 115 D3
Aurora OH, *13556* 91 E2
Aurora OR, *655* 20 C2
Aurora SD, *500* 27 F3
Aurora TX, *853* 59 E2
Aurora UT, *947* 39 E2
Aurora Co. SD, *3058* 27 D4
Au Sable MI, *1533* 76 C1
Au Sable Forks NY, *670* 81 D2
Austin AR, *605* 117 E2
Austin IN, *4724* 99 F3
Austin MN, *23314* 73 D2
Austin NV, *600* 37 F1
Austin PA, *623* 92 C2
Austin TX, *656562* 61 E1
Austin Co. TX, *23590* 61 F2
Austintown OH, *31627* 91 F3
Autauga Co. AL, *43671* 127 F2
Autaugaville AL, *820* 127 F2
Auxvasse MO, *901* 97 E2
Ava IL, *662* 98 B4
Ava MO, *3021* 107 E2
Avalon CA, *3127* 52 C3
Avalon NJ, *2143* 105 D4
Avalon PA, *5294* 250 A1
Avawam KY, *450* 111 D2
Avella PA, *750* 91 F4
Avenal CA, *14674* 44 C3
Avenel NJ, *17552* 147 E1
Aventura FL, *25267* 143 F2
Averill Park NY, *1517* 94 B1
Avery CA, *672* 37 D3
Avery Co. NC, *17167* 111 F4
Avery Creek NC, *1405* 121 E1
Avilla IN, *2049* 90 A2
Avis PA, *1492* 93 D2
Aviston IL, *1231* 98 B3
Avoca AR, *423* 106 C3
Avoca IA, *1610* 86 A2
Avoca NY, *1008* 78 C4
Avoca PA, *2851* 261 C2
Avoca WI, *608* 74 A3
Avon AL, *466* 137 D1
Avon CO, *5561* 40 C1
Avon CT, *1500* 94 C3
Avon IL, *915* 88 A3
Avon IN, *6248* 99 F1
Avon MN, *1242* 66 C2
Avon NY, *2977* 78 C3
Avon NC, *550* 115 F3
Avon OH, *11446* 91 D2
Avon PA, *2856* 146 A2
Avon SD, *561* 35 E1
Avon-by-the-Sea NJ, *2244* 147 F2
Avondale AZ, *35883* 54 C1
Avondale CO, *754* 41 E3
Avondale LA, *5441* 134 B3
Avondale MO, *529* 224 C2
Avondale PA, *1108* 146 B3
Avondale RI, *425* 149 F2
Avondale Estates GA, *2609* 190 E3
Avonia PA, *1331* 91 F1
Avon Lake OH, *18145* 91 D2
Avonmore PA, *820* 92 A4
Avon Park FL, *8542* 141 D3
Avoyelles Par. LA, *41481* 125 F4
Awendaw SC, *1195* 131 D1
Axtell KS, *445* 43 F1
Axtell NE, *696* 35 D4
Ayden NC, *4622* 115 D3
Ayer MA, *2960* 95 D1
Aynor SC, *587* 122 C3
Azalea Park FL, *11073* 246 D2
Azle TX, *9600* 59 E2
Aztec NM, *6378* 48 B1
Azusa CA, *44712* 228 E2

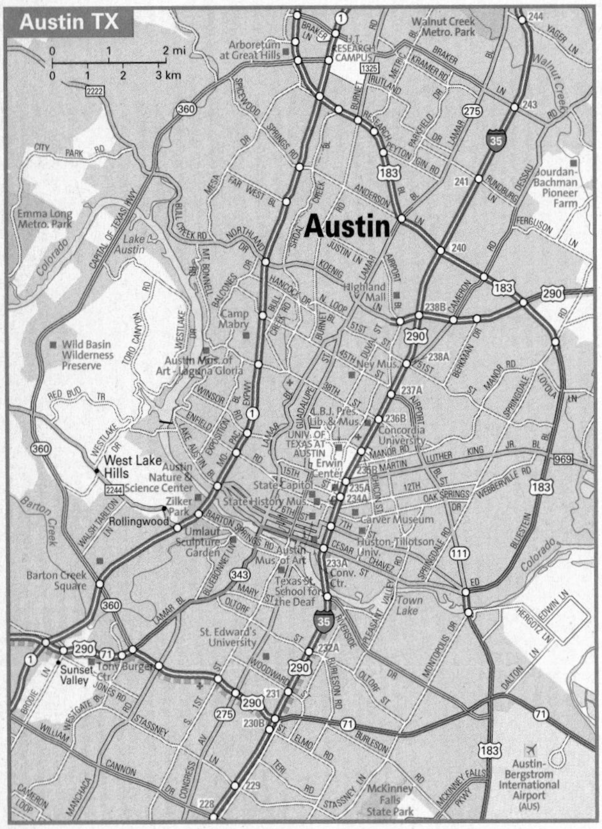

B
Babbie AL, *627* 128 A4
Babbitt MN, *1670* 64 C3
Babson Park FL, *1182* 141 D3
Babylon NY, *12615* 148 C4
Baca Co. CO, *4517* 42 A4
Bacon Co. GA, *10103* 129 F4
Baconton GA, *804* 129 D4
Bad Axe MI, *3462* 76 C2
Baden PA, *4377* 92 A3
Badger IA, *610* 72 C4
Badin NC, *1154* 122 B1
Bagdad AZ, *1578* 46 C4
Bagdad FL, *1490* 135 F2
Baggs WY, *348* 32 C3
Bagley MN, *1235* 64 A3
Bahama NC, *550* 112 C4
Bailey NC, *670* 113 D4
Bailey Co. TX, *6594* 49 F4
Bailey Island ME, *400* 82 B3
Baileys Crossroads VA, *23166* 270 B4
Bailey's Prairie TX, *694* 132 A4
Baileyton AL, *684* 119 F3
Baileyton TN, *504* 111 D3
Bainbridge GA, *11722* 137 D1
Bainbridge IN, *743* 99 E1
Bainbridge NY, *1365* 79 E4
Bainbridge OH, *1012* 101 D2
Bainbridge Island WA, *20308* 12 C3
Baird TX, *1623* 58 C3
Baiting Hollow NY, *1449* 149 E3
Baker LA, *13793* 134 A2
Baker MT, *1695* 17 F4
Baker City OR, *9860* 21 F2
Baker Co. FL, *22259* 138 C2
Baker Co. GA, *4074* 128 C4
Baker Co. OR, *16741* 21 F3
Bakersfield CA, *247057* 45 D4
Bakersville NC, *357* 111 E4
Bala-Cynwyd PA, *10300* 146 C3
Balaton MN, *637* 72 A1
Balch Sprs. TX, *19375* 207 E3
Balcones Hts. TX, *3016* 257 E2
Bald Knob AR, *3210* 117 F1
Baldwin FL, *1634* 139 D2
Baldwin GA, *2425* 121 D3
Baldwin IL, *3627* 98 B4
Baldwin LA, *2497* 133 F3
Baldwin MD, *850* 144 C1
Baldwin MI, *1107* 75 F2
Baldwin PA, *19999* 250 C3
Baldwin WI, *2667* 67 E4
Baldwin City KS, *3400* 96 A3
Baldwin Co. AL, *140415* 135 E1
Baldwin Co. GA, *44700* 129 E1
Baldwin Harbor NY, *8147* 147 F1
Baldwin Park CA, *75837* 228 E2
Baldwinsville NY, *7053* 79 D3
Baldwinville MA, *1852* 95 D1
Baldwyn MS, *3321* 119 D3
Balfour NC, *1200* 121 E1
Bal Harbour FL, *3305* 233 B3
Ball LA, *3681* 125 E4
Ballantine MT, *346* 24 C1
Ballard UT, *566* 32 A4
Ballard Co. KY, *8286* 108 C2
Ballentine SC, *850* 122 A3
Ball Ground GA, *730* 120 C3
Ballinger TX, *4243* 58 C4
Ballouville CT, *950* 150 B3
Ballston Spa NY, *5556* 80 C4
Ballville OH, *3255* 90 C2
Ballwin MO, *31283* 98 A3
Bally PA, *1062* 146 B1
Balmorhea TX, *527* 62 B2
Balmville NY, *3339* 148 B1
Balsam Lake WI, *950* 67 E3
Baltic CT, *1500* 149 F1
Baltic OH, *743* 91 E4
Baltic SD, *811* 27 F4
Baltimore MD, *651154* 144 C2
Baltimore OH, *2881* 101 D1
Baltimore Co. MD, *754292* 144 C1
Baltimore Highlands MD, *15724* 193 C4
Bamberg SC, *3733* 130 C1
Bamberg Co. SC, *16658* 130 B1
Bancroft ID, *382* 31 E1
Bancroft IA, *808* 72 B3
Bancroft KY, *536* 230 F1
Bancroft MI, *616* 76 B3
Bancroft NE, *520* 35 F2
Bancroft WV, *367* 101 E3
Bandera TX, *957* 61 D2
Bandera Co. TX, *17645* 60 C2
Bandon OR, *2833* 28 A1
Bangor ME, *31473* 83 D1
Bangor MI, *1933* 75 E4
Bangor PA, *5319* 93 F3
Bangor WI, *1400* 73 F2
Bangs TX, *1620* 59 D4
Banks OR, *1286* 20 B1
Banks Co. GA, *14422* 121 D3
Banner Co. NE, *819* 33 F3
Banner Elk NC, *811* 111 E4
Banner Hill TN, *1053* 111 E4
Bannertown NC, *950* 112 A3
Banning CA, *23562* 53 D2
Bannockburn IL, *1429* 203 C2
Bannock Co. ID, *75565* 31 E1
Banquete TX, *850* 63 F2
Bantam CT, *802* 94 C3
Baraboo WI, *10711* 74 A2
Baraga MI, *1285* 65 F4
Baraga Co. MI, *8746* 65 F4
Barataria LA, *1333* 134 B3
Barber Co. KS, *5307* 43 D4
Barberton OH, *27899* 91 E3
Barbour Co. AL, *29038* 128 B3
Barbour Co. WV, *15557* 102 A2
Barbourmeade KY, *1260* 230 F1
Barboursville WV, *3183* 101 E4
Barbourville KY, *3589* 110 C2
Bardstown KY, *10374* 110 A1
Bardwell KY, *799* 108 C2
Bardwell TX, *583* 59 F3
Bareville PA, *6625* 146 A2
Bargersville IN, *2120* 99 F1
Bar Harbor ME, *2680* 83 D2
Barker NY, *577* 78 B3
Barling AR, *4176* 116 C1
Barlow KY, *715* 108 C2
Bar Mills ME, *800* 82 B3
Barnegat NJ, *1690* 147 E4
Barnegat Light NJ, *764* 147 E4
Barnegat Pines NJ, *1300* 147 E3
Barnes Co. ND, *11775* 19 D4
Barnesville GA, *5972* 129 D1
Barnesville MN, *2173* 19 F4
Barnesville OH, *4225* 101 F1
Barneveld WI, *1088* 74 A3
Barnhart MO, *6108* 98 A4
Barnsboro NJ, *2500* 146 C4
Barnsdall OK, *1325* 51 F1
Barnstable MA, *47821* 151 F3
Barnstable Co. MA, *222230* 151 E4
Barnum MN, *525* 64 C4
Bar Nunn WY, *936* 33 D1
Barnwell SC, *5035* 130 B1
Barnwell Co. SC, *23478* 130 B1
Baroda MI, *858* 89 E1
Barrackville WV, *1288* 102 A1
Barre MA, *1150* 150 B1
Barre VT, *9291* 81 E2
Barren Co. KY, *38033* 110 A2
Barre Plains MA, *1200* 150 B1
Barrett TX, *2872* 132 B3
Barrington IL, *10168* 203 B2
Barrington NH, *600* 81 F4
Barrington NJ, *7084* 248 D4
Barrington RI, *16819* 151 D3
Barrington Hills IL, *3915* 203 A2
Barron WI, *3248* 67 E3
Barron Co. WI, *44963* 67 E3
Barrow AK, *4581* 154 C1
Barrow Co. GA, *46144* 121 D3
Barry IL, *1368* 97 F1
Barry Co. MI, *56755* 75 F4
Barry Co. MO, *34010* 106 C2
Barstow CA, *21119* 53 D1
Barstow MD, *750* 144 C4
Bartelso IL, *593* 98 B3
Bartholomew Co. IN, *71435* 99 F2
Bartlesville OK, *34748* 51 F1
Bartlett IL, *36706* 203 A3
Bartlett NE, *128* 35 D2
Bartlett NH, *500* 81 F2
Bartlett TN, *40543* 118 B1
Bartlett TX, *1675* 61 E1
Barton MD, *478* 102 C1
Barton VT, *742* 81 E1
Barton Co. KS, *28205* 43 D3
Barton Co. MO, *12541* 106 B1
Bartonsville MD, *12529* 144 A1
Bartonville IL, *6310* 88 B3
Bartow FL, *15340* 140 C2
Bartow Co. GA, *76019* 120 B3
Barview OR, *1872* 20 A4
Basalt CO, *2681* 40 C2
Basalt ID, *419* 23 E4
Basehor KS, *2238* 96 B2
Basile LA, *1660* 133 E2
Basin MT, *255* 15 E4
Basin WY, *1238* 24 C3
Basin City WA, *968* 13 E4
Baskett KY, *550* 99 E4
Basking Ridge NJ, *3600* 148 A4
Bassett NE, *743* 35 D1
Bassett VA, *1338* 112 B2
Bass Harbor ME, *600* 83 D2
Bass Lake IN, *1249* 89 E2
Bastrop LA, *12988* 125 F2
Bastrop TX, *5340* 61 E2
Bastrop Co. TX, *57733* 61 E2
Basye VA, *986* 102 C3
Batavia IL, *23866* 88 C1
Batavia IA, *500* 87 E3
Batavia NY, *16256* 78 B3
Batavia OH, *1617* 100 B2
Batesburg-Leesville SC, *5517* 122 A4
Bates Co. MO, *16653* 96 B4
Batesville AR, *9445* 107 F4
Batesville IN, *6033* 100 A2
Batesville MS, *7113* 118 B3
Batesville TX, *1298* 60 C3
Bath ME, *9266* 82 C3
Bath MI, *1200* 76 A3
Bath NY, *5641* 78 C4
Bath PA, *2678* 93 F3
Bath Co. KY, *11085* 100 C4
Bath Co. VA, *5048* 102 B4
Baton Rouge LA, *227818* 134 A2
Battle Creek IA, *743* 72 A4
Battle Creek MI, *53364* 75 F4
Battle Creek NE, *1158* 35 E2
Battlefield MO, *2385* 107 D2
Battle Ground IN, *1323* 89 E4
Battle Ground WA, *9296* 20 C1
Battle Lake MN, *686* 19 F4
Battlement Mesa CO, *3497* 40 B2
Battle Mtn. NV, *2871* 30 A4
Baudette MN, *1104* 64 A1
Baumstown PA, *1000* 146 B2
Bauxite AR, *432* 117 E2
Bawcomville LA, *7616* 125 E2
Baxley GA, *4150* 129 F3
Baxter IA, *1052* 87 D1
Baxter MN, *5555* 64 A4
Baxter TN, *1279* 110 A4
Baxter Co. AR, *38386* 107 E4
Baxter Estates NY, *1006* 241 G2
Baxter Sprs. KS, *4602* 106 B2
Bay AR, *1800* 108 A4
Bayard IA, *536* 86 B1
Bayard NE, *1247* 33 F2
Bayard NM, *2534* 55 F2
Bayboro NC, *741* 115 D3
Bay City MI, *36817* 76 B2
Bay City OR, *1149* 20 B2
Bay City TX, *18667* 61 F3
Bay Co. FL, *148217* 136 C2
Bay Co. MI, *110157* 76 B2
Bayfield CO, *1549* 40 C4
Bayfield WI, *611* 65 D4
Bayfield Co. WI, *15013* 65 D4
Bay Harbor Islands FL, *5146* 233 C4
Bay Head NJ, *1238* 147 E3
Bay Hill FL, *5177* 246 B3
Baylor Co. TX, *4093* 59 D1
Bay Minette AL, *7820* 135 E1
Bayonet Pt. FL, *23577* 140 B2
Bayonne NJ, *61842* 148 B4
Bayou Cane LA, *17046* 134 A3
Bayou George FL, *800* 136 C2
Bayou Goula LA, *750* 134 A2
Bayou La Batre AL, *2313* 135 E2
Bayou Vista LA, *4351* 134 A3
Bayou Vista TX, *1644* 132 B4
Bay Park NY, *2300* 241 G5
Bay Pines FL, *3065* 266 A3
Bay Pt. CA, *21534* 259 D1
Bayport MN, *3162* 67 D4
Bayport NY, *8662* 149 D4
Bay Ridge MD, *2300* 144 C3
Bay St. Louis MS, *8209* 134 C2
Bayshore FL, *750* 142 C1
Bay Shore NY, *23852* 149 D4
Bayshore Gardens FL, *17350* 266 B5
Bay Side NJ, *1800* 147 E4
Bayside WI, *4518* 234 D1
Bay Sprs. MS, *2097* 126 C3
Baytown TX, *66430* 132 B3
Bay View OH, *692* 91 D2
Bay Vil. OH, *16087* 204 D2
Bayville NJ, *4700* 147 E3
Bayville NY, *7135* 148 C3
Beach ND, *1116* 17 F4
Beach City OH, *1137* 91 E3
Beach City TX, *1645* 132 B3
Beach Haven NJ, *1278* 147 E4
Beach Haven Gardens
NJ, *1200* 147 E4
Beach Haven Terrace NJ, *1100* 147 E4
Beachwood NJ, *10375* 147 E3
Beachwood OH, *12186* 204 G2
Beacon IA, *518* 87 D2
Beacon NY, *13808* 148 B1
Beacon Falls CT, *1500* 149 D1
Beadle Co. SD, *17023* 27 D3
Bealeton VA, *2100* 103 D3
Beals ME, *750* 83 E2
Bean Sta. TN, *2634* 111 D3
Bear DE, *17593* 145 E1
Bear Creek AL, *1053* 119 E3
Bearden AR, *1125* 117 E4
Beardstown IL, *5766* 98 A1
Bear Lake Co. ID, *6411* 31 F2
Bear River City UT, *750* 31 E3
Beasley TX, *590* 132 A4
Beatrice AL, *412* 127 F4
Beatrice NE, *12496* 35 F4
Beatty NV, *1154* 45 F2
Beattyville KY, *1193* 110 C1
Beatyestown NJ, *3223* 94 A4
Beaufort NC, *3771* 115 E4
Beaufort SC, *12950* 130 C2
Beaufort Co. NC, *44958* 113 F4
Beaufort Co. SC, *120937* 130 C3
Beaumont CA, *11384* 53 D2
Beaumont MS, *977* 135 D1
Beaumont TX, *113866* 132 C3
Beaumont Place TX, *4500* 220 D2
Beauregard Par. LA, *32986* 133 D2
Beaver OK, *1570* 50 C1
Beaver PA, *4775* 91 F3
Beaver UT, *2454* 39 D3
Beaver WV, *1378* 111 F1
Beaver City NE, *641* 42 C1
Beaver Co. OK, *5857* 50 C1
Beaver Co. PA, *181412* 91 F3
Beaver Co. UT, *6005* 39 D3
Beavercreek OH, *37984* 100 C1
Beaver Crossing NE, *457* 35 E4
Beaverdale PA, *1230* 92 B4
Beaver Dam KY, *3033* 109 E1
Beaver Dam WI, *15169* 74 B2
Beaver Falls PA, *9920* 91 F3
Beaverhead Co. MT, *9202* 23 D2
Beaver Meadows PA, *968* 93 E3
Beaver Sprs. PA, *634* 93 D3
Beaverton MI, *1106* 76 A2
Beaverton OR, *76129* 20 C2
Beavertown PA, *870* 93 D3
Bechtelsville PA, *931* 146 B1
Beckemeyer IL, *1043* 98 B3
Becker MN, *2673* 66 C3
Becker Co. MN, *30000* 19 F3
Beckett NJ, *4726* 146 C4
Beckham Co. OK, *19799* 50 C3
Beckley WV, *17254* 111 F1
Beckville TX, *752* 124 C3
Bedford IN, *13768* 99 F3
Bedford IA, *1620* 86 B3
Bedford KY, *677* 100 A3
Bedford MA, *12595* 151 D1
Bedford NH, *1300* 95 D1
Bedford NY, *1724* 148 C2
Bedford OH, *14214* 204 G3
Bedford PA, *3141* 102 C1
Bedford TX, *47152* 207 B2
Bedford VA, *6299* 112 B1
Bedford Co. PA, *49984* 92 C4
Bedford Co. TN, *37586* 120 A1
Bedford Co. VA, *60371* 112 B1
Bedford Hts. OH, *11375* 204 G3
Bedford Hills NY, *5500* 148 C2
Bedford Park IL, *574* 203 D5
Beebe AR, *4930* 117 F2
Bee Cave TX, *656* 61 E1
Beech Bottom WV, *606* 91 F4
Beech Creek PA, *717* 93 D3
Beecher IL, *2033* 89 D2
Beech Grove IN, *14880* 99 F1
Beechwood Vil. KY, *1173* 230 E1
Bee Co. TX, *32359* 61 E4
Beemer NE, *773* 35 F2
Bee Ridge FL, *8744* 140 B4
Beersheba Sprs. TN, *553* 120 A1
Beesleys Pt. NJ, *1400* 147 F4
Beeville TX, *13129* 61 E4
Beggs OK, *1364* 51 F2
Bel Air MD, *10080* 145 D1
Belcamp MD, *1900* 145 D1
Belchertown MA, *2626* 150 A1
Belcourt ND, *2440* 18 C1
Belding MI, *5877* 75 F3
Belen NM, *6901* 48 C4
Belfair WA, *700* 12 C3
Belfast ME, *6381* 82 C2
Belfast NY, *800* 78 B4
Belfast PA, *1301* 93 F3
Belfield ND, *866* 18 A4
Belford NJ, *1340* 147 E1
Belfry MT, *219* 24 B2
Belgium WI, *1678* 75 D2
Belgrade MN, *750* 66 B3
Belgrade MT, *5728* 23 F1
Belgrade Lakes ME, *350* 82 B2
Belhaven NC, *1968* 115 E3
Belinda City TN, *2100* 109 F4
Belington WV, *1788* 102 A2
Belknap Co. NH, *56325* 81 F4
Bell CA, *36664* 228 D3
Bellair FL, *16539* 222 C4
Bellaire MI, *1164* 69 F4
Bellaire OH, *4892* 101 F1
Bellaire TX, *15642* 132 A3
Bellamy AL, *600* 127 E2
Bella Villa MO, *687* 256 B3
Bella Vista AR, *16582* 106 C3
Bella Vista CA, *550* 28 C4
Bellbrook OH, *7009* 100 C1
Bell Buckle TN, *391* 119 F1
Bell Co. KY, *30060* 110 C3
Bell Co. TX, *237974* 61 E1
Belle MO, *1344* 97 F4
Belle WV, *1259* 101 F4
Belleair FL, *4067* 140 B2
Belleair Beach FL, *1751* 140 B2
Belleair Bluffs FL, *2243* 266 A2
Belle Ctr. OH, *807* 90 C4
Belle Chasse LA, *9848* 134 B3
Bellefontaine OH, *13069* 90 B4
Bellefontaine Neighbors
MO, *11271* 256 C1
Bellefonte AR, *400* 107 D3
Bellefonte DE, *1249* 146 B4
Bellefonte KY, *837* 101 D3
Bellefonte PA, *6395* 92 C3
Belle Fourche SD, *4565* 25 F3
Belle Glade FL, *14906* 143 E1
Belle Haven VA, *480* 114 B3
Belle Isle FL, *5531* 141 D1
Bellemeade KY, *871* 230 F2
Belle Plaine IA, *2878* 87 E1
Belle Plaine KS, *1708* 43 E4
Belle Plaine MN, *3789* 66 C4
Belle Rose LA, *1944* 134 A3
Bellerose NY, *1173* 241 G3
Bellerose Terrace NY, *2157* 241 G3
Belle Terre NY, *832* 149 D3
Belle Vernon PA, *1211* 92 A4
Belleview FL, *3478* 139 D4
Belleville IL, *41410* 98 B3
Belleville KS, *2239* 43 E1
Belleville MI, *3997* 90 C1
Belleville NJ, *35928* 148 B4
Belleville PA, *1386* 92 C3
Belleville WI, *1908* 74 B3
Bellevue ID, *1876* 22 C4
Bellevue IL, *1887* 88 B3
Bellevue IA, *2350* 74 A4
Bellevue KY, *6480* 204 B3
Bellevue MI, *1365* 76 A4
Bellevue NE, *44382* 86 A2
Bellevue OH, *8193* 204 B3
Bellevue PA, *8770* 92 A4
Bellevue WA, *109569* 12 C3
Bellevue WI, *11828* 74 C1
Bellflower CA, *72878* 228 D3
Bell Gardens CA, *44054* 228 D3
Bellingham MA, *4497* 150 C2
Bellingham WA, *67171* 12 C1
Bellmawr NJ, *11262* 146 C3
Bellmead TX, *9214* 59 E4
Bellows Falls VT, *3165* 81 E4
Bellport NY, *2363* 149 D4
Bells TN, *2171* 108 C4
Bells TX, *1190* 59 F1
Bellview FL, *21201* 247 A1
Bellville OH, *1773* 91 D3
Bellville TX, *3794* 61 F2
Bellwood IL, *20535* 203 C4
Bellwood NE, *446* 35 E3
Bellwood PA, *2016* 92 C4
Bellwood VA, *5974* 254 B3
Belmar NJ, *6045* 147 F2
Belmond IA, *2560* 72 C3
Belmont CA, *25123* 259 B5
Belmont MA, *24194* 151 D1

Entries in **bold black** indicate counties or parishes.
Entries in **bold color** indicate cities with detailed inset maps.

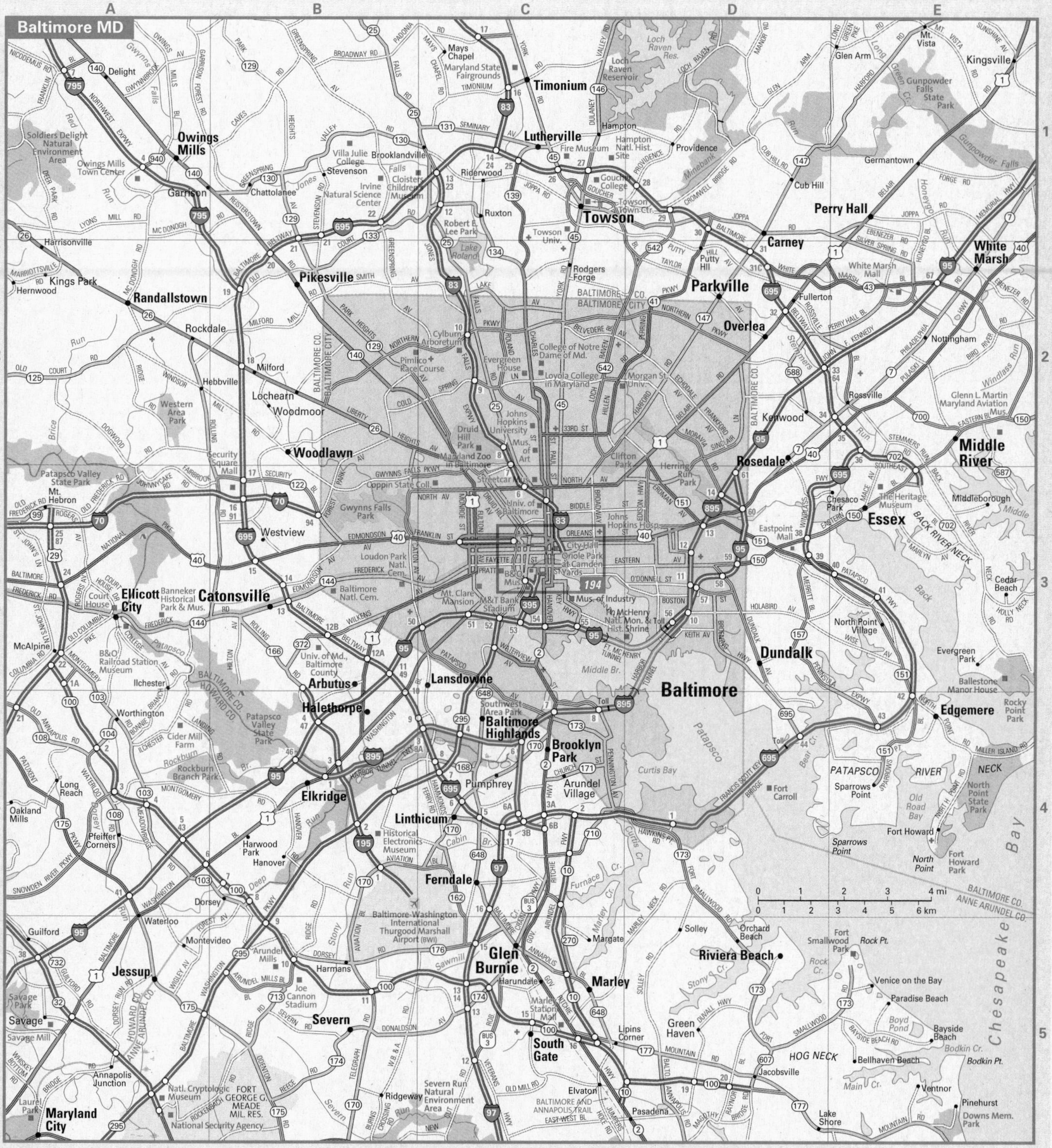

Figures after entries indicate population, page number, and grid reference.

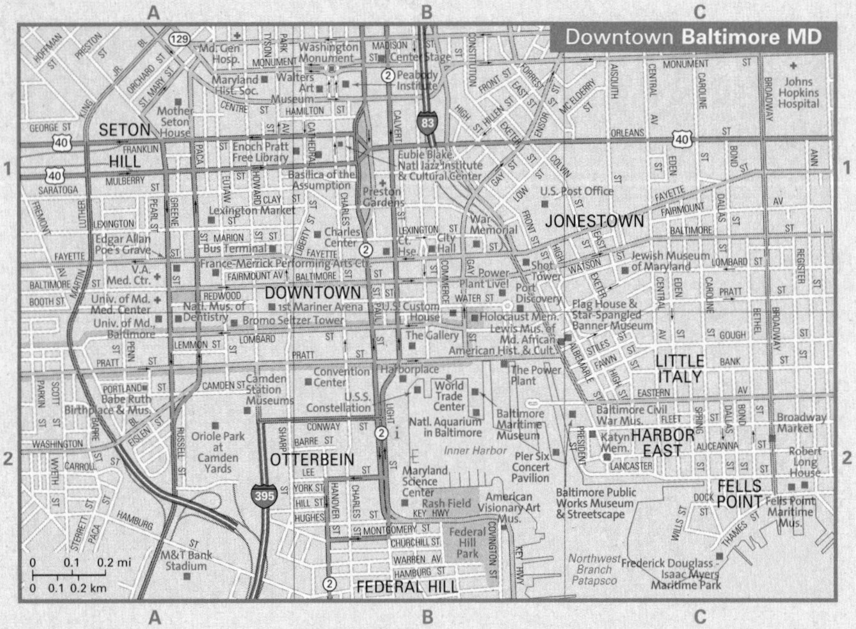

POINTS OF INTEREST

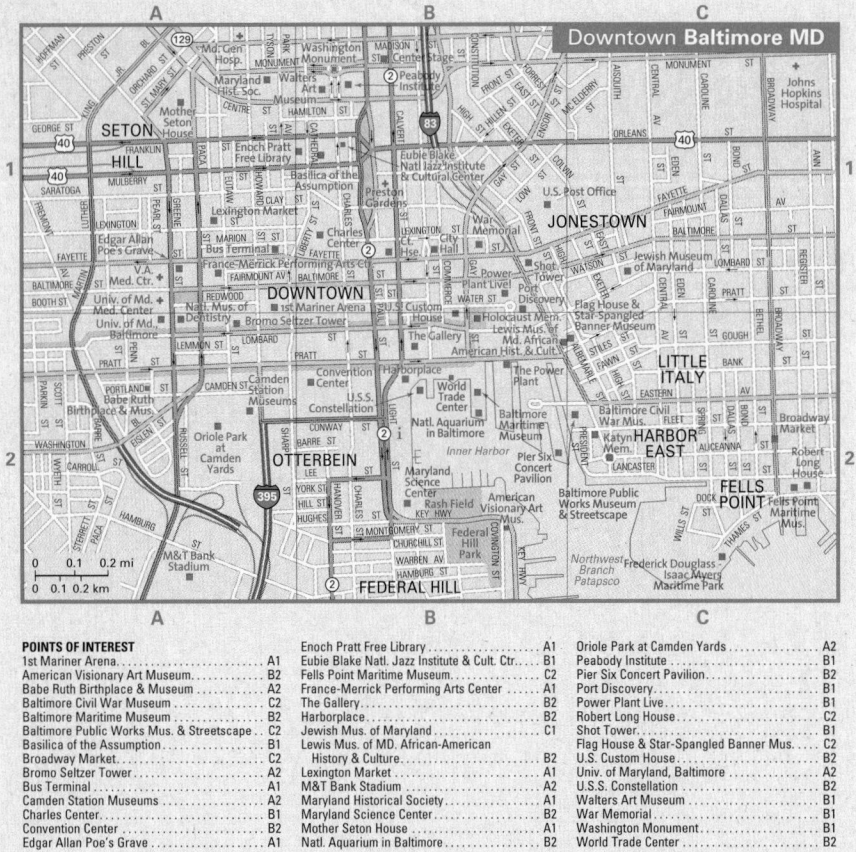

1st Mariner Arena	A1
American Visionary Art Museum	B2
Babe Ruth Birthplace & Museum	A2
Baltimore Civil War Museum	C2
Baltimore Maritime Museum	B2
Baltimore Public Works Mus. & Streetscape	C2
Basilica of the Assumption	B1
Broadway Market	C2
Bromo Seltzer Tower	A2
Bus Terminal	A1
Camden Station Museums	A2
Charles Center	B1
Convention Center	B2
Edgar Allan Poe's Grave	A1
Enoch Pratt Free Library	A1
Eubie Blake Natl. Jazz Institute & Cult. Ctr.	B1
Fells Point Maritime Museum	C2
France-Merrick Performing Arts Center	A1
The Gallery	B2
Harborplace	B2
Jewish Mus. of Maryland	C1
Lewis Mus. of MD. African-American History & Culture	B2
Lexington Market	A1
M&T Bank Stadium	A2
Maryland Historical Society	A1
Maryland Science Center	B2
Mother Seton House	A1
Natl. Aquarium in Baltimore	B2
Oriole Park at Camden Yards	A2
Peabody Institute	B1
Pier Six Concert Pavilion	B2
Port Discovery	B1
Power Plant Live	B1
Robert Long House	C2
Shot Tower	B1
Flag House & Star-Spangled Banner Mus.	C2
U.S. Custom House	B2
Univ. of Maryland, Baltimore	A2
U.S.S. Constellation	B2
Walters Art Museum	B1
War Memorial	B1
Washington Monument	B1
World Trade Center	B2

Belmont MS, *1961* ... **119** D3
Belmont NH, *950* ... **81** F4
Belmont NY, *952* ... **92** C1
Belmont NC, *8705* ... **122** A1
Belmont WV, *1036* ... **101** F2
Belmont WI, *871* ... **74** A4
Belmont Corner ME, *375* ... **82** C2
Belmont Co. OH, *70226* ... **101** F1
Bel-Nor MO, *1598* ... **256** B2
Beloit KS, *4019* ... **43** E1
Beloit OH, *1024* ... **91** F3
Beloit WI, *35775* ... **74** B4
Belpre OH, *6660* ... **101** E2
Bel-Ridge MO, *3082* ... **256** B2
Belt MT, *633* ... **15** F3
Belton KY, *500* ... **109** E2
Belton MO, *21730* ... **96** B3
Belton SC, *4461* ... **121** E3
Belton TX, *14623* ... **59** E4
Beltrami Co. MN, *39650* ... **64** A2
Beltsville MD, *15690* ... **144** B3
Belvedere CA, *2125* ... **259** B2
Belvedere GA, *11100* ... **190** E4
Belvedere SC, *5631* ... **121** F4
Belvidere IL, *20820* ... **74** B4
Belvidere NJ, *2771* ... **93** F3
Belwood NC, *962* ... **121** F1
Belzoni MS, *2663* ... **126** B1
Bement IL, *1784* ... **98** C1
Bemidji MN, *11917* ... **64** A3
Bemiss GA, *1500* ... **137** F1
Benavides TX, *1686* ... **63** E2
Ben Avon PA, *1917* ... **250** A1
Benbrook TX, *20208* ... **207** A3
Bend OR, *52029* ... **21** D3
Bendersville PA, *576* ... **103** E1
Benewah Co. ID, *9171* ... **14** B3
Benham KY, *599* ... **111** D2
Ben Hill Co. GA, *17484* ... **129** E3
Benicia CA, *26865* ... **36** B3
Benjamin TX, *264* ... **58** C1
Benkelman NE, *1006* ... **42** B1
Benld IL, *1541* ... **98** B2
Ben Lomond CA, *2364* ... **44** A2
Bennet NE, *570* ... **35** F4
Bennett CO, *2021* ... **41** E1
Bennett Co. SD, *3574* ... **26** B4
Bennettsville SC, *9425* ... **122** C2
Bennington KS, *623* ... **43** E2
Bennington NE, *937* ... **35** F3
Bennington NH, *425* ... **81** E4
Bennington VT, *9168* ... **94** C1
Bennington Co. VT, *36994* ... **81** D4
Benoit MS, *611* ... **118** A4

Bensenville IL, *20703* ... **203** C4
Bensley VA, *5435* ... **113** E1
Benson AZ, *4711* ... **55** D3
Benson MD, *950* ... **144** C1
Benson MN, *3376* ... **66** A3
Benson NC, *2923* ... **123** D1
Benson Co. ND, *6964* ... **19** D2
Bent Co. CO, *5998* ... **41** F3
Bent Creek NC, *1389* ... **121** E1
Bentleyville OH, *947* ... **204** G3
Bentleyville PA, *2502* ... **92** A4
Benton AR, *21906* ... **117** E2
Benton IL, *6880* ... **98** C4
Benton KS, *827* ... **43** F4
Benton KY, *4197* ... **109** D2
Benton LA, *2035* ... **124** C2
Benton ME, *500* ... **82** C2
Benton MO, *732* ... **108** B2
Benton PA, *955* ... **93** E2
Benton TN, *1138* ... **120** C1
Benton WI, *976* ... **74** A4
Benton City WA, *2624* ... **21** E1
Benton Co. AR, *153406* ... **106** B3
Benton Co. IN, *9421* ... **89** D3
Benton Co. IA, *25308* ... **87** E1
Benton Co. MN, *34226* ... **66** C2
Benton Co. MS, *8026* ... **118** C2
Benton Co. MO, *17180* ... **97** D4
Benton Co. OR, *78153* ... **20** B3
Benton Co. TN, *16537* ... **109** D4
Benton Co. WA, *142475* ... **21** E1
Benton Harbor MI, *11182* ... **89** E1
Bentonia MS, *500* ... **126** B2
Bentonville AR, *19730* ... **106** C3
Benzie Co. MI, *15998* ... **69** F4
Benzonia MI, *519* ... **69** E4
Berea KY, *9851* ... **110** C1
Berea OH, *18970* ... **91** E2
Berea SC, *14158* ... **217** A2
Beresford SD, *2006* ... **35** F1
Bergen NY, *1240* ... **78** B3
Bergen Co. NJ, *884118* ... **148** B3
Bergenfield NJ, *26247* ... **148** B3
Bergholz OH, *769* ... **91** F4
Bergman AR, *407* ... **107** D3
Berino NM, *900* ... **56** C3
Berkeley CA, *102743* ... **36** B4
Berkeley IL, *5245* ... **203** C4
Berkeley MO, *10063* ... **256** B1
Berkeley RI, *2800* ... **150** C3
Berkeley Co. SC, *142651* ... **131** D1
Berkeley Co. WV, *75905* ... **103** D2
Berkeley Hts. NJ, *13407* ... **148** A4
Berkeley Sprs. WV, *663* ... **102** C1
Berkley MA, *850* ... **151** D3
Berkley MI, *15531* ... **210** B2
Berks Co. PA, *373638* ... **146** B1
Berkshire CT, *950* ... **149** D2
Berkshire Co. MA, *134953* ... **94** C2
Berlin CT, *1000* ... **149** E1
Berlin GA, *595* ... **137** F1
Berlin MD, *3491* ... **114** C2
Berlin MA, *650* ... **150** C1
Berlin NH, *10331* ... **81** F2
Berlin NJ, *6149* ... **147** D3
Berlin OH, *1300* ... **91** E4
Berlin PA, *2192* ... **102** C1
Berlin WI, *5305* ... **74** B1
Berlin Hts. OH, *685* ... **91** D2
Bermuda Run NC, *1431* ... **112** A4
Bernalillo NM, *6611* ... **48** C3
Bernalillo Co. NM, *556678* ... **48** C3
Bernardston MA, *1000* ... **94** C1
Bernardsville NJ, *7345* ... **148** A4
Berne IN, *4150* ... **90** A3
Bernice LA, *1809* ... **125** E2
Bernice OK, *504* ... **106** B3
Bernie MO, *1777* ... **108** B3
Bernstadt KY, *475* ... **110** C2
Bernville PA, *865* ... **146** A1
Berrien Co. GA, *16235* ... **129** E4
Berrien Co. MI, *162453* ... **89** E1
Berrien Sprs. MI, *1862* ... **89** E1
Berry AL, *1238* ... **119** E4
Berryville AR, *4433* ... **106** C3
Berryville TX, *891* ... **124** A3
Berryville VA, *2963* ... **103** D2
Berthold ND, *466* ... **18** B2
Berthoud CO, *4839* ... **33** E4
Bertie Co. NC, *19773* ... **113** F4
Bertram IA, *681* ... **87** F1
Bertram TX, *1122* ... **61** D1
Bertrand MI, *1700* ... **89** F1
Bertrand MO, *740* ... **108** B2
Bertrand NE, *786* ... **35** D4
Berwick LA, *4418* ... **134** A3
Berwick ME, *1993* ... **82** A4
Berwick PA, *10774* ... **93** E3
Berwyn IL, *54016* ... **89** D1
Berwyn PA, *5067* ... **146** B3
Berwyn Hts. MD, *2942* ... **270** E2
Bessemer AL, *29672* ... **127** F1
Bessemer MI, *2148* ... **65** E4
Bessemer PA, *1172* ... **91** F3
Bessemer City NC, *5119* ... **122** A1
Bethalto IL, *9454* ... **98** B3
Bethany CT, *900* ... **149** D1
Bethany IL, *1287* ... **98** C1
Bethany MO, *3087* ... **86** C4
Bethany OK, *20307* ... **51** E3
Bethany WV, *985* ... **91** F4
Bethany Beach DE, *903* ... **145** F4
Bethel AK, *5471* ... **154** B3
Bethel CT, *9137* ... **148** C2
Bethel DE, *184* ... **145** E4
Bethel ME, *475* ... **82** B2
Bethel NC, *1681* ... **113** E4
Bethel OH, *2637* ... **100** C2
Bethel VT, *800* ... **81** E3
Bethel VA, *500* ... **103** D3
Bethel Acres OK, *2735* ... **51** F3
Bethel Hts. AR, *714* ... **106** C3
Bethel Park PA, *33556* ... **92** A4
Bethel Sprs. TN, *763* ... **119** D1
Bethesda MD, *55277* ... **144** B3
Bethesda OH, *1413* ... **101** F1
Bethlehem CT, *2022* ... **149** D1
Bethlehem GA, *716* ... **121** D4
Bethlehem MD, *600* ... **145** D4
Bethlehem NH, *700* ... **81** F2
Bethlehem NC, *3713* ... **111** F4
Bethlehem PA, *71329* ... **146** C1
Bethlehem Ctr. NY, *2500* ... **188** D3
Bethpage NY, *16543* ... **148** C4
Betmar Acres FL, *4000* ... **140** C2
Bettendorf IA, *31275* ... **88** A2
Bettsville OH, *784* ... **90** C2
Beulah CO, *1164* ... **41** E3
Beulah MI, *363* ... **69** E4
Beulah MS, *473* ... **118** A4
Beulah ND, *3152* ... **18** B3
Beulaville NC, *1067* ... **123** E2
Beverly MA, *39862* ... **151** F1
Beverly NJ, *2661* ... **147** D3
Beverly OH, *1282* ... **101** E1
Beverly WV, *651* ... **102** B3
Beverly Beach FL, *547* ... **139** E4
Beverly Beach MD, *1600* ... **144** C3
Beverly Hills CA, *33784* ... **52** C2
Beverly Hills FL, *8317* ... **140** B1
Beverly Hills MI, *10437* ... **210** B2
Beverly Hills MO, *603* ... **256** B2
Beverly Shores IN, *708* ... **89** E2
Bevier MO, *723* ... **97** E1
Bevil Oaks TX, *1346* ... **132** C2
Bevis OH, *5700* ... **204** A1
Bexar Co. TX, *1392931* ... **61** D2
Bexley OH, *13203* ... **206** C2
Bibb Co. AL, *20826* ... **127** F1
Bibb Co. GA, *153887* ... **129** D2
Bicknell IN, *3378* ... **99** E3
Bicknell UT, *353* ... **39** E3
Biddeford ME, *20942* ... **82** B4
Bienville Par. LA, *15752* ... **125** D2
Big Bear City CA, *5779* ... **53** D2
Big Bear Lake CA, *5438* ... **53** D2
Big Beaver PA, *2186* ... **91** F3
Big Bend WI, *1278* ... **74** C3
Big Chimney WV, *600* ... **101** F3
Big Coppitt Key FL, *2595* ... **143** D4
Big Delta AK, *749* ... **154** C2
Big Flats NY, *2482* ... **93** D1
Bigfork MT, *1421* ... **15** D2
Biggs CA, *1793* ... **36** B2
Big Horn Co. MT, *12671* ... **24** C2
Big Horn Co. WY, *11461* ... **24** C3
Big Lake MN, *6063* ... **66** C3
Big Lake TX, *2885* ... **58** A4
Big Lake WA, *1153* ... **12** C2
Biglerville PA, *1101* ... **103** E1
Big Oak Flat CA, *3388* ... **37** D4
Big Pine CA, *1350* ... **37** E4
Big Pine Key FL, *5032* ... **143** D4
Big Piney WY, *408* ... **32** A1
Big Rapids MI, *10849* ... **75** F2
Big River CA, *1266* ... **46** B4
Big Run PA, *686* ... **92** B3
Big Sandy MT, *703* ... **16** B2
Big Sandy TN, *518* ... **109** D3
Big Sandy TX, *1288* ... **124** B2
Big Sky MT, *1221* ... **23** F2
Big Spr. TX, *25233* ... **58** A3
Big Sprs. NE, *418* ... **34** B3
Big Stone City SD, *605* ... **27** F2
Big Stone Co. MN, *5820* ... **27** F2
Big Stone Gap VA, *4856* ... **111** E2
Big Timber MT, *1650* ... **24** A1
Big Water UT, *417* ... **47** D1
Big Wells TX, *704* ... **60** C4
Billerica MA, *38981* ... **95** E1
Billings MO, *1091* ... **106** C2
Billings MT, *89847* ... **24** C1
Billings NY, *800* ... **148** B1
Billings OK, *436* ... **51** E1
Billings Co. ND, *888* ... **18** A3
Billington Hts. NY, *1691* ... **78** B4
Biloxi MS, *50644* ... **135** D2
Biltmore Forest NC, *1440* ... **121** E1
Bingen PA, *1300* ... **189** B2
Binger OK, *708* ... **51** D3
Bingham ME, *856* ... **82** B1
Bingham Co. ID, *41735* ... **23** E4
Bingham Farms MI, *1030* ... **210** B2
Binghamton NY, *47380* ... **93** E1
Biola CA, *1037* ... **44** C2
Birch Bay WA, *4961* ... **12** C1
Birch Run MI, *1653* ... **76** B3
Birch Tree MO, *634* ... **107** F2
Birchwood Vil. MN, *968* ... **235** E1
Bird City KS, *482* ... **42** B1
Bird Island MN, *1195* ... **66** B4
Birdsboro PA, *5064* ... **146** B2
Birmingham AL, *242820* ... **119** F4
Birmingham MI, *19291* ... **76** C4
Birnamwood WI, *795* ... **68** B4
Biron WI, *915* ... **74** A1
Bisbee AZ, *6090* ... **55** E4
Biscayne Park FL, *3269* ... **233** B4
Biscoe AR, *476* ... **117** F2
Biscoe NC, *1700* ... **122** C1
Bishop CA, *3575* ... **37** E4
Bishop TX, *3305* ... **63** F2
Bishopville SC, *3670* ... **122** B3
Bismarck MO, *1470* ... **108** A1
Bismarck ND, *55532* ... **18** C4
Bison SD, *373* ... **26** A2
Bithlo FL, *4626* ... **141** D1
Bitter Sprs. AZ, *547* ... **47** D1
Biwabik MN, *954* ... **64** C3
Bixby OK, *13336* ... **106** A4
Blackbird DE, *700* ... **145** E1
Black Canyon City AZ, *2697* ... **47** D4
Black Creek NC, *714* ... **123** E1
Black Creek WI, *1192* ... **68** C4
Black Diamond WA, *3970* ... **12** C3
Blackduck MN, *696* ... **64** A3
Black Eagle MT, *914* ... **15** F3
Black Earth WI, *1320* ... **74** A3
Blackfoot ID, *10419* ... **31** E1
Blackford Co. IN, *14048* ... **90** A4
Black Forest CO, *13247* ... **41** E2
Black Hawk SD, *2432* ... **26** A3
Black Hawk Co. IA, *128012* ... **73** E4
Black Jack MO, *6792* ... **256** B1
Black Lick PA, *1438* ... **92** B4
Blacklick Estates OH, *9518* ... **206** C3
Black Mtn. NC, *7511* ... **121** E1
Black River NY, *1285* ... **79** E1
Black River Falls WI, *3618* ... **73** F1
Black Rock AR, *717* ... **107** F4
Black Rock NM, *1252* ... **48** A3
Blacksburg SC, *1880* ... **122** A1
Blacksburg VA, *39573* ... **112** A2
Blackshear GA, *3283* ... **129** F4
Blackstone MA, *2900* ... **150** C2
Blackstone VA, *3675* ... **113** D2
Blackville SC, *2973* ... **130** B1
Blackwell OK, *7668* ... **51** E1
Blackwells GA, *2200* ... **120** C3
Blackwood NJ, *4692* ... **146** C3
Bladenboro NC, *1718* ... **123** D3
Bladen Co. NC, *32278* ... **123** D2
Bladensburg MD, *7661* ... **270** E3
Blades DE, *956* ... **145** E4
Blaine ME, *1428* ... **85** E2
Blaine MN, *44942* ... **67** D3
Blaine TN, *1585* ... **110** C4
Blaine WA, *3770* ... **12** C1
Blaine Co. ID, *18991* ... **23** D4
Blaine Co. MT, *7009* ... **16** B2
Blaine Co. NE, *583* ... **34** C2
Blaine Co. OK, *11976* ... **51** D2
Blair NE, *7512* ... **35** F3
Blair OK, *894* ... **51** D4
Blair WI, *1273* ... **73** F1
Blair Co. PA, *129144* ... **92** C4

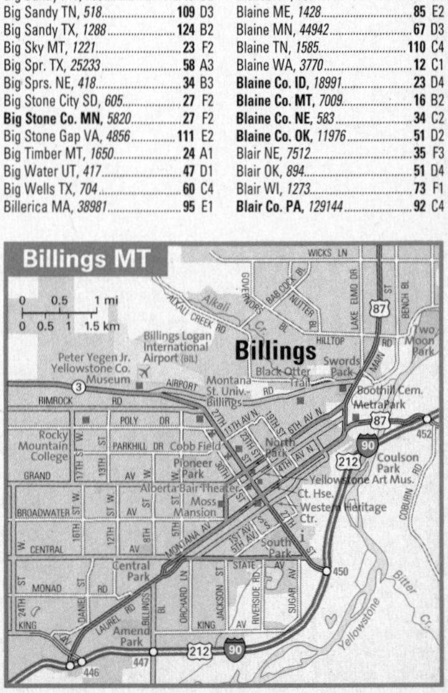

Entries in **bold black** indicate counties or parishes.
Entries in **bold color** indicate cities with detailed inset maps.

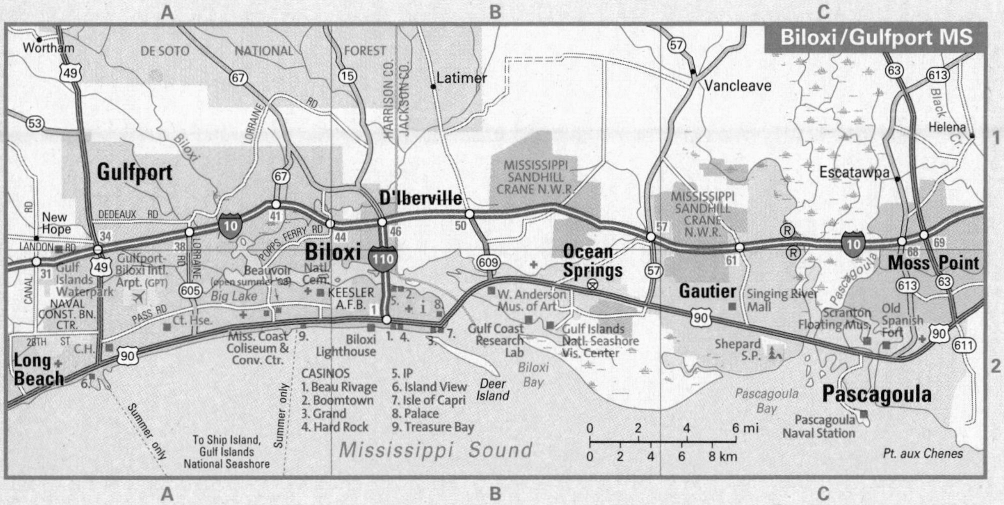

Biloxi A2
D'Iberville B1
Escatawpa C1
Fontainebleau B2
Gautier C2
Gulfport. A1
Helena. C1
Latimer B1
Long Beach A2
Moss Pt. C2
New Hope A1
Ocean Sprs. B2
Pascagoula C2
Vancleave. C1
Wortham. A1

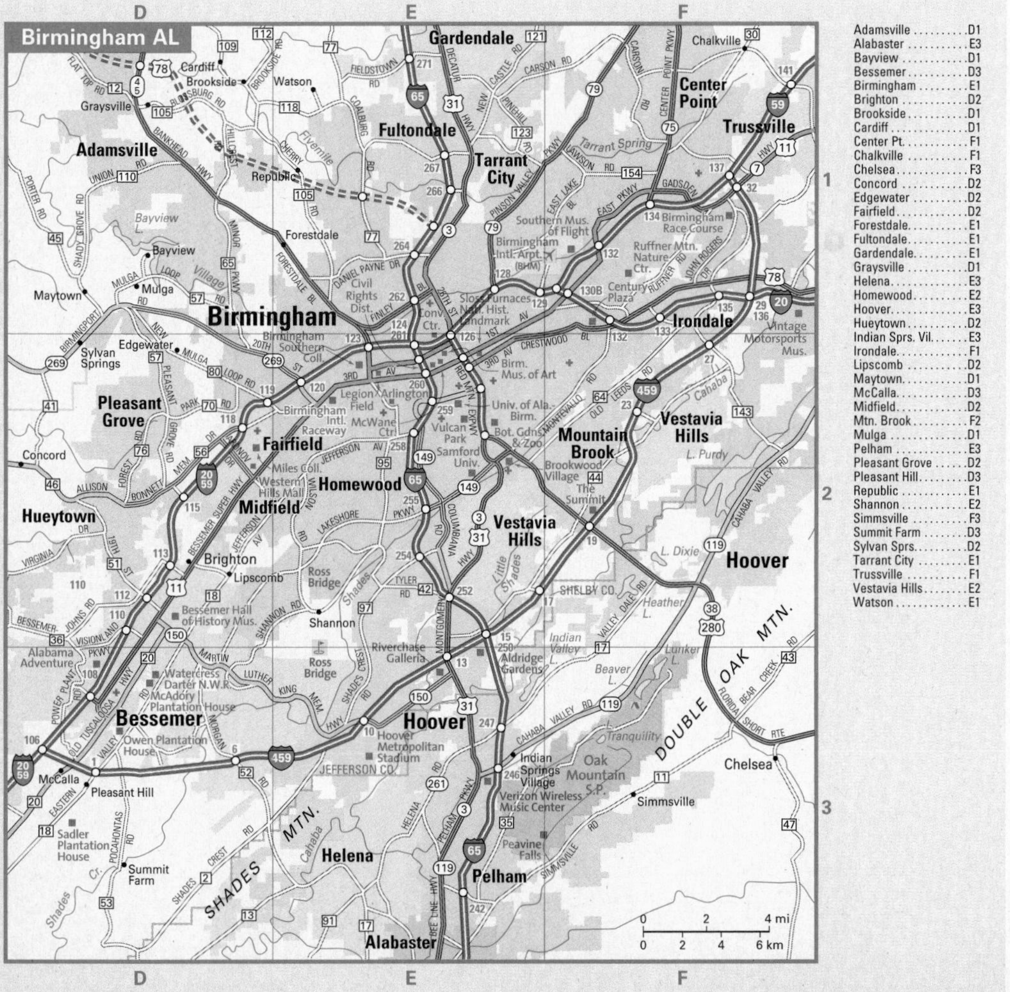

Adamsville D1
Alabaster E3
Bayview D1
Bessemer D3
Birmingham E1
Brighton D2
Brookside D1
Cardiff D1
Center Pt. F1
Chalkville F1
Chelsea F3
Concord D2
Edgewater D2
Fairfield D2
Forestdale E1
Fultondale E1
Gardendale E1
Graysville D1
Helena E3
Homewood E2
Hoover E3
Hueytown D2
Indian Sprs. Vil. E3
Irondale F1
Lipscomb D2
Maytown D1
McCalla D3
Midfield D2
Mtn. Brook F2
Mulga D1
Pelham E3
Pleasant Grove D2
Pleasant Hill D3
Republic E1
Shannon E2
Simmsville F3
Summit Farm D3
Sylvan Sprs. D2
Tarrant City E1
Trussville F1
Vestavia Hills E2
Watson E1

Figures after entries indicate population, page number, and grid reference.

Bowdoinham ME, 600 ... 82 B3
Bowdon GA, 1959 ... 120 B4
Bowen IL, 535 ... 87 F4
Bowers Beach DE, 305 ... 145 F3
Bowie AZ, 550 ... 55 E3
Bowie MD, 50269 ... 144 C3
Bowie TX, 5219 ... 59 E1
Bowie Co. TX, 89306 ... 116 B4
Bowleys Quarters MD, 6314 ... 144 C2
Bowling Green FL, 2892 ... 140 C3
Bowling Green KY, 49296 ... 109 F2
Bowling Green MO, 3260 ... 97 F2
Bowling Green OH, 29636 ... 90 C2
Bowling Green VA, 936 ... 103 D4
Bowman GA, 898 ... 121 E3
Bowman ND, 1600 ... 25 F1
Bowman SC, 1198 ... 130 C1
Bowman Co. ND, 3242 ... 25 F1
Bow Mar CO, 847 ... 209 B4
Boxborough MA, 1400 ... 150 C1
Box Butte Co. NE, 12158 ... 34 A2
Box Elder MT, 794 ... 16 B2
Box Elder SD, 2841 ... 26 A3
Box Elder Co. UT, 42745 ... 31 D3
Boxford MA, 2340 ... 151 F1
Boyce LA, 1190 ... 125 E4
Boyce VA, 426 ... 103 D2
Boyceville WI, 1043 ... 67 E3
Boyd TX, 1099 ... 59 E2
Boyd WI, 680 ... 67 F4
Boyd Co. KY, 49752 ... 101 D4
Boyd Co. NE, 2438 ... 35 D1
Boyden IA, 672 ... 35 F1
Boydton VA, 454 ... 113 D3
Boyertown PA, 3940 ... 146 B2
Boyette FL, 5895 ... 140 C3
Boykins VA, 620 ... 113 E3
Boyle MS, 720 ... 118 A4
Boyle Co. KY, 27697 ... 110 B1
Boyne City MI, 3503 ... 70 B3
Boynton Beach FL, 60389 ... 143 F1
Boys Town NE, 818 ... 245 A2
Bozeman MT, 27509 ... 23 F1
Braceville IL, 792 ... 88 C2
Bracken Co. KY, 8279 ... 100 C3
Brackettville TX, 1876 ... 60 B3
Bradbury CA, 855 ... 228 E2
Braddock PA, 2912 ... 250 C2
Braddock Hts. MD, 4627 ... 144 A1
Braddock Hills PA, 1998 ... 250 C2
Bradenton FL, 49504 ... 140 B3
Bradenton Beach FL, 1482 ... 140 B3
Bradford AR, 800 ... 117 F1
Bradford IL, 787 ... 88 B2
Bradford NH, 600 ... 81 E4
Bradford OH, 1859 ... 90 B4
Bradford PA, 9175 ... 92 B1
Bradford RI, 1497 ... 150 C4
Bradford TN, 1113 ... 108 C4
Bradford VT, 815 ... 81 E3
Bradford Co. FL, 26088 ... 138 C3
Bradford Co. PA, 62761 ... 93 E2
Bradfordville FL, 1100 ... 137 E2
Bradford Woods PA, 1149 ... 92 A3
Bradley AR, 563 ... 125 D1
Bradley IL, 12784 ... 89 D3
Bradley ME, 650 ... 83 D1
Bradley WV, 2371 ... 101 F4
Bradley Beach NJ, 4793 ... 147 F2
Bradley Co. AR, 12600 ... 117 E4
Bradley Co. TN, 87965 ... 120 C1
Bradley Jct. FL, 850 ... 140 C3
Bradner OH, 1171 ... 90 C2
Brady TX, 5523 ... 58 C4
Braham MN, 1276 ... 67 D2
Braidwood IL, 5203 ... 88 C2
Brainerd MN, 13178 ... 64 B4
Braintree MA, 33698 ... 151 D2
Bramwell WV, 426 ... 111 F1
Branch Co. MI, 45787 ... 90 A1
Branchville AL, 825 ... 119 F4
Branchville NJ, 845 ... 94 A4
Branchville SC, 1083 ... 130 C1
Brandenburg KY, 2049 ... 99 F4
Brandon FL, 77895 ... 140 C2
Brandon MS, 16436 ... 126 B3
Brandon SD, 5693 ... 27 F4
Brandon VT, 1684 ... 81 D3
Brandon WI, 912 ... 74 C2
Brandywine MD, 1410 ... 144 B4
Brandywine Manor PA, 1200 ... 146 B3
Branford CT, 5735 ... 149 D2
Branford FL, 695 ... 138 B3
Branson MO, 6050 ... 107 D3
Brant Beach NJ, 800 ... 147 E4
Brantley AL, 920 ... 128 A4
Brantley Co. GA, 14629 ... 129 F4
Brant Rock MA, 5100 ... 151 E2
Braselton GA, 1206 ... 121 D3
Brasher Falls NY, 1140 ... 80 B1
Bratenahl OH, 1337 ... 204 F1
Brattleboro VT, 8289 ... 94 C1
Brawley CA, 22052 ... 53 E4
Braxton Co. WV, 14702 ... 101 F3
Bray OK, 1035 ... 51 E4
Braymer MO, 910 ... 96 C1
Brazil IN, 8188 ... 99 E1
Brazoria TX, 2787 ... 132 A4
Brazoria Co. TX, 241767 ... 132 A4
Brazos Co. TX, 152415 ... 61 F1
Brea CA, 35410 ... 229 F3
Breathitt Co. KY, 16100 ... 111 D1
Breaux Bridge LA, 7281 ... 133 F2
Breckenridge CO, 2408 ... 41 D1
Breckenridge MI, 1339 ... 76 A2
Breckenridge MN, 3559 ... 27 F1
Breckenridge TX, 5868 ... 59 D2
Breckenridge Hills MO, 4817 ... 256 B2
Breckinridge Co. KY, 18648 ... 99 F4
Brecksville OH, 13382 ... 204 F3
Breese IL, 4048 ... 98 B3
Breezy Pt. MD, 800 ... 144 C4
Breezy Pt. MN, 979 ... 64 B4
Breinigsville PA, 1700 ... 146 B1
Bremen GA, 4579 ... 120 B4
Bremen IN, 4486 ... 89 F2
Bremen KY, 365 ... 109 E1
Bremen OH, 1265 ... 101 D1
Bremer Co. IA, 23325 ... 73 E3
Bremerton WA, 37259 ... 12 C3
Bremond TX, 876 ... 59 F4
Brenham TX, 13507 ... 61 F2
Brent AL, 4024 ... 127 F1
Brent FL, 22257 ... 135 F2
Brentsville VA, 650 ... 144 A4
Brentwood CA, 23302 ... 36 B3
Brentwood MD, 2844 ... 270 E2
Brentwood MO, 7693 ... 256 B2
Brentwood NY, 53917 ... 149 D4
Brentwood PA, 10466 ... 250 B3
Brentwood TN, 23445 ... 109 F4
Bressler PA, 2809 ... 218 C2
Brevard NC, 6789 ... 121 E1
Brevard Co. FL, 476230 ... 141 E2
Brewer ME, 8987 ... 83 D1
Brewerton NY, 3453 ... 79 D3
Brewster MA, 2212 ... 151 F3
Brewster MN, 502 ... 72 A2
Brewster NE, 29 ... 34 C2
Brewster NY, 2162 ... 148 C2
Brewster OH, 2324 ... 91 E3
Brewster WA, 2189 ... 13 E2
Brewster Co. TX, 8866 ... 62 C3
Brewster Hill NY, 2226 ... 148 C1
Brewton AL, 5498 ... 135 F1
Briar TX, 5350 ... 59 E2
Briarcliff TX, 895 ... 61 D1
Briarcliffe Acres SC, 470 ... 123 D4
Briarcliff Manor NY, 7696 ... 148 B2
Briar Creek PA, 651 ... 93 E3
Briarwood KY, 554 ... 230 F1
Briceville TN, 650 ... 110 C4
Brickerville PA, 1287 ... 146 A2
Bridge City LA, 8323 ... 239 B2
Bridge City TX, 8651 ... 132 C3
Bridgehampton NY, 1381 ... 149 F3
Bridgeport AL, 2728 ... 120 A2
Bridgeport CA, 200 ... 37 E3
Bridgeport CT, 139529 ... 149 D2
Bridgeport IL, 2168 ... 99 D3
Bridgeport MD, 2700 ... 144 A1
Bridgeport MI, 7849 ... 76 B3
Bridgeport NE, 1594 ... 34 A3
Bridgeport NY, 1665 ... 79 E3
Bridgeport PA, 4371 ... 248 A1
Bridgeport TX, 4309 ... 59 E2
Bridgeport WA, 2059 ... 13 E2
Bridgeport WV, 7306 ... 102 A2
Bridger MT, 745 ... 24 B2
Bridgeton MO, 15550 ... 256 B1
Bridgeton NJ, 22771 ... 145 F1
Bridgetown OH, 12569 ... 204 A2
Bridgeview IL, 15335 ... 203 D5
Bridgeville DE, 1436 ... 145 E4
Bridgeville PA, 5341 ... 250 A3
Bridgewater MA, 6664 ... 151 D2
Bridgewater NJ, 3200 ... 147 D1
Bridgewater NY, 579 ... 79 E3
Bridgewater SD, 607 ... 27 E4
Bridgewater VA, 5203 ... 102 C4
Bridgman MI, 2428 ... 89 E1
Bridgton ME, 2359 ... 82 B3
Brielle NJ, 4893 ... 147 E2
Brier WA, 6383 ... 262 B2
Brigantine NJ, 12594 ... 147 F4
Brigham City UT, 17411 ... 31 E3
Bright IN, 5405 ... 100 B2
Brighton AL, 3640 ... 195 D2
Brighton CO, 20905 ... 41 E1
Brighton IL, 2196 ... 98 A2
Brighton IA, 687 ... 87 E2
Brighton MI, 6701 ... 76 B4
Brighton NY, 35584 ... 78 C3
Brighton TN, 1719 ... 118 B1
Brightwaters NY, 3248 ... 149 D4
Brightwood VA, 500 ... 102 C3
Brilliant AL, 762 ... 119 E3
Brilliant OH, 1600 ... 91 F4
Brillion WI, 2937 ... 74 C1
Brimfield IL, 933 ... 88 B3
Brinckerhoff NY, 2734 ... 148 B1
Brinkley AR, 3940 ... 117 F2
Brinnon WA, 803 ... 12 C3
Brisbane CA, 3597 ... 259 B3
Briscoe Co. TX, 1790 ... 50 B4
Bristol CT, 60062 ... 149 D1
Bristol FL, 845 ... 137 D2
Bristol IN, 1382 ... 89 F1
Bristol NH, 1670 ... 81 F3
Bristol PA, 9923 ... 147 D2
Bristol RI, 22469 ... 151 D3
Bristol SD, 377 ... 27 E2
Bristol TN, 24821 ... 111 E3
Bristol VT, 1800 ... 81 D2
Bristol VA, 17367 ... 111 E3
Bristol WI, 1100 ... 74 C4
Bristol Co. MA, 534678 ... 151 D3
Bristol Co. RI, 50648 ... 151 D3
Bristow OK, 4325 ... 51 F2
Britt IA, 2052 ... 72 C3
Brittany Farms PA, 3268 ... 146 C2
Britton MI, 699 ... 90 B1
Britton SD, 1328 ... 27 E1
Broadalbin NY, 1411 ... 80 C4
Broad Brook CT, 3469 ... 150 A3
Broadmoor CA, 4026 ... 259 B3
Broadus MT, 451 ... 25 E2
Broadview IL, 8264 ... 203 C4
Broadview Hts. OH, 15967 ... 204 F3
Broadwater Co. MT, 4385 ... 15 F4
Broadway NC, 1015 ... 123 D1
Broadway VA, 2192 ... 102 C3
Brock Hall MD, 1200 ... 144 C3
Brockport NY, 8103 ... 78 C3
Brockton MA, 94304 ... 151 D2
Brockton MT, 245 ... 17 E2
Brockway PA, 2182 ... 92 B2
Brocton NY, 1547 ... 78 A4
Brodhead KY, 1193 ... 110 C1
Brodhead WI, 3180 ... 74 B4
Brodheadsville PA, 1637 ... 93 F3
Brogden NC, 2907 ... 123 E1
Broken Arrow OK, 75427 ... 106 A4
Broken Bow NE, 3491 ... 35 D3
Broken Bow OK, 4230 ... 116 B3
Bromley KY, 838 ... 204 A3
Bronson FL, 964 ... 138 C4
Bronson MI, 2421 ... 90 A1
Bronte TX, 1076 ... 58 C3
Bronwood GA, 513 ... 128 C3
Bronx Co. NY, 1332650 ... 148 B4
Brook IN, 1062 ... 89 D3
Brookdale SC, 4724 ... 122 A4
Brooke VA, 650 ... 103 D4
Brooke Co. WV, 25447 ... 91 F4
Brookfield CT, 2700 ... 148 C1
Brookfield IL, 19085 ... 203 C5
Brookfield MA, 1200 ... 150 B2
Brookfield MO, 4769 ... 97 D1
Brookfield OH, 1288 ... 276 C1
Brookfield WI, 38649 ... 234 B2
Brookfield Ctr. CT, 1800 ... 148 C1
Brookhaven MS, 9861 ... 126 B4
Brookhaven NY, 3570 ... 149 D4
Brookhaven PA, 7985 ... 248 A4
Brookings OR, 5447 ... 28 A2
Brookings SD, 18504 ... 27 F3
Brookings Co. SD, 28220 ... 27 F3
Brookland AR, 1332 ... 108 A4
Brooklandville MD, 2200 ... 193 C1
Brooklawn NJ, 2354 ... 248 C4
Brooklet GA, 1113 ... 130 B2
Brookline MA, 57107 ... 151 D1
Brookline NH, 650 ... 95 D1
Brooklyn CT, 1100 ... 150 B3
Brooklyn IL, 676 ... 256 C2
Brooklyn IN, 1545 ... 99 F1
Brooklyn IA, 1367 ... 87 E1
Brooklyn MI, 1176 ... 90 B1
Brooklyn OH, 11586 ... 204 E2
Brooklyn WI, 916 ... 74 B3
Brooklyn Ctr. MN, 29172 ... 235 B1
Brooklyn Hts. OH, 1558 ... 204 F2
Brooklyn Park MD, 10938 ... 193 C4
Brooklyn Park MN, 67388 ... 235 B1
Brookneal VA, 1259 ... 112 C2
Brook Park OH, 21218 ... 204 E3
Brookport IL, 1054 ... 108 C2
Brooks GA, 553 ... 128 C1
Brooks KY, 2678 ... 100 A4
Brooks ME, 550 ... 82 C2
Brooks Co. GA, 16450 ... 137 F1
Brooks Co. TX, 7976 ... 63 E3
Brookshire TX, 3450 ... 61 F2
Brookside AL, 1393 ... 195 D1
Brookside DE, 14806 ... 146 B4
Brookside OH, 644 ... 91 F4
Brookside Vil. TX, 1960 ... 220 C4
Brookston IN, 1717 ... 89 E3
Brooksville FL, 7264 ... 140 C1
Brooksville KY, 589 ... 100 C3
Brooksville MS, 1182 ... 127 D1
Brookville IN, 2652 ... 100 A2
Brookville NY, 2126 ... 148 C4
Brookville OH, 5289 ... 100 B1
Brookville PA, 4230 ... 92 B2
Brookwood AL, 1483 ... 127 F1
Broomall PA, 11046 ... 146 C3
Broome Co. NY, 200536 ... 93 F1
Broomfield CO, 38272 ... 41 E1
Broomfield Co. CO, 38272 ... 41 E1
Brooten MN, 649 ... 66 B3
Broussard LA, 5874 ... 133 F2
Broward Co. FL, 1623018 ... 143 E1
Browerville MN, 735 ... 66 B2
Brown City MI, 1334 ... 76 C3
Brown Co. IL, 6950 ... 87 F4
Brown Co. IN, 14957 ... 99 F2
Brown Co. KS, 10724 ... 96 A1
Brown Co. MN, 26911 ... 72 B1
Brown Co. NE, 3525 ... 34 C1
Brown Co. OH, 42285 ... 100 C3
Brown Co. SD, 35460 ... 27 E1
Brown Co. TX, 37674 ... 59 D3
Brown Co. WI, 226778 ... 74 C1
Brown Deer WI, 12170 ... 234 C1
Brownfield TX, 9488 ... 58 A2
Browning MT, 1065 ... 15 E2
Brownsboro TX, 796 ... 124 A3
Brownsburg IN, 14520 ... 99 F1
Brownsdale MN, 718 ... 73 D2
Browns Mills NJ, 11257 ... 147 D3
Brownstown IL, 705 ... 98 C2
Brownstown IN, 2978 ... 99 F3
Browns Valley MN, 690 ... 27 F1
Brownsville CA, 1069 ... 36 C1
Brownsville KY, 921 ... 109 F2
Brownsville MN, 517 ... 73 F2
Brownsville OR, 1449 ... 20 B3
Brownsville PA, 2804 ... 102 B1
Brownsville TN, 10748 ... 118 C1
Brownsville TX, 139722 ... 63 F4
Brownsville WI, 570 ... 74 C2
Brownton MN, 807 ... 66 C4
Brownville NY, 1022 ... 79 E1
Brownville Jct. ME, 750 ... 84 C4
Brownwood TX, 18813 ... 59 D4
Broxton GA, 1428 ... 129 E4
Broyhill Park VA, 17000 ... 270 B4
Bruce MS, 2097 ... 118 C3
Bruce SD, 272 ... 27 F3
Bruce WI, 787 ... 67 F3
Bruceton TN, 1554 ... 109 D4
Bruceville-Eddy TX, 1490 ... 59 E4
Brule Co. SD, 5364 ... 27 D4
Brundidge AL, 2341 ... 128 B4
Brunson SC, 589 ... 130 B1
Brunswick GA, 15600 ... 139 D1
Brunswick ME, 14816 ... 82 B3
Brunswick MD, 4894 ... 144 A2
Brunswick MO, 925 ... 97 D2
Brunswick OH, 33388 ... 91 E2
Brunswick Co. NC, 73143 ... 123 E3
Brunswick Co. VA, 18419 ... 113 D3
Brush CO, 5117 ... 33 F4
Brush Prairie WA, 2384 ... 20 C1
Brushy OK, 787 ... 116 B1
Brusly LA, 2020 ... 134 A2
Bryan OH, 8333 ... 90 B2
Bryan TX, 65660 ... 61 F1
Bryan Co. GA, 23417 ... 130 B3
Bryan Co. OK, 36534 ... 59 F1
Bryans Road MD, 4912 ... 144 B4
Bryant AR, 9764 ... 117 E2
Bryant SD, 396 ... 27 E3
Bryantville MA, 2600 ... 151 E2
Bryn Athyn PA, 1351 ... 248 D1
Bryn Mawr PA, 4382 ... 146 C3
Bryson TX, 528 ... 59 D2
Bryson City NC, 1411 ... 121 D1
Buchanan GA, 941 ... 120 B4
Buchanan MI, 4681 ... 89 E1
Buchanan NY, 2189 ... 148 B2
Buchanan VA, 1233 ... 112 B1
Buchanan Co. IA, 21093 ... 73 E4
Buchanan Co. MO, 85998 ... 96 B1
Buchanan Co. VA, 26978 ... 111 F2
Buchanan Dam TX, 1688 ... 61 D1
Buchtel OH, 574 ... 101 E2
Buckeye AZ, 6537 ... 54 B1
Buckeye Lake OH, 3049 ... 101 D1
Buckfield ME, 325 ... 82 B2
Buckhannon WV, 5725 ... 102 A2
Buckhead Ridge FL, 1390 ... 141 E4
Buckingham PA, 1400 ... 146 C2
Buckingham VA, 175 ... 113 D1
Buckingham Co. VA, 15623 ... 113 D1
Buckland AK, 406 ... 154 B2
Buckley IL, 593 ... 89 D4
Buckley MI, 550 ... 69 F4
Buckley WA, 4145 ... 12 C3
Bucklin KS, 725 ... 42 C4
Bucklin MO, 524 ... 97 D1
Buckner KY, 4000 ... 100 A4
Buckner MO, 2725 ... 96 C2
Bucks Co. PA, 597635 ... 146 C1
Bucksport ME, 2970 ... 83 D2
Bucksport SC, 1117 ... 123 D4
Bucoda WA, 628 ... 12 C4
Bucyrus OH, 13224 ... 90 C3
Buda IL, 592 ... 88 B2
Buda TX, 2404 ... 61 E2
Budd Lake NJ, 8100 ... 94 A4
Bude MS, 1037 ... 126 A4
Buellton CA, 3828 ... 52 A2
Buena NJ, 3873 ... 147 D4
Buena WA, 950 ... 13 D4
Buena Park CA, 78282 ... 228 E3
Buenaventura Lakes FL, 14100 ... 246 C5
Buena Vista CO, 2195 ... 41 D2
Buena Vista GA, 1664 ... 128 C2
Buena Vista MI, 7845 ... 76 B2
Buena Vista VA, 6349 ... 112 C1
Buena Vista Co. IA, 20411 ... 72 B4
Buffalo IN, 672 ... 89 E3
Buffalo IA, 1321 ... 87 F2
Buffalo KY, 475 ... 110 A1
Buffalo MN, 10097 ... 66 C3
Buffalo MO, 2781 ... 107 D1
Buffalo NY, 292648 ... 78 B3
Buffalo ND, 209 ... 19 E4
Buffalo OK, 1200 ... 50 C1
Buffalo SC, 1426 ... 121 F2
Buffalo SD, 380 ... 25 F1
Buffalo TX, 1804 ... 59 F4
Buffalo WV, 1171 ... 101 E3
Buffalo WI, 1040 ... 73 E1
Buffalo WY, 3900 ... 25 D3
Buffalo Ctr. IA, 963 ... 72 C2
Buffalo Co. NE, 42259 ... 35 D4
Buffalo Co. SD, 2032 ... 27 D3
Buffalo Co. WI, 13804 ... 67 E4
Buffalo Grove IL, 42909 ... 203 C2
Buffalo Lake MN, 768 ... 66 B4
Buford GA, 10668 ... 120 C3
Buhl ID, 3985 ... 30 C1
Buhl MN, 983 ... 64 C3
Buhler KS, 1358 ... 43 E3
Buies Creek NC, 2215 ... 123 D1
Bullard TX, 1150 ... 124 A3
Bullhead SD, 308 ... 26 C1
Bullhead City AZ, 33769 ... 46 B3
Bullitt Co. KY, 61236 ... 99 F4
Bulloch Co. GA, 55983 ... 130 B2
Bullock Co. AL, 11714 ... 128 B3
Bulls Gap TN, 714 ... 111 D3
Bull Shoals AR, 2000 ... 107 E3
Bull Valley IL, 726 ... 74 C4
Bulverde TX, 3761 ... 61 D2
Buna TX, 2269 ... 132 C2
Buncombe Co. NC, 206330 ... 111 E4
Bunker Hill IL, 1801 ... 98 B2
Bunker Hill IN, 987 ... 89 F3
Bunker Hill OR, 1462 ... 20 A4
Bunker Hill WV, 700 ... 103 D2
Bunkerville NV, 1014 ... 46 B1
Bunkie LA, 4662 ... 133 E1
Bunnell FL, 2122 ... 139 E4
Buras LA, 3358 ... 134 C4
Burbank CA, 100316 ... 52 C2
Burbank IL, 27902 ... 203 D5
Burbank WA, 3303 ... 21 E1
Burden KS, 564 ... 43 F4

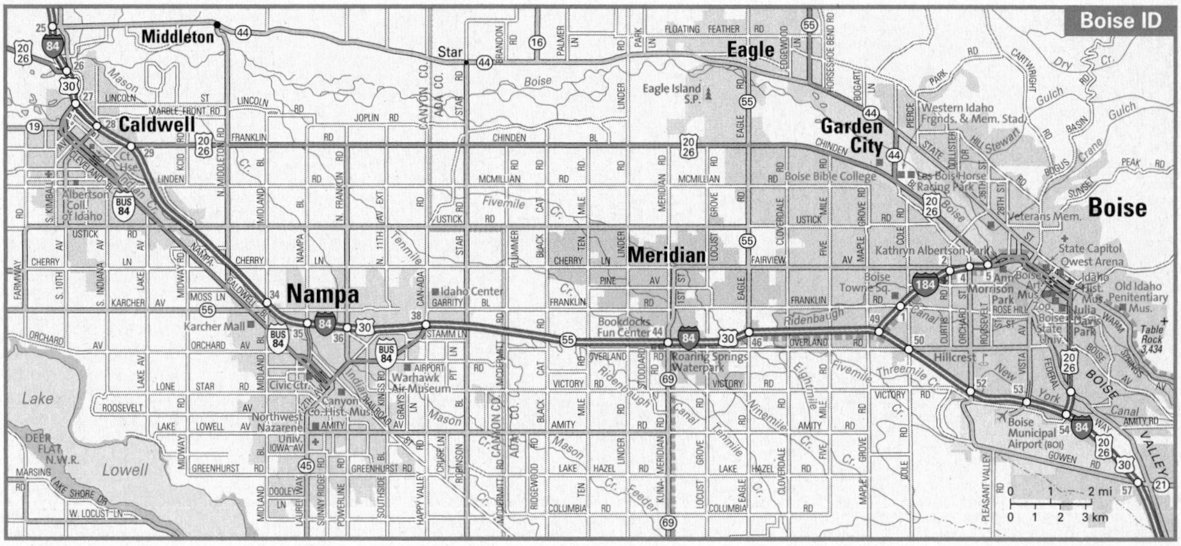

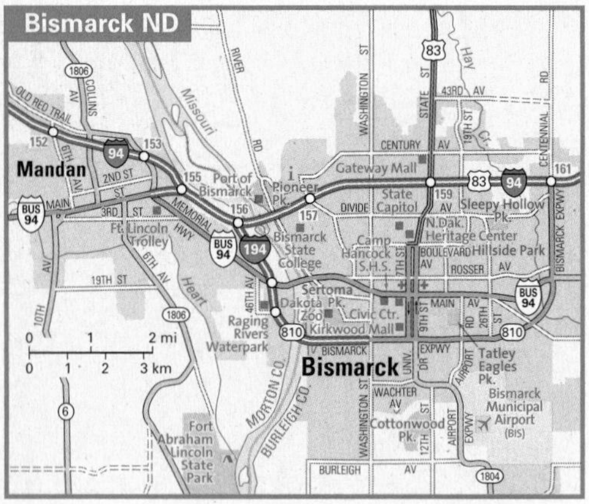

Entries in **bold black** indicate counties or parishes.
Entries in **bold color** indicate cities with detailed inset maps.

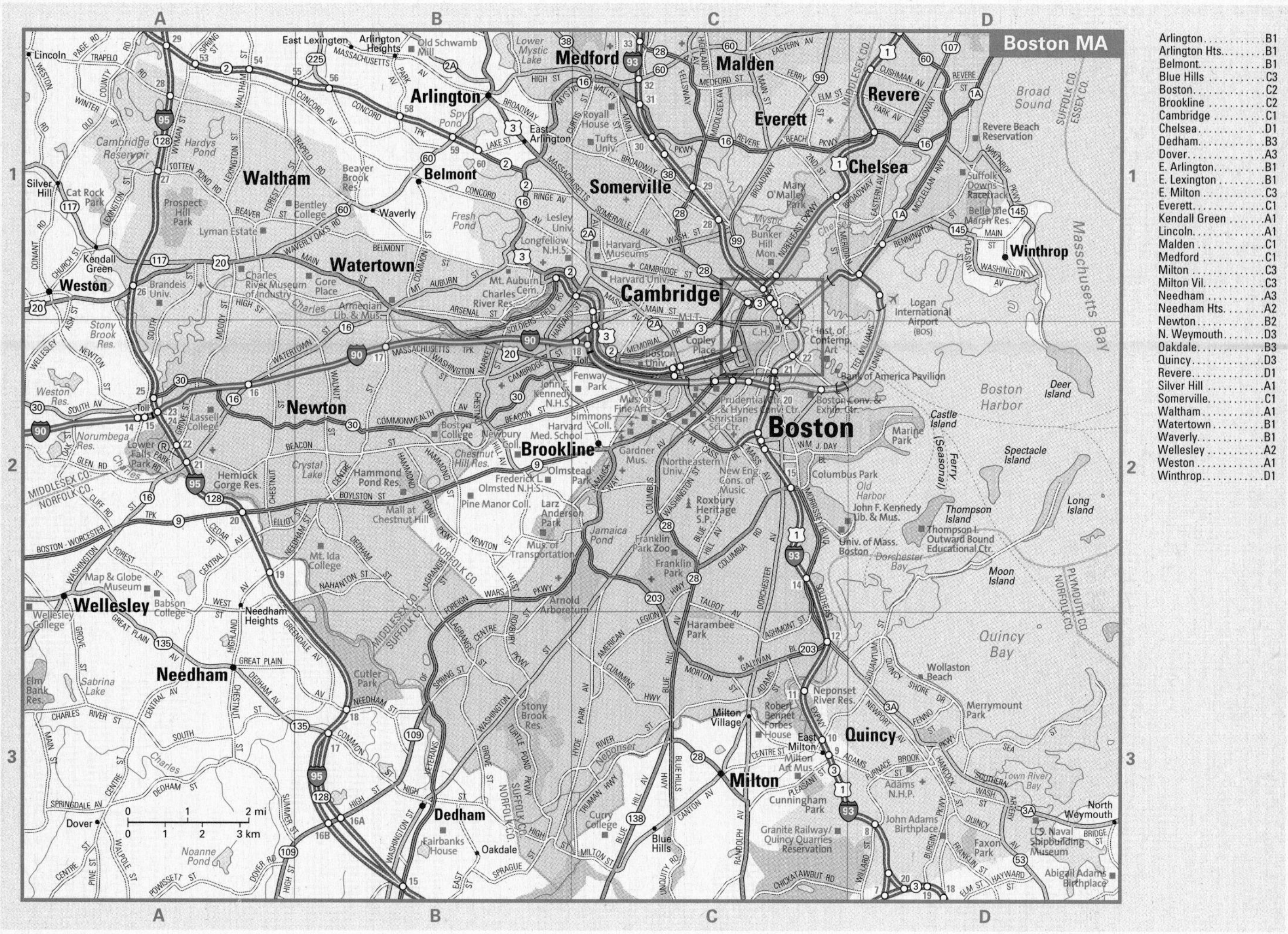

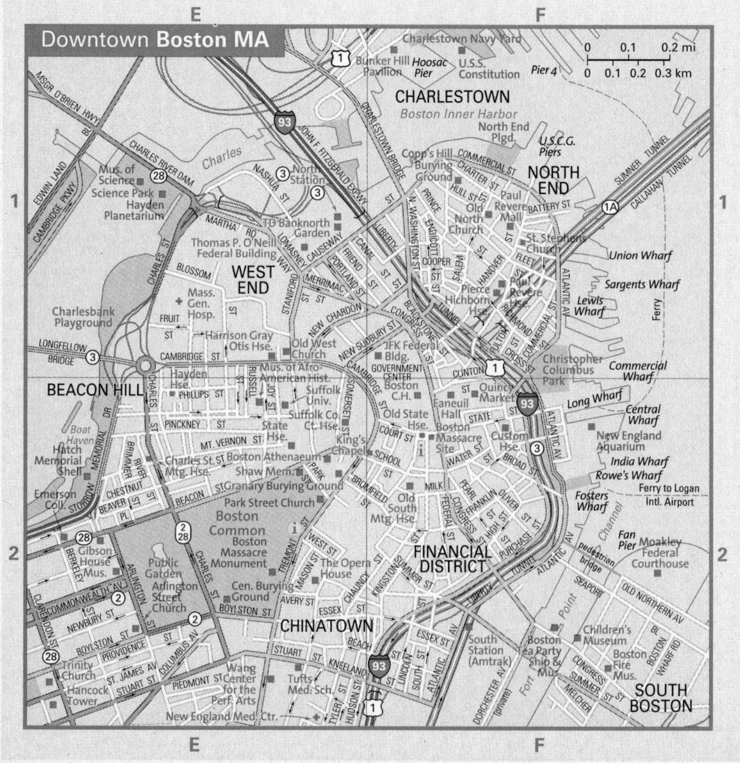

POINTS OF INTEREST

Figures after entries indicate population, page number, and grid reference.

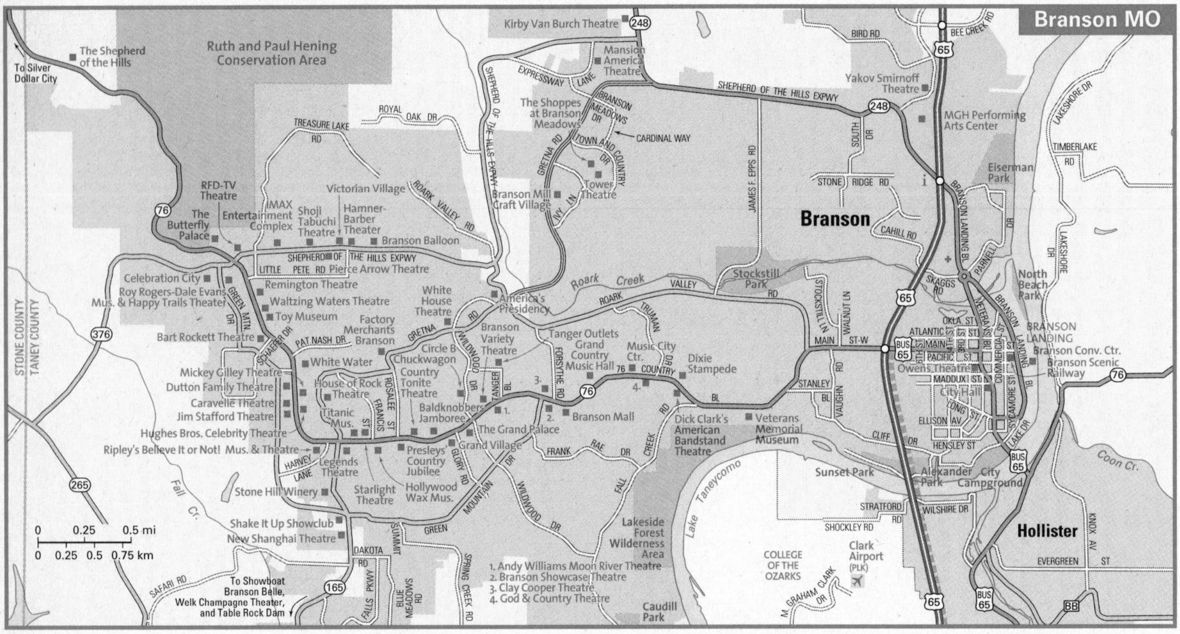

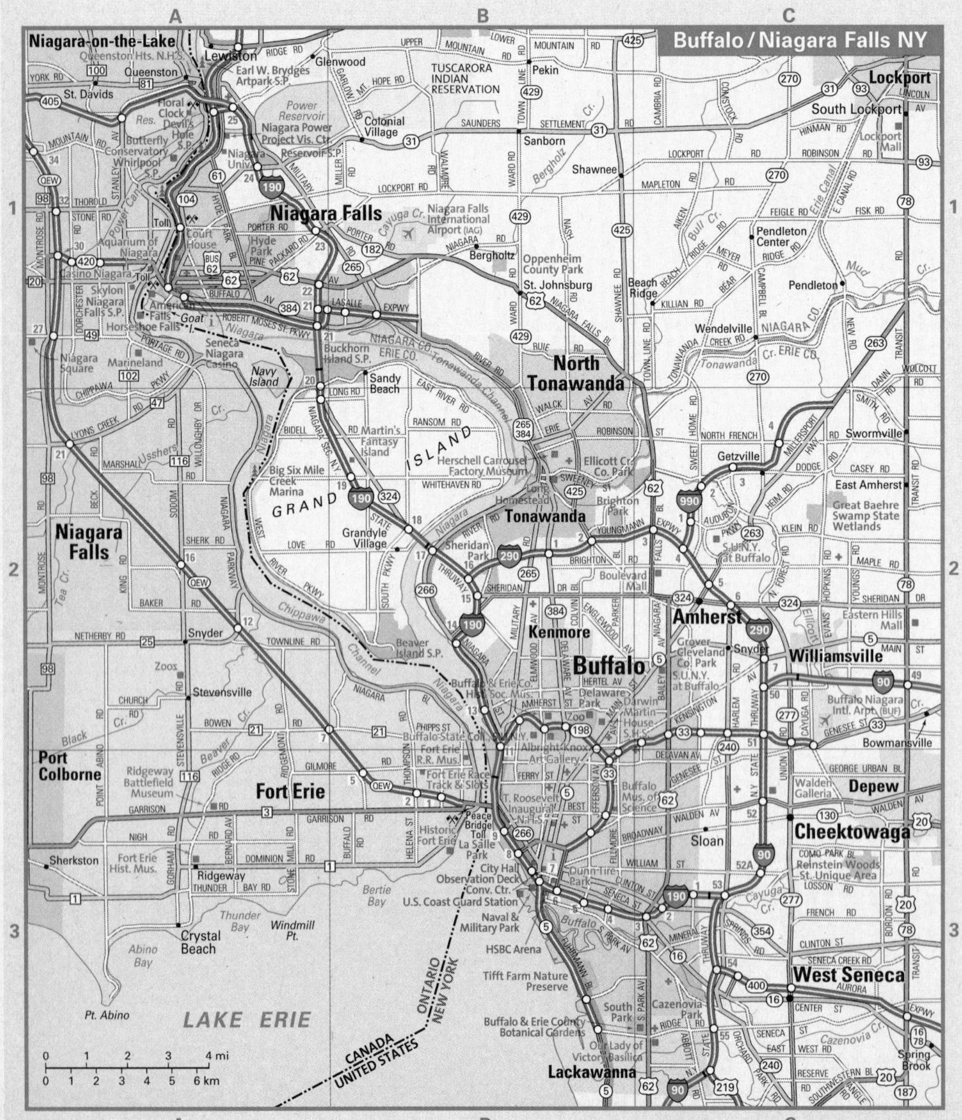

Amherst C2
Beach Ridge C1
Bergholtz B1
Bowmansville C2
Buffalo B2
Cheektowaga C3
Colonial Vil. B1
Crystal Beach A3
Depew C3
E. Amherst C2
Ft. Erie A3
Getzville C2
Glenwood A1
Grandyle Vil. B2
Kenmore B2
Lackawanna B3
Lewiston A1
Lockport C1
Niagara Falls, NY A1
Niagara Falls, ON A2
Niagara-on-the-Lake . . A1
N. Tonawanda B1
Pekin B1
Pendleton C1
Pendleton Ctr. C1
Port Colborne A3
Queenston A1
Ridgeway A3
St. Davids A1
St. Johnsburg B1
Sanborn B1
Sandy Beach B1
Shawnee B1
Sherkston A3
Sloan C3
Snyder, NY C2
Snyder, ON A2
S. Lockport C1
Spring Brook C3
Stevensville A2
Swormville C2
Tonawanda B2
Wendelville C1
W. Seneca C3
Williamsville C2

Entries in **bold black** indicate counties or parishes.
Entries in **bold color** indicate cities with detailed inset maps.

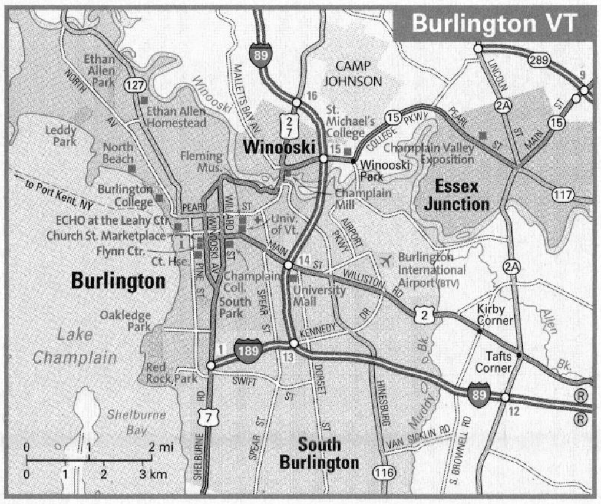

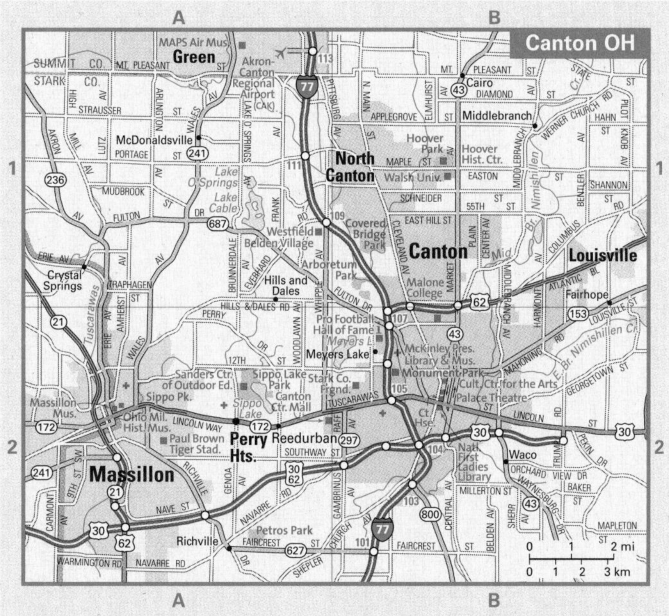

Cairo B1	Green A1	McDonaldsville A1	Perry Hts. A2
Canton B1	Hills and Dales A1	Meyers Lake B2	Reedurban A2
Crystal Sprs. A1	Louisville B1	Middlebranch B1	Richville A2
Fairhope B1	Massillon A2	N. Canton B1	Waco B2

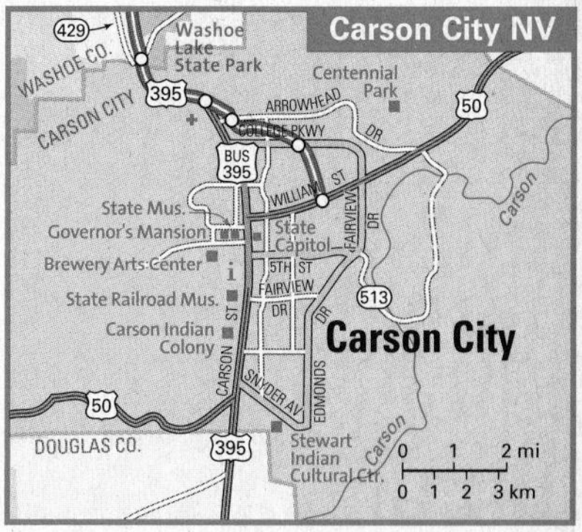

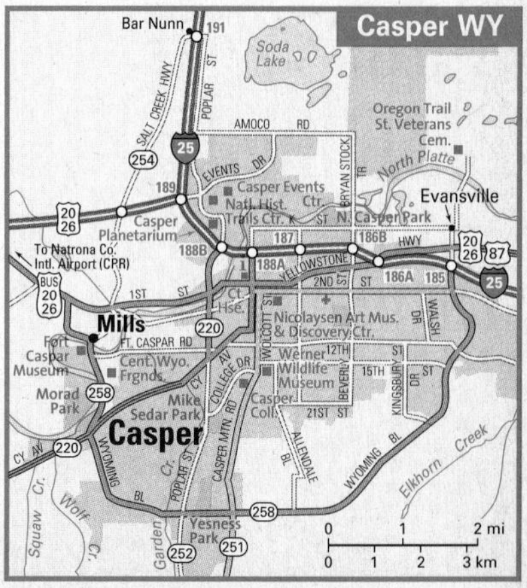

Figures after entries indicate population, page number, and grid reference.

Cattaraugus Co. NY, 83955 78 A4
Cavalier ND, 1537 19 E1
Cavalier Co. ND, 4831 19 D1
Cave City AR, 1946 107 F4
Cave City KY, 1880 110 A2
Cave Creek AZ, 3728 54 C1
Cave Jct. OR, 1363 28 B2
Cave Spr. GA, 975 120 B3
Cave Spr. VA, 24941 112 B2
Cave Sprs. AR, 1103 106 C3
Cavetown MD, 1486 144 A1
Cawker City KS, 521 43 D1
Cawood KY, 800 111 D2
Cayce SC, 12150 122 A3
Cayucos CA, 2943 44 B4
Cayuga IN, 1109 99 E1
Cayuga NY, 509 79 D3
Cayuga Co. NY, 81963 79 D4
Cayuga Hts. NY, 3273 79 D4
Cazenovia NY, 2614 79 E3
Cecil PA, 2585 92 A4
Cecil Co. MD, 85951 145 E1
Cecilia KY, 600 110 A1
Cecilia LA, 1505 133 F2
Cecilton MD, 474 145 E1
Cedar Bluff AL, 1467 120 A3
Cedar Bluff VA, 1085 111 F2
Cedar Bluffs NE, 615 35 F3
Cedar Brook NJ, 1100 147 D4
Cedarburg WI, 10908 74 C3
Cedar City UT, 20527 39 D4
Cedar Co. IA, 18187 87 F1
Cedar Co. MO, 13733 106 C1
Cedar Co. NE, 9615 35 E1
Cedar Creek NE, 396 35 F3
Cedar Crest NM, 1060 48 C3
Cedaredge CO, 1854 40 B2
Cedar Falls IA, 36145 73 D4
Cedar Fort UT, 341 31 E4
Cedar Grove FL, 5367 136 C2
Cedar Grove MD, 950 144 B2
Cedar Grove NJ, 12300 148 A3
Cedar Grove NM, 599 48 C3
Cedar Grove WV, 862 101 F4
Cedar Grove WI, 1887 75 D2
Cedar Hill MO, 1703 98 A4
Cedar Hill TX, 32093 207 C3
Cedar Hills OR, 8949 251 C2
Cedarhurst NY, 6164 241 G5
Cedar Key FL, 790 138 B4
Cedar Lake IN, 9279 89 D2
Cedar Park TX, 26049 61 E1
Cedar Pt. NC, 929 115 D4
Cedar Rapids IA, 120758 87 E1
Cedar Rapids NE, 407 35 E3
Cedar Sprs. MI, 3112 75 F3
Cedartown GA, 9470 120 B3
Cedar Vale KS, 723 51 F1
Cedarville AR, 1133 116 C1
Cedarville IL, 719 74 B4
Cedarville NJ, 793 145 F2
Cedarville OH, 3828 100 C1
Celebration FL, 2736 141 D1
Celeste TX, 817 59 F2
Celina OH, 10303 90 B4
Celina TN, 1379 110 A3
Celina TX, 1861 59 F2
Celoron NY, 1295 92 B1
Cement OK, 530 51 E3
Cement City MI, 452 90 B1
Centennial CO, 102821 209 C4
Center CO, 2392 41 D4
Center MO, 644 97 F1
Center NE, 90 35 E1
Center ND, 678 18 B3
Center TX, 5678 124 C3
Center Barnstead NH, 500 81 F4
Centerbrook CT, 950 149 E2
Center Brunswick NY, 900 94 B1
Centerburg OH, 1432 91 D4
Center City MN, 582 67 D3
Centereach NY, 27285 149 D3
Centerfield UT, 1048 39 E2
Center Harbor NH, 600 81 F3
Center Hill FL, 910 140 C1
Center Line MI, 8531 210 C2
Center Moriches NY, 6655 149 D4
Center Ossipee NH, 650 81 F3
Center Pt. AL, 22784 119 F4
Center Pt. IA, 2007 87 E1
Center Pt. TX, 750 61 D2
Centerport NY, 5446 148 C3
Centerton AR, 2146 106 C3
Centerton NJ, 2000 147 D3
Centertown KY, 416 109 E1
Center Valley PA, 1600 146 C1
Centerville GA, 4278 129 D2
Centerville IN, 2427 100 B1
Centerville IA, 5924 87 D3
Centerville LA, 800 133 F3
Centerville MA, 9200 151 F3
Centerville MO, 171 108 A1
Centerville OH, 23024 100 B1
Centerville PA, 3390 92 A4
Centerville SC, 5181 121 E3
Centerville SD, 910 35 F1
Centerville TN, 3793 109 E4
Centerville TX, 903 124 A4
Centerville UT, 14585 31 E4
Central AZ, 425 55 E2
Central SC, 3522 121 E2
Central TN, 2717 111 E3
Central Bridge NY, 900 79 F4
Central City AR, 531 116 C1
Central City CO, 515 41 D1
Central City IL, 1371 98 C3
Central City IA, 1157 87 F1
Central City KY, 5893 109 E1
Central City NE, 2998 35 E3
Central City PA, 1258 92 B4
Central Falls RI, 18928 150 C3
Central High OK, 954 51 E4
Centralia IL, 14136 98 C3
Centralia KS, 534 43 F1
Centralia MO, 3774 97 E2
Centralia WA, 14742 12 B4
Central Islip NY, 31950 149 D4
Central Lake MI, 990 69 F3
Central Park WA, 2558 12 B4
Central Pt. OR, 12493 28 B2
Central Square NY, 1646 79 D3
Central Valley NY, 1857 148 B2
Central Vil. CT, 1400 150 B3
Central Vil. MA, 600 151 D4
Centre AL, 3216 120 A3
Centre Co. PA, 135758 92 C3
Centre Hall PA, 1079 92 C3
Centreville AL, 2466 127 F1
Centreville IL, 5951 98 A3
Centreville MD, 1970 145 D3
Centreville MI, 1579 89 F1
Centreville MS, 1680 134 A1
Centreville VA, 48661 144 A3
Centuria WI, 865 67 E3
Century FL, 1714 135 F1
Ceres CA, 34609 36 C4
Ceresco NE, 920 35 F3
Cerritos CA, 51488 228 E4
Cerro Gordo IL, 1436 98 C1
Cerro Gordo Co. IA, 46447 73 D3
Chackbay LA, 4018 134 A3
Chadbourn NC, 2129 123 D3
Chadron NE, 5634 34 A1
Chaffee MO, 3044 108 B2
Chaffee Co. CO, 16242 41 D2
Chaffinville MA, 3100 150 B1
Chagrin Falls OH, 4024 91 E2
Chalco NE, 10736 35 F3
Chalfant PA, 870 250 D2
Chalfont PA, 3900 146 C2
Chalkville AL, 3829 195 F1
Challenge CA, 1069 36 C1
Challis ID, 909 23 D3
Chalmers IN, 513 89 E3
Chalmette LA, 32069 134 B3
Chama NM, 1199 48 C1
Chamberino NM, 425 56 C3
Chamberlain SD, 2338 27 D4
Chamberlayne Farms VA, 4700 254 B1
Chambersburg PA, 17862 103 D1
Chambers Co. AL, 36583 128 B1
Chambers Co. TX, 26031 132 B3
Chamblee GA, 9552 120 C4
Chamisal NM, 301 49 D2
Champaign IL, 67518 88 C4
Champaign Co. IL, 179669 89 D4
Champaign Co. OH, 38890 90 C4
Champion OH, 4727 91 F2
Champlain NY, 1173 81 D1
Chancellor SD, 328 27 F4
Chandler AZ, 176581 54 C2
Chandler IN, 3094 99 E4
Chandler OK, 2842 51 F2
Chandler TX, 2099 124 A2
Chandler Hts. AZ, 950 54 C2
Chandlerville IL, 704 88 A4
Chanhassen MN, 20321 66 C4
Channahon IL, 7344 88 C2
Channel Lake IL, 1785 74 C4
Channelview TX, 29685 132 B3
Channing TX, 356 50 A2
Chantilly VA, 41041 144 A3
Chanute KS, 9411 106 A1
Chaparral NM, 6117 56 C3
Chapel Hill NC, 48715 112 C4
Chapel Hill TN, 943 119 F1
Chapin IL, 592 98 A1
Chapin SC, 628 122 A3
Chaplin KY, 500 100 A4
Chapman KS, 1241 43 F2
Chapmanville WV, 1211 101 E4
Chappaqua NY, 9468 148 B2
Chappell NE, 983 34 A3
Chardon OH, 5156 91 E2
Charenton LA, 1944 133 F3
Chariton IA, 4573 87 D3
Chariton Co. MO, 8438 97 D1
Charlack MO, 1431 256 B2
Charleroi PA, 4871 92 A4
Charles City IA, 7812 73 D3
Charles City VA, 175 113 E1
Charles City Co. VA, 6926 113 E1
Charles Co. MD, 120546 144 B4
Charles Mix Co. SD, 9350 27 D4
Charleston AR, 2965 116 C1
Charleston IL, 21039 99 D2
Charleston ME, 300 82 C1
Charleston MS, 2198 118 B3
Charleston MO, 4732 108 C2
Charleston SC, 96650 131 D2
Charleston TN, 630 120 C1
Charleston UT, 378 31 F4
Charleston WV, 53421 101 E4
Charleston Co. SC, 309969 131 D2
Charlestown IN, 5993 100 A3
Charlestown MD, 1019 145 D1
Charlestown NH, 1145 81 E4
Charlestown RI, 950 150 C4
Charles Town WV, 2907 103 D2
Charlevoix MI, 2994 69 F3
Charlevoix Co. MI, 26090 70 B3
Charlo MT, 439 15 D3
Charlotte MI, 8389 76 A4
Charlotte NC, 540828 122 A1
Charlotte TN, 1153 109 E4
Charlotte TX, 1637 61 D3
Charlotte Beach FL, 1600 140 C4
Charlotte Co. FL, 141627 140 C4
Charlotte Co. VA, 12472 113 D2
Charlotte C.H. VA, 404 113 D2
Charlotte Hall MD, 1214 144 C4
Charlotte Harbor FL, 3647 140 C4
Charlottesville VA, 45049 102 C4
Charlton MA, 1100 150 B2
Charlton NY, 700 94 B1
Charlton City MA, 1400 150 B2
Charlton Co. GA, 10282 138 C1
Charlton Depot MA, 1200 150 B2
Charter Oak CA, 9027 229 F2
Charter Oak IA, 530 86 A1
Chartley MA, 1600 151 D2
Chase KS, 490 43 D3
Chase City VA, 2457 113 D2
Chase Co. KS, 3030 43 F3
Chase Co. NE, 4068 34 B4
Chaska MN, 17449 66 C4
Chassahowitzka FL, 700 140 B1
Chassell MI, 800 65 F4
Chateaugay NY, 798 80 C1
Chatfield MN, 2394 73 E2
Chatham IL, 8583 98 B1
Chatham LA, 623 125 E2
Chatham MA, 1667 151 F3
Chatham NJ, 8460 148 A4
Chatham NY, 1758 94 B2
Chatham VA, 1338 112 C2
Chatham Co. GA, 232048 130 B3
Chatham Co. NC, 49329 112 C4
Chatom AL, 1193 127 D4
Chatsworth GA, 3531 120 C2
Chatsworth IL, 1265 88 C3
Chattahoochee FL, 3287 137 D1
Chattahoochee Co. GA, 14882 128 C2
Chattanooga OK, 432 51 D4
Chattanooga TN, 155554 120 B2
Chattaroy WV, 1136 111 E1
Chattooga Co. GA, 25470 120 B3
Chatwood PA, 3600 146 B3
Chaumont NY, 592 79 D1
Chauncey OH, 1067 101 E2
Chautauqua Co. KS, 4359 43 F4
Chautauqua Co. NY, 139750 78 A4
Chauvin LA, 3229 134 B4
Chaves Co. NM, 61382 57 E2
Chazy NY, 600 81 D1
Cheatham Co. TN, 35912 109 E3
Cheat Neck WV, 950 102 B1
Chebanse IL, 1148 89 D3
Cheboygan MI, 5295 70 C2
Cheboygan Co. MI, 26448 70 C3
Checotah OK, 3481 116 A1
Cheektowaga NY, 79988 78 B3
Chefornak AK, 394 154 B3
Chehalis WA, 7057 12 B4
Chelan WA, 3522 13 E3
Chelan Co. WA, 66616 13 D2
Chelmsford MA, 32400 95 E1
Chelsea AL, 2949 127 F1
Chelsea MA, 35080 151 D1
Chelsea MI, 4398 76 B4
Chelsea NY, 2300 148 B1
Chelsea OK, 2136 106 A3
Chelsea VT, 275 81 E3
Cheltenham MD, 650 144 C4
Cheltenham PA, 5500 248 C2
Chelyan WV, 1400 101 F4
Chemung Co. NY, 91070 93 E1
Chenango Bridge NY, 4100 93 E1
Chenango Co. NY, 51401 79 E4
Chenequa WI, 583 74 C3
Cheney KS, 1783 43 E4
Cheney WA, 8832 13 F3
Cheneyville LA, 901 133 E1
Chenoa IL, 1845 88 C3
Chenoweth OR, 3412 21 D1
Chepachet RI, 900 150 C3
Cheraw SC, 5524 122 C2
Cheriton VA, 499 114 B3
Cherokee AL, 1237 119 D2
Cherokee IA, 5369 72 A4
Cherokee KS, 722 106 B1
Cherokee OK, 1630 51 D1
Cherokee Co. AL, 23988 120 A3
Cherokee Co. GA, 141903 120 C3
Cherokee Co. IA, 13035 72 A3
Cherokee Co. KS, 22605 106 B2
Cherokee Co. NC, 24298 121 D1
Cherokee Co. OK, 42521 106 B4
Cherokee Co. SC, 52537 121 F2
Cherokee Co. TX, 46659 124 A3
Cherokee Forest SC, 8000 217 A1
Cherokee Vil. AR, 4648 107 F3
Cherry Co. NE, 6148 34 B1
Cherry Creek NY, 551 78 A4
Cherryfield ME, 375 83 E2
Cherry Grove OH, 4555 204 C3
Cherry Hill NJ, 69965 146 C3
Cherry Hills Vil. CO, 5958 209 C4
Cherryvale KS, 2386 106 A2
Cherryvale SC, 2461 122 B4
Cherry Valley AR, 704 118 A1
Cherry Valley NY, 592 79 F4
Cherryville NC, 5361 122 A1
Chesaning MI, 2548 76 B3
Chesapeake OH, 842 101 D3
Chesapeake VA, 199184 113 F3
Chesapeake WV, 1643 101 F4
Chesapeake Beach MD, 3180 144 C4
Chesapeake City MD, 787 145 E1
Chesapeake Ranch Estates MD, 11503 103 F4
Cheshire CT, 5789 149 D1
Cheshire MA, 1200 94 C1
Cheshire Co. NH, 73825 81 E4
Chesilhurst NJ, 1520 147 D4
Chesnee SC, 1003 121 F1
Chester CA, 2316 29 D4
Chester CT, 1546 149 E2
Chester IL, 5185 98 B4
Chester MD, 3723 145 D3
Chester MA, 800 94 C2
Chester MT, 871 15 F2
Chester NH, 550 81 F4
Chester NJ, 1635 94 A4
Chester NY, 3445 148 A2
Chester PA, 36854 146 C3
Chester SC, 6476 122 A2
Chester VT, 999 81 E4
Chester VA, 17890 113 E1
Chester WV, 2592 91 F3
Chester Co. PA, 433501 146 B3
Chester Co. SC, 34068 122 A2
Chester Co. TN, 15540 119 D1
Chester Depot VT, 850 81 E4
Chesterfield IN, 2969 89 F4
Chesterfield MO, 46802 98 A3
Chesterfield SC, 1318 122 B2
Chesterfield VA, 3558 113 E1
Chesterfield Co. SC, 42768 122 B2
Chesterfield Co. VA, 259903 113 E1
Chester Hts. PA, 2481 146 B3
Chester Hill PA, 918 92 C3
Chesterland OH, 2646 91 E2
Chesterton IN, 10488 89 E2
Chestertown MD, 4746 145 D2
Chestnut Mtn. GA, 650 121 D3
Chestnut Ridge NY, 7829 148 B3
Cheswick PA, 1899 250 D1
Cheswold DE, 313 145 E2
Chetek WI, 2180 67 E3
Chetopa KS, 1281 106 B2
Chevak AK, 765 154 B3
Cheverly MD, 6433 144 B3
Cheviot OH, 9015 100 B2
Chevy Chase MD, 2726 270 C2
Chevy Chase View MD, 863 270 C1
Chewelah WA, 2186 13 F2
Cheyenne OK, 778 50 C3
Cheyenne WY, 53011 33 E3
Cheyenne Co. CO, 2231 42 A2
Cheyenne Co. KS, 3165 42 B1
Cheyenne Co. NE, 9830 34 A3
Cheyenne Wells CO, 1010 42 A2
Cheyney PA, 1600 146 B3
Chicago IL, 2896016 89 D1
Chicago Hts. IL, 32776 89 D2
Chicago Ridge IL, 14127 203 D5
Chichester NH, 500 81 F4
Chickamauga GA, 2245 120 B2
Chickasaw AL, 6364 135 E1
Chickasaw Co. IA, 13095 73 E3
Chickasaw Co. MS, 19440 118 C4
Chickasha OK, 15850 51 E3
Chico CA, 59954 36 B1
Chico TX, 947 59 E2
Chicopee MA, 54653 150 A2
Chicora PA, 1021 92 A3
Chicot Co. AR, 14117 125 F1
Chiefland FL, 1993 138 B4
Chilchinbito AZ, 462 47 F1
Chilcoot CA, 387 37 D1
Childersburg AL, 4927 128 A1
Childress TX, 6778 50 C4
Childress Co. TX, 7688 50 C4
Chilhowie VA, 1827 111 F2
Chillicothe IL, 5996 88 B3
Chillicothe MO, 8968 96 C1
Chillicothe OH, 21796 101 D2
Chillicothe TX, 798 50 C4
Chillum MD, 34252 270 D2
Chiloquin OR, 716 28 C1
Chilton WI, 3708 74 C1
Chilton Co. AL, 39593 127 F2
Chimayo NM, 2924 49 D2
China TX, 1112 132 C3
China Grove NC, 3616 122 B1
China Grove TX, 1247 61 D3
Chinchilla PA, 2300 261 E1
Chincoteague VA, 4317 114 C2
Chinle AZ, 5366 47 F2
Chino CA, 67168 229 G3
Chino Hills CA, 66787 229 G3
Chinook MT, 1386 16 B2
Chinook WA, 457 20 B1
Chino Valley AZ, 7835 47 D4
Chipita Park CO, 1709 205 C1
Chipley FL, 3592 136 C1
Chippewa Co. MI, 38543 70 B1
Chippewa Co. MN, 13088 66 A3
Chippewa Co. WI, 55195 67 F3
Chippewa Falls WI, 12925 67 F4

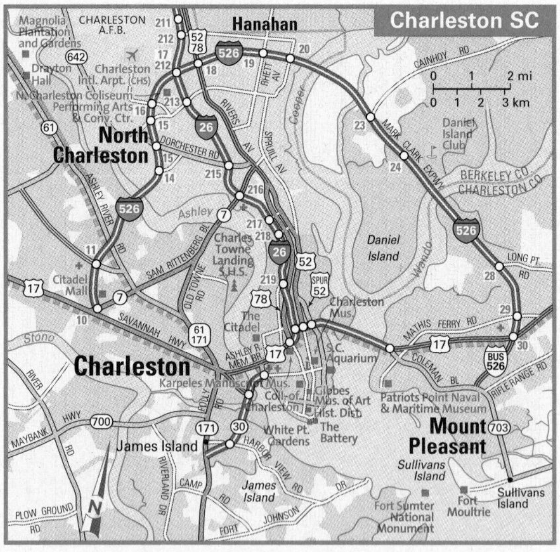

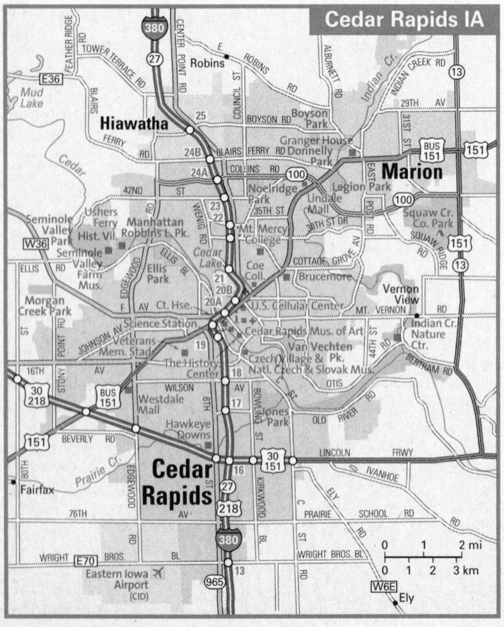

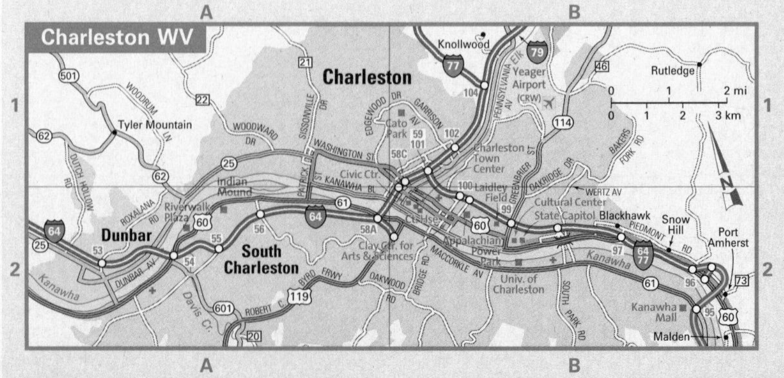

Blackhawk .... B2 Dunbar .... A2 Malden .... B2 Rutledge .... B1 S. Charleston A2
Charleston .... A1 Knollwood .... B1 Port Amherst .... B2 Snow Hill .... B2 Tyler Mtn. A1

Entries in **bold black** indicate counties or parishes.
Entries in **bold color** indicate cities with detailed inset maps.

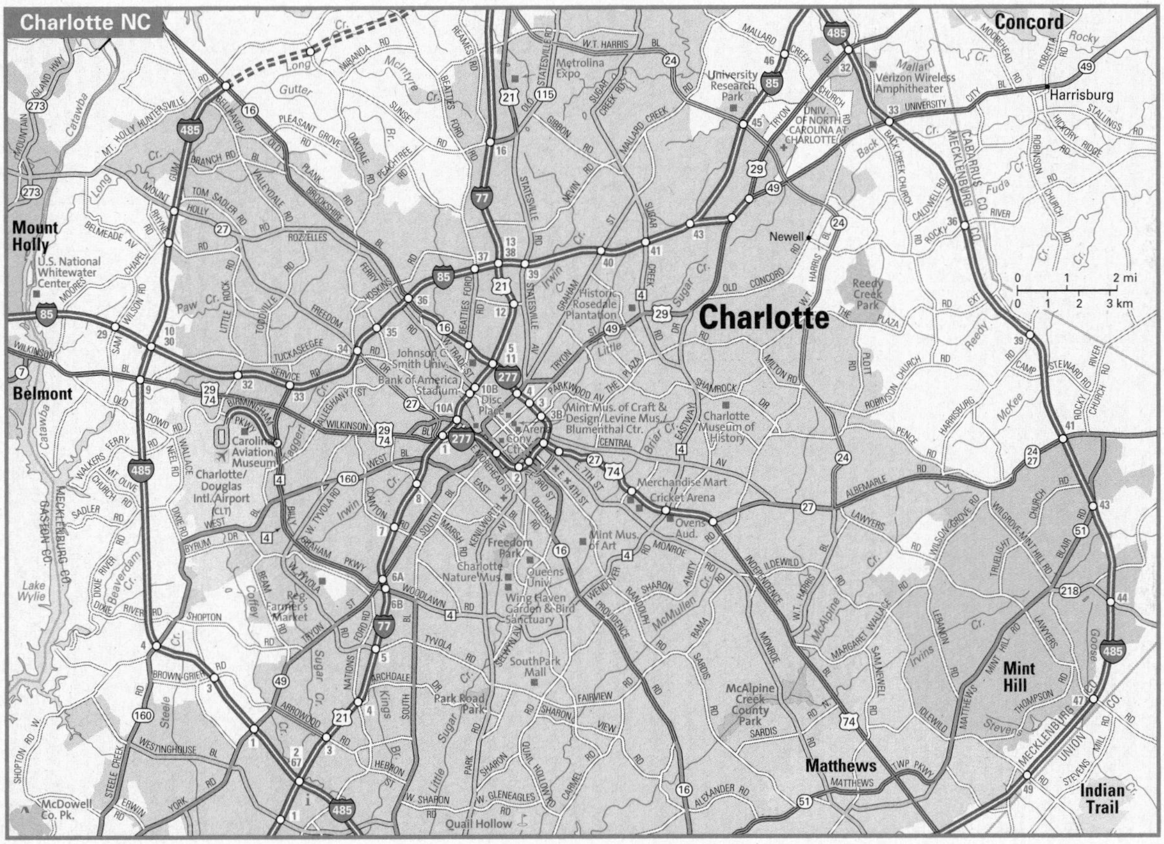

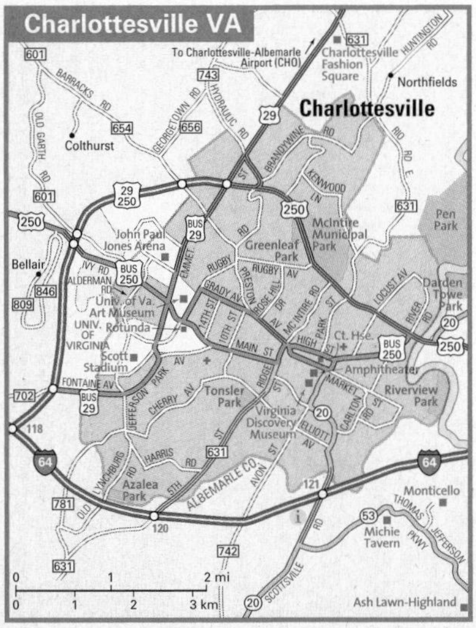

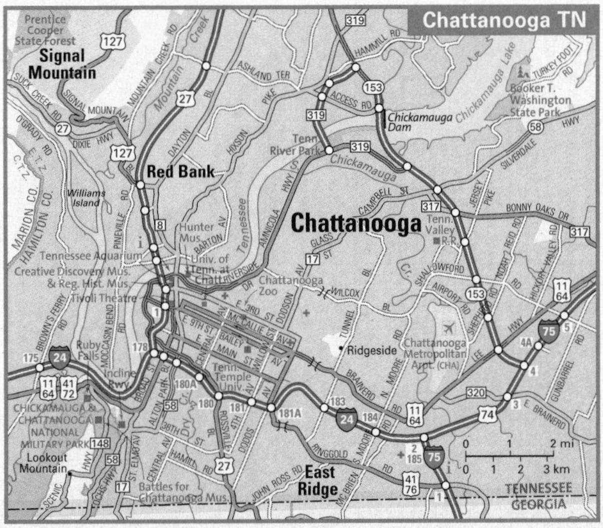

Figures after entries indicate population, page number, and grid reference.

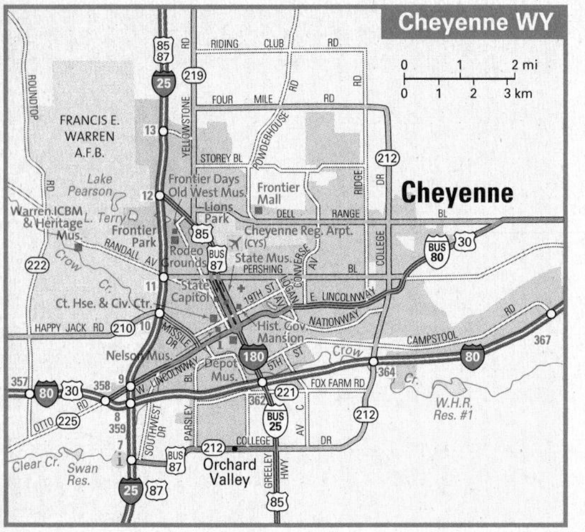

Clinchco VA, *424* **111** E2
Clinch Co. GA, *6878* **138** C1
Clint TX, *980* **56** C4
Clinton AR, *2283* **117** E1
Clinton CT, *3516* **149** E2
Clinton IL, *7485* **88** C4
Clinton IN, *5126* **99** E1
Clinton IA, *27772* **88** A1
Clinton KY, *1415* **108** C3
Clinton LA, *1998* **134** A1
Clinton ME, *1305* **82** C2
Clinton MD, *26064* **144** B4
Clinton MA, *7884* **150** C1
Clinton MI, *2293* **90** B1
Clinton MS, *23347* **126** B2
Clinton MO, *9311* **96** C4
Clinton MT, *549* **15** D4
Clinton NJ, *2632* **147** D1
Clinton NY, *1952* **79** E3
Clinton NC, *8600* **123** E2
Clinton OH, *1337* **91** E3
Clinton OK, *8833* **51** D3
Clinton SC, *8091* **121** F3
Clinton TN, *9409* **110** C4
Clinton UT, *12585* **244** A2
Clinton WA, *868* **12** C2
Clinton WI, *2162* **74** B4
Clinton Co. IL, *35535* **98** B3
Clinton Co. IN, *33866* **89** E4
Clinton Co. IA, *50149* **88** A1
Clinton Co. KY, *9634* **110** B2
Clinton Co. MI, *64753* **76** A3
Clinton Co. MO, *18979* **96** B1
Clinton Co. NY, *79894* **80** C2
Clinton Co. OH, *40543* **100** C2
Clinton Co. PA, *37914* **92** C2
Clintondale NY, *1424* **148** B1
Clintonville WI, *4736* **68** C4
Clintwood VA, *1549* **111** E2
Clio AL, *2206* **128** B4
Clio MI, *2483* **76** B3
Clio SC, *774* **122** C2
Cloquet MN, *11201* **64** C4
Closter NJ, *8383* **148** B3
Cloud Co. KS, *10268* **43** E2
Cloudcroft NM, *749* **56** C2
Clover SC, *4014* **122** A1
Cloverdale CA, *6831* **36** A2
Cloverdale IN, *2243* **99** E1
Cloverdale VA, *2986* **112** B1
Cloverleaf TX, *23508* **220** D2
Cloverport KY, *1256* **109** F1
Clovis CA, *68468* **44** C2
Clovis NM, *32667* **49** F4
Clute TX, *10424* **132** A4
Clyde CA, *694* **259** D1
Clyde KS, *740* **43** E1
Clyde NY, *2269* **79** D3
Clyde NC, *1324* **121** E1
Clyde OH, *6064* **90** C2
Clyde TX, *3345* **58** C3
Clyde Hill WA, *2890* **262** B3
Clyde Park MT, *310* **23** F1
Clymer PA, *1547* **92** B3
Coachella CA, *22724* **53** E3
Coahoma TX, *932* **58** B3
Coahoma Co. MS, *30622* **118** A3
Coal City IL, *4797* **88** C2
Coal City WV, *1905* **111** F1
Coal Co. OK, *6031* **51** F4
Coal Creek CO, *303* **41** E3
Coaldale PA, *2295* **93** E3
Coal Fork WV, *1350* **101** F4
Coalgate OK, *2005* **51** F4
Coal Grove OH, *2027* **101** D3
Coal Hill AR, *1001* **116** C1
Coaling AL, *1115* **127** F1
Coalinga CA, *11668* **44** C3
Coalmont TN, *948* **120** A1
Coal Mtn. GA, *700* **120** C3
Coal Run Vil. KY, *577* **111** E1

Coal Valley IL, *3606* **88** A2
Coalville UT, *1382* **31** F4
Coatesville IN, *516* **99** E1
Coatesville PA, *10838* **146** B3
Coats NC, *1845* **123** D1
Cobb CA, *1638* **36** B2
Cobb Co. GA, *607751* **120** C4
Cobb Island MD, *750* **103** E4
Cobden IL, *1116* **108** C1
Cobleskill NY, *4533* **79** F4
Coburg OR, *969* **20** B3
Cochise Co. AZ, *117755* **55** E3
Cochiti NM, *507* **48** C3
Cochituate MA, *6768* **150** C1
Cochran GA, *4455* **129** E2
Cochran Co. TX, *3730* **57** F1
Cochranton PA, *1148* **92** A2
Cochranville PA, *950* **146** A3
Cocke Co. TN, *33565* **111** D4
Cockeysville MD, *19388* **144** C1
Cockrell Hill TX, *4443* **207** D2
Cocoa FL, *16412* **141** E2
Cocoa Beach FL, *12482* **141** E2
Cocoa West FL, *5921* **232** A2
Coconino Co. AZ, *116320* **47** D3
Coconut Creek FL, *43566* **233** B1
Codington Co. SD, *25897* **27** E2
Cody WY, *8835* **24** B3
Coeburn VA, *1996* **111** E2
Coeur d'Alene ID, *34514* **14** B3
Coffee Co. AL, *43615* **128** A4
Coffee Co. GA, *37413* **129** E4
Coffee Co. TN, *48014* **120** A1
Coffeen IL, *709* **98** B2
Coffeeville MS, *930* **118** B4
Coffey Co. KS, *8865* **96** A4
Coffeyville KS, *11021* **106** A2
Coggon IA, *745* **73** E4
Cohasset MA, *3300* **151** E1
Cohasset MN, *2481* **64** B3
Cohocton NY, *854* **78** C4
Cohoes NY, *15521* **94** B1
Cohutta GA, *582* **120** B2
Cokato MN, *2727* **66** C3
Cokeburg PA, *705* **92** A4
Coke Co. TX, *3864* **58** B4
Coker AL, *808* **127** E1
Cokeville WY, *506* **31** F2
Colbert GA, *488* **121** D3
Colbert OK, *1065* **59** F1
Colbert Co. AL, *54984* **119** E2
Colby KS, *5450* **42** B2
Colby WI, *1616* **68** A4
Colchester CT, *3200* **149** E1
Colchester IL, *1493* **87** F4
Colchester VT, *1000* **81** D2
Colcord OK, *819* **106** B3
Colden NY, *700* **78** B4
Cold Spr. KY, *3806* **204** B3
Cold Spr. MN, *2975* **66** C3
Cold Spr. NY, *1983* **148** B1
Coldspring TX, *691* **132** B2
Coldwater KS, *792* **43** D4
Coldwater MI, *12697* **90** A1
Coldwater MS, *1674* **118** B2
Coldwater OH, *4482* **90** B4
Cole OK, *473* **51** E3
Colebrook NH, *1400* **81** F1
Cole Camp MO, *1028* **97** D3
Cole Co. MO, *71397* **97** E3
Coleman FL, *647* **140** C1
Coleman MI, *1296* **76** A2
Coleman TX, *5127* **58** C3
Coleman WI, *716* **68** C3
Coleman Co. TX, *9235* **58** C3
Coleraine MN, *1110* **64** B3
Coleridge NE, *541* **35** E1
Coles Co. IL, *53196* **99** D1
Colesville MD, *19810* **144** B2
Colfax CA, *1496* **36** C2
Colfax IL, *989* **88** C4

Colfax IN, *768* **89** E4
Colfax IA, *2223* **87** D2
Colfax LA, *1659* **125** E4
Colfax WA, *2844* **14** A4
Colfax WI, *1136* **67** E4
Colfax Co. NE, *10441* **35** F3
Colfax Co. NM, *14189* **49** E1
Collbran CO, *388* **40** B2
College AK, *11402* **154** C2
Collegedale TN, *6514* **120** B2
College Park GA, *20382* **120** C4
College Park MD, *24657* **270** E1
College Place WA, *7818* **21** F1
College Sta. TX, *67890* **61** F1
Collegeville IN, *865* **89** E3
Collegeville PA, *8032* **146** B2
Colleton Co. SC, *38264* **130** C2
Colleyville TX, *19636* **59** E2
Collier Co. FL, *251377* **143** D1
Collierville TN, *31872* **118** B2
Collin Co. TX, *491675* **59** F2
Collingdale PA, *8664* **248** B4

Collingswood NJ, *14326* **146** C3
Collingsworth Co. TX, *3206* **50** C3
Collins GA, *528* **129** F3
Collins IA, *499* **87** D1
Collins MS, *2683* **126** C4
Collins Park DE, *8300* **274** D3
Collinsville AL, *1644* **120** A3
Collinsville IL, *24707* **98** B3
Collinsville MS, *1823* **127** D2
Collinsville OK, *4077* **106** A3
Collinsville TX, *1235* **59** F1
Collinsville VA, *7777* **112** B3
Collinwood TN, *1024* **119** E1
Colma CA, *1191* **259** B3
Colman SD, *572* **27** F4
Colmar PA, *1800* **146** C2
Colmar Manor MD, *1257* **270** E3
Colmesneil TX, *638* **132** C1
Colo IA, *868* **87** D1
Cologne MN, *1012* **66** C4
Cologne NJ, *800* **147** D4
Coloma MI, *1595* **89** F1
Colome SD, *340* **34** C1
Colon MI, *1227* **90** A1
Colona IL, *5173* **88** A2
Colonia NJ, *17811* **147** E1
Colonial Beach VA, *3228* **103** E4
Colonial Hts. TN, *7067* **111** E3
Colonial Hts. VA, *16897* **113** E2
Colonial Park PA, *13259* **218** C1
Colonie NY, *7916* **94** B1
Colony KS, *397* **96** A4
Colorado City AZ, *3334* **46** C1
Colorado City CO, *2018* **41** E3
Colorado City TX, *4281* **58** B3
Colorado Co. TX, *20390* **61** F2
Colorado Sprs. CO, *360890* **41** E2
Colquitt GA, *1939* **137** D1
Colquitt Co. GA, *42053* **137** F1
Colstrip MT, *2346* **25** D1
Colton CA, *47662* **229** J2
Colton SD, *662* **27** F4
Columbia AL, *804* **128** B4
Columbia CA, *2405* **37** D3
Columbia CT, *800* **149** F1
Columbia FL, *550* **138** C3
Columbia IL, *7922* **98** A3
Columbia KY, *4014* **110** B2
Columbia LA, *477* **125** F3
Columbia MD, *88254* **144** B2
Columbia MS, *6603* **126** B4
Columbia MO, *84531* **97** E3
Columbia NC, *819* **113** F4
Columbia PA, *10311* **103** E1

CHICAGO MAP INDEX

Addison C4
Algonquin A2
Alsip D6
Arlington Hts. C2
Aurora A5
Bannockburn C2
Barrington B2
Barrington Hills A2
Bartlett A3
Batavia A4
Bedford Park D5
Bellwood C4
Bensenville C4
Berkeley C4
Berwyn D4
Bloomingdale B3
Blue Island D6
Bolingbrook B6
Bridgeview D5
Broadview C4
Brookfield C5
Buffalo Grove C2
Burbank D5
Burnham E6
Burr Ridge C5
Burtons Bridge A1
Calumet City E6
Calumet Park E6
Carol Stream B4
Carpentersville A2
Cary A2
Chicago E4
Chicago Ridge D5
Cicero D4
Clarendon Hills C5
Countryside C5
Crest Hill B6
Crestwood D6
Crystal Lake A1
Darien C5
Deerfield C2
Deer Park B2
Des Plaines C3
Diamond Lake B1
Dixmoor E6
Dolton E6
Downers Grove B5
E. Dundee A2
Elgin A3
Elk Grove Vil. C3
Elmhurst C4
Elmwood Park D4
Evanston D3
Evergreen Park D5
Forest Lake B1
Forest Park D4
Forest View D5
Fox River Grove A2
Franklin Park C4
Geneva A4
Glencoe D2
Glendale Hts. B4
Glen Ellyn B4
Glenview D2
Golf D3
Grayslake B1
Green Oaks C1
Hanover Park B3
Harvey E6
Harwood Hts. D3
Hawthorn Woods B1
Hazel Crest E6
Hickory Hills D5
Highland Park D2
Highwood D1
Hillside C4
Hinsdale C5
Hodgkins C5
Hoffman Estates B3
Holiday Hills A1
Homer Glen C6
Hometown D5
Indian Creek C1
Indian Head Park C5
Inverness B2
Island Lake A1
Itasca B3
Ivanhoe B1
Justice D5
Kenilworth D2
Kildeer B2
Knollwood C1
La Grange C5
La Grange Park C4
Lake Barrington B1
Lake Bluff C1
Lake Forest C1
Lake in the Hills A2
Lakemoor A1
Lake Zurich B2
Lemont B6
Libertyville C1
Lincolnshire C2
Lincolnwood D3
Lisle B5
Lockport B6
Lombard B4
Long Grove C2
Lyons D5
Markham D6
Maywood C4
McCook D5
McHenry A1
Medinah B3
Melrose Park C4
Merrionette Park D6
Mettawa C1
Midlothian D6
Montgomery A5
Morton Grove D3
Mount Prospect C3
Mundelein B1
Naperville B5
Niles D3
Norridge D3
N. Aurora A5
N. Barrington B2
Northbrook D2
N. Chicago C1
Northfield D2
Northlake C4
N. Riverside D4
Oak Brook C4
Oakbrook Terrace C4
Oak Forest D6
Oak Lawn D5
Oak Park D4
Oakwood Hills A1
Orland Hills C6
Orland Park C6
Oswego A5
Palatine B2
Palos Hts. D6
Palos Hills D5
Palos Park D6
Park Ridge C3
Phoenix E6
Plainfield A6
Port Barrington A1
Posen D6
Prairie Grove A1
Prospect Hts. C2
Riverdale E6
River Forest D4
River Grove D4
Riverside D4
Riverwoods C2
Robbins D6
Rolling Meadows B3
Romeoville B6
Rondout C1
Roselle B3
Rosemont C3
St. Charles A4
Schaumburg B3
Schiller Park C3
Skokie D3
Sleepy Hollow A2
S. Barrington B2
S. Elgin A3
S. Holland E6
Stickney D5
Stone Park C4
Streamwood B3
Summit D5
Tinley Park D6
Tower Lakes B1
Trout Valley A1
Valley View A3
Vernon Hills C1
Villa Park C4
Volo B1
Warrenville A4
Wauconda B1
Wayne A4
Westchester C4
W. Chicago A4
W. Dundee A2
Western Sprs. C5
Westmont C5
Wheaton B4
Wheeling C2
Willowbrook C5
Willow Sprs. C5
Wilmette D3
Winfield B4
Winnetka D2
Wood Dale C3
Woodridge B5
Worth D6

Columbia SC, *116278* **122** A3
Columbia TN, *33055* **119** F1
Columbia Beach MD, *700* **144** C3
Columbia City IN, *7077* **89** F2
Columbia City OR, *1571* **20** C1
Columbia Co. AR, *25603* **125** D1
Columbia Co. FL, *56513* **138** C2
Columbia Co. GA, *89288* **121** F4
Columbia Co. NY, *63094* **94** B2
Columbia Co. OR, *43560* **20** B1
Columbia Co. PA, *64151* **93** E3
Columbia Co. WA, *4064* **13** F4
Columbia Co. WI, *52468* **74** B3
Columbia Falls MT, *3645* **15** D2
Columbia Hts. MN, *18520* **235** C2
Columbiana AL, *3316* **127** F1
Columbiana OH, *5635* **91** F3
Columbiana Co. OH, *112075* **91** F3

POINTS OF INTEREST

900 North Michigan Avenue Shops B1
Adler Planetarium & Astronomy Museum C3
Art Institute of Chicago B2
Auditorium Building B2
Buckingham Fountain B2
Cadillac Palace Theatre B2
Centennial Fountain & Arc B2
Chicago Architecture Foundation B2
Chicago Board of Trade B2
Chicago Ctr. for the Perform. Arts A1
Chicago Children's Museum C1
Chicago Cultural Center B2
Chicago Fire Marker A3
Chicago Mercantile Exchange A2
Chicago Place B1
Chicago Stock Exchange B2
Chicago Theatre B2
Chicago Yacht Club C2
City Hall B2
Civic Opera House A2
Cloud Gate B2
Crown Fountain B2
Daley Plaza B2
Dearborn Station B3
The Field Museum C3
Ford Center for the Performing Arts B2
Goodman Theatre B2
Harris Theater B2
Harold Washington Library Center B2
James R. Thompson Center B2
Jane Addams' Hull House Mus. A3
John G. Shedd Aquarium C3
John Hancock Center B1
LaSalle Bank Theatre B2
Merchandise Mart A2
Monadnock Building B2
Moody Bible Institute B1
Mus. of Broadcast Communications B2
Mus. of Contemporary Art B1
Mus. of Contemporary Photography B3
Navy Pier C1
Newberry Library B1
New Maxwell Street Market A3
Northwestern University (Chicago Campus) C1
Petrillo Band Shell B2
Pritzker Pavillion B2
River East Plaza C1
Sears Tower A2
Smith Museum of Stained Glass Windows C1
Soldier Field C3
Spertus Museum B3
Symphony Center B2
Tribune Tower B1
Univ. of Illinois at Chicago A3
Water Tower B1
Water Tower Place B1
Wrigley Building B1

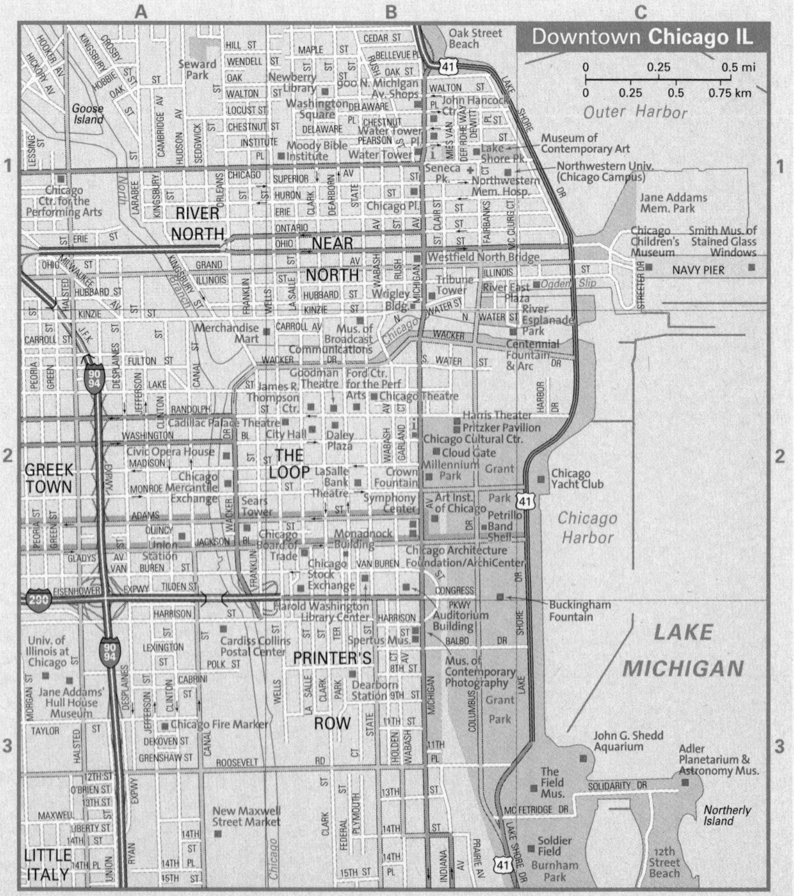

Entries in **bold black** indicate counties or parishes.
Entries in **bold color** indicate cities with detailed inset maps.

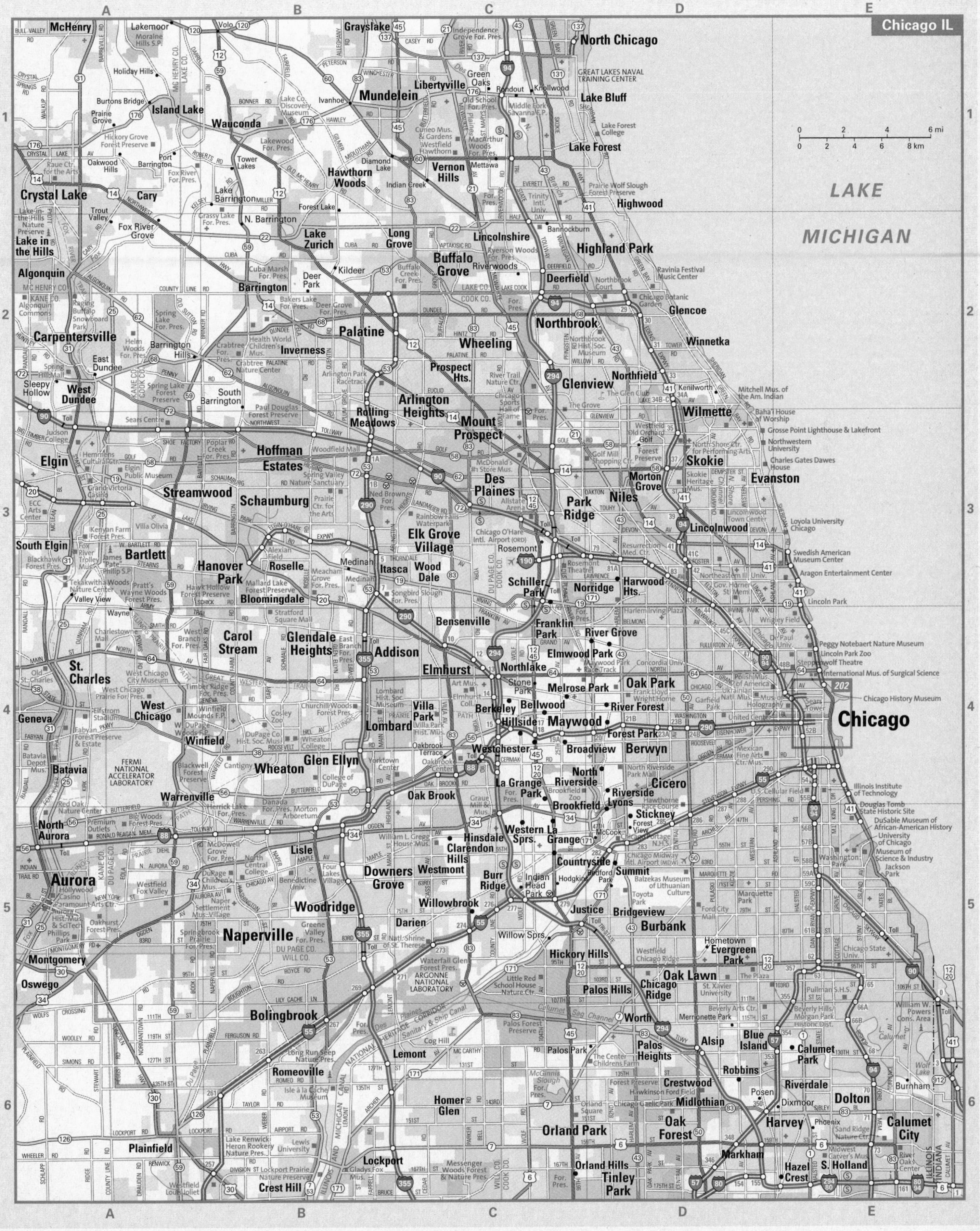

Figures after entries indicate population, page number, and grid reference.

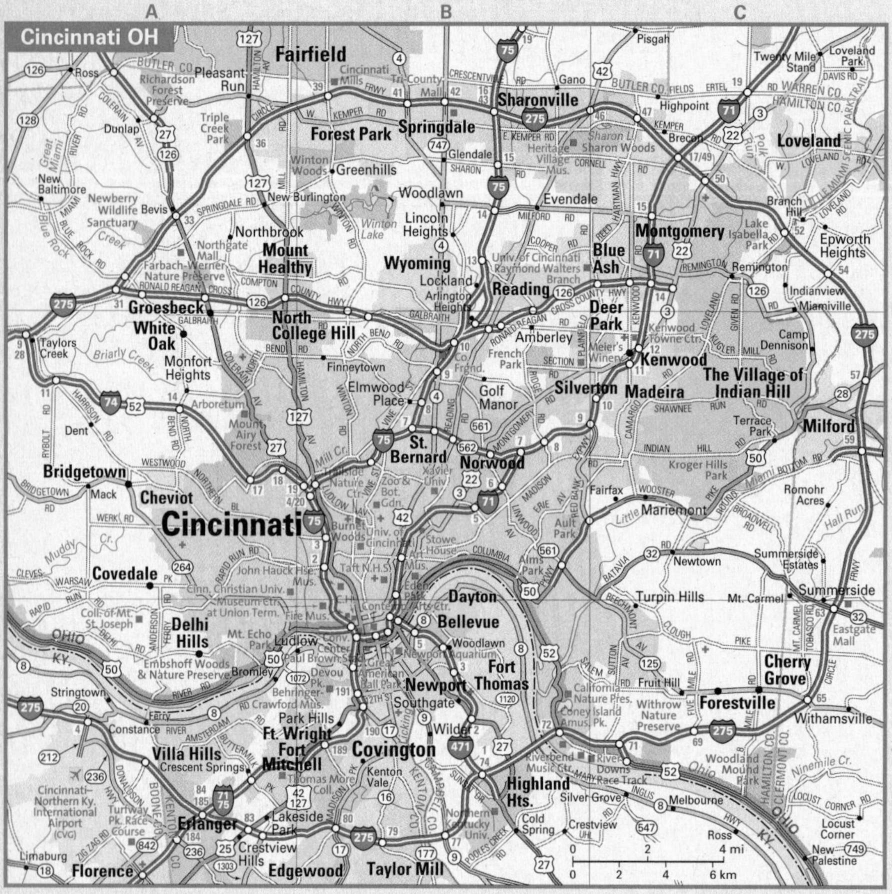

Amberley B2
Arlington Hts. B2
Bellevue B3
Bevis A1
Blue Ash C1
Branch Hill C1
Brecon C1
Bridgetown A2
Bromley A3
Camp Dennison C2
Cherry Grove C3
Cheviot A2
Cincinnati A2
Cold Spr. B3
Constance A3
Covedale A2
Covington B3
Crescent Sprs. A3
Crestview B3
Crestview Hills A3
Dayton B2
Deer Park C2
Delhi Hills A3
Dent A2
Dunlap A1
Edgewood B3
Elmwood Place B2
Epworth Hts. C1
Erlanger A3
Evendale B1
Fairfax C2
Fairfield B1
Finneytown B2
Florence A3
Forest Park B1
Forestville C3
Ft. Mitchell A3
Ft. Thomas B3
Ft. Wright B3
Fruit Hill C3
Gano B1
Glendale B1
Golf Manor B2
Greenhills B1
Groesbeck A2
Highland Hts. B3
Highpoint C1
Indianview C1
Kenton Vale B3
Kenwood C2
Lakeside Park A3
Limaburg A3
Lincoln Hts. B1
Lockland B1
Locust Corner C3
Loveland C1
Loveland Park C1
Ludlow A3
Mack A2
Madeira C2
Mariemont C2
Melbourne C3
Miamiville C1
Milford C2
Monfort Hts. A2
Montgomery C1
Mt. Carmel C2
Mt. Healthy A1
New Baltimore A1
New Burlington A1
New Palestine C3
Newport B3
Newtown C2
Northbrook A1
N. College Hill A2
Norwood B2
Park Hills B3
Pisgah C1
Pleasant Run A1
Reading B1
Remington C1
Romohr Acres C2
Ross, KY C3
Ross, OH A1
St. Bernard B2
Sharonville B1
Silver Grove C3
Silverton B2
Southgate B3
Springdale B1
Stringtown A3
Summerside C2
Summerside Estates C2
Taylor Mill B3
Taylors Creek A2
Terrace Park C2
The Vil. of Indian Hill C2
Turpin Hills C3
Twenty Mile Stand C1
Villa Hills A3
White Oak A2
Wilder B3
Withamsville C3
Woodlawn, KY B3
Woodlawn, OH B1
Wyoming B1

Avon D2
Avon Lake D2
Bay Vil. D2
Beachwood G2
Bedford G3
Bedford Hts. G3
Bentleyville G3
Berea D3
Bratenahl F1
Brecksville F3
Broadview Hts. F3
Brooklyn E2
Brooklyn Hts. F2
Brook Park E3
Cleveland E1
Cleveland Hts. F2
Cuyahoga Hts. F2
E. Cleveland F1
Euclid G1
Fairview Park D2
Garfield Hts. F3
Gates Mills G1
Glenwillow G3
Highland Hts. G1
Highland Hills G2
Hunting Valley G2
Independence F3
Lakewood E2
Linndale E2
Lyndhurst G2
Macedonia G3
Maple Hts. F3
Mayfield G1
Mayfield Hts. G2
Middleburg Hts. E3
Moreland Hills G2
Newburgh Hts. F2
Northfield G3
N. Olmsted D2
N. Randall G2
N. Ridgeville D3
N. Royalton E3
Oakwood G3
Olmsted Falls D3
Orange G2
Parma E3
Parma Hts. E3
Pepper Pike G2
Richmond Hts. G1
Rocky River E2
Sagamore Hills F3
Seven Hills F3
Shaker Hts. G2
Solon G3
S. Euclid G1
Strongsville E3
Twinsburg G3
University Hts. G2
Valley View F3
Walton Hills G3
Warrensville Hts. G2
Westlake D2
Wickliffe G1
Willoughby Hills G1
Woodmere G2

Entries in **bold black** indicate counties or parishes.
Entries in **bold color** indicate cities with detailed inset maps.

Coweta OK, 7139	106 A4	**Craven Co. NC**, 91436	115 D3
Coweta Co. GA, 89215	128 C1	Crawford CO, 366	40 C3
Cowley WY, 560	24 C2	Crawford GA, 807	121 E4
Cowley Co. KS, 36291	43 F4	Crawford MS, 655	127 D1
Cowlitz Co. WA, 92948	20 C1	Crawford NE, 1107	33 F1
Coxsackie NY, 2895	94 B2	Crawford TX, 705	59 E4
Coyanosa TX, 2279	121 F2	**Crawford Co. AR**, 53247	116 C1
Cozad NE, 4163	34 C4	**Crawford Co. GA**, 12495	129 D2
Crab Orchard KY, 842	110 B1	**Crawford Co. IL**, 20452	99 D3
Crab Orchard TN, 838	110 B4	**Crawford Co. IN**, 10743	99 F4
Crab Orchard WV, 2761	111 F1	**Crawford Co. IA**, 16942	86 A1
Crafton PA, 6706	250 A2	**Crawford Co. KS**, 38242	106 B1
Craig AK, 1397	155 E4	**Crawford Co. MI**, 14273	70 C4
Craig CO, 9189	32 C4	**Crawford Co. MO**, 22804	97 F4
Craig Beach OH, 1254	91 F3	**Crawford Co. OH**, 46966	90 C3
Craig Co. OK, 14950	106 A2	**Crawford Co. PA**, 90366	91 F1
Craig Co. VA, 5190	112 B1	**Crawford Co. WI**, 17243	73 F3
Craighead Co. AR, 82148	108 A4	Crawfordsville AR, 514	118 B1
Craigmont ID, 556	22 B1	Crawfordsville IN, 15243	89 E4
Craigsville VA, 979	102 B4	Crawfordville FL, 750	137 E2
Craigsville WV, 2204	102 A4	Crawfordville GA, 572	121 E4
Crainville IL, 1432	108 C1	Creal Sprs. IL, 702	108 C1
Cramerton NC, 2976	122 A1	Creede CO, 377	40 C4
Cranberry PA, 2650	147 D2	Creedmoor NC, 2232	112 C4
Cranbury NJ, 1800	147 E3	**Creek Co. OK**, 67367	51 F2
Crandall TX, 2774	59 F2	Creekside KY, 336	230 F1
Crandon WI, 1961	68 B3	Creighton NE, 1270	35 E2
Crane MO, 1390	106 C2	Crenshaw MS, 916	118 B3
Crane TX, 3191	58 A4	**Crenshaw Co. AL**, 13665	128 A4
Crane Co. TX, 3996	57 F4	Creola AL, 2002	135 E1
Cranesville PA, 600	91 F1	Cresaptown MD, 5884	102 C1
Cranford NJ, 22578	147 E1	Crescent IA, 537	86 A2
Cranston RI, 79269	150 C3		

Black Forest	D1
Cascade	C1
Chipita Park	C1
Colorado Sprs.	D1
Crystola	C1
Fountain	D2
Green Mtn. Falls	C1
Manitou Sprs.	C2
Security	D2
Stratmoor Hills	D2
Widefield	D2

Colorado Springs CO

Columbia SC

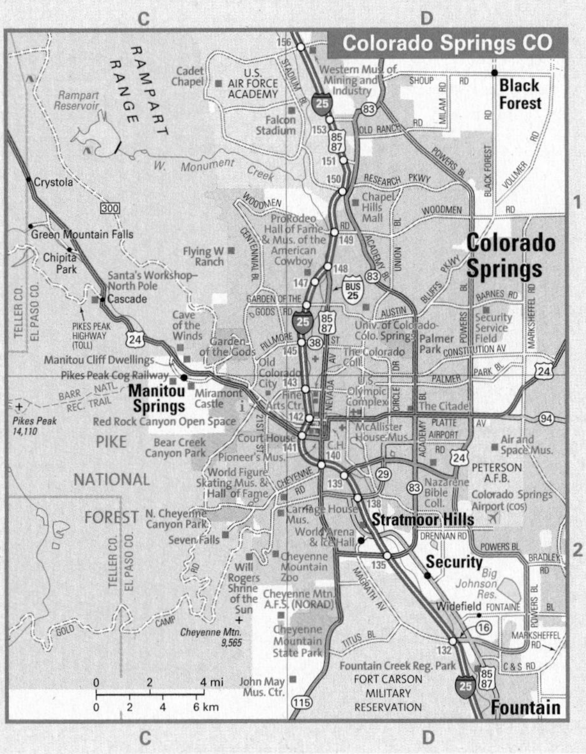

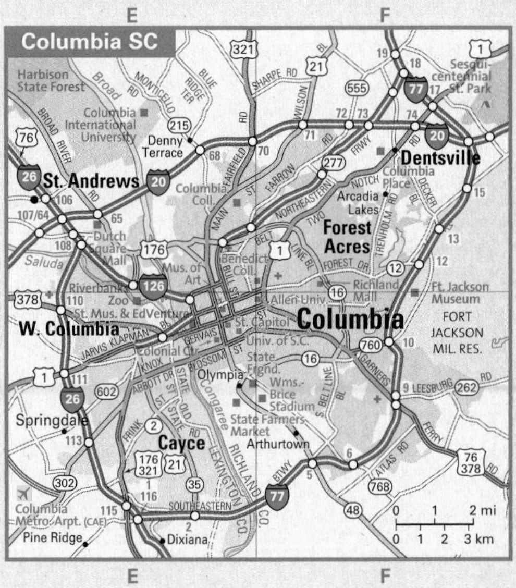

Arcadia Lakes	F1	Dentsville	F1	St. Andrews	E1
Arthurtown	F2	Dixiana	E2	Springdale	E2
Cayce	F2	Forest Acres	E2	W. Columbia	E2
Columbia	F2	Olympia	E2		
Denny Terrace	F1	Pineridge	E2		

Downtown Cleveland OH

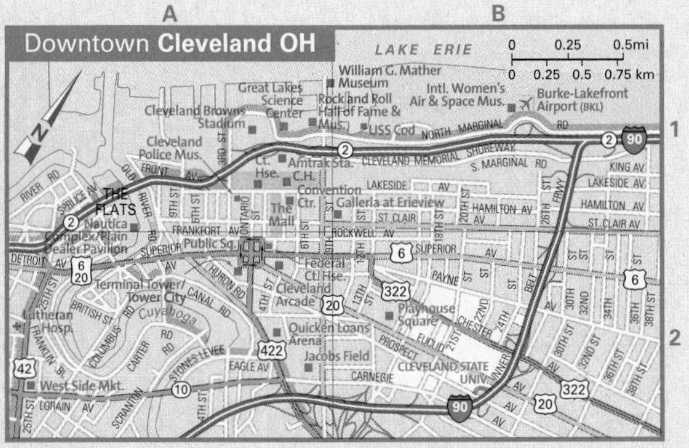

POINTS OF INTEREST

Amtrak Station	A1
Burke-Lakefront Airport	B1
Cleveland Arcade	A2
Cleveland Browns Stadium	A1
Cleveland Police Museum	A1
Cleveland State University	B2
Convention Center	A1
Galleria at Erieview	B1
Great Lakes Science Center	A1
International Women's Air & Space Museum	B1
Jacobs Field	A2
Nautica Complex/Plain Dealer Pavilion	A1
Playhouse Square	B2
Quicken Loans Arena	A2
Rock and Roll Hall of Fame & Museum	A1
Terminal Tower/Tower City	A2
U.S.S. Cod	B1
West Side Market	A2
William G. Mather Museum	B1

Columbus GA

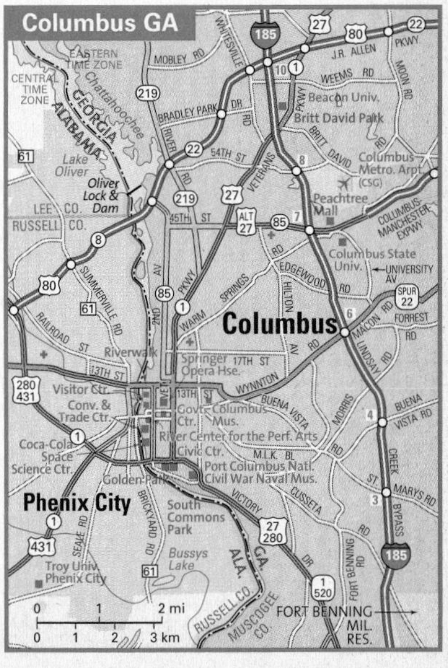

Figures after entries indicate population, page number, and grid reference.

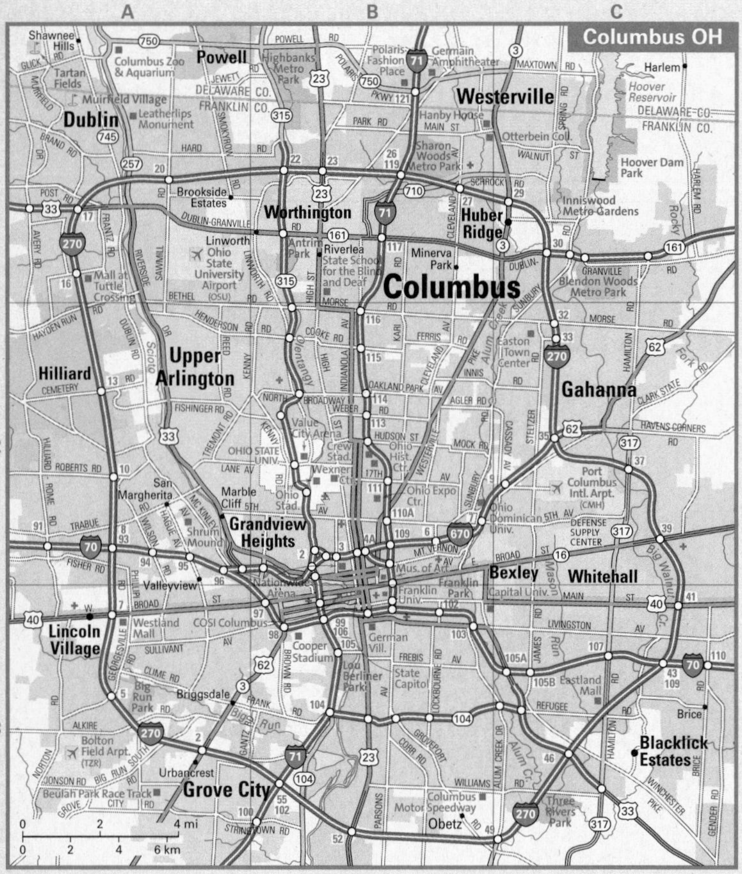

D

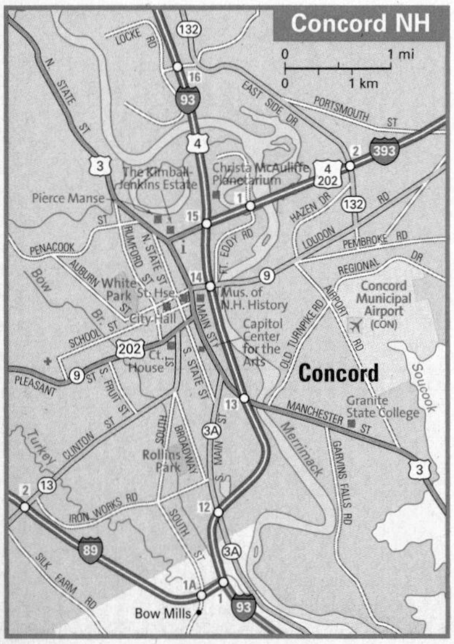

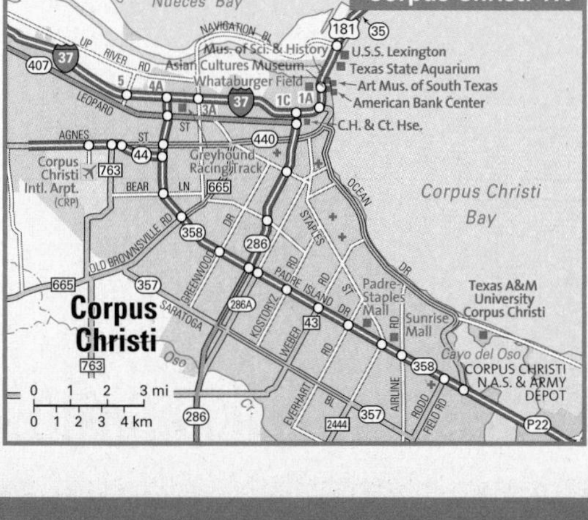

Entries in **bold black** indicate counties or parishes.
Entries in **bold color** indicate cities with detailed inset maps.

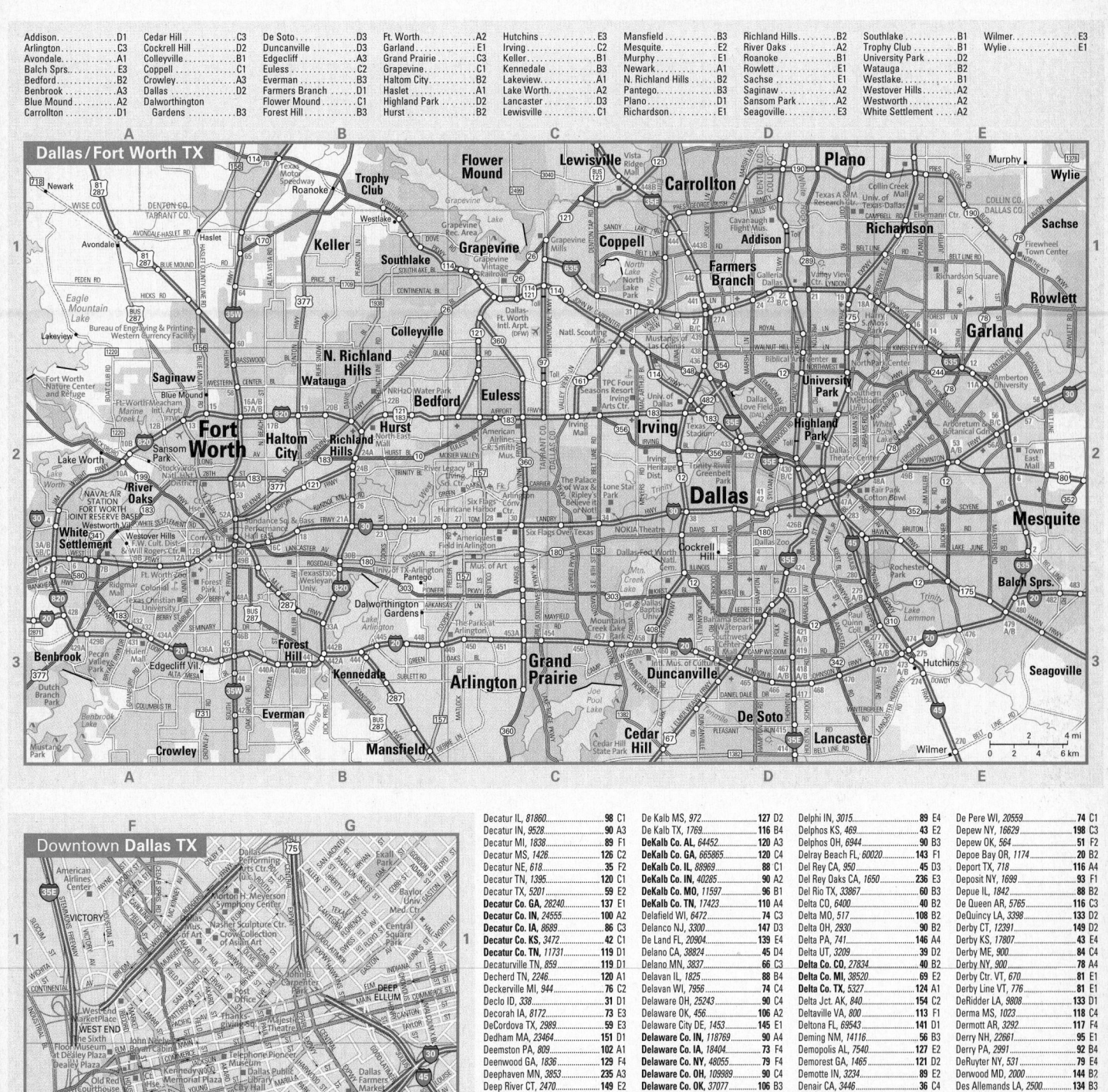

Addison....D1
Arlington....C3
Avondale....A1
Balch Sprs....E3
Bedford....B2
Benbrook....A3
Blue Mound....A2
Carrollton....D1
Cedar Hill....C3
Cockrell Hill....D2
Colleyville....B1
Coppell....C1
Crowley....A3
Dallas....D2
Dalworthington Gardens....B3
De Soto....D3
Duncanville....D3
Edgecliff....A3
Euless....C2
Everman....B3
Farmers Branch....D1
Flower Mound....C1
Forest Hill....B3
Ft. Worth....A2
Garland....E1
Grand Prairie....C3
Grapevine....C1
Haltom City....B2
Haslet....A1
Highland Park....D2
Hurst....B2
Hutchins....E3
Irving....C2
Keller....B1
Kennedale....B3
Lakeview....A1
Lake Worth....A2
Lancaster....D3
Lewisville....C1
Mansfield....B3
Mesquite....E2
Murphy....E1
Newark....A1
N. Richland Hills....B2
Pantego....B3
Plano....D1
Richardson....E1
Richland Hills....B2
River Oaks....A2
Roanoke....B1
Rowlett....E1
Sachse....E1
Saginaw....A2
Sansom Park....A2
Seagoville....E3
Southlake....B1
Trophy Club....B1
University Park....D2
Watauga....B2
Westlake....B1
Westover Hills....A2
Westworth....A2
White Settlement....A2
Wilmer....E3
Wylie....E1

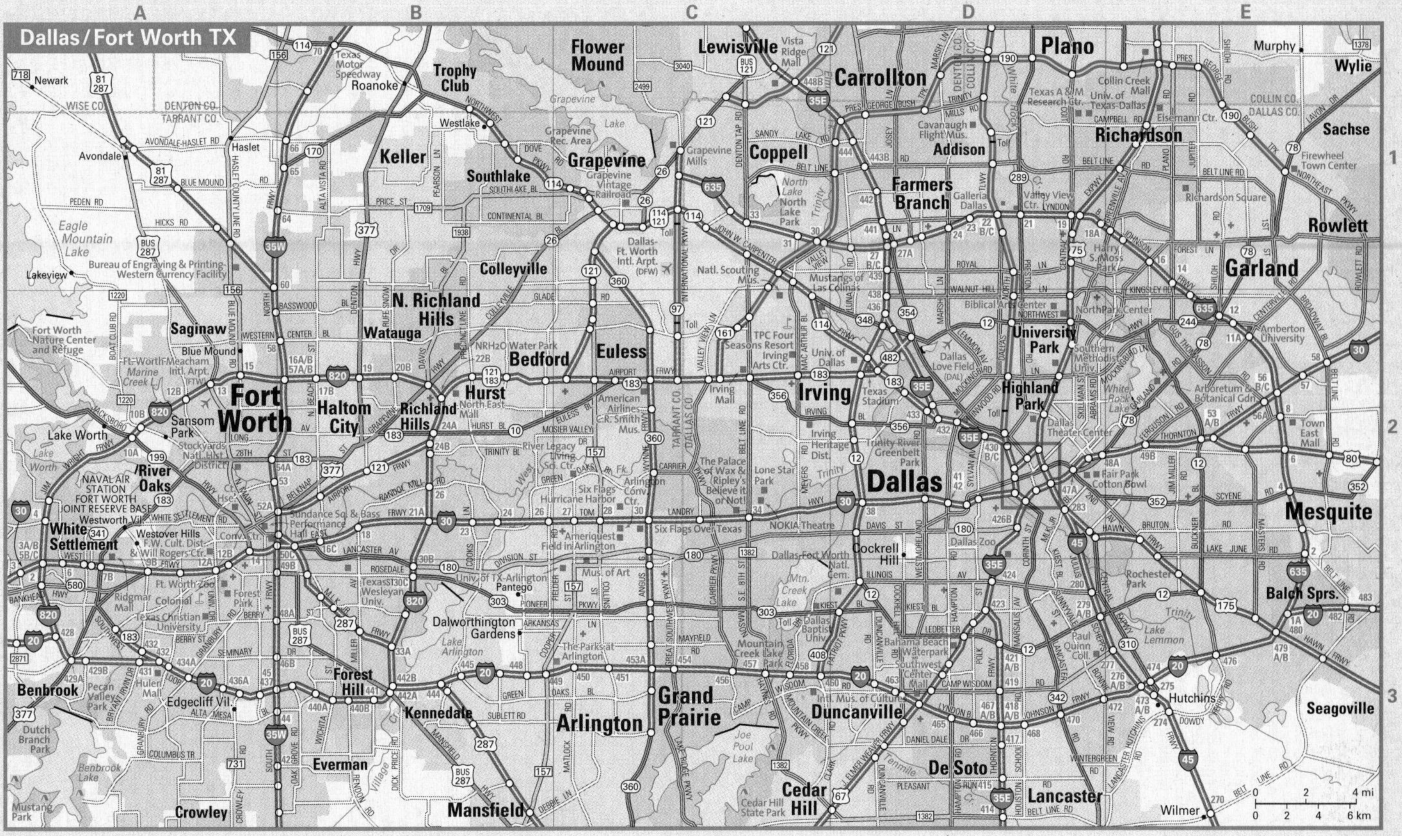

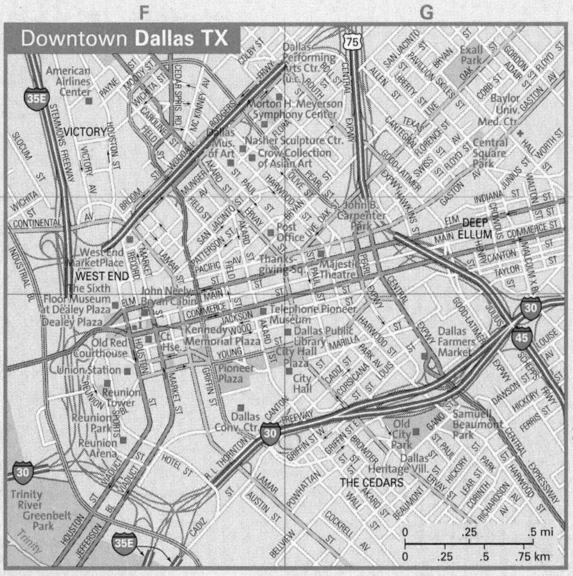

POINTS OF INTEREST
AmericanAirlines Center....F1
City Hall....G2
Crow Collection of Asian Art....F1
Dallas Convention Center....F2
Dallas Heritage Village....G2
Dallas Museum of Art....F1
Dallas Public Library....G2
Farmers Market....G2
John Neely Bryan Cabin....F2
Kennedy Memorial Plaza....F2
Majestic Theatre....G1
Morton H. Meyerson Symphony Center....F1
Nasher Sculpture Center....F1
Old Red Courthouse....F2
Reunion Arena....F2
Reunion Tower....F2
The Sixth Floor Museum at Dealey Plaza....F2
Telephone Pioneer Museum....F2
West End MarketPlace....F1

Figures after entries indicate population, page number, and grid reference.

Barstow C2	Cleveland C2	Davenport A1	Le Claire C1	Panorama Park B1	Riverdale B1
Bettendorf B1	Coal Valley B2	E. Moline B1	Milan B2	Port Byron C1	Rock Island A2
Carbon Cliff C2	Colona C2	Hampton C1	Moline B2	Rapids City C1	Silvis C2

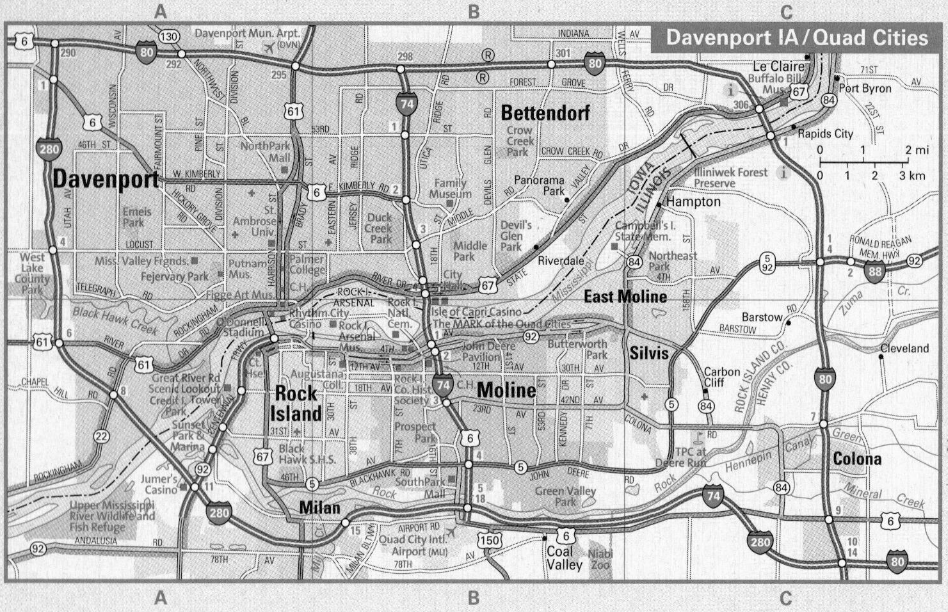

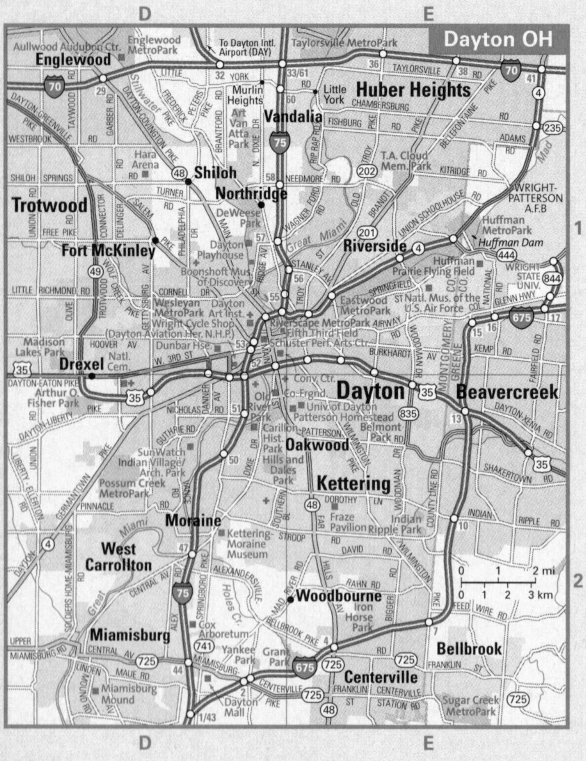

Beavercreek E2	Ft. McKinley D1	Murlin Hts. D1	Vandalia E1
Bellbrook E2	Huber Hts. E1	Northridge D1	W. Carrollton D2
Centerville E2	Kettering E2	Oakwood E2	Woodbourne E2
Dayton E2	Little York E1	Riverside E1	
Drexel D1	Miamisburg D2	Shiloh D1	
Englewood D1	Moraine D2	Trotwood D1	

Des Plaines IL, *58720* **89** D1
Destin FL, *11119* **136** B2
Destrehan LA, *11260* **134** B3
Detroit MI, *951270* 76 C4
Detroit TX, *776* **116** A4
Detroit Beach MI, *2289* **90** C1
Detroit Lakes MN, *7348* **19** F4
Deuel Co. NE, *2098* **34** A3
Deuel Co. SD, *4498* **27** F2
De Valls Bluff AR, *783* **117** F2
Deville LA, *1007* **125** E4
Devils Lake ND, *7222* **19** D2
Devine TX, *4140* **61** D3
Devola OH, *2771* **101** F2
Devon PA, *5067* **248** A2
Dewar OK, *919* **116** A1
Dewey AZ, *6295* **47** D4
Dewey OK, *3179* **51** F1
Dewey Beach DE, *301* **145** F4
Dewey Co. OK, *4743* **51** D2
Dewey Co. SD, *5972* **26** C2
Deweyville TX, *1190* **132** C2
Deweyville UT, *278* **31** E3
De Witt AR, *3552* **117** F3
De Witt IA, *5049* **88** A1
De Witt MI, *4702* **76** A3
De Witt NE, *572* **35** F4
De Witt NY, *8200* **79** E3
De Witt Co. IL, *16798* **88** B4
DeWitt Co. TX, *20013* **61** E3
Dexter GA, *509* **129** E2
Dexter IA, *689* **86** C2
Dexter ME, *2201* **82** C1
Dexter MI, *2338* **76** B4
Dexter MO, *7356* **108** B2
Dexter NM, *1235* **57** E2
Dexter NY, *1120* **79** E1
Dexter OR, *750* **20** C4
Diablo CA, *988* **259** D2
Diamond IL, *1393* **88** C2
Diamond MO, *807* **106** C2
Diamond Bar CA, *56287* **229** F3
Diamond City AR, *730* **107** D3
Diamondhead MS, *5912* **134** C2
Diamond Hill RI, *1100* **150** C2
Diamond Sprs. CA, *4888* **36** C2
Diamondville WY, *716* **31** F2
Diaz AR, *1284* **107** F4
D'Iberville MS, *7608* **135** D2
Diboll TX, *5470* **132** B1
Dickens TX, *332* **58** B1
Dickens Co. TX, *2762* **58** B1
Dickenson Co. VA, *16395* **111** E2
Dickey Co. ND, *5757* **27** D1
Dickeyville WI, *1043* **73** F4
Dickinson ND, *16010* **18** A4
Dickinson TX, *17093* **132** B4
Dickinson Co. IA, *16424* **72** B3
Dickinson Co. KS, *19344* **43** E2
Dickinson Co. MI, *27472* **69** D2
Dickson OK, *1139* **51** F4
Dickson TN, *12244* **109** E4
Dickson City PA, *6205* **93** F2
Dickson Co. TN, *43156* **109** E4
Dierks AR, *1230* **116** C3
Dieterich IL, *591* **98** C2
Dighton KS, *1261* **42** C3
Dighton MA, *1200* **151** D3
Dike IA, *944* **73** D4
Dilkon AZ, *1265* **47** F3
Dill City OK, *526* **51** D3
Dilley TX, *3674* **60** C4
Dillingham AK, *2466* **154** B3
Dillon CO, *802* **41** D1
Dillon MT, *3752* **23** E2
Dillon SC, *6316* **122** C3
Dillon Co. SC, *30722* **122** C3
Dillonvale OH, *781* **91** F4
Dillsboro IN, *1436* **100** A2
Dillsburg PA, *2063* **93** D4
Dillwyn VA, *447* **113** D1

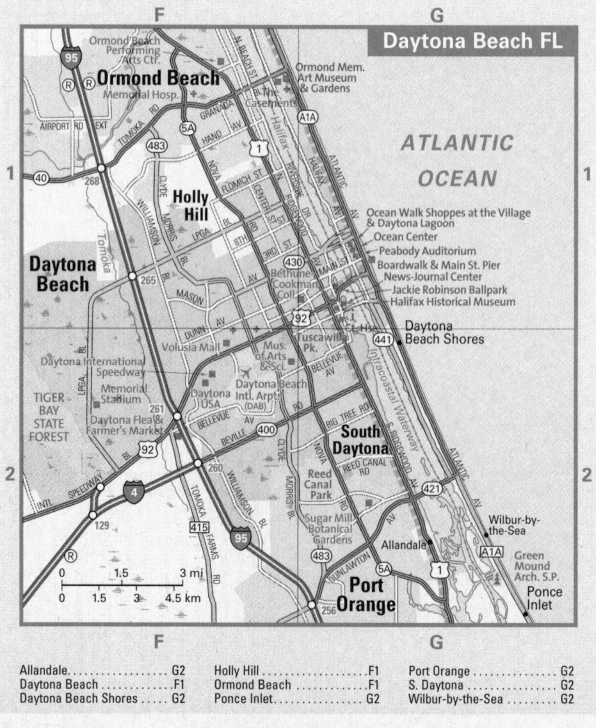

Allandale G2	Holly Hill F1	Port Orange G2
Daytona Beach F1	Ormond Beach F1	S. Daytona G2
Daytona Beach Shores G2	Ponce Inlet G2	Wilbur-by-the-Sea G2

Dilworth MN, *3001* **19** F4
Dimmit Co. TX, *10248* **60** C4
Dimmitt TX, *4375* **50** A4
Dimondale MI, *1342* **76** A4
Dinosaur CO, *319* **32** B4
Dinuba CA, *16844* **45** D3
Dinwiddie VA, *350* **113** E2
Dinwiddie Co. VA, *24533* **113** E2
District Hts. MD, *5958* **271** F4
Divernon IL, *1201* **98** B1
Divide Co. ND, *2283* **17** F1
Dixfield ME, *1137* **82** B2
Dixie Co. FL, *13827* **137** F3
Dixmoor IL, *3934* **203** E6
Dixon CA, *16103* **36** B3
Dixon IL, *15941* **88** B1
Dixon KY, *632* **109** E1
Dixon MO, *1570* **97** E4
Dixon MT, *216* **15** D3
Dixon Co. NE, *6339* **35** F1
D'Lo MS, *394* **126** B3
Dobbins Hts. NC, *936* **122** C2
Dobbs Ferry NY, *10622* **148** B3
Dobson NC, *1457* **112** A3
Dock Jct. GA, *6951* **139** D1
Doctor Phillips FL, *9548* **246** B3
Doctors Inlet FL, *1400* **139** D2
Doddridge Co. WV, *7403* **102** A2
Dodge NE, *700* **35** F2
Dodge Ctr. MN, *2226* **73** D1
Dodge City AL, *612* **119** F3
Dodge City KS, *25176* **42** C4
Dodge Co. GA, *19171* **129** E3
Dodge Co. MN, *17731* **73** D1
Dodge Co. NE, *36160* **35** F3
Dodge Co. WI, *85897* **74** C2
Dodgeville WI, *4220* **74** A3
Doerun GA, *828* **129** D4
Doland SD, *297* **27** E2
Dolan Sprs. AZ, *1867* **46** B3
Dolgeville NY, *2166* **79** F3
Dollar Bay MI, *950* **65** F3
Dolores CO, *857* **40** B4
Dolores Co. CO, *1844* **40** B4
Dolton IL, *25614* **203** E6
Doña Ana NM, *1379* **56** B3
Dona Ana Co. NM, *174682* **56** B3
Donald OR, *608* **20** B2
Donaldsonville LA, *7605* **134** A2
Donalsonville GA, *2796* **137** D1
Dongola IL, *806* **108** C1
Doniphan MO, *1932* **108** A3
Doniphan NE, *763* **35** E4
Doniphan Co. KS, *8249* **96** B1
Donley Co. TX, *3828* **50** B3
Donna TX, *14768* **63** E4
Donnellson IA, *963* **87** F3
Donora PA, *5653* **92** A4
Doolittle MO, *644* **97** E4
Dooly Co. GA, *11525* **129** D3
Doon IA, *533* **27** F4
Door Co. WI, *27961* **69** D4
Dora AL, *2413* **119** F4
Doraville GA, *9862* **120** C4
Dorchester NE, *615* **35** F4
Dorchester WI, *827* **68** A4
Dorchester Co. MD, *30674* **103** F4
Dorchester Co. SC, *96413* **130** C1
Dormont PA, *9305* **250** B3
Dorr MI, *2800* **75** F4
Dorris CA, *886* **28** C2
Dorset VT, *400* **81** D4
Dorsey MD, *1000* **193** B4
Dortches NC, *809* **113** D4
Dos Palos CA, *4581* **44** C2
Dothan AL, *57737* **128** B4
Double Sprs. AL, *1003* **119** E3
Doubs MD, *750* **144** A2
Dougherty Co. GA, *96065* **128** C4
Douglas AL, *530* **120** A3
Douglas AZ, *14312* **55** E4
Douglas GA, *10639* **129** E4
Douglas MA, *800* **150** C2
Douglas MI, *1214* **75** E4
Douglas WY, *5288* **33** E1
Douglas Co. CO, *175766* **41** E2
Douglas Co. GA, *92174* **120** C4
Douglas Co. IL, *19922* **99** D1
Douglas Co. KS, *99962* **96** A3
Douglas Co. MN, *32821* **66** B2
Douglas Co. MO, *13084* **107** E2
Douglas Co. NE, *463585* **35** F3
Douglas Co. NV, *41259* **37** D2
Douglas Co. OR, *100399* **20** B4
Douglas Co. SD, *3458* **27** E4
Douglas Co. WA, *32603* **13** E3
Douglas Co. WI, *43287* **64** C4
Douglass KS, *1813* **43** F4
Douglass Hills KY, *5718* **230** F2
Douglassville PA, *1300* **146** B2
Douglasville GA, *20065* **120** C4
Dousman WI, *1584* **74** C3
Dove Creek CO, *698* **40** A4
Dover AR, *1329* **117** D1
Dover DE, *32135* 145 E2
Dover FL, *2798* **140** C2
Dover ID, *342* **14** B2
Dover MA, *2216* **151** D1
Dover NH, *26884* **82** A4
Dover NJ, *18188* **148** A3
Dover OH, *12210* **91** E4
Dover PA, *1815* **103** E1
Dover TN, *1442* **109** D3
Dover-Foxcroft ME, *2592* **82** C1
Dover Plains NY, *1996* **94** B3
Dowagiac MI, *6147* **89** F1
Dow City IA, *503* **86** A1
Dowling Park FL, *650* **137** F2
Downers Grove IL, *48724* **89** D1
Downey CA, *107323* **52** C2
Downey ID, *613* **31** E2
Downieville CA, *80* **36** C1
Downingtown PA, *7589* **146** B3
Downs IL, *776* **88** C4
Downs KS, *1038* **43** D1
Dows IA, *675* **72** C4
Doyle TN, *525* **110** A4
Doylestown OH, *2799* **91** E3
Doylestown PA, *8227* **146** C2
Doyline LA, *841* **125** D2
Drain OR, *1021* **20** B4
Drake ND, *322* **18** C2
Drakesboro KY, *627* **109** E2
Drakes Branch VA, *504* **113** D2
Draper UT, *25220* **31** E4
Dravosburg PA, *2015* **250** C3
Drayton ND, *913* **19** E2
Dresden OH, *1423* **91** D4
Dresden TN, *2855* **108** C3
Dresser WI, *732* **67** D3
Drew MS, *2434* **118** A4
Drew Co. AR, *18723* **117** F4
Drexel MO, *1090* **96** B3
Drexel NC, *1938* **111** F4
Drexel OH, *2057* **208** D1
Drexel Hill PA, *29364* **146** C3
Driggs ID, *1100* **23** F4
Dripping Sprs. TX, *1548* **61** D2
Driscoll TX, *825* **63** F2
Druid Hills GA, *12741* **190** D3
Drummond MT, *318* **15** E4
Drumright OK, *2905* **51** F2
Dryden ME, *1100* **82** B2
Dryden MI, *815* **76** C3
Dryden NY, *1832* **79** D4
Dryden VA, *1253* **111** D2
Dry Mills ME, *700* **82** B3
Dry Prong LA, *421* **125** E4
Dry Ridge KY, *1995* **100** B3
Duarte CA, *21486* **228** E2
Dubach LA, *800* **125** E2
Dublin CA, *29973* **259** E3
Dublin GA, *15857* **129** E2
Dublin IN, *697* **100** A1
Dublin MD, *650* **145** D1
Dublin OH, *31392* **90** C4
Dublin PA, *2083* **146** C2
Dublin TX, *3754* **59** D3
Dublin VA, *2288* **112** A2
Dubois ID, *647* **23** E3
DuBois PA, *8123* **92** B3
Dubois WY, *962* **24** B4
Dubois Co. IN, *39674* **99** E3
Duboistown PA, *1280* **93** D2
Dubuque IA, *57686* **73** F4
Dubuque Co. IA, *89143* **73** F4
Duchesne UT, *1408* **32** A4
Duchesne Co. UT, *14371* **39** F1
Duck Hill MS, *746* **118** B4

Entries in **bold bla**
Entries in **bold col**

Entries in **bold black** indicate counties or parishes.
Entries in **bold color** indicate cities with detailed inset maps.

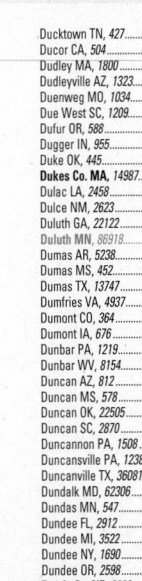

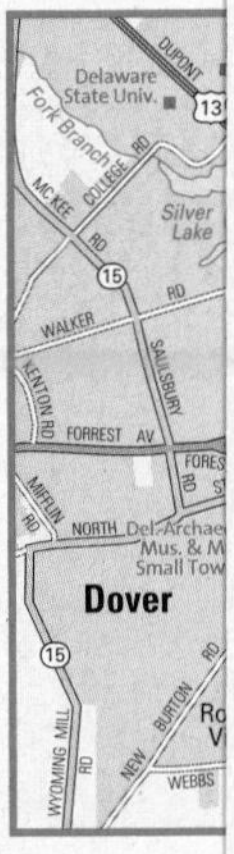

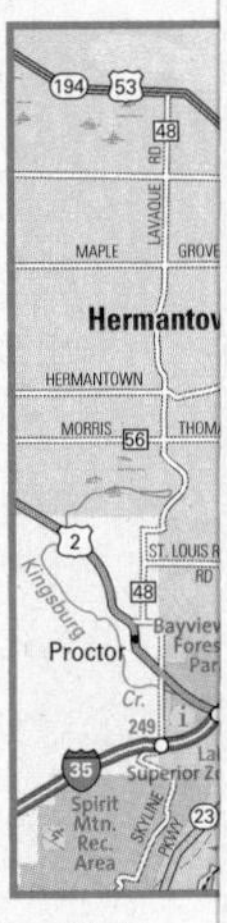

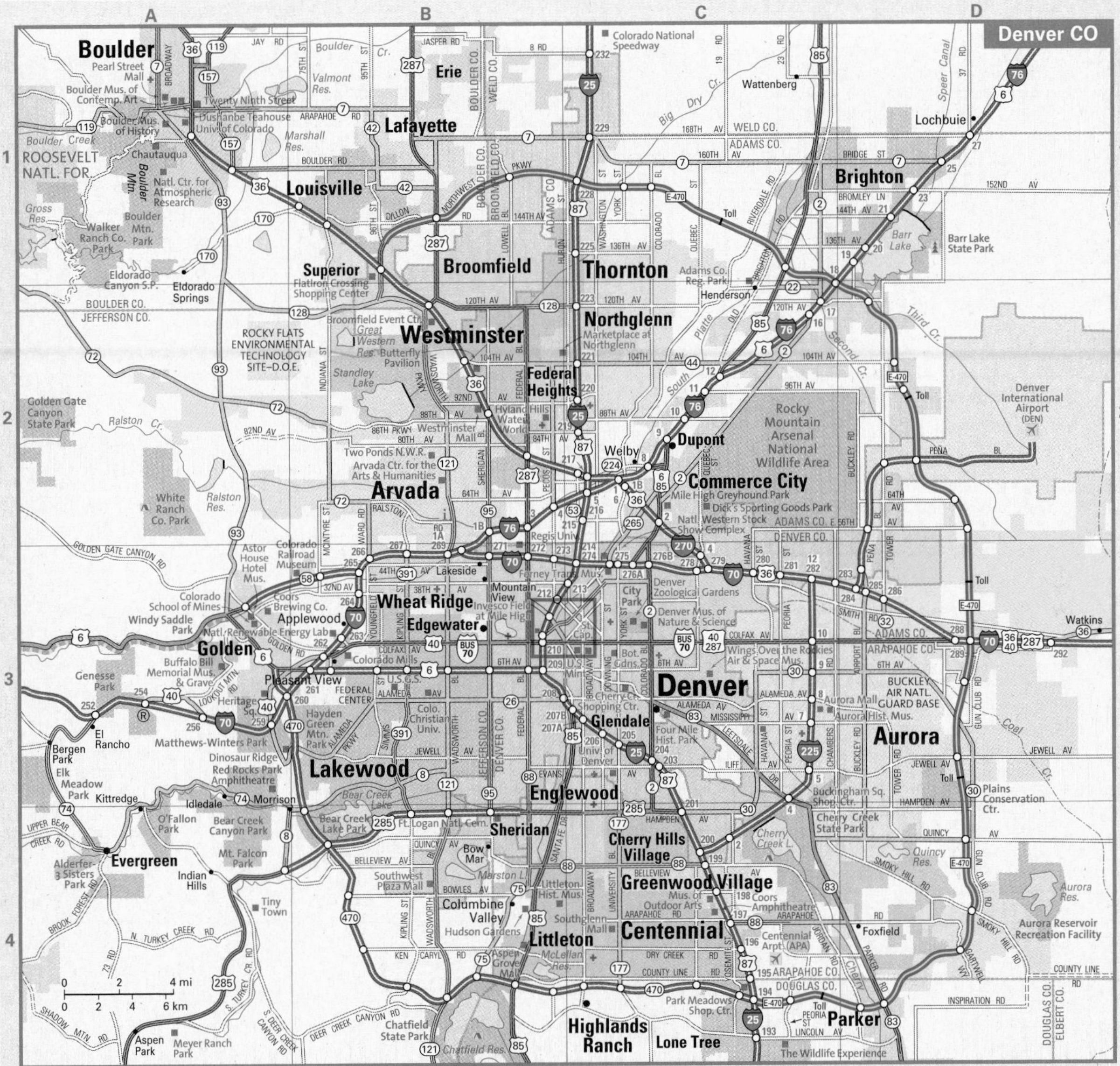

Applewood	B3
Arvada	B2
Aspen Park	A4
Aurora	D3
Bergen Park	A3
Boulder	A1
Bow Mar	B4
Brighton	D1
Broomfield	B1
Centennial	C4
Cherry Hills Vil.	C4
Columbine Valley	B4
Commerce City	C2
Denver	C3
Dupont	C2
Edgewater	B3
Eldorado Sprs.	A1
El Rancho	A3
Englewood	C3
Erie	B1
Evergreen	A4
Federal Hts.	B2
Foxfield	D4
Glendale	C3
Golden	A3
Greenwood Vil.	C4
Henderson	C2
Highlands Ranch	C4
Idledale	A3
Indian Hills	A4
Kittredge	A4
Lafayette	B1
Lakeside	B3
Lakewood	B3
Littleton	B4
Lochbuie	D1
Lone Tree	C4
Louisville	B1
Morrison	A4
Mtn. View	B3
Northglenn	C2
Parker	D4
Pleasant View	B3
Sheridan	B4
Superior	B1
Thornton	C1
Watkins	D3
Wattenberg	C1
Welby	C2
Westminster	B2
Wheat Ridge	B3

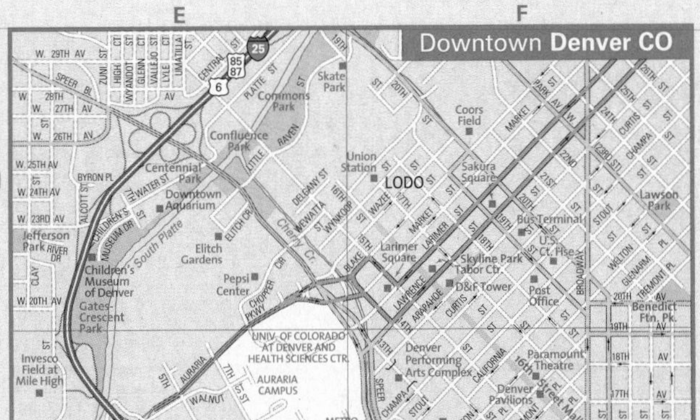

POINTS OF INTEREST

16th Street Mall	E1
Auraria Campus	E2
Bus Terminal	F1
Byers-Evans House	F2
Children's Museum of Denver	E1
Colorado Convention Center	F2
Colorado History Museum	F2
Coors Field	F1
D&F Tower	F1
Denver Art Museum	F2
Denver Pavilions	F2
Denver Performing Arts Complex	F2
Downtown Aquarium	E1
Elitch Gardens	E1
Firefighters Museum	F2
Invesco Field at Mile High	E2
Larimer Square	F1
LoDo	F1
Metropolitan State Coll. of Denver	E2
Paramount Theatre	F2
Pepsi Center	E1
Post Office	F1
Public Library	F2
Sakura Square	F1
Skate Park	E1
State Capitol	F2
Tabor Center	F1
Union Station	F1
U.S. Court House	F1
U.S. Mint	F2
Univ. of Colorado at Denver and Health Sciences Center	E2

Figures after entries indicate population, page number, and grid reference.

Elizabethton TN, 13372 ..... 111 E3
Elizabethtown IL, 348 ..... 109 D1
Elizabethtown KY, 22542 ..... 110 A1
Elizabethtown NY, 750 ..... 81 D2
Elizabethtown NC, 3698 ..... 123 D2
Elizabethtown PA, 11887 ..... 93 E4
Elizabethville PA, 1344 ..... 93 D4
El Jebel CO, 4488 ..... 40 C2
Elkader IA, 1465 ..... 73 F3
Elk City OK, 10510 ..... 50 C3
Elk Co. KS, 3261 ..... 43 F4
Elk Co. PA, 35112 ..... 92 B2
Elk Grove CA, 75175 ..... 36 C3
Elk Grove Vil. IL, 34727 ..... 203 C3
Elkhart IN, 51874 ..... 89 F2
Elkhart KS, 2233 ..... 50 A1
Elkhart TX, 1215 ..... 124 A4
Elkhart Co. IN, 182791 ..... 89 F2
Elkhart Lake WI, 1021 ..... 74 C2
Elkhorn CA, 1591 ..... 236 E2
Elk Horn IA, 649 ..... 86 B2
Elkhorn NE, 6062 ..... 35 F3
Elkhorn WI, 7305 ..... 74 C4
Elkhorn City KY, 1060 ..... 111 E1
Elkin NC, 4109 ..... 112 A3
Elkins AR, 1251 ..... 106 C4
Elkins WV, 7032 ..... 102 A3
Elkland PA, 1786 ..... 93 D1
Elkmont AL, 470 ..... 119 F2
Elk Mound WI, 785 ..... 67 E4
Elko NV, 16708 ..... 30 B4
Elko Co. NV, 45291 ..... 30 C3
Elk Pt. SD, 1714 ..... 35 F1
Elk Rapids MI, 1700 ..... 69 F4
Elkridge MD, 22042 ..... 144 C2
Elk Ridge UT, 1838 ..... 39 E1
Elk River MN, 16447 ..... 66 C3
Elk Run Hts. IA, 1052 ..... 73 E4
Elkton KY, 1984 ..... 109 E2
Elkton MD, 11893 ..... 145 E1
Elkton MI, 863 ..... 76 C2
Elkton SD, 677 ..... 27 F3
Elkton TN, 510 ..... 119 F2
Elkton VA, 2042 ..... 102 C3
Elkview WV, 1182 ..... 101 F3
Elkville IL, 1001 ..... 98 B4
El Lago TX, 3075 ..... 132 B3
Ellaville GA, 1609 ..... 129 D3
Ellenboro NC, 479 ..... 121 F1
Ellenboro WV, 373 ..... 101 F2
Ellendale DE, 327 ..... 145 F3
Ellendale MN, 590 ..... 72 C2
Ellendale ND, 1559 ..... 27 D1
Ellensburg WA, 15414 ..... 13 D4
Ellenton FL, 3142 ..... 266 B4
Ellenville NY, 4130 ..... 94 A3
Ellerbe NC, 1021 ..... 122 C2
Ellerslie MD, 600 ..... 102 C1
Ellettsville IN, 5078 ..... 99 F2
Ellicott NY, 2200 ..... 78 B4
Ellicott City MD, 56397 ..... 144 C2
Ellijay GA, 1584 ..... 120 C2
Ellington CT, 1300 ..... 150 A3
Ellington MO, 1045 ..... 108 A2
Ellinwood KS, 2164 ..... 43 D3
Elliott Co. KY, 6748 ..... 101 D4
Ellis KS, 1873 ..... 42 C2
Ellis Co. KS, 27507 ..... 43 D2
Ellis Co. OK, 4075 ..... 50 C2
Ellis Co. TX, 111360 ..... 59 F3
Ellisport WA, 1200 ..... 262 A4
Elliston MT, 225 ..... 15 E4
Elliston VA, 1241 ..... 112 B2
Ellisville MS, 3465 ..... 126 C4
Ellisville MO, 9104 ..... 256 A2
Elloree SC, 742 ..... 122 B4
Ellsworth IA, 531 ..... 72 C4
Ellsworth KS, 2965 ..... 43 E2
Ellsworth ME, 6456 ..... 83 D2
Ellsworth MI, 483 ..... 69 F3
Ellsworth MN, 540 ..... 27 F4
Ellsworth PA, 1083 ..... 92 A4
Ellsworth WI, 2909 ..... 67 E4
Ellsworth Co. KS, 6525 ..... 43 E3
Ellwood City PA, 8688 ..... 91 F3
Elma IA, 598 ..... 73 D3
Elma NY, 2491 ..... 78 B3
Elma WA, 3049 ..... 12 B4
Elm City NC, 1165 ..... 113 D4
Elm Creek NE, 894 ..... 35 D4
Elmendorf TX, 664 ..... 61 D3
Elmer NJ, 1384 ..... 145 F1
Elm Grove WI, 6249 ..... 234 B2
Elmhurst IL, 42762 ..... 89 D1
Elmira NY, 30940 ..... 93 D1
El Mirage AZ, 7609 ..... 249 A1
Elmira Hts. NY, 4170 ..... 93 D1
Elm Mott TX, 1200 ..... 59 E4
Elmo UT, 368 ..... 39 F2
Elmont NY, 32657 ..... 148 C4
Elmont VA, 500 ..... 113 E1
El Monte CA, 115965 ..... 228 E2
Elmore MN, 735 ..... 72 C2
Elmore OH, 1426 ..... 90 C2
Elmore City OK, 756 ..... 51 E4
Elmore Co. AL, 65874 ..... 128 A2
Elmore Co. ID, 29130 ..... 22 B4
Elm Sprs. AR, 1044 ..... 106 C3
Elmville CT, 1300 ..... 150 B3
Elmwood IL, 1945 ..... 88 A3
Elmwood NE, 668 ..... 35 F4
Elmwood WI, 841 ..... 67 E4
Elmwood Park IL, 25405 ..... 203 D4
Elmwood Park NJ, 18925 ..... 240 C1
Elmwood Place OH, 2681 ..... 204 B2
Elnora IN, 721 ..... 99 E3
Elnora NY, 2700 ..... 94 B1
Elon NC, 6738 ..... 112 C4
Eloy AZ, 10375 ..... 54 C2
El Paso IL, 2695 ..... 88 B3
El Paso TX, 563662 ..... 56 C4
El Paso Co. CO, 516929 ..... 41 E2
El Paso Co. TX, 679622 ..... 56 C4
El Portal FL, 2505 ..... 233 B4
El Prado NM, 400 ..... 49 D1
El Reno OK, 16212 ..... 51 E3
El Rio CA, 6193 ..... 52 B2
El Rito NM, 425 ..... 48 C2
Elroy WI, 1578 ..... 74 A2
Elsa TX, 5549 ..... 63 E4
Elsah IL, 635 ..... 98 A2
Elsberry MO, 2047 ..... 98 A2
El Segundo CA, 16033 ..... 228 C3
Elsie MI, 1055 ..... 76 A3
Elsinore UT, 733 ..... 39 E3
Elsmere DE, 5800 ..... 146 B4
Elsmere KY, 8139 ..... 100 B2
Elsmere NY, 3200 ..... 188 D3
El Sobrante CA, 12260 ..... 259 C1
Elton LA, 1261 ..... 133 E2
Elvaton MD, 3500 ..... 193 C5
Elverson PA, 959 ..... 146 B2
Elwood IL, 1620 ..... 89 D2
Elwood IN, 9737 ..... 89 F4
Elwood KS, 1145 ..... 96 B1
Elwood NE, 761 ..... 34 C4
Elwood NJ, 1392 ..... 147 D4
Elwood UT, 678 ..... 31 E3
Ely IA, 1149 ..... 87 E1
Ely MN, 3724 ..... 64 C2
Ely NV, 4041 ..... 38 B2
Elyria OH, 55953 ..... 91 D2
Elysburg PA, 2067 ..... 93 E3
Elysian MN, 486 ..... 72 C1
Emanuel Co. GA, 21837 ..... 129 F2
Emerado ND, 510 ..... 19 E2
Emerald Isle NC, 3488 ..... 115 D4
Emerson GA, 1092 ..... 120 C3
Emerson NE, 817 ..... 35 F2
Emerson NJ, 7197 ..... 148 B3
Emery SD, 439 ..... 27 E4
Emery UT, 308 ..... 39 E2
Emery Co. UT, 10860 ..... 39 F2
Emery Mills ME, 350 ..... 82 A4
Emeryville CA, 6882 ..... 259 C2
Emigsville PA, 2467 ..... 103 E1
Emily MN, 847 ..... 64 B4
Eminence KY, 2231 ..... 100 A4
Eminence MO, 548 ..... 107 F2
Emlenton PA, 784 ..... 92 A2
Emmaus PA, 11313 ..... 146 B1
Emmet AR, 506 ..... 117 D4
Emmet Co. IA, 11027 ..... 72 B3
Emmet Co. MI, 31437 ..... 70 B3
Emmetsburg IA, 3958 ..... 72 B3
Emmett ID, 5490 ..... 22 B4
Emmitsburg MD, 2290 ..... 103 D1
Emmonak AK, 767 ..... 154 B2
Emmons Co. ND, 4331 ..... 18 C4
Emmorton MD, 4000 ..... 145 D1
Emory TX, 1021 ..... 124 A1
Emory VA, 2266 ..... 111 F2
Empire CO, 355 ..... 41 D1
Empire LA, 2211 ..... 134 C4
Empire NV, 499 ..... 29 E4
Empire City OK, 734 ..... 51 E4
Emporia KS, 26760 ..... 43 F3
Emporia VA, 5665 ..... 113 E3
Emporium PA, 2526 ..... 92 C2
Emsworth PA, 2598 ..... 250 A1
Encampment WY, 443 ..... 33 D3
Encinal TX, 629 ..... 60 C4
Encinitas CA, 58014 ..... 53 D4
Enderlin ND, 947 ..... 19 E4
Endicott NY, 13038 ..... 93 E1
Endicott WA, 621 ..... 13 F4
Endwell NY, 11706 ..... 93 E1
Energy IL, 1175 ..... 108 C1
Enfield CT, 8125 ..... 150 A2
Enfield IL, 625 ..... 99 D4
Enfield NH, 1698 ..... 81 E3
Enfield NC, 2347 ..... 113 E4
Enfield Ctr. NH, 600 ..... 81 E3
England AR, 2972 ..... 117 E2
Englewood CO, 31727 ..... 41 E1
Englewood FL, 16196 ..... 140 C4
Englewood NJ, 26203 ..... 148 B3
Englewood OH, 12235 ..... 100 B1
Englewood TN, 1590 ..... 120 C1
Englewood Beach FL, 1000 ..... 140 C4
Englewood Cliffs NJ, 5322 ..... 240 D1
English IN, 673 ..... 99 F4
Englishtown NJ, 1764 ..... 147 E2
Enhaut PA, 2809 ..... 218 C2
Enid OK, 47045 ..... 51 E1
Enigma GA, 869 ..... 129 E4
Enka NC, 1500 ..... 121 E1
Ennis MT, 840 ..... 23 E2
Ennis TX, 16045 ..... 59 F3
Enoch UT, 3467 ..... 39 D4
Enochville NC, 2851 ..... 122 B1
Enola PA, 5627 ..... 218 A1
Enon OH, 2638 ..... 100 C1
Enoree SC, 700 ..... 121 F2
Enosburg Falls VT, 1473 ..... 81 D1
Ensley FL, 18752 ..... 135 F2
Ensor KY, 500 ..... 109 E1
Enterprise AL, 21178 ..... 128 B4
Enterprise KS, 836 ..... 43 F2
Enterprise MS, 474 ..... 127 D3
Enterprise OR, 1895 ..... 22 A2
Enterprise UT, 1285 ..... 38 C4
Enterprise WV, 939 ..... 102 A2
Entiat WA, 957 ..... 13 D3
Enumclaw WA, 11116 ..... 12 C3
Ephraim UT, 4505 ..... 39 E2
Ephrata PA, 13213 ..... 146 A2
Ephrata WA, 6808 ..... 13 E3
Epping NH, 1673 ..... 81 F4
Epps LA, 1153 ..... 125 F2
Epworth IA, 1428 ..... 73 F4
Epworth Hts. OH, 3300 ..... 204 C1
Equality IL, 721 ..... 109 D1
Erath LA, 2187 ..... 133 F3
Erath Co. TX, 33001 ..... 59 D3
Erda UT, 2473 ..... 31 E4
Erial NJ, 6200 ..... 146 C4
Erick OK, 1023 ..... 50 C3
Erie CO, 6291 ..... 209 B1
Erie IL, 1589 ..... 88 A2
Erie KS, 1211 ..... 106 A1
Erie PA, 103717 ..... 92 A1
Erie Co. NY, 950265 ..... 78 B4
Erie Co. OH, 79551 ..... 91 D2
Erie Co. PA, 280843 ..... 92 A1
Erin TN, 1490 ..... 109 E3
Erlanger KY, 16676 ..... 100 B2
Erwin NC, 4537 ..... 123 D1
Erwin TN, 5610 ..... 111 E4
Erwinville LA, 700 ..... 134 A2
Escalante UT, 818 ..... 39 E4
Escalon CA, 5963 ..... 36 C4
Escambia Co. AL, 38440 ..... 136 A1
Escambia Co. FL, 294410 ..... 135 F1
Escanaba MI, 13140 ..... 69 D2
Escatawpa MS, 3566 ..... 195 C1
Escobares TX, 1954 ..... 63 D4
Escondido CA, 133559 ..... 53 D4
Esko MN, 1300 ..... 64 C4
Eskridge KS, 589 ..... 43 F2
Esmeralda Co. NV, 971 ..... 37 F4
Espanola NM, 9688 ..... 49 D2
Espanong NJ, 2700 ..... 148 A3
Esparto CA, 1858 ..... 36 B2
Espy PA, 1428 ..... 93 E3
Essex CT, 2573 ..... 149 E2
Essex IA, 884 ..... 86 A3
Essex MD, 39078 ..... 144 C2
Essex MA, 1426 ..... 151 F1
Essex MO, 524 ..... 108 B2
Essex Co. MA, 723419 ..... 151 F1
Essex Co. NJ, 793633 ..... 148 A3
Essex Co. NY, 38851 ..... 80 C3
Essex Co. VT, 6459 ..... 81 F1
Essex Co. VA, 9989 ..... 103 E4
Essex Fells NJ, 2162 ..... 240 A2
Essex Jct. VT, 8591 ..... 81 D2
Essexville MI, 3766 ..... 76 B2
Estacada OR, 2371 ..... 20 C2
Estancia NM, 1584 ..... 49 D4
Estelle LA, 15880 ..... 239 C2
Estelline SD, 675 ..... 27 F3
Estell Manor NJ, 1585 ..... 104 C3
Ester AK, 1680 ..... 154 C2
Estero FL, 9503 ..... 142 C1
Estes Park CO, 5413 ..... 33 E4
Estherville IA, 6656 ..... 72 B2
Estherwood LA, 807 ..... 133 E2
Estill SC, 2425 ..... 130 B2
Estill Co. KY, 15307 ..... 110 C1
Estill Sprs. TN, 2152 ..... 120 A1
Estral Beach MI, 486 ..... 90 C1
Ethan SD, 330 ..... 27 E4
Ethel MS, 452 ..... 126 C1
Ethete WY, 1455 ..... 32 B1
Ethridge TN, 536 ..... 119 E1
Etna CA, 781 ..... 28 B3
Etna PA, 3924 ..... 250 C1
Etna Green IN, 663 ..... 89 F2
Etowah NC, 2766 ..... 121 E1
Etowah TN, 3663 ..... 120 C1
Etowah Co. AL, 103459 ..... 120 A3
Ettrick VA, 5627 ..... 113 E2
Ettrick WI, 521 ..... 73 F1
Eubank KY, 358 ..... 110 B2
Euclid OH, 52717 ..... 91 E2
Eudora AR, 2819 ..... 126 A1
Eudora KS, 4307 ..... 96 B3
Eufaula AL, 13908 ..... 128 B3
Eufaula OK, 2639 ..... 116 A1
Eugene OR, 137893 ..... 20 B4
Euharlee GA, 3208 ..... 120 B3
Euless TX, 46005 ..... 207 C2
Eunice LA, 11499 ..... 133 E2
Eunice NM, 2562 ..... 57 F3
Eupora MS, 2326 ..... 118 C4
Eureka CA, 26128 ..... 28 A4
Eureka IL, 4871 ..... 88 B3
Eureka KS, 2914 ..... 43 F4
Eureka MO, 7676 ..... 98 A3
Eureka MT, 1017 ..... 14 C1
Eureka NV, 550 ..... 38 A1
Eureka SD, 1101 ..... 27 D1
Eureka UT, 766 ..... 39 E1
Eureka Co. NV, 1651 ..... 30 B4
Eureka Mill SC, 1737 ..... 122 A2
Eureka Sprs. AR, 2278 ..... 106 C3
Eustace TX, 798 ..... 59 F3
Eustis FL, 15106 ..... 140 C1
Eustis NE, 464 ..... 34 C4
Eutaw AL, 1878 ..... 127 E2
Eva AL, 491 ..... 119 F3
Evadale TX, 1430 ..... 132 C2
Evangeline Par. LA, 35434 ..... 133 E1
Evans CO, 9514 ..... 33 E4
Evans GA, 17727 ..... 121 F4
Evans WV, 750 ..... 101 E3
Evans City PA, 2009 ..... 92 A3
Evans Co. GA, 10495 ..... 130 B3
Evansdale IA, 4526 ..... 73 E4
Evans Mills NY, 605 ..... 79 E1
Evanston IL, 74239 ..... 89 D1
Evanston WY, 11507 ..... 31 F3
Evansville IL, 724 ..... 98 B4
Evansville IN, 121582 ..... 99 D4
Evansville MN, 566 ..... 66 A2
Evansville WI, 4039 ..... 74 B4
Evansville WY, 2255 ..... 33 D1
Evaro MT, 329 ..... 15 D4
Evart MI, 1738 ..... 75 F1
Evarts KY, 1101 ..... 111 D2
Eveleth MN, 3865 ..... 64 C3
Evendale OH, 3090 ..... 204 B1
Evening Shade AR, 465 ..... 107 F4
Everett MA, 38037 ..... 197 C1
Everett PA, 1905 ..... 102 C1
Everett WA, 91488 ..... 12 C2
Everglades City FL, 479 ..... 143 D2
Evergreen AL, 3630 ..... 127 F4
Evergreen CO, 9216 ..... 41 D1
Evergreen MT, 6215 ..... 15 D2
Evergreen Park IL, 20821 ..... 203 D5
Everly IA, 647 ..... 72 A3
Everman TX, 5836 ..... 207 B3
Everson PA, 842 ..... 92 A4
Everson WA, 2035 ..... 12 C1
Evesboro NJ, 2400 ..... 147 D3
Ewa Beach HI, 14650 ..... 152 A3

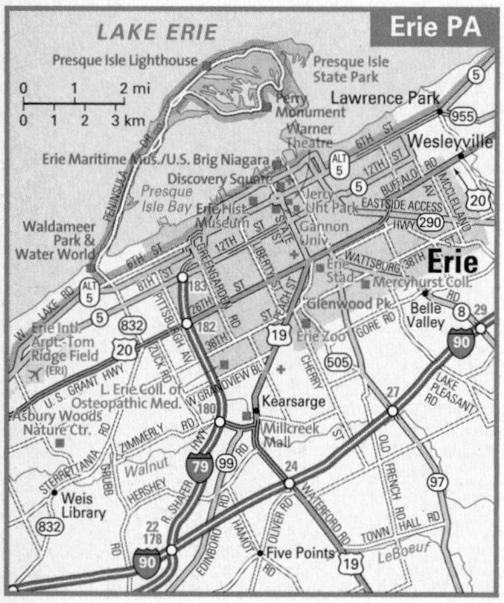

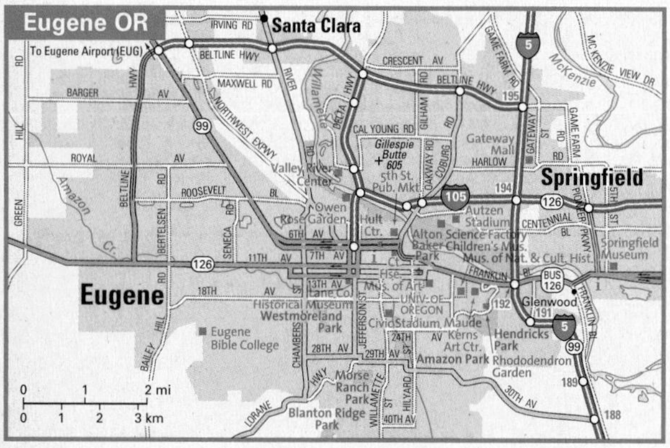

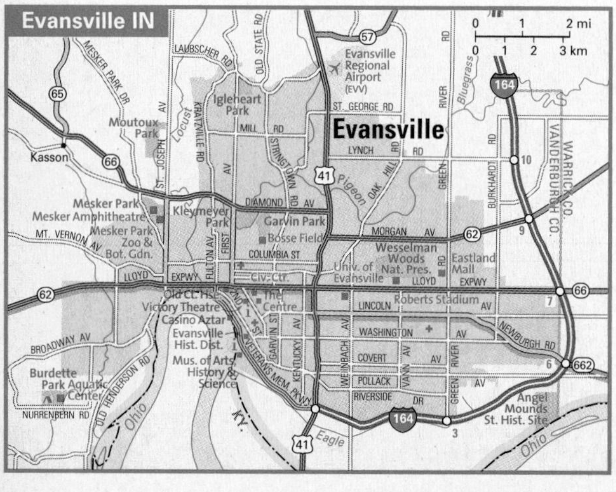

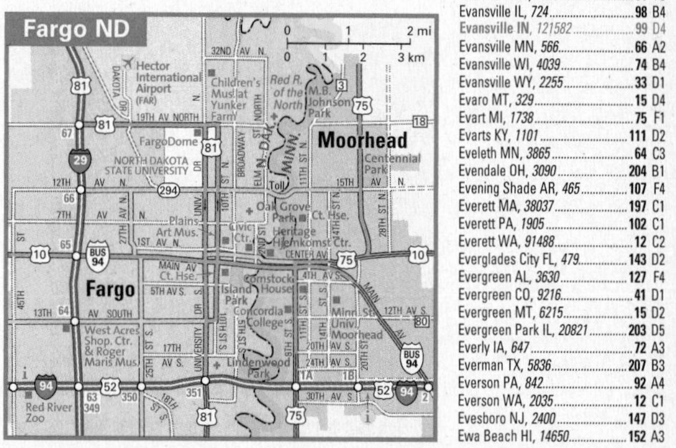

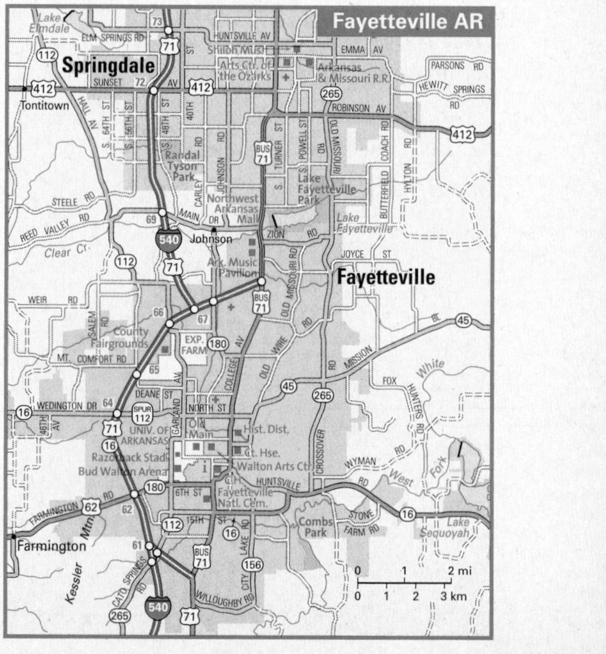

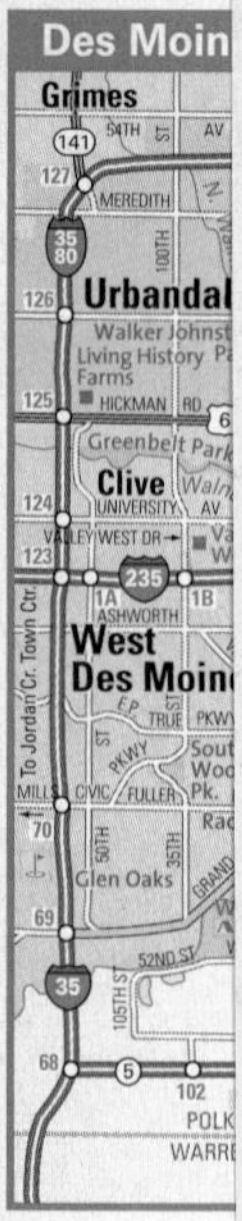

Entries in **bold black** indicate counties or parishes.
Entries in **bold color** indicate cities with detailed inset maps.

Entries in **bold black** indicate counties or parishes.
Entries in **bold color** indicate cities with detailed inset maps.

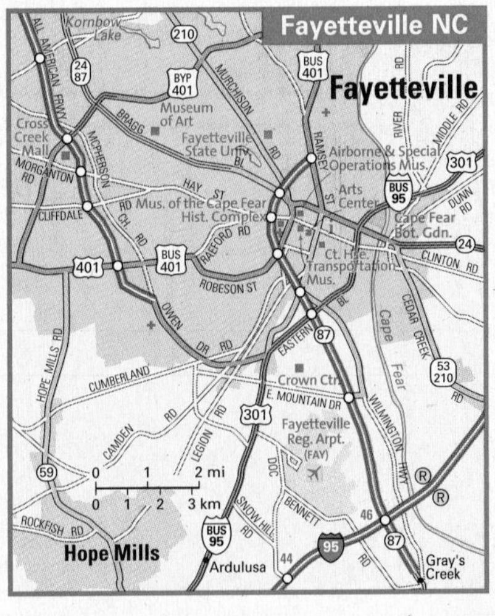

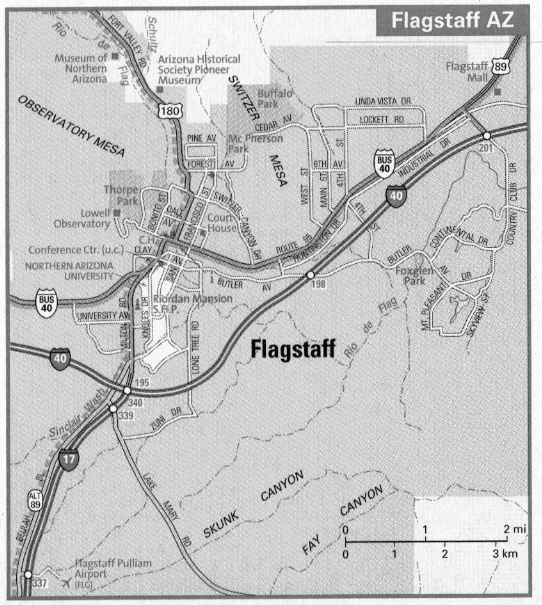

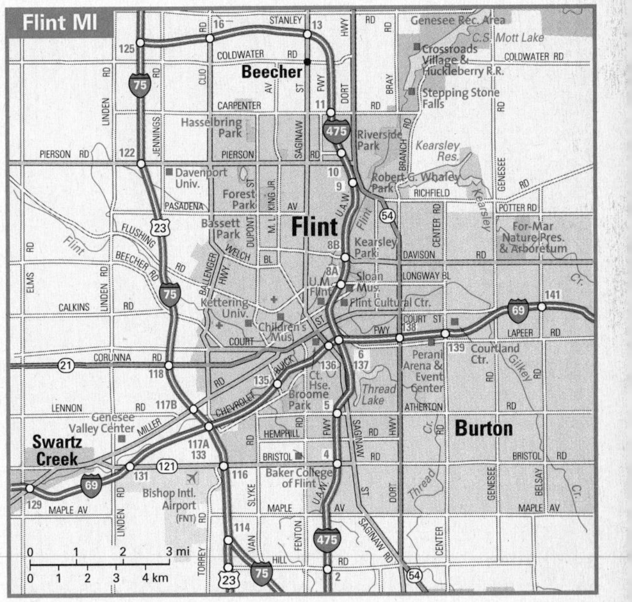

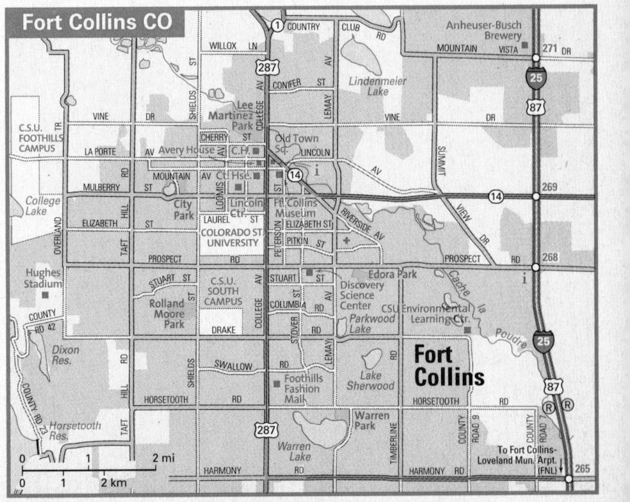

Folcroft PA, *6978* ... **248** B4
Foley AL, *7590* ... **135** E2
Foley MN, *2154* ... **66** C2
Folkston GA, *2178* ... **139** D1
Follansbee WV, *3115* ... **91** F4
Folly Beach SC, *2116* ... **131** D2
Folsom CA, *51884* ... **36** C3
Folsom LA, *525* ... **134** B2
Folsom NJ, *1972* ... **147** D4
Folsom PA, *8072* ... **248** A4
Fonda IA, *648* ... **72** B4
Fonda NY, *810* ... **79** F3
Fond du Lac WI, *42203* ... **74** C2
Fond du Lac Co. WI, *97296* ... **74** C2
Fontana CA, *128929* ... **229** H2
Fontana WI, *1754* ... **74** C4
Fontanelle IA, *692* ... **86** B2
Foothill Farms CA, *17426* ... **255** C1
Footville WI, *788* ... **74** B4
Ford City CA, *3512* ... **52** B1
Ford City PA, *3451* ... **92** A3
Ford Co. IL, *14241* ... **88** C4
Ford Co. KS, *32458* ... **42** C4
Fordland MO, *684* ... **107** D2
Fordoche LA, *933* ... **133** F2
Fords NJ, *15032* ... **240** A6
Fords Prairie WA, *1961* ... **12** B4
Fordsville KY, *531* ... **109** F1
Fordville ND, *266* ... **19** E2
Fordyce AR, *4799* ... **117** E4
Foreman AR, *1125* ... **116** C4
Forest MS, *5987* ... **126** C2
Forest OH, *1488* ... **90** C3
Forest VA, *8006* ... **112** C1
Forest Acres SC, *10558* ... **205** F1
Forestbrook SC, *3391* ... **123** D4
Forest City FL, *12612* ... **246** B1
Forest City IA, *4362* ... **72** C3
Forest City NC, *7549* ... **121** F1
Forest City PA, *1855* ... **93** F2
Forest Co. PA, *4946* ... **92** B2
Forest Co. WI, *10024* ... **68** C2
Forestdale AL, *10509* ... **195** E1
Forestdale MA, *3992* ... **151** E3
Forest Dale VT, *800* ... **81** D3
Forest Grove OR, *17708* ... **20** B2
Forest Grove PA, *3800* ... **250** A1
Forest Hts. MD, *2585* ... **270** D5
Foresthill CA, *1791* ... **36** C2
Forest Hill LA, *456* ... **133** E1
Forest Hill TX, *12949* ... **59** E2
Forest Hills FL, *989* ... **139** D4
Forest Hills KY, *494* ... **230** F2
Forest Hills PA, *6831* ... **250** C2
Forest Lake IL, *1530* ... **203** B1
Forest Lake MN, *15098* ... **67** D3
Forest Park GA, *21447* ... **190** D5
Forest Park IL, *15688* ... **203** D4
Forest Park OH, *19463* ... **100** B2
Forest Park OK, *1066* ... **244** E2
Forest View IL, *778* ... **203** D5
Forestville CA, *2370* ... **36** A3
Forestville MD, *12707* ... **271** F4
Forestville NY, *770* ... **78** A4
Forestville OH, *10978* ... **100** B2
Forgan OK, *532* ... **50** C1
Forked River NJ, *4914* ... **147** E3
Forkland AL, *629* ... **127** E2
Forks WA, *3120* ... **12** A2
Forman ND, *506* ... **27** E1
Forney TX, *5588* ... **59** F2
Forrest IL, *1225* ... **88** C3
Forrest City AR, *14774* ... **118** A2
Forrest Co. MS, *72604* ... **126** C4
Forreston IL, *1469* ... **88** B1
Forsyth GA, *3776* ... **129** D1
Forsyth IL, *2434* ... **98** C1
Forsyth MO, *1686* ... **107** D3
Forsyth MT, *1944* ... **25** D1
Forsyth Co. GA, *98407* ... **120** C3

Bayshore ... B1
Bokeelia ... A1
Bonita Sprs. ... B2
Cape Coral ... A1
Captiva ... A2
Estero ... B2
Flamingo Bay ... A1
Ft. Myers ... B1
Ft. Myers Beach ... B2
Ft. Myers Shores ... B1
Ft. Myers Villas ... B1
Matlacha ... A1
N. Ft. Myers ... B1
Pine Island Ctr. ... A1
Pineland ... A1
Punta Rassa ... A2
St. James City ... A2
San Carlos Park ... B2
Sanibel ... A2
Tice ... B1
Truckland ... B2

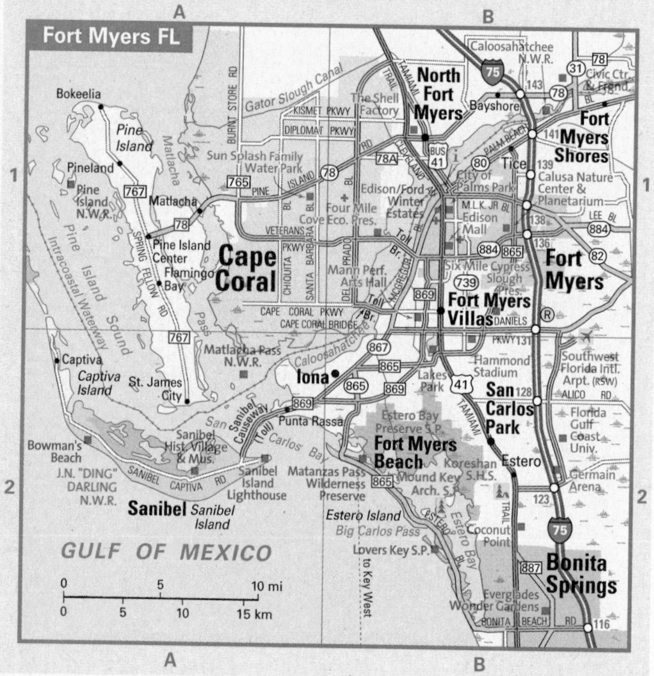

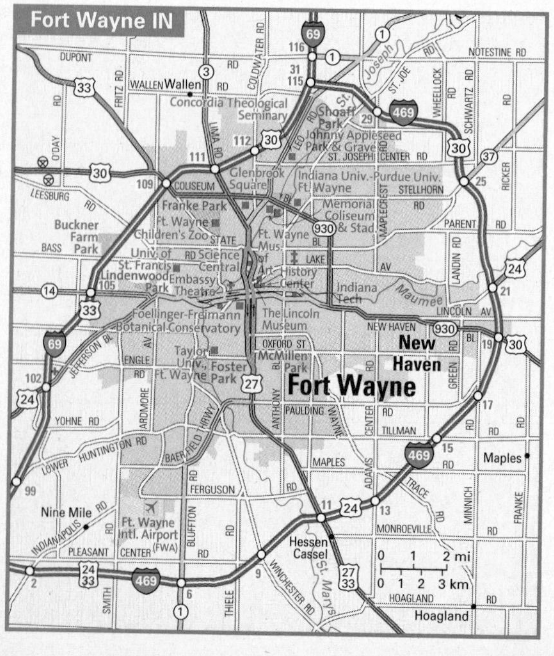

Forsyth Co. NC, *306067* ... **112** A4
Ft. Ann NY, *471* ... **81** D4
Ft. Ashby WV, *1354* ... **102** C2
Ft. Atkinson WI, *11621* ... **74** B3
Ft. Belknap Agency MT, *1262* ... **16** C2
Ft. Bend Co. TX, *354452* ... **132** A4
Ft. Benton MT, *1594* ... **15** F2
Ft. Bragg CA, *7026* ... **36** A1
Ft. Branch IN, *2320* ... **99** D4
Ft. Bridger WY, *400* ... **32** A3
Ft. Calhoun NE, *856* ... **35** F3
Ft. Chiswell VA, *911* ... **112** A2
Ft. Cobb OK, *667* ... **51** D3
Ft. Collins CO, *118652* ... **33** E4
Ft. Covington NY, *700* ... **80** C1
Ft. Davis TX, *1050* ... **62** B2
Ft. Defiance AZ, *4061* ... **48** A2
Ft. Deposit AL, *1270* ... **128** A3
Ft. Dodge IA, *25136* ... **72** C4
Ft. Dodge KS, *550* ... **42** C4
Ft. Duchesne UT, *621* ... **32** A4
Ft. Edward NY, *3141* ... **81** D4
Ft. Fairfield ME, *1600* ... **85** E2
Ft. Gaines GA, *1110* ... **128** C4
Ft. Garland CO, *432* ... **41** D4
Ft. Gates TX, *800* ... **59** E4
Ft. Gay WV, *819* ... **101** D4
Ft. Gibson OK, *4054* ... **106** A4
Ft. Grant AZ, *800* ... **55** E2
Ft. Hall ID, *3193* ... **31** E1
Ft. Hancock TX, *1713* ... **56** C4
Ft. Johnson NY, *491* ... **80** C4
Ft. Jones CA, *660* ... **28** B3
Ft. Kent ME, *1978* ... **85** D1
Ft. Kent Mills ME, *325* ... **85** D1
Ft. Laramie WY, *243* ... **33** E2
Ft. Lauderdale FL, *152397* ... **143** F2
Ft. Lawn SC, *864* ... **122** A2
Ft. Lee NJ, *35461* ... **148** B3
Ft. Loramie OH, *1344* ... **90** B4

Figures after entries indicate population, page number, and grid referenc

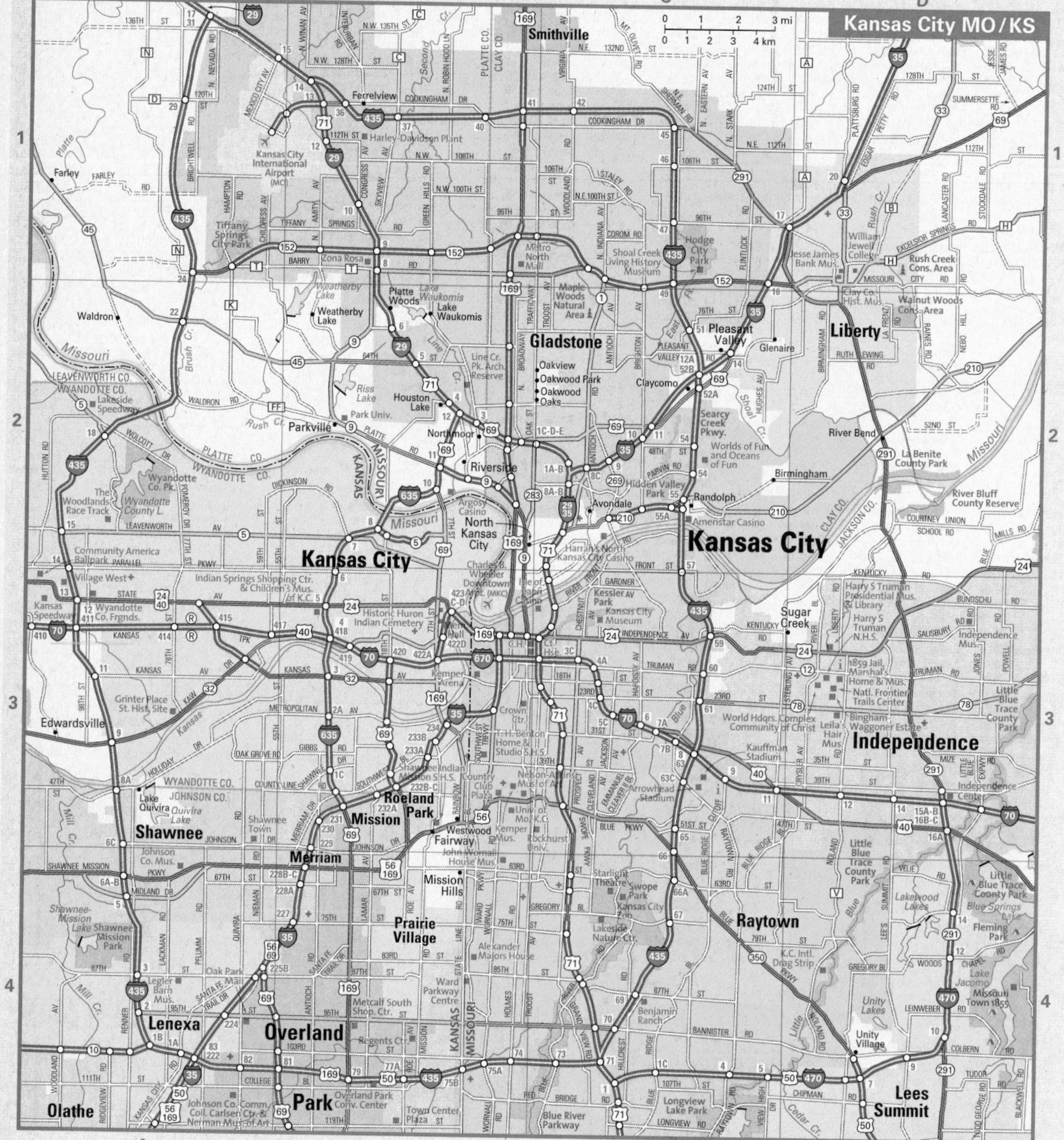

Avondale ... C2
Birmingham ... C2
Claycomo ... C2
Countryside ... B4
Edwardsville ... A3
Fairway ... B3
Farley ... A1
Ferrelview ... B1
Gladstone ... C2
Glenaire ... C2
Houston Lake ... B2
Independence ... D3
Kansas City, KS ... B2
Kansas City, MO ... C2
Lake Quivira ... A3
Lake Waukomis ... B2
Lees Summit ... D4
Lenexa ... A4
Liberty ... D2
Merriam ... B4
Mission ... B3
Mission Hills ... B4
N. Kansas City ... B2
Northmoor ... B2
Oaks ... C2
Oakview ... C2
Oakwood ... C2
Oakwood Park ... C2
Olathe ... A4
Overland Park ... B4
Parkville ... B2
Platte Woods ... B2
Pleasant Valley ... C2
Prairie Vil. ... B4
Randolph ... C2
Raytown ... C4
River Bend ... D2
Riverside ... B2
Roeland Park ... B3
Shawnee ... A3
Smithville ... C1
Sugar Creek ... D3
Unity Vil. ... D4
Waldron ... A2
Weatherby Lake ... B2
Westwood ... B3

Keysville VA, *817* ... **113** D2
Keytesville MO, *533* ... **97** D2
Key West FL, *25478* ... **142** C4
Kezar Falls ME, *750* ... **82** A3
Kiamesha Lake NY, *700* ... **148** A1
Kiawah Island SC, *1163* ... **131** D2
Kibler AR, *969* ... **116** C1
Kidder Co. ND, *2753* ... **18** C4
Kiefer OK, *1026* ... **51** F2
Kiel WI, *3450* ... **74** C2
Kiester MN, *540* ... **72** C2
Kihei HI, *16749* ... **153** D1
Kilauea HI, *2092* ... **152** B1
Kilbourne LA, *436* ... **125** F1
Kildeer IL, *3460* ... **203** B2
Kilgore TX, *11301* ... **124** B2
Killbuck OH, *839* ... **91** E4
Killdeer ND, *713* ... **18** A3
Kill Devil Hills NC, *5897* ... **115** F2
Killeen TX, *86911* ... **59** E4
Killen AL, *1119* ... **119** E2
Killian LA, *1053* ... **134** B2
Killona LA, *797* ... **239** A1
Kilmarnock VA, *1244* ... **113** F1
Kilmichael MS, *830* ... **118** C4
Kiln MS, *2040* ... **134** C2
Kimball MN, *635* ... **66** C3
Kimball NE, *2559* ... **33** F3
Kimball SD, *745* ... **27** D4
Kimball TN, *1312* ... **120** A2
Kimball WV, *411* ... **111** F1
Kimball Co. NE, *4089* ... **33** F3
Kimberling City MO, *2253* ... **107** D3
Kimberly AL, *1801* ... **119** F4
Kimberly ID, *2614* ... **30** C1
Kimberly WI, *6146* ... **74** C1
Kimberton PA, *2500* ... **146** B2
Kimble Co. TX, *4468* ... **60** C1
Kincaid IL, *1441* ... **98** B1
Kinde MI, *534* ... **76** C1
Kinder LA, *2148* ... **133** E2
Kinderhook NY, *1275* ... **94** B2
Kindred ND, *614* ... **19** E4
King NC, *5952* ... **112** A3
King WI, *2215* ... **74** B1
King and Queen Co. VA, *6630* ... **113** F1
King and Queen C.H. VA, *50* ... **113** F1
King City CA, *11094* ... **44** B3
King City MO, *1012* ... **86** B4
King City OR, *1949* ... **251** C3
King Co. TX, *356* ... **58** C1
King Co. WA, *1737034* ... **12** C3
King Cove AK, *792* ... **154** B4
Kingfield ME, *600* ... **82** B1
Kingfisher OK, *4380* ... **51** E2
Kingfisher Co. OK, *13926* ... **51** E2
King George VA, *375* ... **103** E4
King George Co. VA, *16803* ... **103** E4
Kingman AZ, *20069* ... **46** B3
Kingman IN, *538* ... **99** E1
Kingman KS, *3387* ... **43** E4
Kingman Co. KS, *8673* ... **43** E4
King of Prussia PA, *18511* ... **146** C2
King Salmon AK, *442* ... **154** B3
Kings Beach CA, *4037* ... **37** D2
Kingsburg CA, *9199* ... **45** D3
Kingsbury TX, *652* ... **61** E2
Kingsbury Co. SD, *5815* ... **27** E3
Kings Co. CA, *129461* ... **44** C3
Kings Co. NY, *2465326* ... **147** F1
Kingsford MI, *5549* ... **68** C2
Kingsford Hts. IN, *1453* ... **89** E2
Kingsland AR, *449* ... **117** E4
Kingsland GA, *10506* ... **139** D1
Kingsland TX, *4584* ... **61** D1
Kingsley IA, *1245* ... **35** F1
Kingsley MI, *1469* ... **69** F4
Kings Mtn. NC, *9693* ... **122** A1
Kings Park NY, *16146* ... **149** D3
Kings Pt. NY, *5076* ... **148** B4
Kingsport TN, *44905* ... **111** E3
Kingston GA, *659* ... **120** B3
Kingston ID, *650* ... **14** B3
Kingston IL, *980* ... **88** C1
Kingston KY, *600* ... **110** C1
Kingston MA, *5380* ... **151** E2
Kingston MI, *450* ... **76** C2
Kingston MO, *287* ... **96** C1
Kingston NH, *1700* ... **95** E1
Kingston NJ, *1292* ... **147** D1
Kingston NY, *23456* ... **94** B2
Kingston OH, *1032* ... **101** D2
Kingston OK, *1390* ... **59** F1
Kingston PA, *13855* ... **93** E2
Kingston RI, *5446* ... **150** C4
Kingston TN, *5264* ... **110** B4

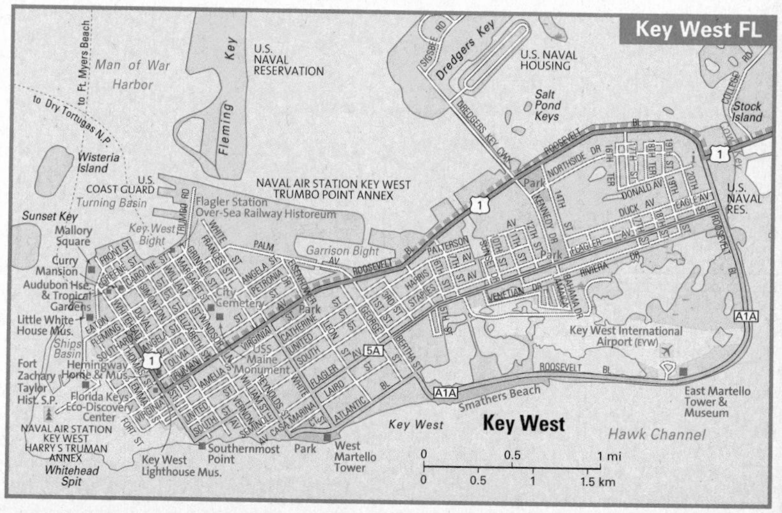

Entries in **bold black** indicate counties or parishes.
Entries in **bold color** indicate cities with detailed inset maps.

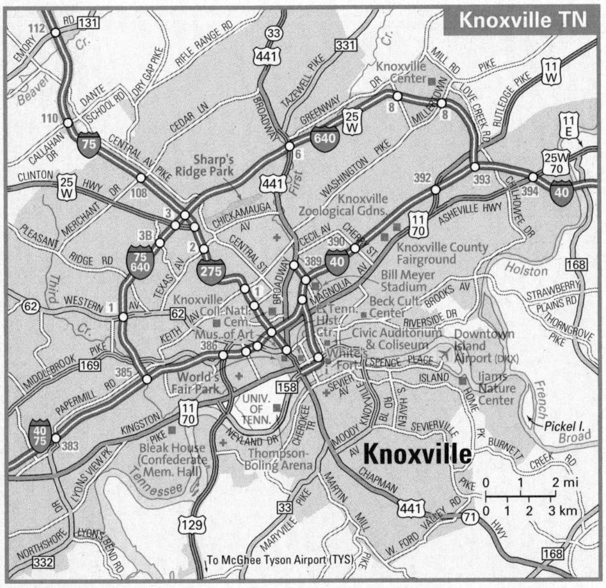

Bird in Hand	C2
E. Petersburg	A1
Lancaster	B2
Landisville	A1
Leacock	C1
Leola	C1
Millersville	A2
Mountville	A2
Neffsville	B1
Rohrerstown	A2
Salunga	A1
Smoketown	C2

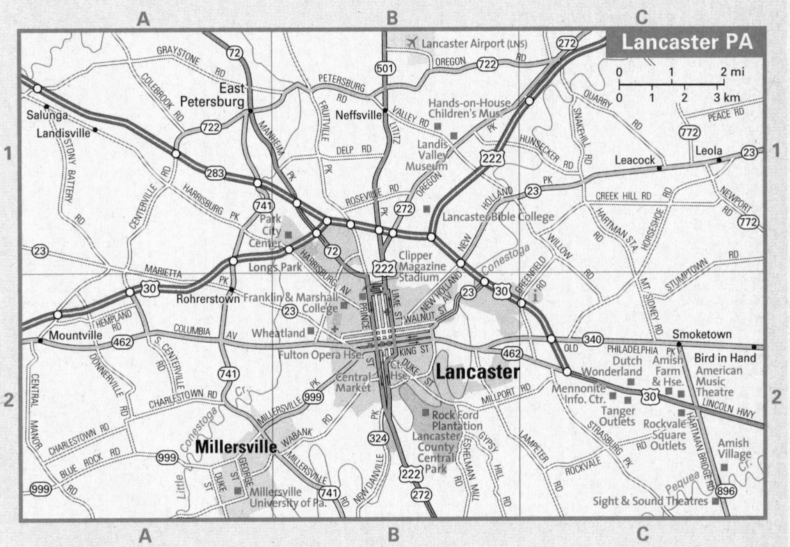

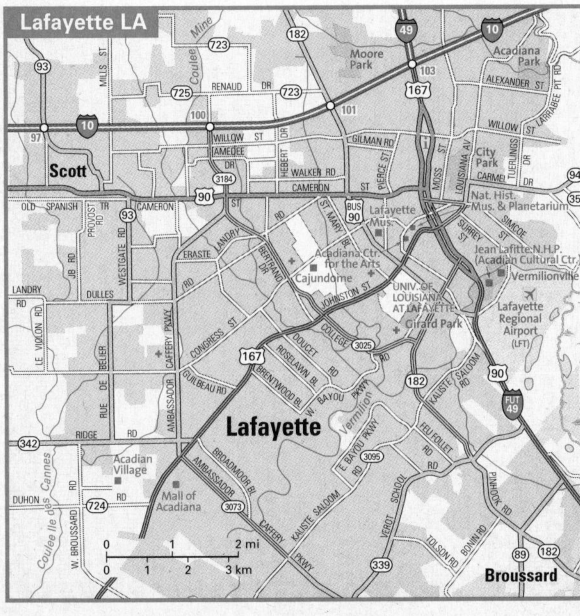

Figures after entries indicate population, page number, and grid reference.

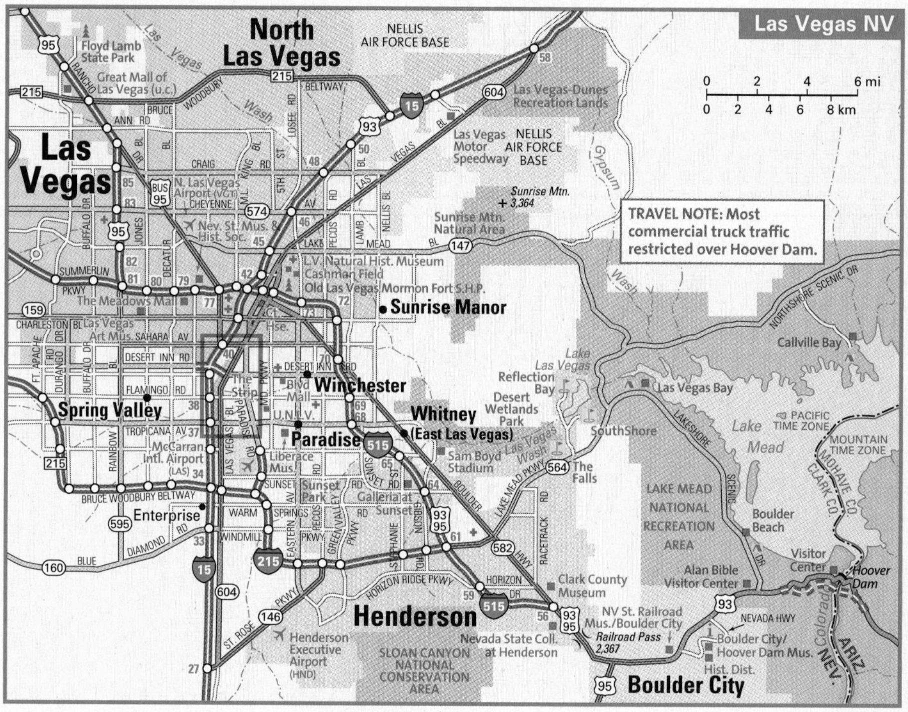

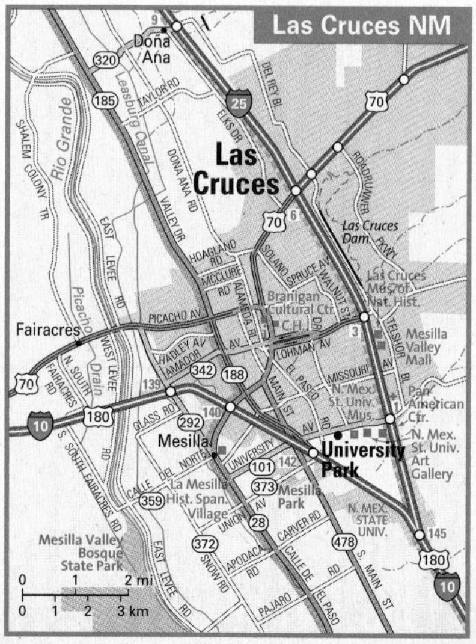

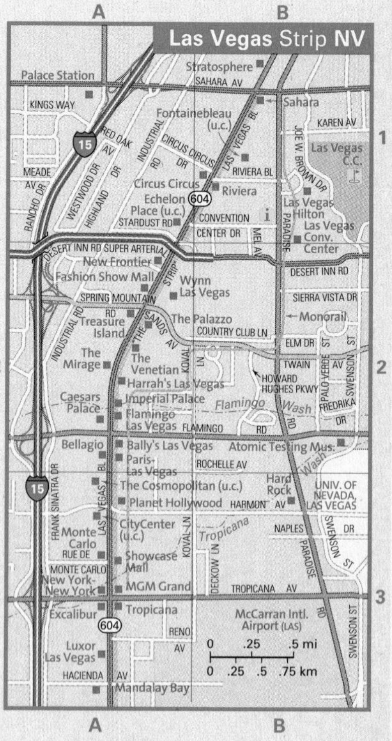

POINTS OF INTEREST

Entries in **bold black** indicate counties or parishes.
Entries in **bold color** indicate cities with detailed inset maps.

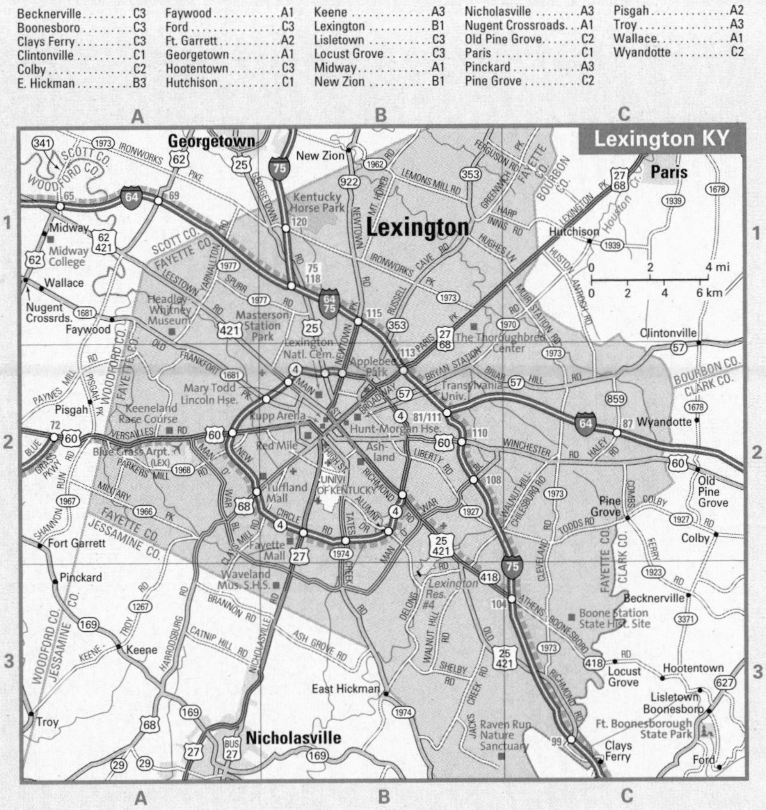

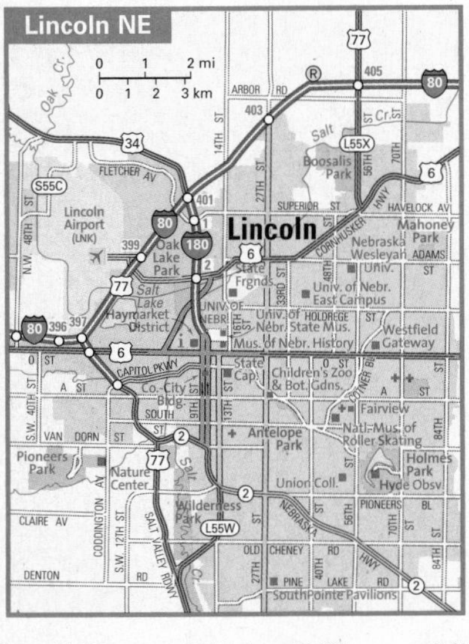

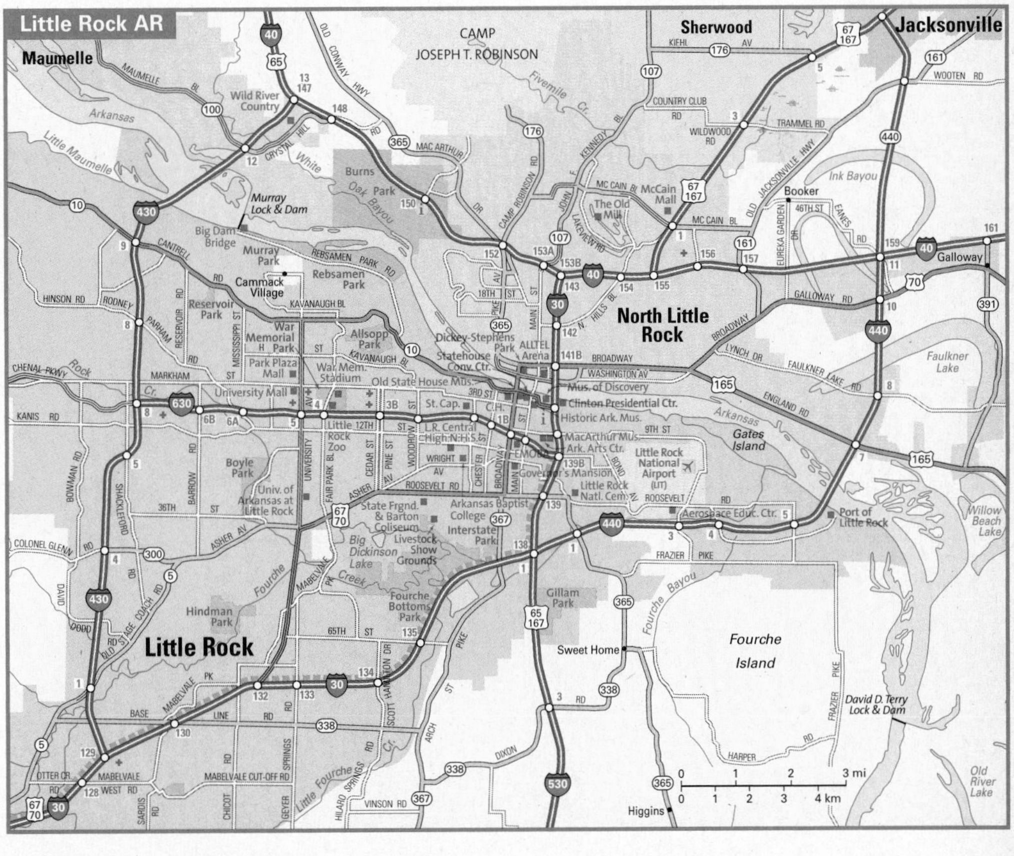

Figures after entries indicate population, page number, and grid reference.

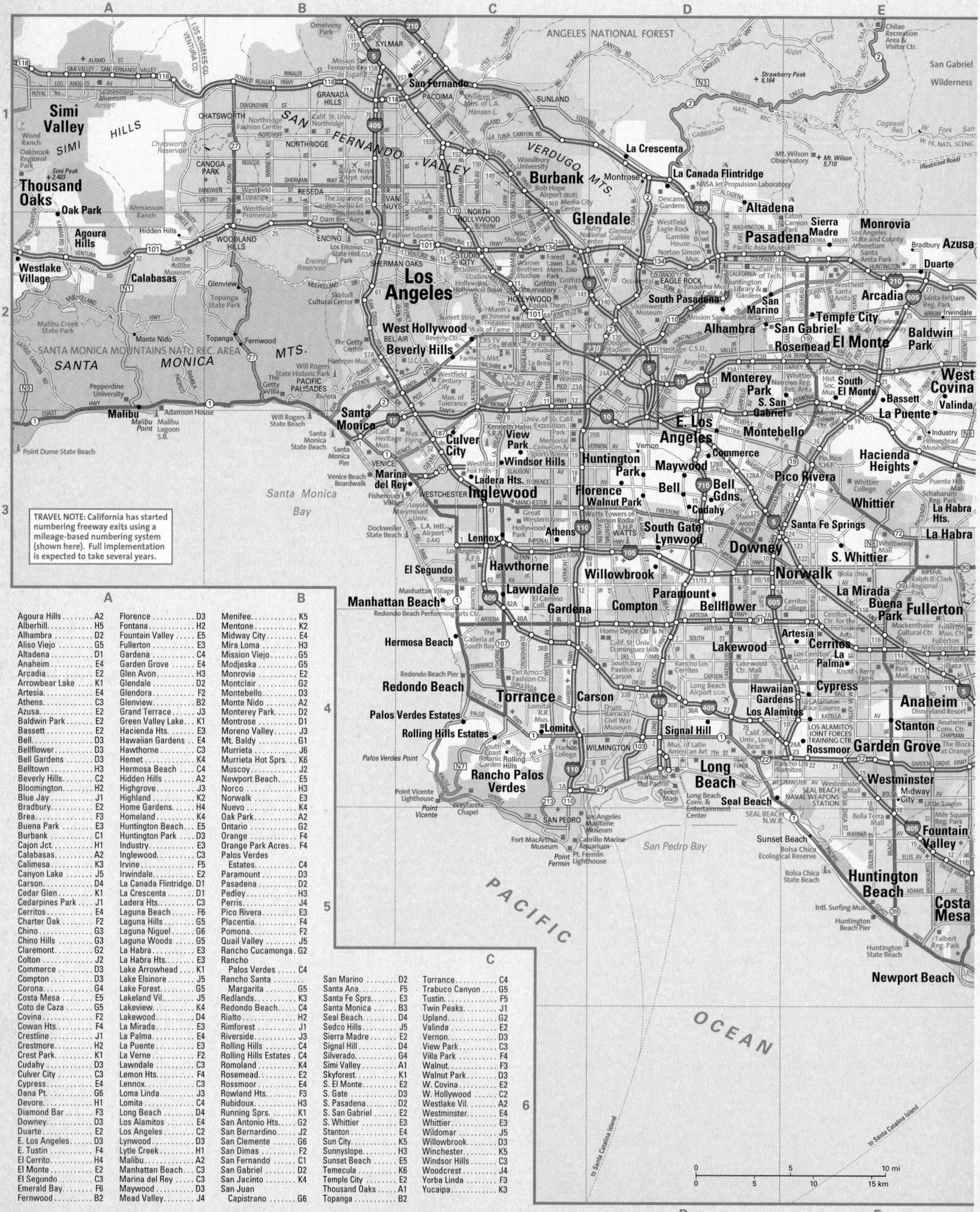

Entries in **bold black** indicate counties or parishes.
Entries in **bold color** indicate cities with detailed inset maps.

Figures after entries indicate population, page number, and grid reference.

POINTS OF INTEREST

Angels Flight A1
Bradbury Building A1
Bus Terminal B2
California Plaza A1
Cathedral of Our Lady of the Angels .. A1
Chinese American Museum B1
City Hall B1
Convention Center A2
Court House A1
Dodger Stadium B1
El Pueblo de Los Angeles Historical Monument B1
Flower District A2
Japanese American Natl. Museum ... B1
Jewelry District A2
L.A. Center Studios A1
L.A. Live A2
Library A1
MOCA at the Geffen Contemporary ... B1
Mt. St. Mary's College A2
Museum of Contemporary Art (MOCA) A1
Museum of Neon Art A2
Music Center A1
Olvera Street B1
Post Office B1
STAPLES Center A2
Union Station B1
Walt Disney Concert Hall A1

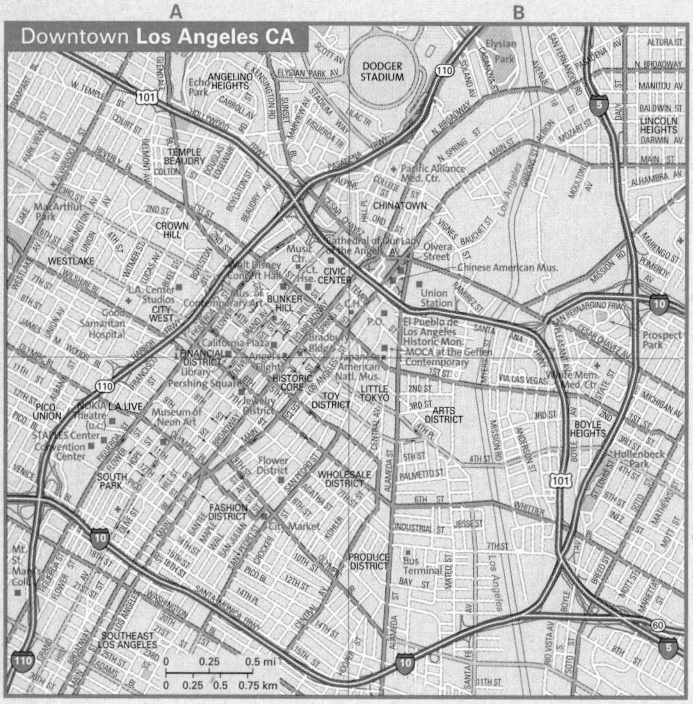

Lexington IL, *1912* **88** C3
Lexington KY, *260512* **100** B4
Lexington MA, *30355* **151** D1
Lexington MI, *1104* **76** C3
Lexington MN, *2214* **235** C1
Lexington MS, *2025* **126** B1
Lexington MO, *4453* **96** C2
Lexington NE, *10011* **34** C4
Lexington NC, *19953* **112** B4
Lexington OH, *4165* **91** D3
Lexington OK, *2086* **51** E3
Lexington SC, *9793* **122** A3
Lexington TN, *7393* **109** D4
Lexington TX, *1178* **61** E1
Lexington VA, *6867* **112** C1
Lexington Co. SC, *216014* **122** A4
Lexington Park MD, *11021* **103** F4
Libby MT, *2626* **14** C2
Liberal KS, *19666* **50** B1
Liberal MO, *779* **106** B1
Liberty IL, *519* **97** F1
Liberty IN, *2061* **100** B1
Liberty KY, *1850* **110** B1
Liberty ME, *300* **82** C2
Liberty MS, *633* **134** A1
Liberty MO, *26232* **96** B2
Liberty NY, *3975* **94** A3
Liberty NC, *2661* **112** B4
Liberty PA, *2670* **250** C3
Liberty SC, *3009* **121** E2
Liberty TN, *367* **110** A4
Liberty TX, *8033* **132** B3
Liberty Ctr. OH, *1109* **90** B2
Liberty City TX, *1935* **124** B2
Liberty Corner NJ, *1700* **147** D1
Liberty Co. FL, *7021* **137** D2
Liberty Co. GA, *61610* **130** B3
Liberty Co. MT, *2158* **15** F1
Liberty Co. TX, *70154* **132** B2
Liberty Hill TX, *1409* **61** E1
Liberty Lake WA, *3076* **14** B3
Libertyville IL, *20742* **74** C4
Licking MO, *1471* **107** E1
Licking Co. OH, *145491* **91** D4
Lidgerwood ND, *738* **27** E1
Lido Beach NY, *2825* **147** F1
Lighthouse Pt. FL, *10767* **143** F1
Ligonier IN, *4357* **89** F2
Ligonier PA, *1695* **92** B4
Lihue HI, *5674* **152** B1
Lilbourn MO, *1303* **108** B3
Lilburn GA, *11307* **120** C4
Lillington NC, *2915* **123** D1
Lilly PA, *948* **92** B4
Lily KY, *1200* **110** C2
Lilydale MN, *552* **235** D3
Lily Lake IL, *825* **88** C1
Lima MT, *242* **23** E3
Lima NY, *2459* **78** C3
Lima OH, *40081* **90** B3
Lima PA, *3225* **248** A4
Lime Lake NY, *1422* **78** B4
Limeport PA, *1100* **146** C1
Limerick ME, *425* **82** A3
Limerick PA, *850* **146** B2
Lime Sprs. IA, *496* **73** E2
Limestone ME, *1453* **85** E1
Limestone Co. AL, *65676* **119** F2
Limestone Co. TX, *22051* **59** F4
Limon CO, *2071* **41** F2
Lincoln AL, *4577* **120** A4
Lincoln AR, *1752* **106** B4
Lincoln CA, *11205* **36** C2
Lincoln DE, *950* **145** F3
Lincoln ID, *2800* **23** E4
Lincoln IL, *15369* **88** B4
Lincoln KS, *1349* **43** E2
Lincoln ME, *2933* **85** D4
Lincoln MA, *650* **197** A1
Lincoln MO, *1026* **97** D4
Lincoln MT, *1100* **15** E4
Lincoln NE, *225581* **35** F4
Lincoln NH, *750* **81** F3
Lincoln ND, *1730* **18** C4
Lincoln Beach OR, *2078* **20** B2
Lincoln City OR, *7437* **20** B2
Lincoln Co. AR, *14492* **117** F3
Lincoln Co. CO, *6087* **41** F2
Lincoln Co. GA, *8348* **121** E4
Lincoln Co. ID, *4044* **31** D1
Lincoln Co. KS, *3578* **43** E2
Lincoln Co. KY, *23361* **110** B1
Lincoln Co. ME, *33616* **82** C2
Lincoln Co. MN, *6429* **27** F3
Lincoln Co. MS, *33166* **126** B4
Lincoln Co. MO, *38944* **97** F2
Lincoln Co. MT, *18837* **14** C2
Lincoln Co. NE, *34632* **34** C3
Lincoln Co. NV, *4165* **38** B4
Lincoln Co. NM, *19411* **49** D4
Lincoln Co. NC, *63780* **122** A1
Lincoln Co. OK, *32080* **51** F2
Lincoln Co. OR, *44479* **20** B3
Lincoln Co. SD, *24131* **35** F1
Lincoln Co. TN, *31340* **119** F1
Lincoln Co. WA, *10184* **13** F3
Lincoln Co. WV, *22108* **101** E4
Lincoln Co. WI, *29641* **68** A3
Lincoln Co. WY, *14573* **32** A2
Lincolndale NY, *2018* **148** C2
Lincoln Hts. OH, *4113* **204** B1
Lincolnia VA, *15788* **270** B4
Lincoln Par. LA, *42509* **125** E2
Lincoln Park CO, *3904* **41** E3
Lincoln Park MI, *40008* **90** C1
Lincoln Park NJ, *10930* **148** A3
Lincolnshire IL, *6108* **203** C2
Lincolnton GA, *1595* **121** E4
Lincolnton NC, *9965* **122** A1
Lincoln Vil. OH, *9482* **206** A3
Lincolnville SC, *904* **131** D1
Lincolnville Ctr. ME, *325* **82** C2
Lincolnwood IL, *12359* **203** D3
Lincroft NJ, *6255* **147** E2
Lind WA, *582* **13** F4
Linda CA, *13474* **36** C2
Lindale GA, *4088* **120** B3
Lindale TX, *2954* **124** A2
Lindcove CA, *650* **45** D3
Linden AL, *2424* **127** E2
Linden CA, *1103* **36** C3
Linden IN, *700* **89** E4
Linden MI, *2861* **76** B3
Linden NJ, *39394* **147** E1
Linden TN, *1015* **119** E1
Linden TX, *2256* **124** C1
Linden WI, *615* **74** A3
Lindenhurst IL, *12539* **74** C4
Lindenhurst NY, *27819* **148** C4
Lindenwold NJ, *17414* **146** C3
Lindon UT, *8363* **31** F4
Lindsay CA, *10297* **45** D3
Lindsay OK, *2889* **51** E4
Lindsay TX, *788* **59** E1
Lindsborg KS, *3321* **43** E3
Lindstrom MN, *3015* **67** D3
Linesville PA, *1155* **91** F2
Lineville AL, *2401* **128** B1
Lingle WY, *510* **33** F2
Linglestown PA, *6414* **93** D4
Linn KS, *425* **43** F1
Linn MO, *1354* **97** E3
Linn TX, *958* **63** E3
Linn Co. IA, *191701* **87** E1
Linn Co. KS, *9570* **96** B4
Linn Co. MO, *13754* **97** D1
Linn Co. OR, *103069* **20** C3
Linneus MO, *369* **97** D1
Linn Valley KS, *562* **96** B4
Lino Lakes MN, *16791* **235** D7
Linthicum MD, *7539* **193** C4
Linton IN, *5774* **99** E2
Linton ND, *1321* **26** C1
Linwood IN, *700* **89** F4
Linwood KS, *374* **96** B2
Linwood MI, *1200* **76** B2
Linwood NJ, *7172* **147** F4
Lionville PA, *6298* **146** B3
Lipscomb AL, *2458* **195** D2
Lipscomb TX, *44* **50** C2
Lipscomb Co. TX, *3057* **50** C2
Lisbon IA, *1898* **87** F1
Lisbon ME, *1800* **82** B3
Lisbon NH, *1070* **81** E2
Lisbon ND, *2292* **19** E4
Lisbon OH, *2788* **91** F3
Lisbon Falls ME, *4420* **82** B3
Lisle IL, *21182* **203** B5
Lisman AL, *653* **127** D3
Litchfield CT, *1328* **94** C3
Litchfield IL, *6815* **98** B2
Litchfield ME, *325* **82** B2
Litchfield MI, *1458* **90** A1
Litchfield MN, *6562* **66** B3
Litchfield Co. CT, *182193* **94** B3
Litchfield Park AZ, *3810* **249** A2
Lithia Sprs. GA, *2072* **120** C4
Lithonia GA, *2187* **120** C4
Lithopolis OH, *600* **101** D1
Lititz PA, *9029* **146** A2
Little Canada MN, *9771* **235** D2
Little Chute WI, *10476* **74** C1
Little Compton RI, *400* **151** D4
Little Creek DE, *195* **145** E2
Little Cypress TX, *1800* **132** C2
Little Eagle SD, *370* **26** C1
Little Elm TX, *3646* **59** F2
Little Falls MN, *7719* **66** C2
Little Falls NJ, *10855* **148** A3
Little Falls NY, *5188* **79** F3
Little Ferry NJ, *10800* **240** C2
Littlefield TX, *6507* **58** A1
Little Flock AR, *2585* **106** C3
Littlefork MN, *680* **64** B2
Little Heaven DE, *1400* **145** E3
Little River KS, *536* **43** E3
Little River SC, *7027* **123** D4
Little River-Academy TX, *1645* **61** E1
Little River Co. AR, *13628* **116** C4
Little Rock AR, *183133* **117** E2
Littlerock CA, *1402* **52** C2
Little Silver NJ, *6170* **147** E2
Littlestown PA, *3947* **103** E1
Littleton CO, *40340* **41** E1
Littleton NH, *4431* **81** F2
Littleton NC, *692* **113** D3
Little Valley NY, *1130* **92** B1
Littleville AL, *978* **119** E2
Live Oak CA, *6229* **36** C2
Live Oak CA, *16628* **236** D1
Live Oak FL, *6480* **138** B2
Live Oak TX, *9156* **61** D2
Live Oak Co. TX, *12309* **61** D4
Livermore CA, *73345* **36** B4
Livermore KY, *1482* **109** E1
Livermore Falls ME, *1626* **82** B2
Liverpool NY, *2505* **265** A1
Liverpool PA, *876* **93** D4
Livingston AL, *3297* **127** D2
Livingston CA, *10473* **36** C4
Livingston IL, *825* **98** B2
Livingston LA, *1342* **134** A2
Livingston MT, *6851* **23** F1
Livingston NJ, *27391* **148** A4
Livingston TN, *3498* **110** A3
Livingston TX, *5433* **132** B1
Livingston WI, *597* **74** A3
Livingston Co. IL, *39678* **88** C3
Livingston Co. KY, *9804* **109** D2
Livingston Co. MI, *156951* **76** B4
Livingston Co. MO, *14558* **96** C1
Livingston Co. NY, *64328* **78** C4
Livingston Manor NY, *1355* **94** A2
Livingston Par. LA, *91814* **134** B2
Livonia LA, *1339* **133** F2
Livonia MI, *100545* **76** B4
Livonia NY, *1373* **78** C3
Llangollen Estates DE, *5600* **145** E1
Llano TX, *3325* **61** D1
Llano Co. TX, *17044* **61** D1
Lloyd Harbor NY, *3675* **148** C3
Loa UT, *525* **39** E3
Loami IL, *804* **98** B1
Lobelville TN, *915* **109** D4
Lochbuie CO, *2049* **209** D1
Lochearn MD, *25269* **144** C2
Loch Lynn Hts. MD, *469* **102** B2
Loch Sheldrake NY, *800* **94** A3
Lockeford CA, *3179* **36** C3
Lockesburg AR, *711* **116** C4
Lockhart AL, *548* **136** B1
Lockhart FL, *12944* **246** B1
Lockhart TX, *11615* **61** E2
Lock Haven PA, *9149* **93** D3
Lockland OH, *3707* **204** B1
Lockney TX, *2056* **50** B4
Lockport IL, *15191* **89** D2
Lockport LA, *2624* **134** B3
Lockport NY, *22279* **78** B3
Lockwood MO, *989* **106** C1
Locust NC, *2416* **122** B1
Locust Fork AL, *1016* **119** F4
Locust Grove GA, *2322* **129** D1
Locust Grove OK, *1366* **106** B3
Locust Valley NY, *3521* **148** C3
Lodge Grass MT, *510* **24** C2
Lodgepole MT, *214* **16** C2
Lodi CA, *56999* **36** C3
Lodi NJ, *23971* **240** C1
Lodi OH, *3061* **91** D3
Lodi WI, *2882* **74** B3

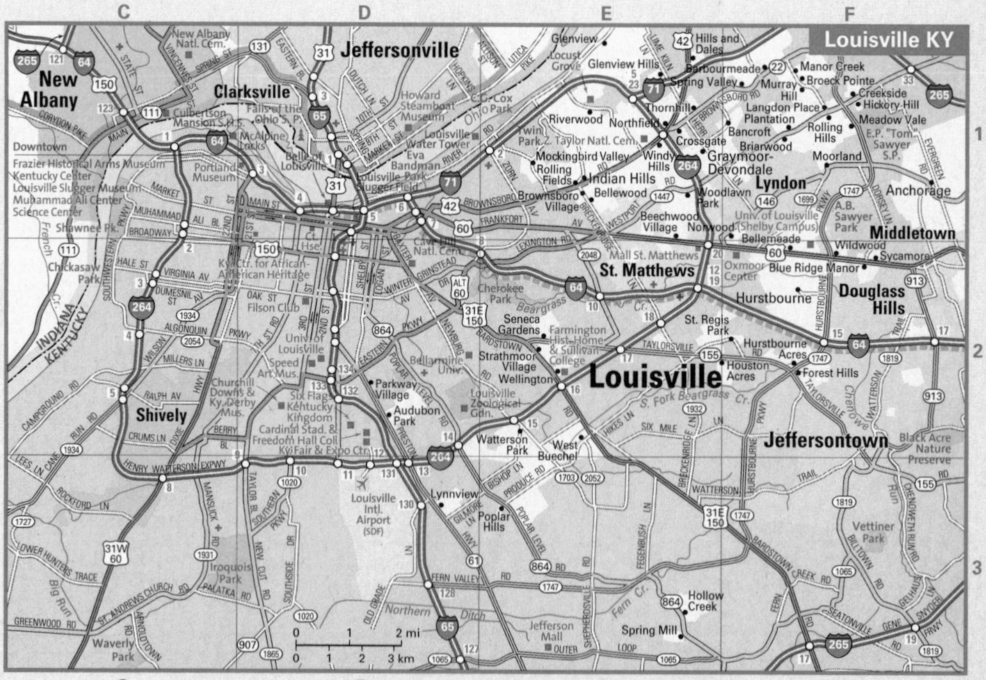

Bancroft F1
Barbourmeade F1
Beechwood Vil. E1
Bellemeade F2
Bellewood E1
Blue Ridge Manor F2
Briarwood F1
Broeck Pointe F1
Brownsboro Vil. E1
Clarksville D1
Creekside F1
Crossgate E1
Douglass Hills F2
Fincastle F1
Forest Hills F2
Glenview E1
Glenview Hills E1
Graymoor-Devondale E1
Hickory Hill F1
Hills and Dales E1
Hollow Creek E3
Houston Acres F2
Hurstbourne F2
Hurstbourne Acres F2
Indian Hills E1
Jeffersontown F2
Jeffersonville D1
Langdon Place F1
Louisville E2
Lyndon F1
Lynnview D3
Manor Creek F1
Meadow Vale F1
Middletown F1
Mockingbird Valley E1
Moorland F1
Murray Hill F1
New Albany C1
Northfield E1
Norwood F1
Parkway Vil. D2
Plantation F1
Poplar Hills E3
Riverwood E1
Rolling Fields E1
Rolling Hills F1
St. Matthews E1
St. Regis Park F2
Seneca Gardens E2
Shively C2
Spring Mill E3
Spring Valley F1
Strathmoor Vil. E2
Sycamore F2
Thornhill E1
Watterson Park E2
Wellington E2
W. Buechel E2
Wildwood F2
Windy Hills E1
Woodlawn Park E1

Entries in **bold black** indicate counties or parishes.
Entries in **bold color** indicate cities with detailed inset maps.

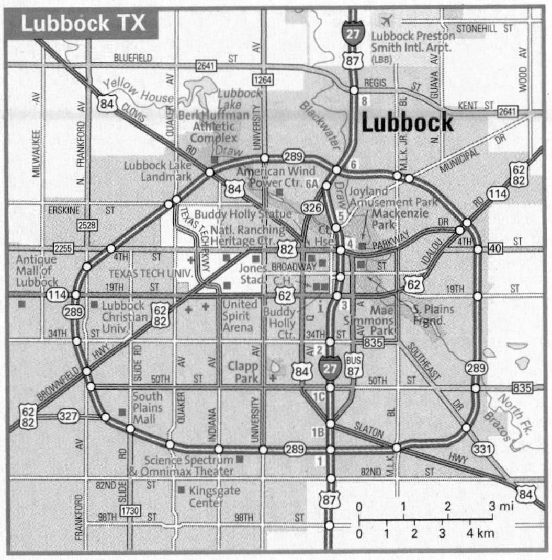

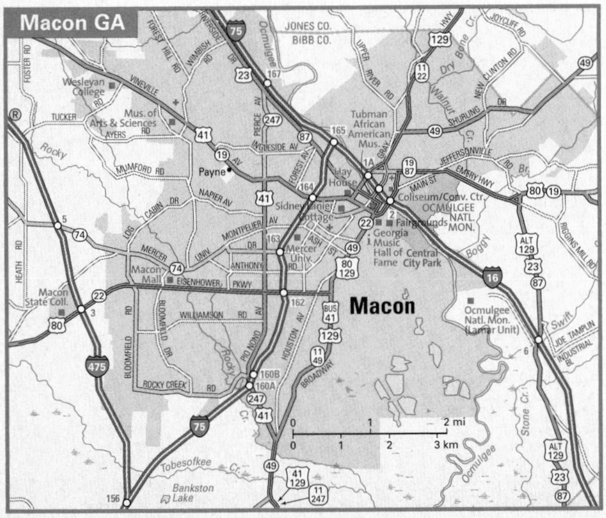

Fitchburg	A2	McFarland	B2	Shorewood Hills	A2
Madison	A1	Middleton	A1	Sun Prairie	B1
Maple Bluff	B1	Monona	B2	Waunakee	A1

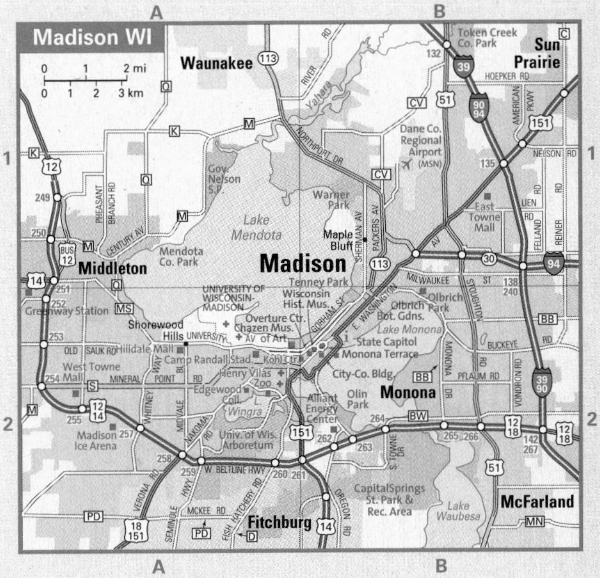

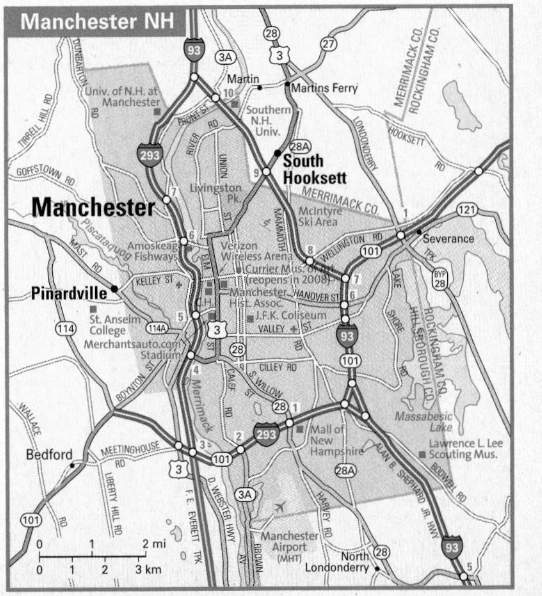

Figures after entries indicate population, page number, and grid reference.

Macon Co. AL, *24105* **128** B3
Macon Co. GA, *14074* **129** D2
Macon Co. IL, *114706* **98** C1
Macon Co. MO, *15762* **97** D1
Macon Co. NC, *29811* **121** D1
Macon Co. TN, *20386* **110** A3
Macoupin Co. IL, *49019* **98** B2
Macungie PA, *3039* **146** B1
Macy NE, *956* **35** F2
Madawaska ME, *3326* **85** D1
Maddock ND, *498* **19** D2
Madeira OH, *8923* **204** C2
Madeira Beach FL, *4511* **140** B3
Madelia MN, *2340* **72** B1
Madera CA, *43207* **44** C2
Madera Co. CA, *123109* **44** C2

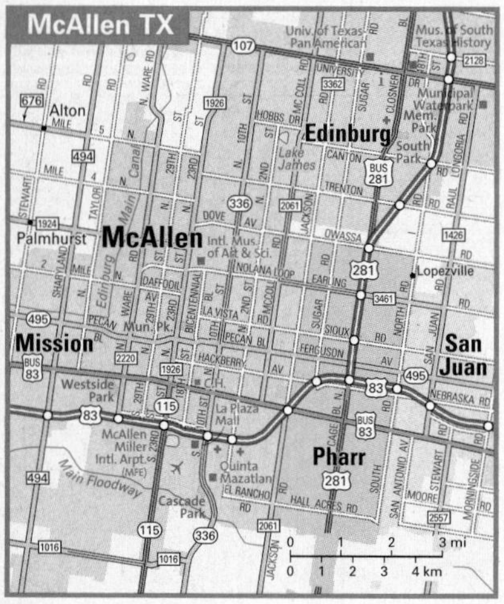

Madill OK, *3410* **51** F4
Madison AL, *29329* **119** F2
Madison AR, *987* **118** A2
Madison CT, *2222* **149** E2
Madison FL, *3061* **137** F2
Madison GA, *3636* **121** D4
Madison IL, *4545* **98** A3
Madison IN, *12004* **100** A3
Madison KS, *857* **43** F3
Madison ME, *2733* **82** B1
Madison MN, *1768* **27** F2
Madison MS, *14692* **126** B2
Madison MO, *586* **97** E2
Madison NE, *2367* **35** E2
Madison NJ, *16530* **148** A4
Madison NC, *2262* **112** B3
Madison OH, *2921* **91** F1
Madison SD, *6540* **27** F3
Madison VA, *210* **102** C4
Madison WV, *2677* **101** E4
Madison WI, *208054* **74** B3
Madison Co. AL, *276700* **119** F2
Madison Co. AR, *14243* **106** C4
Madison Co. FL, *18733* **137** F2
Madison Co. GA, *25730* **121** E3
Madison Co. ID, *27467* **23** F4
Madison Co. IL, *258941* **98** B3
Madison Co. IN, *133358* **89** F4
Madison Co. IA, *14019* **86** C2
Madison Co. KY, *70872* **110** C1
Madison Co. MS, *74674* **126** B2
Madison Co. MO, *11800* **108** A1
Madison Co. MT, *6851* **23** E2
Madison Co. NE, *35226* **35** E2
Madison Co. NY, *69441* **79** E4
Madison Co. NC, *19635* **111** E4
Madison Co. OH, *40213* **100** C1
Madison Co. TN, *91837* **108** C4
Madison Co. TX, *12940* **61** F1
Madison Co. VA, *12520* **102** C3
Madison Hts. MI, *31101* **210** C2
Madison Hts. VA, *11584* **112** C1
Madison Lake MN, *837* **72** C1
Madison Par. LA, *13728* **126** A3
Madisonville KY, *19307* **109** E1
Madisonville LA, *677* **134** B2
Madisonville TN, *3939* **120** C1
Madisonville TX, *4159* **61** F1
Madras OR, *5078* **21** D3
Madrid IA, *2264* **86** C1
Madrid NY, *650* **80** B1
Maeser UT, *2855* **32** A4
Magalia CA, *10569* **36** C1
Magazine AR, *915* **116** C1
Magdalena NM, *913* **48** B4
Magee MS, *4200* **126** C3
Maggie Valley NC, *607* **121** E1
Magna UT, *22770* **257** A2
Magnolia AR, *10858* **125** D1
Magnolia DE, *226* **145** E2
Magnolia MS, *2071* **134** B1
Magnolia NJ, *4409* **248** D4
Magnolia NC, *932* **123** E2
Magnolia OH, *931* **91** E3
Magnolia TX, *1111* **132** A2
Magoffin Co. KY, *13332* **111** D1
Mahanoy City PA, *4647* **93** E3
Mahaska Co. IA, *22335* **87** D2
Mahnomen MN, *1202* **19** F3
Mahnomen Co. MN, *5190* **19** F3
Mahomet IL, *4877* **88** C4
Mahoning Co. OH, *257555* **91** F3
Mahopac NY, *8478* **148** C2
Mahtomedi MN, *7563* **235** E1
Mahwah NJ, *5200* **148** B3
Maiden NC, *3282* **122** A1
Maili HI, *5943* **152** A3
Maine NY, *1000* **93** E1
Maineville OH, *885* **100** B2
Maitland FL, *12019* **141** D1
Maize KS, *1868* **43** E4
Majestic KY, *600* **111** E1
Major Co. OK, *7545* **51** D2
Makaha HI, *7753* **152** A3
Makakilo City HI, *13156* **152** A3
Makawao HI, *6327* **153** D1
Makena HI, *5671* **153** D1
Malabar FL, *2622* **141** E2
Malad City ID, *2158* **31** E2
Malaga CA, *1400* **44** C3
Malaga NJ, *1700* **145** F1
Malakoff TX, *2257* **59** F3
Malcolm NE, *413* **35** F4
Malden MA, *56340* **151** D1
Malden MO, *4782* **108** B3
Malden WV, *750* **200** B2
Malheur Co. OR, *31615* **22** A4
Malibu CA, *12575* **52** B2
Malin OR, *638* **29** D2
Mallory WV, *1143* **111** F1
Malone FL, *2007* **137** D1
Malone NY, *6075* **80** C1
Malone WA, *473* **12** B4
Malta ID, *177* **31** D2
Malta IL, *969* **88** C1
Malta MT, *2120* **16** C2
Malta NY, *2100* **94** B1
Malta OH, *696* **101** E1
Maltby WA, *8267* **262** B2
Malvern AL, *1215* **136** C1
Malvern AR, *9021* **117** D3
Malvern IA, *1256* **86** A3
Malvern OH, *1218* **91** E3
Malvern PA, *3059* **146** B3
Malverne NY, *8934* **241** G4
Mamaroneck NY, *18752* **148** C3
Mammoth AZ, *1762* **55** D2
Mammoth Hot Sprs. WY, *275* **23** F2
Mammoth Lakes CA, *7093* **37** E4
Mammoth Spr. AR, *1147* **107** F3
Mamou LA, *3566* **133** E2
Man WV, *770* **111** F1
Manahawkin NJ, *2004* **147** E4
Manasquan NJ, *6310* **147** E2
Manassa CO, *1042* **41** D4
Manassas VA, *35135* **144** A4
Manassas Park VA, *10290* **144** A3
Manatee Co. FL, *264002* **140** C3
Manawa WI, *1330* **74** B1
Mancelona MI, *1408* **70** B4
Manchaca TX, *1200* **61** E2
Manchaug MA, *850* **150** C2
Manchester CT, *30595* **150** A3
Manchester GA, *3988* **128** C2
Manchester IA, *5257* **73** F4
Manchester KY, *1738* **110** C2
Manchester ME, *600* **82** B2
Manchester MD, *3329* **103** E1
Manchester MI, *2160* **90** B1
Manchester MO, *19161* **256** A3
Manchester NH, *107006* **81** F4
Manchester NY, *1475* **78** C3
Manchester OH, *2043* **100** C3
Manchester PA, *2350* **93** E4
Manchester TN, *8294* **120** A1
Manchester VT, *602* **81** D4
Manchester WA, *4958* **262** A3
Manchester-by-the-Sea MA, *3600* **151** F1
Manchester Ctr. VT, *2065* **81** D4
Mancos CO, *1119* **40** B4
Mandan ND, *16718* **18** C4
Mandaree ND, *558* **18** A3
Manderson SD, *626* **26** A4
Mandeville AR, *700* **116** C4
Mandeville LA, *10489* **134** B2
Mangham LA, *595* **125** F2
Mango FL, *8842* **266** C2
Mangonia Park FL, *1283* **141** F4
Mangum OK, *2924* **50** C3
Manhasset NY, *8362* **148** C4
Manhasset Hills NY, *3661* **241** G3
Manhattan IL, *3330* **89** D2
Manhattan KS, *44831* **43** F2
Manhattan MT, *1396* **23** F1
Manhattan Beach CA, *33852* **228** C3
Manheim PA, *4784* **93** E4
Manila AR, *3055* **108** B4
Manila UT, *308* **32** A3
Manilla IA, *839* **86** A1
Manistee MI, *6586* **75** E1
Manistee Co. MI, *24527* **70** A4
Manistique MI, *3583* **69** E2
Manito IL, *1733* **88** B4
Manitou Beach MI, *2080* **90** B1
Manitou Sprs. CO, *4980* **41** E2
Manitowoc WI, *34053* **75** D1
Manitowoc Co. WI, *82887* **75** D1
Mankato KS, *976* **43** E1
Mankato MN, *32427* **72** C1
Manlius NY, *4819* **79** E3
Manly IA, *1342* **73** D3
Mannford OK, *2095* **51** F2
Manning IA, *1490* **86** B1
Manning ND, *30* **18** A3
Manning SC, *4025* **122** B4
Mannington WV, *2124* **102** A1
Mannsville OK, *587* **51** F4
Manokotak AK, *399* **154** B3
Manomet MA, *2900* **151** E3
Manor PA, *2796* **92** A4
Manor TX, *1204* **61** E1
Manorhaven NY, *6138* **241** G2
Manorville NY, *11131* **149** D3
Mansfield AR, *1097* **116** C2
Mansfield IL, *949* **88** C4
Mansfield LA, *5582* **124** C3
Mansfield MA, *7320* **151** D2
Mansfield MO, *1349* **107** E2
Mansfield OH, *49346* **91** D3
Mansfield PA, *3411* **93** D1
Mansfield TX, *28031* **59** E3
Mansfield Ctr. CT, *973* **150** B3
Mansfield Four Corners CT, *700* . **150** B3
Manson IA, *1893* **72** B4
Manson WA, *900* **13** E2
Mansura LA, *1573* **133** F1
Mantachie MS, *1107* **119** D3
Manteca CA, *49258* **36** C4
Manteno IL, *6414* **89** D2
Manteo NC, *1052* **115** F2
Manti UT, *3040* **39** E2
Manton MI, *1221* **75** F1
Mantorville MN, *1054* **73** D1
Mantua OH, *1046* **91** E2
Mantua UT, *791* **31** E3
Mantua VA, *7485* **270** A4
Manvel ND, *370* **19** E2
Manvel TX, *3046* **132** A4
Manville NJ, *10343* **147** D1
Manville RI, *3800* **150** C2
Many LA, *2889* **125** D4
Many Farms AZ, *1548* **47** F2

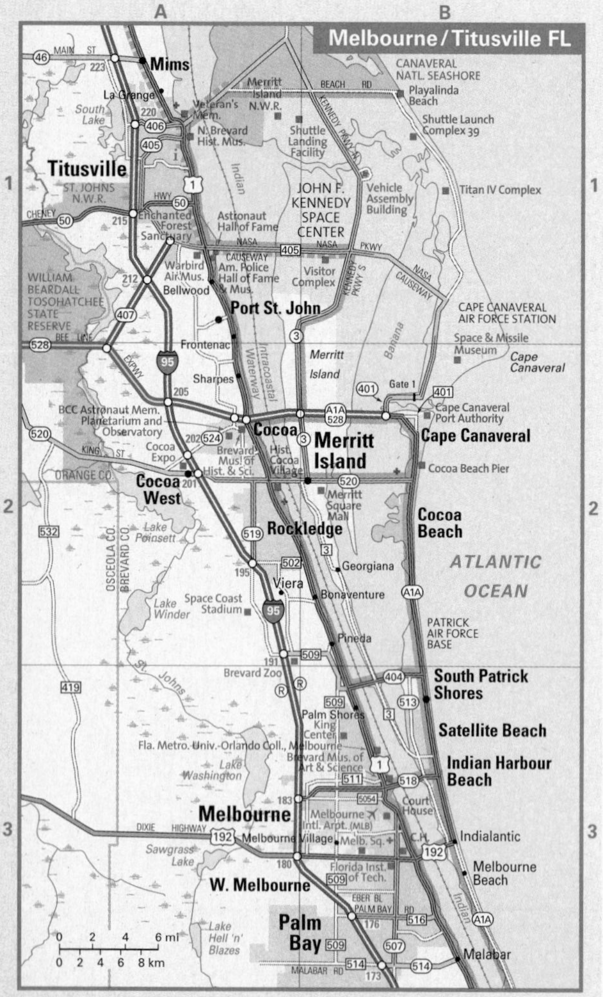

Bellwood A1	Indialantic. B3	Merritt Island B2	Sharpes. A2
Bonaventure B2	Indian Harbour	Mims A1	S. Patrick Shores . . . B3
Cape Canaveral. B2	Beach B3	Palm Bay B3	Titusville A1
Cocoa A2	La Grange. A1	Palm Shores. B3	Viera A2
Cocoa Beach. B2	Malabar B3	Pineda B2	W. Melbourne B3
Cocoa West. A2	Melbourne A3	Port St. John A1	
Frontenac A1	Melbourne Beach. . . . B3	Rockledge. A2	
Georgiana. B2	Melbourne Vil. B3	Satellite Beach B3	

Entries in **bold black** indicate counties or parishes.
Entries in **bold color** indicate cities with detailed inset maps.

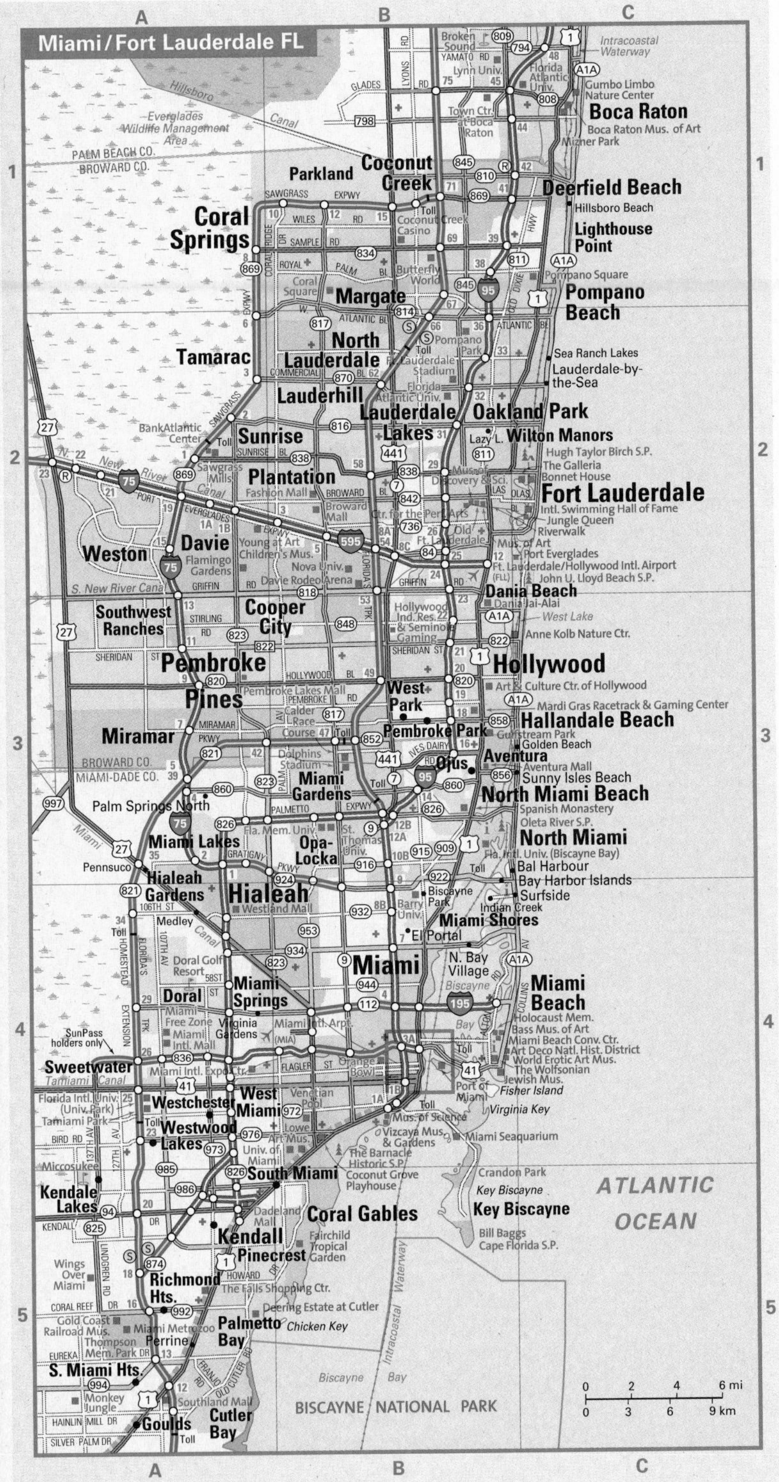

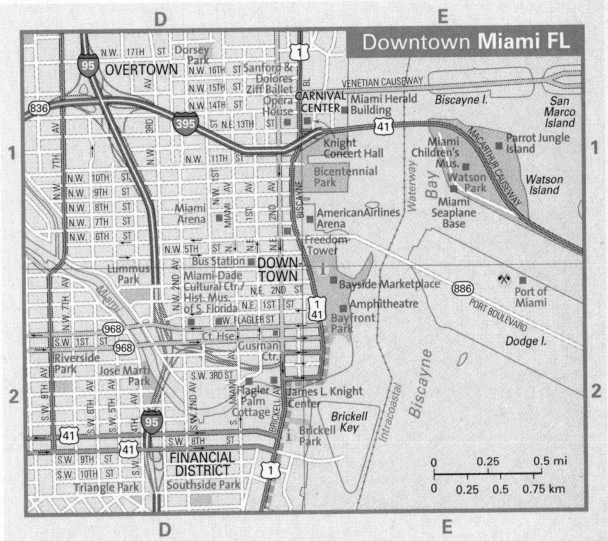

Figures after entries indicate population, page number, and grid reference.

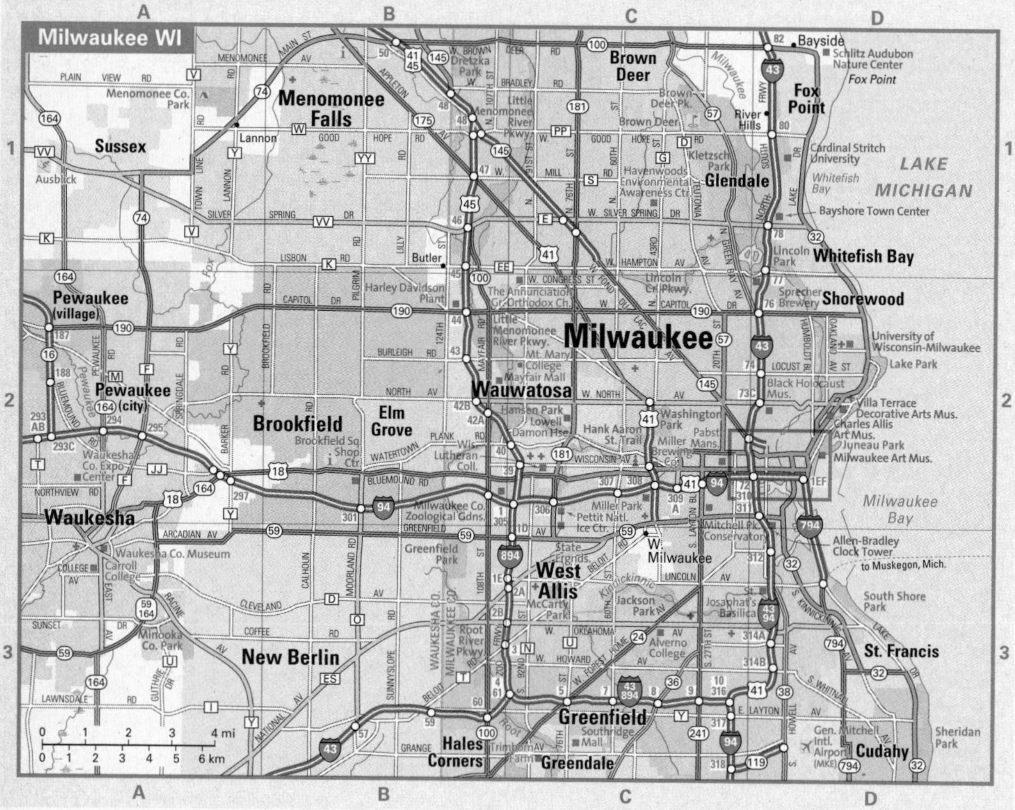

POINTS OF INTEREST

Amtrak Station F2
Betty Brinn Children's Museum G2
Bradley Center F1
Broadway Theatre Center F2
City Hall F1
Court House E1
Cudahy Gardens G2
Discovery World at Pier Wisconsin G2
The Eisner American Museum of Advertising & Design F2
Federal Plaza F2
Grain Exchange F2
Haggerty Museum of Art E2
Helfaer Theatre E2
Historic Third Ward F2
IMAX E1
Intercity Bus Depot F2
Lakeshore State Park G2
Maier Festival Park G2
Marcus Ctr. for the Performing Arts F1
Marquette University E2
Midwest Airlines Center F2
Milwaukee Art Museum & War Memorial Center G1
Milwaukee County Hist. Center F1
Milwaukee Institute of Art & Design F2
Milwaukee Public Market F2
Milwaukee Public Museum E1
Milwaukee School of Engineering F1
Milwaukee Theatre F1
Pabst Theater F1
Potawatomi Bingo & Casino E2
St. Joan of Arc Chapel E2
The Shops of Grand Avenue F2
State Office Building F1
U.S. Cellular Arena F1
Wisconsin Conservatory of Music G1

Entries in **bold black** indicate counties or parishes.
Entries in **bold color** indicate cities with detailed inset maps.

Medinah IL, *2500* **203** B3
Medley FL, *1098* **233** A4
Medora IN, *565* **99** F3
Medora ND, *100* **18** A4
Medulla FL, *6637* **140** C2
Medway ME, *650* **85** D4
Medway MA, *4000* **150** C2
Meeker CO, *2242* **40** B1
Meeker OK, *978* **51** F3
Meeker Co. MN, *22644* **66** B3
Meenon WI, *250* **67** E2
Meeteetse WY, *351* **24** B3
Meggett SC, *1230* **131** D2
Mehlville MO, *28822* **98** A3
Meigs GA, *1090* **137** E1
Meigs Co. OH, *23072* **101** E2
Meigs Co. TN, *11086* **110** B4
Meiners Oaks CA, *3750* **52** B2
Melba ID, *439* **22** B4
Melbourne AR, *1673* **107** E4
Melbourne FL, *71382* **141** E2
Melbourne IA, *794* **87** D1
Melbourne KY, *457* **204** C3
Melbourne Beach FL, *3335* **141** E2
Melbourne Vil. FL, *706* **141** E2
Melcher-Dallas IA, *1298* **87** D2
Melfa VA, *450* **114** C3
Melissa TX, *1350* **59** F2
Mellen WI, *845* **65** D4
Mellette SD, *248* **27** E2
Mellette Co. SD, *2083* **26** C4
Melrose MN, *3091* **66** B2
Melrose NM, *736* **49** F4
Melrose NY, *800* **94** B1
Melrose WI, *529* **73** F1
Melrose Park IL, *23171* **203** C4
Melrose Park NY, *2359* **79** D3
Melvern KS, *429* **96** A3
Melville LA, *1376* **133** F1
Melvindale MI, *10735* **210** B4
Memphis FL, *7264* **140** B3
Memphis MI, *1129* **76** C3
Memphis MO, *2061* **87** E4
Memphis TN, *650100* **118** B1
Memphis TX, *2479* **50** B4
Memphis Jct. KY, *550* **109** F2
Mena AR, *5637* **116** C2
Menahga MN, *1220* **64** A4
Menan ID, *707* **23** E4
Menands NY, *3910* **188** E3
Menard TX, *1653* **60** C1
Menard Co. IL, *12486* **88** B4
Menard Co. TX, *2360* **60** C1
Menasha WI, *16331* **74** C1
Mendenhall MS, *2555* **126** B3
Mendenhall PA, *1600* **146** B3
Mendham NJ, *5097* **148** A4
Mendocino CA, *824* **36** A2
Mendocino Co. CA, *86265* **36** A1
Mendon IL, *883* **87** F4
Mendon MA, *1100* **150** C2
Mendon MI, *917* **89** F1
Mendon OH, *697* **90** B3
Mendon UT, *898* **31** E3
Mendota CA, *7890* **44** C3
Mendota IL, *7272* **88** B2
Mendota Hts. MN, *11434* **235** D3
Menifee Co. KY, *6556* **100** C4
Menlo GA, *485* **120** B3
Menlo Park CA, *30785* **259** C5
Menno SD, *729* **27** E4
Menominee MI, *9131* **69** D3
Menominee Co. MI, *25326* **69** D3
Menominee Co. WI, *4562* **68** C3
Menomonee Falls WI, *32647* **74** C3
Menomonie WI, *14937* **67** E4
Mentone AL, *451* **120** B2
Mentone CA, *7803* **53** D2
Mentone IN, *898* **89** F2
Mentone TX, *20* **57** E4
Mentor OH, *50278* **91** E2
Mentor TN, *1500* **110** C4
Mentor-on-the-Lake OH, *8127* **91** E1
Mequon WI, *21823* **74** C3
Meraux LA, *10192* **134** C3
Merced CA, *63893* **37** D4
Merced Co. CA, *210554* **44** C2
Mercedes TX, *13649* **63** E4
Mercer PA, *2391* **91** F2
Mercer Co. IL, *16957* **87** F2
Mercer Co. KY, *20817* **110** B1
Mercer Co. MO, *3757* **86** C4
Mercer Co. NJ, *350761* **147** D2
Mercer Co. ND, *8644* **18** B3
Mercer Co. OH, *40924* **90** A4
Mercer Co. PA, *120293* **91** F2
Mercer Co. WV, *62980* **112** A1
Mercer Island WA, *22036* **262** B3
Mercersburg PA, *1540* **103** D1
Mercerville NJ, *26419* **147** D2
Merchantville NJ, *3801* **248** D3
Meredith NH, *1739* **81** F3
Meredosia IL, *1041* **98** A1
Meriden CT, *58244* **149** D1
Meriden KS, *706* **96** A2
Meridian ID, *34919* **22** B4
Meridian MS, *39968* **127** D2
Meridian PA, *3794* **92** A3
Meridian TX, *1491* **59** E3
Meridian Hills IN, *1713* **99** F1
Meridianville AL, *4117* **119** F2
Merigold MS, *664* **118** A4
Meriwether Co. GA, *22534* **128** C1
Merkel TX, *2637* **58** C3
Merlin OR, *650* **28** B1
Mermentau LA, *721* **133** E2
Merriam KS, *11008* **224** B4
Merriam Woods MO, *1142* **107** D3
Merrick NY, *22764* **148** C4
Merrick Co. NE, *8204* **35** E3
Merrifield VA, *11170* **270** A3
Merrill IA, *754* **35** F1
Merrill MI, *782* **76** A2
Merrill OR, *897* **28** C2
Merrill WI, *10146* **68** B3
Merrillan WI, *585* **73** F1
Merrillville IN, *30560* **89** D2
Merrimac MA, *2900* **151** F1
Merrimack NH, *1600* **95** D1
Merrimack Co. NH, *136225* **81** F4
Merrionette Park IL, *1999* **203** D6
Merritt Island FL, *36090* **141** E2
Mer Rouge LA, *721* **125** F2
Merrow CT, *1000* **150** A3
Merryville LA, *1126* **133** D1
Merton WI, *1926* **74** C3
Mertzon TX, *839* **58** B4
Mertztown PA, *1200* **146** B1
Mesa AZ, *396375* **54** C1
Mesa WA, *425* **13** E4
Mesa Co. CO, *116255* **40** B2
Mescalero NM, *1233* **56** C2
Mesilla NM, *2180* **56** C3
Mesita NM, *776* **48** B3
Mesquite NV, *9389* **46** B1
Mesquite NM, *948* **56** C3
Mesquite TX, *124523* **59** F2
Metairie LA, *146136* **134** B3
Metamora IL, *2700* **88** B3
Metamora MI, *507* **76** C3
Metamora OH, *563* **90** B2
Metcalfe MS, *1109* **118** A4
Metcalfe Co. KY, *10037* **110** A2
Methuen MA, *43789* **95** E1
Metlakatla AK, *1375* **155** E4
Metolius OR, *635* **21** D3
Metropolis IL, *6482* **108** C2
Metter GA, *3879* **129** F2
Metuchen NJ, *12840* **147** E1
Metzger OR, *3354* **251** C2
Mexia TX, *6563* **59** F4

Apple Valley C4
Arden Hills C1
Bald Eagle E1
Birchwood Vil. E1
Blaine C1
Bloomington B4
Brooklyn Ctr. B1
Brooklyn Park B1
Burnsville B4
Circle Pines C1
Columbia Hts. C2
Coon Rapids B1
Cottage Grove E4
Crystal B2
Deephaven A3
Dellwood E1
Eagan D4
Eden Prairie A4
Edina B3
Falcon Hts. C2
Fridley C1
Gem Lake D1
Golden Valley B2
Grant E1
Hilltop C1
Hopkins A3
Hugo E1
Inver Grove Hts. D4
Lake Elmo E2
Landfall E3
Lauderdale C2
Lexington C1
Lilydale D3
Lino Lakes D1
Little Canada D2
Mahtomedi E1
Maple Grove A1
Maplewood D2
Medicine Lake A2
Mendota C3
Mendota Hts. D3
Minneapolis B3
Minnetonka A3
Mounds View C1
New Brighton C1
New Hope B2
Newport D3
Nininger E4
N. Oaks D1
N. St. Paul D2
Oakdale E2
Osseo A1
Pine Sprs. E2
Plymouth A2
Richfield B3
Robbinsdale B2
Rosemount D4
Roseville C2
St. Anthony C2
St. Louis Park B2
St. Paul D2
St. Paul Park E4
Savage B4
Shakopee A4
Shoreview C1
S. St. Paul D3
Spring Lake Park C1
Sunfish Lake D3
Vadnais Hts. D1
Wayzata A2
W. St. Paul D3
White Bear Lake D1
Willernie E1
Withrow E1
Woodbury E3
Woodland A3

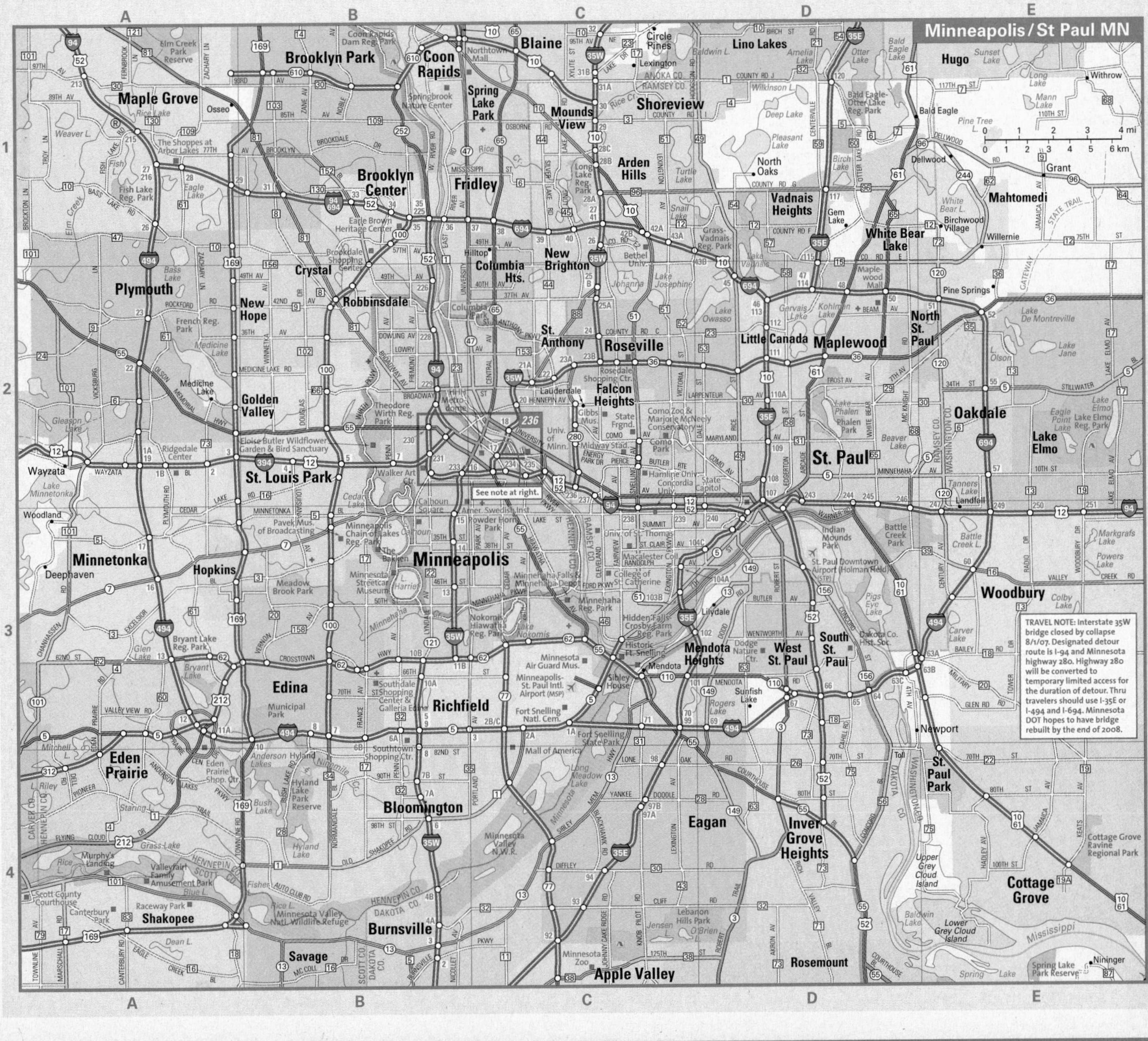

Figures after entries indicate population, page number, and grid reference.

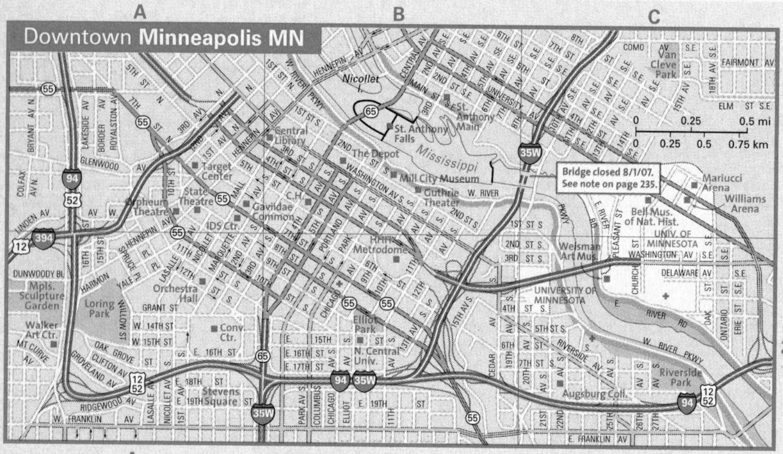

POINTS OF INTEREST

Augsburg College C2
Bell Museum of Natural History C1
Central Library B1
City Hall B1
Convention Center A2
The Depot B1
Gaviidae Common A1
Guthrie Theater B1
HHH Metrodome B2
IDS Center A1
Mariucci Arena C1
Mill City Museum B1
Minneapolis Sculpture Garden A2
North Central University B2
Orchestra Hall A2
Orpheum Theatre A1
St. Anthony Falls B1
St. Anthony Main B1
State Theatre A1
Target Center A1
University of Minnesota C1, C2
Walker Art Center A2
Weisman Art Museum C2
Williams Arena C1

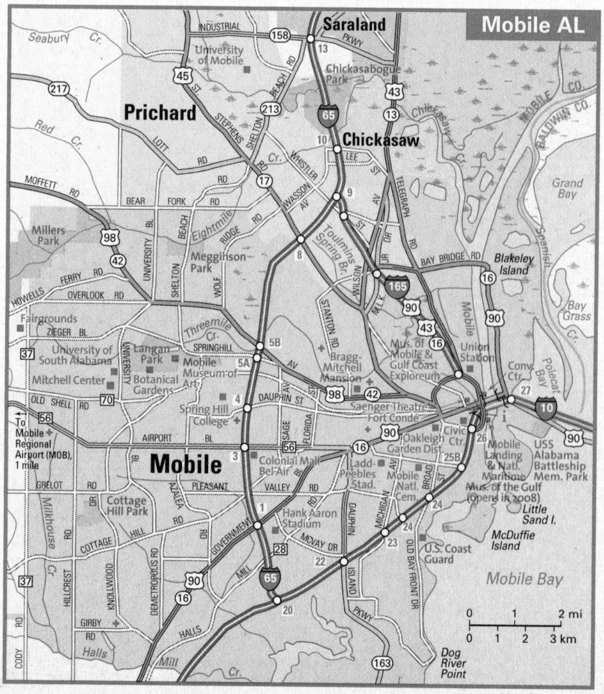

Aptos D1
Aromas E2
Ben Lomond D1
Bolsa Knolls E2
Capitola D1
Carmel-by-the-Sea . . D3
Carmel Highlands . . . D3
Carmel Valley E3
Castroville E2
Corralitos E1
Del Rey Oaks E3
Elkhorn E2
Felton D1
Freedom E1
Gabilan Acres E2
Gilroy E1
La Selva Beach D1
Las Lomas E2
Live Oak D1
Marina E2
Monterey D3
Moss Landing E2
Mt. Hermon D1
Opal Cliffs D1
Pacific Grove D3
Pajaro E1
Paradise Park D1
Pebble Beach D3
Prunedale E2
Rio del Mar D1
Salinas E3
Sand City E3
Santa Cruz D1
Scotts Valley D1
Seaside E3
Soquel D1
Spreckels E3
Twin Lakes D1
Watsonville E1
Zayante D1

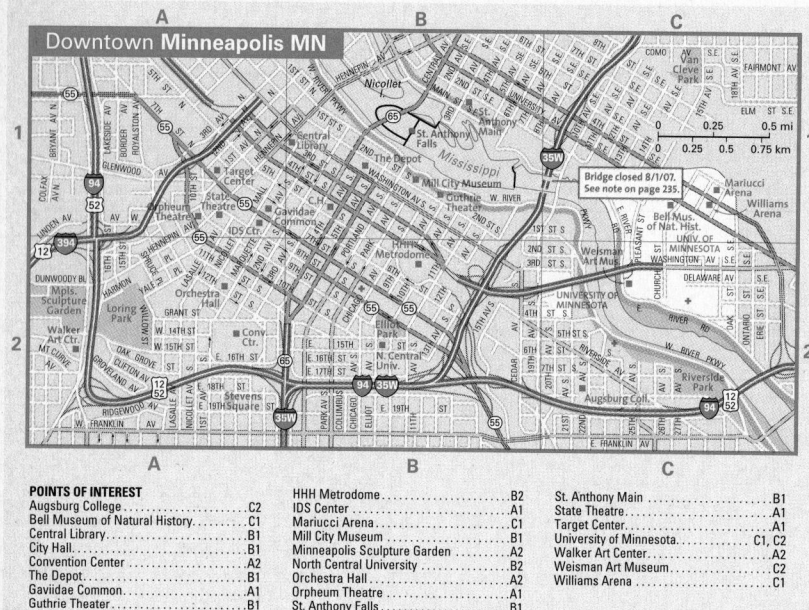

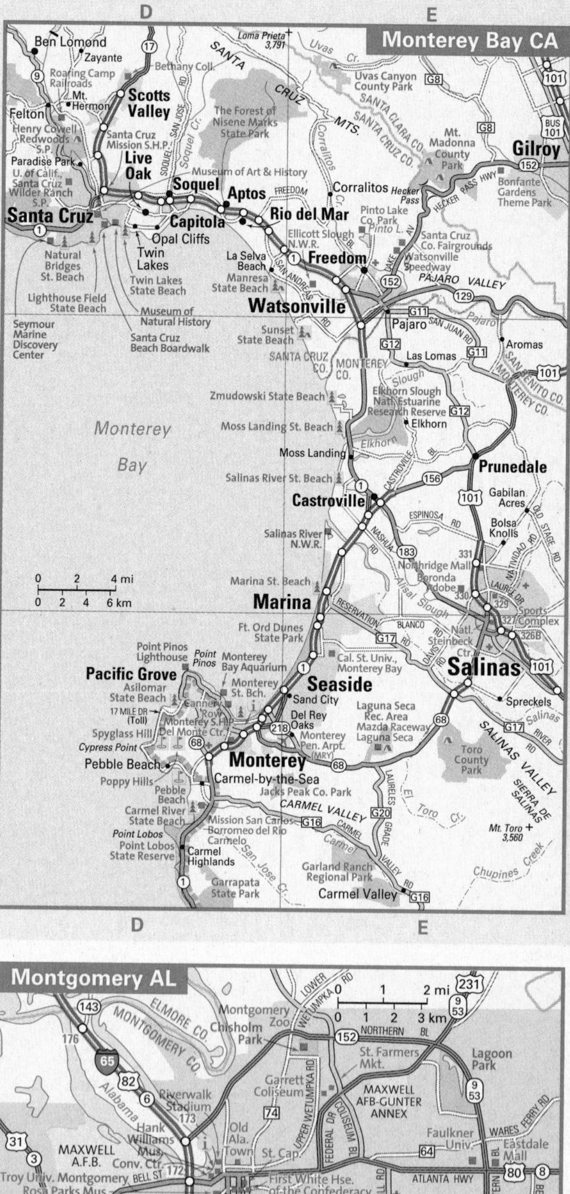

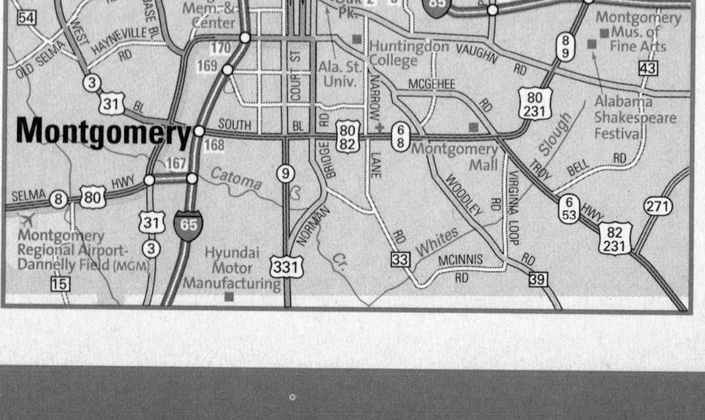

Entries in **bold black** indicate counties or parishes.
Entries in **bold color** indicate cities with detailed inset maps.

Miles City MT, 8487 ... 17 E4
Milford CT, 52305 ... 149 D2
Milford DE, 6732 ... 145 F3
Milford IL, 1369 ... 89 D3
Milford IN, 1550 ... 89 F2
Milford IA, 2474 ... 72 A3
Milford KS, 502 ... 43 F2
Milford ME, 2197 ... 83 D1
Milford MD, 26527 ... 193 B2
Milford MA, 24230 ... 150 C2
Milford MI, 6272 ... 76 B4
Milford NE, 2070 ... 35 F4
Milford NH, 8293 ... 95 D1
Milford NJ, 1195 ... 146 C1
Milford NY, 511 ... 79 F4
Milford OH, 6284 ... 100 B2
Milford PA, 1104 ... 94 A3
Milford TX, 685 ... 59 F3
Milford UT, 1451 ... 39 D3
Milford Ctr. OH, 626 ... 90 C4
Milford Square PA, 1100 ... 146 C1
Mililani Town HI, 28608 ... 152 A3
Millard Co. UT, 12405 ... 39 D2
Millbourne PA, 943 ... 248 B3
Millbrae CA, 20718 ... 259 B4
Millbrook AL, 10386 ... 128 A2
Millbrook NY, 1429 ... 94 B3
Millburn NJ, 19765 ... 240 A3
Millbury MA, 4700 ... 150 C2
Millbury OH, 1161 ... 90 C2
Mill City NV, 300 ... 29 F4
Mill City OR, 1537 ... 20 C3
Mill Creek WA, 11525 ... 12 C3
Mill Creek WV, 662 ... 102 A3
Milledgeville GA, 18757 ... 129 E1
Milledgeville IL, 1016 ... 88 B1
Mille Lacs Co. MN, 22330 ... 66 C2
Millen GA, 3492 ... 129 F2
Miller MO, 754 ... 106 C2
Miller SD, 1530 ... 27 D3
Miller Co. AR, 40443 ... 124 C1
Miller Co. GA, 6383 ... 128 C4
Miller Co. MO, 23564 ... 97 E4
Miller Place NY, 10580 ... 149 D3
Millersburg IN, 868 ... 89 F2
Millersburg KY, 842 ... 100 B4
Millersburg OH, 3326 ... 91 E4
Millersburg OR, 651 ... 20 B3
Millersburg PA, 2562 ... 93 D4
Millers Creek NC, 2071 ... 111 F3
Millers Falls MA, 1072 ... 94 C1
Millersport OH, 963 ... 101 D1
Millerstown PA, 679 ... 93 D4
Millersville PA, 7774 ... 146 A3
Millersville TN, 5308 ... 109 F3
Millerton NY, 925 ... 94 B2
Mill Hall PA, 1568 ... 93 D3
Milliken CO, 2888 ... 33 E4
Millington MI, 1137 ... 76 B3
Millington NJ, 3500 ... 148 A4
Millington TN, 10433 ... 118 B1
Millinocket ME, 5190 ... 85 D4
Millis MA, 4607 ... 150 C2
Mill Neck NY, 825 ... 148 C3
Millport AL, 1160 ... 119 D4
Millry AL, 615 ... 127 D4
Mills WY, 2591 ... 33 D1
Millsboro DE, 2360 ... 145 F4
Mills Co. IA, 14547 ... 86 A3
Mills Co. TX, 5151 ... 59 D4
Millstadt IL, 2794 ... 98 A3
Millstone KY, 650 ... 111 E2
Milltown IN, 932 ... 99 F4
Milltown NJ, 7000 ... 147 E1
Milltown WI, 888 ... 67 E3
Millvale PA, 4028 ... 250 B1
Mill Valley CA, 13600 ... 36 B4
Millville CA, 610 ... 28 C4
Millville DE, 259 ... 145 F4
Millville NJ, 26847 ... 145 F1
Millville OH, 817 ... 100 B2
Millville PA, 991 ... 93 E3
Millville UT, 1507 ... 31 E3
Millwood NY, 2300 ... 148 B2
Millwood WA, 1649 ... 14 A3
Milner GA, 522 ... 129 D1
Milnor ND, 711 ... 27 E1
Milo IA, 839 ... 86 C2
Milo ME, 1898 ... 84 C4
Milpitas CA, 62698 ... 36 B4
Milroy IN, 800 ... 100 A2
Milroy PA, 1386 ... 93 D3
Milton DE, 1657 ... 145 F3
Milton FL, 7045 ... 135 F1
Milton IN, 611 ... 100 A1
Milton IA, 550 ... 87 E3
Milton KY, 525 ... 100 A3
Milton MA, 26062 ... 151 D1
Milton NH, 950 ... 81 F4
Milton NY, 1251 ... 148 B1
Milton PA, 6650 ... 93 D3
Milton VT, 1537 ... 81 D1
Milton WA, 5795 ... 262 B5
Milton WV, 2206 ... 101 E3
Milton WI, 5132 ... 74 B4
Milton-Freewater OR, 6470 ... 21 F1
Miltonvale KS, 523 ... 43 E2
Milwaukee WI, 596974 ... 74 C3
Milwaukee Co. WI, 940164 ... 75 D3
Milwaukie OR, 20490 ... 251 D2
Mimosa Park LA, 4500 ... 239 A2
Mims FL, 9147 ... 141 E1
Mina NV, 275 ... 37 F3
Minatare NE, 810 ... 33 F2
Minco OK, 1672 ... 51 E3
Minden IA, 564 ... 86 A2
Minden LA, 13027 ... 125 D2
Minden NE, 2964 ... 35 D4
Minden NV, 2836 ... 37 D2
Mineola NY, 19234 ... 148 C4
Mineola TX, 4550 ... 124 A2
Miner MO, 1056 ... 108 B2
Mineral VA, 424 ... 103 D4
Mineral City OH, 841 ... 91 E4
Mineral Co. CO, 831 ... 40 C4
Mineral Co. MT, 3884 ... 14 C3
Mineral Co. NV, 5071 ... 37 E3
Mineral Co. WV, 27078 ... 102 C2
Mineral Pt. WI, 2617 ... 74 A3
Mineral Ridge OH, 3900 ... 276 B2
Mineral Sprs. AR, 1264 ... 116 C4
Mineral Sprs. NC, 1370 ... 122 B2
Mineral Wells TX, 16946 ... 59 E2
Mineral Wells WV, 1860 ... 101 E2
Miner Co. SD, 2884 ... 27 E3
Minersville PA, 4552 ... 93 E3
Minersville UT, 817 ... 39 D3
Minerva OH, 3934 ... 91 F3
Minerva Park OH, 1288 ... 206 B1
Minetto NY, 1086 ... 79 D2
Mineville NY, 1747 ... 81 D2
Mingo Co. WV, 28253 ... 101 E4
Mingo Jct. OH, 3631 ... 91 F4
Minidoka Co. ID, 20174 ... 31 D1
Minier IL, 1244 ... 88 B4
Minneapolis KS, 2046 ... 43 E2
Minneapolis MN, 382618 ... 67 D4
Minnehaha Co. SD, 148281 ... 27 F4
Minneola FL, 5435 ... 140 C1
Minneola KS, 717 ... 42 C4
Minneota MN, 1449 ... 27 F3
Minnesota Lake MN, 681 ... 72 C2
Minnetonka MN, 51301 ... 235 A3
Minnewaukan ND, 318 ... 19 D2
Minoa NY, 3348 ... 79 E3
Minocqua WI, 750 ... 68 B2
Minong WI, 531 ... 67 E2
Minonk IL, 2168 ... 88 B3
Minooka IL, 3971 ... 88 C2
Minor Hill TN, 437 ... 119 E2
Minot ME, 400 ... 82 B3
Minot MA, 1100 ... 151 E2
Minot ND, 36567 ... 18 B2
Minster OH, 2794 ... 90 B4
Mint Hill NC, 14922 ... 122 B1
Minto ND, 657 ... 19 E2
Minturn CO, 1068 ... 41 D1
Mio MI, 2016 ... 70 C4
Mira Loma CA, 17617 ... 229 H3
Miramar FL, 72739 ... 143 E2
Miramar Beach FL, 2435 ... 136 B2
Misenheimer NC, 750 ... 122 B1
Mishawaka IN, 46557 ... 89 F2
Mishicot WI, 1422 ... 75 D1
Missaukee Co. MI, 14478 ... 75 F1
Mission KS, 9727 ... 224 B3
Mission OR, 1019 ... 21 F1
Mission SD, 904 ... 26 C4
Mission TX, 45408 ... 63 E4
Mission Bend TX, 30831 ... 220 A3
Mission Hills KS, 3593 ... 224 B4
Mission Viejo CA, 93102 ... 52 C3
Mississippi Co. AR, 51979 ... 108 B4
Mississippi Co. MO, 13427 ... 108 C2
Mississippi State MS, 3500 ... 119 D4
Missoula MT, 57053 ... 15 D4
Missoula Co. MT, 95802 ... 15 D4
Missouri City TX, 52913 ... 132 A3
Missouri Valley IA, 2992 ... 86 A2
Mitchell IN, 4567 ... 99 F3
Mitchell NE, 1831 ... 33 F2
Mitchell SD, 14558 ... 27 E4
Mitchell Co. GA, 23932 ... 137 E1
Mitchell Co. IA, 10874 ... 73 D2
Mitchell Co. KS, 6932 ... 43 E2
Mitchell Co. NC, 15687 ... 111 E4
Mitchell Co. TX, 9698 ... 58 B3
Mitchellville AR, 497 ... 117 F4
Mitchellville IA, 1715 ... 87 D2
Mi-Wuk Vil. CA, 1485 ... 37 D3
Moab UT, 4779 ... 40 A3
Moapa NV, 928 ... 46 B1
Moberly MO, 11945 ... 97 E2
Mobile AL, 198915 ... 135 E1
Mobile Co. AL, 399843 ... 135 E1
Mobridge SD, 3574 ... 26 C2
Mocksville NC, 4178 ... 112 A4
Moclips WA, 615 ... 12 B3
Modena NY, 1100 ... 148 B1
Modena PA, 610 ... 146 B3
Modesto CA, 188856 ... 36 C4
Modoc Co. CA, 9449 ... 29 D3
Moenkopi AZ, 901 ... 47 E2
Moffat Co. CO, 13184 ... 32 B4
Mogadore OH, 3893 ... 188 B2
Mohall ND, 812 ... 18 B1
Mohave Co. AZ, 155032 ... 46 B3
Mohave Valley AZ, 13694 ... 46 B3
Mohawk MI, 750 ... 65 F3
Mohawk NY, 2660 ... 79 F3
Mohegan CT, 3500 ... 149 F1
Mohegan Lake NY, 5700 ... 148 B2
Mohnton PA, 2963 ... 146 A2
Mohrsville PA, 800 ... 146 A1
Mojave CA, 3836 ... 52 C1
Mokelumne Hill CA, 774 ... 36 C3
Mokuleia HI, 1839 ... 152 A2
Molalla OR, 5647 ... 20 C2
Molena GA, 475 ... 128 C1
Moline IL, 43768 ... 88 A2
Moline KS, 457 ... 43 F4
Moline Acres MO, 2662 ... 256 C1
Molino FL, 1312 ... 135 F1
Momence IL, 3171 ... 89 D2
Mona UT, 850 ... 39 E1
Monaca PA, 6286 ... 91 F3
Monahans TX, 6821 ... 57 F4
Monarch Mills SC, 1930 ... 121 F2
Moncks Corner SC, 5952 ... 131 D1
Mondovi WI, 2634 ... 67 E4
Monee IL, 2924 ... 89 D2
Monessen PA, 8669 ... 92 A4
Monett MO, 7396 ... 106 C2
Monette AR, 1179 ... 108 A4
Monfort Hts. OH, 3880 ... 204 A2
Moniteau Co. MO, 14827 ... 97 D3
Monmouth IL, 9841 ... 88 A3
Monmouth ME, 500 ... 82 B2
Monmouth OR, 7741 ... 20 B2
Monmouth Beach NJ, 3595 ... 147 F2
Monmouth Co. NJ, 615301 ... 147 E2
Monmouth Jct. NJ, 2721 ... 147 D1
Mono Co. CA, 12853 ... 37 E4
Monon IN, 1733 ... 89 E3
Monona IA, 1550 ... 73 F3
Monona WI, 8018 ... 74 B3
Monona Co. IA, 10020 ... 86 A1
Monongah WV, 939 ... 102 A2
Monongahela PA, 4761 ... 92 A4
Monongalia Co. WV, 81866 ... 102 A1
Monponsett MA, 1700 ... 151 E2
Monroe CT, 3000 ... 149 D2
Monroe GA, 11407 ... 121 D4
Monroe IN, 734 ... 90 A3
Monroe IA, 1808 ... 87 D2
Monroe LA, 53107 ... 125 E2
Monroe MI, 22076 ... 90 C1
Monroe NY, 7780 ... 148 B2
Monroe NC, 26228 ... 122 B2
Monroe OH, 7133 ... 100 B2
Monroe OR, 607 ... 20 B3
Monroe UT, 1845 ... 39 E3
Monroe VA, 1200 ... 112 C1
Monroe WA, 13795 ... 12 C3
Monroe WI, 10843 ... 74 B4
Monroe City IN, 548 ... 99 E3
Monroe City MO, 2588 ... 97 E1
Monroe Co. AL, 24324 ... 127 F4
Monroe Co. AR, 10254 ... 118 A2
Monroe Co. FL, 79589 ... 143 D3
Monroe Co. GA, 21757 ... 129 D1
Monroe Co. IL, 27619 ... 98 A4
Monroe Co. IN, 120563 ... 99 F2
Monroe Co. IA, 8016 ... 87 D3
Monroe Co. KY, 11756 ... 110 A2
Monroe Co. MI, 145945 ... 90 C1
Monroe Co. MS, 38014 ... 119 D4
Monroe Co. MO, 9311 ... 97 E2
Monroe Co. NY, 735343 ... 78 C3
Monroe Co. OH, 15180 ... 101 F1
Monroe Co. PA, 138687 ... 93 F3
Monroe Co. TN, 38961 ... 120 C1
Monroe Co. WV, 14583 ... 112 A1
Monroe Co. WI, 40899 ... 73 F1
Monroeville AL, 6862 ... 127 F4
Monroeville IN, 1236 ... 90 A3
Monroeville OH, 1433 ... 91 D2
Monroeville PA, 29349 ... 92 A4
Monrovia CA, 36929 ... 228 E2
Monrovia IN, 628 ... 99 F1
Monsey NY, 14504 ... 148 B2
Monson MA, 2103 ... 150 A2
Montague CA, 1456 ... 28 C3
Montague MA, 800 ... 94 C1
Montague MI, 2407 ... 75 E2
Montague TX, 225 ... 59 E1
Montague Co. TX, 19117 ... 59 E1
Mont Alto PA, 1357 ... 103 D1
Montana City MT, 2094 ... 15 E4
Montara CA, 2950 ... 259 A4
Montauk NY, 3851 ... 149 F3
Mont Belvieu TX, 2324 ... 132 B3
Montcalm WV, 885 ... 111 F1
Montcalm Co. MI, 61266 ... 75 F2
Montclair CA, 33049 ... 229 G2
Montclair NJ, 38977 ... 148 A3
Mont Clare PA, 1900 ... 146 B2
Monteagle TN, 1238 ... 120 A1
Monte Alto TX, 1611 ... 63 E4
Montebello CA, 62150 ... 228 D3
Montecito CA, 10000 ... 52 B2
Montegut LA, 1803 ... 134 B4

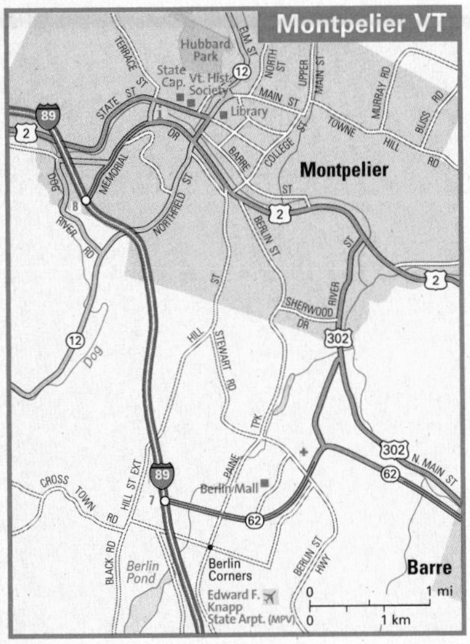

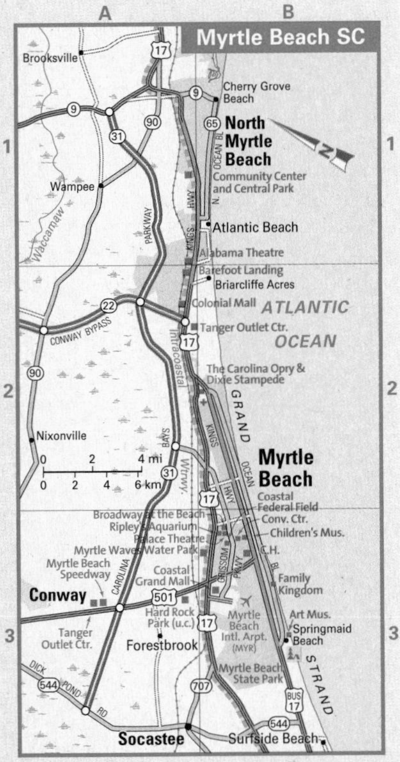

Atlantic Beach... B1	Conway......... A3	Socastee........ A3
Briarcliffe Acres . B2	Forestbrook A3	Springmaid Beach B3
Brooksville....... A1	Myrtle Beach.... B2	Surfside Beach .. B3
Cherry Grove Beach B1	Nixonville A2	Wampee........ A1
	N. Myrtle Beach . B1	

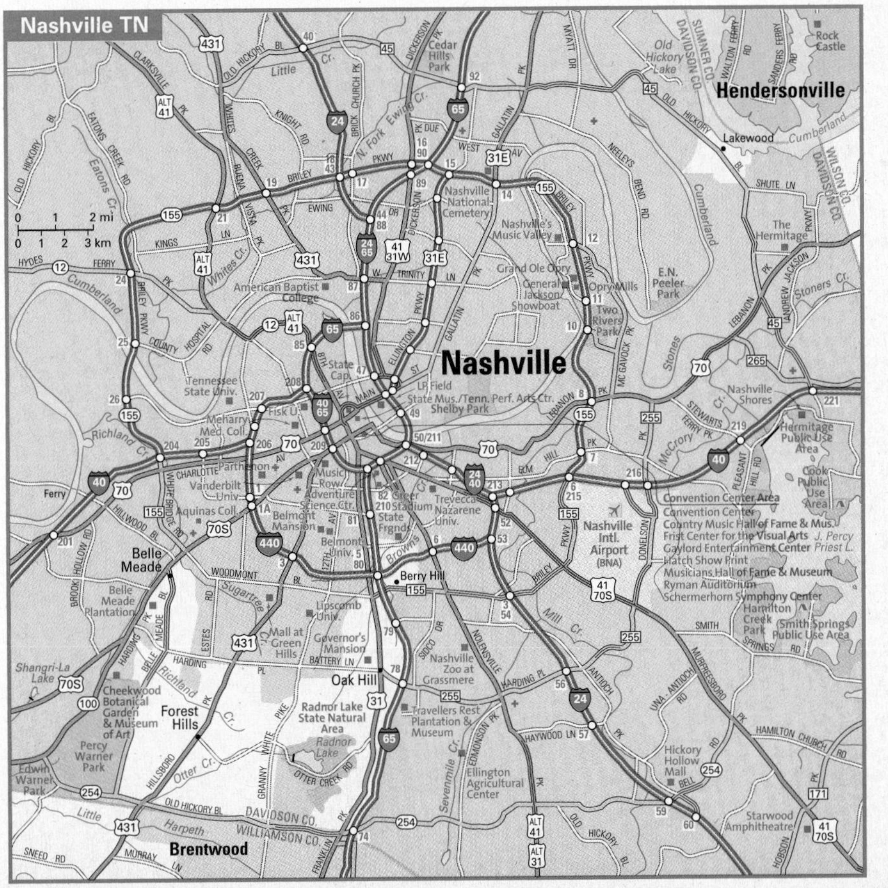

Figures after entries indicate population, page number, and grid reference.

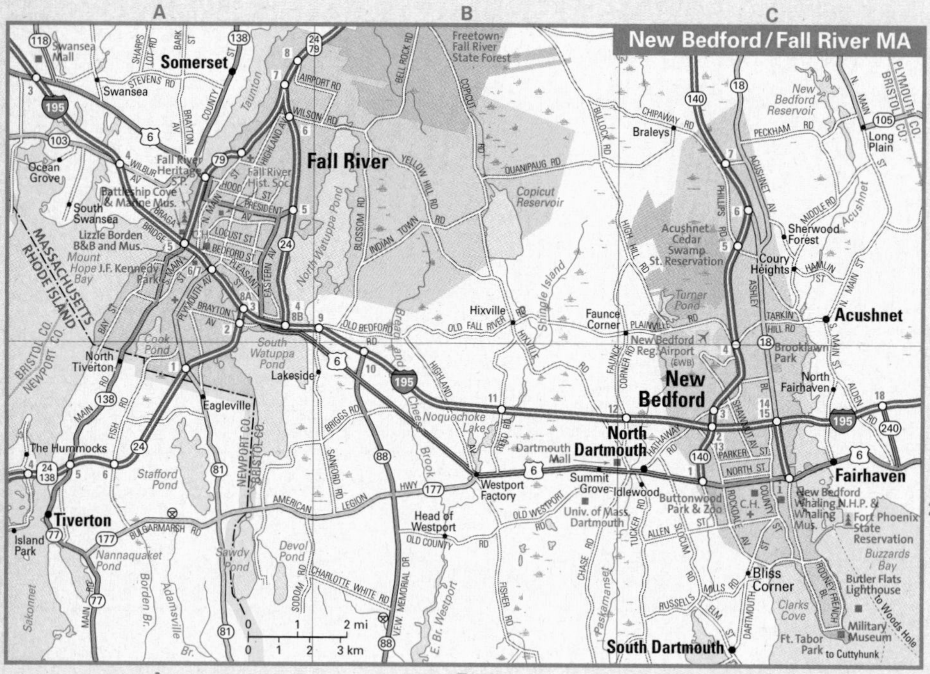

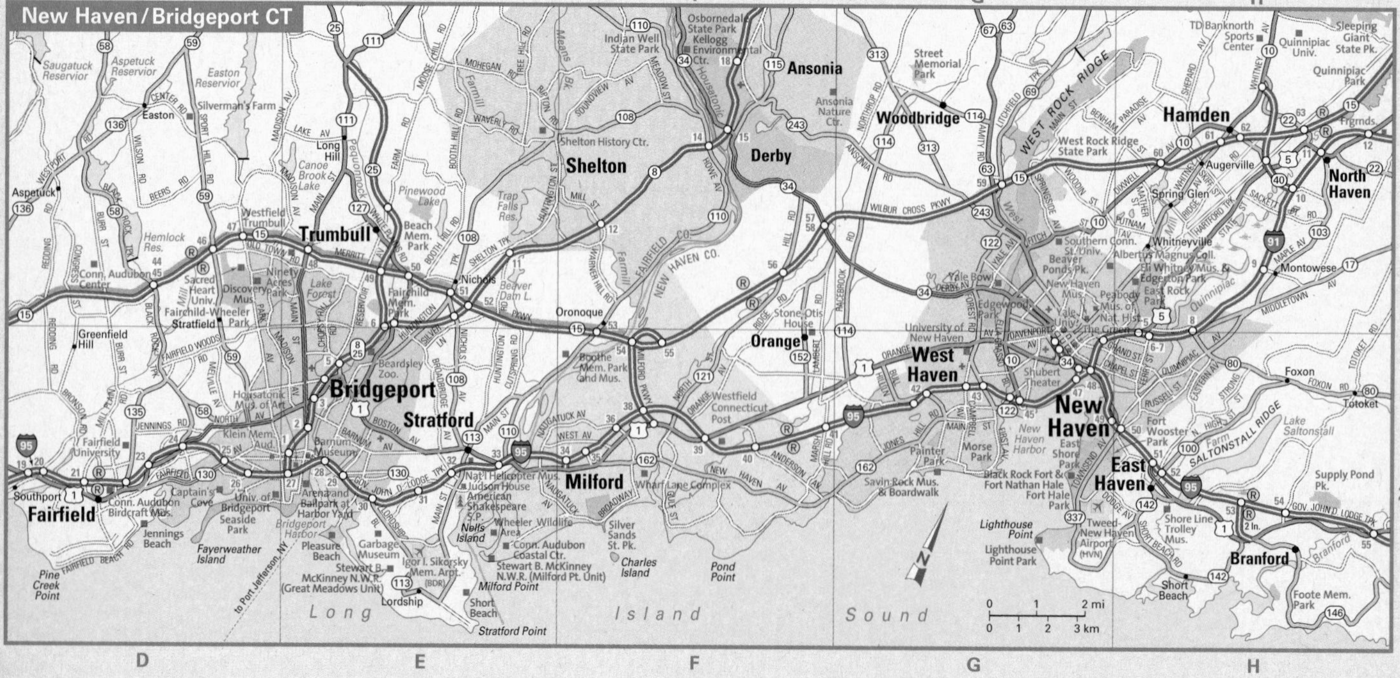

Entries in **bold black** indicate counties or parishes.
Entries in **bold color** indicate cities with detailed inset maps.

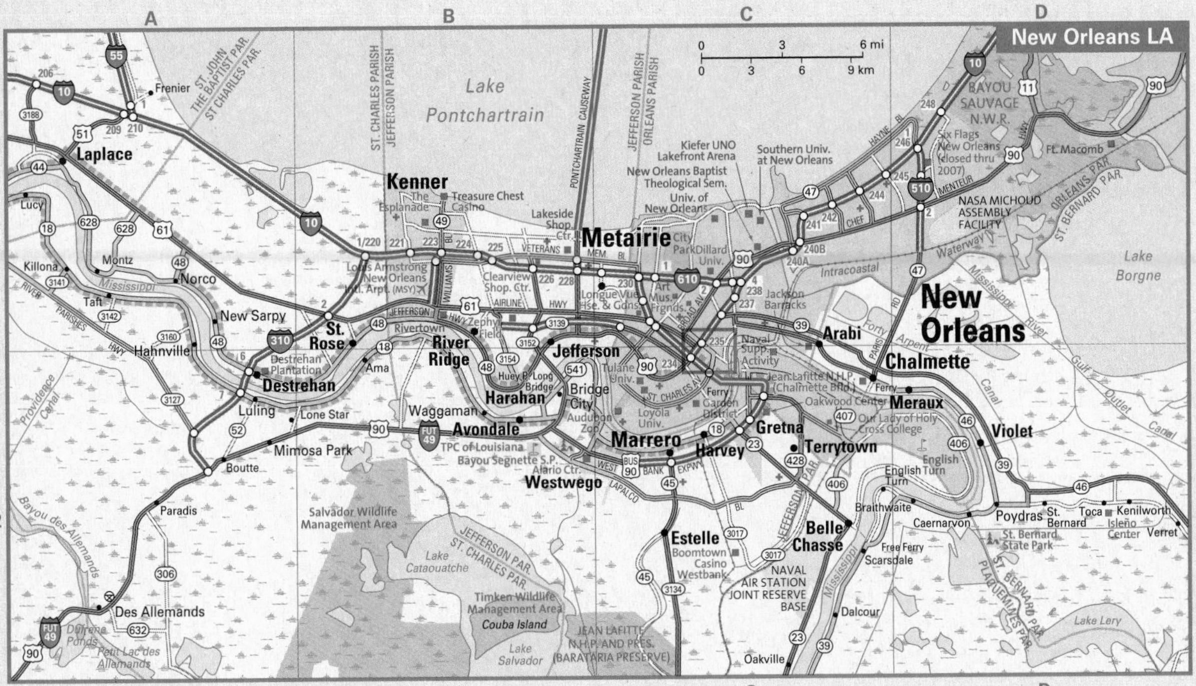

POINTS OF INTEREST

Aquarium of the Americas F2
The Cabildo F1
Confederate Museum F2
Contemporary Arts Center F2
Creole Queen F2
Ernest N. Morial Convention Center F2
French Quarter (Vieux Carré) F1
Harrah's F2
Jackson Square F1
Jean Lafitte Natl. Hist. Park (Visitor Center) F1
Louisiana Children's Museum F2
Louisiana Superdome E2
Mahalia Jackson Theatre of the Performing Arts. F1
Morris Jeff Municipal Auditorium F1
Musée Conti Wax Museum F1
National World War II Museum F2
New Orleans Arena E2
New Orleans Centre E2
New Orleans Jazz N.H.P. Visitor Center F1
Ogden Museum F2
Old U.S. Mint F1
Pontalba Buildings F1
The Presbytère F1
Public Library E2
St. Charles Avenue Streetcar F2
St. Louis Cathedral F1
U.S. Customs House F2
Woldenberg Riverfront Park F2
World Trade Center F2

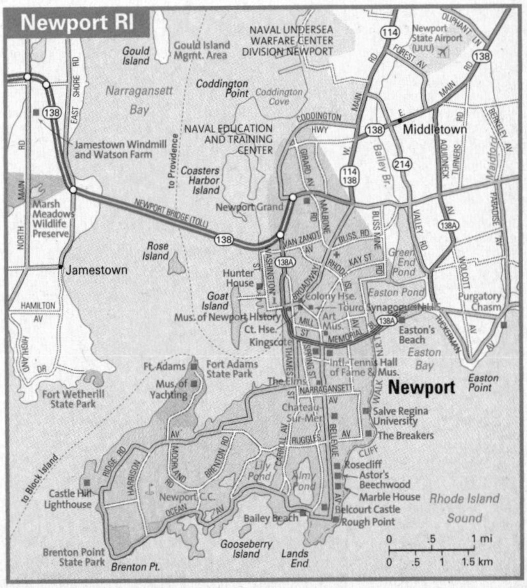

Figures after entries indicate population, page number, and grid reference.

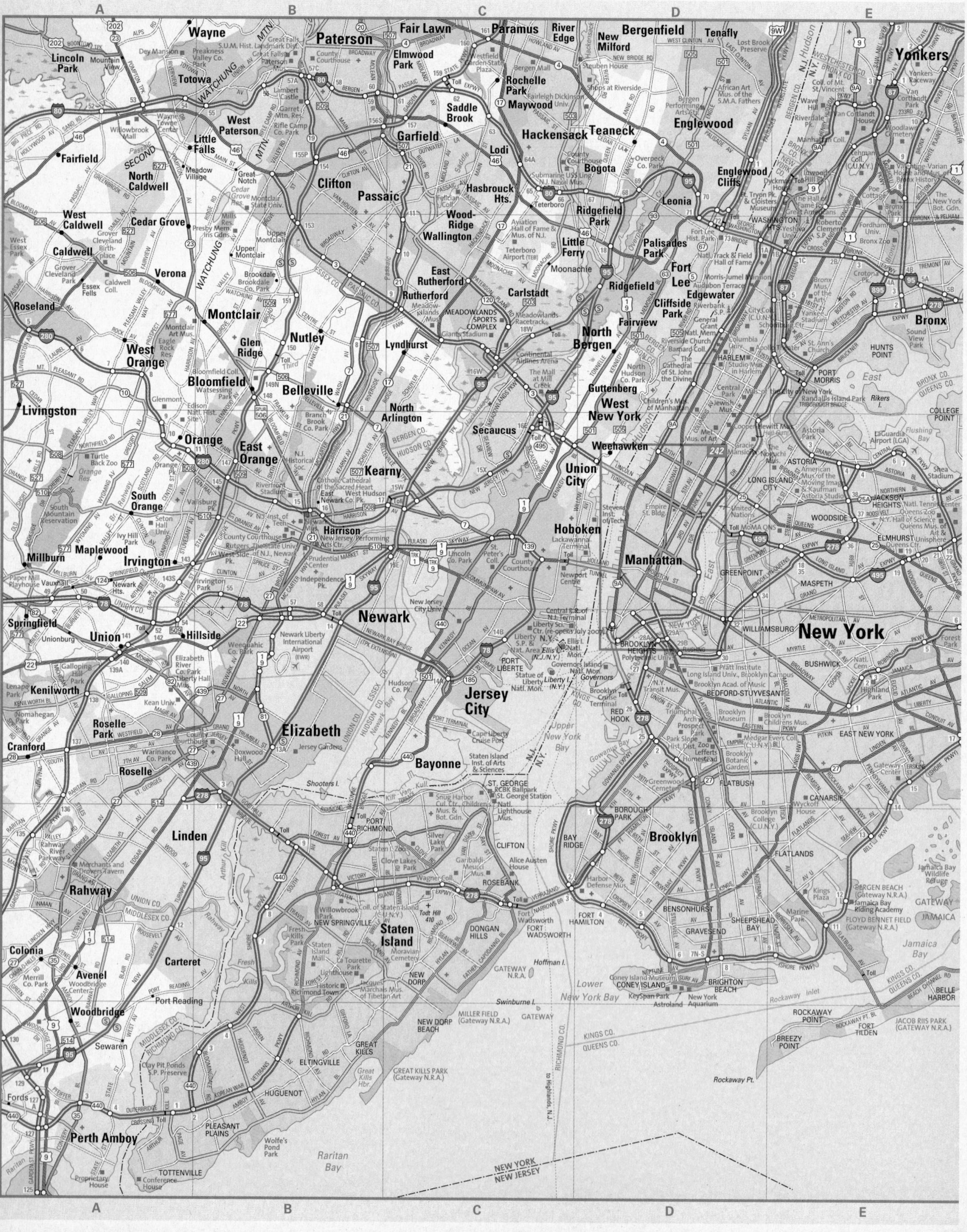

Entries in **bold black** indicate counties or parishes.
Entries in **bold color** indicate cities with detailed inset maps.

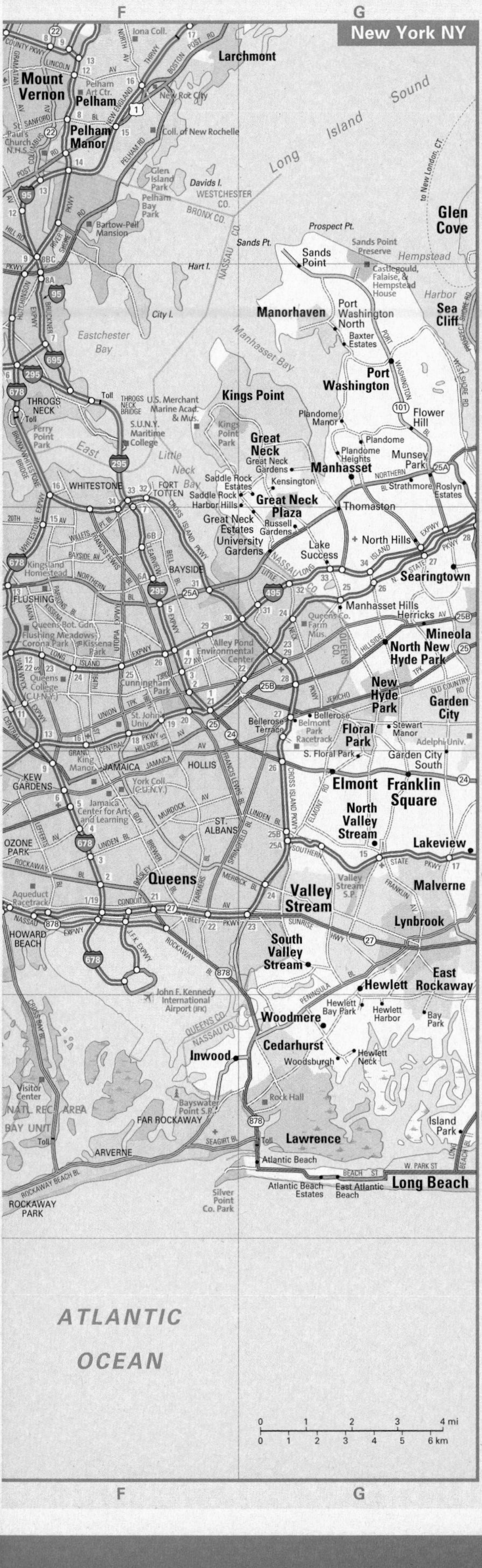

Atlantic Beach . . . G5
Atlantic Beach Estates . . G5
Avenel . . . A5
Baxter Estates . . . G2
Bayonne . . . C4
Bay Park . . . G5
Bellerose . . . G3
Bellerose Terrace . . . G3
Belleville . . . B2
Bergenfield . . . D1
Bloomfield . . . B2
Bogota . . . D1
Brookdale . . . B2
Caldwell . . . A2
Carlstadt . . . C2
Carteret . . . A5
Cedar Grove . . . A2
Cedarhurst . . . G5
Cliffside Park . . . D2
Clifton . . . B1
Colonia . . . A5
Cranford . . . A4
E. Atlantic Beach . . . G5
E. Newark . . . B3
E. Orange . . . B3
E. Rockaway . . . G4
E. Rutherford . . . C2
Edgewater . . . D2
Elizabeth . . . B4
Elmont . . . G4
Elmwood Park . . . C1
Englewood . . . D1
Englewood Cliffs . . . D1
Essex Fells . . . A2
Fairfield . . . A1
Fair Lawn . . . C1
Fairview . . . D2
Floral Park . . . G3
Flower Hill . . . G2
Fords . . . A6
Ft. Lee . . . D2
Franklin Square . . . G4
Garden City . . . G3
Garden City South . . . G4
Garfield . . . C1
Glen Cove . . . G1
Glen Ridge . . . B2
Great Neck . . . G2
Great Neck Estates . . . G3
Great Neck Gardens . . . G2
Great Neck Plaza . . . G2
Great Notch . . . B1
Guttenberg . . . D2
Hackensack . . . C1
Harbor Hills . . . G2
Harrison . . . B3
Hasbrouck Hts. . . . C1
Herricks . . . G3
Hewlett . . . G4
Hewlett Bay Park . . . G5
Hewlett Harbor . . . G5
Hewlett Neck . . . G5
Hillside . . . A4
Hoboken . . . C3
Inwood . . . F5
Irvington . . . A3
Island Park . . . G5
Jersey City . . . C4
Kearny . . . B3
Kenilworth . . . A4
Kensington . . . G2
Kings Pt. . . . G2
Lake Success . . . G3
Lakeview . . . G4
Larchmont . . . F1
Lawrence . . . G5
Leonia . . . D1
Lincoln Park . . . A1
Linden . . . A5
Little Falls . . . A1
Little Ferry . . . C2
Livingston . . . A2
Lodi . . . C1
Long Beach . . . G5
Lynbrook . . . G4
Lyndhurst . . . C2
Malverne . . . G4
Manhasset . . . G2
Manhasset Hills . . . G3
Manorhaven . . . G2
Maplewood . . . A3
Maywood . . . C1
Meadow Vil. . . . A1
Millburn . . . A3
Mineola . . . G3
Montclair . . . B2
Moonachie . . . C2
Mtn. View . . . A1
Mt. Vernon . . . F1
Munsey Park . . . G2
Newark . . . B4
Newark Hts. . . . A3
New Hyde Park . . . G3
New Milford . . . D1
New York . . . E4
N. Arlington . . . C2
N. Bergen . . . D2
N. Caldwell . . . A1
N. Hills . . . G3
N. New Hyde Park . . . G3
N. Valley Stream . . . G4
Nutley . . . B2
Orange . . . A3
Palisades Park . . . D2
Paramus . . . C1
Passaic . . . B1
Paterson . . . B1
Pelham . . . F1
Pelham Manor . . . F1
Perth Amboy . . . A6
Plandome . . . G2
Plandome Hts. . . . G2
Plandome Manor . . . G2
Port Reading . . . A5
Port Washington . . . G2
Port Washington North . . G2
Rahway . . . A5
Ridgefield . . . D2
Ridgefield Park . . . D1
River Edge . . . C1
Rochelle Park . . . C1
Roseland . . . A2
Roselle . . . A4
Roselle Park . . . A4
Roslyn Estates . . . G2
Russell Gardens . . . G3
Rutherford . . . C2
Saddle Brook . . . C1
Saddle Rock . . . G2
Saddle Rock Estates . . . G2
Sands Pt. . . . G2
Sea Cliff . . . G2
Searingtown . . . G3
Secaucus . . . C3
Sewaren . . . A6
S. Floral Park . . . G4
S. Orange . . . A3
S. Valley Stream . . . G4
Springfield . . . A4
Stewart Manor . . . G3
Strathmore . . . G2
Teaneck . . . D1
Tenafly . . . D1
Teterboro . . . C1
Thomaston . . . G3
Totowa . . . A1
Union . . . A4
Unionburg . . . A4
Union City . . . C3
University Gardens . . . G3
Upper Montclair . . . B2
Valley Stream . . . G4
Vauxhall . . . A3
Verona . . . A2
Wallington . . . C2
Wayne . . . A1
Weehawken . . . D3
W. Caldwell . . . A2
W. New York . . . D2
W. Orange . . . A2
W. Paterson . . . B1
Woodbridge . . . A6
Woodmere . . . G5
Wood-Ridge . . . C2
Woodsburgh . . . G5
Yonkers . . . E1

Mukilteo WA, *18019* . . . 12 C2
Mukwonago WI, *6162* . . . 74 C3
Mulberry AR, *1627* . . . 116 C1
Mulberry FL, *3230* . . . 140 C2
Mulberry IN, *1387* . . . 89 E4
Mulberry KS, *577* . . . 106 B1
Mulberry NC, *2269* . . . 111 F3
Mulberry OH, *3139* . . . 100 B2
Mulberry Grove IL, *671* . . . 98 B3
Muldraugh KY, *1298* . . . 99 F4
Muldrow OK, *3104* . . . 116 B1
Muleshoe TX, *4530* . . . 49 F4
Mulga AL, *973* . . . 195 D1
Mullan ID, *840* . . . 14 C3
Mullen NE, *491* . . . 34 B2
Mullens WV, *1769* . . . 111 F1
Mullica Hill NJ, *1658* . . . 146 C4
Mulliken MI, *557* . . . 76 A4
Mullins SC, *5029* . . . 122 C3
Multnomah Co. OR, *660486* . . . 20 C2
Mulvane KS, *5155* . . . 43 E4
Muncie IN, *67430* . . . 90 A4
Muncy PA, *2663* . . . 93 D2
Munday TX, *1527* . . . 58 C1
Mundelein IL, *30935* . . . 74 C4
Munds Park AZ, *1250* . . . 47 E3
Munford AL, *2446* . . . 120 A4
Munford TN, *4708* . . . 118 B1
Munfordville KY, *1563* . . . 110 A2
Munhall PA, *12264* . . . 250 C3
Munich ND, *268* . . . 19 D1
Munising MI, *2539* . . . 69 E1
Munroe Falls OH, *5314* . . . 188 B1
Munsey Park NY, *2632* . . . 241 G2
Munsons Corners NY, *2426* . . . 79 D4
Munster IN, *21511* . . . 89 D2
Murchison TX, *592* . . . 124 A3
Murdo SD, *612* . . . 26 C4
Murfreesboro AR, *1764* . . . 116 C3
Murfreesboro NC, *2045* . . . 113 E3
Murfreesboro TN, *68816* . . . 109 F4
Murphy ID, *40* . . . 30 B1
Murphy MO, *9048* . . . 98 A3
Murphy NC, *1568* . . . 120 C2
Murphy TX, *3099* . . . 207 E1
Murphys CA, *2061* . . . 37 D3
Murphysboro IL, *13295* . . . 108 C1
Murray IA, *766* . . . 86 C3
Murray KY, *14950* . . . 109 D3
Murray NE, *481* . . . 86 A3
Murray UT, *34024* . . . 31 E4
Murray Co. GA, *36506* . . . 120 C2
Murray Co. MN, *9165* . . . 72 A1
Murray Co. OK, *12623* . . . 51 F4
Murray Hill KY, *616* . . . 230 F1
Murrayville IL, *644* . . . 98 A1
Murrells Inlet SC, *5519* . . . 123 D4
Murrieta CA, *44282* . . . 53 D3
Murrieta Hot Sprs. CA, *2948* . . . 229 K6
Murrysville PA, *18872* . . . 92 A4
Muscatine IA, *22697* . . . 87 F2
Muscatine Co. IA, *41722* . . . 87 F2
Muscle Shoals AL, *11924* . . . 119 E2
Muscoda WI, *1453* . . . 74 A3
Muscogee Co. GA, *186291* . . . 128 C2
Muscoy CA, *8919* . . . 229 J2
Muskego WI, *21397* . . . 74 C3
Muskegon MI, *40105* . . . 75 E3
Muskegon Co. MI, *170200* . . . 75 F3
Muskegon Hts. MI, *12049* . . . 75 E3
Muskingum Co. OH, *84585* . . . 91 E4
Muskogee OK, *38310* . . . 106 A4
Muskogee Co. OK, *69451* . . . 106 A4
Musselshell Co. MT, *4497* . . . 16 C4
Mustang OK, *13156* . . . 51 E3
Mustang Ridge TX, *785* . . . 61 E2
Muttontown NY, *3412* . . . 148 C3
Myers Corner NY, *5546* . . . 148 B1
Myerstown PA, *3171* . . . 146 A1
Myersville MD, *1382* . . . 144 A1
Myricks MA, *600* . . . 151 D3
Myrtle MS, *407* . . . 118 C2
Myrtle Beach SC, *22759* . . . 123 D4
Myrtle Creek OR, *3419* . . . 28 B1
Myrtle Grove FL, *17211* . . . 135 F2
Myrtle Grove NC, *7125* . . . 123 E3
Myrtle Pt. OR, *2451* . . . 28 A1
Mystic CT, *4001* . . . 149 F2
Mystic IA, *588* . . . 87 D3
Mystic Island NJ, *8694* . . . 147 E4
Myton UT, *539* . . . 32 A4

N

Naalehu HI, *919* . . . 153 E4
Naches WA, *643* . . . 13 D4
Naco AZ, *833* . . . 55 E4
Nacogdoches TX, *29914* . . . 124 B4
Nacogdoches Co. TX, *59203* . . . 124 B4
Nageezi NM, *296* . . . 48 B2
Nags Head NC, *2700* . . . 115 F2
Nahant MA, *3632* . . . 151 D1
Nahunta GA, *930* . . . 129 F4
Nain VA, *700* . . . 102 C2
Naknek AK, *678* . . . 154 B3
Nambe NM, *1200* . . . 49 D2
Nameloc Hts. MA, *1500* . . . 151 E3
Nampa ID, *51867* . . . 22 B4
Nanakuli HI, *10814* . . . 152 A3
Nance Co. NE, *4038* . . . 35 E3
Nanticoke PA, *10955* . . . 93 E2
Nantucket MA, *3830* . . . 151 F4
Nantucket Co. MA, *9520* . . . 151 F4
Nanty Glo PA, *3054* . . . 92 B4
Nanuet NY, *16707* . . . 148 B3
Napa CA, *72585* . . . 36 B3
Napa Co. CA, *124279* . . . 36 B2
Napanoch NY, *1168* . . . 94 A3
Napaskiak AK, *390* . . . 154 B3
Napavine WA, *1361* . . . 12 B4
Naperville IL, *128358* . . . 88 C1
Naples FL, *20976* . . . 142 C2
Naples NY, *1072* . . . 78 C4
Naples TX, *1410* . . . 124 B1
Naples UT, *1300* . . . 32 A4
Naples Manor FL, *5186* . . . 142 C2
Naples Park FL, *6741* . . . 142 C1
Napoleon MI, *1254* . . . 90 B1
Napoleon ND, *857* . . . 18 C4
Napoleon OH, *9318* . . . 90 B2
Napoleonville LA, *686* . . . 134 A3
Nappanee IN, *6710* . . . 89 F2
Naranja FL, *4034* . . . 143 E3
Narberth PA, *4233* . . . 248 B3
Narragansett Pier RI, *3671* . . . 150 C4
Narrows VA, *2111* . . . 112 A1
Naschitti NM, *360* . . . 48 A2
Nash TX, *2169* . . . 116 C4
Nash Co. NC, *87420* . . . 113 D4
Nashotah WI, *1266* . . . 74 C3
Nashua IA, *1618* . . . 73 D3
Nashua MT, *325* . . . 17 D2
Nashua NH, *86605* . . . 95 D1
Nashville AR, *4878* . . . 116 C4
Nashville GA, *4697* . . . 129 E4
Nashville IL, *3147* . . . 98 B4
Nashville IN, *825* . . . 99 F2
Nashville MI, *1684* . . . 76 A4
Nashville NC, *4309* . . . 113 D4
Nashville TN, *569891* . . . 109 F4
Nashwauk MN, *935* . . . 64 B3
Nassau DE, *1600* . . . 145 F4
Nassau NY, *1161* . . . 94 B1
Nassau Bay TX, *4170* . . . 132 B3
Nassau Co. FL, *57663* . . . 139 D2
Nassau Co. NY, *1334544* . . . 148 C4
Nassau Vil. FL, *1900* . . . 222 C1
Nassawadox VA, *572* . . . 114 B3
Natalbany LA, *1739* . . . 134 B2
Natalia TX, *1663* . . . 61 D3
Natchez LA, *583* . . . 125 D4
Natchez MS, *18464* . . . 125 F4
Natchitoches LA, *17865* . . . 125 D3
Natchitoches Par. LA, *39080* . . . 125 D4
Natick MA, *32170* . . . 150 C1
National City CA, *54260* . . . 258 B2
National Park NJ, *3205* . . . 146 C3
Natrona Co. WY, *66533* . . . 33 D1
Natrona Hts. PA, *10934* . . . 92 A3
Naturita CO, *635* . . . 40 B3
Naugatuck CT, *30989* . . . 149 D1
Nauvoo IL, *1063* . . . 87 F4
Navajo NM, *2097* . . . 48 A2
Navajo Co. AZ, *97470* . . . 47 F3
Navarre OH, *1440* . . . 91 E3
Navarro Co. TX, *45124* . . . 59 F3
Navasota TX, *6789* . . . 61 F1
Navesink NJ, *1962* . . . 147 F1
Naylor MO, *610* . . . 108 A3
Naytahwaush MN, *583* . . . 19 F3
Nazareth KY, *1000* . . . 100 A4
Nazareth PA, *6023* . . . 93 F3
Nazlini AZ, *397* . . . 47 F2
Neah Bay WA, *794* . . . 12 A2
Neapolis OH, *1000* . . . 90 B2
Nebraska City NE, *7228* . . . 86 A3
Necedah WI, *888* . . . 74 A1
Neche ND, *437* . . . 19 E1
Nederland CO, *1394* . . . 41 D1
Nederland TX, *17422* . . . 132 C3
Nedrow NY, *2265* . . . 79 D3
Needham MA, *28911* . . . 151 D1
Needles CA, *4830* . . . 46 B4
Needville TX, *2609* . . . 132 A4
Neelyville MO, *487* . . . 108 A3
Neenah WI, *24507* . . . 74 C1
Neffs OH, *1138* . . . 101 F1
Negaunee MI, *4576* . . . 65 F4
Neillsville WI, *2731* . . . 68 A4
Nekoosa WI, *2590* . . . 74 A1
Neligh NE, *1651* . . . 35 E2
Nelliston NY, *622* . . . 79 F3
Nelson GA, *626* . . . 120 C3
Nelson NE, *587* . . . 35 E4
Nelson Co. KY, *37477* . . . 100 A4
Nelson Co. ND, *3715* . . . 19 E2
Nelson Co. VA, *14445* . . . 112 C1
Nelsonville NY, *565* . . . 148 B1
Nelsonville OH, *5230* . . . 101 E2
Nemacolin PA, *1034* . . . 102 A1
Nemaha Co. KS, *10717* . . . 96 A1
Nemaha Co. NE, *7576* . . . 86 A4
Nenana AK, *402* . . . 154 C2
Neodesha KS, *2848* . . . 106 A1
Neoga IL, *1854* . . . 98 C2
Neola IA, *845* . . . 86 A2
Neola UT, *533* . . . 32 A4
Neopit WI, *839* . . . 68 C4
Neosho MO, *10505* . . . 106 B2
Neosho WI, *593* . . . 74 C3
Neosho Co. KS, *16997* . . . 106 A1
Neotsu OR, *650* . . . 20 B2
Nephi UT, *4733* . . . 39 E1
Neptune NJ, *9000* . . . 147 F2
Neptune Beach FL, *7270* . . . 139 D2
Neptune City NJ, *5218* . . . 147 F2
Nesbit MS, *700* . . . 118 B2
Nesconset NY, *11992* . . . 149 D3
Nescopeck PA, *1528* . . . 93 E3
Neshoba Co. MS, *28684* . . . 126 C2
Nesquehoning PA, *3288* . . . 93 F3
Ness City KS, *1534* . . . 42 C3
Ness Co. KS, *3454* . . . 42 C3
Netarts OR, *744* . . . 20 B2
Netcong NJ, *2580* . . . 94 A4
Nettleton MS, *1932* . . . 119 D3
Nevada IA, *6658* . . . 86 C1
Nevada MO, *8607* . . . 106 C1
Nevada OH, *814* . . . 90 C3
Nevada City CA, *3001* . . . 36 C2
Nevada Co. AR, *9955* . . . 117 D4
Nevada Co. CA, *92033* . . . 36 C2
New Albany IN, *37603* . . . 99 F4
New Albany MS, *7607* . . . 118 C3
New Albin IA, *527* . . . 73 F2
New Alexandria PA, *595* . . . 92 A4
Newark AR, *1219* . . . 107 F4
Newark CA, *42471* . . . 259 D5
Newark DE, *28547* . . . 146 B4
Newark IL, *887* . . . 88 C2
Newark NJ, *273546* . . . 148 B4
Newark NY, *9682* . . . 78 C3
Newark OH, *46279* . . . 91 D4
Newark TX, *887* . . . 59 E2
Newark Valley NY, *1071* . . . 93 E1
New Athens IL, *1981* . . . 98 B4
New Auburn MN, *488* . . . 66 C4
New Auburn WI, *562* . . . 67 F3
New Augusta MS, *715* . . . 126 C4
Newaygo MI, *1670* . . . 75 F2
Newaygo Co. MI, *47874* . . . 75 F2
New Baden IL, *3001* . . . 98 B3
New Baltimore MI, *7405* . . . 76 C4
New Baltimore NY, *800* . . . 94 B1
New Baltimore VA, *800* . . . 144 A3
New Beaver PA, *1677* . . . 91 F3
New Bedford MA, *93768* . . . 151 D4
Newberg OR, *18064* . . . 20 B2
New Berlin IL, *1030* . . . 98 B1
New Berlin NY, *1129* . . . 79 E4
New Berlin PA, *838* . . . 93 D3
New Berlin WI, *38220* . . . 74 C3
New Bern NC, *23128* . . . 115 D3
Newbern TN, *2988* . . . 108 C4
Newberry FL, *3316* . . . 138 C3
Newberry MI, *2686* . . . 69 F1
Newberry SC, *10580* . . . 121 F3
Newberry Co. SC, *36108* . . . 121 F3
New Bethlehem PA, *1057* . . . 92 B3
New Bloomfield MO, *599* . . . 97 E3
New Bloomfield PA, *1077* . . . 93 D4
Newborn GA, *520* . . . 121 D4
New Boston IL, *632* . . . 87 F2
New Boston NH, *550* . . . 81 F4
New Boston OH, *2340* . . . 101 D3
New Boston TX, *4808* . . . 116 B4
New Braunfels TX, *36494* . . . 61 D2
New Bremen OH, *2909* . . . 90 B4
New Brighton MN, *22206* . . . 235 C1
New Brighton PA, *6641* . . . 91 F3
New Britain CT, *71538* . . . 149 E1
New Britain PA, *3125* . . . 146 C2
New Brockton AL, *1250* . . . 128 A4
New Brunswick NJ, *48573* . . . 147 E1
New Buffalo MI, *2200* . . . 89 E1
Newburg WV, *360* . . . 102 B2
Newburg WI, *1119* . . . 74 C2
Newburgh IN, *3088* . . . 99 E4
Newburgh NY, *28259* . . . 148 B1
Newburgh Hts. OH, *2389* . . . 204 F2
Newbury VT, *396* . . . 81 E2
Newburyport MA, *17189* . . . 151 F1
New Canaan CT, *6600* . . . 148 C2
New Carlisle IN, *1505* . . . 89 E1
New Carlisle OH, *5735* . . . 100 C1
New Carrollton MD, *12589* . . . 144 B3
New Castle CO, *1984* . . . 40 C1
New Castle DE, *4862* . . . 146 B4
New Castle IN, *17780* . . . 100 A1

Figures after entries indicate population, page number, and grid reference.

POINTS OF INTEREST

Entries in **bold black** indicate counties or parishes.
Entries in **bold color** indicate cities with detailed inset maps.

Norfolk VA/Hampton Roads

Bartlett A3	Grafton A1	Newport News B2	Rescue A2
Battery Park A2	Hampton C2	Norfolk C3	Suffolk A4
Carrollton A3	Hobson A3	Poquoson B1	Tabb A1
Chesapeake B4	Kiptopeke E1	Portsmouth B3	Virginia Beach E3

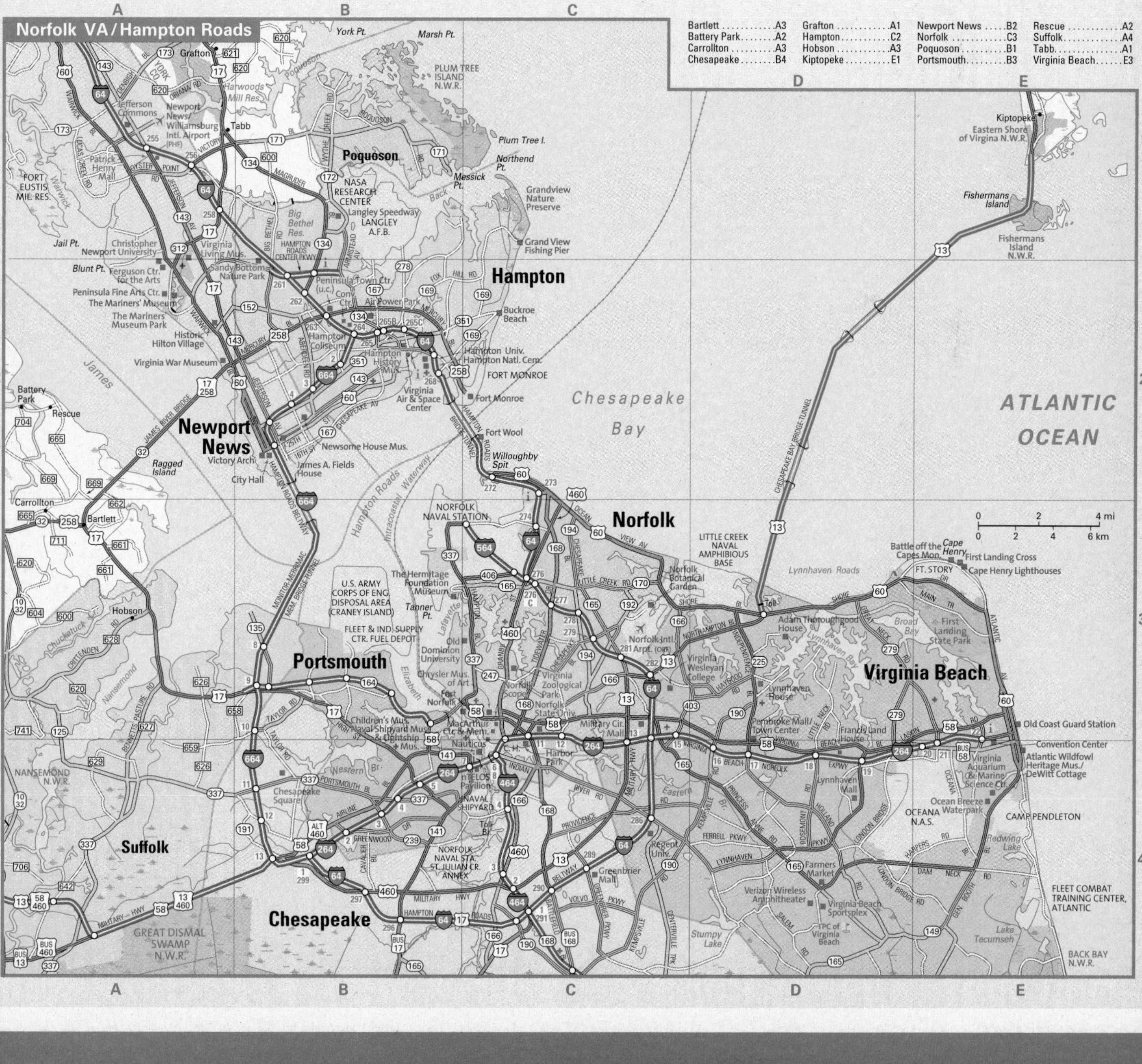

Figures after entries indicate population, page number, and grid reference.

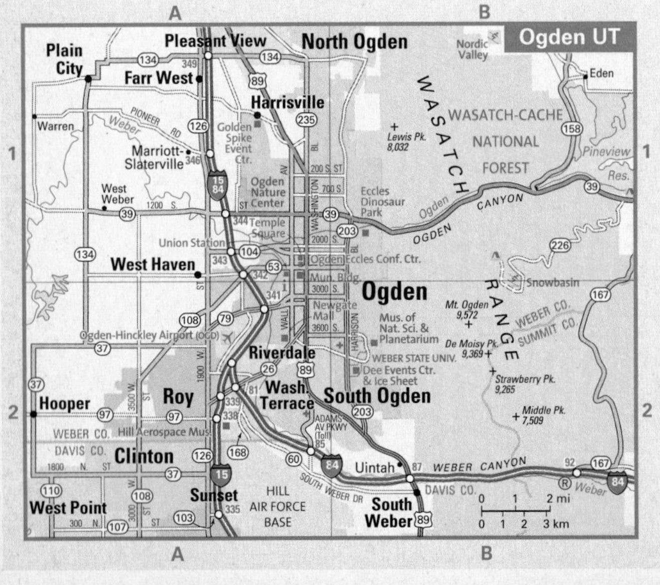

Entries in **bold black** indicate counties or parishes.
Entries in **bold color** indicate cities with detailed inset maps.

Northwood NH, 500 81 F4
Northwood ND, 959 19 E3
Northwood OH, 5471 90 C2
Northwoods MO, 4643 256 B2
N. Woodstock NH, 750 81 F3
Northwye MO, 550 97 E4
N. York PA, 1689 275 E1
Norton KS, 3012 42 C1
Norton MA, 2618 151 D2
Norton OH, 11523 91 E3
Norton VA, 3904 111 E2
Norton Co. KS, 5953 42 C1
Norton Shores MI, 22527 75 E3
Nortonville KS, 620 96 A2
Nortonville KY, 1264 109 E2
Norwalk CA, 103298 52 C3
Norwalk CT, 82951 148 C2
Norwalk IA, 6884 86 C2
Norwalk OH, 16238 91 D2
Norwalk WI, 653 73 F2
Norway IA, 601 87 E1
Norway ME, 2623 82 B2
Norway MI, 2959 69 D2
Norwell MA, 1300 151 E2
Norwich CT, 36117 149 F1
Norwich KS, 551 43 E4
Norwich NY, 7355 79 E4
Norwich VT, 1200 81 E3
Norwood CO, 438 40 B3
Norwood KY, 395 230 F1
Norwood MA, 28587 151 D2
Norwood MO, 552 107 E2
Norwood NY, 1685 80 B1
Norwood NC, 2216 122 B1
Norwood OH, 21675 100 B2
Norwood PA, 5985 248 B4
Norwood Young America MN, 3108 66 C4
Notasulga AL, 916 128 B2
Nottingham NH, 550 81 F4
Nottoway VA, 100 113 D2
Nottoway Co. VA, 15725 113 D2
Notus ID, 458 22 B4
Novato CA, 47630 36 B3
Novi MI, 47386 76 B4
Novinger MO, 534 87 D4
Nowata OK, 3971 106 A3
Nowata Co. OK, 10569 106 A2
Noxapater MS, 419 126 C1
Noxon MT, 230 14 C2
Noxubee Co. MS, 12548 127 D1
Noyack NY, 2696 149 E3
Nuangola PA, 671 93 E3
Nuckolls Co. NE, 5057 35 E4
Nucla CO, 734 40 B3
Nueces Co. TX, 313645 63 F2
Nuevo CA, 4135 53 D3
Nuiqsut AK, 433 154 C1
Nunda NY, 1330 78 B4
Nunn CO, 471 33 E4
Nutley NJ, 27362 148 B3
Nutter Fort WV, 1686 102 A2
Nyack NY, 6737 148 B3
Nye Co. NV, 32485 37 F3
Nyssa OR, 3163 22 A4

O

Oacoma SD, 390 27 D4
Oak Bluffs MA, 1700 151 E4
Oakboro NC, 1198 122 B1
Oak Brook IL, 8702 203 C4
Oakbrook Terrace IL, 2300 203 C4
Oak City UT, 650 39 D2
Oak Creek CO, 849 32 C4
Oak Creek WI, 28456 74 C3
Oakdale CA, 15503 36 C4
Oakdale CT, 1100 149 F1
Oakdale KY, 4937 109 D2
Oakdale LA, 8137 133 E1
Oakdale MA, 1100 150 C1
Oakdale MN, 26653 235 E2
Oakdale NY, 8075 149 D4
Oakdale PA, 1551 92 A4
Oakes ND, 1979 27 E1
Oakesdale WA, 420 14 A3
Oakfield ME, 475 85 D3
Oakfield NY, 1805 78 B3
Oakfield WI, 1012 74 C2
Oak Forest IL, 28051 89 D2
Oak Grove AL, 457 128 A1
Oak Grove IL, 1318 88 A2
Oak Grove KY, 7064 109 E3
Oak Grove LA, 2174 125 F1
Oak Grove MI, 700 70 C4
Oak Grove MN, 6903 67 D3
Oak Grove MS, 1400 126 C4
Oak Grove MO, 5535 96 C2
Oak Grove OR, 12808 251 D3
Oak Grove SC, 8183 122 A3
Oak Grove TN, 4072 111 E3
Oak Grove Hts. AR, 727 108 A4
Oak Harbor OH, 2841 90 C2
Oak Harbor WA, 19795 12 C2
Oak Hill FL, 1378 141 E1
Oak Hill OH, 1685 101 D3
Oak Hill WV, 7589 101 F4
Oak Hills IL, 2400 256 D2
Oakhurst CA, 2868 37 D4
Oakhurst NJ, 4152 147 F2
Oakhurst OK, 2731 51 F2
Oak Island NC, 6571 123 E4
Oakland CA, 399484 36 B4
Oakland FL, 936 141 D1
Oakland IL, 996 99 D1
Oakland IA, 1487 86 A2
Oakland ME, 2758 82 C2
Oakland MD, 1930 102 B2
Oakland MD, 2100 144 B1
Oakland MS, 586 118 B3
Oakland MO, 1540 256 B3
Oakland NE, 1367 35 F2
Oakland NJ, 12466 148 A3
Oakland OK, 674 51 F4
Oakland OR, 954 20 B4
Oakland PA, 622 93 F1
Oakland TN, 1279 118 C1
Oakland City IN, 2588 99 E4
Oakland Co. MI, 1194156 76 B4
Oakland Park FL, 30966 143 F1
Oak Lawn IL, 55245 89 D1
Oakley ID, 668 31 D2
Oakley KS, 2173 42 B2
Oakley UT, 948 31 F4
Oaklyn NJ, 4188 248 D4
Oakman AL, 944 119 E4
Oakmont PA, 6911 92 A4
Oak Orchard DE, 750 145 F4
Oak Park CA, 2320 228 A2
Oak Park IL, 52524 89 D1
Oak Park MI, 29793 76 C4
Oak Park Hts. MN, 3957 67 D3
Oak Ridge NJ, 750 148 A3
Oak Ridge NC, 3988 112 B4
Oakridge OR, 3148 20 C4
Oak Ridge TN, 27387 110 C4
Oak Ridge North TX, 2991 132 A2
Oak Shade NJ, 1500 147 D3
Oakton VA, 29348 144 A3
Oaktown IN, 633 99 E3
Oak Trail Shores TX, 2475 59 E3
Oak View CA, 4199 52 B2
Oakville CT, 8618 149 D1
Oakville MO, 35309 98 A3
Oakville WA, 675 12 B4
Oakwood GA, 2689 121 D3
Oakwood IL, 1502 89 D4
Oakwood OH, 607 90 B3
Oakwood OH, 3667 204 G3
Oakwood OH, 9215 208 E2
Oakwood Beach NJ, 700 145 E1
Oakwood Hills IL, 2194 203 A1
Oasis NV, 200 30 C3
Oberlin KS, 1994 42 C1
Oberlin LA, 1853 133 E2
Oberlin OH, 8195 91 D2
Oberlin PA, 2809 218 C2
Obetz OH, 3977 206 B3
Obion TN, 1134 108 C3
Obion Co. TN, 32450 108 C3
Oblong IL, 1580 99 D2
O'Brien Co. IA, 15102 72 A3
Ocala FL, 45943 138 C4
Occidental CA, 1272 36 A3
Occoquan VA, 759 144 B4
Oceana WV, 1550 111 F1
Oceana Co. MI, 26873 75 E2
Ocean Bluff MA, 5100 151 E2
Ocean Breeze Park FL, 463 141 E4
Ocean City FL, 5594 136 B2
Ocean City MD, 7173 114 C1
Ocean City NJ, 15378 147 F4
Ocean Co. NJ, 510916 147 E3
Ocean Gate NJ, 2076 147 E3
Ocean Grove MA, 3012 151 D3
Ocean Grove NJ, 4256 147 F2
Oceano CA, 7260 52 A1
Ocean Park WA, 1459 12 B4
Ocean Pines MD, 10496 114 C1
Oceanport NJ, 5807 147 F2
Ocean Reef Club FL, 1000 143 E3
Ocean Ridge FL, 1636 143 F1
Ocean Shores WA, 3836 12 B4
Oceanside CA, 161029 53 D3
Oceanside NY, 32733 147 F1
Oceanside OR, 326 20 B2
Ocean Sprs. MS, 17225 135 D2
Ocean View DE, 1006 145 F4
Ochelata OK, 494 51 F1
Ocheyedan IA, 536 72 A2
Ochiltree Co. TX, 9006 50 B2
Ochlocknee GA, 605 137 E1
Ocilla GA, 3270 129 E4
Ocoee FL, 24391 141 D1
Oconee Co. GA, 26225 121 D4
Oconee Co. SC, 66215 121 E2
Oconomowoc WI, 12382 74 C3
Oconto WI, 4708 69 D4
Oconto Co. WI, 35634 68 C4
Oconto Falls WI, 2843 68 C4
Ocracoke NC, 769 115 E3
Odebolt IA, 1153 72 A4
Odell IL, 1014 88 C3
Odell OR, 1849 20 C1
Odem TX, 2499 61 E4
Odenton MD, 20534 144 C2
Odenville AL, 1131 119 F4
Odessa DE, 286 145 E1
Odessa FL, 3173 140 B2
Odessa MO, 4818 96 C2
Odessa NY, 617 79 D4
Odessa TX, 90943 58 A3
Odessa WA, 957 13 F3
Odin IL, 1122 98 C3
Odon IN, 1376 99 E3
O'Donnell TX, 1011 58 A2
Oelwein IA, 6692 73 E4
O'Fallon IL, 21910 98 B3
O'Fallon MO, 46169 98 A3
Ogallala NE, 4930 34 B3
Ogden IL, 743 89 D4
Ogden IA, 2023 86 C1
Ogden KS, 1762 43 F2
Ogden NC, 5481 123 E3
Ogden UT, 77226 31 E3
Ogden Dunes IN, 1313 89 E2
Ogdensburg NJ, 2638 148 A3
Ogdensburg NY, 12364 80 B1
Ogemaw Co. MI, 21645 76 B1
Oglala SD, 1229 34 A1
Ogle Co. IL, 51032 88 B1
Oglesby IL, 3647 88 B2
Oglethorpe GA, 1200 129 D3
Oglethorpe Co. GA, 12635 121 E4
Ogunquit ME, 800 82 B4
Ohatchee AL, 1215 120 A4
Ohio City OH, 784 90 B3
Ohio Co. IN, 5623 100 A2
Ohio Co. KY, 22916 109 F1
Ohio Co. WV, 47427 91 F4
Ohioville PA, 3759 91 F3
Oil City LA, 1219 124 C2
Oil City PA, 11504 92 A2
Oildale CA, 27885 45 D4
Oilton OK, 1099 51 F2
Ojai CA, 7862 52 B2
Okaloosa Co. FL, 170498 136 B2
Okanogan WA, 2484 13 E2
Okanogan Co. WA, 39564 13 E1
Okarche OK, 1110 51 E2
Okauchee WI, 2100 74 C3
Okawville IL, 1355 98 B3
Okay OK, 597 106 A4
Okeechobee FL, 5376 141 E4
Okeechobee Co. FL, 35910 141 E3
Okeene OK, 1240 51 D2
Okemah OK, 3038 51 F3
Okemos MI, 22805 76 A4
Okfuskee Co. OK, 11814 51 F3
Oklahoma City OK, 506132 51 E3
Oklahoma Co. OK, 660448 51 E2
Okmulgee OK, 13022 51 F2
Okmulgee Co. OK, 39685 51 F2
Okoboji IA, 820 72 A2
Okolona MS, 3056 119 D3
Oktibbeha Co. MS, 42902 118 C4
Ola AR, 1204 117 D2
Olanta SC, 613 122 C4
Olathe CO, 1573 40 B3
Olathe KS, 92962 96 B3
Olcott NY, 1156 78 B3
Old Bennington VT, 232 94 C1
Old Bridge NJ, 22833 147 E1
Oldenburg IN, 647 100 A2
Old Field NY, 947 149 D3
Old Forge NY, 800 79 F2
Old Forge PA, 8798 93 F2
Old Fort NC, 963 111 E4
Oldham SD, 206 27 E3
Oldham Co. KY, 46178 100 A4
Oldham Co. TX, 2185 49 F3
Old Lyme CT, 850 149 E2
Old Mystic CT, 3205 149 F2
Old Orchard Beach ME, 8856 82 B4
Old River-Winfree TX, 1364 132 B3
Old Saybrook CT, 1962 149 E2
Oldsmar FL, 11910 140 B2
Oldtown ID, 190 14 B2
Old Town ME, 8130 83 D1
Old Zionsville PA, 950 146 B1
Olean NY, 15347 92 C1
Oley PA, 1100 146 B1
Olin IA, 716 87 F1
Olive Branch MS, 21054 118 B2
Olive Hill KY, 1813 101 D4
Olivehurst CA, 11061 36 C2
Oliver PA, 2925 102 B1
Oliver Co. ND, 2065 18 B4
Oliver Sprs. TN, 3303 110 C4
Olivet MI, 1758 76 A4
Olivet NJ, 1420 145 F1
Olivet SD, 70 35 E1
Olivette MO, 7438 256 B2
Olivia MN, 2570 66 B4
Olla LA, 1417 125 E3
Olmos Park TX, 2343 257 E2
Olmsted Co. MN, 124277 73 E1
Olmsted Falls OH, 7962 204 D3
Olney IL, 8631 99 D3
Olney MD, 31438 144 B2
Olney TX, 3396 59 D1
Olney Sprs. CO, 389 41 F3
Olpe KS, 504 43 F3
Olton TX, 2288 50 A4
Olustee OK, 680 50 C4
Olympia WA, 42514 12 B4
Olympian Vil. MO, 669 98 A4
Olyphant PA, 4978 261 E2
Omaha NE, 390007 86 A2
Omaha TX, 999 124 B1
Omak WA, 4721 13 E2
Omao HI, 1221 152 B1
Omega GA, 1340 129 D4
Omro WI, 3177 74 C1
Onaga KS, 704 43 F1
Onalaska TX, 1174 132 B1
Onalaska WI, 14839 73 F2
Onamia MN, 847 66 C2
Onancock VA, 1525 114 C3
Onarga IL, 1438 89 D3
Onawa IA, 3091 35 F2
Onaway ID, 230 14 B4
Onaway MI, 993 70 C3
Oneco FL, 7500 140 B3
Oneida IL, 752 88 A3
Oneida NY, 10987 79 E3
Oneida TN, 3615 110 B3
Oneida WI, 1070 68 C4
Oneida Co. ID, 4125 31 E2
Oneida Co. NY, 235469 79 E2
Oneida Co. WI, 36776 68 B2
O'Neill NE, 3733 35 D2
Onekama MI, 647 75 E1
Oneonta AL, 5576 119 F4
Oneonta NY, 13292 79 F4
Onida SD, 740 26 C3
Onley VA, 496 114 C3
Onondaga Co. NY, 458336 79 E3
Onset MA, 1292 151 E3
Onslow Co. NC, 150355 115 D4
Onsted MI, 813 90 B1
Ontario CA, 158007 52 C2
Ontario OH, 5303 91 D3
Ontario OR, 10985 22 A4
Ontario Co. NY, 100224 78 C4
Ontonagon MI, 1769 65 E4
Ontonagon Co. MI, 7818 65 E4
Onyx CA, 476 45 E4
Oolitic IN, 1152 99 F3
Oologah OK, 883 106 A3
Ooltewah TN, 5681 120 B2
Oostburg WI, 2660 75 D2
Opal Cliffs CA, 6458 236 D1
Opa-Locka FL, 14951 143 E2
Opelika AL, 23498 128 B2
Opelousas LA, 22860 133 F2
Opp AL, 6607 128 A4
Oppelo AR, 725 117 E1
Oquawka IL, 1539 87 F3
Oracle AZ, 3563 55 D2
Oracle Jct. AZ, 550 55 D3
Oradell NJ, 8047 148 B3
Oran MO, 1264 108 B2
Orange CA, 128821 52 C3
Orange CT, 13233 149 D2
Orange MA, 3945 95 D1
Orange NJ, 32868 240 A3
Orange OH, 3236 204 G2
Orange TX, 18643 132 C2
Orange VA, 4123 103 D4
Orange Beach AL, 3784 135 F2
Orangeburg NY, 3388 148 B3
Orangeburg SC, 12765 122 A4
Orangeburg Co. SC, 91582 122 A4
Orange City FL, 6604 141 D1
Orange City IA, 5582 35 F1
Orange Co. CA, 2846289 52 C3
Orange Co. FL, 896344 141 D1
Orange Co. IN, 19306 99 F3
Orange Co. NY, 341367 148 A2
Orange Co. NC, 118227 112 C4
Orange Co. TX, 84966 132 C2
Orange Co. VT, 28226 81 E3
Orange Co. VA, 25881 103 D4
Orange Cove CA, 7722 45 D3
Orange Grove TX, 1288 61 E4
Orange Lake FL, 650 138 C4
Orange Lake NY, 6085 148 B1
Orange Park FL, 9081 139 D2
Orangeville IL, 751 74 B4
Orangeville UT, 1398 39 F2
Orchard Beach MD, 4400 193 D5
Orchard City CO, 2880 40 B2
Orchard Lake MI, 2215 210 A1
Orchard Mesa CO, 6456 40 B2
Orchard Park NY, 3294 78 B4
Orchards WA, 17852 20 C1
Orcutt CA, 28830 52 A1
Ord NE, 2269 35 D3
Orderville UT, 596 39 D4
Ordway CO, 1248 41 F3
Oreana IL, 892 98 C1
Ore City TX, 1106 124 B2
Orefield PA, 1600 146 B1
Oregon IL, 4060 88 B1
Oregon MO, 935 96 B1
Oregon OH, 19355 90 C2
Oregon WI, 7514 74 B3
Oregon City OR, 25754 20 C2
Oregon Co. MO, 10344 107 F3
Oreland PA, 5509 248 C1
Orem UT, 84324 31 F4
Orfordville WI, 1272 74 B4
Orient NY, 709 149 F2
Oriental NC, 875 115 E3
Orient Park FL, 5703 266 C2
Orinda CA, 17599 259 C2
Orion IL, 1713 88 A2
Oriskany NY, 1459 79 E3
Oriskany Falls NY, 698 79 E3
Orland CA, 6281 36 B1
Orland ME, 500 83 D2
Orland Hills IL, 6779 203 C6
Orlando FL, 185951 141 D1
Orland Park IL, 51077 89 D2
Orleans IN, 2273 99 F3
Orleans IA, 583 72 B2
Orleans MA, 1716 151 F3
Orleans NE, 425 43 D1
Orleans VT, 826 81 E1
Orleans Co. NY, 44171 78 B3
Orleans Co. VT, 26277 81 E1
Orleans Par. LA, 484674 134 B2
Orlinda TN, 594 109 F3
Orlovista FL, 6047 246 B2
Ormond Beach FL, 36301 139 E4
Ormond-by-the-Sea FL, 8430 139 E4
Orofino ID, 3247 14 B4
Orono ME, 8253 83 D1
Orono MN, 7538 66 C4
Oronoco MN, 883 73 D1
Oronogo MO, 976 106 B2
Orosi CA, 7318 45 D3
Oro Valley AZ, 29700 55 D3
Oroville CA, 13004 36 C1
Oroville WA, 1653 13 E1
Orrick MO, 889 96 C2
Orrington ME, 325 83 D1
Orrs Island ME, 550 82 B3
Orrville OH, 8551 91 E3
Orting WA, 3760 12 C3
Ortonville MI, 1535 76 B3
Ortonville MN, 2158 27 F2
Orwell OH, 1519 91 F2
Orwigsburg PA, 3106 146 A1
Osage IA, 3451 73 D3
Osage WY, 215 25 F4
Osage Beach MO, 3662 97 D4
Osage City KS, 3034 96 A3
Osage Co. KS, 16712 96 A3
Osage Co. MO, 13062 97 F4
Osage Co. OK, 44437 51 F1
Osakis MN, 1567 66 B2
Osawatomie KS, 4645 96 B3
Osborne KS, 1607 43 D2
Osborne PA, 566 250 A1
Osborne Co. KS, 4452 43 D2
Osburn ID, 1545 14 B3
Oscawana Corners NY, 1500 148 B2
Osceola AR, 8875 108 B4
Osceola IN, 1859 89 F2
Osceola IA, 4659 86 C3
Osceola MO, 835 96 C4
Osceola NE, 921 35 E3
Osceola WI, 2421 67 D3
Osceola Co. FL, 172493 141 D2
Osceola Co. IA, 7003 72 A2
Osceola Co. MI, 23197 75 F1
Osceola Mills PA, 1249 92 C3
Oscoda MI, 992 76 C1
Oscoda Co. MI, 9418 70 C4
Osgood IN, 1669 100 A2
Oshkosh NE, 887 34 A3
Oshkosh WI, 62916 74 C1
Oskaloosa IA, 10938 87 D2
Oskaloosa KS, 1165 96 A2
Oslo FL, 1900 141 E3
Osmond NE, 796 35 E2
Osprey FL, 4143 140 B4
Osseo MN, 2434 235 A1
Osseo WI, 1669 67 F4
Ossian IN, 2943 90 A3
Ossian IA, 853 73 E3
Ossineke MI, 1059 71 D4
Ossining NY, 24010 148 B2

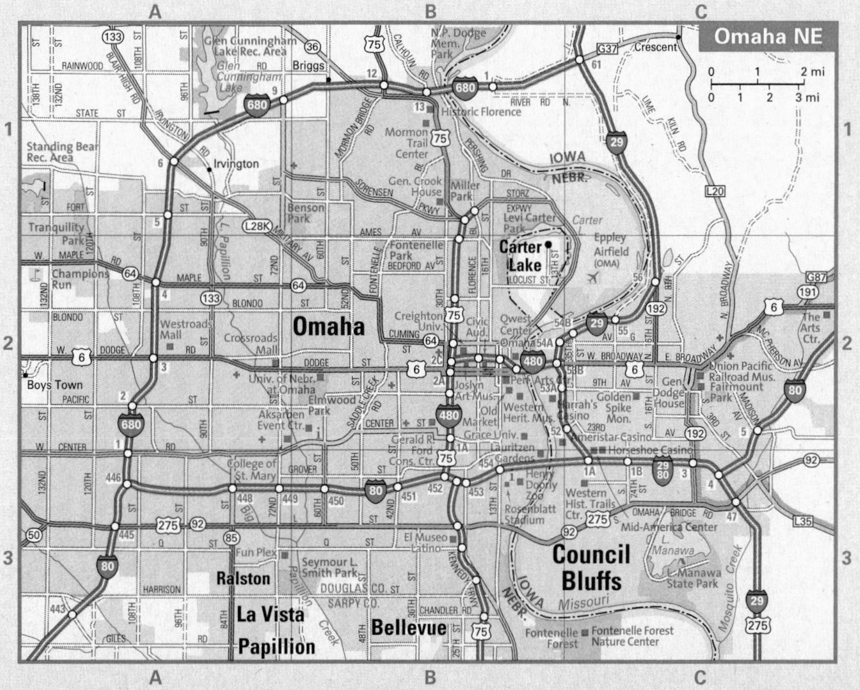

Bellevue B3
Boys Town A2
Briggs B1
Carter Lake B2
Council Bluffs C3
Crescent C1
Irvington A1
La Vista A3
Omaha B2
Papillion A3
Ralston A3

Figures after entries indicate population, page number, and grid reference.

P

Entries in **bold black** indicate counties or parishes.
Entries in **bold color** indicate cities with detailed inset maps.

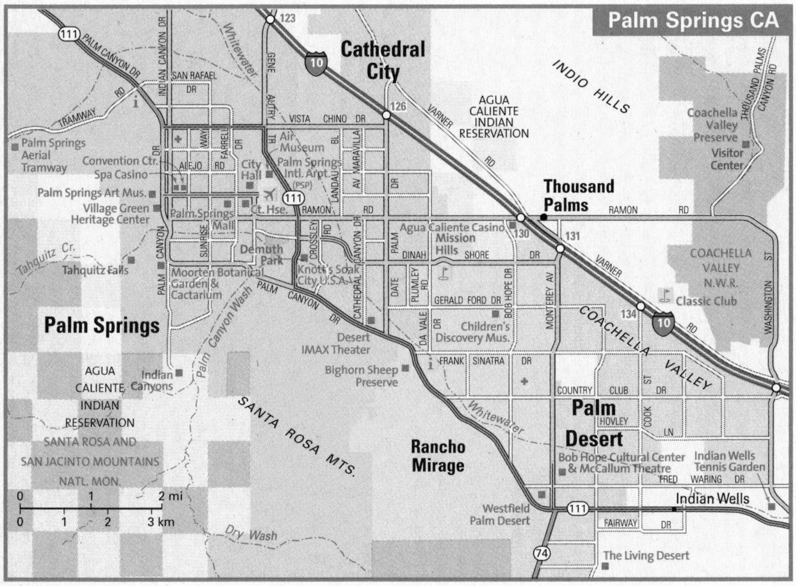

Bellview A1	Ferry Pass B1	Pensacola B1
Brent B1	Goulding B1	Pleasant Grove A2
Brownsville B2	Gulf Breeze B2	Warrington A2
Ensley A1	Myrtle Grove A2	W. Pensacola A1

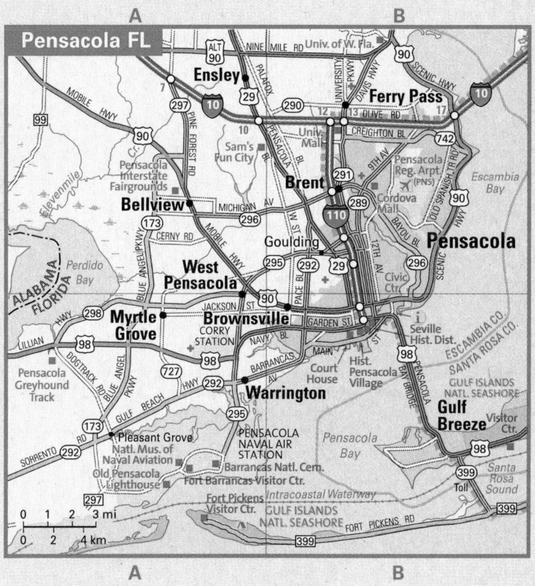

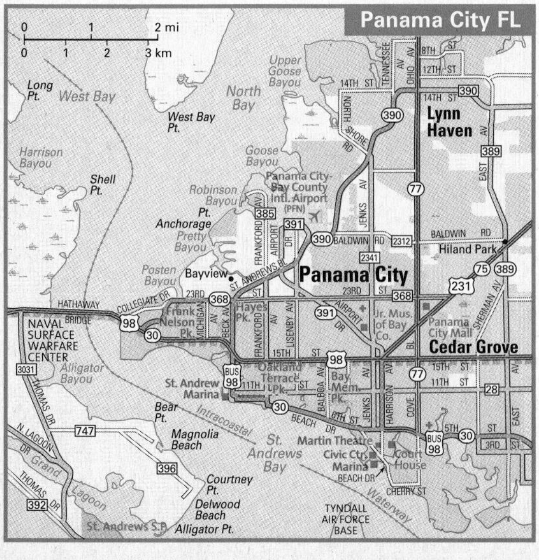

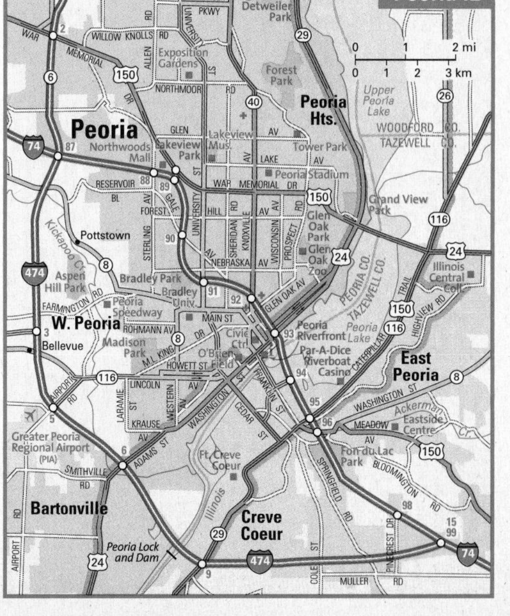

Figures after entries indicate population, page number, and grid reference.

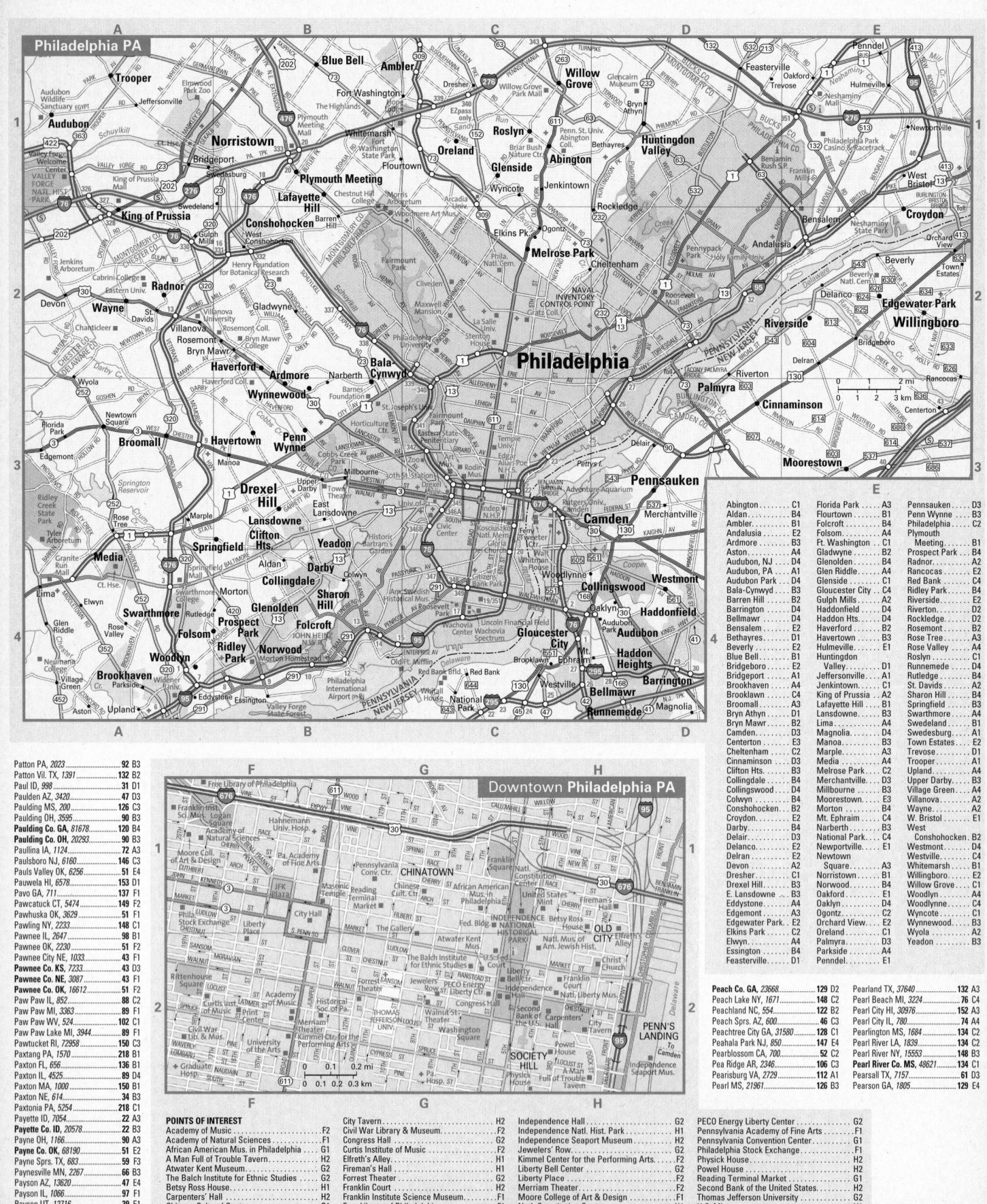

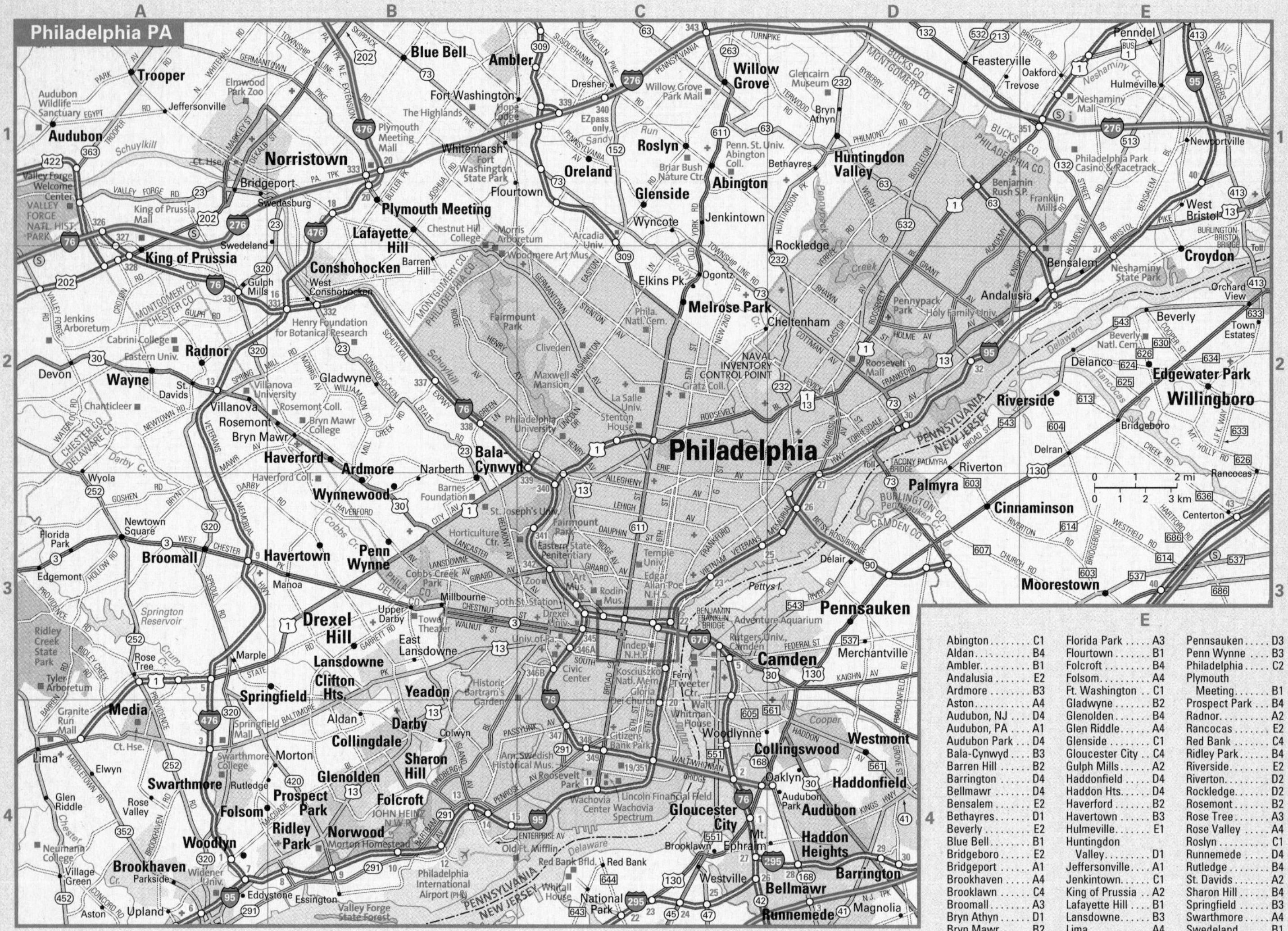

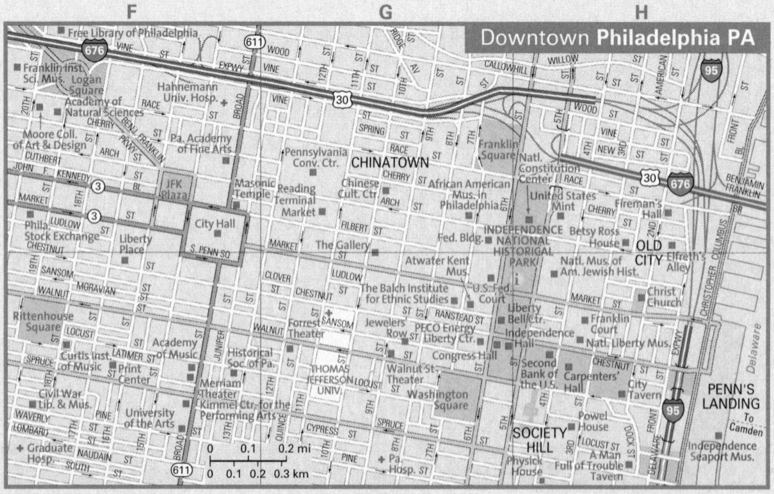

POINTS OF INTEREST

Entries in **bold black** indicate counties or parishes.
Entries in **bold color** indicate cities with detailed inset maps.

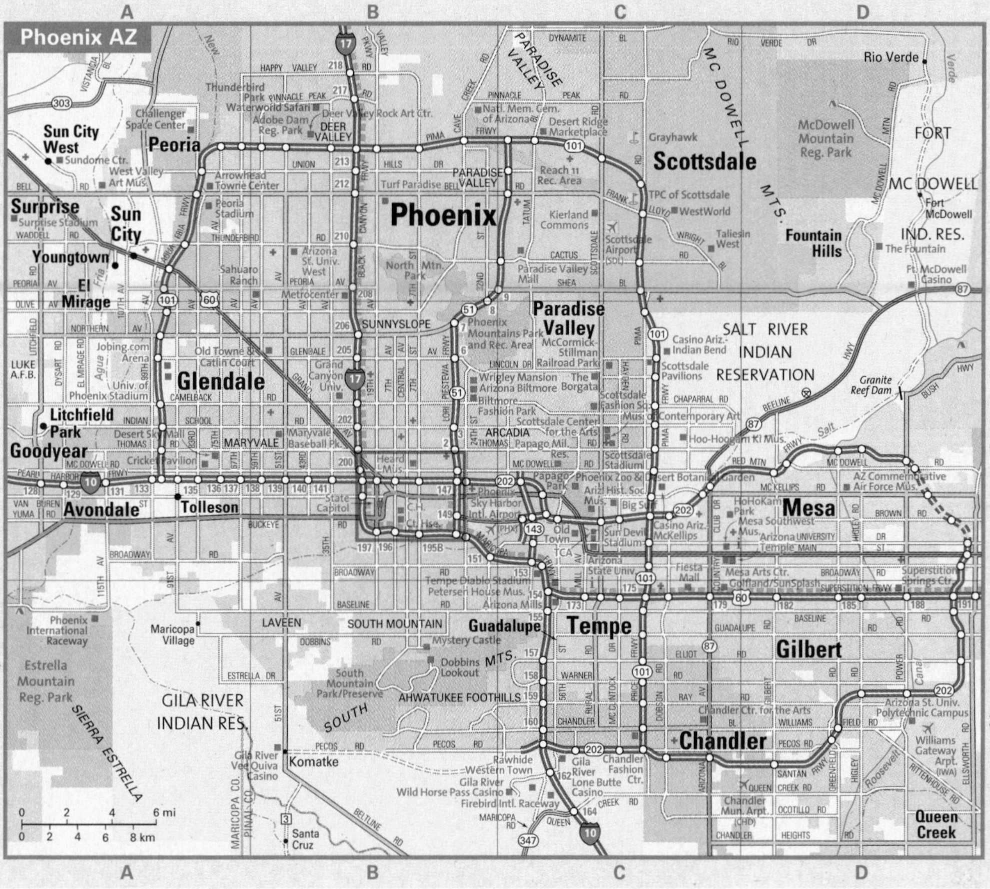

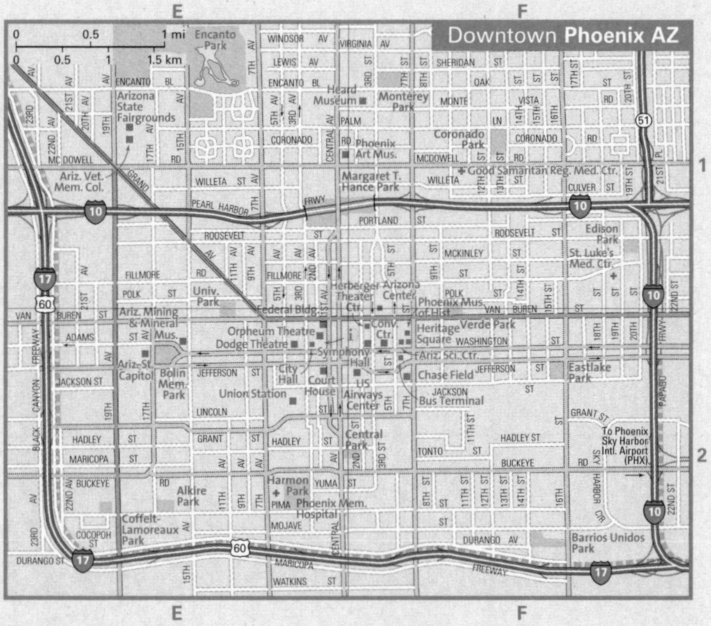

POINTS OF INTEREST

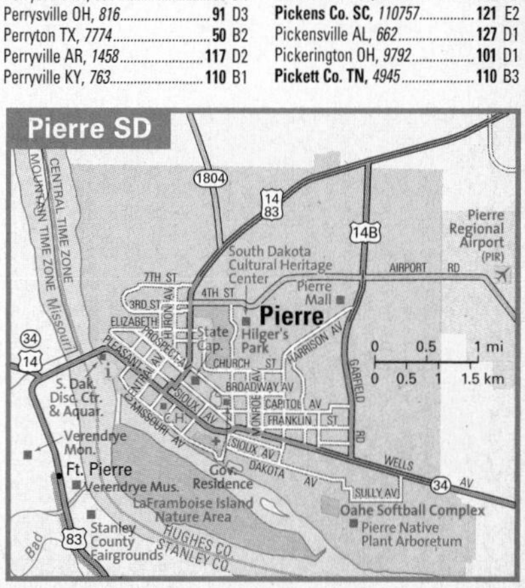

Figures after entries indicate population, page number, and grid reference.

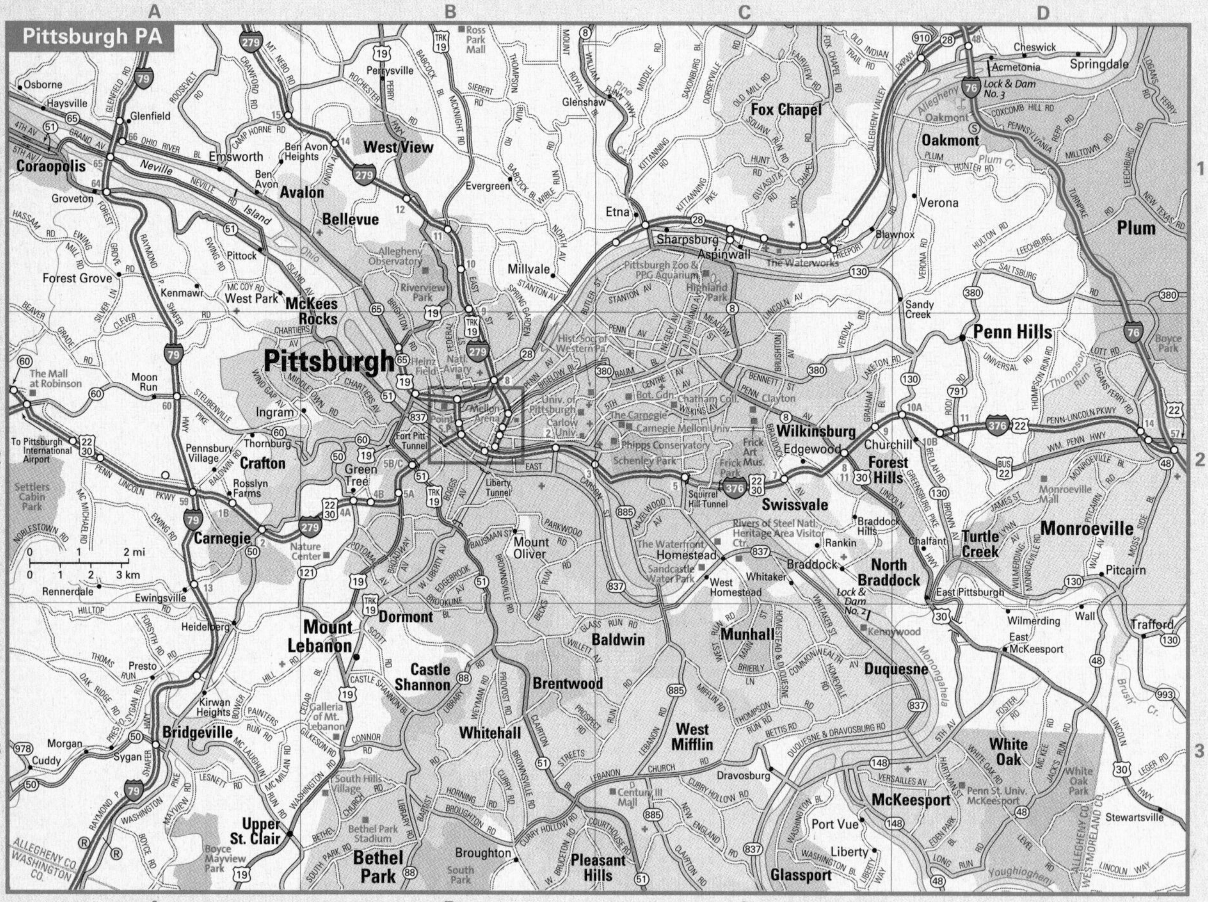

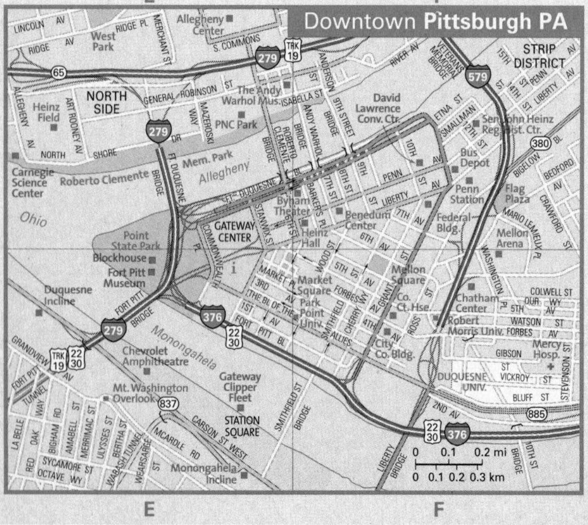

POINTS OF INTEREST

Allegheny Center........E1
The Andy Warhol Museum........E1
Benedum Center........F1
Blockhouse........E2
Bus Depot........F1
Byham Theater........F1
Carnegie Science Center........E1
Chatham Center........F2
Chevrolet Amphitheatre........E2
City County Building........F2
County Court House........F2
David Lawrence Convention Center........F1
Duquesne Incline........E2
Duquesne University........F2
Federal Building........F1
Fort Pitt Museum........E2
Gateway Center........E1
Gateway Clipper Fleet........E2
Heinz Field........E1
Heinz Hall........F1
Mellon Arena........F2
Monongahela Incline........E2
Mt. Washington Overlook........E2
Penn Station........F1
PNC Park........E1
Point Park University........F2
Point State Park........E1
Robert Morris University........F2
Senator John Heinz Regional Hist. Center........F1
Station Square........E2

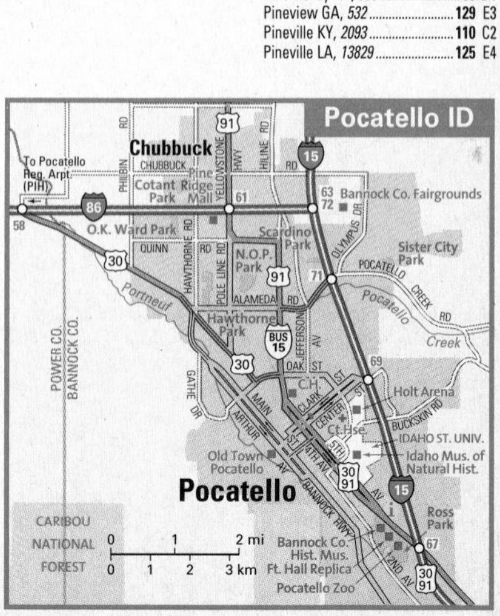

Entries in **bold black** indicate counties or parishes.
Entries in **bold color** indicate cities with detailed inset maps.

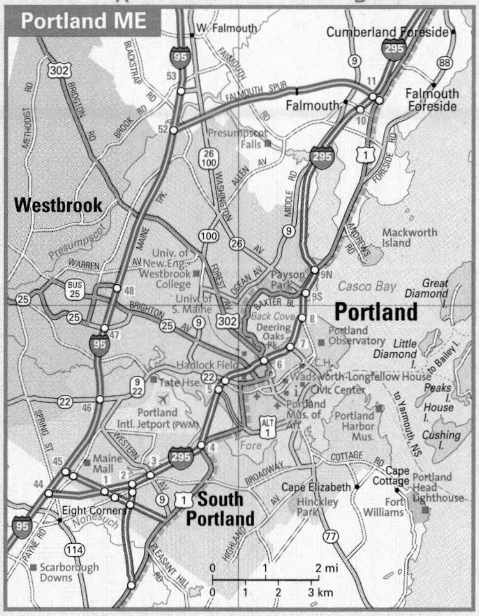

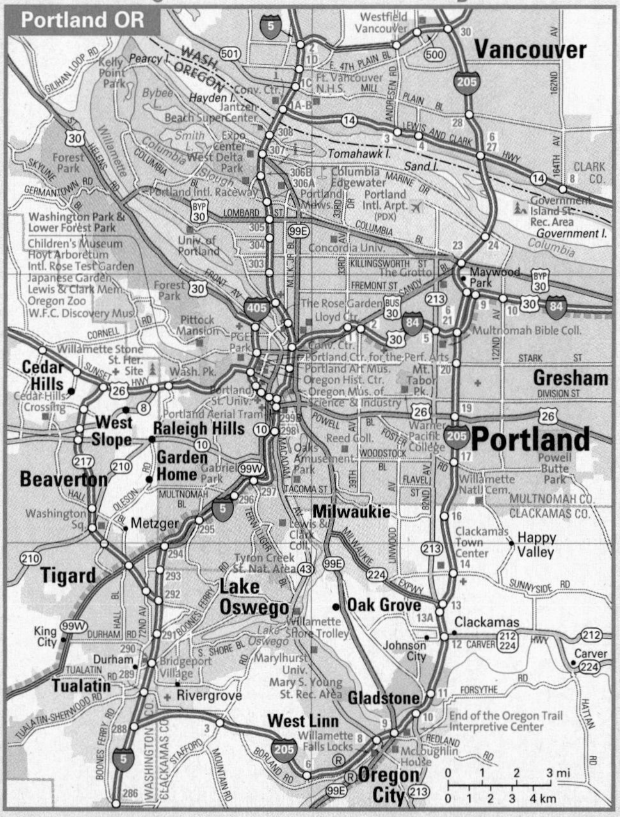

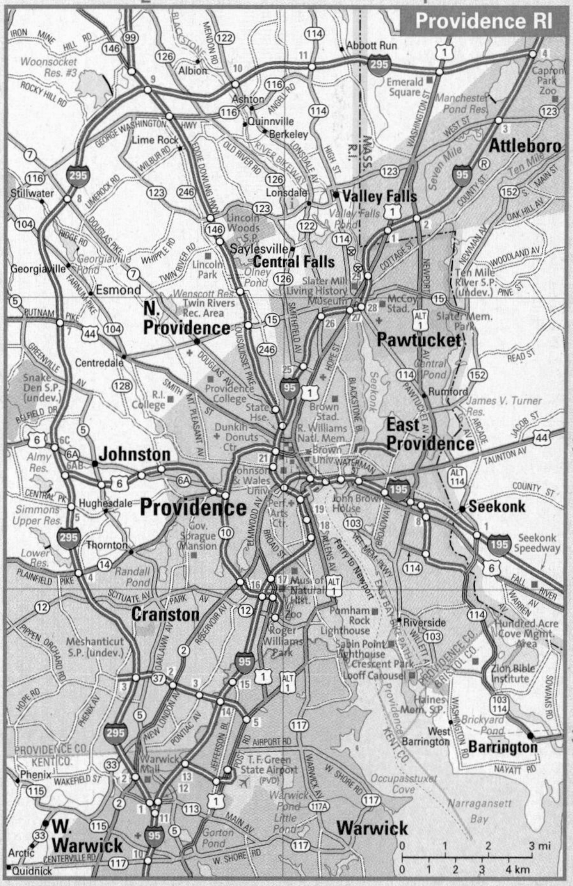

Figures after entries indicate population, page number, and grid reference.

Pt. Pleasant NJ, 19306 ... 147 E2
Pt. Pleasant WV, 4637 ... 101 E3
Pt. Pleasant Beach NJ, 5314 ... 147 E2
Pt. Reyes Sta. CA, 818 ... 36 B3
Poipu HI, 1075 ... 152 B1
Pojoaque NM, 1261 ... 49 D2
Polacca AZ, 1100 ... 47 F2
Poland OH, 2866 ... 276 C3
Polk PA, 1031 ... 92 A2
Polk City FL, 1516 ... 140 C2
Polk City IA, 2344 ... 86 C1
Polk Co. AR, 20229 ... 116 C3
Polk Co. FL, 483924 ... 140 C3
Polk Co. GA, 38127 ... 120 B4
Polk Co. IA, 374601 ... 86 C1
Polk Co. MN, 31369 ... 19 F2
Polk Co. MO, 26992 ... 107 D1
Polk Co. NE, 5639 ... 35 E3
Polk Co. NC, 18324 ... 121 F1
Polk Co. OR, 62380 ... 20 B2
Polk Co. TN, 16050 ... 120 C1
Polk Co. TX, 41133 ... 132 B1
Polk Co. WI, 41319 ... 67 E3
Polkton NC, 1195 ... 122 B2
Polkville NC, 535 ... 121 F1
Pollock SD, 339 ... 26 C1
Pollock Pines CA, 4728 ... 36 C2
Polo IL, 2477 ... 88 B1
Polo MO, 582 ... 96 C1
Polson MT, 4041 ... 15 D3
Pomeroy IA, 710 ... 72 B4
Pomeroy OH, 1966 ... 101 E2
Pomeroy WA, 1517 ... 13 F4
Pomona CA, 149473 ... 52 C2
Pomona KS, 923 ... 96 A3
Pomona NJ, 4019 ... 105 D3
Pomona NY, 2726 ... 148 B2
Pomona Park FL, 789 ... 139 D4
Pomonkey MD, 750 ... 144 B4
Pompano Beach FL, 78191 ... 143 F1
Pompton Lakes NJ, 10640 ... 148 A3
Pompton Plains NJ, 6500 ... 148 A3
Ponca NE, 1062 ... 35 F1
Ponca City OK, 25919 ... 51 E1
Ponce de Leon FL, 457 ... 136 C1
Ponce Inlet FL, 2513 ... 139 E4
Poncha Sprs. CO, 466 ... 41 D3
Ponchatoula LA, 5180 ... 134 B2
Pond Creek OK, 896 ... 51 E1
Pondera Co. MT, 6424 ... 15 E2
Ponderay ID, 638 ... 14 B2
Ponderosa NM, 310 ... 48 C2
Ponemah MN, 874 ... 64 A2
Ponte Vedra Beach FL, 1100 ... 139 D2
Pontiac IL, 11864 ... 88 C3
Pontiac MI, 66337 ... 76 C4
Pontoon Beach IL, 5620 ... 256 D1
Pontotoc MS, 5253 ... 118 C3
Pontotoc Co. MS, 26726 ... 118 C3
Pontotoc Co. OK, 35143 ... 51 F4
Pooler GA, 6239 ... 130 B3
Poolesville MD, 5151 ... 144 A2
Pope Co. AR, 54469 ... 117 D1
Pope Co. IL, 4413 ... 109 D1
Pope Co. MN, 11236 ... 66 A3
Poplar CA, 1496 ... 45 D3
Poplar MT, 911 ... 17 F2
Poplar WI, 552 ... 64 C4
Poplar Bluff MO, 16651 ... 108 A2
Poplar Grove IL, 1368 ... 74 B4
Poplar Hills KY, 396 ... 230 E3
Poplarville MS, 2601 ... 134 C1
Poquetanuck CT, 1100 ... 149 F1
Poquonock CT, 2200 ... 150 A3
Poquonock Bridge CT, 1592 ... 149 F2
Poquoson VA, 11566 ... 113 F2
Poquott NY, 975 ... 149 D3
Porcupine SD, 407 ... 26 B4
Portage IN, 33496 ... 89 E2
Portage MI, 44897 ... 89 F1
Portage PA, 2837 ... 92 B4
Portage WI, 9728 ... 74 B2
Portage Co. OH, 161261 ... 91 E3
Portage Co. WI, 67182 ... 74 B1
Portage Lakes OH, 9870 ... 188 A2
Portageville MO, 3295 ... 108 B3
Portal GA, 597 ... 129 F2
Portales NM, 11131 ... 49 F4
Port Allegany PA, 2355 ... 92 C1
Port Allen LA, 5278 ... 134 A2
Port Angeles WA, 18397 ... 12 B2
Port Aransas TX, 3370 ... 63 F2
Port Arthur TX, 57755 ... 132 C3
Port Austin MI, 737 ... 76 C1
Port Barre LA, 2287 ... 133 F2
Port Barrington IL, 1788 ... 203 A1
Port Byron IL, 1535 ... 88 A2
Port Byron NY, 1297 ... 79 D3
Port Carbon PA, 2019 ... 93 E3
Port Charlotte FL, 46451 ... 140 C4
Port Chester NY, 27867 ... 148 C3
Port Clinton OH, 6391 ... 90 C2
Port Deposit MD, 676 ... 145 D1
Port Edwards WI, 1944 ... 74 A1
Porter IN, 4972 ... 89 E2
Porter OK, 574 ... 106 A4
Porter WA, 473 ... 12 B4
Porter Corners NY, 800 ... 80 C4
Porter Co. IN, 164343 ... 89 E2
Porterdale GA, 1281 ... 121 D4
Porter Hts. TX, 1490 ... 132 A2
Porterville CA, 39615 ... 45 D3
Port Ewen NY, 3650 ... 94 B2
Port Gibson MS, 1840 ... 126 A3
Port Hadlock WA, 3476 ... 12 C2
Port Henry NY, 1152 ... 81 D3
Port Hueneme CA, 21845 ... 52 B2
Port Huron MI, 32338 ... 77 D3
Portia AR, 483 ... 107 F4
Port Isabel TX, 4865 ... 63 F4
Port Jefferson NY, 7837 ... 149 D3
Port Jefferson Sta. NY, 7527 ... 149 D3
Port Jervis NY, 8860 ... 148 A2
Port La Belle FL, 3050 ... 141 D4
Portland AR, 552 ... 125 F1
Portland CT, 5534 ... 149 E1
Portland IN, 6437 ... 90 A4
Portland ME, 64249 ... 82 B3
Portland MI, 3789 ... 76 A3
Portland ND, 604 ... 19 E3
Portland OR, 529121 ... 20 C2
Portland PA, 579 ... 93 F3
Portland TN, 8458 ... 109 F3
Portland TX, 14827 ... 63 F2
Port Lavaca TX, 12035 ... 61 F4
Port Leyden NY, 665 ... 79 E2
Port Ludlow WA, 1968 ... 12 C2
Port Matilda PA, 638 ... 92 C3
Port Monmouth NJ, 3742 ... 147 E1
Port Neches TX, 13601 ... 132 C3
Port Norris NJ, 1507 ... 145 F2
Portola CA, 2227 ... 37 D1
Portola Valley CA, 4462 ... 259 C6
Port Orange FL, 45823 ... 139 E4
Port Orchard WA, 7693 ... 12 C3
Port Orford OR, 1153 ... 28 A1
Port Reading NJ, 3829 ... 240 A5
Port Republic NJ, 1037 ... 147 E4
Port Richey FL, 3021 ... 140 B2
Port Royal PA, 977 ... 93 D4
Port Royal SC, 3950 ... 130 C2
Port St. Joe FL, 3644 ... 137 D3
Port St. John FL, 12112 ... 141 E1
Port St. Lucie FL, 88769 ... 141 E4
Port Salerno FL, 10141 ... 141 E4
Port Sanilac MI, 658 ... 76 C2
Portsmouth NH, 20784 ... 82 A4
Portsmouth OH, 20909 ... 101 D3
Portsmouth RI, 2700 ... 151 D4
Portsmouth VA, 100565 ... 113 F2
Port Sulphur LA, 3115 ... 134 C4
Port Townsend WA, 8334 ... 12 C2
Portville NY, 1024 ... 92 C1
Port Vincent LA, 463 ... 134 A2
Port Vue PA, 4228 ... 250 C3
Port Washington NY, 15215 ... 148 C3
Port Washington WI, 10467 ... 75 D2
Port Washington North NY, 2700 ... 241 G2
Port Wentworth GA, 3276 ... 130 B3
Porum OK, 725 ... 116 A1
Posen IL, 4730 ... 203 D6
Posen MI, 234 ... 70 C3
Poston AZ, 389 ... 53 F2
Postville IA, 2273 ... 73 E3
Poteau OK, 7939 ... 116 B2
Poteet TX, 3305 ... 61 D3
Poth TX, 1850 ... 61 D3
Potlatch ID, 791 ... 14 B4
Potomac IL, 681 ... 89 D4
Potomac MD, 44822 ... 144 B3
Potomac Hts. MD, 1154 ... 144 B4
Potosi MO, 2662 ... 98 A4
Potosi TX, 1664 ... 58 C3
Potosi WI, 688 ... 73 F4
Potsdam NY, 9425 ... 80 B1
Pottawatomie Co. KS, 18209 ... 43 F2
Pottawatomie Co. OK, 65521 ... 51 F3
Pottawattamie Co. IA, 87704 ... 86 A2
Potter Co. PA, 18080 ... 92 C2
Potter Co. SD, 2693 ... 27 D2
Potter Co. TX, 113546 ... 50 A3
Pottersville NY, 750 ... 80 C3
Potter Valley CA, 650 ... 36 A2
Potterville MI, 2168 ... 76 A4
Pottsboro TX, 1579 ... 59 F1
Potts Camp MS, 494 ... 118 C2
Pottstown PA, 21859 ... 146 B2
Pottsville AR, 1271 ... 117 D1
Pottsville PA, 15549 ... 93 E3
Potwin KS, 457 ... 43 F3
Poughkeepsie NY, 29871 ... 148 B1
Poughquag NY, 900 ... 148 C1
Poulan GA, 946 ... 129 D4
Poulsbo WA, 6813 ... 12 C3
Poultney VT, 1575 ... 81 D4
Pound VA, 1089 ... 111 E2
Pound Ridge NY, 800 ... 148 C2
Poway CA, 48044 ... 53 D4
Powderly KY, 846 ... 109 E2
Powder River Co. MT, 1858 ... 25 E2
Powder Sprs. GA, 12481 ... 120 C4
Powell AL, 926 ... 120 A2
Powell OH, 6247 ... 90 C4
Powell TN, 7500 ... 110 C4
Powell WY, 5373 ... 24 B3
Powell Co. KY, 13237 ... 100 C4
Powell Co. MT, 7180 ... 15 D3
Powells Crossroads TN, 1286 ... 120 B1
Powellton WV, 1796 ... 101 F4
Power Co. ID, 7538 ... 31 E1
Powers OR, 734 ... 28 A1
Powers Lake ND, 309 ... 18 A2
Powers Lake WI, 1500 ... 113 E4
Poweshiek Co. IA, 18815 ... 87 D1
Powhatan VA, 550 ... 113 D1
Powhatan Co. VA, 22377 ... 113 E1
Powhatan Pt. OH, 1744 ... 101 F1
Pownal VT, 700 ... 94 C1
Poydras LA, 3886 ... 134 C3
Poynette WI, 2266 ... 74 B2
Prague OK, 2138 ... 51 F3
Prairie City IA, 1365 ... 87 D2
Prairie City OR, 1080 ... 21 F3
Prairie Co. AR, 9539 ... 117 F2
Prairie Co. MT, 1199 ... 17 E4
Prairie Creek AR, 1849 ... 106 C3
Prairie du Chien WI, 6018 ... 73 F3
Prairie du Rocher IL, 613 ... 98 A4
Prairie du Sac WI, 3231 ... 74 A3
Prairie Grove AR, 2540 ... 106 C4
Prairie Grove IL, 960 ... 203 A1
Prairie View TX, 4410 ... 61 F2
Prairie Vil. KS, 22072 ... 224 B4
Prairieville LA, 950 ... 134 A2
Pratt KS, 6570 ... 43 D4
Pratt WV, 551 ... 101 F4
Pratt Co. KS, 9647 ... 43 D4
Prattsburgh NY, 650 ... 78 C4
Prattville AL, 24303 ... 128 A2
Preble Co. OH, 42337 ... 100 B1
Premont TX, 2772 ... 63 E2
Prentice WI, 626 ... 68 A3
Prentiss MS, 1158 ... 126 B4
Prentiss Co. MS, 25556 ... 119 D2
Prescott AZ, 39938 ... 47 D4
Prescott AR, 3686 ... 117 D4
Prescott WI, 3764 ... 67 D4
Prescott Valley AZ, 23535 ... 47 D4
Presho SD, 588 ... 26 C4
Presidential Lakes Estates NJ, 2332 ... 147 D3
Presidio TX, 4167 ... 62 B4
Presidio Co. TX, 7304 ... 62 B3
Presque Isle ME, 9511 ... 85 D2
Presque Isle Co. MI, 14411 ... 70 C3
Preston GA, 453 ... 128 C3
Preston ID, 4682 ... 31 E2
Preston IA, 949 ... 88 A1
Preston MD, 566 ... 145 E4
Preston MN, 1426 ... 73 E2
Preston Co. WV, 29334 ... 102 B2
Prestonsburg KY, 3612 ... 111 E1
Pretty Prairie KS, 615 ... 43 E4
Price UT, 8402 ... 39 F1
Price Co. WI, 15822 ... 68 A2
Priceville AL, 1631 ... 119 F3
Prichard AL, 28633 ... 135 E1
Priest River ID, 1754 ... 14 B2
Primghar IA, 891 ... 72 A3
Prince Edward Co. VA, 19720 ... 113 D2
Prince Frederick MD, 1432 ... 144 C4
Prince George VA, 750 ... 113 E2
Prince George Co. VA, 33047 ... 113 E2
Prince George's Co. MD, 801515 ... 144 C3
Princes Lakes IN, 1506 ... 99 F2
Princess Anne MD, 2313 ... 103 F4
Princeton FL, 10090 ... 143 E3
Princeton IL, 7501 ... 88 B2
Princeton IN, 8175 ... 99 D4
Princeton IA, 946 ... 88 A2
Princeton KY, 6536 ... 109 D2
Princeton MN, 3933 ... 66 C3
Princeton MO, 1047 ... 86 C4
Princeton NJ, 14203 ... 147 D2
Princeton NC, 1066 ... 123 E1
Princeton TX, 3477 ... 59 F2
Princeton WV, 6347 ... 111 F1
Princeton WI, 1504 ... 74 B2
Princeton Jct. NJ, 2382 ... 147 D2
Princeville HI, 1698 ... 152 B1
Princeville IL, 1621 ... 88 B3
Princeville NC, 940 ... 113 E4
Prince William Co. VA, 280813 ... 144 A4
Prineville OR, 7356 ... 21 D3
Pringle PA, 991 ... 261 B1
Prior Lake MN, 15917 ... 67 D4
Proctor MN, 2852 ... 64 C4
Proctor VT, 1700 ... 81 D3
Proctorville OH, 620 ... 101 E3
Progreso TX, 4851 ... 63 E4
Progress PA, 9647 ... 218 B1
Progress Vil. FL, 2482 ... 266 C2
Prophetstown IL, 2023 ... 88 A2
Prospect CT, 2200 ... 149 D1
Prospect KY, 4657 ... 100 A4
Prospect OH, 1191 ... 90 C4
Prospect PA, 1234 ... 92 A3
Prospect Hts. IL, 17081 ... 203 C2
Prospect Park PA, 6594 ... 248 B4
Prosper TX, 2097 ... 59 F2
Prosperity SC, 1047 ... 122 A3
Prosser WA, 4838 ... 21 E1
Protection KS, 558 ... 43 D4
Provencal LA, 708 ... 125 D4
Providence FL, 1100 ... 140 C2
Providence KY, 3611 ... 109 D1
Providence RI, 173618 ... 150 C3
Providence UT, 4377 ... 31 E3
Providence Co. RI, 621602 ... 150 C3
Provincetown MA, 3192 ... 151 F2
Provo UT, 105166 ... 31 F4
Prowers Co. CO, 14483 ... 42 A3
Prudenville MI, 1737 ... 76 A1
Prue OK, 433 ... 51 F2
Prunedale CA, 16432 ... 44 B2
Pruntytown WV, 600 ... 102 A2
Pryor MT, 628 ... 24 C2
Pryor OK, 8659 ... 106 A3
Pueblo CO, 102121 ... 41 E3
Pueblo Co. CO, 141472 ... 41 E3
Pueblo Pintado NM, 247 ... 48 B2
Pueblo West CO, 16899 ... 41 E3
Puhi HI, 1186 ... 152 B1
Pukalani HI, 7380 ... 153 D1
Pukwana SD, 287 ... 27 D4
Pulaski NY, 2398 ... 79 D2
Pulaski TN, 7871 ... 119 F1
Pulaski VA, 9473 ... 112 A2
Pulaski WI, 3060 ... 68 C4
Pulaski Co. AR, 361474 ... 117 E2
Pulaski Co. GA, 9588 ... 129 E3
Pulaski Co. IL, 7348 ... 108 C2
Pulaski Co. IN, 13755 ... 89 E3
Pulaski Co. KY, 56217 ... 110 B2
Pulaski Co. MO, 41165 ... 97 E4
Pulaski Co. VA, 35127 ... 112 A2
Pullman MI, 700 ... 75 E4
Pullman WA, 24675 ... 14 B4
Pumphrey MD, 5317 ... 193 C4
Pumpkin Ctr. CA, 700 ... 45 D4
Pumpkin Ctr. NC, 2228 ... 115 D4
Punaluu HI, 881 ... 152 A2
Punta Gorda FL, 14344 ... 140 C4
Punta Rassa FL, 1731 ... 142 C1
Punxsutawney PA, 6271 ... 92 B3
Pupukea HI, 4250 ... 152 A2
Purcell OK, 5571 ... 51 E3
Purcellville VA, 3584 ... 144 A2
Purdy MO, 1103 ... 106 C2
Purdys NY, 1200 ... 148 C2
Purvis MS, 2164 ... 134 C1
Puryear TN, 667 ... 109 D3
Pushmataha Co. OK, 11667 ... 116 A3
Putnam CT, 6746 ... 150 B3
Putnam Co. FL, 70423 ... 139 D3
Putnam Co. GA, 18812 ... 129 E1
Putnam Co. IL, 6086 ... 88 B2
Putnam Co. IN, 36019 ... 99 E1
Putnam Co. MO, 5223 ... 87 D4
Putnam Co. NY, 95745 ... 148 B2
Putnam Co. OH, 34726 ... 90 B3
Putnam Co. TN, 62315 ... 110 A3
Putnam Co. WV, 51589 ... 101 E3
Putnam Lake NY, 3855 ... 148 C1
Putney GA, 2998 ... 129 D4
Putney KY, 600 ... 111 D2
Putney VT, 800 ... 81 E4
Puxico MO, 1145 ... 108 B2
Puyallup WA, 33011 ... 12 C3
Pyatt AR, 253 ... 107 D3

Q

Quail Valley CA, 1639 ... 229 J5
Quaker City OH, 563 ... 101 F1
Quaker Hill CT, 4200 ... 149 F2
Quakertown PA, 8931 ... 146 C1
Quanah TX, 3022 ... 50 C4
Quantico VA, 561 ... 144 A4
Quapaw OK, 984 ... 106 B2
Quarryville CT, 2500 ... 150 A3
Quarryville PA, 1994 ... 146 A3
Quartz Hill CA, 9890 ... 52 C2
Quartzsite AZ, 3354 ... 54 A1
Quasqueton IA, 574 ... 73 E4
Quay Co. NM, 10155 ... 49 F3
Quechee VT, 550 ... 81 E3
Queen Anne's Co. MD, 40563 ... 145 D3
Queen City MO, 638 ... 87 D4
Queen City TX, 1613 ... 124 C1
Queen Creek AZ, 4316 ... 54 C2
Queens Co. NY, 2229379 ... 148 B4
Queenstown MD, 617 ... 145 D3
Quenemo KS, 468 ... 96 A3
Questa NM, 1864 ... 49 D1
Quidnessett RI, 3600 ... 150 C4
Quidnick RI, 6300 ... 150 C3
Quilcene WA, 591 ... 12 C3
Quimby IA, 368 ... 72 A4
Quinby SC, 842 ... 122 C3
Quince Orchard MD, 23044 ... 144 B2
Quincy CA, 1879 ... 36 C1
Quincy FL, 6982 ... 137 D2
Quincy IL, 40366 ... 97 F1
Quincy MA, 88025 ... 151 D1
Quincy MI, 1701 ... 90 A1
Quincy OH, 734 ... 90 B4

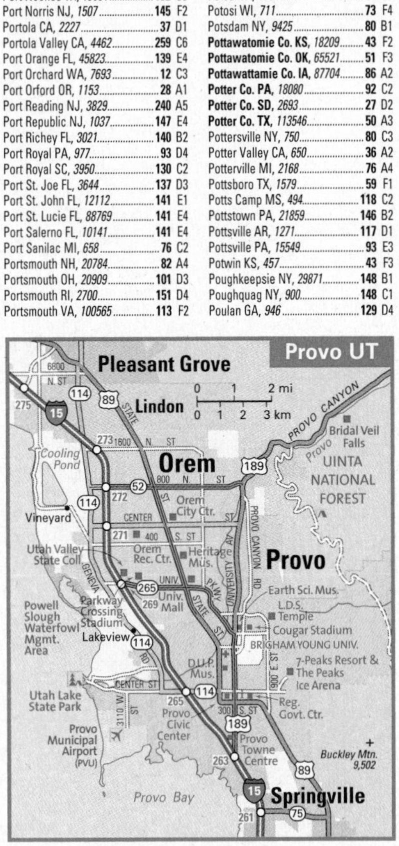

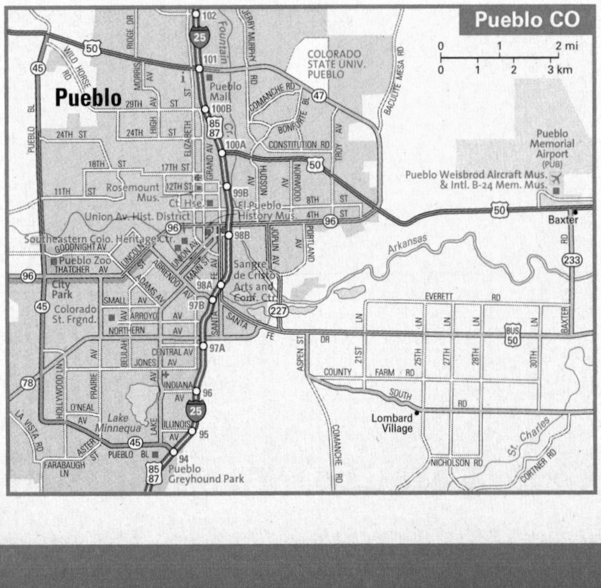

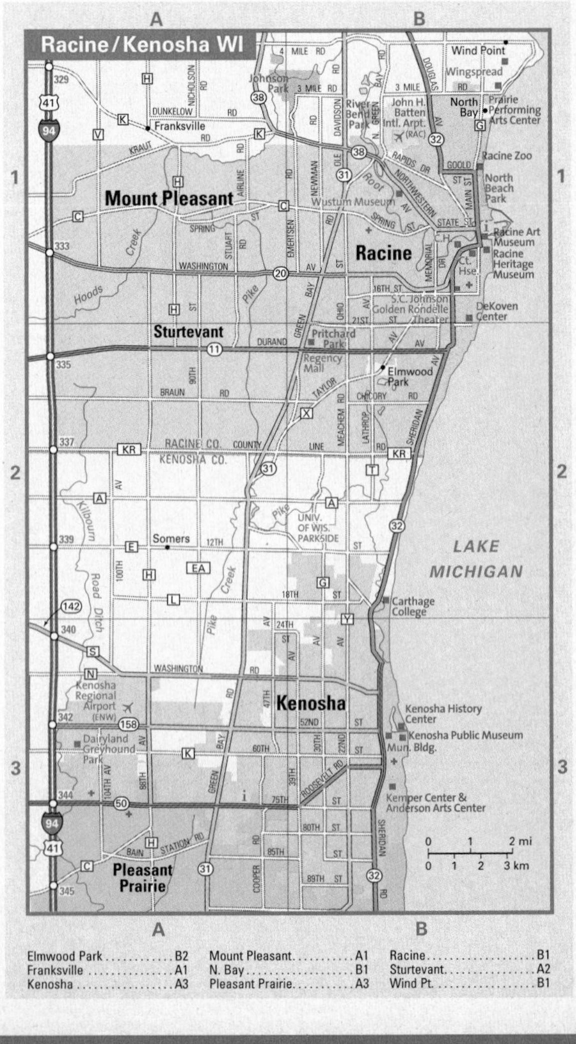

Elmwood Park	B2	Mount Pleasant	A1	Racine	B1
Franksville	A1	N. Bay	B1	Sturtevant	A2
Kenosha	A3	Pleasant Prairie	A3	Wind Pt.	B1

Entries in **bold black** indicate counties or parishes.
Entries in **bold color** indicate cities with detailed inset maps.

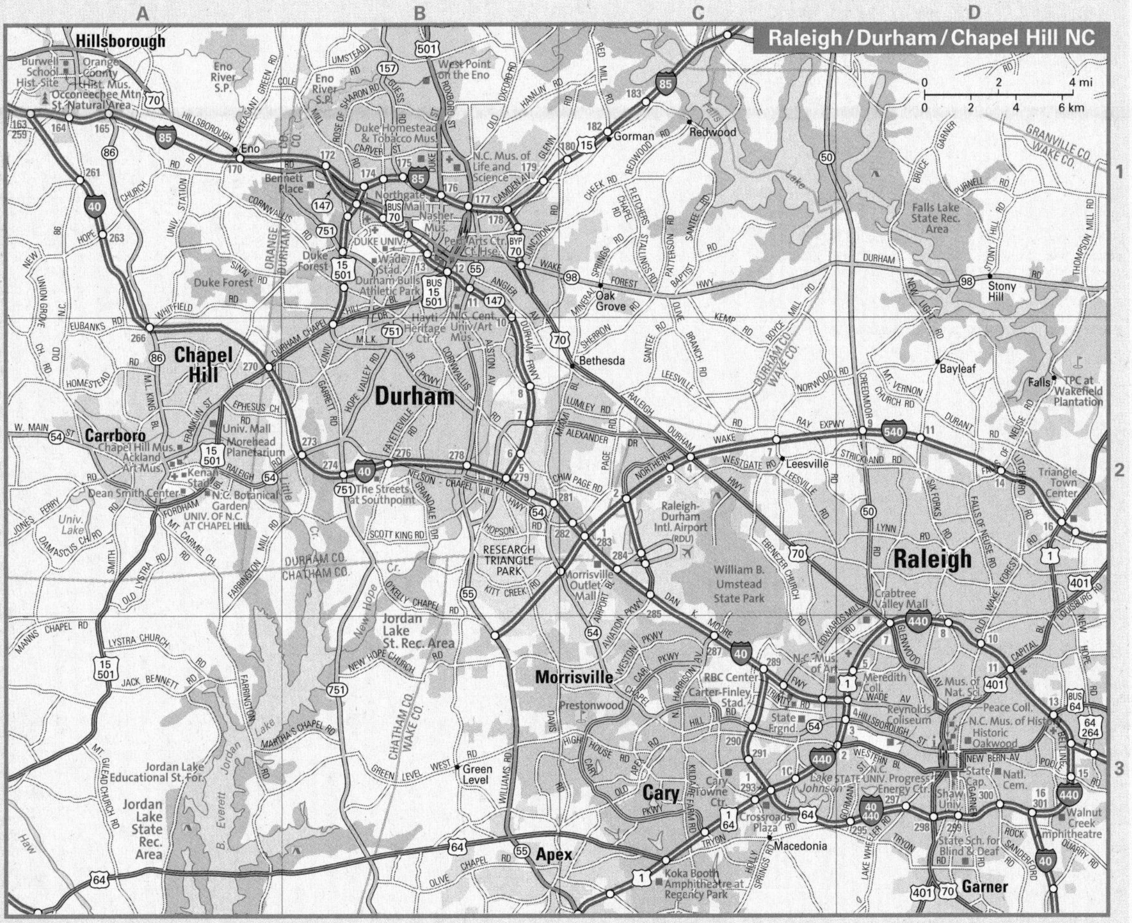

Apex	B3	Carrboro	A2	Durham	B2	Garner	D3	Hillsborough	A1	Morrisville	C3	Redwood	C1
Bayleaf	D2	Cary	C3	Eno	A1	Gorman	C1	Leesville	C2	Oak Grove	C1	Stony Hill	D1
Bethesda	C2	Chapel Hill	A2	Falls	D2	Green Level	B3	Macedonia	C3	Raleigh	D2		

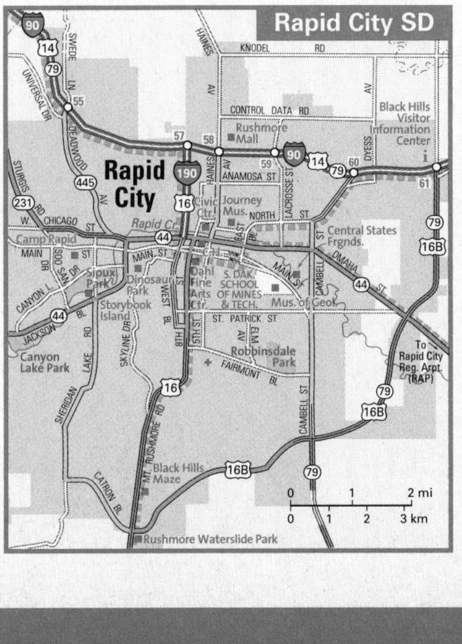

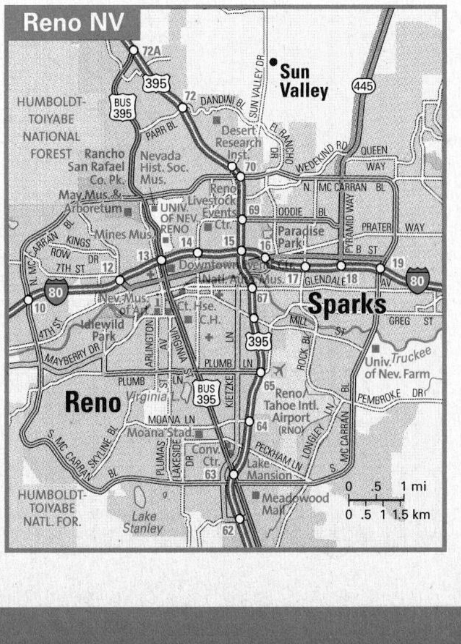

Figures after entries indicate population, page number, and grid reference.

Atlee	B1	Chamberlayne Farms	B1	Glen Allen	B1	Mechanicsville	C1	Rivermont	C3
Bellwood	B3	Chester	B3	Greendale	B1	Midlothian	A2	Sandston	C2
Bensley	B3	Chesterfield	B3	Highland Sprs.	C2	Newmans	C1	Seven Pines	C2
Bon Air	A2	E. Highland Park	B1	Hunton	B1	Old Cold Harbor	C1	Short Pump	A1
Borkeys Store	C1	Fair Oaks	C2	Lakeside	B1	Pearsons Corner	C1	Studley	C1
Centralia	B3			Laurel	B1	Richmond	B2	Varina	C2

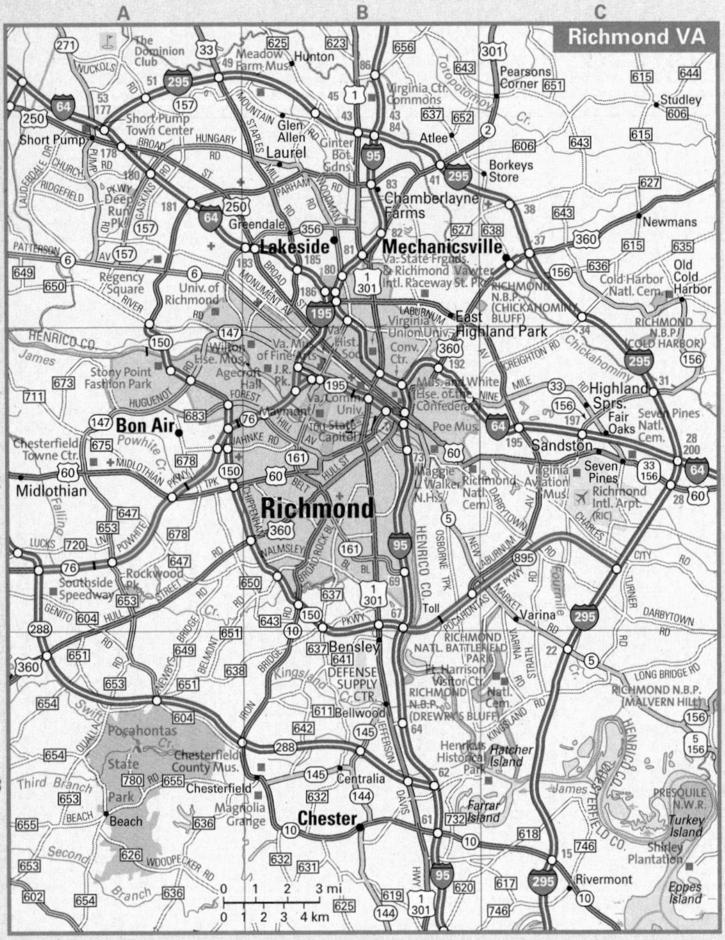

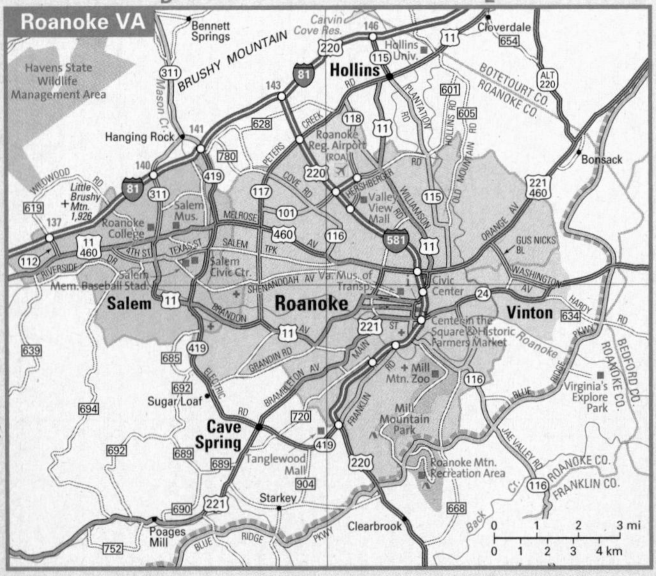

Bennett Sprs.	D1	Clearbrook	E2	Hollins	E1	Salem	D2	Vinton	E2
Bonsack	E1	Cloverdale	E1	Poages Mill	D2	Starkey	D2		
Cave Spr.	D2	Hanging Rock	D1	Roanoke	D2	Sugar Loaf	D2		

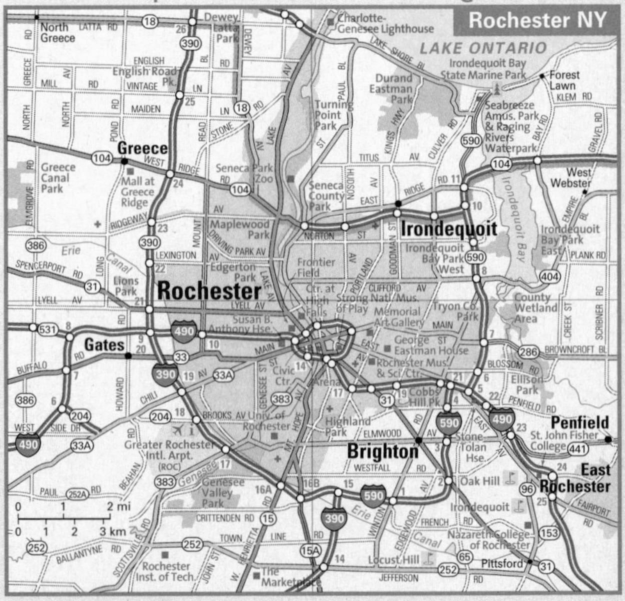

Brighton	G2	Gates	F2	N. Greece	F1	Rochester	F1		
E. Rochester	G2	Greece	F1	Penfield	G2				
Forest Lawn	G1	Irondequoit	G1	Pittsford	G2				

Entries in **bold black** indicate counties or parishes.
Entries in **bold color** indicate cities with detailed inset maps.

River Edge NJ, *10946* 240 C1
River Falls AL, *616* 128 A4
River Falls WI, *12560* 67 D4
River Forest IL, *11635* 203 D4
River Grove IL, *10668* 203 D4
Rivergrove OR, *324* 251 C3
Riverhead NY, *10513* 149 E3
River Hills WI, *1631* 234 D1
River Oaks TX, *6985* 207 A2
River Ridge LA, *14588* 239 B1
River Rouge MI, *9917* 210 C4
Riverside AL, *1564* 120 A4
Riverside CA, *255166* 53 D2
Riverside IL, *8895* 203 D4
Riverside IA, *928* 87 F2
Riverside MO, *2979* 224 B2
Riverside NJ, *8000* 147 D3
Riverside NY, *2875* 149 E3
Riverside OH, *23545* 100 C1
Riverside PA, *1861* 93 E3
Riverside Co. CA, *1545387* 53 E3
Riverton IL, *3048* 98 B1
Riverton KS, *850* 106 B2
Riverton NJ, *2759* 146 C3
Riverton UT, *25011* 31 E4
Riverton WY, *9310* 32 B1
Riverview FL, *12035* 140 C2
Riverview MO, *3146* 256 C1
Riverwood KY, *469* 230 E1
Riverwoods IL, *3843* 203 C2
Rives Jct. MI, *650* 76 A4
Rivesville WV, *913* 102 A1
Riviera Beach FL, *29884* 141 F4
Riviera Beach MD, *12695* 144 C2
Roachdale IN, *975* 99 E1
Roaming Shores OH, *1239* 91 F2
Roane Co. TN, *51910* 110 B4
Roane Co. WV, *15446* 101 F3
Roan Mtn. TN, *1160* 111 E3
Roanoke AL, *6563* 128 B1
Roanoke IL, *1994* 88 B3
Roanoke IN, *1495* 90 A3
Roanoke TX, *2810* 207 B1
Roanoke VA, *94911* 112 B2
Roanoke Co. VA, *85778* 112 B1
Roanoke Rapids NC, *16957* 113 E3
Roaring Spr. PA, *2418* 92 C4
Robards KY, *564* 109 E1
Robbins IL, *6635* 203 D6
Robbins NC, *1195* 122 C1
Robbinsdale MN, *14123* 235 B2
Robbinsville NJ, *1900* 147 D2
Robbinsville NC, *747* 121 D1
Robersonville NC, *1731* 113 E4
Roberta GA, *808* 129 D2
Robert Lee TX, *1171* 58 B3
Roberts ID, *647* 23 E4
Roberts WI, *969* 67 E4
Roberts Co. SD, *10016* 27 F1
Roberts Co. TX, *887* 50 B2
Robertsdale AL, *3782* 135 E2
Robertson Co. KY, *2266* 100 C3
Robertson Co. TN, *54433* 109 F3
Robertson Co. TX, *16000* 61 F1
Robertsville NJ, *9800* 147 E2
Robeson Co. NC, *123339* 123 D2
Robesonia PA, *2036* 146 A2
Robins IA, *1806* 87 E1
Robinson IL, *6822* 99 D2
Robinson TX, *7845* 59 E4
Robstown TX, *12727* 63 F2
Roby TX, *673* 58 B2
Rochdale MA, *1400* 150 B2
Rochelle GA, *1415* 129 E3
Rochelle IL, *9424* 88 B1
Rochelle Park NJ, *5528* 240 C1
Rochester IL, *2893* 98 B1
Rochester IN, *6414* 89 F3
Rochester MN, *85806* 73 D1
Rochester NH, *28461* 81 F4
Rochester NY, *219773* 78 C3
Rochester PA, *4014* 91 F3
Rochester VT, *300* 81 D3
Rochester WA, *1829* 12 B4
Rochester WI, *1149* 74 C4
Rochester Hills MI, *68825* 76 C4
Rock MA, *850* 151 D3
Rockaway NJ, *6473* 148 A3
Rockaway Beach MO, *577* 107 D3
Rockaway Beach OR, *1267* 20 B1
Rockbridge Co. VA, *20808* 102 B4
Rockcastle Co. KY, *16582* 110 C2
Rock Co. MN, *9721* 27 F4
Rock Co. NE, *1756* 35 D2
Rock Co. WI, *152307* 74 B4
Rock Creek MN, *1119* 67 D2
Rock Creek OH, *584* 91 F2
Rockdale MD, *16100* 144 C2
Rockdale TX, *5439* 61 E1
Rockdale Co. GA, *70111* 120 C4
Rockfall CT, *1500* 149 E1
Rock Falls IL, *9580* 88 B1
Rockford AL, *428* 128 A1
Rockford IL, *150115* 74 B4
Rockford IA, *907* 73 D3
Rockford MI, *4626* 75 F3
Rockford MN, *3484* 66 C3
Rockford OH, *1126* 90 A3
Rockford TN, *798* 110 C4
Rockford WA, *413* 14 B3
Rock Hall MD, *1396* 145 D2
Rock Hill MO, *4765* 256 B2
Rock Hill NY, *1056* 148 A1
Rock Hill SC, *49765* 122 A2
Rockingham NC, *9672* 122 C2
Rockingham Co. NH, *277359* 81 F4
Rockingham Co. NC, *91928* 112 B3
Rockingham Co. VA, *67725* 102 C3
Rock Island IL, *39684* 88 A2
Rock Island OK, *709* 116 B1
Rock Island WA, *863* 13 E3
Rock Island Co. IL, *149374* 88 A2
Rockland ID, *316* 31 E1
Rockland ME, *7609* 82 C2
Rockland MA, *17670* 151 D2
Rockland WI, *628* 73 F1
Rockland Co. NY, *286753* 148 B2
Rockledge AL, *600* 120 A3
Rockledge FL, *20170* 141 E2
Rockledge PA, *2577* 248 D2
Rocklin CA, *36330* 36 C2
Rockmart GA, *3870* 120 B3
Rock Mills AL, *676* 128 B1
Rock Pt. AZ, *724* 47 F1
Rockport AR, *792* 117 D3
Rockport IN, *2160* 99 E4
Rockport KY, *334* 109 E1
Rockport ME, *900* 82 C2
Rockport MA, *5606* 151 F1
Rock Port MO, *1395* 86 A4
Rockport TX, *7385* 61 E4
Rock Rapids IA, *2573* 27 F4
Rock River WY, *235* 33 D2
Rocksprings TX, *1285* 60 B2
Rock Sprs. WY, *18708* 32 B3
Rockton IL, *5296* 74 B4
Rockvale CO, *426* 41 E3
Rock Valley IA, *2702* 35 F1
Rockville CT, *7708* 150 A3
Rockville IN, *2765* 99 E1
Rockville MD, *47388* 144 B2
Rockville MN, *749* 66 C3
Rockville RI, *425* 150 C4
Rockwall TX, *17976* 59 F2
Rockwall Co. TX, *43080* 59 F2
Rockwell AR, *3024* 117 D3
Rockwell IA, *989* 73 D3
Rockwell NC, *1971* 122 B1
Rockwell City IA, *2264* 72 B4
Rockwood MI, *3442* 90 C1
Rockwood PA, *954* 102 B1
Rockwood TN, *5774* 110 B4
Rocky Ford CO, *4286* 41 F3
Rocky Hill CT, *17966* 149 E1
Rocky Hill NJ, *662* 147 D1
Rocky Mount NC, *55893* 113 E4
Rocky Mount VA, *4066* 112 B2
Rocky Pt. NY, *10185* 149 D3
Rocky Ridge UT, *403* 39 E1
Rocky Ripple IN, *712* 221 B2
Rocky River OH, *20735* 204 E2
Rodarte NM, *350* 49 D2
Rodeo CA, *8717* 259 C1
Roebuck SC, *1725* 121 F2
Roeland Park KS, *6817* 224 B3
Roessleville NY, *10800* 188 D2
Roff OK, *734* 51 F4
Roger Mills Co. OK, *3436* 50 C2
Rogers AR, *38829* 106 C3
Rogers MN, *3588* 66 C3
Rogers TX, *1117* 61 E1
Rogers City MI, *3322* 71 D3
Rogers Co. OK, *70641* 106 A3
Rogersville AL, *1199* 119 E2
Rogersville MO, *1508* 107 D2
Rogersville TN, *4240* 111 D3
Rogue River OR, *1847* 28 B2
Rohnert Park CA, *42236* 36 B3
Roland IA, *1324* 86 C1
Roland OK, *2842* 116 B1
Rolesville NC, *907* 113 D4
Rolette ND, *538* 18 C1
Rolette Co. ND, *13674* 18 C1
Rolfe IA, *675* 72 B3
Rolla KS, *482* 50 B1
Rolla MO, *16367* 97 F4
Rolla ND, *1417* 18 C1
Rolling Fields KY, *648* 230 E1
Rolling Fork MS, *2486* 126 A1
Rolling Hills CA, *1871* 228 C4
Rolling Hills KY, *907* 230 F1
Rolling Hills WY, *449* 33 D1
Rolling Hills Estates CA, *7676* 228 C4
Rolling Meadows IL, *24604* 203 B3
Rolling Prairie IN, *800* 89 E2
Rollingstone MN, *697* 73 E1
Rollinsford NH, *1500* 82 A4
Roma TX, *9617* 63 D4
Romancoke MD, *800* 145 D3
Roman Forest TX, *1279* 132 B2
Rome GA, *34980* 120 B3
Rome IL, *1776* 88 B3
Rome NY, *34950* 79 E3
Rome City IN, *1615* 90 A2
Romeo CO, *375* 41 D4
Romeo MI, *3721* 76 C4
Romeoville IL, *21153* 89 D2
Romney WV, *1940* 102 C2
Romoland CA, *2764* 229 K4
Romulus MI, *22979* 90 C1
Ronan MT, *1812* 15 D3
Ronceverte WV, *1557* 112 A1
Ronkonkoma NY, *20029* 149 D3
Roodhouse IL, *2214* 98 A2
Rooks Co. KS, *5685* 43 D2
Roosevelt NJ, *933* 147 E2
Roosevelt UT, *4299* 32 A4
Roosevelt Co. MT, *10620* 17 E2
Roosevelt Co. NM, *18018* 49 F4
Roosevelt Park MI, *3890* 75 E3
Roper NC, *613* 113 F4
Ropesville TX, *517* 58 A1
Rosalia WA, *648* 14 A3
Rosamond CA, *14349* 52 C1
Rosaryville MD, *12322* 144 C4
Rosburg WA, *40* 12 B4
Roscoe IL, *6244* 74 B4
Roscoe NY, *597* 94 A2
Roscoe SD, *324* 27 D2
Roscoe TX, *1378* 58 B3
Roscommon MI, *1133* 70 C4
Roscommon Co. MI, *25469* 76 A1
Roseau MN, *2756* 19 F1
Roseau Co. MN, *16338* 19 F1
Roseboro NC, *1267* 123 D2
Rose Bud AR, *429* 117 E1
Rosebud SD, *1557* 34 C1
Rosebud TX, *1493* 59 F4
Rosebud Co. MT, *9383* 17 D4
Roseburg OR, *20017* 28 B1
Rose City MI, *721* 76 B1
Rose City TX, *519* 132 C2
Rosedale IN, *750* 99 E1
Rosedale LA, *753* 133 F2
Rosedale MD, *19199* 144 C2
Rosedale MS, *2414* 118 A4
Rosedale Beach DE, *750* 145 F4
Rose Haven MD, *1400* 144 C4
Rose Hill KS, *3432* 43 F4
Rose Hill NC, *1330* 123 E2
Rose Hill VA, *714* 111 D3
Rose Hill VA, *15058* 270 C5
Roseland FL, *1775* 141 E3
Roseland LA, *1162* 134 B1
Roseland NJ, *5298* 240 A2
Roselawn IN, *3933* 89 D3
Roselle IL, *23115* 203 B3
Roselle NJ, *21274* 147 E1
Roselle Park NJ, *13281* 147 E1
Rose Lodge OR, *1708* 20 B2
Rosemead CA, *53505* 228 E2
Rosemont CA, *22904* 255 C3
Rosemont IL, *4224* 203 C3
Rosemount MN, *14619* 235 D4
Rosemount OH, *2043* 101 D3
Rosenberg TX, *24043* 132 A3
Rosendale NY, *1374* 94 B3
Rosendale WI, *923* 74 C2
Rosenhayn NJ, *1099* 145 F1
Rosepine LA, *1390* 133 D1
Roseto PA, *1653* 93 F3
Rose Valley PA, *944* 248 A4
Roseville CA, *79921* 36 C2
Roseville IL, *1083* 88 A3
Roseville MI, *48129* 210 D2
Roseville MN, *33690* 235 C2
Roseville OH, *1936* 101 E1
Roseville Park DE, *6200* 146 B4
Rosholt SD, *419* 27 F1
Rosiclare IL, *1213* 109 D1
Roslyn PA, *16900* 248 C1
Roslyn SD, *225* 27 E2
Roslyn WA, *1017* 13 D3
Roslyn Estates NY, *1210* 241 G2
Rosman NC, *490* 121 E1
Ross CA, *2329* 259 A1
Ross OH, *1971* 100 B2
Ross Co. OH, *73345* 101 D2
Rossford OH, *6406* 267 B2
Rossiter PA, *790* 92 B3
Rossmoor CA, *10298* 228 E4
Rossmoor NJ, *3129* 147 E2
Rossville GA, *3511* 120 B2
Rossville IL, *1217* 89 D4
Rossville IN, *1513* 89 E4
Rossville KS, *1014* 43 F2
Rossville MD, *11515* 193 E2
Roswell GA, *79334* 120 C3
Roswell NM, *45293* 57 E1
Rotan TX, *1611* 58 B2
Rothsay MN, *497* 19 F4
Rothschild WI, *4970* 68 B4
Rothsville PA, *3017* 146 A2
Rotonda FL, *6574* 140 C4
Rotterdam NY, *20536* 94 B1
Rotterdam Jct. NY, *918* 94 B1
Rougemont NC, *600* 112 C3
Rough Rock AZ, *469* 47 F1
Round Hill VA, *500* 103 D2
Round Lake NY, *604* 94 B1
Round Mtn. NV, *550* 37 F2
Round Pond ME, *325* 82 C3
Round Rock AZ, *601* 47 F1
Round Rock TX, *61136* 61 E1
Roundup MT, *1931* 16 C4
Rouses Pt. NY, *2277* 81 D1
Routt Co. CO, *19690* 32 C4
Rouzerville PA, *862* 103 D1
Rowan Co. KY, *22094* 100 C4
Rowan Co. NC, *130340* 112 A4
Rowland NC, *1146* 122 C3
Rowland Hts. CA, *48553* 229 F3
Rowlesburg WV, *613* 102 B2
Rowlett TX, *44503* 207 E1
Rowley MA, *1434* 151 F1
Roxana DE, *375* 145 F4
Roxboro NC, *8696* 112 C3
Roxie MS, *569* 126 A4
Roxton TX, *694* 116 A4
Roy NM, *304* 49 E2
Roy UT, *32885* 31 E3
Royal Ctr. IN, *832* 89 E3
Royal City WA, *1823* 13 E4
Royal Oak MD, *750* 145 D4
Royal Oak MI, *60062* 76 C4
Royal Palm Beach FL, *21523* 143 E1
Royal Pines NC, *5334* 121 E1
Royalton IL, *1130* 98 C4
Royalton MN, *816* 66 C2
Royersford PA, *4246* 146 B2
Royse City TX, *2957* 59 F2
Royston GA, *2493* 121 E3
Rubidoux CA, *29180* 229 H3
Rubonia FL, *1700* 266 B4
Ruch OR, *700* 28 B2
Rudyard MT, *275* 15 F2
Rugby ND, *2939* 18 C2
Ruidoso NM, *7698* 57 D2
Ruidoso Downs NM, *1824* 57 D2
Rule TX, *698* 58 C2
Ruleville MS, *3234* 118 A4
Rumford ME, *4795* 82 B2
Rumson NJ, *7137* 147 F2
Runaway Bay TX, *1104* 59 E2
Runge TX, *1080* 61 E3
Runnels Co. TX, *11495* 58 C3
Runnemede NJ, *8533* 146 C3
Running Sprs. CA, *5125* 229 K1
Rupert ID, *5645* 31 D1
Rupert WV, *940* 102 A4
Rural Hall NC, *2464* 112 B3
Rural Retreat VA, *1350* 111 F2
Rural Valley PA, *922* 92 B3
Rush City MN, *2102* 67 D2
Rush Co. IN, *18261* 100 A1
Rush Co. KS, *3551* 43 D3
Rushford MN, *1696* 73 E2
Rushford Vil. MN, *714* 73 E2
Rushmere VA, *1083* 113 F2
Rush Sprs. OK, *1278* 51 E4
Rush Valley UT, *453* 31 E4
Rushville IL, *3212* 88 A4
Rushville IN, *5995* 100 A1
Rushville NE, *999* 34 A1
Rushville NY, *621* 78 C4
Rusk TX, *5085* 124 B3

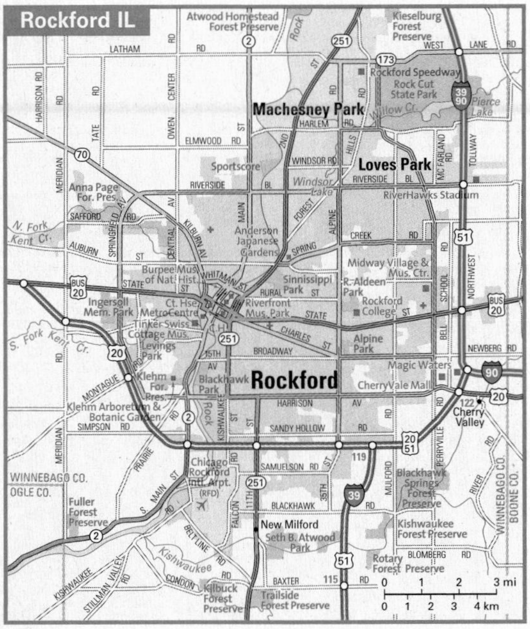

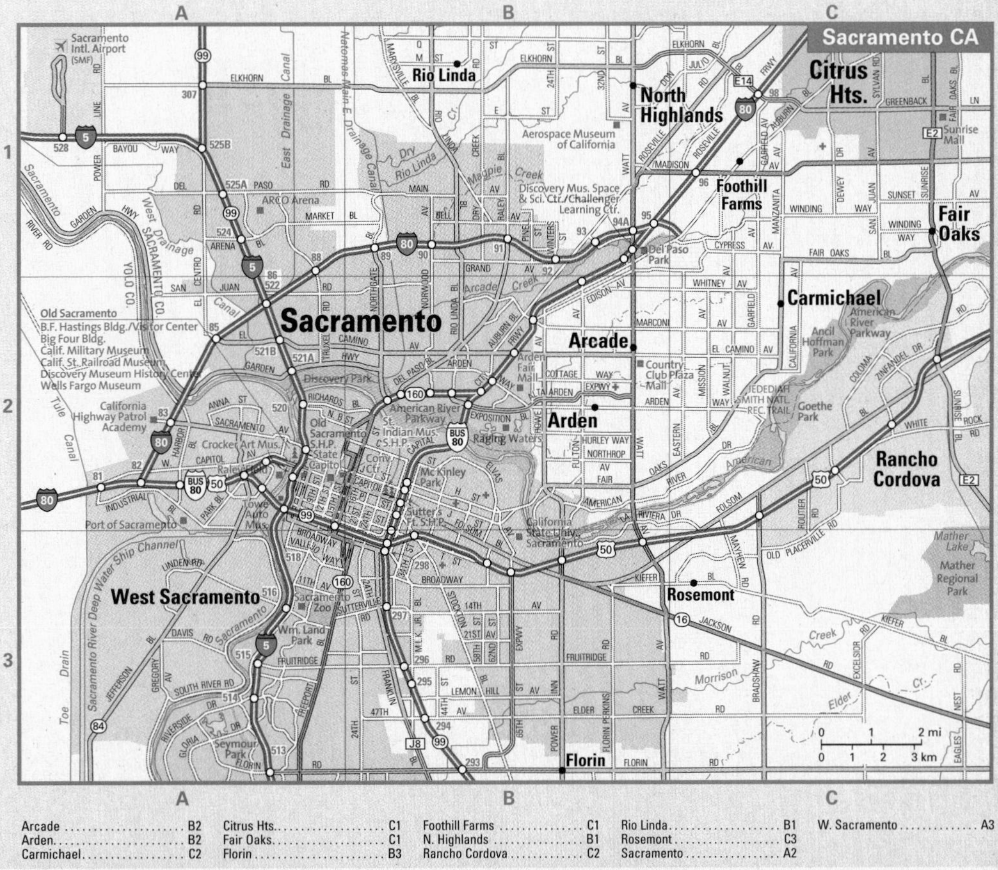

Arcade B2	Citrus Hts. C1	Foothill Farms C1	Rio Linda B1	W. Sacramento A3
Arden B2	Fair Oaks C1	N. Highlands B1	Rosemont C3	
Carmichael C2	Florin B3	Rancho Cordova C2	Sacramento A2	

Figures after entries indicate population, page number, and grid reference.

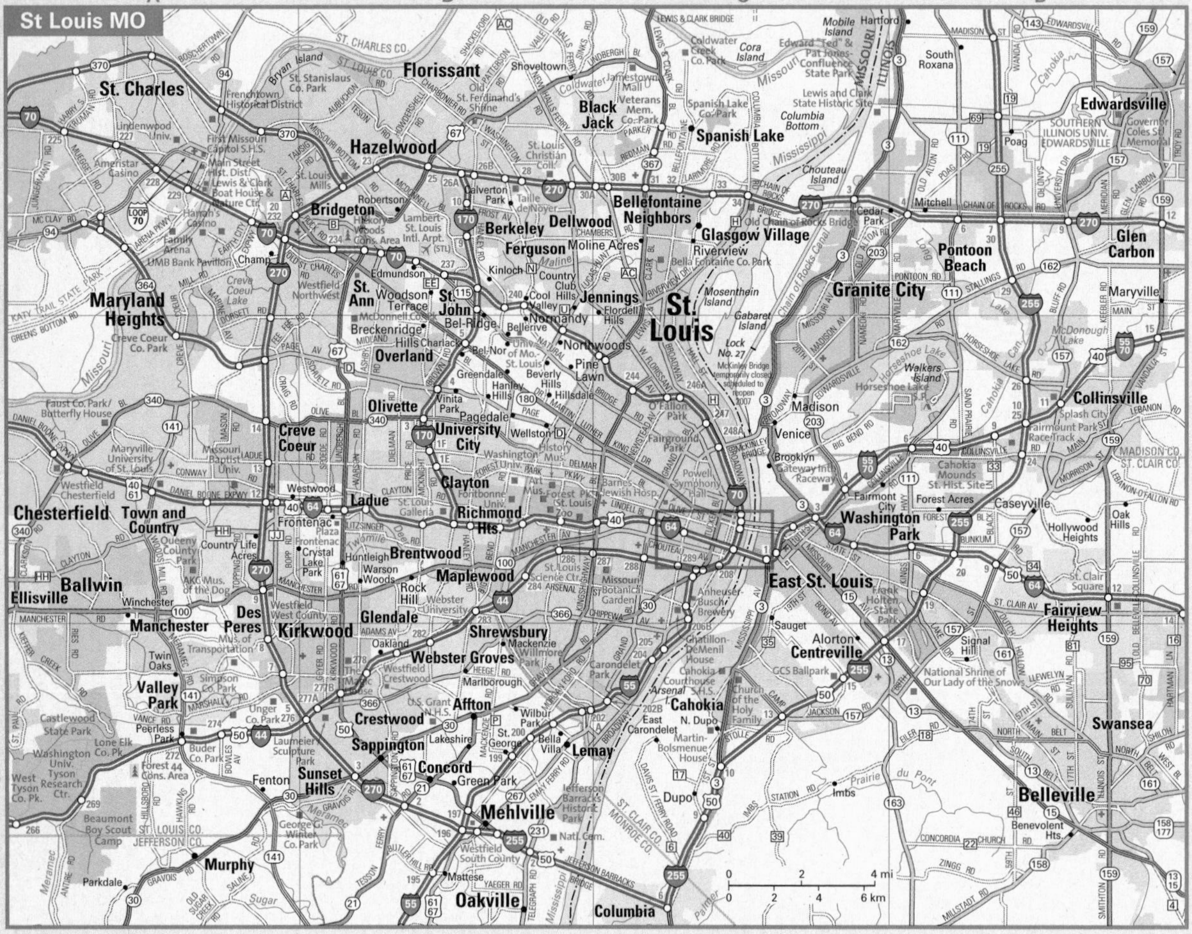

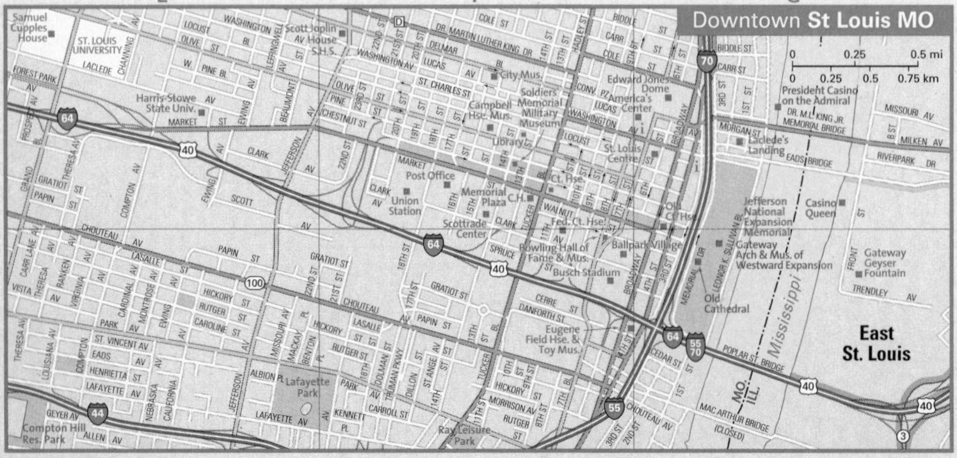

POINTS OF INTEREST

Entries in **bold black** indicate counties or parishes.
Entries in **bold color** indicate cities with detailed inset maps.

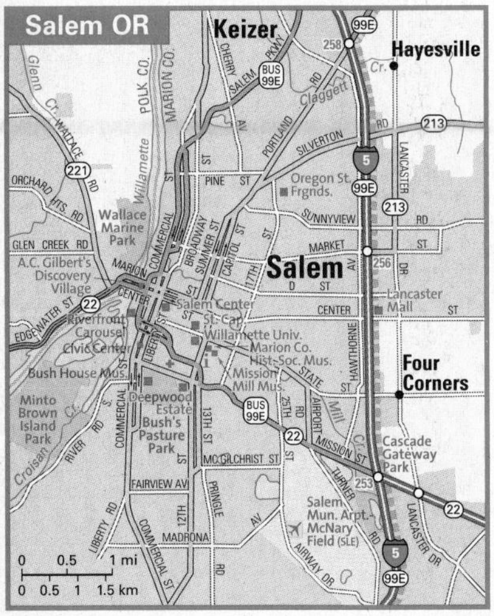

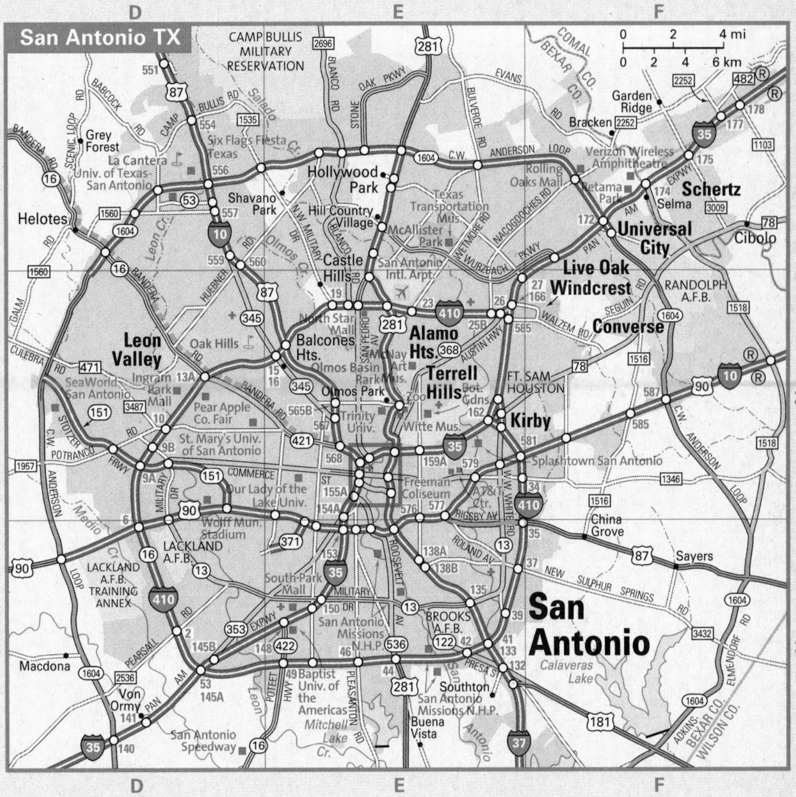

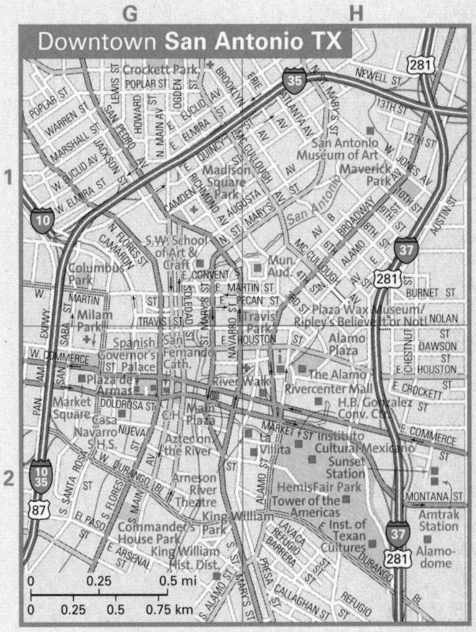

POINTS OF INTEREST

Figures after entries indicate population, page number, and grid reference.

Bonita . . . C2
Chula Vista . . . B3
Coronado . . . A2
El Cajon . . . C1
Imperial Beach . . . B3
Lakeside . . . C1
La Mesa . . . C2
Lemon Grove . . . C2
National City . . . B2
San Diego . . . C2
Santee . . . C1
Spring Valley . . . C2
Sunnyside . . . C2
Tijuana, MX . . . C3

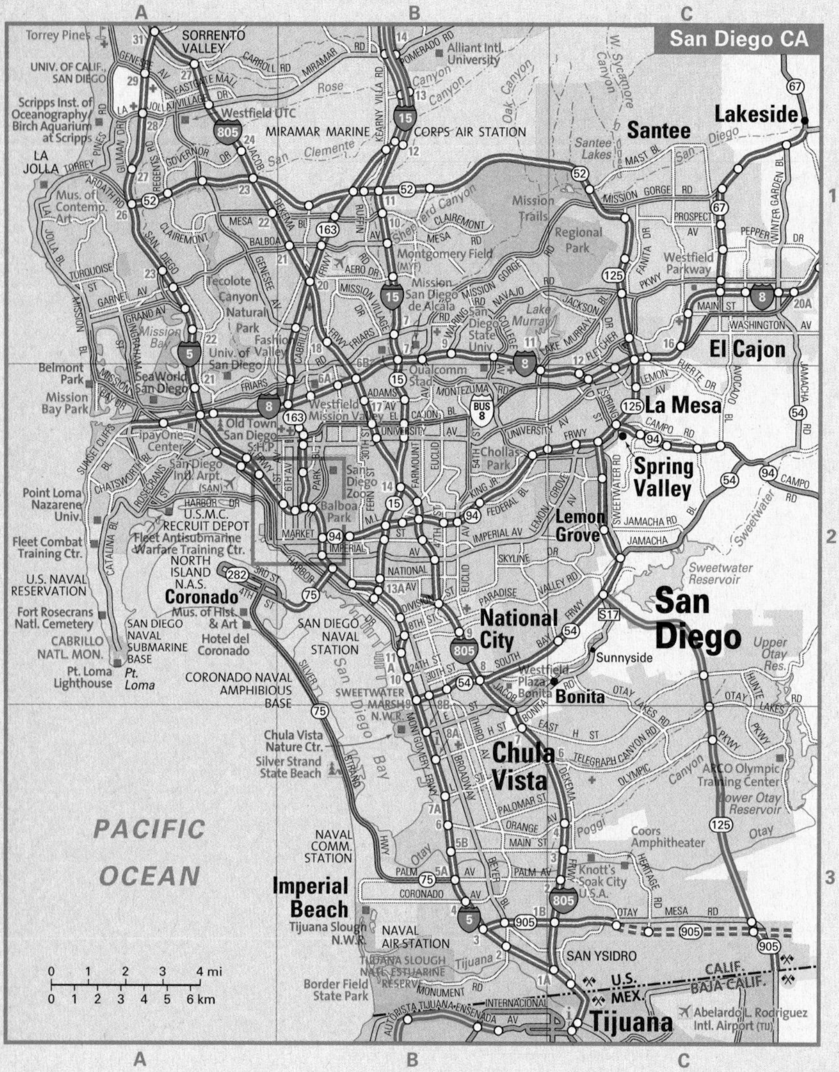

POINTS OF INTEREST

Automotive Museum . . . E1
Balboa Park . . . E1
Balboa Stadium . . . E2
Casa del Prado . . . E1
Children's Museum . . . D2
Civic Center . . . D2
Copley Symphony Hall . . . E2
County Court House . . . D2
Firehouse Museum . . . D2
Gaslamp Quarter & W. H. Davis House . . . E2
The Globe Theatres . . . E1
House of Hospitality . . . E1
Maritime Museum . . . D2
Museum of Contemporary Art, San Diego . . D2
PETCO Park . . . E2
Reuben H. Fleet Science Center . . . E1
San Diego Aerospace Museum . . . E1
San Diego Aircraft Carrier Museum . . . D2
San Diego Convention Center . . . D2
San Diego Hall of Champions . . . E1
San Diego International Airport . . . D1
San Diego Museum of Art . . . E1
San Diego Museum of Man . . . E1
San Diego Natural History Museum . . . E1
San Diego Zoo . . . E1
Santa Fe Depot . . . D2
Seaport Village . . . D2
Spanish Village Art Center . . . E1
Spreckels Organ Pavilion . . . E1
Spreckels Theatre . . . D2
Starlight Bowl . . . E1
Timken Museum of Art . . . E1
Veterans Museum & Memorial Center . . . E1
Villa Montezuma . . . E2
Westfield Horton Plaza . . . D2

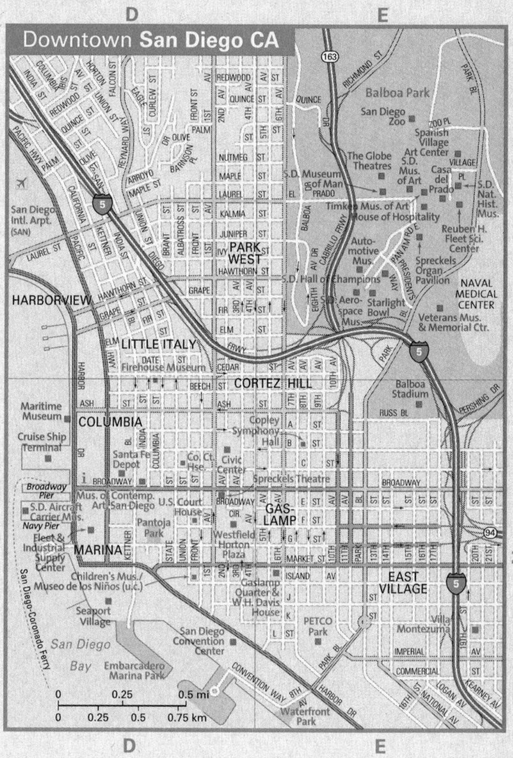

St. Clair MI, *5802* . . . 76 C3
St. Clair MN, *827* . . . 72 C1
St. Clair MO, *4390* . . . 97 F4
St. Clair PA, *3254* . . . 93 E3
St. Clair Co. AL, *64742* . . . 120 A4
St. Clair Co. IL, *256082* . . . 98 B4
St. Clair Co. MI, *164235* . . . 76 C3
St. Clair Co. MO, *9652* . . . 96 C4
St. Clair Shores MI, *63096* . . . 76 C4
St. Clairsville OH, *5057* . . . 91 F4
St. Cloud FL, *20074* . . . 141 D2
St. Cloud MN, *59107* . . . 66 C3
St. Croix Co. WI, *63155* . . . 67 E3
St. Croix Falls WI, *2033* . . . 67 D3
St. David AZ, *1744* . . . 55 E3
St. David IL, *587* . . . 88 A4
St. Edward NE, *796* . . . 35 E3
Ste. Genevieve MO, *4476* . . . 98 B4
Ste. Genevieve Co. MO, *17842* . . . 98 A4
St. Elmo IL, *1456* . . . 98 C2
St. Francis KS, *1497* . . . 42 B1
St. Francis MN, *4910* . . . 67 D3
St. Francis SD, *675* . . . 34 C1
St. Francis WI, *8662* . . . 234 D3
St. Francis Co. AR, *29329* . . . 118 A2
St. Francisville IL, *759* . . . 99 D3
St. Francisville LA, *1712* . . . 134 A1
St. Francois Co. MO, *55641* . . . 98 A4
St. Gabriel LA, *5514* . . . 134 A2
St. George KS, *434* . . . 43 F2
St. George ME, *350* . . . 82 C3
St. George MO, *1288* . . . 256 B3
St. George SC, *2092* . . . 130 C1
St. George UT, *49663* . . . 46 C1
St. Hedwig TX, *1875* . . . 61 D3
St. Helen MI, *2993* . . . 76 A1
St. Helena CA, *5950* . . . 36 B3
St. Helena Par. LA, *10525* . . . 134 A1
St. Helens OR, *10019* . . . 20 C1
St. Henry OH, *2271* . . . 90 A4
St. Ignace MI, *2678* . . . 70 C2
St. Ignatius MT, *788* . . . 15 D3
St. Jacob IL, *801* . . . 98 B3
St. James MD, *1657* . . . 144 A1
St. James MN, *4695* . . . 72 B1
St. James MO, *3704* . . . 97 F4
St. James NY, *13268* . . . 149 D3
St. James NC, *804* . . . 123 E4
St. James City FL, *4105* . . . 142 C1
St. James Par. LA, *21216* . . . 134 A3
St. Jo TX, *977* . . . 59 E1
St. John IN, *8382* . . . 89 D2
St. John KS, *1318* . . . 43 D3
St. John MO, *6871* . . . 256 B1
St. John ND, *358* . . . 18 C1
St. John WA, *548* . . . 13 F3
St. Johns AZ, *3269* . . . 48 A4
St. Johns MI, *7485* . . . 76 A3
St. Johnsbury VT, *6319* . . . 81 E2
St. Johnsbury Ctr. VT, *1000* . . . 81 E2
St. Johns Co. FL, *123135* . . . 139 D3
St. Johnsville NY, *1685* . . . 79 F3
St. John the Baptist Par. LA, *43044* . . . 134 B2
St. Joseph IL, *2912* . . . 89 D4
St. Joseph LA, *1340* . . . 126 A3
St. Joseph MI, *8789* . . . 89 E1
St. Joseph MN, *4681* . . . 66 C3
St. Joseph MO, *73990* . . . 96 B1
St. Joseph TN, *829* . . . 119 E2
St. Joseph Co. IN, *265559* . . . 89 F2
St. Joseph Co. MI, *62422* . . . 89 F1
St. Landry Par. LA, *87700* . . . 133 F1
St. Lawrence PA, *1812* . . . 146 B2
St. Lawrence SD, *210* . . . 27 D3
St. Lawrence Co. NY, *111931* . . . 79 F1
St. Leo FL, *595* . . . 140 C2
St. Libory IL, *583* . . . 98 B4
St. Louis MI, *4494* . . . 76 A2
St. Louis MO, *348189* . . . 98 A3
St. Louis Co. MN, *200528* . . . 64 C3
St. Louis Co. MO, *1016315* . . . 98 A3
St. Louis Park MN, *44126* . . . 235 B2
St. Lucie FL, *604* . . . 141 E3
St. Lucie Co. FL, *192695* . . . 141 E3
St. Maries ID, *2652* . . . 14 B3
St. Martin Par. LA, *48583* . . . 133 F2
St. Martins MO, *1023* . . . 97 E3
St. Martinville LA, *6989* . . . 133 F2
St. Mary Par. LA, *53500* . . . 134 A3
St. Marys AK, *500* . . . 154 B2
St. Marys GA, *13761* . . . 139 D1
St. Marys KS, *2198* . . . 43 F2
St. Marys OH, *8342* . . . 90 B4
St. Marys PA, *14502* . . . 92 B2
St. Marys WV, *2017* . . . 101 F2
St. Mary's City MD, *900* . . . 103 F4
St. Mary's Co. MD, *86211* . . . 103 E3
St. Matthews KY, *16855* . . . 230 E1
St. Matthews SC, *2107* . . . 122 B4
St. Meinrad IN, *700* . . . 99 E4
St. Michael MN, *9099* . . . 66 C3
St. Michaels AZ, *1295* . . . 48 A2
St. Michaels MD, *1193* . . . 145 D3
St. Nazianz WI, *749* . . . 74 C1
St. Paris OH, *1998* . . . 90 B4
St. Paul AK, *532* . . . 154 A3
St. Paul IN, *1022* . . . 100 A2
St. Paul KS, *646* . . . 106 B1
St. Paul MN, *287151* . . . 67 D4
St. Paul MO, *1634* . . . 98 A3
St. Paul NE, *2218* . . . 35 D3
St. Paul OR, *354* . . . 20 B2
St. Paul TX, *630* . . . 59 F2
St. Paul TX, *542* . . . 61 E4
St. Paul VA, *1000* . . . 111 E2
St. Paul Park MN, *5070* . . . 235 E4
St. Pauls NC, *2137* . . . 123 D2
St. Pete Beach FL, *9929* . . . 140 B3
St. Peter MN, *9747* . . . 72 C1
St. Peters MO, *51381* . . . 98 A3
St. Petersburg FL, *248232* . . . 140 B3
St. Regis MT, *315* . . . 14 C3
St. Regis Falls NY, *750* . . . 80 C1
St. Regis Park KY, *1520* . . . 230 F2
St. Robert MO, *2760* . . . 107 E1
St. Rose LA, *6540* . . . 134 B3
St. Simons Island GA, *13381* . . . 139 D1
St. Stephen MN, *860* . . . 66 C2
St. Stephen SC, *1776* . . . 131 D1
St. Stephens NC, *9439* . . . 111 F4
St. Tammany Par. LA, *191268* . . . 134 B2
St. Thomas ND, *447* . . . 19 E2
Saks AL, *10698* . . . 120 A4
Salado TX, *1951* . . . 61 E1
Salamanca NY, *6097* . . . 92 B1
Sale Creek TN, *750* . . . 120 B1
Salem AR, *1591* . . . 107 F3
Salem AR, *2789* . . . 117 E2
Salem IL, *7909* . . . 98 C3
Salem IN, *6172* . . . 99 F3
Salem KY, *769* . . . 109 D2
Salem MA, *40407* . . . 151 D1
Salem MO, *4854* . . . 107 F1
Salem NH, *28112* . . . 95 E1
Salem NJ, *5857* . . . 145 E1
Salem NM, *795* . . . 56 B2
Salem NY, *964* . . . 81 D4
Salem NC, *2923* . . . 111 F4
Salem OH, *12197* . . . 91 F3
Salem OR, *136924* . . . 20 B2
Salem SD, *1371* . . . 27 E4
Salem UT, *4372* . . . 39 E1
Salem VA, *24747* . . . 112 B2
Salem WV, *2006* . . . 102 A2
Salem Co. NJ, *64285* . . . 145 F1
Salesville AR, *437* . . . 107 E3
Salida CA, *12560* . . . 36 C4
Salida CO, *5504* . . . 41 D3
Salina KS, *45679* . . . 43 E2
Salina OK, *1422* . . . 106 B3
Salina UT, *2393* . . . 39 E2
Salinas CA, *151060* . . . 44 B3
Saline MI, *8034* . . . 90 B1
Saline Co. AR, *83529* . . . 117 D2
Saline Co. IL, *26733* . . . 98 C4
Saline Co. KS, *53597* . . . 43 E2
Saline Co. MO, *23756* . . . 97 D2
Saline Co. NE, *13843* . . . 35 E4
Salineville OH, *1397* . . . 91 F3
Salisbury CT, *700* . . . 94 B2
Salisbury MD, *23743* . . . 114 C1
Salisbury MA, *4484* . . . 151 F1
Salisbury MO, *1726* . . . 97 D2
Salisbury NC, *26462* . . . 112 A4
Salisbury PA, *878* . . . 102 B1
Salisbury Beach MA, *1300* . . . 151 F1
Salix PA, *1259* . . . 92 B4
Sallisaw OK, *7989* . . . 116 B1
Salmon ID, *3122* . . . 23 D2
Salome AZ, *1690* . . . 54 A1
Salsbury Cove ME, *300* . . . 83 D2
Saltillo MS, *3393* . . . 119 D3
Salt Lake City UT, *181743* . . . 31 E4
Salt Lake Co. UT, *898387* . . . 31 E4
Salt Lick KY, *342* . . . 100 C4
Salton City CA, *978* . . . 53 E3
Saltsburg PA, *955* . . . 92 A4
Salt Sprs. FL, *600* . . . 139 D4
Saltville VA, *2204* . . . 111 F2
Saluda NC, *575* . . . 121 E1
Saluda SC, *3066* . . . 121 F3
Saluda VA, *250* . . . 113 F1
Saluda Co. SC, *19181* . . . 121 F3
Salunga PA, *4771* . . . 225 A1
Salyersville KY, *1604* . . . 111 D1

SAN FRANCISCO BAY MAP INDEX

Alameda . . . C3
Alamo . . . D2
Albany . . . C2
Antioch . . . E1
Ashland . . . D3
Atherton . . . C5
Bay Pt. . . . D1
Belmont . . . C5
Belvedere . . . B2
Benicia . . . C1
Berkeley . . . C2
Brisbane . . . B3
Broadmoor . . . B3
Burlingame . . . B4
Campbell . . . E6
Canyon . . . C2
Castro Valley . . . D3
Clayton . . . E2
Clyde . . . D1
Colma . . . B3
Concord . . . D1
Corte Madera . . . A2
Crockett . . . C1
Cupertino . . . D6
Daly City . . . B3
Danville . . . D2
Diablo . . . D2
Dublin . . . E3
E. Palo Alto . . . C5
El Cerrito . . . B2
El Granada . . . B5
El Sobrante . . . C1
Emeryville . . . C2
Fairfax . . . A1
Foster City . . . C4
Fremont . . . D5
Greenbrae . . . A2
Half Moon Bay . . . B5
Hayward . . . D4
Hercules . . . C1
Hillsborough . . . B4
Homestead Valley . . . A2
Ignacio . . . A1
Kensington . . . C2
Kentfield . . . A2
Ladera . . . C5
Lafayette . . . D2
La Honda . . . C6
Larkspur . . . A2
Livermore . . . E3
Los Altos . . . D6
Los Altos Hills . . . C6
Marinwood . . . A1
Martinez . . . D1
Menlo Park . . . C5
Millbrae . . . B4
Mill Valley . . . A2
Milpitas . . . E5
Montara . . . A4
Moraga . . . D2
Moss Beach . . . A5
Mtn. View . . . D5
Muir Beach . . . A2
Newark . . . D5
N. Fair Oaks . . . C5
N. Richmond . . . B2
Novato . . . A1
Oakland . . . C3
Orinda . . . C2
Pacheco . . . D1
Pacifica . . . A4
Palo Alto . . . D5
Pescadero . . . B6
Piedmont . . . C2
Pinole . . . C1
Pittsburg . . . E1
Pleasant Hill . . . D2
Pleasanton . . . E4
Port Costa . . . C1
Portola Valley . . . C6
Redwood City . . . C5
Richmond . . . B2
Rodeo . . . C1
Ross . . . A1
San Anselmo . . . A1
San Bruno . . . B4
San Carlos . . . C5
San Francisco . . . A3
San Gregorio . . . B6
San Jose . . . E6
San Leandro . . . C3
San Lorenzo . . . D4
San Mateo . . . B4
San Pablo . . . B1
San Quentin . . . A2
San Rafael . . . B1
San Ramon . . . D3
Santa Clara . . . D6
Santa Venetia . . . A1
Saratoga . . . D6
Sausalito . . . B2
S. San Francisco . . . B4
Stinson Beach . . . A2
Sunnyvale . . . D6
Sunol . . . E4
Tamalpais Valley . . . A2
Tara Hills . . . B1
Tiburon . . . B2
Union City . . . D4
Vallejo . . . C1
Vine Hill . . . D1
Walnut Creek . . . D2
Woodacre . . . A1
Woodside . . . C5

Entries in **bold black** indicate counties or parishes.
Entries in **bold color** indicate cities with detailed inset maps.

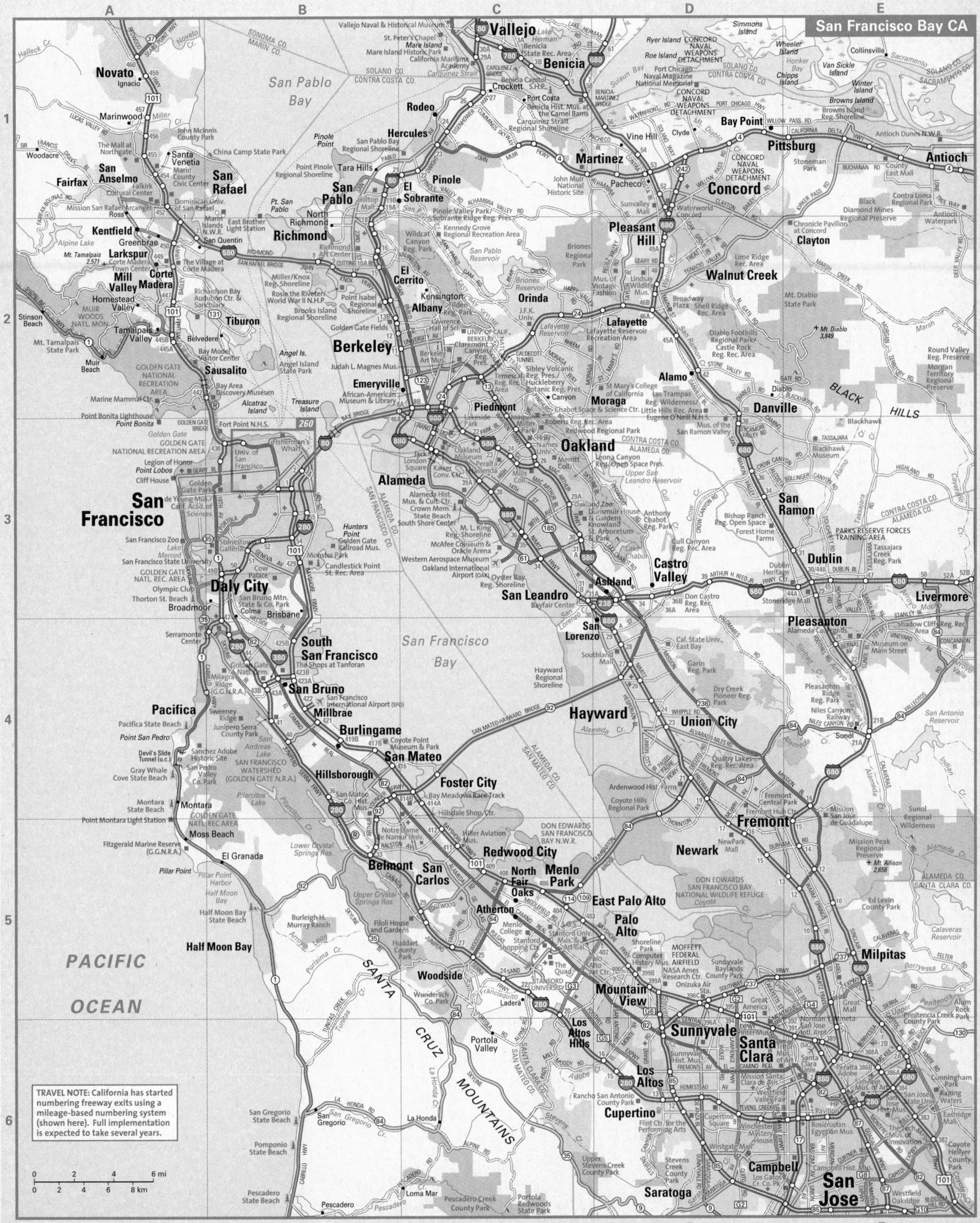

Figures after entries indicate population, page number, and grid reference.

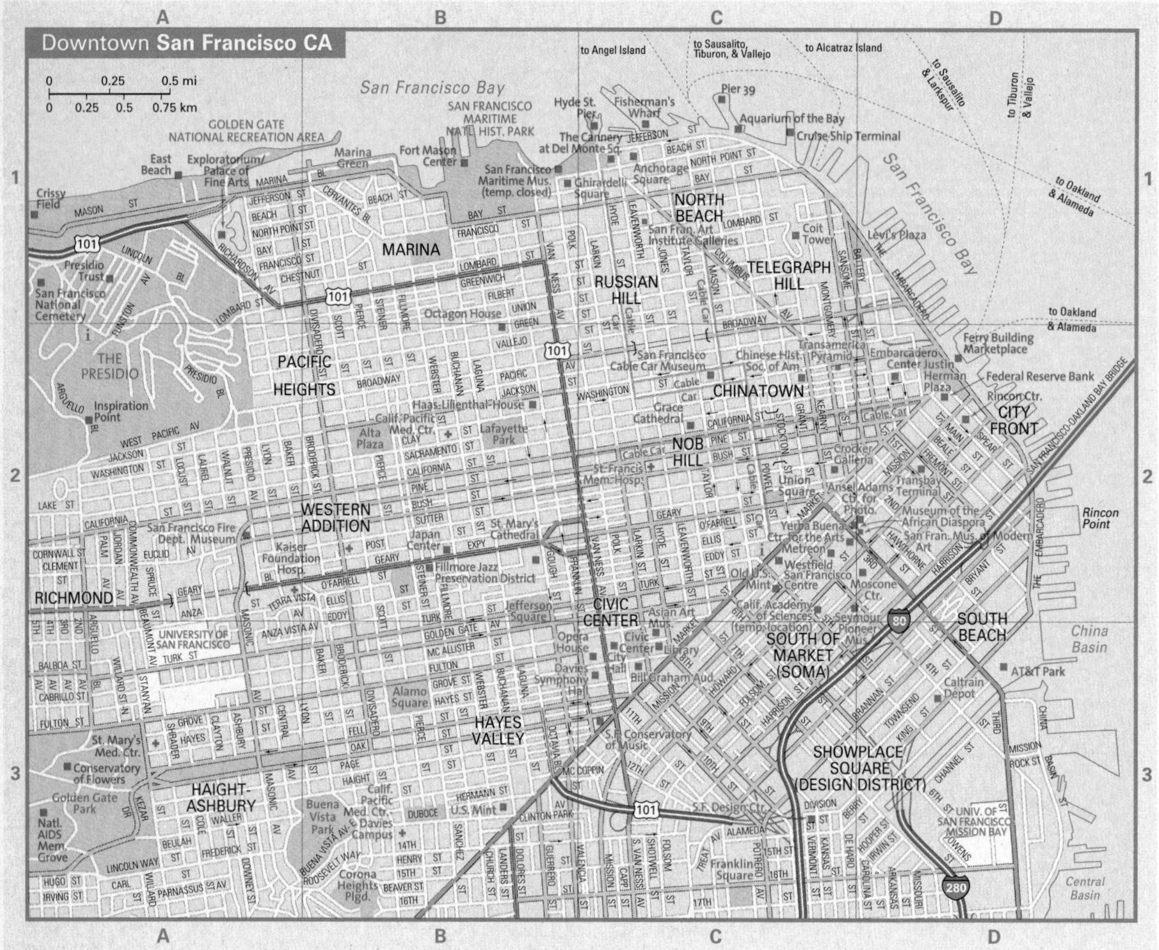

POINTS OF INTEREST

Anchorage Square C1
Ansel Adams Center for Photography D2
Aquarium of the Bay C1
Asian Art Museum C3
AT&T Park D3
Bill Graham Auditorium C3
California Academy of Sciences C2
Caltrain Depot D3
The Cannery at Del Monte Square C1
Chinese Historical Society of America C2
City Hall C3
Coit Tower C1
Conservatory of Flowers A3
Crissy Field A1
Crocker Galleria C2
Cruise Ship Terminal C1
Davies Symphony Hall C3
East Beach A1
Embarcadero Center D2
Exploratorium/ Palace of Fine Arts A1
Fillmore Jazz Preservation District B2
Ferry Building Marketplace D2
Fisherman's Wharf C1
Fort Mason Center B1
Ghirardelli Square B1
Golden Gate Natl. Rec. Area A1
Golden Gate Park A3
Grace Cathedral C2
Haas-Lilienthal House B2
Hyde Street Pier C1
Inspiration Point A2
Japan Center B2
Levi's Plaza D1
Library C3
Metreon C2
Moscone Center D2
Museum of the African Diaspora D2
National AIDS Memorial Grove A3
Octagon House B2
Old U.S. Mint C2
Opera House C3
Pier 39 .... C1
The Presidio A2
Presidio Trust A1
Rincon Center D2
St. Mary's Cathedral B2
San Francisco Art Institute Galleries C1
San Francisco Cable Car Mus. C2
San Francisco Conservatory of Music C3
San Francisco Design Center C3
San Francisco Fire Dept. Mus. A2
San Francisco Maritime Mus. B1
San Francisco Maritime Natl. Hist. Park B1
San Francisco Museum of Modern Art D2
San Francisco Natl. Cemetery A1
Seymour Pioneer Museum C2
Transamerica Pyramid C2
Transbay Terminal D2
U.S. Mint B3
Univ. of San Francisco A3
Univ. of San Francisco- Mission Bay D3
Westfield San Francisco Centre C2
Yerba Buena Center for the Arts C2

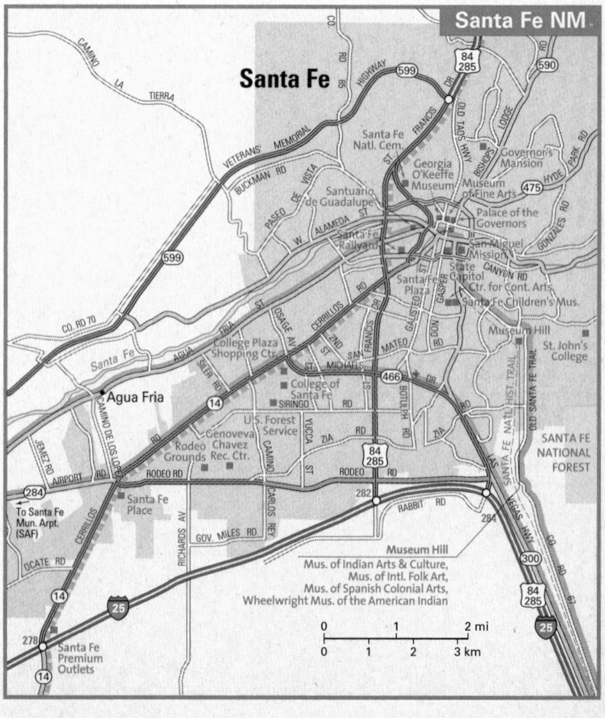

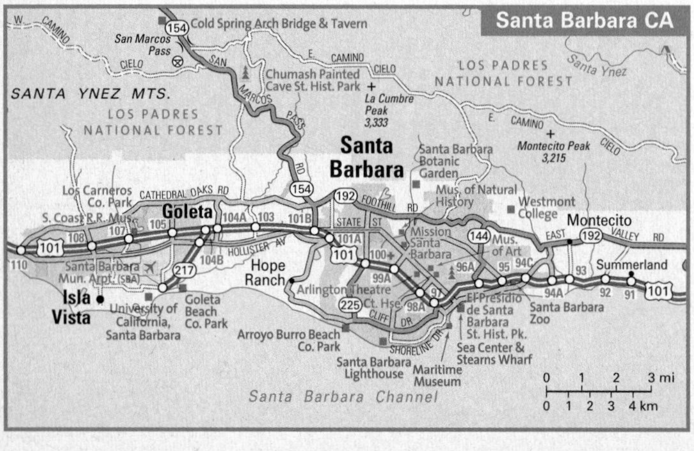

Entries in **bold black** indicate counties or parishes.
Entries in **bold color** indicate cities with detailed inset maps.

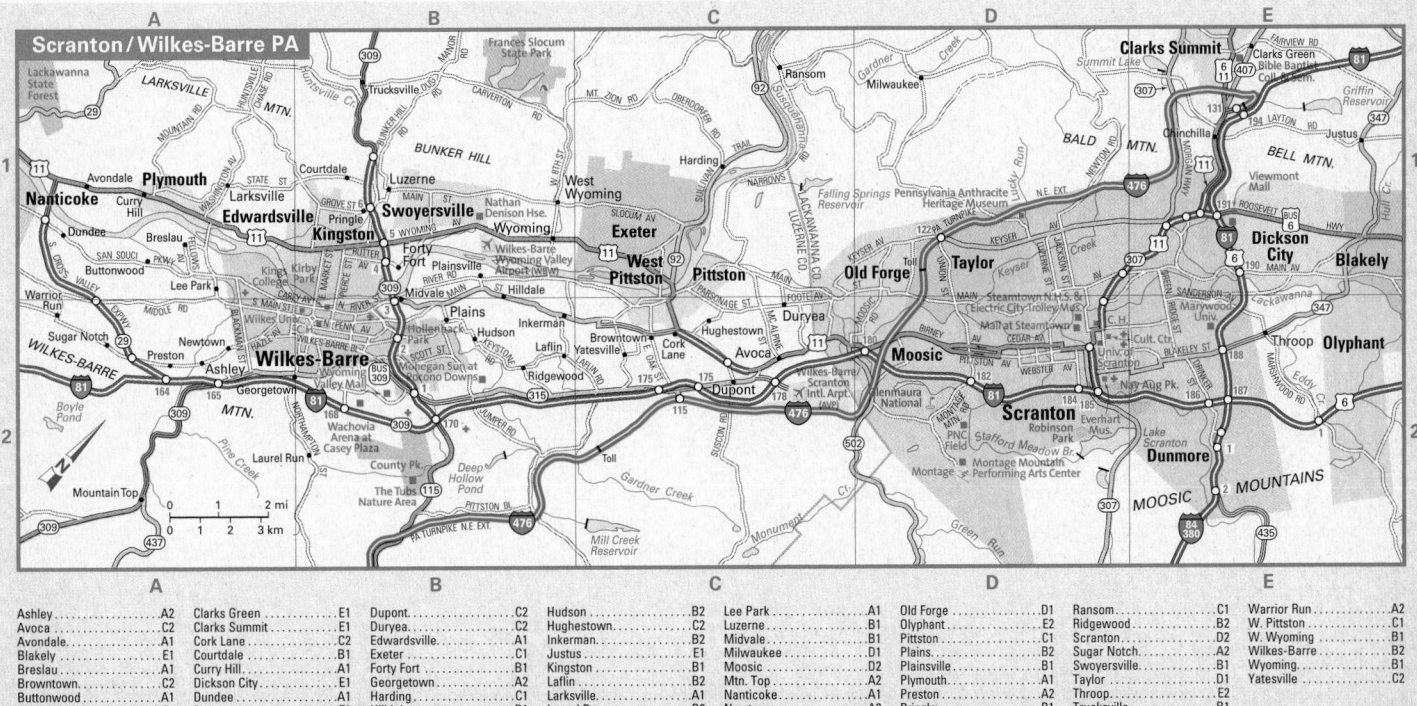

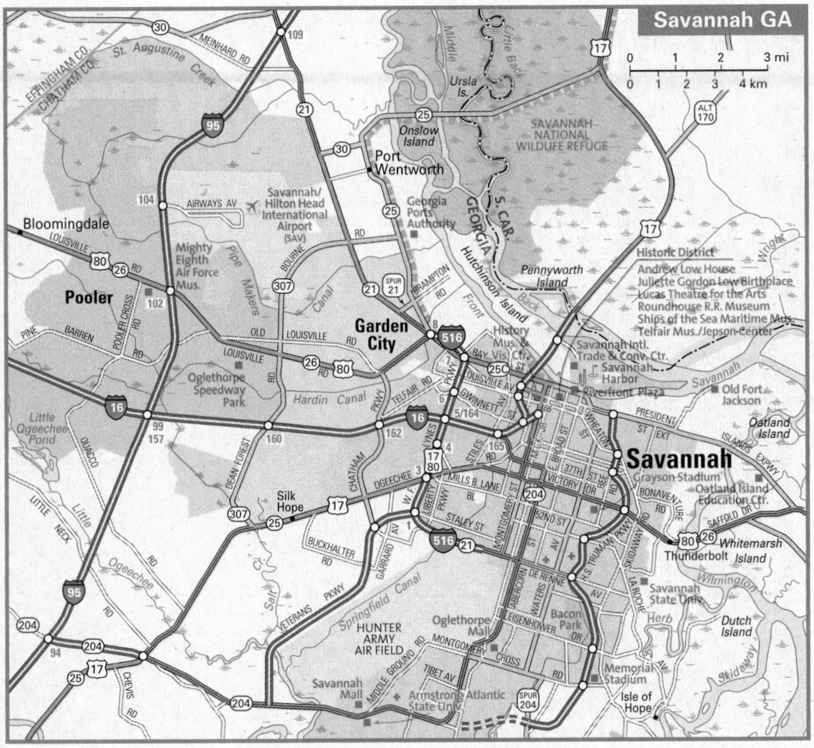

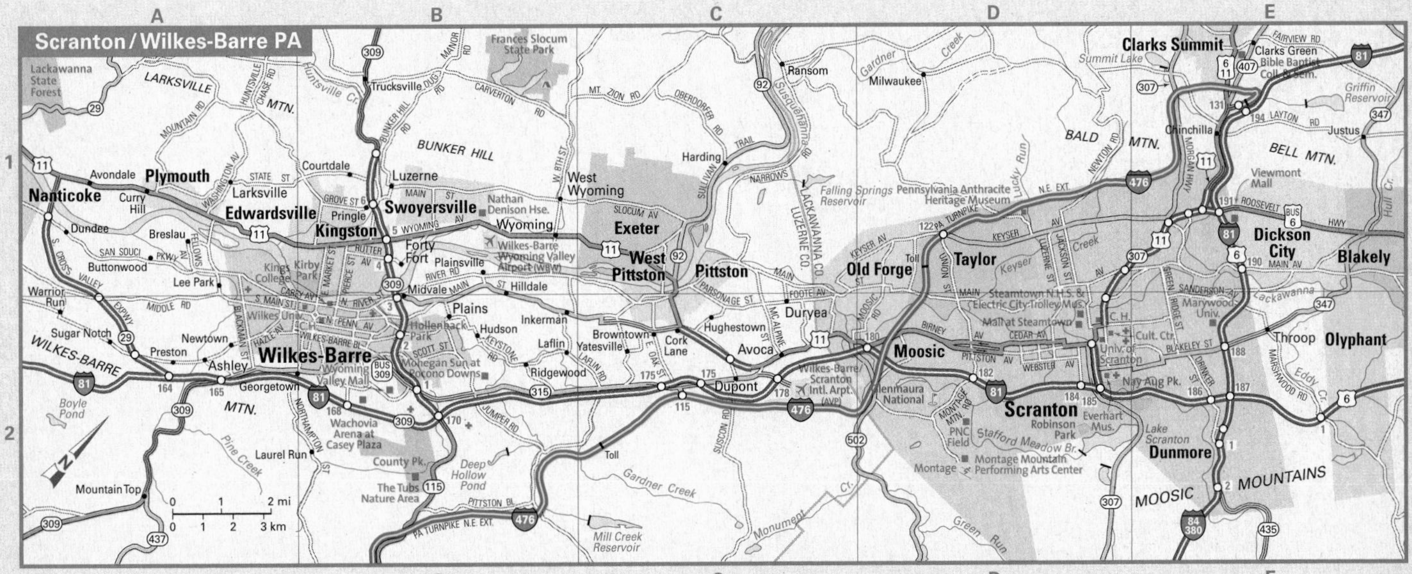

Figures after entries indicate population, page number, and grid reference.

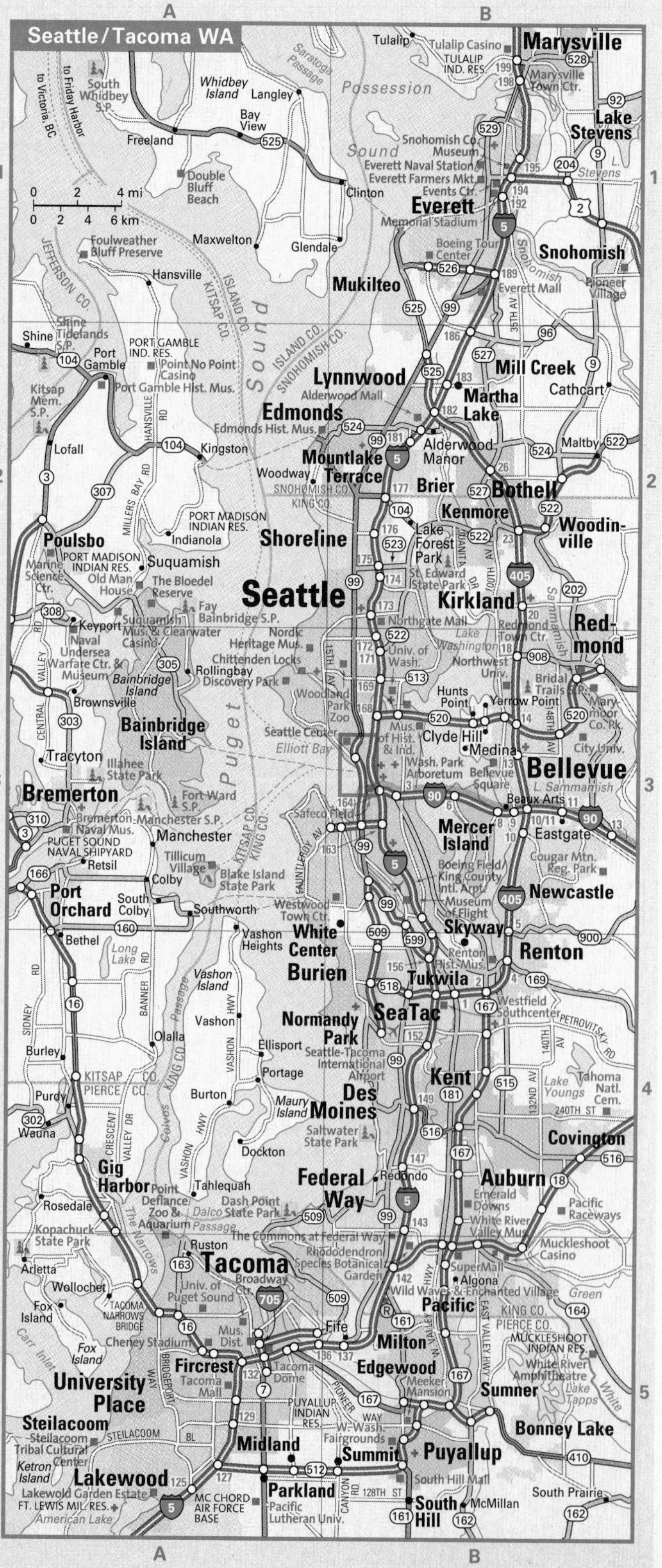

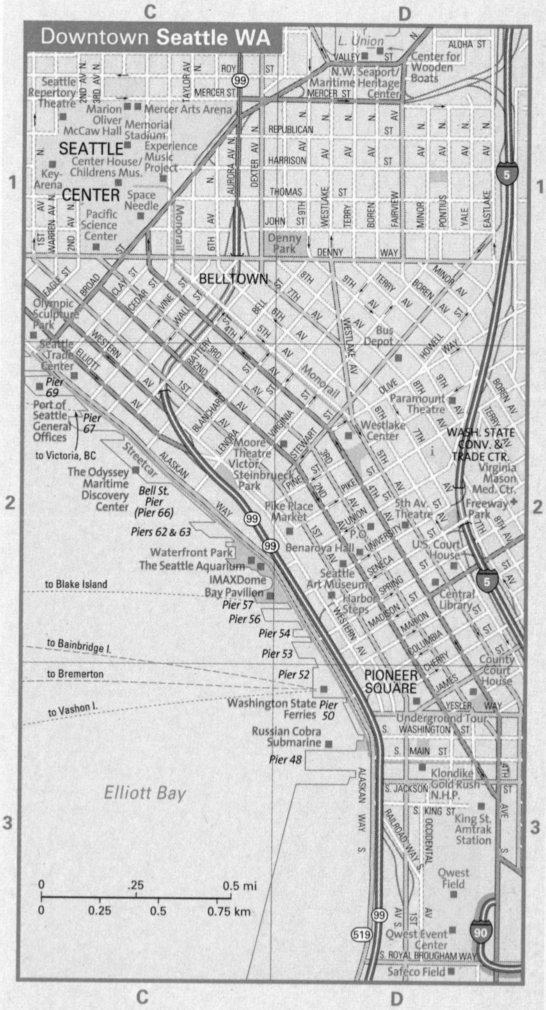

POINTS OF INTEREST

Entries in **bold black** indicate counties or parishes.
Entries in **bold color** indicate cities with detailed inset maps.

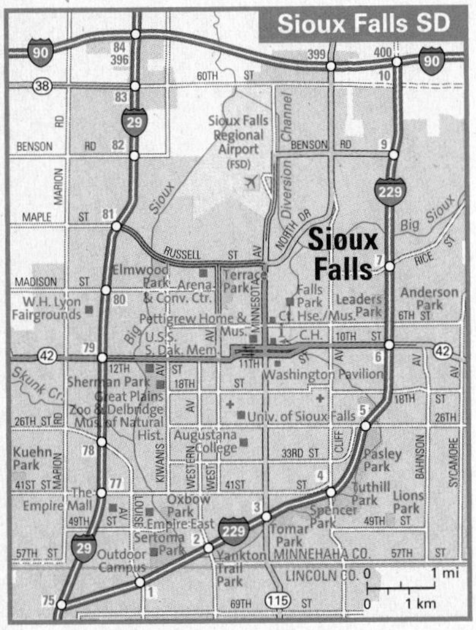

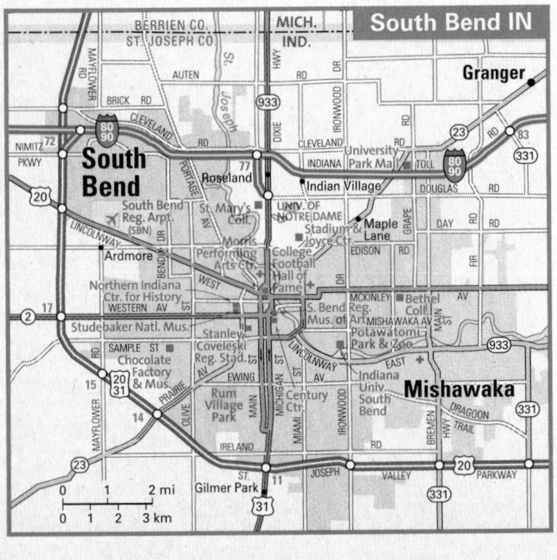

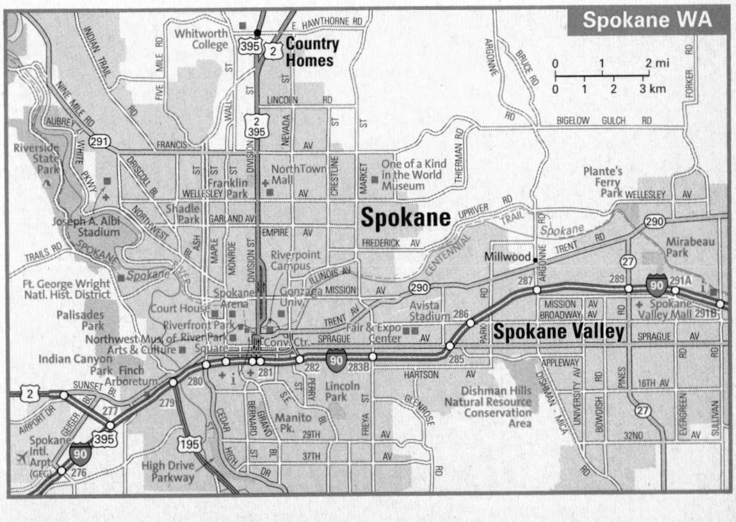

Figures after entries indicate population, page number, and grid reference.

Slatersville RI, *2600* ... **150** C2
Slatington PA, *4434* ... **93** F3
Slaton TX, *6109* ... **58** A1
Slaughter LA, *1011* ... **134** A1
Slaughter Beach DE, *198* ... **145** F3
Slaughterville OK, *3609* ... **51** E3
Slayton MN, *2072* ... **72** A1
Sledge MS, *529* ... **118** B3
Sleepy Eye MN, *3515* ... **72** B1
Sleepy Hollow IL, *3553* ... **203** A2
Sleepy Hollow NY, *9212* ... **148** B3
Slidell LA, *25695* ... **134** C2
Sligo PA, *728* ... **92** A3
Slinger WI, *3901* ... **74** C3
Slingerlands NY, *2900* ... **188** D3
Slippery Rock PA, *3068* ... **92** A3
Sloan IA, *1032* ... **35** F2
Sloan NY, *3775* ... **198** C3
Sloatsburg NY, *3117* ... **148** B2
Slocomb AL, *2052* ... **136** C1
Slocum RI, *350* ... **150** C4
Slope Co. ND, *767* ... **18** A4
Smackover AR, *2005* ... **125** E1
Smallwood MD, *950* ... **144** B1
Smallwood NY, *566* ... **94** A3
Smelterville ID, *651* ... **14** B3
Smethport PA, *1684* ... **92** C1
Smith NV, *1425* ... **37** D2
Smith Ctr. KS, *1931* ... **43** D1
Smith Co. KS, *4536* ... **43** D1
Smith Co. MS, *16182* ... **126** C3
Smith Co. TN, *17712* ... **110** A3
Smith Co. TX, *174706* ... **124** B2
Smithers WV, *904* ... **101** F4
Smithfield NC, *11510* ... **123** E1
Smithfield OH, *867* ... **91** F4
Smithfield PA, *854* ... **102** B1
Smithfield UT, *7261* ... **31** E2
Smithfield VA, *6324* ... **113** F2
Smithland KY, *401* ... **109** D2
Smith River CA, *750* ... **28** A2
Smithsburg MD, *2146* ... **144** A1
Smiths Grove KY, *784* ... **109** F2
Smiths Sta. AL, *4534* ... **128** C2
Smithton IL, *2248* ... **98** B3
Smithton MO, *510* ... **97** D3
Smithtown NY, *26901* ... **149** D3
Smithville GA, *774* ... **129** D3
Smithville IN, *750* ... **99** F2
Smithville MS, *882* ... **119** D3
Smithville MO, *5514* ... **96** B2
Smithville NJ, *3100* ... **147** E4
Smithville OH, *1333* ... **91** E3
Smithville PA, *1000* ... **146** A3
Smithville TN, *3994* ... **110** A4
Smithville TX, *3901* ... **61** E2
Smokey Pt. WA, *1556* ... **12** C2
Smyrna DE, *5679* ... **145** E2
Smyrna GA, *40999* ... **120** C4
Smyrna TN, *25569* ... **109** F4
Smyth Co. VA, *33081* ... **111** F2
Snead AL, *748* ... **119** F3
Sneads FL, *1919* ... **137** D1
Sneads Ferry NC, *2248* ... **115** D4
Sneedville TN, *1257* ... **111** D3
Snellville GA, *15351* ... **120** C4
Snohomish WA, *8494* ... **12** C2
Snohomish Co. WA, *606024* ... **12** C2
Snook TX, *568* ... **61** F1
Snoqualmie WA, *1631* ... **12** C3
Snowflake AZ, *4460* ... **47** F4
Snow Hill MD, *2409* ... **114** C2
Snow Hill NC, *1514* ... **115** C3
Snowmass Vil. CO, *1822* ... **40** C2
Snow Shoe PA, *771* ... **92** C3
Snyder OK, *1509* ... **51** D4
Snyder TX, *10783* ... **58** B2
Snyder Co. PA, *37546* ... **93** D3
Soap Lake WA, *1733* ... **13** E3
Socastee SC, *14295* ... **123** D4
Social Circle GA, *3379* ... **121** D4
Society Hill SC, *700* ... **122** C3
Socorro NM, *8877* ... **56** B1
Socorro TX, *27152* ... **56** C4
Socorro Co. NM, *18078* ... **48** C4
Soda Sprs. ID, *3381* ... **31** F1
Sodaville OR, *290* ... **20** B3
Soddy-Daisy TN, *11530* ... **120** B1
Sodus NY, *1735* ... **78** C3
Sodus Pt. NY, *1160* ... **79** D3
Solana FL, *1011* ... **140** C4
Solana Beach CA, *12979* ... **53** D4
Solano Co. CA, *394542* ... **36** B3
Soldiers Grove WI, *653* ... **73** F2
Soldotna AK, *3759* ... **154** C3
Soledad CA, *11263* ... **44** B3
Solomon AZ, *475* ... **55** E2
Solomon KS, *1072* ... **43** E2
Solomons MD, *1536* ... **103** F4
Solon IA, *1177* ... **87** F1
Solon OH, *21802* ... **91** E2
Solon Sprs. WI, *576* ... **64** C4
Solvang CA, *5332* ... **52** A2
Solvay NY, *6845* ... **265** A2
Solway TN, *1200* ... **110** C4
Somerdale NJ, *5192* ... **146** C3
Somers CT, *1626* ... **150** A2
Somers MT, *556* ... **15** D2
Somers NY, *2000* ... **148** C2
Somers WI, *1100* ... **74** C4
Somerset KY, *11352* ... **110** B2
Somerset MD, *1124* ... **270** C2
Somerset MA, *18234* ... **151** D3
Somerset NJ, *23040* ... **147** E1
Somerset OH, *1549* ... **101** E1
Somerset PA, *6762* ... **102** B1
Somerset TX, *1550* ... **61** D3
Somerset WI, *1556* ... **67** D3
Somerset Co. ME, *50888* ... **82** B1
Somerset Co. MD, *24747* ... **103** F4
Somerset Co. NJ, *297490* ... **147** D1
Somerset Co. PA, *80023* ... **92** B4
Somers Pt. NJ, *11614* ... **147** F4
Somersworth NH, *11477* ... **82** A4
Somerton AZ, *7266* ... **53** F4
Somervell Co. TX, *6809* ... **59** E3
Somerville MA, *77478* ... **151** D1
Somerville NJ, *12423* ... **147** D1
Somerville TN, *2519* ... **118** C1
Somerville TX, *1704* ... **61** F1
Somis CA, *900* ... **52** B2
Somonauk IL, *1295* ... **88** C2
Sonoita AZ, *826* ... **55** D4
Sonoma CA, *9128* ... **36** B3
Sonoma Co. CA, *458614* ... **36** A2
Sonora CA, *4423* ... **37** D3
Sonora KY, *350* ... **110** A1
Sonora TX, *2924* ... **60** B1
Sopchoppy FL, *426* ... **137** E3
Soperton GA, *2824* ... **129** F2
Sophia WV, *1301* ... **111** F1
Soquel CA, *5081* ... **236** D1
Sorento IL, *601* ... **98** B2
Sorrento LA, *1227* ... **134** A2
Souderton PA, *6730* ... **146** C2
Soulsbyville CA, *1729* ... **37** D3
Sound Beach NY, *9807* ... **149** D3
Sound View CT, *1200* ... **149** F2
Sour Lake TX, *1667* ... **132** C2
S. Acton MA, *2700* ... **150** C1
S. Amboy NJ, *7913* ... **147** E1
S. Amherst MA, *5039* ... **150** A1
S. Amherst OH, *1863* ... **91** D2
Southampton NY, *3965* ... **149** E3
Southampton PA, *8000* ... **146** C2
Southampton Co. VA, *17482* ... **113** E3
S. Apopka FL, *5800* ... **246** B1
Southaven MS, *28977* ... **118** B2
S. Barrington IL, *3760* ... **203** B2
S. Bay FL, *3859* ... **143** E1
S. Belmar NJ, *1806* ... **147** F2
S. Beloit IL, *5397* ... **74** B4
S. Bend IN, *107789* ... **89** F2
S. Bend WA, *1807* ... **12** B4
S. Berlin MA, *800* ... **150** C1
S. Berwick ME, *2300* ... **82** A4
S. Bethany DE, *492* ... **145** F4
S. Bloomfield OH, *1179* ... **101** D1
Southborough MA, *1700* ... **150** C1
S. Boston VA, *8491* ... **112** C3
S. Bound Brook NJ, *4492* ... **147** D1
Southbridge MA, *12878* ... **150** B2
S. Bridgewater MA, *2600* ... **151** D2
S. Burlington VT, *15814* ... **81** D2
Southbury CT, *550* ... **149** D1
S. Cairo NY, *700* ... **94** B2
S. Carthage TN, *1302* ... **110** A3
S. Casco ME, *650* ... **82** B3
S. Charleston OH, *1850* ... **100** C1
S. Charleston WV, *13390* ... **101** E4
S. Chatham MA, *800* ... **151** F3
S. China ME, *325* ... **82** C2
S. Cle Elum WA, *457* ... **13** D3
S. Coatesville PA, *997* ... **146** B3
S. Coffeyville OK, *790* ... **106** A2
S. Congaree SC, *2266* ... **122** A4
S. Connellsville PA, *2281* ... **102** B1
S. Corning NY, *1147* ... **93** D1
S. Dartmouth MA, *1900* ... **151** D4
S. Dayton NY, *662* ... **78** A4
S. Daytona FL, *13177* ... **139** E4
S. Deerfield MA, *1868* ... **150** A1
S. Dennis MA, *3679* ... **151** F3
S. Dos Palos CA, *1385* ... **44** C2
S. Dover NY, *750* ... **148** C1
S. Duxbury MA, *3062* ... **151** E2
S. Easton MA, *3600* ... **151** D2
S. Elgin IL, *16100* ... **88** C1
S. Eliot ME, *3445* ... **82** A4
S. El Monte CA, *21144* ... **228** E2
Southern Pines NC, *10918* ... **122** C1
Southern Shores NC, *2201* ... **115** F2
S. Euclid OH, *23537* ... **204** G1
S. Fallsburg NY, *2061* ... **94** A3
Southfield MI, *78296* ... **76** C4
S. Floral Park NY, *1578* ... **241** G4
S. Fork CO, *604* ... **40** C4
S. Fork PA, *1138* ... **92** B4
S. Fulton TN, *2517* ... **108** C3
S. Gastonia NC, *5433* ... **122** A1
S. Gate CA, *96375* ... **228** D3
Southgate KY, *3472* ... **204** B3
S. Gate MD, *28672* ... **193** C5
S. Glastonbury CT, *1900* ... **149** E1
S. Glens Falls NY, *3368* ... **81** D4
S. Greensburg PA, *2280* ... **92** A4
S. Hadley MA, *3900* ... **150** A1
S. Hadley Falls MA, *4300* ... **150** A2
S. Haven KS, *390* ... **51** E1
S. Haven MI, *5021* ... **75** E4
S. Heart ND, *307* ... **18** A4
S. Hill NY, *6003* ... **79** D4
S. Hill VA, *4403* ... **113** D3
S. Hill WA, *31623* ... **262** B5
S. Holland IL, *22147* ... **203** E6
S. Hooksett NH, *5282* ... **81** F4
S. Houston TX, *15833* ... **220** C3
S. Hutchinson KS, *2539* ... **43** E3
Southington CT, *39728* ... **149** D1
S. Jacksonville IL, *3475* ... **98** A1
S. Jamesport NY, *1100* ... **149** E3
S. Jordan UT, *29437* ... **31** E4
S. Killingly CT, *750* ... **150** B3
Southlake TX, *21519* ... **207** B1
S. Lake Tahoe CA, *23609* ... **37** D2
S. Lancaster MA, *1742* ... **150** C1
S. Lebanon ME, *350* ... **81** F4

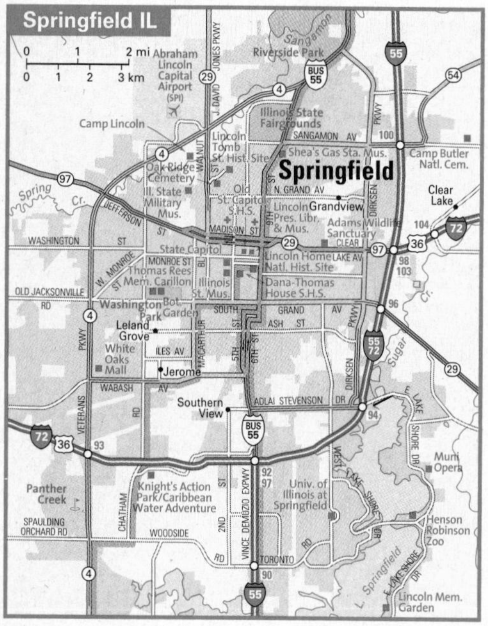

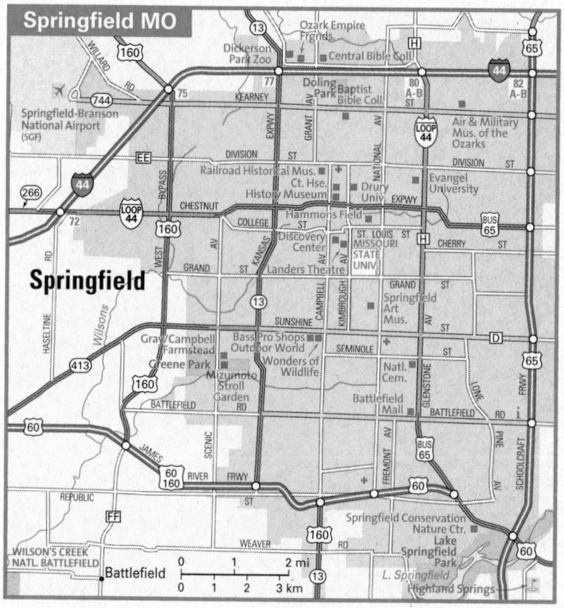

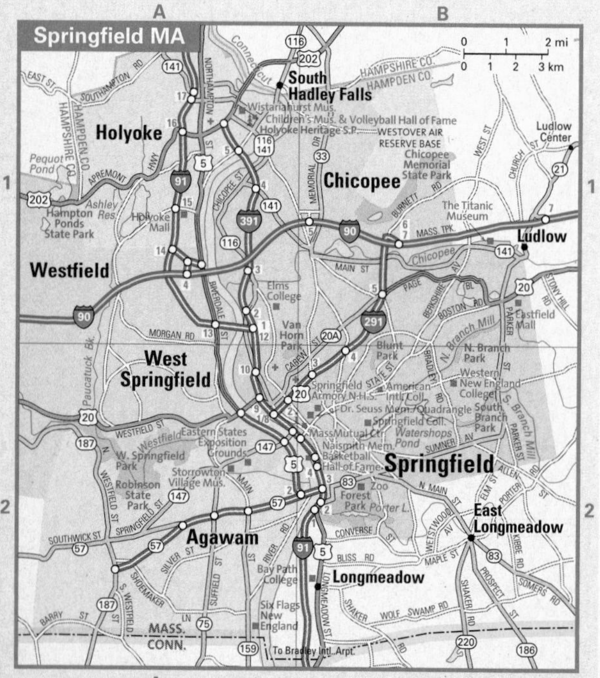

Agawam ... A2
Chicopee ... B1
E. Longmeadow ... B2
Holyoke ... A1
Longmeadow ... B2
Ludlow ... B1
Ludlow Ctr. ... B1
S. Hadley Falls ... A1
Springfield ... B2
Westfield ... A1
W. Springfield ... A2

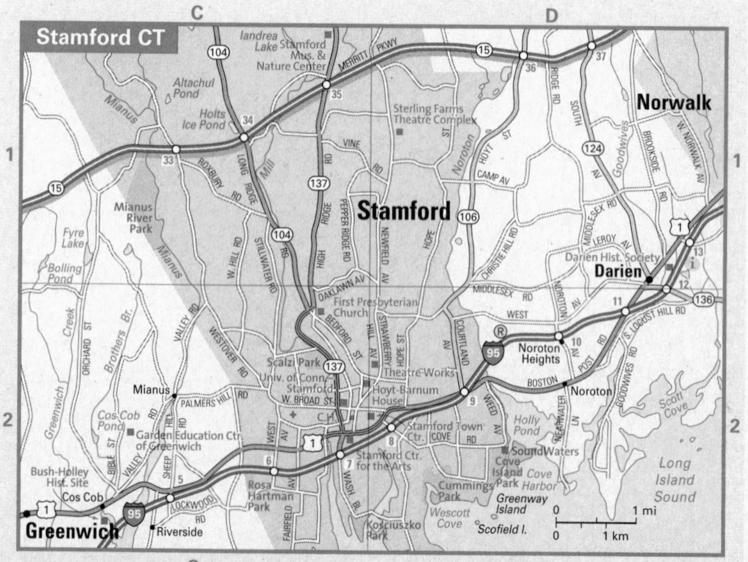

Cos Cob ... C2
Darien ... D1
Greenwich ... C2
Mianus ... C2
Noroton ... D2
Noroton Hts. ... D2
Norwalk ... D1
Riverside ... C2
Stamford ... D1

Entries in **bold black** indicate counties or parishes.
Entries in **bold color** indicate cities with detailed inset maps.

S. Lebanon OH, 2538 ..... 100 B2
S. Lockport NY, 8552 ..... 78 B3
S. Lyon MI, 10036 ..... 76 B4
Southmayd TX, 992 ..... 59 F1
S. Miami FL, 10741 ..... 143 E2
S. Miami Hts. FL, 35522 ..... 233 A5
S. Middleboro MA, 750 ..... 151 E3
S. Mills NC, 700 ..... 113 F3
S. Milwaukee WI, 21256 ..... 75 D3
S. Monroe MI, 6370 ..... 90 C1
Southmont NC, 850 ..... 112 B4
S. Naknek AK, 137 ..... 154 B3
S. Ogden UT, 14377 ..... 244 A2
Southold NY, 5465 ..... 149 E3
S. Orange NJ, 16964 ..... 240 A3
S. Orleans MA, 1100 ..... 151 F3
S. Padre Island TX, 2422 ..... 63 F4
S. Palm Beach FL, 699 ..... 143 F1
S. Paris ME, 2237 ..... 82 B2
S. Pasadena CA, 24292 ..... 228 D2
S. Pasadena FL, 5778 ..... 140 B3
S. Patrick Shores FL, 8913 ..... 141 E2
S. Pekin IL, 1162 ..... 88 B4
S. Pittsburg TN, 3295 ..... 120 A2
S. Plainfield NJ, 21810 ..... 147 E1
S. Point OH, 3742 ..... 101 D3
Southport FL, 1500 ..... 136 C2
Southport IN, 1852 ..... 99 F1
Southport NY, 7396 ..... 93 D1
Southport NC, 2351 ..... 123 E4
S. Portland ME, 23324 ..... 82 B3
S. Pottstown PA, 2135 ..... 146 B2
S. Range MI, 727 ..... 65 F3
S. River NJ, 15322 ..... 147 E1
S. Rockwood MI, 1284 ..... 90 C1
S. Roxana IL, 1888 ..... 256 D1
S. Royalton VT, 800 ..... 81 E3
S. Russell OH, 4022 ..... 91 E2
S. St. Paul MN, 20167 ..... 235 D3
S. Salt Lake UT, 22038 ..... 257 B2
S. Sanford ME, 4173 ..... 82 B2
S. San Francisco CA, 60552 ..... 259 B4
S. San Gabriel CA, 7595 ..... 228 E2
S. Shaftsbury VT, 772 ..... 94 C1
S. Shore KY, 1226 ..... 101 D3
S. Shore SD, 270 ..... 27 F2
Southside AL, 7036 ..... 120 A4
Southside Place TX, 1546 ..... 220 B3
S. Sioux City NE, 11925 ..... 35 F2
S. Toms River NJ, 3634 ..... 147 E3
S. Torrington WY, 550 ..... 33 F2
S. Tucson AZ, 5490 ..... 55 D3
S. Valley Stream NY, 5638 ..... 241 G4
S. Venice FL, 13539 ..... 140 B4
S. Wallins KY, 996 ..... 111 D2
S. Walpole MA, 1900 ..... 151 D2
S. Waverly PA, 987 ..... 93 E1
S. Weber UT, 4260 ..... 244 B2
S. Webster OH, 764 ..... 101 D3
S. Weldon NC, 1414 ..... 113 E3
South West City MO, 855 ..... 106 B3
Southwest Harbor ME, 950 ..... 83 D2
S. Westport MA, 600 ..... 151 D4
S. Weymouth MA, 11100 ..... 151 D2
S. Whitley IN, 1782 ..... 89 F3
S. Whittier CA, 55193 ..... 228 E3
Southwick MA, 2000 ..... 150 A2
S. Williamson KY, 600 ..... 111 E1
S. Williamsport PA, 6412 ..... 93 D2
S. Willington CT, 1700 ..... 150 A3
S. Wilmington IL, 621 ..... 88 C2
S. Windham CT, 1278 ..... 149 F1
S. Windham ME, 1792 ..... 82 B3
S. Windsor CT, 700 ..... 150 A3
Southwood Acres CT, 8067 ..... 150 A2
S. Woodstock CT, 1211 ..... 150 B3
S. Yarmouth MA, 11603 ..... 151 F3
S. Zanesville OH, 1936 ..... 101 E1
Spalding NE, 537 ..... 35 E3
Spalding Co. GA, 58417 ..... 128 C1
Spanaway WA, 21588 ..... 12 C3
Spanish Fork UT, 20246 ..... 39 E1
Spanish Fort AL, 5423 ..... 135 E1
Spanish Lake MO, 21337 ..... 256 C1
Sparkman AR, 586 ..... 117 D4
Sparks GA, 1755 ..... 137 F1
Sparks MD, 1800 ..... 144 C1
Sparks NV, 90264 ..... 37 D1
Sparrow Bush NY, 1100 ..... 94 A3
Sparta GA, 1522 ..... 129 E1
Sparta IL, 4486 ..... 98 B4
Sparta MI, 4159 ..... 75 F3
Sparta MO, 1144 ..... 107 D2
Sparta NJ, 9755 ..... 148 A3
Sparta NC, 1817 ..... 112 A3
Sparta TN, 4599 ..... 110 A4
Sparta WI, 8648 ..... 73 F1
Spartanburg SC, 39673 ..... 121 F2
Spartanburg Co. SC, 253791 ..... 121 F2
Spavinaw OK, 563 ..... 106 B3
Spearfish SD, 8606 ..... 25 F3
Spearman TX, 3021 ..... 50 B2
Spearville KS, 813 ..... 42 C4
Speedway IN, 12881 ..... 99 F1
Speigner AL, 1700 ..... 128 A2
Spencer IN, 2508 ..... 99 E2
Spencer IA, 11317 ..... 72 A3
Spencer MA, 6032 ..... 150 B1
Spencer NE, 541 ..... 35 D1
Spencer NY, 731 ..... 93 E1
Spencer NC, 3355 ..... 112 A4
Spencer OH, 747 ..... 91 D3
Spencer OK, 3746 ..... 51 E3
Spencer TN, 1713 ..... 110 A4
Spencer WV, 2352 ..... 101 F3
Spencer WI, 1932 ..... 68 A4
Spencer Co. IN, 20391 ..... 99 E4
Spencer Co. KY, 11766 ..... 100 A4
Spencerport NY, 3559 ..... 78 C3
Spencerville OH, 2235 ..... 90 B3
Speonk NY, 2675 ..... 149 E3
Sperry OK, 981 ..... 51 F2
Spiceland IN, 807 ..... 100 A1
Spicer MN, 1126 ..... 66 B3
Spindale NC, 4022 ..... 121 F1
Spink Co. SD, 7454 ..... 27 E2
Spinnerstown PA, 1100 ..... 146 C1
Spirit Lake ID, 1376 ..... 14 B2
Spirit Lake IA, 4261 ..... 72 B2
Spiro OK, 2227 ..... 116 B1
Splendora TX, 1275 ..... 132 B2
Spofford NH, 500 ..... 94 C1
Spokane WA, 195629 ..... 14 A3
Spokane Co. WA, 417939 ..... 14 B2
Spokane Valley WA, 80700 ..... 14 B3
Spooner WI, 2653 ..... 67 E2
Spotswood NJ, 7880 ..... 147 E1
Spotsylvania VA, 3833 ..... 103 D4
Spotsylvania Co. VA, 90395 ..... 103 D4
Spottsville KY, 600 ..... 99 E4
Spout Sprs. NC, 550 ..... 123 D1
Sprague WA, 490 ..... 13 F3
Spreckels CA, 485 ..... 236 E3
Spring TX, 36385 ..... 132 A2
Spring Arbor MI, 2188 ..... 90 B1
Springboro OH, 12380 ..... 100 B1
Spring City PA, 3305 ..... 146 B2
Spring City TN, 2025 ..... 110 B4
Spring City UT, 956 ..... 39 E1
Spring Creek NV, 10548 ..... 30 B4
Springdale AR, 45798 ..... 106 C4
Springdale OH, 10563 ..... 204 B1
Springdale PA, 3828 ..... 250 D1
Springdale SC, 2877 ..... 205 E2
Springdale UT, 457 ..... 39 D4
Springer NM, 1285 ..... 49 E2
Springer OK, 577 ..... 51 E4
Springerville AZ, 1972 ..... 48 A4
Springfield CO, 1562 ..... 42 A4
Springfield FL, 8810 ..... 136 C2
Springfield GA, 1821 ..... 130 B2
Springfield IL, 111454 ..... 98 B1
Springfield KY, 2634 ..... 110 B1
Springfield LA, 395 ..... 134 B2
Springfield MA, 152082 ..... 150 A2
Springfield MI, 5189 ..... 75 F4
Springfield MN, 2215 ..... 72 B1
Springfield MO, 151580 ..... 107 D2
Springfield NE, 1450 ..... 35 F3
Springfield NJ, 14429 ..... 148 A4
Springfield OH, 65358 ..... 100 C1
Springfield OR, 52864 ..... 20 B4
Springfield PA, 23677 ..... 146 C3
Springfield SC, 504 ..... 122 A4
Springfield SD, 792 ..... 35 E1
Springfield TN, 14329 ..... 109 F3
Springfield VT, 3938 ..... 81 E4
Springfield VA, 30417 ..... 144 B3
Spring Garden Estates MD, 2400 ..... 144 A1
Spring Glen UT, 1000 ..... 39 F1
Spring Green WI, 1444 ..... 74 A3
Spring Grove IL, 3880 ..... 74 C4
Spring Grove MN, 1304 ..... 73 E2
Spring Grove PA, 2050 ..... 103 E1
Spring Hill FL, 69078 ..... 140 B1
Spring Hill KS, 2727 ..... 96 B3
Springhill LA, 5439 ..... 125 D1
Spring Hill TN, 7715 ..... 109 E4
Spring Hope NC, 1261 ..... 113 D4
Spring Lake MI, 2514 ..... 75 E3
Spring Lake NJ, 3567 ..... 147 F2
Spring Lake NC, 8098 ..... 123 D1
Spring Lake Hts. NJ, 5227 ..... 147 E2
Spring Lake Park MN, 6772 ..... 235 C1
Spring Mill KY, 380 ..... 230 E3
Springport MI, 704 ..... 76 A4
Springs NY, 4950 ..... 149 F3
Springtown TX, 2062 ..... 59 E2
Springvale ME, 3488 ..... 82 A4
Spring Valley CA, 26663 ..... 53 D4
Spring Valley IL, 5398 ..... 88 B2
Spring Valley KY, 668 ..... 230 F1
Spring Valley MN, 2518 ..... 73 D2
Spring Valley NY, 25464 ..... 148 B2
Spring Valley TX, 3611 ..... 220 B2
Spring Valley WI, 1189 ..... 67 E4
Springview NE, 242 ..... 34 C1
Springville AL, 2521 ..... 119 F4
Springville CA, 1109 ..... 45 D3
Springville IA, 1091 ..... 87 F1
Springville NJ, 2400 ..... 147 D3
Springville NY, 4252 ..... 78 B4
Springville UT, 20424 ..... 39 E1
Springwater NY, 700 ..... 78 C4
Spruce Pine NC, 2030 ..... 111 E4
Spry PA, 4903 ..... 275 F2
Spur TX, 1088 ..... 58 B1
Squaw Valley CA, 2691 ..... 45 D3
Staatsburg NY, 911 ..... 94 B3
Stacy MN, 1278 ..... 67 D3
Stafford CT, 900 ..... 150 B2
Stafford KS, 1161 ..... 43 D3
Stafford TX, 15681 ..... 132 A3
Stafford VA, 3000 ..... 103 D3
Stafford Co. KS, 4789 ..... 43 D3
Stafford Co. VA, 92446 ..... 144 A4
Stafford Sprs. CT, 4100 ..... 150 A2
Stallings NC, 3189 ..... 122 B2
Stamford CT, 117083 ..... 148 C3
Stamford NY, 1265 ..... 79 F4
Stamford TX, 3636 ..... 58 C2
Stamping Ground KY, 566 ..... 100 B4
Stamps AR, 2131 ..... 125 D1
Stamardsville VA, 476 ..... 102 C4
Stanberry MO, 1243 ..... 86 B4
Standing Pine MS, 509 ..... 126 C2
Standish ME, 600 ..... 82 B3
Standish MI, 1581 ..... 76 B1
Stanfield AZ, 651 ..... 54 C2
Stanfield NC, 1113 ..... 122 B1
Stanfield OR, 1979 ..... 21 E1
Stanford IL, 670 ..... 88 B4
Stanford KY, 3430 ..... 110 B1
Stanford MT, 454 ..... 16 B3
Stanhope NJ, 3584 ..... 94 A4
Stanislaus Co. CA, 446997 ..... 36 C4
Stanley KY, 400 ..... 109 E1
Stanley NC, 3053 ..... 122 A1
Stanley ND, 1279 ..... 18 A2
Stanley VA, 1326 ..... 102 C3
Stanley WI, 1898 ..... 67 F4
Stanley Co. SD, 2772 ..... 26 C3
Stanleytown VA, 1515 ..... 112 B2
Stanly Co. NC, 58100 ..... 122 B1
Stannards NY, 868 ..... 92 C1
Stanton CA, 37403 ..... 228 E4
Stanton DE, 5000 ..... 146 B4
Stanton IA, 714 ..... 86 B3
Stanton KY, 3029 ..... 110 C1
Stanton MI, 1504 ..... 76 A3
Stanton NE, 1627 ..... 35 E2
Stanton ND, 345 ..... 18 B3
Stanton TN, 615 ..... 118 C1
Stanton TX, 2556 ..... 58 A3
Stanton Co. KS, 2406 ..... 42 B4
Stanton Co. NE, 6455 ..... 35 E2
Stantonsburg NC, 726 ..... 123 E1
Stanwood IA, 680 ..... 87 F1
Stanwood WA, 3923 ..... 12 C2
Staples MN, 3104 ..... 64 A4
Stapleton NE, 301 ..... 34 C3
Star ID, 1795 ..... 22 B4
Star NC, 807 ..... 122 C1
Starbuck MN, 1314 ..... 66 A2
Star City AR, 2471 ..... 117 F4
Star City WV, 1366 ..... 102 A1
Star Cross NJ, 1400 ..... 145 F1
Stark Co. IL, 6332 ..... 88 B3
Stark Co. ND, 22636 ..... 18 A4
Stark Co. OH, 378098 ..... 91 E3
Starke FL, 5593 ..... 138 C3
Starke Co. IN, 23556 ..... 89 E2
Starkey VA, 600 ..... 254 D2
Starks LA, 650 ..... 133 D2
Starkville MS, 21869 ..... 118 C4
Star Lake NY, 860 ..... 79 F1
Star Prairie WI, 574 ..... 67 E3
Starr Co. TX, 53597 ..... 63 E3
Startex SC, 988 ..... 217 C1
Startup WA, 817 ..... 12 C2
Star Valley AZ, 550 ..... 47 E4
Stateburg SC, 1264 ..... 122 B4
State Ctr. IA, 1349 ..... 87 D1
State College PA, 38420 ..... 92 C3
State Line MS, 555 ..... 127 D4
Stateline NV, 1215 ..... 37 D2
State Line PA, 1300 ..... 103 D1
Statenville GA, 50 ..... 137 F1
State Road NC, 900 ..... 112 A3
Statesboro GA, 22698 ..... 130 B2
Statesville NC, 23320 ..... 112 A4
Statham GA, 2040 ..... 121 D4
Staunton IL, 5030 ..... 98 B2
Staunton IN, 550 ..... 99 E2
Staunton VA, 23853 ..... 102 B4
Stayton OR, 6816 ..... 20 C3
Steamboat NV, 1100 ..... 37 D2
Steamboat Canyon AZ, 1100 ..... 47 F2
Steamboat Sprs. CO, 9815 ..... 33 D4
Stearns KY, 1586 ..... 110 B3
Stearns Co. MN, 133166 ..... 66 B3
Stebbins AK, 547 ..... 154 B2
Stedman NC, 664 ..... 123 D2
Steele AL, 1093 ..... 120 A4
Steele MO, 2263 ..... 108 B4
Steele ND, 761 ..... 18 C4
Steele Co. MN, 33680 ..... 73 D1
Steele Co. ND, 2258 ..... 19 E3
Steeleville IL, 2077 ..... 98 B4
Steelton PA, 5858 ..... 218 B2
Steelville MO, 1429 ..... 97 F4
Steilacoom WA, 6049 ..... 12 C3
Steinhatchee FL, 700 ..... 137 F3
Stephen MN, 708 ..... 19 E2
Stephens AR, 1152 ..... 125 D1
Stephens City VA, 1146 ..... 102 C2
Stephens Co. GA, 25435 ..... 121 D2
Stephens Co. OK, 43182 ..... 51 E4
Stephens Co. TX, 9674 ..... 59 D2
Stephenson MI, 875 ..... 69 D3
Stephenson VA, 1100 ..... 103 D2
Stephenson Co. IL, 48979 ..... 74 B4
Stephenville TX, 14921 ..... 59 D3
Sterling CO, 11360 ..... 34 A4
Sterling CT, 650 ..... 150 B3
Sterling IL, 15451 ..... 88 B1
Sterling KS, 2642 ..... 43 E3
Sterling MA, 1300 ..... 150 C1
Sterling MI, 533 ..... 76 B1
Sterling NE, 507 ..... 35 F4
Sterling OK, 762 ..... 51 E4
Sterling VA, 20500 ..... 144 A3
Sterling City TX, 1081 ..... 58 B3
Sterling Co. TX, 1393 ..... 58 B4
Sterling Forest NY, 700 ..... 148 A2
Sterling Hts. MI, 124471 ..... 76 C4
Sterlington LA, 1276 ..... 125 E2
Stetsonville WI, 563 ..... 68 A3
Steuben Co. IN, 33214 ..... 90 A2
Steuben Co. NY, 98726 ..... 78 C4
Steubenville KY, 550 ..... 110 B2
Steubenville OH, 19015 ..... 91 F4
Stevens Co. KS, 5463 ..... 42 B4
Stevens Co. MN, 10053 ..... 27 F2
Stevens Co. WA, 40066 ..... 13 F2
Stevens Creek VA, 550 ..... 112 A2
Stevenson AL, 1770 ..... 120 A2
Stevenson CT, 800 ..... 149 D2
Stevenson WA, 1200 ..... 20 C1
Stevens Pt. WI, 24551 ..... 68 B4
Stevensville MD, 5880 ..... 145 D3
Stevensville MI, 1191 ..... 89 E1
Stevensville MT, 1553 ..... 15 D4
Stewardson IL, 747 ..... 98 C2
Stewart MN, 564 ..... 66 B4
Stewart Co. GA, 5252 ..... 128 C3
Stewart Co. TN, 12370 ..... 109 D3
Stewart Manor NY, 1935 ..... 241 G3
Stewartstown PA, 1752 ..... 103 E1
Stewartsville MO, 759 ..... 96 B1
Stewartville MN, 5411 ..... 73 D2
Stickney IL, 6148 ..... 203 D5
Stickney SD, 334 ..... 27 E4
Stigler OK, 2731 ..... 116 B1
Stillman Valley IL, 1048 ..... 88 B1
Stillmore GA, 730 ..... 129 F2

Bayberry	A1	Galeville	A1	Nedrow	A2	Taunton	A2
Collamer	B1	Jamesville	B2	N. Syracuse	B1	Westvale	A2
DeWitt	B2	Lakeland	A1	Onondaga Hill	A2		
E. Syracuse	B2	Liverpool	A1	Solvay	A2		
Fairmount	A2	Lyndon	B2	Split Rock	A2		
Franklin Park	B1	Mattydale	B1	Syracuse	A2		

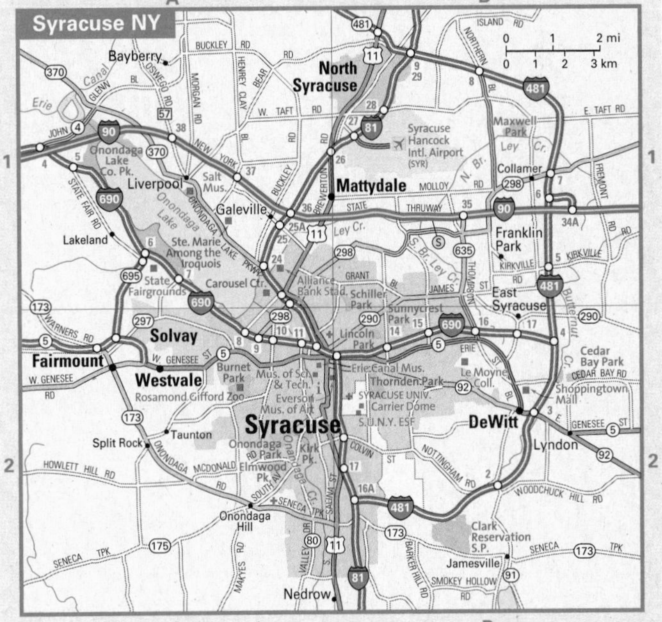

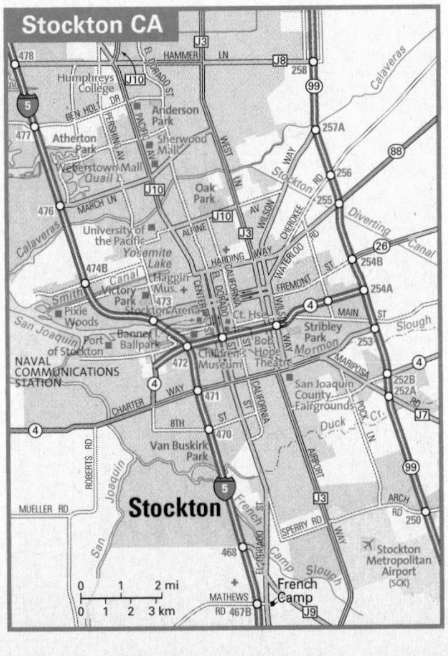

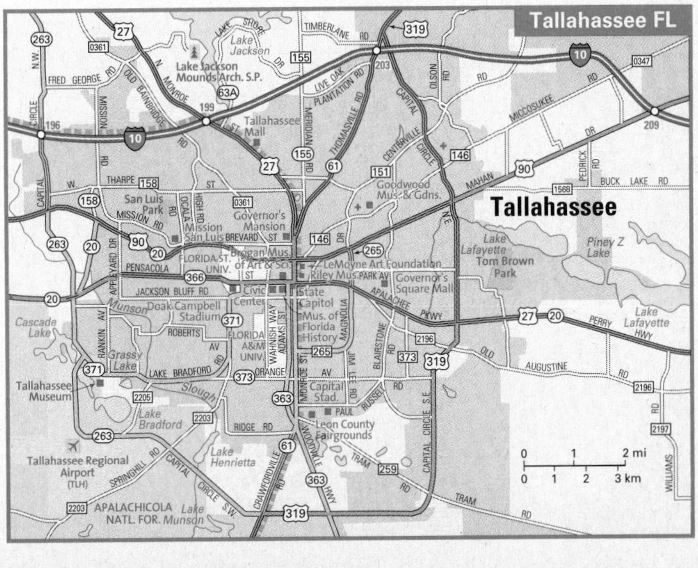

Figures after entries indicate population, page number, and grid reference.

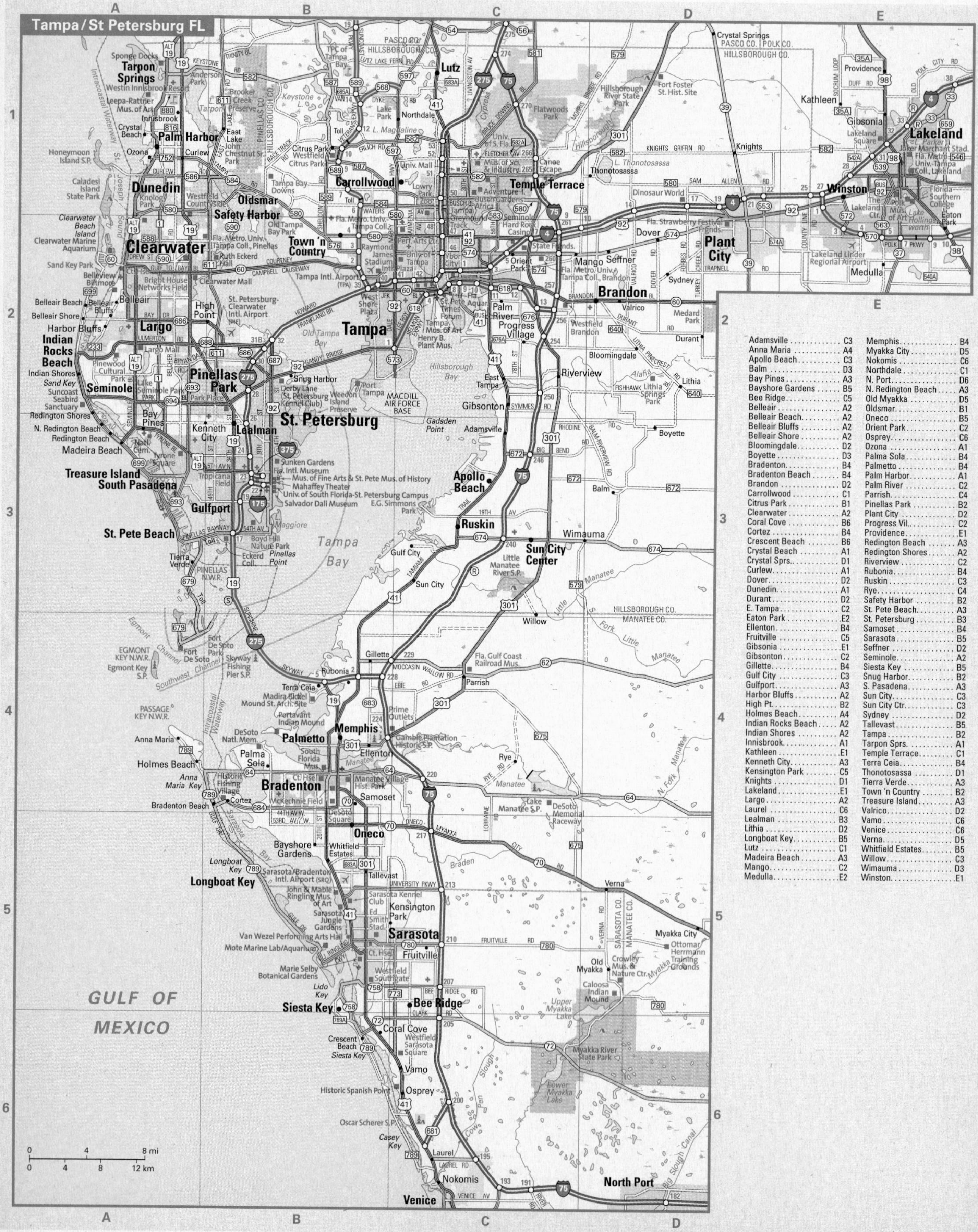

Entries in **bold black** indicate counties or parishes.
Entries in **bold color** indicate cities with detailed inset maps.

Harbor View ... B1
Holland ... A2
Lime City ... B2
Maumee ... A2
Moline ... B2
Northwood ... B2
Oregon ... B1
Ottawa Hills ... A1
Perrysburg ... A2
Rossford ... B2
Stony Ridge ... B2
Sylvania ... A1
Toledo ... A1
Walbridge ... B2

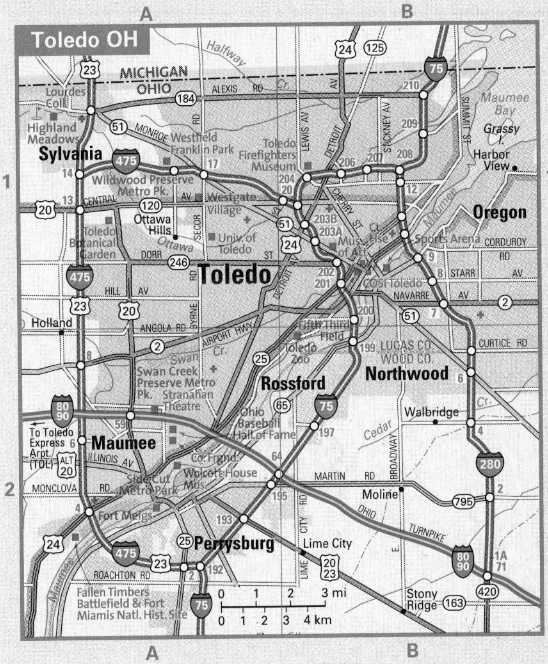

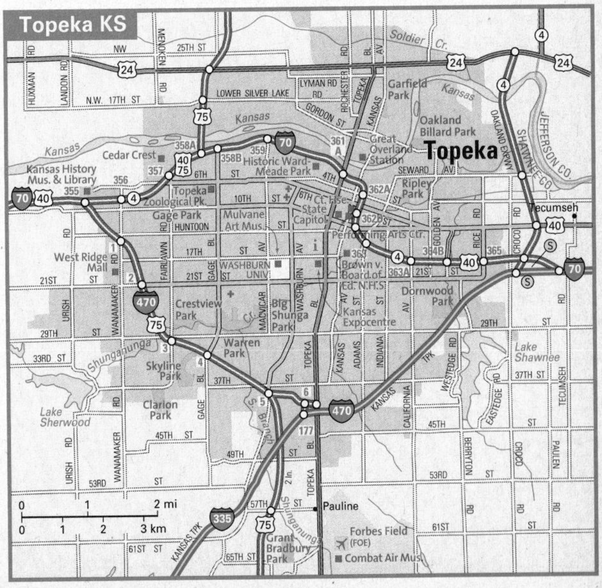

Figures after entries indicate population, page number, and grid reference.

Bakersville	B1	Fallsington	A2	Mercerville	B1	Trenton	B2
Ewing	A1	Lawrenceville	B1	Morrisville	A2	W. Trenton	A1
Ewingville	A1	Lewisville	B1	Slackwood	B1	White Horse	B2

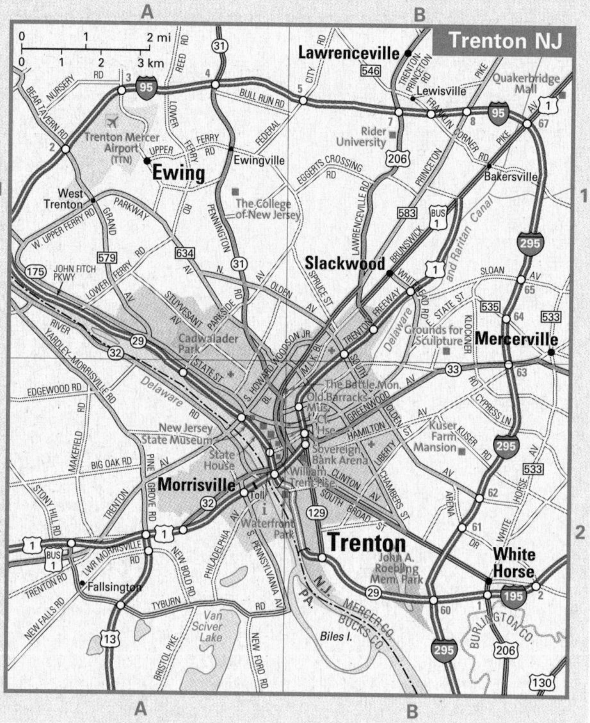

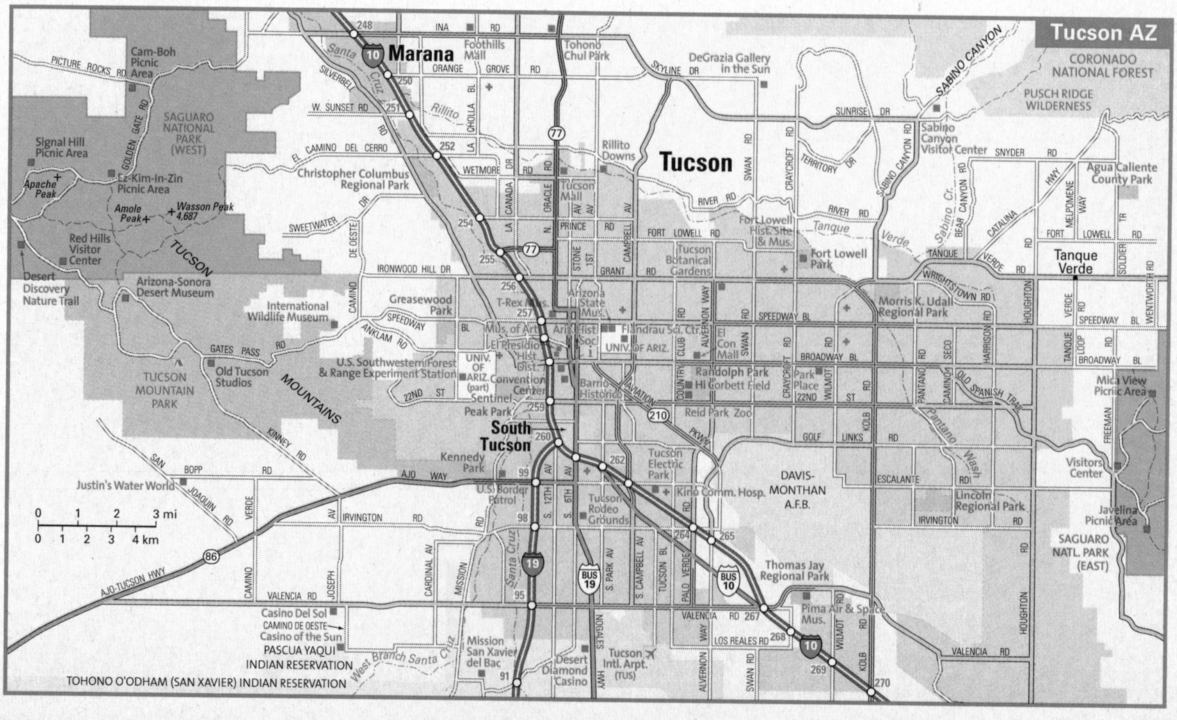

Entries in **bold black** indicate counties or parishes.
Entries in **bold color** indicate cities with detailed inset maps.

Bixby C3
Bowden A3
Broken Arrow C3
Catoosa C1
Jenks B3
Oakhurst A3
Sand Sprs. A2
Sapulpa A3
Tiger C1
Tulsa B2

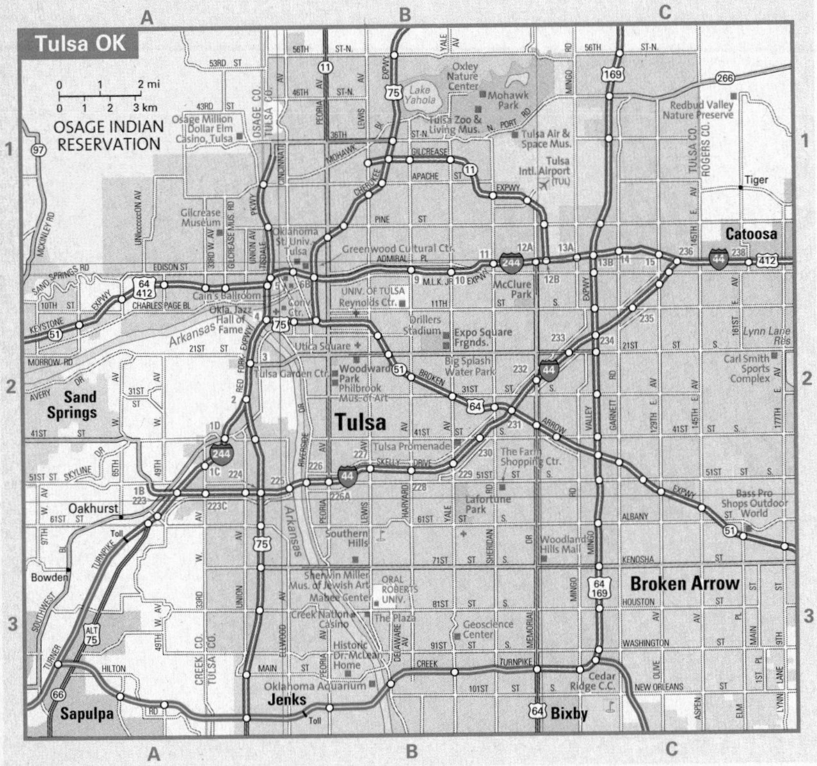

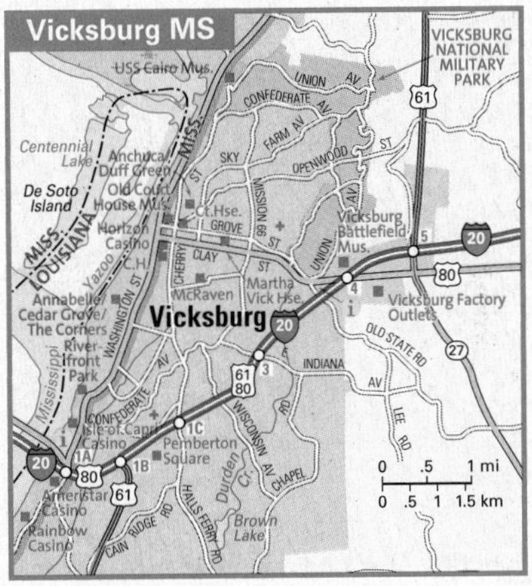

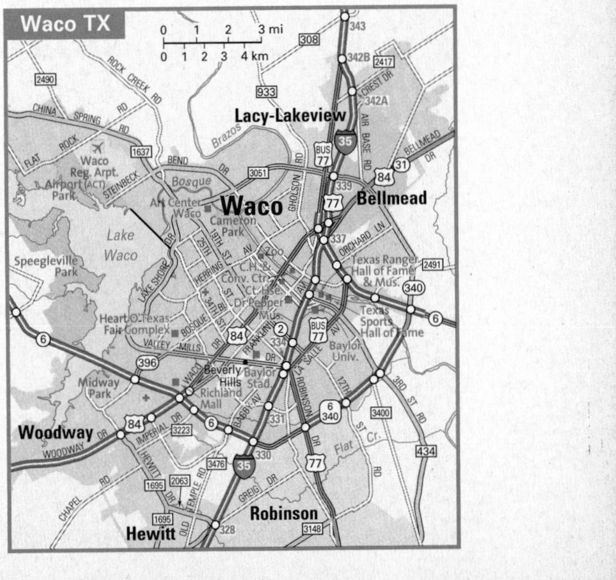

Figures after entries indicate population, page number, and grid reference.

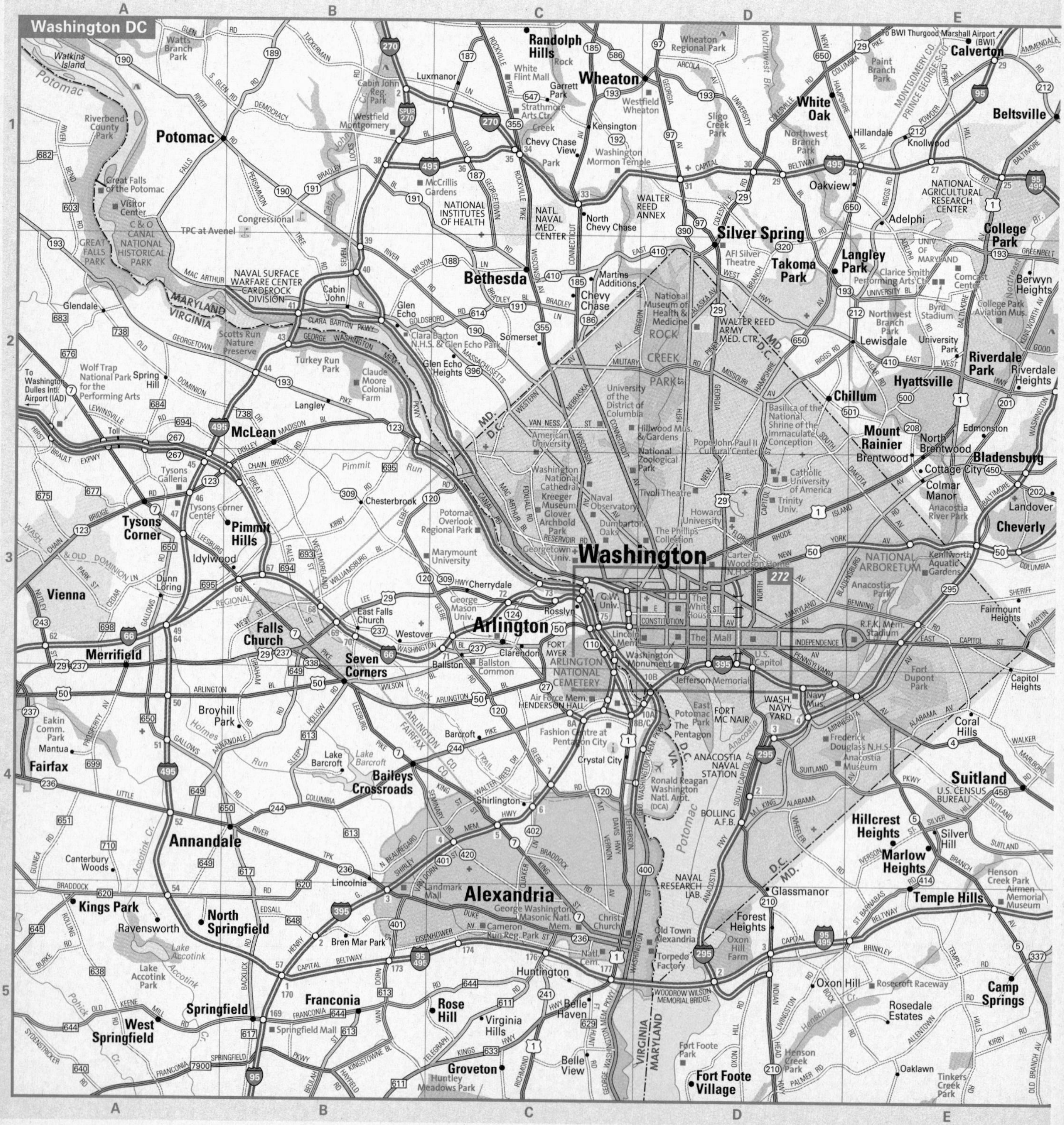

Entries in **bold black** indicate counties or parishes.
Entries in **bold color** indicate cities with detailed inset maps.

Vermilion Par. LA, 53807 133 E3
Vermillion SD, 9765 35 F1
Vermillion Co. IN, 16788 99 E1
Vermont IL, 792 88 A4
Vermontville MI, 789 76 A4
Vernal UT, 7714 32 A4
Verndale MN, 575 64 A4
Vernon AL, 2143 119 D4
Vernon CT, 28063 150 A3
Vernon FL, 743 136 C2
Vernon IN, 330 100 A2
Vernon MI, 847 76 B3
Vernon NY, 1155 79 E3
Vernon TX, 11660 51 D4
Vernon VT, 400 94 C1
Vernon Co. MO, 20454 96 C4
Vernon Co. WI, 28056 73 F2
Vernon Hills IL, 20120 203 C1
Vernonia OR, 2228 20 B1
Vernon Par. LA, 52531 125 D4
Vernon Valley NJ, 1737 148 A2
Vero Beach FL, 17705 141 E3
Verona MS, 3334 119 D3
Verona MO, 714 106 C2
Verona NJ, 13533 148 A3
Verona NC, 700 115 D4
Verona PA, 3124 250 D1
Verona VA, 3638 102 C4
Verona WI, 7052 74 B3
Verplanck NY, 777 148 B2

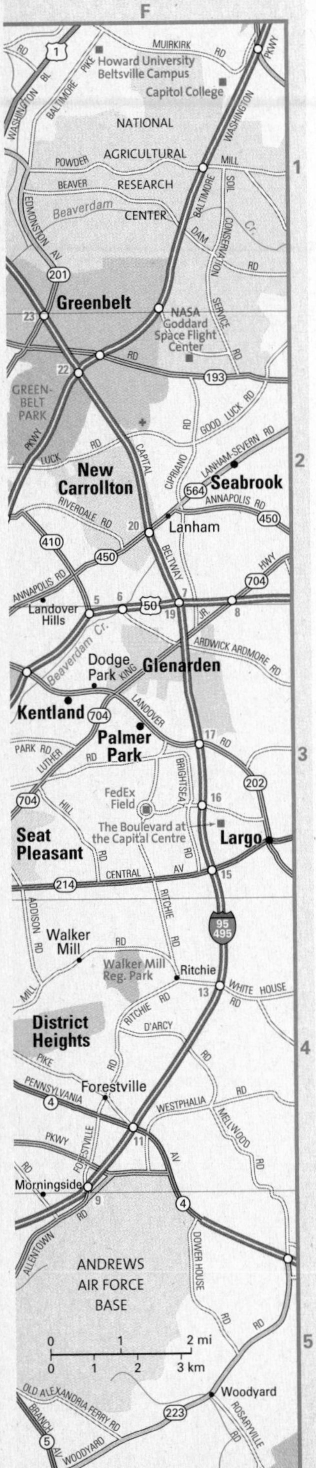

Versailles CT, 750 149 F1
Versailles IN, 1784 100 A2
Versailles KY, 7511 100 B4
Versailles MO, 2565 97 D3
Versailles OH, 2589 90 B4
Vesper WI, 541 68 A4
Vestal NY, 3900 93 E1
Vestal Ctr. NY, 850 93 E1
Vestavia Hills AL, 24476 119 F4
Vevay IN, 1735 100 A3
Vian OK, 1362 116 B1
Viborg SD, 832 35 F1
Viburnum MO, 825 107 F1
Vici OK, 668 51 D2
Vicksburg MI, 2320 89 F1
Vicksburg MS, 26407 126 A2
Victor CO, 445 41 E2
Victor ID, 840 23 F4
Victor IA, 952 87 E1
Victor MT, 859 15 D4
Victor NY, 2433 78 C3
Victoria KS, 1208 43 D2
Victoria MN, 4025 66 C4
Victoria TX, 60603 61 F3
Victoria VA, 1821 113 D2
Victoria Co. TX, 84088 61 F3
Victorville CA, 64029 53 D2
Victory Gardens NJ, 1546 148 A3
Vidalia GA, 10491 129 F3
Vidalia LA, 4543 125 F4
Vidor TX, 11440 132 C2
Vienna GA, 2973 129 D3
Vienna IL, 1234 108 C1
Vienna LA, 424 125 E2
Vienna MO, 628 97 E4
Vienna VA, 14453 144 B3
Vienna WV, 10861 101 E2
View Park CA, 10958 228 C3
Vigo Co. IN, 105848 99 E2
Vilano Beach FL, 2533 139 E3
Vilas Co. WI, 21033 68 B2
Village Green NY, 3945 79 D3
Village Green PA, 8279 248 A4
Villages of Oriole FL, 4758 143 F1
Villa Grove IL, 2553 99 D1
Villa Hills KY, 7948 204 A3
Villa Park CA, 5999 229 F4
Villa Park IL, 22075 203 C4
Villa Rica GA, 4134 120 B4
Villa Ridge MO, 2417 98 A3
Villas NJ, 9064 104 C4
Ville Platte LA, 8145 133 E1
Villisca IA, 1344 86 B3
Vilonia AR, 2106 117 E1
Vimy Ridge AR, 600 117 E2
Vinalhaven ME, 750 83 D3
Vincennes IN, 18701 99 D3
Vincent AL, 1853 128 A1
Vincent CA, 15097 52 C2
Vincentown NJ, 750 147 D3
Vinco PA, 1429 92 B4
Vine Grove KY, 4169 110 A1
Vine Hill CA, 3260 259 D1
Vineland MN, 607 66 C1
Vineland NJ, 56271 145 F1
Vinemont AL, 425 119 F3
Vineyard Haven MA, 2048 151 E4
Vinings GA, 9677 190 C2
Vinita OK, 6472 106 B3
Vinita Park MO, 1924 256 B2
Vinton CA, 387 37 D1
Vinton IA, 5102 87 E1
Vinton LA, 3338 133 D2
Vinton TX, 1892 56 C3
Vinton VA, 7782 112 B2
Vinton Co. OH, 12806 101 E2
Viola DE, 156 145 E3
Viola IL, 956 88 A2
Viola WI, 667 73 F2
Violet LA, 8555 134 C3
Virden IL, 3488 98 B2
Virgin UT, 394 39 D4
Virginia IL, 1728 98 A1
Virginia MN, 9157 64 C3
Virginia Beach VA, 425257 114 B4
Virginia City MT, 130 23 E2
Virginia City NV, 800 37 D2
Virginia Gardens FL, 2348 233 A4
Viroqua WI, 4335 73 F2
Visalia CA, 91565 45 D3
Vista CA, 89857 53 D3
Vivian LA, 4031 124 C1
Volcano HI, 2231 153 F4
Volga SD, 1435 27 F3
Volin SD, 207 35 E1
Voluntown CT, 850 149 F1
Volusia Co. FL, 443343 139 E4
Vonore TN, 1162 120 C1
Voorheesville NY, 2705 94 B1

W

Wabash IN, 11743 89 F3
Wabasha MN, 2599 73 E1
Wabasha Co. MN, 21610 73 E1
Wabash Co. IL, 12937 99 D3
Wabash Co. IN, 34960 89 F3
Wabasso FL, 918 141 E3
Wabasso MN, 643 72 A1
Wabaunsee Co. KS, 6885 43 F2
Waco TX, 113726 59 E4
Waconia MN, 6814 66 C4
Waddington NY, 923 80 B1
Wade MS, 491 135 D2
Wade NC, 480 123 D1
Wadena MN, 4294 64 A4
Wadena Co. MN, 13713 64 A4
Wadesboro NC, 3552 122 B2
Wading River NY, 6668 149 D3
Wadley AL, 640 128 B1
Wadley GA, 2088 129 F1
Wadsworth NV, 881 37 D1
Wadsworth OH, 18437 91 E3
Waelder TX, 947 61 E2
Wagener SC, 863 122 A4
Waggaman LA, 9435 239 B2
Wagner SD, 1675 35 E1
Wagoner OK, 7669 106 A4
Wagoner Co. OK, 57491 106 A4
Wagon Mound NM, 369 49 E2
Wagontown PA, 1100 146 B3
Wagram NC, 801 122 C2
Wahiawa HI, 16151 152 A2
Wahkiakum Co. WA, 3824 12 B4
Wahoo NE, 3942 35 F3
Wahpeton ND, 8586 27 F1
Waialua HI, 3761 152 A2
Waianae HI, 10506 152 A3
Waiehu HI, 7310 153 D1
Waihee HI, 7310 153 D1
Waikane HI, 726 152 A2
Waikapu HI, 1115 153 D1
Waikoloa Vil. HI, 4806 153 E3
Wailua HI, 2083 152 B1
Wailuku HI, 12296 153 D1
Waimalu HI, 29371 152 A3
Waimanalo HI, 3664 152 B3
Waimanalo Beach HI, 4271 152 B3
Waimea HI, 1787 152 B1
Waimea (Kamuela) HI, 7028 153 E2
Wainscott NY, 628 149 F3
Wainwright AK, 546 154 C1
Waipahu HI, 33108 152 A3
Waipio Acres HI, 5298 152 A3
Waite Park MN, 6568 66 C3
Waitsburg WA, 1212 13 F4
Waitsfield VT, 350 81 D2
Wakarusa IN, 1618 89 F2
Wakeby MA, 1600 151 E3
Wake Co. NC, 627846 113 D4
WaKeeney KS, 1924 42 C2
Wakefield KS, 838 43 F2
Wakefield MA, 24804 151 D1
Wakefield MI, 2085 65 E4
Wakefield NE, 1411 35 F2
Wakefield RI, 8468 150 C4
Wakefield VA, 1038 113 F2
Wake Forest NC, 12588 113 D4
Wakeman OH, 951 91 D2
Wake Vil. TX, 5129 124 C1
Wakonda SD, 374 35 F1
Wakulla Co. FL, 22863 137 E2
Walbridge OH, 2546 267 B2
Walcott IA, 1528 87 F2
Walden CO, 734 33 D4
Walden NY, 6164 148 B1
Walden TN, 1960 120 B1
Waldo AR, 1594 125 D1
Waldo FL, 821 138 C3
Waldoboro ME, 1291 82 C2
Waldo Co. ME, 36280 82 C2
Waldorf MD, 22312 144 B4
Waldport OR, 2050 20 B3
Waldron AR, 3508 116 C2
Waldron IN, 1100 100 A2
Waldron MI, 590 90 B1
Wales MA, 850 150 B2
Wales WI, 2523 74 C3
Waleska GA, 616 120 C3
Walford IA, 1224 87 E1
Walhalla ND, 1057 19 E1
Walhalla SC, 3801 121 E2
Walker IA, 750 73 E4
Walker LA, 4801 134 A2
Walker MI, 21842 75 F3
Walker MN, 1069 64 A3
Walker Co. AL, 70713 119 E4
Walker Co. GA, 61053 120 B2
Walker Co. TX, 61758 132 A1
Walker Mill MD, 11104 271 F4
Walkersville MD, 5192 144 A1
Walkerton IN, 2274 89 E2
Walkertown NC, 4009 112 B4
Walker Valley NY, 758 148 A1
Walkerville MT, 714 23 E1
Wall PA, 727 250 D3
Wall SD, 818 26 B3
Wallace ID, 960 14 B3
Wallace NC, 3344 123 E2
Wallace VA, 550 111 E3
Wallace Co. KS, 1749 42 B2
Walla Walla WA, 29686 21 F1
Walla Walla Co. WA, 55180 21 F1
Walled Lake MI, 6713 210 A2
Waller TX, 2092 61 F2
Waller Co. TX, 32663 132 A3
Wallingford CT, 17509 149 D1
Wallingford VT, 948 81 D4
Wallington NJ, 11583 148 B3
Wallis TX, 1172 61 F2
Wallkill NY, 2143 148 B1
Wall Lake IA, 841 72 B4
Wallowa OR, 869 22 A1
Wallowa Co. OR, 7226 22 A1
Walnut CA, 30004 229 F3
Walnut IL, 1461 88 B2
Walnut IA, 778 86 A2
Walnut MS, 754 118 C2
Walnut Cove NC, 1465 112 B3
Walnut Creek CA, 64296 36 B4
Walnut Creek NC, 859 123 E1
Walnut Grove AL, 710 120 A3
Walnut Grove CA, 669 36 C3
Walnut Grove GA, 1241 121 D4
Walnut Grove MN, 599 72 A1
Walnut Grove MS, 488 126 C2
Walnut Grove MO, 630 106 C1
Walnut Grove TN, 677 109 F3
Walnut Hill TN, 2756 111 E3
Walnut Park CA, 16180 228 D3
Walnutport PA, 2043 93 F3
Walnut Ridge AR, 4925 108 A4
Walnut Sprs. TX, 755 59 E3
Walpole MA, 5867 151 D2
Walpole NH, 700 81 E4
Walsenburg CO, 4182 41 E4
Walsh CO, 723 42 A4
Walsh Co. ND, 12389 19 E2
Walterboro SC, 5153 130 C1
Walterhill TN, 1523 109 F4
Walters OK, 2657 51 D4
Walthall MS, 170 118 C4
Walthall Co. MS, 15156 134 B1
Waltham MA, 59226 151 D1
Walthill NE, 909 35 F2
Walthourville GA, 4030 130 B3
Walton IN, 1069 89 F3
Walton KY, 2450 100 B3
Walton NY, 3070 93 F1
Walton Co. FL, 40601 136 B2
Walton Co. GA, 60687 121 D4
Walton Hills OH, 2400 204 G3
Walworth WI, 2304 74 C4
Walworth Co. SD, 5974 26 C2
Walworth Co. WI, 93759 74 C4
Wamac IL, 1378 98 C3
Wamego KS, 4246 43 F2
Wampsville NY, 561 79 E3
Wamsutter WY, 261 32 C3
Wanakah NY, 3200 78 A4
Wanamassa NJ, 4551 147 E2
Wanamingo MN, 1007 73 D1
Wanaque NJ, 10266 148 A3
Wanatah IN, 1013 89 E2
Wanblee SD, 641 26 B4
Wanchese NC, 1527 115 F2
Wapakoneta OH, 9474 90 B4
Wapanucka OK, 445 51 F4
Wapato WA, 4582 13 D4
Wapella IL, 651 88 C4
Wapello IA, 2124 87 F2
Wapello Co. IA, 36051 87 D3
Waples TX, 800 59 E3
Wappingers Falls NY, 4929 148 B1
War WV, 788 111 F1
Ward AR, 2580 117 F2
Ward Co. ND, 58795 18 B2
Ward Co. TX, 10909 57 F4
Warden WA, 2544 13 E4
Wardner ID, 215 14 B3
Wardsville MO, 976 97 E3
Ware MA, 6174 150 B1
Ware Co. GA, 35483 129 F4
Wareham MA, 2874 151 E3
Ware Shoals SC, 2363 121 F3
Waretown NJ, 1582 147 E3
Warm Beach WA, 2040 12 C2
Warm Sprs. GA, 485 128 C1
Warm Sprs. OR, 2431 21 D3
Warm Sprs. VA, 200 102 B4
Warner NH, 650 81 F4
Warner OK, 1430 116 A1
Warner SD, 419 27 D2
Warner Robins GA, 48804 129 D2
Warr Acres OK, 9735 244 D2
Warren AR, 6442 117 E4
Warren IL, 1496 74 A4
Warren IN, 1272 90 A3
Warren ME, 900 82 C2
Warren MA, 1452 150 B2
Warren MI, 138247 76 C4
Warren MN, 1678 19 F2
Warren OH, 46832 91 F2
Warren PA, 10259 92 B1
Warren RI, 11360 151 D3
Warren Co. GA, 6336 129 F1
Warren Co. IL, 18735 88 A3
Warren Co. IN, 8419 89 D4
Warren Co. IA, 40671 86 C2
Warren Co. KY, 92522 109 F2
Warren Co. MS, 49644 126 A2
Warren Co. MO, 24525 97 F3
Warren Co. NJ, 102437 93 F3
Warren Co. NY, 63303 80 C3
Warren Co. NC, 19972 113 D3
Warren Co. OH, 158383 100 C2
Warren Co. PA, 43863 92 B1
Warren Co. TN, 38276 110 A4
Warren Co. VA, 31584 103 D2
Warren Glen NJ, 650 146 C1
Warren Park IN, 1656 99 F1
Warrensburg IL, 1289 98 C1
Warrensburg MO, 16340 96 C3
Warrensburg NY, 3208 80 C4
Warrensville PA, 1100 93 D2
Warrensville Hts. OH, 15109 204 G2
Warrenton GA, 2013 129 F1
Warrenton MO, 5281 97 F3
Warrenton NC, 811 113 D3
Warrenton OR, 4096 20 B1
Warrenton VA, 6670 103 D3
Warrenville IL, 13363 203 A4
Warrick Co. IN, 52383 99 E4
Warrington FL, 15207 135 F2
Warrington PA, 5400 146 C2
Warrior AL, 3169 119 F4
Warrior Run PA, 624 261 A2
Warroad MN, 1722 64 A1
Warsaw IL, 1793 87 F4
Warsaw IN, 12415 89 F2
Warsaw KY, 1811 100 B3
Warsaw MN, 550 72 C1
Warsaw MO, 2070 97 D4
Warsaw NY, 3814 78 B4
Warsaw NC, 3051 123 E2
Warsaw OH, 781 91 D4
Warsaw VA, 1375 103 E4
Warson Woods MO, 1983 256 B2
Wartburg TN, 890 110 B4
Wartrace TN, 548 120 A1
Warwick NY, 6412 148 A2
Warwick RI, 85808 150 C3
Wasatch Co. UT, 15215 31 F4
Wasco CA, 21263 45 D4
Wasco OR, 381 21 D1
Wasco Co. OR, 23791 21 D2
Waseca MN, 8493 72 C1
Waseca Co. MN, 19526 72 C1
Washakie Co. WY, 8289 24 C4
Washburn IL, 1147 88 B3
Washburn IA, 1000 73 E4
Washburn ME, 800 85 D2
Washburn ND, 1389 18 B3
Washburn WI, 2280 65 D4
Washburn Co. WI, 16036 67 E2
Washington CT, 600 148 C1
Washington DC, 572059 144 B3
Washington GA, 4295 121 E4
Washington IL, 10841 88 B3
Washington IN, 11380 99 E3
Washington IA, 7047 87 E2
Washington KS, 1223 43 F1
Washington LA, 1082 133 F2
Washington MO, 13243 97 F3
Washington NJ, 6712 93 F3
Washington NC, 9583 115 D3
Washington OK, 520 51 E3
Washington PA, 15268 91 F4
Washington UT, 8186 46 C1
Washington VT, 350 81 E2
Washington VA, 183 103 D3
Washington WV, 1170 101 E2
Washington Co. AL, 18097 127 E4
Washington Co. AR, 157715 106 B4
Washington Co. CO, 4926 34 A4
Washington Co. FL, 20973 136 C2
Washington Co. GA, 21176 129 E1
Washington Co. ID, 9977 22 A3
Washington Co. IL, 15148 98 B4
Washington Co. IN, 27223 99 F3
Washington Co. IA, 20670 87 E2
Washington Co. KS, 6483 43 E1
Washington Co. KY, 10916 110 B1
Washington Co. ME, 33941 83 E1
Washington Co. MD, 131923 103 D1
Washington Co. MN, 201130 67 D3
Washington Co. MS, 62977 126 A1
Washington Co. MO, 23344 97 F4
Washington Co. NE, 18780 35 F3
Washington Co. NY, 61042 81 D4
Washington Co. NC, 13723 113 F4
Washington Co. OH, 63251 101 E2
Washington Co. OK, 48996 106 A2
Washington Co. OR, 445342 20 B2
Washington Co. PA, 202897 92 A4
Washington Co. RI, 123546 150 C4
Washington Co. TN, 107198 111 E3
Washington Co. TX, 30373 61 F2
Washington Co. UT, 90354 38 C4
Washington Co. VT, 58039 81 E2
Washington Co. VA, 51103 111 F3
Washington Co. WI, 117493 74 C2
Washington C.H. OH, 13524 100 C1
Washington Crossing NJ, 950 147 D2
Washington Crossing PA, 1300 147 D2
Washington Hts. NY, 1318 148 A1
Washington Par. LA, 43926 134 B1
Washington Park IL, 5345 256 D2
Washington Terrace UT, 8551 244 A2
Washingtonville NY, 5851 148 B1
Washingtonville OH, 789 91 F3
Washita Co. OK, 11508 51 D3
Washoe Co. NV, 339486 29 E4
Washougal WA, 8595 20 C2
Washtenaw Co. MI, 322895 90 B1
Wasilla AK, 5469 154 C3
Waskom TX, 2068 124 C2
Wataga IL, 857 88 A3
Watauga TN, 403 111 E3
Watauga TX, 21908 207 B2
Watauga Co. NC, 42695 111 F3
Watchung NJ, 5613 147 E1
Waterboro ME, 800 82 B3
Waterbury CT, 107271 149 D1
Waterbury VT, 1706 81 D2
Waterbury Ctr. VT, 850 81 E2
Waterford CA, 6924 36 C4
Waterford CT, 2935 149 F2
Waterford MI, 73150 76 B4
Waterford NY, 2204 188 E1
Waterford PA, 1449 92 A1
Waterford WI, 4048 74 C4
Waterford Works NJ, 1000 147 D4
Waterloo IL, 7614 98 A4
Waterloo IN, 2200 90 A2
Waterloo IA, 68747 73 E4
Waterloo MD, 900 144 C2
Waterloo NE, 459 35 F3
Waterloo NY, 5111 79 D3
Waterloo WI, 3259 74 B3
Waterman IL, 1224 88 C1
Water Mill NY, 1724 149 E3
Waterproof LA, 834 125 F3
Watertown CT, 5300 149 D1
Watertown FL, 2837 138 C2
Watertown MA, 32986 151 D1
Watertown MN, 3029 66 C4
Watertown NY, 26705 79 E1
Watertown SD, 20237 27 F2
Watertown TN, 1358 109 F4
Watertown WI, 21598 74 C3
Water Valley MS, 3677 118 B3
Waterville KS, 681 43 F1
Waterville ME, 15605 82 C2
Waterville MN, 1833 72 C1
Waterville NY, 1721 79 E3
Waterville OH, 4828 90 C2
Waterville WA, 1163 13 E3
Watervliet MI, 1843 89 F1
Watervliet NY, 10207 188 E2
Watford City ND, 1435 18 A3
Wathena KS, 1348 96 B1
Watkins MN, 880 66 C3
Watkins Glen NY, 2149 79 D4
Watkinsville GA, 2097 121 D4
Watonga OK, 4658 51 D2
Watonwan Co. MN, 11876 72 B1
Watseka IL, 5670 89 D3
Watson IL, 729 98 C2
Watson LA, 1400 134 A2
Watsontown PA, 2255 93 D3
Watsonville CA, 44265 44 B2
Watterson Park KY, 953 230 E2
Waubay SD, 662 27 E2
Wauchula FL, 4368 140 C3
Wauconda IL, 9448 203 B1
Waukee IA, 5126 86 C2
Waukegan IL, 87901 75 D4
Waukesha WI, 64825 74 C3
Waukesha Co. WI, 360767 74 C3
Waukomis OK, 1261 51 E2
Waukon IA, 4131 73 E3
Waunakee WI, 8995 74 B3
Wauneta NE, 625 34 B4
Waupaca WI, 5676 74 B1
Waupaca Co. WI, 51731 68 B4
Waupun WI, 10718 74 C2
Wauregan CT, 1085 150 B3
Waurika OK, 1988 51 E4
Wausa NE, 636 35 E1
Wausau WI, 38426 68 B4
Wausaukee WI, 572 68 C3
Wauseon OH, 7091 90 B2
Waushara Co. WI, 23154 74 B1
Wautoma WI, 1998 74 B1
Wauwatosa WI, 47271 234 C2
Wauzeka WI, 768 73 F3
Waveland MS, 6674 134 C2
Waverly FL, 1927 141 D2
Waverly IL, 1346 98 B1
Waverly IA, 8968 73 D4
Waverly KS, 589 96 A3
Waverly MN, 732 66 C3
Waverly MO, 806 96 C2
Waverly NE, 2448 35 F4
Waverly NY, 4607 93 E1
Waverly OH, 4433 101 D2
Waverly TN, 4028 109 D4
Waverly VA, 2309 113 E2
Waverly Hall GA, 709 128 C2
Waxahachie TX, 21426 59 F3
Waxhaw NC, 2625 122 B2
Waycross GA, 15333 129 F4
Wayland IA, 945 87 E2
Wayland MA, 1300 150 C1
Wayland MI, 3939 75 F4
Wayland NY, 1893 78 C4
Waymart PA, 1429 93 F2
Wayne IL, 2137 203 A4
Wayne MI, 19051 210 A4
Wayne NE, 5583 35 F2
Wayne NJ, 54069 148 A3
Wayne OH, 842 90 C2
Wayne OK, 714 51 E3
Wayne PA, 1000 146 C3
Wayne WV, 1105 101 D4
Wayne City IL, 1089 98 C4
Wayne Co. GA, 26565 130 A4
Wayne Co. IL, 17151 98 C3
Wayne Co. IN, 71097 100 A1
Wayne Co. IA, 6730 87 D3
Wayne Co. KY, 19923 110 B3
Wayne Co. MI, 2061162 76 B4
Wayne Co. MS, 21216 127 D4
Wayne Co. MO, 13259 108 A2
Wayne Co. NE, 9851 35 F2
Wayne Co. NY, 93765 79 D3
Wayne Co. NC, 113329 123 E1
Wayne Co. OH, 111564 91 E3
Wayne Co. PA, 47722 93 F1
Wayne Co. TN, 16842 119 E1
Wayne Co. UT, 2509 39 F3
Wayne Co. WV, 42903 101 E4
Wayne Lakes OH, 684 100 B1
Waynesboro GA, 5813 129 F1
Waynesboro MS, 5197 127 D4

Figures after entries indicate population, page number, and grid reference.

POINTS OF INTEREST

Arena Stage . . . E4
Arlington Natl. Cemetery . . . A4
Arthur M. Sackler Gallery . . . E3
Art Museum of the Americas . . . C2
Arts & Industries Building . . . E3
Blair-Lee House . . . C2
B'nai B'rith Klutznick Natl. Jewish Museum . . . D1
Bureau of Engraving & Printing . . . D3
Corcoran Gallery of Art . . . C2
Daughters of the American Revolution Constitution Hall . . . C2
Decatur House . . . C1
Dept. of Agriculture . . . D3
Dept. of Commerce . . . D2
Dept. of Education . . . E3
Dept. of Energy . . . E3
Dept. of Health & Human Services . . . E3
Dept. of Housing and Urban Development . . . E3
Dept. of Justice . . . E2
Dept. of Labor . . . F2
Dept. of State . . . C2
Dept. of the Interior . . . C2
Dept. of the Treasury . . . D2
Dept. of Transportation . . . E3
Dept. of Veterans Affairs . . . D1
District of Columbia Court House . . . E2
District of Columbia War Memorial . . . C3
Donald W. Reynolds Center for American Art & Portraiture . . . E2
The Ellipse . . . D2
Environmental Protection Agency . . . D2
Fish Wharf . . . D4
Folger Shakespeare Library . . . G3
Ford's Theatre Natl. Hist. Site . . . E2
Franklin Delano Roosevelt Memorial . . . C3
Freer Gallery of Art . . . D3
Friendship Archway . . . E1
Gallaudet Univ. . . . G1
Georgetown Univ. Law Center . . . F2
George Washington Univ. . . . C2
Government Printing Office . . . F1
Hirshhorn Mus. & Sculpture Garden . . . E3
Ice Skating Rink . . . E2
Internal Revenue Service . . . E2
International Spy Museum . . . E2
James Madison Building . . . G3
J. Edgar Hoover FBI Building . . . E2
John Adams Building . . . G3
John F. Kennedy Center for the Performing Arts . . . B2
John F. Kennedy Gravesite . . . A4
Judiciary Square . . . E2
Korean War Veterans Memorial . . . C3
Koshland Science Museum . . . E2
Lafayette Square . . . D1
L'Enfant Plaza . . . E3
Library of Congress . . . G3
Lincoln Memorial . . . B3
Lyndon B. Johnson Memorial Grove . . . B4
The Mall . . . E3
Marine Corps War Memorial (Iwo Jima Memorial) . . . A3
Martin Luther King, Jr. Memorial Library . . . E2
NASA . . . E4
Natl. Air & Space Museum . . . E3
The Natl. Aquarium . . . D2
The Natl. Archives . . . E2
Natl. Building Museum . . . E2
Natl. Gallery of Art East Building . . . E3
Natl. Gallery of Art West Building . . . E2
Natl. Geographic Society & Explorers Hall . . . D1
Natl. Mus. of African Art . . . E3
Natl. Mus. of American Hist. . . . D2
Natl. Mus. of Natural Hist. . . . E2
Natl. Mus. of the American Indian . . . E3
Natl. Mus. of Women in the Arts . . . D1
Natl. Portrait Gallery . . . E2
Natl. Postal Museum . . . F2
Natl. Theatre . . . D2
Natl. WWII Memorial . . . C3
Navy and Marine Memorial . . . C4
The Netherlands Carillon . . . A3
Newseum . . . E2
Octagon House . . . C2
Old Post Office Pavilion . . . D2
Old Stone House . . . B1
Organization of American States . . . C2
Reflecting Pool . . . C3
Renwick Gallery . . . C2
Ronald Reagan Building and Intl. Trade Center . . . D2
St. Matthew's Cathedral . . . C1
Seabees of the United States Navy Memorial . . . A3
Sewall-Belmont House . . . G2
Shakespeare Theatre . . . E2
Shops at Georgetown Park . . . A1
Signers of the Declaration of Independence Memorial . . . C3
Smithsonian American Art Museum . . . E2
Smithsonian Institution Castle . . . E3
Southeastern University . . . E4
The Supreme Court . . . G3
Taft Memorial Carillon . . . F2
Theodore Roosevelt Memorial . . . A2
Thomas Jefferson Building . . . G3
Thomas Jefferson Memorial . . . D4
Union Station . . . F2
United Spanish War Veterans Memorial . . . A3
U.S. Botanic Garden . . . F3
U.S. Capitol . . . F3
U.S. Claims Court . . . D1
U.S. District Court House . . . E2
U.S. Grant Memorial . . . F3
U.S. Holocaust Memorial Museum . . . D3
U.S. Navy Memorial & Naval Heritage Center . . . E2
U.S. Postal Service Headquarters . . . E3
Verizon Center . . . E2
Vietnam Veterans Memorial . . . C2
Vietnam Women's Memorial . . . C3
Warner Theatre . . . D2
Washington Convention Center . . . E1
The Washington Design Center . . . E3
Washington Harbour . . . A1
Washington Monument . . . D3
Washington Post . . . D1
Waterside Mall . . . E4
The White House . . . D2
Women in Military Service for America Memorial . . . A4
Zero Milestone . . . D2

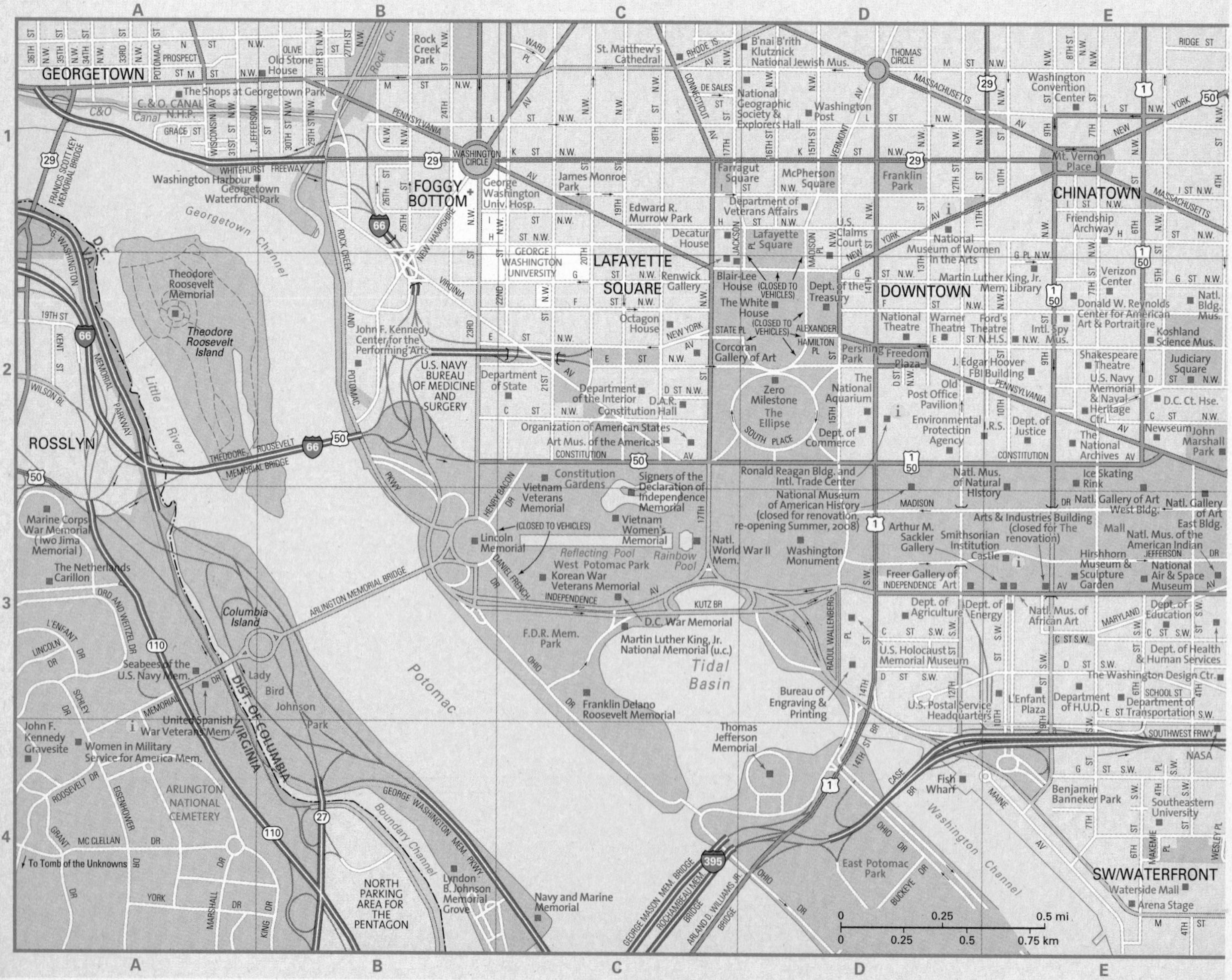

Entries in **bold black** indicate counties or parishes.
Entries in **bold color** indicate cities with detailed inset maps.

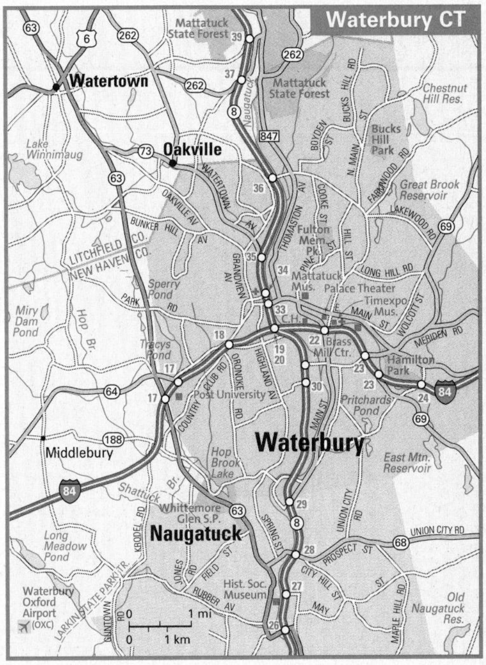

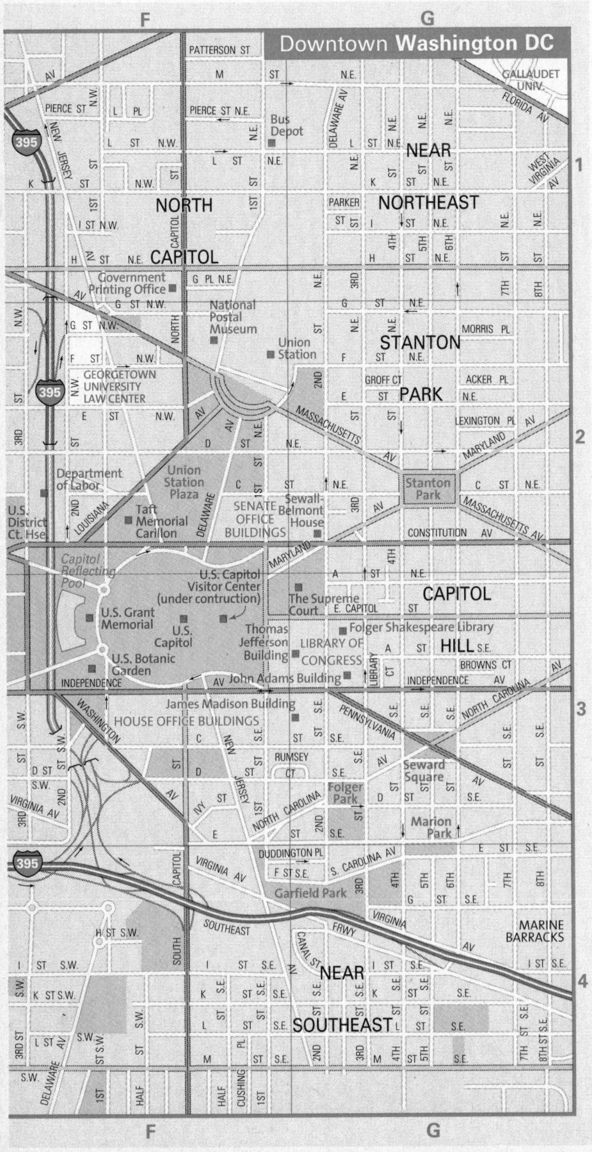

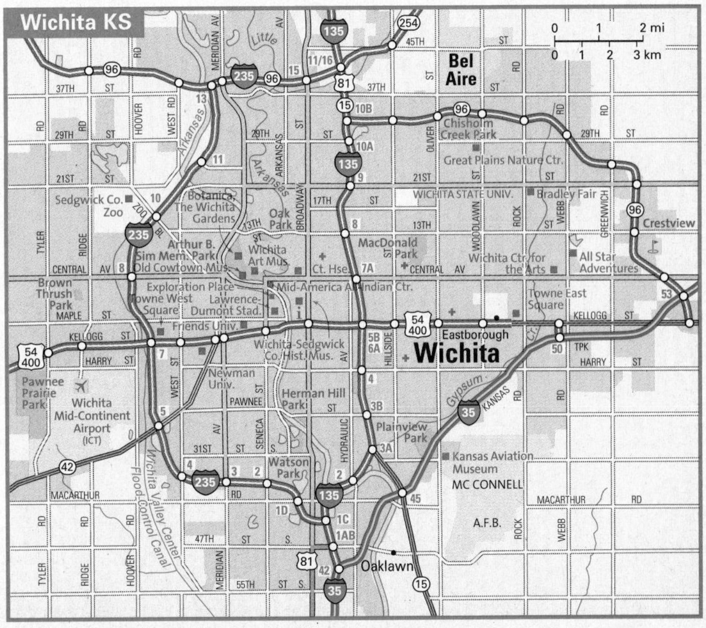

Figures after entries indicate population, page number, and grid reference.

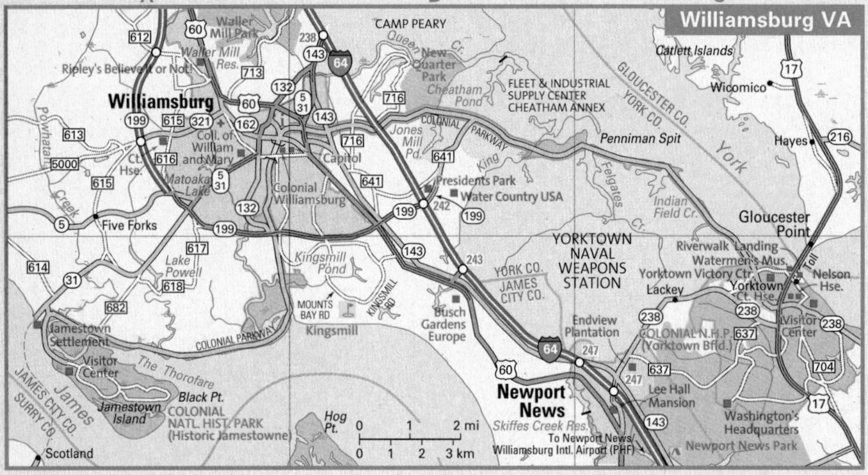

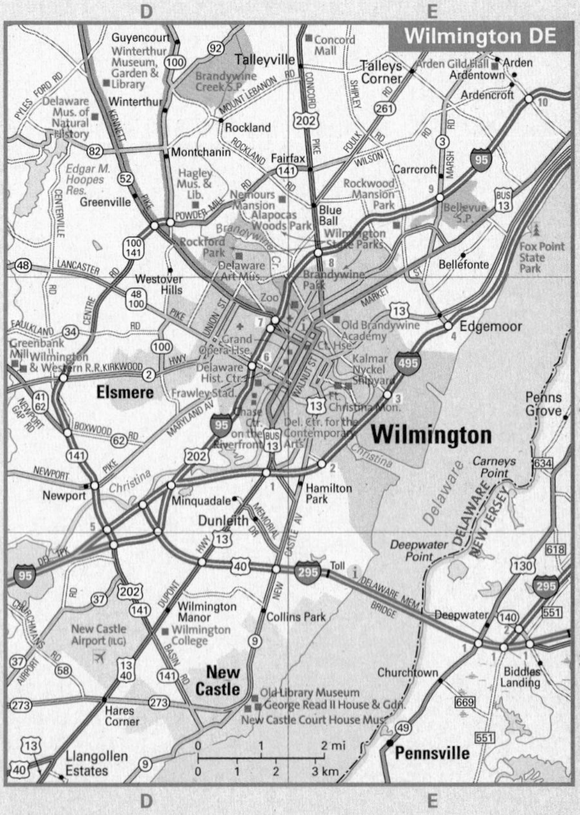

Entries in **bold black** indicate counties or parishes.
Entries in **bold color** indicate cities with detailed inset maps.

Winnebago Co. IA, *11723* **72** C2
Winnebago Co. WI, *156763* **74** C1
Winneconne WI, *2401* **74** C1
Winnecunnet MA, *750* **151** D2
Winnemucca NV, *7174* **30** A3
Winner SD, *3137* **26** C4
Winneshiek Co. IA, *21310* **73** E3
Winnetka IL, *12419* **203** D2
Winnett MT, *185* **16** C4
Winnfield LA, *5749* **125** E3
Winnie TX, *2914* **132** C3
Winnisquam NH, *850* **81** F4
Winn Par. LA, *16894* **125** E3
Winnsboro LA, *5344* **125** F3
Winnsboro SC, *3599* **122** A3
Winnsboro TX, *3584* **124** A1
Winnsboro Mills SC, *2263* **122** A3
Winona MN, *27069* **73** E1
Winona MS, *5482* **118** B4
Winona MO, *1290* **107** F2
Winona OH, *475* **91** F3
Winona TX, *582* **124** B2
Winona Co. MN, *49985* **73** E1
Winona Lake IN, *3987* **89** F2
Winooski VT, *6561* **81** D2
Winside NE, *468* **35** E2
Winslow AZ, *9520* **47** E3
Winslow AR, 399 .... **106** C4
Winslow IN, *881* **99** E4
Winslow ME, *7743* **82** C2
Winslow NJ, *700* **147** D4
Winsted CT, *7321* **94** C2
Winsted MN, *2094* **66** C4
Winston FL, *9024* **140** C2
Winston OR, *4613* **28** B1
Winston Co. AL, *24843* **119** E3
Winston Co. MS, *20160* **127** D1
Winston-Salem NC, *185776* **112** B4
Winstonville MS, *319* **118** A4
Winter Beach FL, *965* **141** E3
Winter Garden FL, *14351* **141** D1
Winter Harbor ME, *250* **83** D2
Winterhaven CA, *529* **53** F4
Winter Haven FL, *26487* **140** C2
Winter Park CO, *662* **41** D1
Winter Park FL, *24090* **141** D1
Winterport ME, *1307* **83** D2
Winters CA, *6125* **36** B3
Winters TX, *2880* **58** C3
Winterset IA, *4768* **86** C2
Winter Sprs. FL, *31666* **141** D1
Winterstown PA, *546* **103** E1
Wintersville OH, *4067* **91** F4
Winterville GA, *1068* **121** D4
Winterville NC, *4791* **115** D3
Winthrop CT, *400* **149** E2
Winthrop IA, *772* **73** E4
Winthrop ME, *2893* **82** B2
Winthrop MA, *18303* **151** D1
Winthrop MN, *1367* **66** C4
Winthrop NY, *1140* **80** B1
Winthrop Harbor IL, *6670* **75** D4
Winton CA, *8832* **36** C4
Winton NC, *956* **113** F3
Wirt Co. WV, *5873* **101** F2
Wiscasset ME, *1203* **82** C3
Wisconsin Dells WI, *2418* **74** A2
Wisconsin Rapids WI, *18435* **74** A1
Wisdom MT, *114* **23** D1
Wise VA, *3255* **111** E2
Wise Co. TX, *48793* **59** E2
Wise Co. VA, *40123* **111** E2
Wishek ND, *1122* **27** D1
Wisner LA, *1140* **125** F3
Wisner NE, *1270* **35** F2
Wister OK, *1002* **116** B2
Withamsville OH, *3145* **204** C3
Withee WI, *508* **68** A4
Witherbee NY, *1747* **81** D2
Witt IL, *991* **98** B2
Witt TN, *500* **111** D4
Wittenberg WI, *1177* **68** B4
Wittmann AZ, *550* **54** B1
Wixom MI, *13263* **76** B4
Woburn MA, *37258* **151** D1
Wofford KY, *375* **110** C2
Wofford Hts. CA, *2276* **45** E4
Wolcott CT, *4000* **149** D1
Wolcott IN, *989* **89** E3
Wolcott NY, *1712* **79** D3
Wolcottville IN, *933* **90** A2
Wolfdale PA, *2873* **91** F4
Wolfeboro NH, *2979* **81** F3
Wolfeboro Falls NH, *1000* **81** F3
Wolfe City TX, *1566* **59** F1
Wolfe Co. KY, *7065* **111** D1
Wolfforth TX, *2554* **58** A1
Wolf Lake MI, *4455* **75** E3
Wolf Pt. MT, *2663* **17** E2
Wolsey SD, *418* **27** E3
Wolverine Lake MI, *4415* **210** A1
Womelsdorf PA, *2599* **146** A2
Wonewoc WI, *834* **74** A2
Wonnie KY, *250* **111** D1
Woodacre CA, *1393* **259** A1
Woodbine GA, *1218* **139** D1
Woodbine IA, *1564* **86** A1
Woodbine KY, *550* **110** C2
Woodbine MD, *425* **144** B2
Woodbine NJ, *2716* **104** C4
Woodbourne NY, *1600* **94** A3
Woodbourne OH, *7910* **208** E2
Woodbranch TX, *1305* **132** B2
Woodbridge CT, *1400* **149** D2
Woodbridge NJ, *18309* **147** E1
Woodbridge VA, *31941* **144** B4
Woodburn IN, *1579* **90** A3
Woodburn KY, *323* **109** F2
Woodburn OR, *20100* **20** B2
Woodbury CT, *1298* **149** D1
Woodbury GA, *1184* **128** C1
Woodbury MN, *46463* **67** D4
Woodbury NJ, *10307* **146** C3
Woodbury TN, *2428* **110** A4
Woodbury VT, *200* **81** E2
Woodbury Co. IA, *103877* **35** F2
Woodbury Hts. NJ, *2988* **146** C3
Wood Co. OH, *121065* **90** C2
Wood Co. TX, *36752* **124** A2
Wood Co. WV, *87986* **101** E2
Wood Co. WI, *75555* **68** A4
Woodcreek TX, *1274* **61** D2
Woodcrest CA, *8342* **229** J4
Wood Dale IL, *13535* **203** C3
Woodfield MD, *2500* **144** B2
Woodfin NC, *3162* **111** E4
Woodford Co. IL, *35469* **88** B3
Woodford Co. KY, *23208* **100** B4
Woodhaven MI, *12530* **90** C1
Woodhull IL, *809* **88** A2
Woodhull NY, *400* **93** D1
Woodinville NY, *600* **148** C1
Woodinville WA, *9194* **12** C3
Woodlake CA, *6651* **45** D3
Wood Lake MN, *436* **66** A4
Woodland CA, *49151* **36** B3
Woodland GA, *432* **128** C2
Woodland MI, *495* **76** A4
Woodland MN, *480* **235** A3
Woodland NC, *833* **113** E3
Woodland WA, *3780* **20** C1
Woodland (Baileyville) ME, *1044* .. **83** E1
Woodland Beach MI, *2179* **90** C1
Woodland Hills UT, *941* **39** E1
Woodland Park CO, *6515* **41** E2
Woodlawn IL, *630* **98** C4
Woodlawn KY, *268* **204** B3
Woodlawn MD, *1600* **145** D1
Woodlawn MD, *36079* **193** B2
Woodlawn OH, *2816* **204** B1
Woodlawn VA, *2249* **112** A3
Woodlawn Park KY, *1033* **230** E1
Woodlyn PA, *10036* **248** A4
Woodlynne NJ, *2796* **248** C4
Woodmere NY, *16447* **241** G5
Woodmere OH, *828* **204** G2
Woodmont CT, *1711* **149** D2
Woodmore MD, *6077* **144** C3
Woodport NJ, *2000* **148** A3
Woodridge IL, *30934* **203** B5
Wood-Ridge NJ, *7644* **240** C2
Woodridge NY, *902* **94** A3
Wood River IL, *11296* **98** A3
Wood River NE, *1204* **35** D4
Woodruff SC, *4229* **121** F2
Woodruff WI, *1300* **68** B2
Woodruff Co. AR, *8741* **117** F1
Woodsboro MD, *846* **144** A1
Woodsboro TX, *1685* **61** E4
Woodsburgh NY, *831* **241** G5
Woods Co. OK, *9089* **51** D1
Woods Cross UT, *6419* **257** B1
Woodsfield OH, *2598* **101** F1
Woods Hts. MO, *742* **96** C2
Woods Hole MA, *925* **151** E4
Woodside CA, *5352* **259** C5
Woodside DE, *184* **145** E2
Woodside MT, *175* **15** D4
Woods Landing WY, *100* **33** D3
Woodson AR, *445* **117** E3
Woodson IL, *559* **98** A1
Woodson Co. KS, *3788* **96** A4
Woodson Terrace MO, *4189* **256** B1
Woodstock CT, *425* **150** B2
Woodstock GA, *10050* **120** C3
Woodstock IL, *20151* **74** C4
Woodstock NY, *2187* **94** B2
Woodstock VT, *977* **81** E3
Woodstock VA, *3952* **102** C3
Woodstock Valley CT, *750* **150** B2
Woodstown NJ, *3136* **145** F1
Woodsville NH, *1081* **81** E2
Woodville AL, *761* **120** A2
Woodville CA, *1678* **45** D3
Woodville FL, *3006* **137** E2
Woodville GA, *400* **121** E4
Woodville MA, *500* **150** C2
Woodville MS, *1192* **134** A1
Woodville OH, *1977* **90** C2
Woodville TX, *2415* **132** C1
Woodville WI, *1104* **67** E4
Woodward IA, *1200* **86** C1
Woodward OK, *11853* **50** C1
Woodward Co. OK, *18486* **51** D2
Woodway TX, *8733* **59** E4
Woodway WA, *936* **262** A2
Woodworth LA, *1080* **133** E1
Woody Creek CO, *275* **40** C2
Woonsocket RI, *43224* **150** C2
Woonsocket SD, *720* **27** E3
Wooster AR, *516* **117** E1
Wooster OH, *24811* **91** E3
Worcester MA, *172648* **150** C1
Worcester NY, *900* **79** F4
Worcester VT, *225* **81** E2
Worcester Co. MD, *46543* **114** C2
Worcester Co. MA, *750963* **150** B1
Worden IL, *905* **98** B3
Worden MT, *506* **24** C1
Workmans Corners DE, *400* **145** F4
Worland WY, *5250* **24** C4
Worley ID, *223* **14** B3
Wormleysburg PA, *2607* **218** B2
Worth IL, *11047* **203** D6
Wortham TX, *1082* **59** F3
Worth Co. GA, *21967* **129** D4
Worth Co. IA, *7909* **73** D2
Worth Co. MO, *2382* **86** B4
Worthing SD, *585* **27** F4
Worthington IN, *1481* **99** E2
Worthington IA, *381* **73** F4
Worthington KY, *1673* **101** D3
Worthington MN, *11283* **72** A2
Worthington OH, *14125* **90** C4
Worthington PA, *778* **92** A3
Wounded Knee SD, *328* **34** A1
Wrangell AK, *2308* **155** E4
Wray CO, *2187* **42** A1
Wrens GA, *2314* **129** F1
Wrentham MA, *2600* **151** D2
Wright FL, *21697* **136** B2
Wright WY, *1347* **25** E4
Wright City MO, *1532* **97** F3
Wright City OK, *848* **116** B3
Wright Co. IA, *14334* **72** C4
Wright Co. MN, *89986* **66** C3
Wright Co. MO, *17955* **107** E2
Wrightsboro NC, *4496* **123** E3
Wrightstown NJ, *748* **147** D3
Wrightstown PA, *650* **147** D2
Wrightstown WI, *1934* **74** C1
Wrightsville AR, *1368* **117** E2
Wrightsville GA, *2223* **129** E2
Wrightsville PA, *2223* **103** E1
Wrightsville Beach NC, *2593* **123** E3
Wrightwood CA, *3837* **52** C2
Wrigley TN, *475* **109** E4
Wurtland KY, *1049* **101** D3
Wurtsboro NY, *1234* **148** A1
Wyalusing PA, *564* **93** E2
Wyandot Co. OH, *22908* **90** C3
Wyandotte MI, *28006* **90** C1
Wyandotte OK, *363* **106** B2
Wyandotte Co. KS, *157882* **96** B2
Wyanet IL, *1028* **88** B2
Wyckoff NJ, *16508* **148** B3

Auburn C2
Bramanville D2
Chaffinville C1
Cherry Valley C2
Dorothy Pond D2
E. Millbury D2
Edgemere D2
Leicester C2
Millbury D2
Morningdale D1
Pondville C2
Rochdale C2
Stoneville C2
Worcester C1

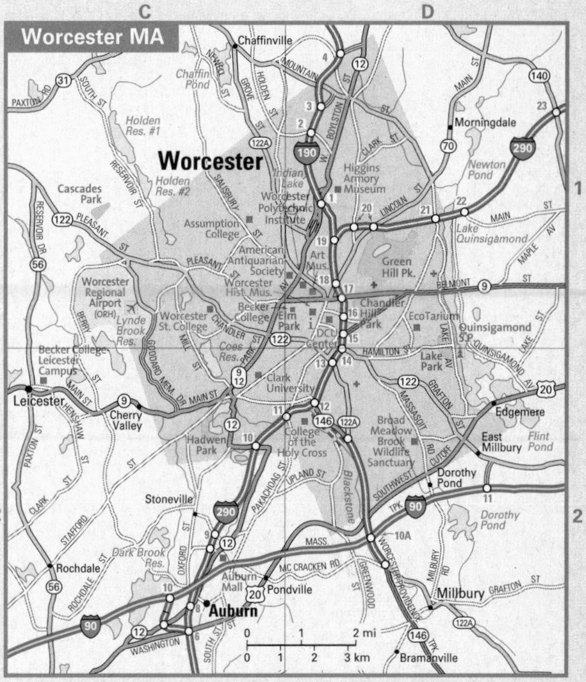

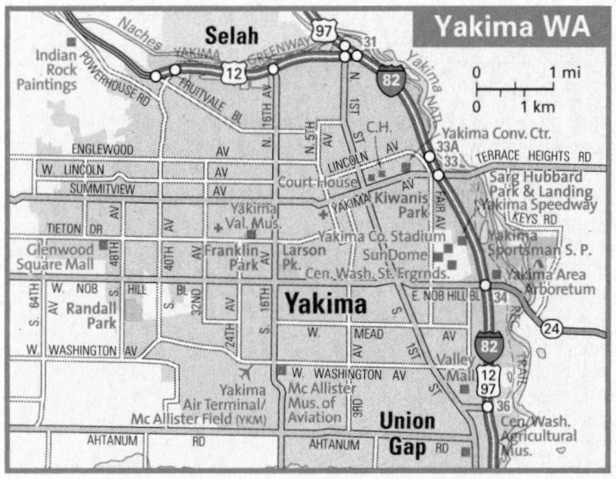

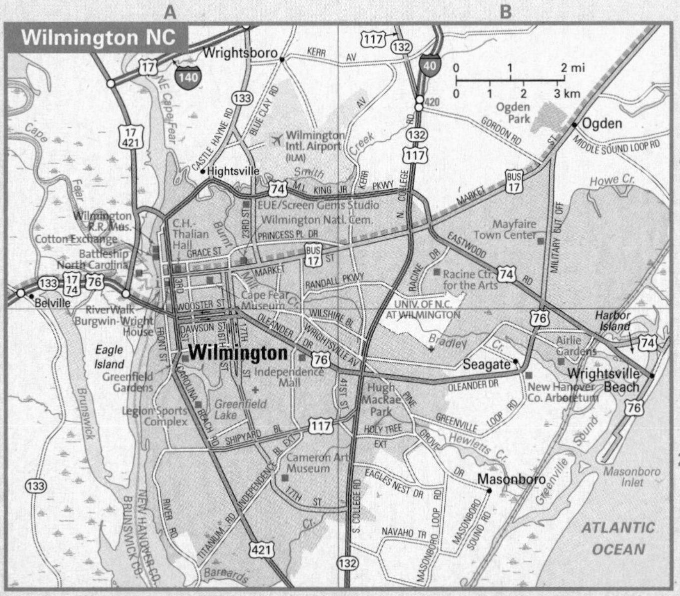

Belville A1
Hightsville A1
Masonboro B2
Ogden B1
Seagate B2
Wilmington A2
Wrightsboro A1
Wrightsville Beach B2

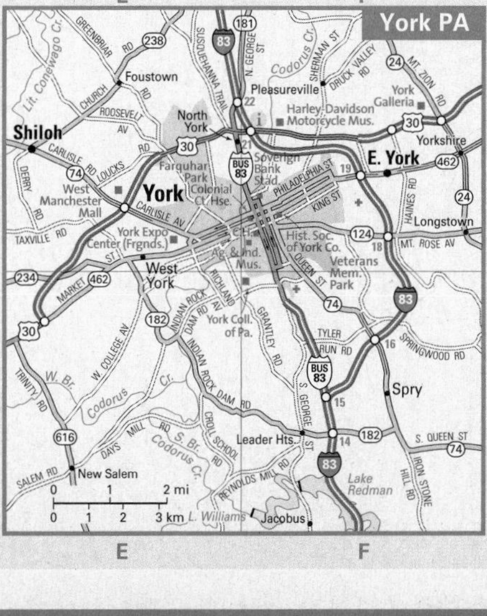

E. York F1
Foustown E1
Jacobus F2
Leader Hts. F2
Longstown F1
New Salem E2
N. York E1
Pleasureville F1
Shiloh E1
Spry F2
W. York E1
York E1
Yorkshire F1

Figures after entries indicate population, page number, and grid reference.

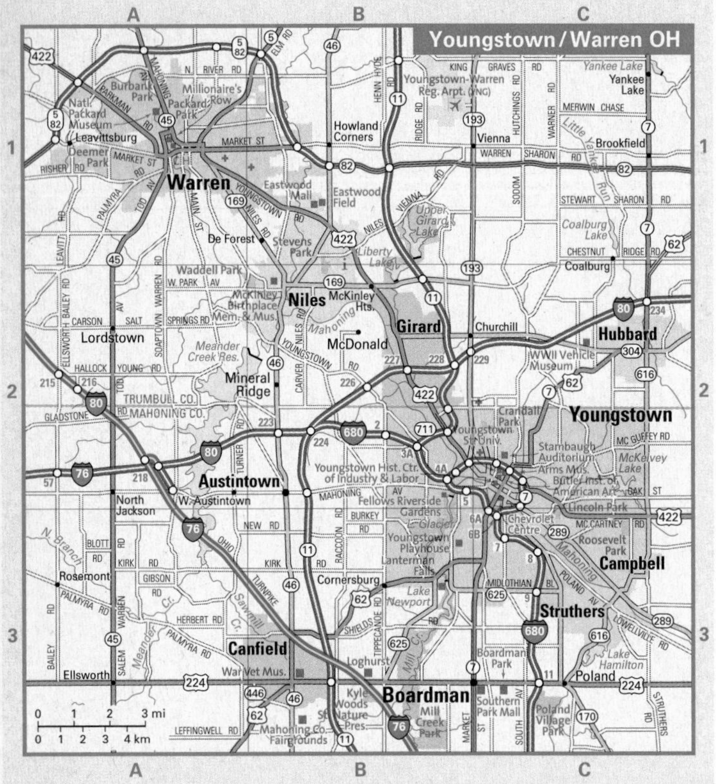

Austintown . . . B2
Boardman . . . B3
Brookfield . . . C1
Campbell . . . C2
Canfield . . . B3
Churchill . . . C2
Coalburg . . . C1
Cornersburg . . . B3
De Forest . . . B1
Ellsworth . . . A3
Girard . . . B2
Howland Corners . . . B1
Hubbard . . . C2
Leavittsburg . . . A1
Lordstown . . . A2
McDonald . . . B2
McKinley Hts. . . . B2
Mineral Ridge . . . B2
Niles . . . B2
N. Jackson . . . A2
Poland . . . C3
Rosemont . . . A3
Struthers . . . C3
Vienna . . . C1
Warren . . . A1
W. Austintown . . . A2
Yankee Lake . . . C1
Youngstown . . . C2

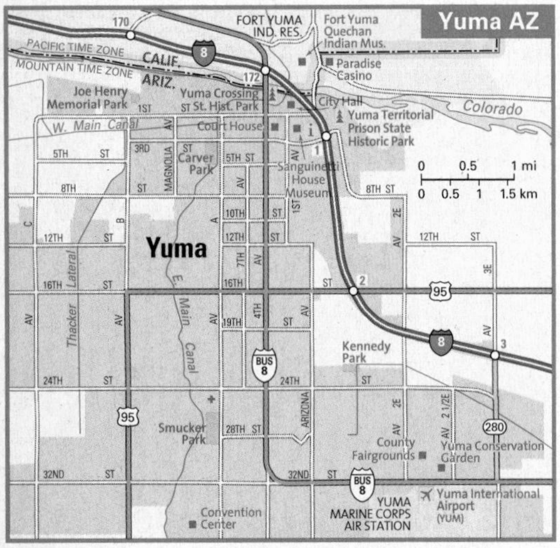

Wycombe PA, *650*...... **146** C2
Wykoff MN, *460*...... **73** E2
Wylie TX, *15132*...... **59** F2
Wymore NE, *1656*...... **43** F1
Wynantskill NY, *3018*...... **188** E2
Wyncote PA, *3046*...... **248** C1
Wyndmere ND, *533*...... **27** F1
Wynne AR, *8615*...... **118** A1
Wynnewood OK, *2367*...... **51** E4
Wynona OK, *531*...... **51** F1
Wyocena WI, *668*...... **74** B2
Wyodak WY, *125*...... **25** E3
Wyola MT, *186*...... **24** C2
Wyoming DE, *1141*...... **145** E2
Wyoming IL, *1424*...... **88** B3
Wyoming IA, *626*...... **87** F1
Wyoming MI, *69368*...... **75** F3
Wyoming MN, *3048*...... **67** D3
Wyoming NY, *513*...... **78** B3
Wyoming OH, *8261*...... **204** B1
Wyoming PA, *3221*...... **261** B1
Wyoming RI, *475*...... **150** C4
Wyoming Co. NY, *43424*...... **78** B4
Wyoming Co. PA, *28080*...... **93** E2
Wyoming Co. WV, *25708*...... **111** F1
Wyomissing PA, *11155*...... **146** A2
Wythe Co. VA, *27599*...... **112** A2
Wytheville VA, *7804*...... **112** A2

X

Xenia IL, *407*...... **98** C3
Xenia OH, *24164*...... **100** C1

Y

Yachats OR, *617*...... **20** B3
Yacolt WA, *1055*...... **20** C1
Yadkin Co. NC, *36348*...... **112** A3
Yadkinville NC, *2818*...... **112** A4
Yah-Tah-Hey NM, *580*...... **48** A2
Yakima WA, *71845*...... 13 D4
Yakima Co. WA, *222581*...... **13** D4
Yakutat AK, *680*...... **155** D3
Yalaha FL, *1175*...... **140** C1
Yale MI, *2063*...... **76** C3
Yale OK, *1342*...... **51** F2
Yalesville CT, *3600*...... **149** D1
Yalobusha Co. MS, *13051*...... **118** B3
Yamhill OR, *794*...... **20** B2
Yamhill Co. OR, *84992*...... **20** B2
Yampa CO, *443*...... **40** C1
Yancey Co. NC, *17774*...... **111** E4
Yanceyville NC, *2091*...... **112** C3
Yankeetown FL, *629*...... **138** C4
Yankton SD, *13528*...... **35** E1
Yankton Co. SD, *21652*...... **35** E1
Yaphank NY, *5025*...... **149** D3
Yardley PA, *2498*...... **147** D2
Yardville NJ, *9208*...... **147** D2
Yarmouth ME, *3560*...... **82** B3
Yarmouth MA, *2100*...... **151** F3
Yarmouth Port MA, *5395*...... **151** F3
Yarnell AZ, *645*...... **47** D4
Yarrow Pt. WA, *1008*...... **262** B3
Yates Ctr. KS, *1599*...... **96** A4
Yates City IL, *725*...... **88** A3
Yates Co. NY, *24621*...... **78** C4
Yatesville GA, *408*...... **129** D1
Yatesville PA, *649*...... **261** C2
Yavapai Co. AZ, *167517*...... **47** D4
Yazoo City MS, *14550*...... **126** B2
Yazoo Co. MS, *28149*...... **126** B2
Yeadon PA, *11762*...... **146** C3
Yeagertown PA, *1035*...... **93** D3
Yell Co. AR, *21139*...... **117** D2
Yellow House PA, 475...... **146** B2
Yellow Medicine Co. MN, *11080*. **66** A4
Yellow Sprs. MD, *1100*...... **144** A1
Yellow Sprs. OH, *3761*...... **100** C1
Yellowstone Co. MT, *129352*...... **24** C1
Yellville AR, *1312*...... **107** E3
Yelm WA, *3289*...... **12** C4
Yemassee SC, *807*...... **130** C2
Yerington NV, *2883*...... **37** E2
Yerkes KY, *500*...... **111** D2
Yermo CA, *900*...... **53** D1
Yoakum TX, *5731*...... **61** E3
Yoakum Co. TX, *7322*...... **57** F2
Yoder WY, *169*...... **33** F2
Yoe PA, *1022*...... **103** E1
Yolo CA, *550*...... **36** B2
Yolo Co. CA, *168660*...... **36** B2
Yoncalla OR, *1052*...... **20** B4
Yonkers NY, *196086*...... **148** B3
Yorba Linda CA, *58918*...... **229** F3
York AL, *2854*...... **127** D2
York NE, *8081*...... **35** E4
York NY, *450*...... **78** C3
York PA, *40862*...... 103 E1
York SC, *6985*...... **122** A2
York Beach ME, *1400*...... **82** B4
York Co. ME, *186742*...... **82** B4
York Co. NE, *14598*...... **35** E4
York Co. PA, *381751*...... **103** E1
York Co. SC, *164614*...... **122** A2
York Co. VA, *56297*...... **113** F2
York Harbor ME, *3321*...... **82** B4
York Haven PA, *809*...... **93** E4
Yorkshire NY, *1403*...... **78** B4
Yorkshire VA, *6732*...... **144** A3
York Sprs. PA, *574*...... **103** E1
Yorktown IN, *4785*...... **89** F4
Yorktown NY, *14891*...... **148** B2
Yorktown TX, *2271*...... **61** E3
Yorktown VA, *203*...... **113** F2
Yorktown Hts. NY, *7972*...... **148** B2
York Vil. ME, *2000*...... **82** B4
Yorkville IL, *6189*...... **88** C2
Yorkville OH, *1230*...... **91** F4
Young AZ, *561*...... **47** E4
Young Co. TX, *17943*...... **59** D2
Young Harris GA, *604*...... **121** D2
Youngstown NY, *1957*...... **78** A3
Youngstown OH, *82026*...... 91 F3
Youngstown PA, *400*...... **92** B4
Youngsville LA, *3992*...... **133** F2
Youngsville NC, *651*...... **113** D4
Youngsville PA, *1834*...... **92** B1
Youngtown AZ, *3010*...... **249** A1
Youngwood PA, *4138*...... **92** A4
Yountville CA, *2916*...... **36** B3
Ypsilanti MI, *22362*...... **90** C1
Yreka CA, *7290*...... **28** C2
Yuba City CA, *36758*...... **36** C2
Yuba Co. CA, *60219*...... **36** C2
Yucaipa CA, *41207*...... **53** D2
Yucca Valley CA, *16865*...... **53** E2
Yukon OK, *21043*...... **51** E3
Yulee FL, *8392*...... **139** D2
Yuma AZ, *77515*...... 53 F4
Yuma CO, *3285*...... **42** A1
Yuma Co. AZ, *160026*...... **54** A2
Yuma Co. CO, *9841*...... **34** A4
Yutan NE, *1216*...... **35** F3

Z

Zacata VA, *450*...... **103** E4
Zachary LA, *11275*...... **134** A1
Zanesville IN, *602*...... **90** A3
Zanesville OH, *25586*...... **101** E1
Zap ND, *231*...... **18** B3
Zapata TX, *4856*...... **63** D3
Zapata Co. TX, *12182*...... **63** D3
Zavala Co. TX, *11600*...... **60** C3
Zavalla TX, *647*...... **132** C1
Zearing IA, *617*...... **87** D1
Zeb OK, *498*...... **106** B4
Zebulon GA, *1181*...... **128** C1
Zebulon KY, *700*...... **111** E1
Zebulon NC, *4046*...... **113** D4
Zeeland MI, *5805*...... **75** F3
Zeigler IL, *1669*...... **98** C4
Zelienople PA, *4123*...... **92** A3
Zephyr Cove NV, *1649*...... **37** D2
Zephyrhills FL, *10833*...... **140** C2
Zia Pueblo NM, *646*...... **48** C3
Ziebach Co. SD, *2519*...... **26** B2
Zillah WA, *2198*...... **13** E4
Zimmerman MN, *2851*...... **66** C3
Zion IL, *22866*...... **75** D4
Zion KY, *550*...... **109** E1
Zion PA, *2054*...... **92** C3
Zion Crossroads VA, 375...... **102** C4
Zionsville IN, *8775*...... **99** F1
Zolfo Sprs. FL, *1641*...... **140** C3
Zumbrota MN, *2789*...... **73** D1
Zuni Pueblo NM, *6367*...... **48** A3
Zwolle LA, *1783*...... **125** D4

PUERTO RICO

Aceitunas PR, *1688*...... **187** D1
Adjuntas PR, *4980*...... **187** D1
Aguada PR, *3871*...... **187** D1
Aguadilla PR, *16776*...... **187** D1
Aguas Buenas PR, *4368*...... **187** E1
Aguilita PR, *4922*...... **187** E1
Aibonito PR, *9269*...... **187** E1
Añasco PR, *5880*...... **187** D1
Arecibo PR, *9318*...... **187** E1
Arroyo PR, *7244*...... **187** E1
Bajadero PR, *3877*...... **187** E1
Barceloneta PR, *4253*...... **187** E1
Barranquitas PR, *2910*...... **187** E1
Bayamón PR, *203499*...... **187** E1
Betances PR, *835*...... **187** D1
Boquerón PR, *1218*...... **187** D1
Cabo Rojo PR, *10610*...... **187** D1
Caguas PR, *88680*...... **187** E1
Camuy PR, *4013*...... **187** D1
Canóvanas PR, *8069*...... **187** F1
Carolina PR, *168164*...... **187** F1
Cataño PR, *30071*...... **187** E1
Cayey PR, *19940*...... **187** E1
Cayuco PR, *1284*...... **187** D1
Ceiba PR, *6277*...... **187** E1
Ceiba PR, *3698*...... **187** E1
Ciales PR, *3082*...... **187** E1
Cidra PR, *4881*...... **187** E1
Coamo PR, *12356*...... **187** E1
Coco PR, *5803*...... **187** E1
Comerío PR, *4478*...... **187** E1
Comunas PR, *2027*...... **187** F1
Coquí PR, *3590*...... **187** E1
Corazón PR, *2925*...... **187** E1
Corozal PR, *11444*...... **187** E1
Coto Norte PR, *1381*...... **187** E1
Daguao PR, *1488*...... **187** F1
Dorado PR, *12747*...... **187** E1
Duque PR, *1529*...... **187** F1
El Mangó PR, *1979*...... **187** F1
Esperanza PR, *1092*...... **187** F1
Fajardo PR, *33286*...... **187** F1
Florida PR, *5652*...... **187** E1
Guánica PR, *9247*...... **187** D1
Guayabal PR, *2377*...... **187** E1
Guayama PR, *21624*...... **187** E1
Guayanilla PR, *5110*...... **187** D1
Guaynabo PR, *78806*...... **187** E1
Gurabo PR, *9046*...... **187** F1
Hatillo PR, *5321*...... **187** D1
Hormigueros PR, *12444*...... **187** D1
Humacao PR, *20682*...... **187** F1
Isabela PR, *12818*...... **187** D1
Jagual PR, *1402*...... **187** E1
Jayuya PR, *3516*...... **187** E1
Jobos PR, *3475*...... **187** E1
Juana Diaz PR, *9505*...... **187** E1
Juncos PR, *8978*...... **187** F1
Lajas PR, *5036*...... **187** D1
La Parguera PR, *1141*...... **187** D1
La Plena PR, *1036*...... **187** E1
Lares PR, *7042*...... **187** D1
Las Marías PR, *1823*...... **187** D1
Las Marías PR, *988*...... **187** D1
Las Piedras PR, *6352*...... **187** F1
Levittown PR, *30140*...... **187** E1
Loíza PR, *4123*...... **187** F1
Los Llanos PR, *2301*...... **187** E1
Luquillo PR, *7947*...... **187** F1
Manatí PR, *16173*...... **187** E1
Maricao PR, *1123*...... **187** D1
Maunabo PR, *2075*...... **187** F1
Mayagüez PR, *78647*...... **187** D1
Moca PR, *4757*...... **187** D1
Mora PR, *1857*...... **187** D1
Morovis PR, *2285*...... **187** E1
Naguabo PR, *4432*...... **187** F1
Naranjito PR, *1931*...... **187** E1
Orocovis PR, *909*...... **187** E1
Palmarejo PR, *1087*...... **187** D1
Palomas PR, *1742*...... **187** D1
Patillas PR, *4091*...... **187** E1
Peñuelas PR, *6712*...... **187** D1
Playita PR, *2192*...... **187** F1
Pole Ojea PR, *1829*...... **187** D1
Ponce PR, *155038*...... **187** E1
Potala Pastillo PR, *3819*...... **187** E1
Puerto Real PR, *6166*...... **187** D1
Punta Santiago PR, *5803*...... **187** F1
Quebrada PR, *1130*...... **187** D1
Quebradillas PR, *5319*...... **187** D1
Rafael Capó PR, *1863*...... **187** D1
Rincón PR, *1436*...... **187** D1
Río Grande PR, *13467*...... **187** F1
Sabana Eneas PR, *1847*...... **187** D1
Sabana Grande PR, *8784*...... **187** D1
Sabana Hoyos PR, *1823*...... **187** E1
Salinas PR, *6141*...... **187** E1
San Antonio PR, *6456*...... **187** D1
San Antonio PR, *2300*...... **187** D1
San Germán PR, *12033*...... **187** D1
San Isidro PR, *8071*...... **187** F1
San Juan PR, *421958*...... 187 E1
San Lorenzo PR, *8947*...... **187** F1
San Sebastián PR, *11598*...... **187** D1
Santa Isabel PR, *6993*...... **187** E1
Santo Domingo PR, *3633*...... **187** D1
Tallaboa PR, *1150*...... **187** D1
Toa Alta PR, *4368*...... **187** E1
Trujillo Alto PR, *50841*...... **187** E1
Utuado PR, *9887*...... **187** E1
Vázquez PR, *2297*...... **187** E1
Vega Alta PR, *11755*...... **187** E1
Vega Baja PR, *28811*...... **187** E1
Vieques PR, *4325*...... **187** F1
Villalba PR, *4388*...... **187** E1
Yabucoa PR, *6636*...... **187** F1
Yauco PR, *19609*...... **187** D1
Yaurel PR, *1468*...... **187** E1

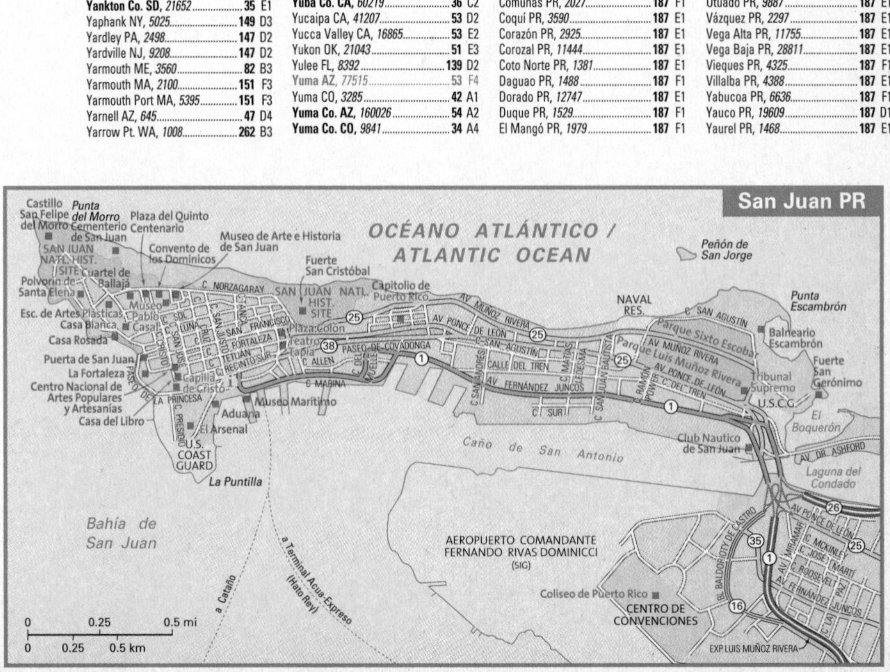

Entries in **bold color** indicate cities with detailed inset maps.

CANADA

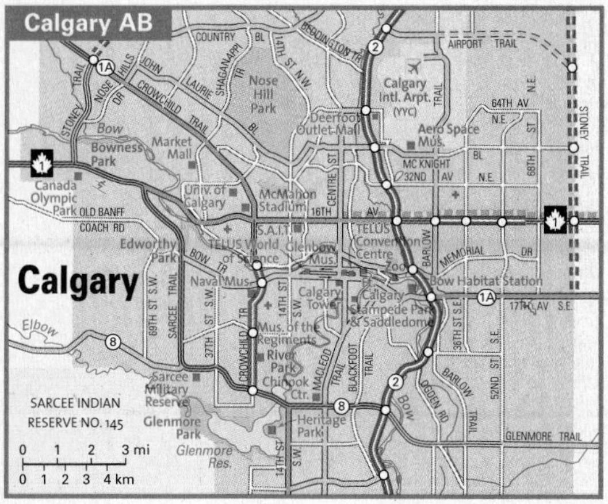

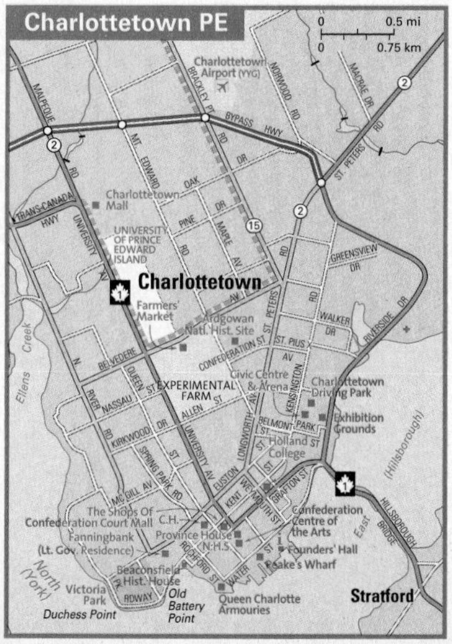

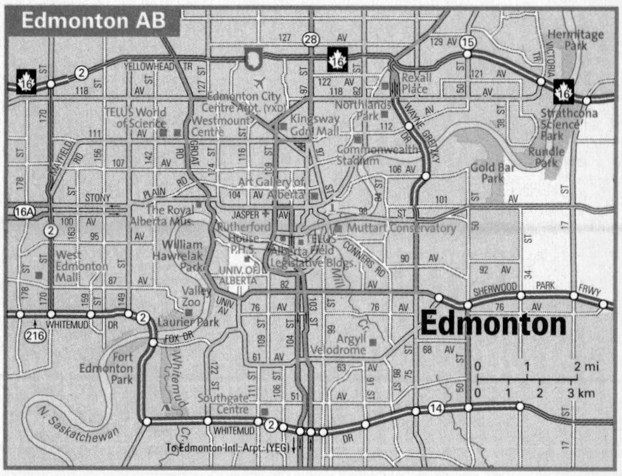

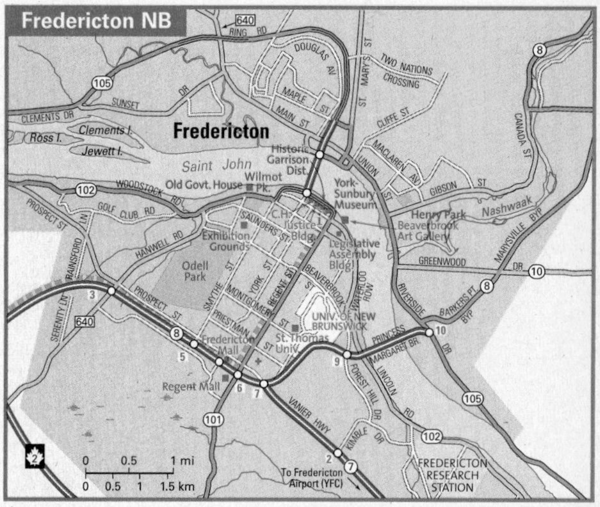

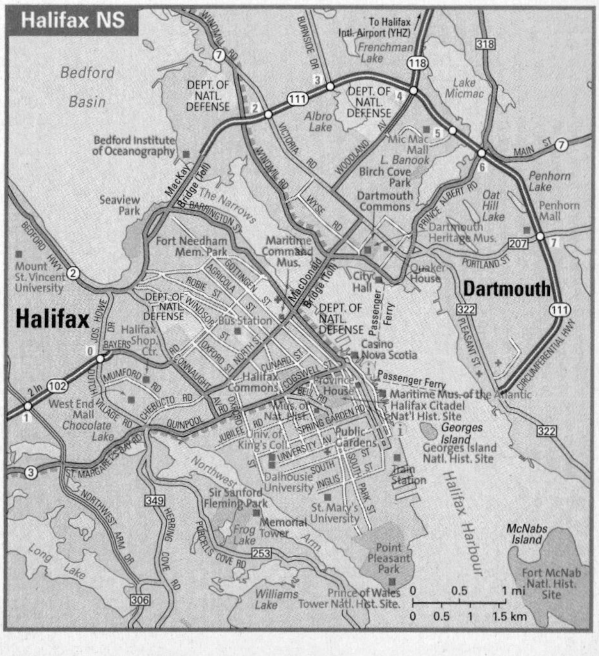

Figures after entries indicate population, page number, and grid reference.

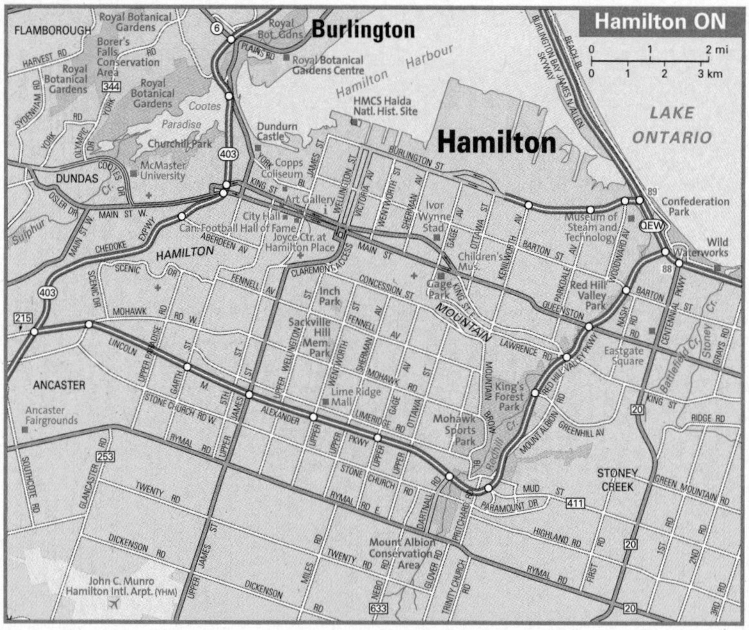

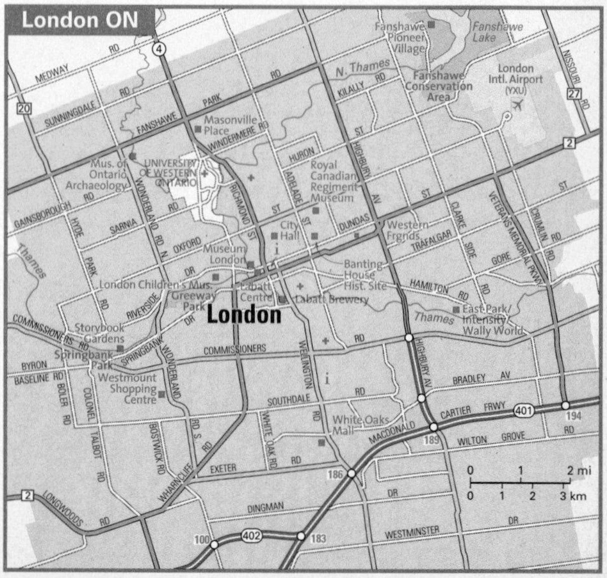

Entries in **bold color** indicate cities with detailed inset maps.

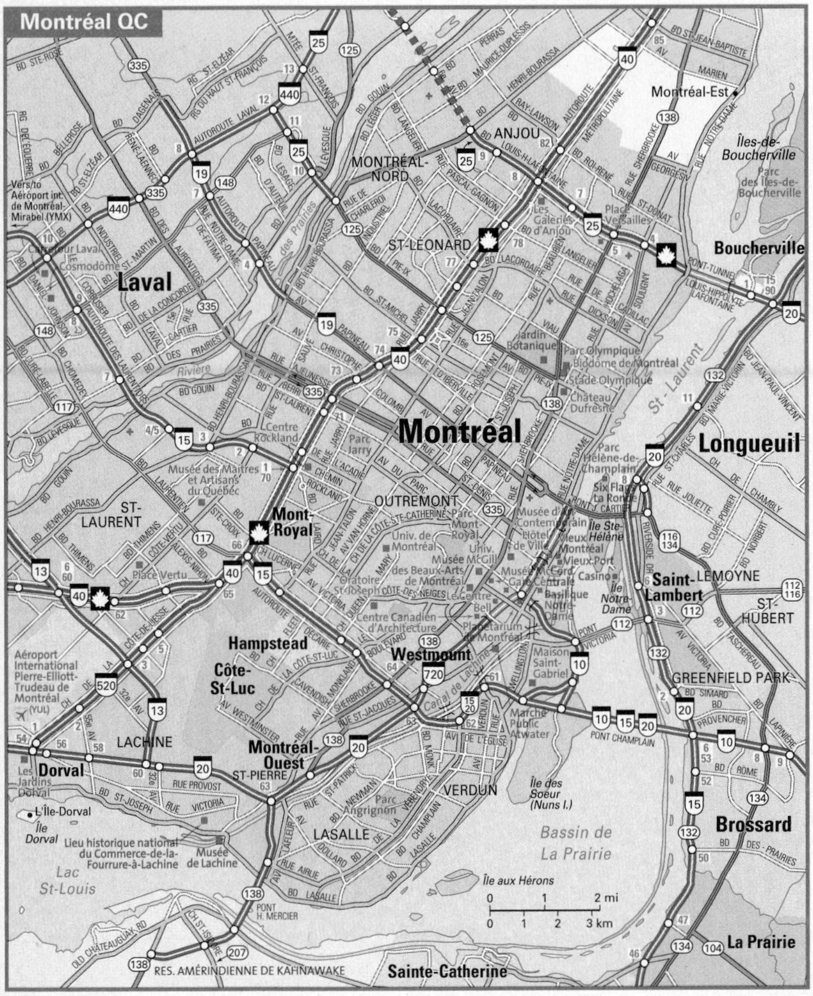

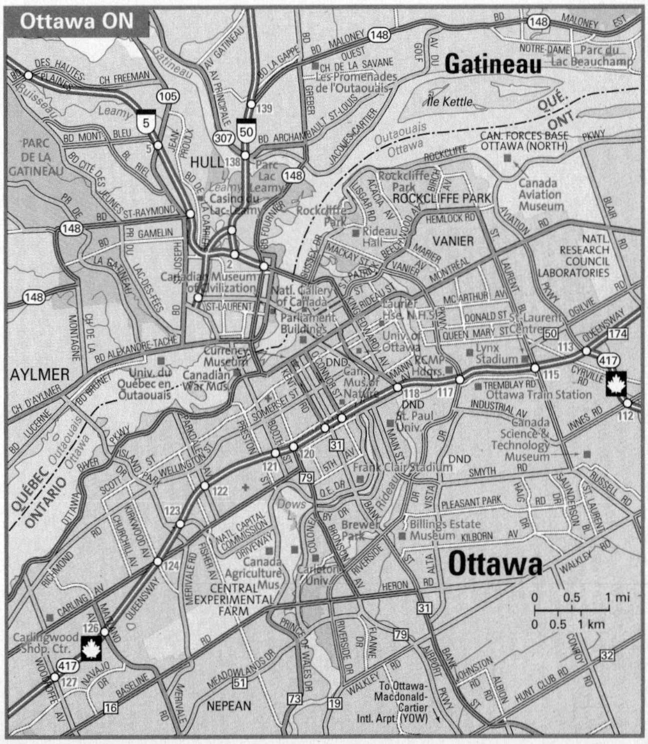

Figures after entries indicate population, page number, and grid reference.

Québec QC

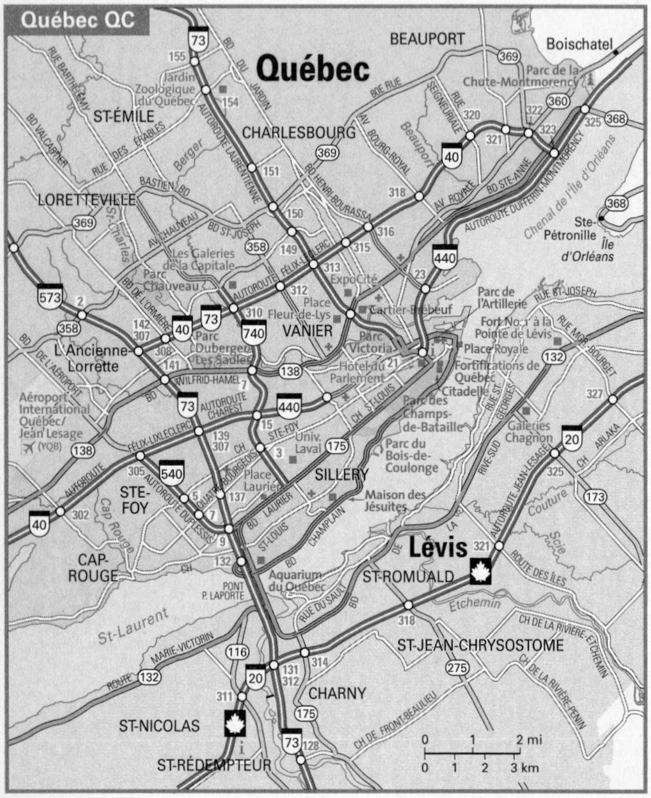

St John's NL

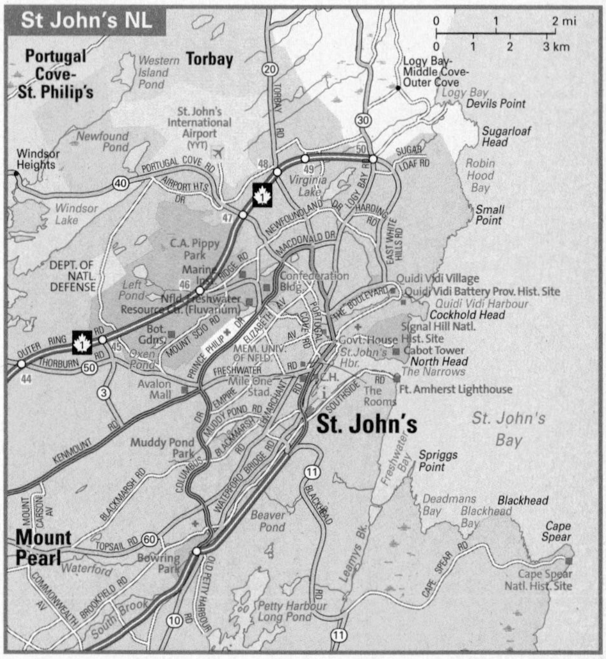

Regina SK

Saint John NB

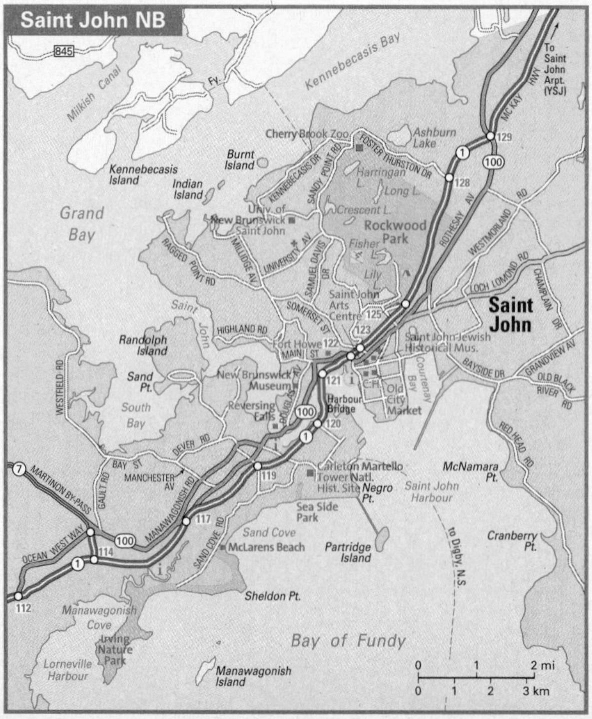

Saskatoon SK

Entries in **bold color** indicate cities with detailed inset maps.

POINTS OF INTEREST
Air Canada Centre....B2
Art Gallery of Ontario....A1
CBC Broadcast Center....A2
CN Tower....A2
Eaton Centre....B1
Four Seasons Centre for the Performing Arts....A1
The Grange....A1
Harbourfront Centre....A2
Hockey Hall of Fame....B2
Hummingbird Centre....B2
MacKenzie House....B1
Metro Convention Center....A2
Old City Hall....B1
Princess of Wales Theatre....A1
Queen's Quay Terminal....A2
Redpath Sugar Museum....B2
Rogers Centre....A2
Royal Alexandra Theatre....A1
Royerson Polytechnic University....B1
Roy Thomson Hall....A2
Saint Lawrence Centre....B2
Saint Lawrence Market....B2
Textile Museum of Canada....A1
Toronto Island Ferry Terminal....B2
Toronto Stock Exchange....A1

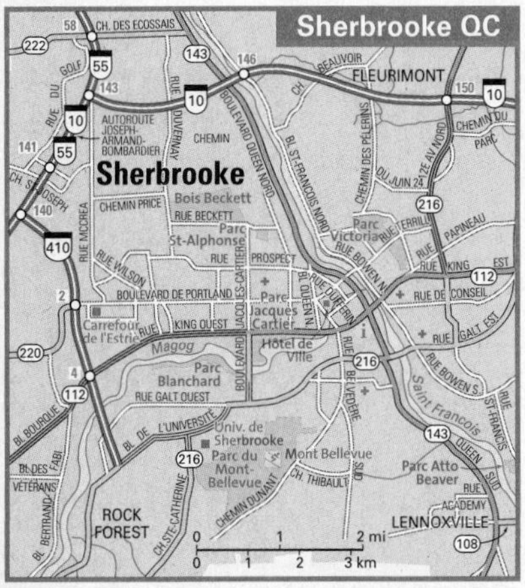

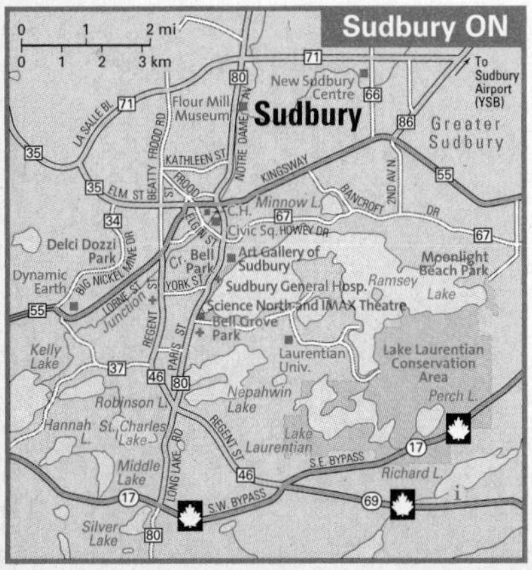

Figures after entries indicate population, page number, and grid reference.

St-Stanislas QC, *1076* **175** D2
St. Stephen NB, *4667* **180** A2
St-Théodore-d'Acton QC, *1544* **175** D3
St. Thomas ON, *33236* **172** B3
St-Timothée QC, *8299* **174** C3
St-Tite QC, *3845* **175** D1
St-Tite-des-Caps QC, *1426* **175** E1
St-Ubalde QC, *1460* **175** D1
St-Ulric QC, *1649* **178** B1
St-Urbain QC, *1430* **176** C4
St-Valère QC, *1308* **175** D2
St-Victor QC, *2460* **175** E2
St. Walburg SK, *672* **159** F4
St-Wenceslas QC, *1132* **175** D2
Scoudouc NB, *1047* **179** D4
Seaforth ON, *2692* **172** B2
Sechelt BC, *7775* **162** C3
Sedgewick AB, *865* **165** D1
Selkirk MB, *9752* **167** E3
Senneterre QC, *3275* **171** E2
Sept-Îles QC, *23791* **177** E2
Sexsmith AB, *1653* **157** F1
Shannon QC, *3668* **175** E1
Shaunavon SK, *1775* **165** F4
Shawinigan QC, *17535* **175** D2
Shawinigan-Sud QC, *11544* **175** D2
Shawville QC, *1582* **174** A3
Shediac NB, *4892* **179** D4
Smoky Lake AB, *1011* **159** D3
Smooth Rock Falls ON, *1830* **170** C1
Snow Lake MB, *1207* **161** D3
Somerset MB, *459* **167** E4
Sooke BC, *8735* **162** C4
Sorel-Tracy QC, *34194* **175** D2
Sorrento BC, *1197* **163** F1
Souris MB, *1683* **167** D4
Souris PE, *1248* **179** F4
Southampton ON, *3360* **172** B1
S. Bruce Peninsula ON, *8090* **172** B1
Southey SK, *693* **166** B3
S. Huron ON, *10019* **172** B3
S. Indian Lake MB, *808* **161** E1
Stewiacke NS, *1388* **181** D2
Stirling AB, *877* **165** D4
Stirling ON, *2149* **173** E1
Stoke QC, *2475* **175** E3
Stonewall MB, *4012* **167** E3
Stoney Pt. ON, *1316* **172** A4
Stony Mtn. MB, *1700* **167** E3
Stony Plain AB, *9589* **159** D4
Stouffville ON, *11073* **173** D2
Stoughton SK, *720* **166** C4
Strasbourg SK, *760* **166** B3
Stratford ON, *29676* **172** C2
Strathmore AB, *7621* **164** C3
Strathroy ON, *12805* **172** B3
Tofield AB, *1818* **159** D4
Tofino BC, *1466* **162** B3
Torbay NL, *5474* **183** F4
Toronto ON, *2481494* 173 D2
Tottenham ON, *4829* **173** D2
Tracadie-Sheila NB, *4724* **179** D3
Trail BC, *7575* **164** A4
Treherne MB, *644* **167** E4
Trent Hills ON, *12569* **173** E1
Trenton NS, *2798* **181** D2
Trenton ON **173** E2
Trepassey NL, *889* **183** E4
Tring-Jonction QC, *1333* **175** E2
Triton NL, *1102* **183** D2
Valemount BC, *1195* **157** F3
Val-Joli QC, *1532* **175** E3
Vallée-Jonction QC, *1882* **175** E2
Valleyview AB, *1856* **158** B3
Val-Morin QC, *2216* **174** C2
Val-Senneville QC, *2479* **171** E2
Vancouver BC, *545671* 163 D3
Vanderhoof BC, *4390* **157** D2
Vankleek Hill ON, *2022* **174** B3
Varennes QC, *19653* **174** C3
Vaudreuil-Dorion QC, *19920* **174** C3
Vauxhall AB, *1112* **165** D4
Vegreville AB, *5376* **159** D4
Venise-en-Québec QC, *1243* **175** D4
Wallaceburg ON, *11114* **172** A3
Warfield BC, *1739* **164** A4
Warman SK, *3481* **165** F2
Warwick QC, *4874* **175** E2
Wasaga Beach ON, *12419* **172** C1
Waterford ON, *2871* **172** C3
Waterloo ON, *86543* **172** C2
Waterloo QC, *3993* **175** D3
Waterville NS, *808* **180** C2
Waterville QC, *1824* **175** E3
Watford ON, *1625* **172** B3
Watrous SK, *1808* **166** B2
Watson SK, *794* **166** B2
Watson Lake YT, *912* **155** E3
Waverley NS, *934* **181** D3
Wawa ON, *3279* **169** F4
Wawanesa MB, *516* **167** D4
Wawota SK, *538* **166** C4
Wedgeport NS, *1217* **180** B4
Weedon QC, *2646* **175** E3
Welland ON, *48402* **173** D3
Wellesley ON, *1666* **172** C2
Wellington ON, *1943* **173** E2
Wellington PE, *382* **179** E4
Wembley AB, *1497* **157** F1
Wentworth-Nord QC, *1121* **174** C3
Westbank BC, *15700* **163** F2
Western Shore NS, *1015* **180** C3
Westlock AB, *4819* **159** D3
W. Lorne ON, *1458* **172** B4
W. Nipissing ON, *13114* **171** D3
W. Vancouver BC, *41421* **163** D3
Westville NS, *3879* **181** D2
Wetaskiwin AB, *11154* **159** D4
Weyburn SK, *9534* **166** B4
Wheatley ON, *1920* **172** A4
Whistler BC, *8896* **163** D2
Whitbourne NL, *930* **183** E4
Whitby ON, *87413* **173** D2
Whitchurch-Stouffville ON, *22008* **173** D2
White City SK, *1013* **166** B3
Whitecourt AB, *8334* **158** C3
Whitehorse YT, *19058* **155** D3
White Rock BC, *18250* **163** D3
Whitewood SK, *947* **166** C3
Wiarton ON, *2349* **172** B1
Wickham QC, *2516* **175** D3
Wikwemikong ON, *1352* **170** C4
Wilkie SK, *1282* **165** E2
Williams Lake BC, *11153* **157** E4
Winchester ON, *2427* **174** B4
Windermere BC, *1060* **164** B3
Windsor NS, *3778* **180** C2
Windsor ON, *208402* **172** A4
Windsor QC, *5321* **175** E3
Wingham ON, *2885* **172** B2
Winkler MB, *7943* **167** E4
Winnipeg MB, *619544* 167 E4
Winnipeg Beach MB, *801* **167** E3
Winnipegosis MB, *621* **167** D2
Witless Bay NL, *1056* **183** F4
Wolfville NS, *3658* **180** C2
Wolseley SK, *766* **166** C3

Anmore D1
Belcarra D1
Burnaby D2
Coquitlam D2
New Westminster D2
N. Vancouver B1
N. Vancouver (DM) C1
Port Moody D1
Richmond C2
Surrey D2
Vancouver B2
W. Vancouver A1

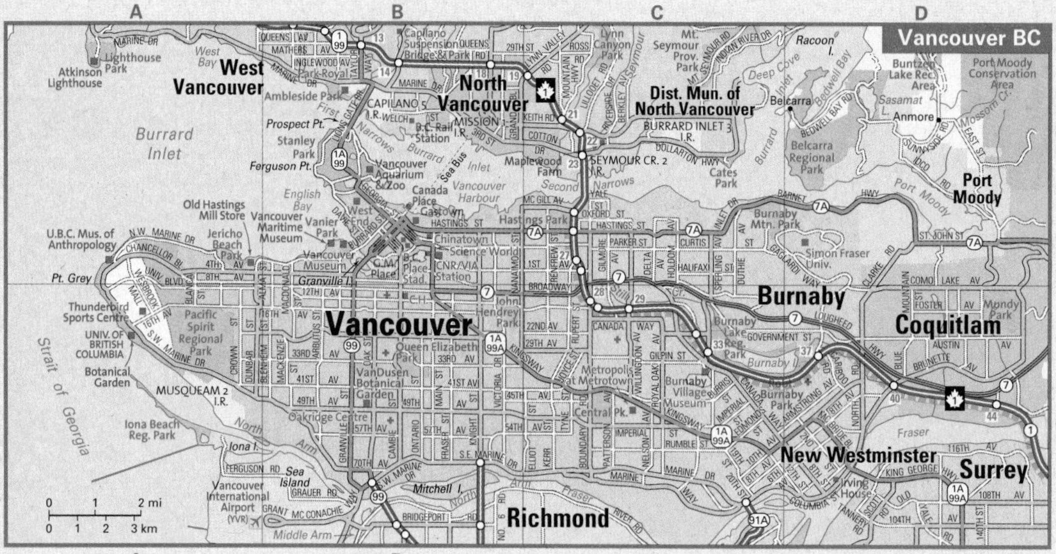

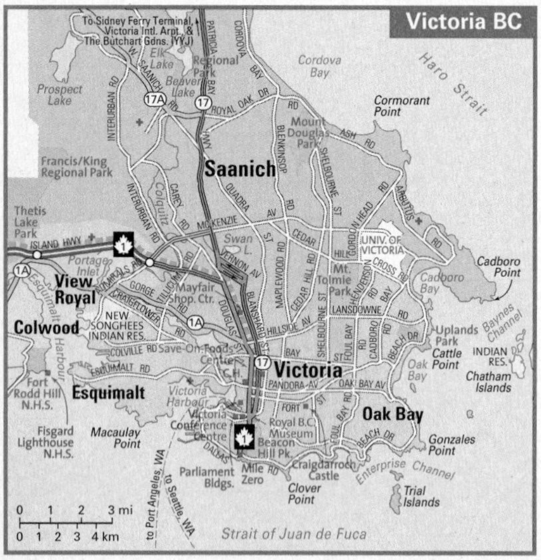

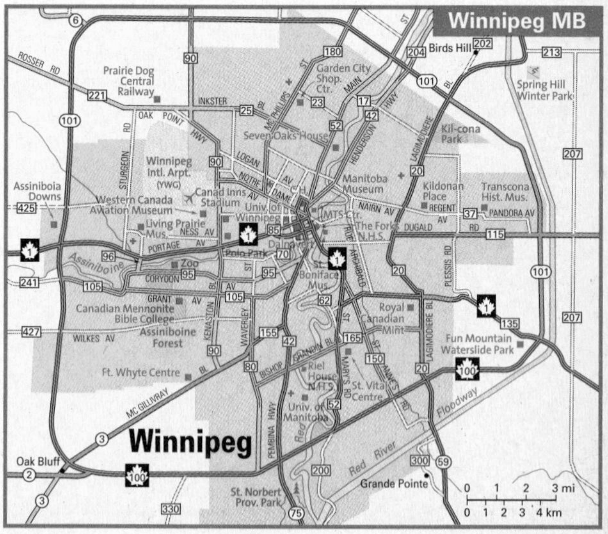

Salmon Arm BC, *15210* **163** F1
Salmon Cove NL, *746* **183** E4
Salmon River NS, *2259* **180** B4
Sandy Bay SK, *1092* **160** C2
Saratoga Beach BC, *1627* **162** B2
Sarnia ON, *70876* **172** A3
Saskatoon SK, *196811* 165 F2
Saugeen Shores ON, *11388* **172** B1
Sault Ste. Marie ON, *74566* **170** B3
Sayabec QC, *1999* **178** B1
Schomberg ON, *1216* **173** D2
Scott QC, *1705* **175** E2
Shippagan NB, *2872* **179** D2
Shoal Lake MB, *801* **167** D3
Shubenacadie NS, *906* **181** D2
Sicamous BC, *2720* **164** A3
Sidney BC, *10929* **163** D4
Silver Creek BC, *1062* **163** F1
Simcoe ON, *14175* **172** C3
Sioux Lookout ON, *5336* **168** C2
Slave Lake AB, *6600* **158** C3
Smithers BC, *5414* **156** C2
Smiths Falls ON, *9140* **174** A4
Smithville ON, *3317* **173** D3
Springhill NS, *4091* **180** C1
Springside SK, *525* **166** C2
Sproat Lake BC, *1888* **162** C3
Spruce Grove AB, *15983* **159** D4
Squamish BC, *14247* **163** D2
Stanstead QC, *2995* **175** D4
Stayner ON, *3885* **172** C1
Steinbach MB, *9227* **167** E4
Stellarton NS, *4809* **181** D2
Stephenville NL, *7109* **182** C3
Stephenville Crossing NL, *1993* **182** C3
Stettler AB, *5215* **165** D2
Tweed ON, *1539* **173** E1
Twillingate NL, *2611* **183** E2
Two Hills AB, *1091* **159** E4
Ucluelet BC, *1559* **162** B3
Union Bay BC, *1167* **162** C3
Unity SK, *2243* **165** E2
Upton QC, *1986* **175** D3
Uxbridge ON, *8540* **173** D2
Val-Comeau NB, *823* **179** D3
Valcourt QC, *2411* **175** D3
Val-David QC, *3819* **174** C2
Val-des-Monts QC, *7842* **174** B3
Val-d'Or QC, *22748* **171** E2
Viking AB, *1052* **159** E4
Ville-Marie QC, *2770* **171** D2
Virden MB, *3109* **167** D4
Vulcan AB, *1762* **164** C3
Wabana NL, *2679* **183** F4
Wabasca AB, *1114* **159** D2
Wabowden MB, *497* **161** E3
Wabush NL, *1894* **183** E1
Wadena SK, *1412* **166** B2
Wainwright AB, *5117* **159** E4
Wakaw SK, *884* **165** F1
Waldheim SK, *889* **165** F1
Walkerton ON, *4970* **172** B1
Wood Buffalo AB, *41466* **159** E1
Woodstock NB, *5198* **178** B4
Woodstock ON, *33061* **172** C3
Woodville ON, *871* **173** D1
Wotton QC, *1568* **175** E3
Wright QC, *1137* **174** A2
Wynyard SK, *1919* **166** B2
Wyoming ON, *2200* **172** B3
Yamachiche QC, *2631* **175** D2
Yarmouth NS, *7561* **180** B4
Yellowknife NT, *16541* **155** F2
Yorkton SK, *15107* **166** C3
Youbou BC, *727* **162** C3

Entries in **bold color** indicate cities with detailed inset maps.

Cancún MX

Isla Mujeres
Laguna Chakmochuk
El Meco
Ruinas El Cerrito
Aeropista
Isla Mujeres
Punta Sam
Punta Sam
Bahía Garrafón
Playa Garrafón
Punta Sur
Ruinas El Meco
Puerto Juárez
Palacio de Gobierno
Cancún
Centro de Convenciones
Laguna Nichupté
ZONA HOTELERA
Alfredo V. Bonfil
Punta Cancún
Ruinas El Rey
Playa San Miguelito
Aeropuerto Internacional de Cancún (CUN)
MAR CARIBE / CARIBBEAN SEA
Parque Nizuc
Punta Nizuc

0 2 4 mi
0 2 4 6 km

México MX

C — D — E

Santo Tomás Chiconautla
Acolman
Xometla
Cuautitlán Izcalli
Tultitlán
Coacalco
Santa Catarina
San Pedro Tepetitlán
Nicolás Romero
Guadalupe
Peaje
Tepexpan
Buenavista
Ecatepec de Morelos
Tequisistlán
Tezoyuca
Ciudad López Mateos
Santa Clara
Nexquipayac
Chiconcuac
Tlalnepantla
San Salvador Atenco
Chiautla
Texcoco
Naucalpan
México
San Bernardino
Montecillo
Santiago Cuautlalpan
Magdalena Chichicaspa
Chimalhuacán
San Vicente Chicoloapan
Netzahualcóyotl
Cuajimalpa
Los Reyes
San Lorenzo Acopilco
Ixtapaluca
Xochimilco
Tláhuac
Xico
Chalco

0 2 4 mi
0 2 4 6 km

Guadalajara MX

Nuevo México
Mascuala
Base Aérea Militar Emilio Carranza
Parque El Centinela
Barranca de Oblatos
Museo Trompo Mágico
Zapopan
Estadio Jalisco
Parque Huentitán y Zoológico
Planetario
El Aguacate
Colimilla
Parque de los Colomos
Fuente de Minerva
Guadalajara
Instituto Cultural Cabañas
SAN GASPAR
Bosque La Primavera
Plaza del Sol
Tlaquepaque
Coyula
Matatlán
Cerro de la Reina
Casa del Rey
Tonalá
El Vado
San Pedrito
Santa Ana Tepetitlán
Santa María Tequepexpan
San Sebastianito
La Tijera
Los Gavilanes
La Calerilla
San Martín de las Flores
Santa Paula
El Verde
San Agustín
Santa Anita
Santa Cruz del Valle
Las Pintitas
San Francisco de la Soledad
La Punta
San Antonio Juanacaxtle
El Quince
Aeropuerto Internacional Miguel Hidalgo (GDL)
San Sebastián El Grande
Puente Grande

0 2 4 mi
0 2 4 6 km

Monterrey MX

General Escobedo
Apodaca
Universidad Autónoma de Nuevo León
San Nicolás de los Garza
Monterrey
Cerro del Topo
Cerro de las Mitras
Parque Niños Héroes
Guadalupe
Garza García
Parque Nacional Cumbres de Monterrey
Ciudad Benito Juárez
Aeropuerto Internacional General Mariano Escobedo (MTY)
Cerro de la Silla

0 2 4 mi
0 2 4 6 km

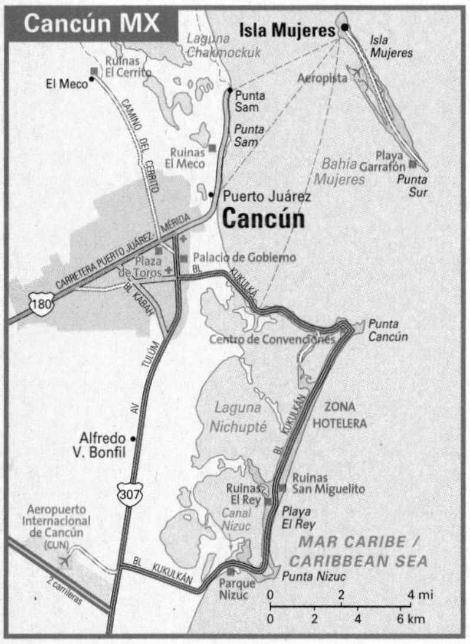

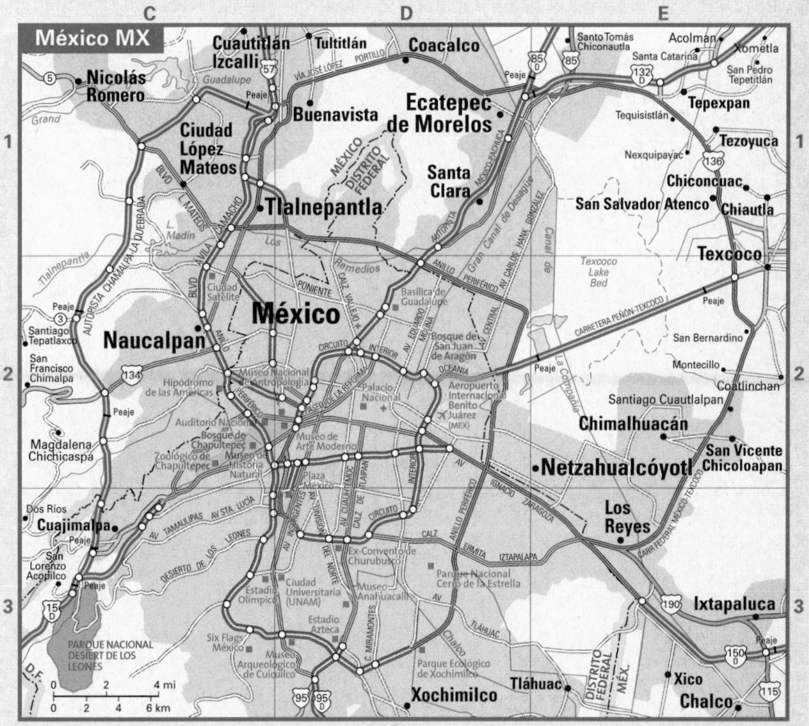

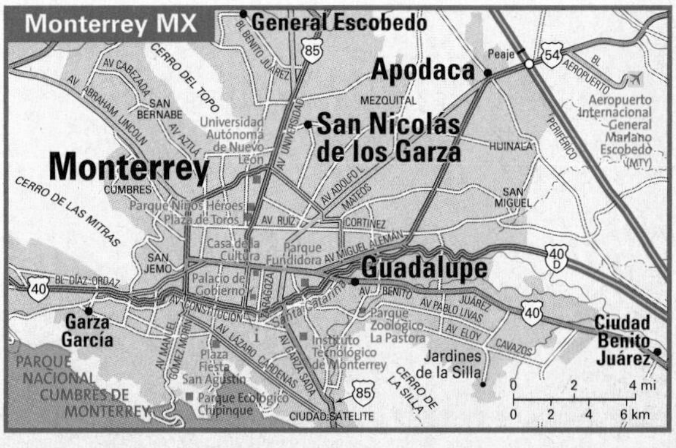

City to City Distance Chart

Miles

	Albuquerque, NM	Amarillo, TX	Anchorage, AK	Atlanta, GA	Baltimore, MD	Billings, MT	Birmingham, AL	Bismarck, ND	Boise, ID	Boston, MA	Buffalo, NY	Calgary, AB	Charleston, SC	Charleston, WV	Charlotte, NC	Cheyenne, WY	Chicago, IL	Cincinnati, OH	Cleveland, OH	Columbus, OH	Dallas, TX	Denver, CO	Des Moines, IA	Detroit, MI	El Paso, TX	Halifax, NS	Houston, TX	Indianapolis, IN	Jackson, MS	Jacksonville, FL	Kansas City, MO	Las Vegas, NV	Little Rock, AR	Los Angeles, CA	Louisville, KY
Albany, NY	2095	1811	4421	1010	333	2083	1093	1675	2526	172	292	2512	913	634	771	1789	832	730	484	621	1680	1833	1155	571	2326	877	1768	795	1331	1094	1282	2586	1354	2859	832
Albuquerque, NM		286	3563	1490	1902	991	1274	1333	966	2240	1808	1498	1793	1568	1649	538	1352	1409	1619	1476	754	438	1091	1608	263	2945	994	1298	1157	1837	894	578	900	806	1320
Amarillo, TX			3734	1206	1618	988	991	1398	1266	1957	1524	1669	1510	1285	1365	534	1069	1126	1335	1192	470	434	808	1324	438	2662	711	1014	874	1517	610	864	617	1092	1036
Anchorage, AK				4304	4297	2601	4253	2724	2745	4592	4133	2065	4495	4093	4348	3056	3584	3890	3935	3946	4087	3300	3421	3872	4002	4821	4328	3771	4294	4652	3547	3356	3929	3403	3886
Atlanta, GA					679	1889	150	1559	2218	1100	910	2395	317	503	238	1482	717	476	726	577	792	1403	967	735	1437	1805	800	531	386	344	801	2067	528	2237	419
Baltimore, MD						1959	795	1551	2401	422	370	2388	583	352	441	1665	708	521	377	420	1399	1690	1031	532	2045	1128	1470	600	1032	763	1087	2445	1072	2705	602
Billings, MT							1839	413	626	2254	1796	536	2157	1755	2012	455	1246	1552	1597	1608	1433	554	1007	1534	1255	2806	1673	1432	1836	2237	1088	965	1530	1239	1547
Birmingham, AL								1509	2170	1215	909	2346	466	578	389	1434	667	475	725	576	647	1356	919	734	1292	1921	678	481	241	494	753	1852	381	2092	369
Bismarck, ND									1039	1846	1388	794	1749	1347	1604	594	838	1144	1189	1200	1342	693	675	1126	1597	2398	1582	1024	1548	1906	801	1378	1183	1702	1139
Boise, ID										2697	2239	735	2520	2182	2375	737	1708	1969	2040	2036	1711	833	1369	1977	1206	3249	1952	1852	2115	2566	1376	760	1808	1033	1933
Boston, MA											462	2683	1003	741	861	1961	1003	862	654	760	1819	2004	1326	741	2465	714	1890	940	1453	1184	1427	2757	1493	3046	964
Buffalo, NY												2224	899	431	695	1502	545	442	197	333	1393	1546	868	277	2039	1167	1513	508	1134	1080	995	2299	1066	2572	545
Calgary, AB													2586	2184	2441	991	1675	1981	2026	2037	2114	1234	1512	1963	1936	2912	2355	1862	2385	2743	1638	1291	2020	1565	1977
Charleston, SC														468	204	1783	907	622	724	637	1109	1705	1204	879	1754	1708	1110	721	703	238	1102	2371	900	2554	610
Charleston, WV															265	1445	506	209	255	168	1072	1367	802	410	1718	1446	1192	320	816	649	764	2122	745	2374	251
Charlotte, NC																1637	761	476	520	433	1031	1559	1057	675	1677	1566	1041	575	625	385	956	2225	754	2453	464
Cheyenne, WY																	972	1233	1304	1300	979	100	633	1241	801	2513	1220	1115	1382	1829	640	843	1076	1116	1197
Chicago, IL																		302	346	359	936	1015	337	283	1543	1555	1108	184	750	1065	532	1768	662	2042	299
Cincinnati, OH																			253	105	958	1200	599	261	1605	1567	1079	116	700	803	597	1955	632	2215	106
Cleveland, OH																				144	1208	1347	669	171	1854	1359	1328	319	950	904	806	2100	882	2374	356
Columbus, OH																					1059	1266	665	192	1706	1465	1179	176	801	818	663	2021	733	2281	207
Dallas, TX																						887	752	1218	647	2524	241	913	406	1049	554	1331	327	1446	852
Denver, CO																							676	1284	701	2556	1127	1088	1290	1751	603	756	984	1029	1118
Des Moines, IA																								606	1283	1878	992	481	931	1315	194	1429	567	1703	595
Detroit, MI																									1799	1278	1338	318	960	1060	795	2037	891	2310	366
El Paso, TX																										3171	758	1489	1051	1642	1085	717	974	801	1499
Halifax, NS																											2595	1646	2158	1889	2133	3309	2198	3583	1669
Houston, TX																												839	445	884	795	1474	447	1558	972
Indianapolis, IN																													675	879	485	1843	587	2104	112
Jackson, MS																														598	747	1735	269	1851	594
Jacksonville, FL																															1148	2415	873	2441	766
Kansas City, MO																																1358	382	1632	516
Las Vegas, NV																																	1478	274	1874
Little Rock, AR																																		1706	526
Los Angeles, CA																																			2126

Kilometers

	Albany, NY	Albuquerque, NM	Amarillo, TX	Anchorage, AK	Atlanta, GA	Baltimore, MD	Billings, MT	Birmingham, AL	Bismarck, ND	Boise, ID	Boston, MA	Buffalo, NY	Calgary, AB	Charleston, SC	Charleston, WV	Charlotte, NC	Cheyenne, WY	Chicago, IL	Cincinnati, OH	Cleveland, OH	Columbus, OH	Dallas, TX	Denver, CO	Des Moines, IA	Detroit, MI	El Paso, TX	Halifax, NS	Houston, TX	Indianapolis, IN	Jackson, MS	Jacksonville, FL	Kansas City, MO	Las Vegas, NV	Little Rock, AR	Los Angeles, CA	Louisville, KY
Albuquerque, NM	3371																																			
Amarillo, TX	2914	460																																		
Anchorage, AK	7113	5733	6008																																	
Atlanta, GA	1625	2397	1940	6925																																
Baltimore, MD	536	3060	2603	6914	1093																															
Billings, MT	3352	1595	1590	4185	3039	3152																														
Birmingham, AL	1759	2050	1595	6843	241	1279	2959																													
Bismarck, ND	2695	2145	2249	4383	2508	2496	665	2428																												
Boise, ID	4064	1554	2037	4417	3569	3863	1007	3492	1672																											
Boston, MA	277	3604	3149	7389	1770	679	3627	1955	2970	4339																										
Buffalo, NY	470	2909	2452	6650	1464	595	2890	1463	2233	3603	743																									
Calgary, AB	4042	2410	2685	3323	3854	3842	862	3775	1278	1183	4317	3578																								
Charleston, SC	1469	2885	2430	7232	510	938	3471	750	2814	4055	1614	1446	4161																							
Charleston, WV	1020	2523	2068	6586	809	566	2824	930	2167	3511	1192	693	3514	753																						
Charlotte, NC	1241	2653	2196	6996	383	710	3237	626	2581	3821	1385	1118	3928	328	426																					
Cheyenne, WY	2879	866	859	4917	2385	2679	732	2307	956	1186	3155	2417	1595	2869	2325	2634																				
Chicago, IL	1339	2175	1720	5767	1154	1139	2005	1073	1348	2748	1614	877	2695	1459	814	1224	1564																			
Cincinnati, OH	1175	2267	1812	6259	766	838	2497	764	1841	3168	1387	711	3187	1001	336	766	1984	486																		
Cleveland, OH	779	2605	2148	6331	1168	607	2570	1167	1913	3282	1052	317	3260	1165	410	837	2098	557	407																	
Columbus, OH	999	2375	1918	6349	928	676	2587	927	1931	3276	1223	536	3278	1025	270	697	2092	578	169	232																
Dallas, TX	2703	1213	756	6576	1274	2251	2306	1041	2159	2753	2927	2241	3401	1784	1725	1659	1575	1506	1541	1944	1704															
Denver, CO	2949	705	698	5310	2257	2719	891	2182	1115	1340	3224	2488	1986	2743	2200	2508	161	1633	1931	2167	2037	1427														
Des Moines, IA	1858	1755	1300	5504	1556	1659	1620	1479	1086	2203	2134	1397	2433	1937	1290	1701	1018	542	964	1076	1070	1210	1088													
Detroit, MI	919	2587	2130	6230	1183	856	2468	1181	1812	3181	1192	446	3158	1414	660	1086	1997	455	420	275	309	1960	2066	975												
El Paso, TX	3743	423	705	6439	2312	3290	2019	2079	2570	1940	3966	3281	3115	2822	2764	2698	1289	2483	2582	2983	2745	1041	1128	2064	2895											
Halifax, NS	1411	4739	4283	7757	2904	1815	4515	3091	3858	5228	1149	1878	4685	2748	2327	2520	4043	2502	2521	2187	2357	4061	4113	3022	2056	5102										
Houston, TX	2845	1599	1144	6964	1287	2365	2692	1091	2545	3141	3041	2434	3789	1786	1918	1675	1963	1783	1736	2137	1897	388	1813	1596	2153	1220	4175									
Indianapolis, IN	1279	2088	1632	6068	854	965	2304	774	1648	2980	1512	817	2996	1160	515	925	1794	296	187	513	283	1469	1751	774	512	2396	2648	1350								
Jackson, MS	2142	1862	1406	6909	621	1660	2954	388	2491	3403	2338	1825	3837	1131	1313	1006	2224	1207	1126	1529	1289	653	2076	1498	1545	1691	3472	716	1086							
Jacksonville, FL	1760	2956	2441	7485	553	1228	3599	795	3067	4129	1905	1738	4413	383	1044	619	2943	1714	1292	1455	1316	1688	2817	2116	1706	2642	3039	1422	1414	962						
Kansas City, MO	2063	1438	981	5707	1289	1749	1751	1212	1289	2214	2296	1601	2636	1773	1229	1538	1030	856	961	1297	1067	891	970	312	1279	1746	3432	1279	780	1202	1847					
Las Vegas, NV	4161	930	1390	5400	3326	3934	1553	2980	2217	1223	4436	3699	2077	3815	3414	3580	1356	2845	3146	3379	3252	2142	1216	2299	3278	1154	5324	2372	2965	2792	3886	2185				
Little Rock, AR	2179	1448	993	6322	850	1725	2462	613	1903	2909	2402	1715	3250	1448	1199	1213	1731	1065	1017	1419	1179	526	1583	912	1434	1567	3537	719	944	433	1405	615	2378			
Los Angeles, CA	4600	1297	1757	5475	3599	4352	1994	3366	2739	1662	4901	4138	2518	4109	3820	3947	1796	3286	3564	3820	3670	2327	1656	2740	3717	1289	5765	2507	3385	2978	3928	2626	441	2745		
Louisville, KY	1339	2124	1667	6253	674	969	2489	594	1833	3110	1551	877	3181	981	404	747	1926	481	171	573	333	1371	1799	957	589	2412	2685	1564	180	956	1232	830	3015	846	3421	
Memphis, TN	1953	1662	1207	6570	626	1501	2615	388	2151	3144	2177	1492	3498	1223	975	988	1958	867	793	1194	956	750	1796	1158	1210	1789	3311	943	747	339	1179	862	2592	225	2959	621
México, MX	4520	2352	2051	8061	2821	3899	3641	2624	3952	3985	4574	4058	4737	3319	3541	3208	2911	3421	3360	3760	3522	1815	2750	3002	3776	1926	5709	1535	3287	2249	2956	2684	2846	2344	2981	3187
Miami, FL	2315	3467	2951	7997	1064	1784	4109	1307	3578	4639	2460	2293	4925	938	1599	1175	3455	2224	1836	2011	1871	2200	3329	2626	2254	3152	3595	1932	1924	1472	555	2359	4397	1915	4439	1744
Milwaukee, WI	1495	2294	1837	5651	1308	1295	1891	1228	1234	2813	1770	1033	2579	1614	967	1379	1628	143	640	713	730	1625	1697	608	611	2602	2658	1920	449	1344	1866	922	2909	1202	3350	634
Minneapolis, MN	2003	2154	1697	5110	1817	1804	1350	1736	693	2357	2280	1541	2039	2122	1477	1887	1418	658	1149	1223	1241	1607	1487	396	1121	2462	3168	1995	959	1852	2376	710	2698	1310	3139	1144
Mobile, AL	2162	2162	1780	7258	534	1630	3249	415	2840	3704	2306	1874	4187	1033	1347	920	2526	1485	1176	1578	1339	1028	2378	1794	1595	1981	1981	761	1186	301	660	1496	3092	735	3268	1006
Montréal, QC	370	3495	3038	6607	1997	907	3368	2074	2711	4079	504	639	3535	1842	1323	1614	2895	1353	1311	946	1167	2851	2965	1874	907	3802	1150	3044	1403	2436	2132	2187	4177	2327	4616	1480
Nashville, TN	1614	2008	1553	6534	389	1152	2652	312	2116	3179	1828	1152	3463	874	636	639	1995	763	452	854	615	1096	1870	1167	870	2137	2962	1289	462	681	948	899	2938	571	3305	282
New Orleans, LA	2317	2053	1598	7207	761	1837	3146	565	2790	3595	2515	2018	4135	1260	1490	1147	2417	1504	1319	1722	1482	845	2267	1797	1736	1799	3649	579	1329	298	895	1500	2983	732	3084	1149
New York, NY	243	3242	2785	7062	1398	309	3297	1585	2640	4008	346	644	3990	1244	829	1015	2824	1282	1023	750	861	2557	2895	1804	1001	3596	1480	2671	1150	1968	1533	1934	4106	2031	4537	1189
Oklahoma City, OK	2492	879	422	6245	1519	2179	1974	1173	1828	2423	2726	2031	3070	2008	1644	1773	1244	1298	1389	1726	1496	336	1096	879	1709	1186	3862	722	1210	985	2077	560	1809	571	2175	1245
Omaha, NE	2079	1566	1168	5409	1591	1879	1455	1514	991	1986	2354	1617	2338	2076	1532	1841	800	763	1184	1297	1290	1076	870	219	1195	1989	3242	1464	994	1504	2150	302	2082	917	2521	1133
Orlando, FL	1987	3112	2595	7641	708	1455	3754	951	3223	4283	2130	1965	4570	610	1271	845	3099	1868	1480	1681	1541	1844	2972	2270	1899	2796	3266	1577	1569	1117	227	2003	4042	1559	4084	1389
Ottawa, ON	486	3392	2936	6455	1866	842	3265	1971	2608	3977	665	536	3384	1780	1221	1483	2793	1252	1208	845	1064	2748	2862	1772	805	3701	1324	2941	1302	2333	2069	2085	4074	2224	4515	1377
Philadelphia, PA	359	3144	2689	7010	1258	167	3249	1443	2592	3961	516	666	3939	1102	730	874	2776	1236	927	703	763	2415	2806	1755	953	3455	1651	2529	1054	1826	1393	1836	4023	1891	4441	1091
Phoenix, AZ	4121	750	1212	5776	3006	3807	1929	2772	2674	1598	4354	3659	2454	3514	3274	3390	1615	2927	3018	3355	3125	1733	1455	2507	3337	695	5490	1911	2838	2385	3334	2188	459	2200	594	2874
Pittsburgh, PA	780	2687	2230	6526	1088	396	2766	1228	2109	3477	953	349	3455	1033	349	705	2293	751	470	219	306	2005	2349	1273	470	3046	2087	2198	595	1590	1323	1379	3564	1480	3984	634
Portland, ME	434	3762	3305	7546	1926	837	3784	2113	3128	4497	172	901	4475	1772	1350	1543	3313	1772	1545	1208	1381	3084	3382	2291	1348	4124	872	3199	1670	2494	2061	2454	4594	2558	5059	1709
Portland, OR	4753	2245	2727	3902	4259	4553	1430	4182	2093	695	5030	4291	1371	4743	4199	4508	1876	3438	3858	3973	3965	3443	2029	2893	3870	2843	5918	3831	3669	4093	4817	2904	1911	3599	1562	3800
Québec, QC	582	3734	3279	6846	2209	1120	3607	2314	2951	4320	624	879	3775	2055	1564	1826	3136	1595	1564	1187	1406	3091	3205	2114	1147	4043	940	3284	1644	2676	2344	2428	4417	2566	4858	1720
Raleigh, NC	1028	2867	2412	7157	637	497	3395	880	2739	4014	1173	1033	4085	449	504	254	2829	1385	840	914	776	1913	2703	1862	1165	2951	2307	1928	1028	1260	740	1733	3797	1430	4164	907
Rapid City, SD	2816	1353	1347	4795	2431	2616	610	2354	515	1496	3091	2354	1472	2935	2288	2700	491	1469	1961	2034	2051	1733	650	1012	1932	1778	3979	2121	1772	2346	2991	1142	1665	1759	2106	1955
Reno, NV	4420	1641	2101	4843	3926	4220	1545	3849	2208	692	4697	3958	2069	4410	3866	4175	1543	3105	3525	3640	3632	3110	1696	2560	3537	2116	5585	3334	3335	3760	4484	2571	711	3266	835	3467
Richmond, VA	776	3018	2563	7065	848	245	3303	1091	2647	4016	920	780	3994	689	518	465	2832	1290	853	758	832	2106	2716	1812	1009	3146	2055	2140	1031	1471	980	1746	3932	1582	4315	920
St. Louis, MO	1667	1691	1234	6113	883	1353	2158	806	1694	2619	1900	1205	3041	1368	824	1133	1435	473	563	901	671	1022	1376	702	883	1998	3036	1389	385	813	1442	405	2590	669	2986	425
Salt Lake City, UT	3578	1004	1551	4729	3083	3379	882	3006	1545	550	3854	3115	1406	3569	3025	3334	702	2262	2682	2796	2790	2269	854	1717	2695	1390	4742	2655	2492	2917	3643	1728	671	2425	1112	2624
San Antonio, TX	3142	1316	825	6833	1609	2689	2414	1413	2573	2833	3366	2679	3511	2108	2162	1997	1683	2043	1981	2383	2143	436	1522	1623	2397	895	4500	322	1908	1036	1744	1307	2047	965	2182	1810
San Diego, CA	4697	1327	1788	5673	3485	4383	2095	3252	2840	1763	4932	4235	2619	3995	3850	3870	1897	3387	3595	3921	3701	2212	1757	2841	3818	1175	5866	2393	3414	2864	3813	2727	542	2740	200	3450
San Francisco, CA	4769	1788	2248	4940	4212	4570	1892	3977	2814	1039	5044	4307	2409	4721	4216	4439	1892	3453	3873	3987	3981	2940	2045	2907	3886	1900	5932	3118	3685	3591	4541	2919	925	3237	619	3817
Seattle, WA	4664	2354	2837	3623	4352	4465	1313	4275	1977	805	4940	4203	1093	4784	4137	4549	1986	3318	3810	3883	3900	3553	2138	2932	3781	3128	5828	3940	3619	4203	4911	3012	2021	3709	1847	3804
Tampa, FL	2076	3136	2619	7664	732	1545	3778	975	3247	4307	2220	2053	4592	698	1360	935	3123	1892	1504	1772	1667	1868	2996	2294	1921	2821	3355	1601	1593	1141	315	2026	4064	1583	4108	1413
Toronto, ON	644	2962	2505	6595	1541	909	2835	1541	2179	3546	917	171	3524	1619	864	1290	2362	821	779	488	708	2319	2433	1342	375	3269	1681	2512	870	1903	1910	1654	3644	1794	4084	948
Vancouver, BC	4878	2570	3052	3430	4566	4679	1527	4491	2191	1018	5155	4417	899	4998	4352	4763	2201	3533	4024	4098	4116	3768	2354	3147	3995	3358	6043	4156	3834	4418	5126	3229	2237	3924	2077	4018
Washington, DC	594	3051	2594	6903	1023	61	3142	1220	2486	3854	737	618	3831	867	557	639	2669	1128	832	595	669	2191	2713	1649	846	3231	1873	2306	959	1603	1158	1743	3928	1667	4348	959
Wichita, KS	2367	1138	681	5921	1591	2053	1717	1348	1503	2166	2600	1905	2814	2077	1533	1842	986	1171	1263	1601	1371	591	838	628	1583	1445	3736	978	1084	1241	2151	309	2053	747	2434	1134
Winnipeg, MB	2730	2587	2285	4385	2542	2531	1324	2463	668	2336	3006	2269	1313	2850	2203	2615	1821	1384	1876	1948	1966	2193	1892	1121	1847	3010	3361	2581	1685	2526	3102	1324	3012	1939	3453	1870

285
Tableau des distances

Milles

	Memphis, TN	México, MX	Miami, FL	Milwaukee, WI	Minneapolis, MN	Mobile, AL	Montréal, QC	Nashville, TN	New Orleans, LA	New York, NY	Oklahoma City, OK	Omaha, NE	Orlando, FL	Ottawa, ON	Philadelphia, PA	Phoenix, AZ	Pittsburgh, PA	Portland, ME	Portland, OR	Québec, QC	Raleigh, NC	Rapid City, SD	Reno, NV	Richmond, VA	St. Louis, MO	Salt Lake City, UT	San Antonio, TX	San Diego, CA	San Francisco, CA	Seattle, WA	Tampa, FL	Toronto, ON	Vancouver, BC	Washington, DC	Wichita, KS	Winnipeg, MB
Albany, NY	1214	2809	1439	929	1245	1344	230	1003	1440	151	1549	1292	1235	302	223	2561	485	270	2954	362	639	1750	2747	482	1036	2224	1953	2919	2964	2899	1290	400	3032	369	1471	1697
Albuquerque, NM	1033	1462	2155	1426	1339	1344	2172	1248	1276	2015	546	973	1934	2108	1954	466	1670	2338	1395	2321	1782	841	1020	1876	1051	624	818	825	1111	1463	1949	1841	1597	1896	707	1608
Amarillo, TX	750	1275	1834	1142	1055	1106	1888	965	993	1731	262	726	1613	1825	1671	753	1386	2054	1695	2038	1499	837	1306	1593	767	964	513	1111	1397	1763	1628	1557	1897	1612	423	1420
Anchorage, AK	4083	5010	4970	3512	3176	4511	4106	4061	4479	4389	3881	3362	4749	4012	4357	3590	4056	4690	2425	4255	4448	2980	3010	4391	3799	2939	4247	3526	3070	2252	4763	4099	2132	4290	3680	2725
Atlanta, GA	389	1753	661	813	1129	332	1241	242	473	869	944	989	440	1160	782	1868	676	1197	2647	1373	396	1511	2440	527	549	1916	1000	2166	2618	2705	455	958	2838	636	989	1580
Baltimore, MD	933	2423	1109	805	1121	1013	564	716	1142	192	1354	1168	904	523	104	2366	246	520	2830	696	309	1626	2623	152	841	2100	1671	2724	2840	2775	960	565	2908	38	1276	1573
Billings, MT	1625	2263	2554	1175	839	2019	2093	1648	1955	2049	1227	904	2333	2029	2019	1199	1719	2352	889	2242	2110	379	960	2053	1341	548	1500	1302	1176	816	2348	1762	949	1953	1067	823
Birmingham, AL	241	1631	812	763	1079	258	1289	194	351	985	729	941	591	1225	897	1723	763	1313	2599	1438	547	1463	2392	678	501	1868	878	2021	2472	2657	606	958	2791	758	838	1531
Bismarck, ND	1337	2456	2224	767	431	1765	1685	1315	1734	1641	1136	616	2003	1621	1611	1662	1311	1944	1301	1834	1702	320	1372	1645	1053	960	1599	1765	1749	1229	2018	1354	1362	1545	934	415
Boise, ID	1954	2477	2883	1748	1465	2302	2535	1976	2234	2491	1506	1234	2662	2472	2462	993	2161	2795	432	2685	2495	930	430	2496	1628	342	1761	1096	646	500	2677	2204	633	2395	1346	1452
Boston, MA	1353	2843	1529	1100	1417	1433	313	1136	1563	215	1694	1463	1324	413	321	2706	592	107	3126	388	729	1921	2919	572	1181	2395	2092	3065	3135	3070	1380	570	3204	458	1616	1868
Buffalo, NY	927	2522	1425	642	958	1165	397	716	1254	400	1262	1005	1221	333	414	2274	217	560	2667	546	642	1463	2460	485	749	1936	1665	2632	2677	2612	1276	106	2745	384	1184	1410
Calgary, AB	2174	2944	3061	1603	1267	2602	2197	2152	2570	2480	1908	1453	2840	2103	2448	1525	2147	2781	852	2346	2539	915	1286	2482	1890	874	2182	1628	1497	679	2854	2190	559	2381	1749	816
Charleston, SC	760	2063	583	1003	1319	642	1145	543	783	773	1248	1290	379	1106	685	2184	642	1101	2948	1277	279	1824	2741	428	850	2218	1310	2483	2934	2973	434	1006	3106	539	1291	1771
Charleston, WV	606	2201	994	601	918	837	822	395	926	515	1022	952	790	759	454	2035	217	839	2610	972	313	1422	2403	322	512	1880	1344	2393	2620	2571	845	537	2705	346	953	1369
Charlotte, NC	614	1994	730	857	1173	572	1003	397	713	631	1102	1144	525	922	543	2107	438	959	2802	1135	158	1678	2595	289	704	2072	1241	2405	2759	2827	581	802	2960	397	1145	1625
Cheyenne, WY	1217	1809	2147	1012	881	1570	1799	1240	1502	1755	773	497	1926	1736	1725	1004	1425	2059	1166	1949	1758	305	959	1760	892	436	1046	1179	1176	1234	1941	1468	1368	1659	613	1132
Chicago, IL	539	2126	1382	89	409	923	841	474	935	797	807	474	1161	778	768	1819	467	1101	2137	991	861	913	1930	802	294	1406	1270	2105	2146	2062	1176	510	2196	701	728	860
Cincinnati, OH	493	2088	1141	398	714	731	815	281	820	636	863	736	920	751	576	1876	292	960	2398	972	522	1219	2191	530	350	1667	1231	2234	2407	2368	935	484	2501	517	785	1166
Cleveland, OH	742	2337	1250	443	760	981	588	531	1070	466	1073	806	1045	525	437	2085	136	751	2469	738	568	1264	2262	471	560	1738	1481	2437	2478	2413	1101	303	2547	370	995	1211
Columbus, OH	594	2189	1163	454	771	832	725	382	921	535	930	802	958	661	474	1942	190	858	2464	874	482	1275	2257	517	417	1734	1332	2300	2474	2424	1036	440	2558	416	852	1222
Dallas, TX	466	1128	1367	1010	999	639	1772	681	525	1589	209	669	1146	1708	1501	1077	1246	1917	2140	1921	1189	1077	1933	1309	635	1410	271	1375	1827	2208	1161	1441	2342	1362	367	1363
Denver, CO	1116	1709	2069	1055	924	1478	1843	1162	1409	1799	681	541	1847	1779	1744	904	1460	2102	1261	1992	1680	404	1054	1688	855	531	946	1092	1271	1329	1862	1512	1463	1686	521	1176
Des Moines, IA	720	1866	1632	378	246	1115	1165	725	1117	1121	546	136	1411	1101	1091	1558	791	1424	1798	1314	1157	629	1591	1126	436	1067	1009	1766	1807	1822	1426	834	1956	1025	390	697
Detroit, MI	752	2347	1401	380	697	991	564	541	1079	622	1062	743	1180	500	592	2074	292	838	2405	713	724	1201	2198	627	549	1675	1490	2373	2415	2350	1194	233	2483	526	984	1148
El Paso, TX	1112	1197	1959	1617	1530	1231	2363	1328	1118	2235	737	1236	1738	2300	2147	432	1893	2563	1767	2513	1834	1105	1315	1955	1242	864	556	730	1181	1944	1753	2032	2087	2008	898	1871
Halifax, NS	2058	3548	2234	1652	1969	1231	715	1841	2268	920	2400	2015	2030	823	1026	3412	1297	542	3678	584	1434	2473	3471	1277	1887	2947	2797	3646	3687	3622	2085	1045	3756	1164	2322	2089
Houston, TX	586	954	1201	1193	1240	473	1892	801	360	1660	449	910	980	1828	1572	1188	1366	1988	2381	2041	1198	1318	2072	1330	863	1650	200	1487	1938	2449	995	1561	2583	1433	608	1604
Indianapolis, IN	464	2043	1196	279	596	737	872	287	826	715	752	618	975	809	655	1764	370	1038	2280	1022	639	1101	2073	641	239	1549	1186	2122	2290	2249	990	541	2383	596	674	1047
Jackson, MS	211	1398	915	835	1151	187	1514	423	185	1223	612	935	694	1450	1135	1482	988	1550	2544	1663	783	1458	2337	914	505	1813	644	1780	2232	2612	709	1183	2746	996	771	1570
Jacksonville, FL	733	1837	345	1160	1477	410	1325	589	556	953	1291	1336	141	1286	866	2072	822	1281	2994	1457	460	1859	2787	609	896	2264	1084	2370	2822	3052	196	1187	3186	720	1337	1928
Kansas City, MO	536	1668	1466	573	441	930	1359	559	932	1202	348	188	1245	1296	1141	1360	857	1525	1805	1509	1077	710	1598	1085	252	1074	812	1695	1814	1872	1259	1028	2007	1083	192	823
Las Vegas, NV	1611	1769	2733	1808	1677	1922	2596	1826	1854	2552	1124	1294	2512	2532	2500	285	2215	2855	1188	2745	2360	1035	442	2444	1610	417	1272	337	575	1256	2526	2265	1390	2441	1276	1872
Little Rock, AR	140	1457	1190	747	814	457	1446	355	455	1262	355	570	969	1382	1175	1367	920	1590	2237	1595	889	1093	2030	983	416	1507	600	1703	2012	2305	984	1115	2439	1036	464	1205
Los Angeles, CA	1839	1853	2759	2082	1951	2031	2869	2054	1917	2820	1352	1567	2538	2806	2760	369	2476	3144	971	3019	2588	1309	519	2682	1856	691	1356	124	385	1148	2553	2538	1291	2702	1513	2146
Louisville, KY	386	1981	1084	394	711	625	920	175	714	739	774	704	863	856	678	1786	394	1062	2362	1069	564	1215	2155	572	264	1631	1125	2144	2372	2364	878	589	2497	596	705	1162
Memphis, TN		1595	1051	624	940	395	1306	215	396	1123	487	724	830	1243	1035	1500	780	1451	2382	1456	749	1247	2175	843	294	1652	739	1841	2144	2440	845	975	2574	896	597	1359
México, MX			2154	2200	2113	1426	2900	1810	1313	2619	1323	1783	1933	2838	2525	1484	2375	2941	2819	3051	2151	2365	2367	2283	1825	2135	853	1683	2233	2996	1948	2570	3139	2386	1481	2477
Miami, FL				1478	1794	727	1671	907	874	1299	1609	1654	232	1631	1211	2390	1167	1627	3312	1803	805	2176	3105	954	1214	2581	1401	2688	3140	3370	274	1532	3504	1065	1655	2246
Milwaukee, WI					337	1019	939	569	1020	894	880	514	1257	875	865	1892	564	1198	2063	1088	956	842	1970	899	367	1446	1343	2145	2186	1991	1272	607	2124	799	769	789
Minneapolis, MN						1335	1255	886	1337	1211	793	383	1573	1192	1181	1805	881	1515	1727	1405	1273	606	1839	1216	621	1315	1257	2014	2055	1654	1588	924	1788	1115	637	452
Mobile, AL							1575	450	146	1203	799	1119	506	1481	1115	1662	1019	1531	2731	1707	730	1641	2545	861	688	2000	673	1960	2411	2799	521	1214	2933	970	958	1787
Montréal, QC								1094	1632	383	1625	1300	1466	121	454	2637	607	282	2963	155	871	1758	2756	714	1112	2232	2043	2931	2972	2907	1522	330	3041	600	1547	1374
Nashville, TN									539	906	703	747	686	1031	818	1715	569	1234	2405	1244	532	1269	2198	626	307	1675	954	2056	2360	2463	701	764	2597	679	748	1337
New Orleans, LA										1332	731	1121	653	1570	1245	1548	1108	1660	2663	1783	871	1643	2431	1002	690	1932	560	1846	2298	2731	668	1302	2865	1106	890	1755
New York, NY											1469	1258	1094	439	91	2481	367	313	2920	515	499	1716	2713	342	956	2189	1861	2839	2929	2864	1150	507	2998	228	1391	1665
Oklahoma City, OK												463	1388	1563	1408	1012	1124	1792	1934	1776	1237	871	1727	1331	505	1204	466	1370	1657	2002	1403	1295	2136	1350	161	1158
Omaha, NE													1433	1238	1228	1440	928	1561	1662	1451	1265	525	1455	1263	440	932	927	1630	1672	1719	1448	971	1853	1162	307	638
Orlando, FL														1427	1006	2169	963	1422	3091	1598	601	1955	2884	750	993	2360	1180	2467	2918	3149	82	1327	3283	860	1434	2025
Ottawa, ON															451	2575	545	382	2901	257	831	1696	2694	675	1050	2170	1981	2869	2910	2845	1483	268	2978	562	1485	1280
Philadelphia, PA																2420	306	419	2890	586	411	1686	2683	254	895	2160	1774	2779	2900	2835	1062	522	2968	140	1330	1633
Phoenix, AZ																	2136	2804	1335	2788	2249	1308	883	2343	1517	651	987	358	750	1513	2184	2307	1655	2362	1173	2075
Pittsburgh, PA																		690	2590	758	497	1386	2383	341	611	1859	1519	2494	2599	2534	1019	321	2668	240	1046	1332
Portland, ME																			3223	264	827	2019	3016	670	1279	2493	2189	3162	3233	3168	1478	668	3301	556	1714	1966
Portland, OR																				3114	2923	1268	578	2925	2057	771	2322	1093	638	170	3106	2633	313	2824	1775	1463
Québec, QC																					1003	1908	2905	846	1261	2381	2193	3080	3122	3057	1654	479	3190	732	1696	1523
Raleigh, NC																						1777	2716	157	825	2193	1398	2563	2894	2926	656	820	3060	265	1266	1724
Rapid City, SD																							1151	1720	963	628	1335	1372	1368	1195	1970	1429	1328	1620	712	792
Reno, NV																								2718	1850	524	1870	642	217	755	2899	2426	898	2617	1568	1867
Richmond, VA																									834	2194	1530	2684	2934	2869	805	660	3003	108	1274	1667
St. Louis, MO																										1326	968	1875	2066	2125	1008	782	2259	837	441	1075
Salt Lake City, UT																											1419	754	740	839	2375	1902	973	2094	1044	1455
San Antonio, TX																												1285	1737	2275	1195	1714	2410	1635	624	1621
San Diego, CA																													508	1271	2481	2601	1414	2720	1531	2209
San Francisco, CA																														816	2933	2643	958	2834	1784	2193
Seattle, WA																															3164	2577	140	2769	1843	1390
Tampa, FL																																1383	3297	916	1448	2039
Toronto, ON																																	2711	563	1217	1375
Vancouver, BC																																		2902	1977	1375
Washington, DC																																			1272	1566
Wichita, KS																																				956

Kilomètres

	Memphis, TN	México, MX	Miami, FL	Milwaukee, WI	Minneapolis, MN	Mobile, AL	Montréal, QC	Nashville, TN	New Orleans, LA	New York, NY	Oklahoma City, OK	Omaha, NE	Orlando, FL	Ottawa, ON	Philadelphia, PA	Phoenix, AZ	Pittsburgh, PA	Portland, ME	Portland, OR	Québec, QC	Raleigh, NC	Rapid City, SD	Reno, NV	Richmond, VA	St. Louis, MO	Salt Lake City, UT	San Antonio, TX	San Diego, CA	San Francisco, CA	Seattle, WA	Tampa, FL	Toronto, ON	Vancouver, BC	Washington, DC	Wichita, KS
México, MX	2566																																		
Miami, FL	1691	3466																																	
Milwaukee, WI	1004	3540	2378																																
Minneapolis, MN	1512	3400	2887	542																															
Mobile, AL	636	2294	1170	1640	2148																														
Montréal, QC	2101	4666	2689	1511	2019	2534																													
Nashville, TN	346	2912	1459	916	1426	724	1760																												
New Orleans, LA	637	2113	1406	1641	2151	235	2626	867																											
New York, NY	1807	4214	2090	1438	1948	1936	616	1458	2143																										
Oklahoma City, OK	784	2129	2589	1416	1276	1286	2615	1131	1176	2364																									
Omaha, NE	1165	2869	2661	827	616	1800	2092	1202	1804	2024	745																								
Orlando, FL	1335	3110	373	2023	2531	814	2359	1104	1051	1760	2233	2306																							
Ottawa, ON	2000	4566	2624	1408	1918	2383	194	1659	2526	706	2515	1992	2296																						
Philadelphia, PA	1665	4063	1948	1392	1900	1794	730	1316	2003	146	2265	1976	1619	726																					
Phoenix, AZ	2414	2388	3846	3044	2904	2674	4243	2759	2491	3992	1628	2317	3490	4143	3894																				
Pittsburgh, PA	1255	3821	1878	907	1418	1640	977	916	1783	591	1809	1493	1549	877	492	3437																			
Portland, ME	2335	4732	2618	1928	2438	2463	454	1986	2671	504	2883	2512	2288	615	674	4512	1110																		
Portland, OR	3833	4536	5329	3319	2779	4394	4767	3870	4285	4698	3112	2674	4973	4668	4650	2148	4167	5186																	
Québec, QC	2343	4909	2901	1751	2261	2747	249	2002	2869	829	2858	2335	2571	414	943	4486	1220	425	5010																
Raleigh, NC	1205	3461	1295	1538	2048	1175	1401	856	1401	803	1990	2035	967	1337	661	3619	800	1331	4703	1614															
Rapid City, SD	2006	3805	3501	1355	975	2640	2829	2042	2644	2761	1401	845	3146	2729	2713	2105	2230	3249	2040	3070	2859														
Reno, NV	3500	3809	4996	3170	2959	4095	4434	3537	3911	4365	2779	2341	4640	4335	4317	1421	3834	4853	930	4674	4370	1852													
Richmond, VA	1356	3673	1535	1446	1957	1385	1149	1007	1612	550	2142	2032	1207	1086	409	3770	549	1078	4706	1361	253	2767	4373												
St. Louis, MO	473	2936	1953	591	999	1107	1789	494	1110	1538	813	708	1598	1689	1440	2441	983	2058	3310	2029	1327	1549	2977	1342											
Salt Lake City, UT	2658	3435	4153	2327	2116	3218	3591	2695	3109	3522	1937	1500	3797	3492	3475	1047	2991	4011	1241	3831	3529	1010	843	3530	2134										
San Antonio, TX	1189	1372	2254	2161	2023	1083	3287	1535	901	2994	750	1492	1899	3187	2854	1588	2444	3522	3736	3529	2249	2148	3009	2462	1558	2283									
San Diego, CA	2962	2708	4325	3451	3241	3154	4716	3308	2970	4568	2204	2623	3969	4616	4471	576	4013	5088	1759	4956	4124	2208	1033	4319	3017	1213	2068								
San Francisco, CA	3450	3593	5052	3517	3306	3879	4782	3797	3697	4713	2666	2690	4695	4682	4666	1207	4182	5202	1027	5023	4656	2201	349	4721	3324	1191	2795	817							
Seattle, WA	3926	4821	5422	3204	2661	4504	4677	3963	4394	4608	3221	2766	5067	4578	4562	2434	4077	5097	274	4919	4708	1923	1215	4616	3419	1350	3660	2045	1313						
Tampa, FL	1360	3134	441	2047	2555	838	2449	1128	1075	1850	2257	2330	132	2386	1709	3514	1640	2378	4998	2661	1056	3170	4664	1295	1622	3821	1923	3992	4719	5091					
Toronto, ON	1569	4135	2465	977	1487	1953	531	1229	2095	816	2084	1562	2135	431	840	3712	516	1075	4236	771	1319	2299	3903	1062	1258	3060	2758	4185	4253	4146	2225				
Vancouver, BC	4142	5051	5638	3418	2877	4719	4893	4179	4610	4824	3437	2981	5282	4792	4776	2663	4293	5311	504	5133	4924	2137	1445	4832	3635	1566	3878	2275	1541	225	5305	4362			
Washington, DC	1442	3839	1714	1286	1794	1561	965	1093	1780	367	2172	1870	1384	904	225	3800	386	895	4544	1178	426	2607	4211	174	1347	3369	2631	4376	4560	4455	1474	906	4669		
Wichita, KS	961	2383	2663	1237	1025	1541	2489	1204	1432	2238	259	494	2307	2389	2140	1887	1683	2758	2856	2729	2037	1146	2523	2050	710	1680	1004	2463	2870	2965	2330	1958	3181	2047	
Winnipeg, MB	2187	3985	3614	1270	727	2875	2211	2151	2824	2679	1863	1027	3258	2060	2627	3339	2143	3163	2354	2451	2774	1274	3004	2682	1730	2341	2608	3554	3529	2237	3281	2212	2212	2520	1538

Distances and Driving Times across North America

TEMPERATURE CONVERSIONS

°F	°C	°C	°F
110	43.3	40	104
100	37.8	35	95
90	32.2	30	86
80	26.7	25	77
70	21.1	20	68
60	15.6	15	59
50	10.0	10	50
40	4.4	5	41
32	0	0	32
30	-1.1	-5	23
20	-6.7	-10	14
10	-12.2	-15	5
0	-17.8	-20	-4
-10	-23.3	-25	-13
-20	-28.9	-30	-22
-30	-34.4	-35	-31
-40	-40.0	-40	-40
-50	-45.6	-45	-49

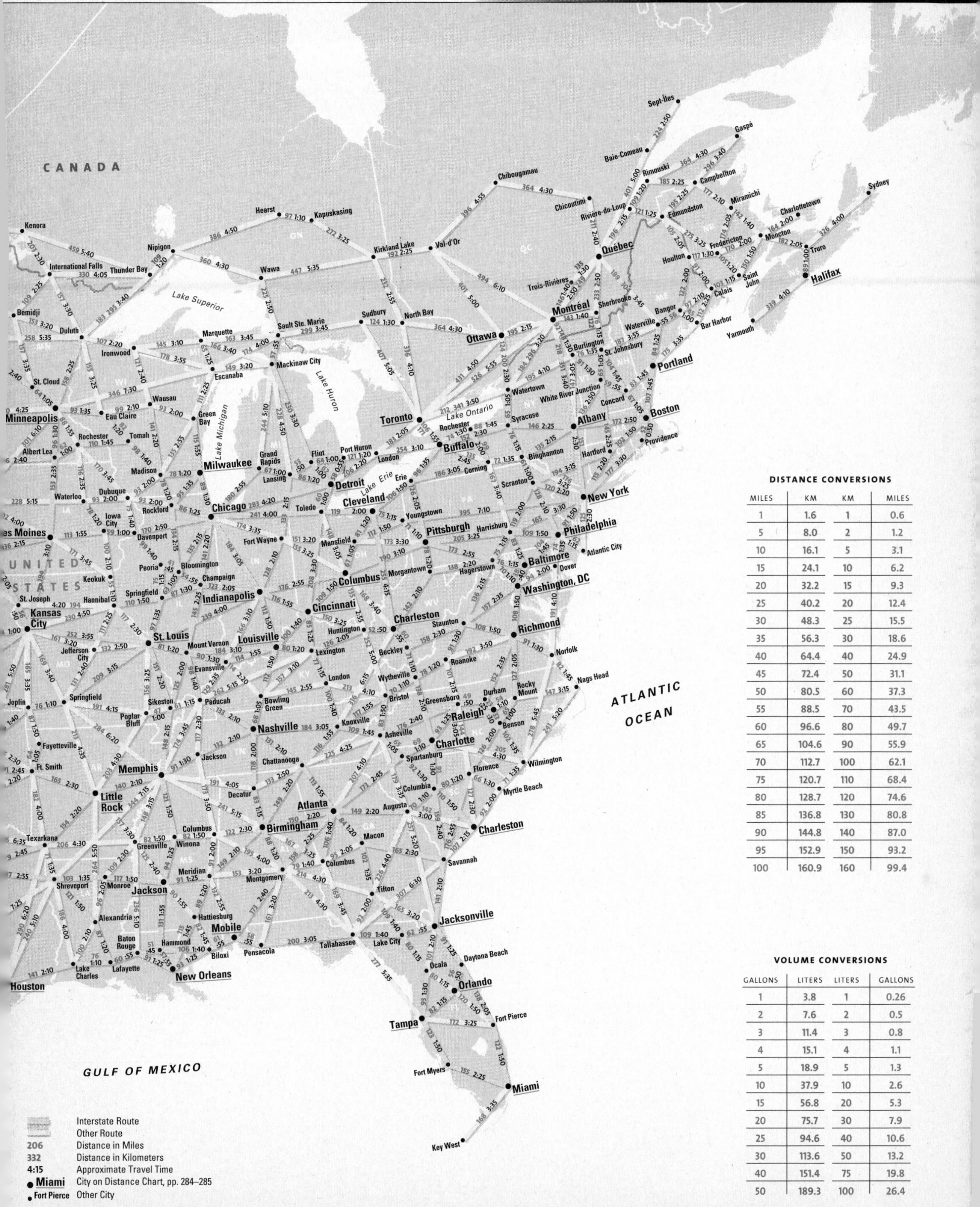

Distances and driving times may vary depending on actual route traveled and driving conditions.

DISTANCE CONVERSIONS

MILES	KM	KM	MILES
1	1.6	1	0.6
5	8.0	2	1.2
10	16.1	5	3.1
15	24.1	10	6.2
20	32.2	15	9.3
25	40.2	20	12.4
30	48.3	25	15.5
35	56.3	30	18.6
40	64.4	40	24.9
45	72.4	50	31.1
50	80.5	60	37.3
55	88.5	70	43.5
60	96.6	80	49.7
65	104.6	90	55.9
70	112.7	100	62.1
75	120.7	110	68.4
80	128.7	120	74.6
85	136.8	130	80.8
90	144.8	140	87.0
95	152.9	150	93.2
100	160.9	160	99.4

VOLUME CONVERSIONS

GALLONS	LITERS	LITERS	GALLONS
1	3.8	1	0.26
2	7.6	2	0.5
3	11.4	3	0.8
4	15.1	4	1.1
5	18.9	5	1.3
10	37.9	10	2.6
15	56.8	20	5.3
20	75.7	30	7.9
25	94.6	40	10.6
30	113.6	50	13.2
40	151.4	75	19.8
50	189.3	100	26.4

TOURISM INFORMATION

UNITED STATES

Alabama
Alabama Bureau of Tourism & Travel
800.252.2262, 334.242.4169
www.800alabama.org

Alaska
Alaska Travel Industry Association
www.travelalaska.com

Arizona
Arizona Office of Tourism
866.275.5816, 602.364.3700
www.arizonaguide.com

Arkansas
Arkansas Dept. of Parks & Tourism
800.628.8725, 501.682.7777
www.1800natural.com

California
California Tourism
800.862.2543, 916.444.4429
www.visitcalifornia.com

Colorado
Colorado Tourism Office
800.265.6723, 303.892.3885
www.colorado.com

Connecticut
Connecticut Commission on Culture & Tourism
888.288.4748, 860.256.2800
www.ctvisit.com

Delaware
Delaware Tourism Office
866.284.7483
www.visitdelaware.com

District of Columbia
DC Convention and Tourism Corp.
800.422.8644, 202.789.7000 or 7030
www.washington.org

Florida
Visit Florida
850.488.5607, 888.735.2872
www.visitflorida.com

Georgia
Georgia Dept. of Industry, Trade & Tourism
800.847.4842
www.georgia.org/travel

Hawaii
Hawaii Visitors & Conv. Bureau
800.464.2924, 808.923.1811
www.gohawaii.com

Idaho
Idaho Div. of Tourism Development
800.847.4843, 208.334.2470
www.visitidaho.org

Illinois
Illinois Bureau of Tourism
800.406.6418
www.enjoyillinois.com

Indiana
Indiana Office of Tourism Development
800.677.9800, 317.232.8860
www.enjoyindiana.com

Iowa
Iowa Division of Tourism
888.472.6035, 515.242.4705
www.traveliowa.com

Kansas
Kansas Travel & Tourism
800.252.6727, 785.296.2009
www.travelks.com

Kentucky
Kentucky Department of Tourism
800.225.8747, 502.564.4930
www.kentuckytourism.com

Louisiana
Louisiana Office of Tourism
225.342.8100, 800.334.8626
www.louisianatravel.com

Maine
Maine Office of Tourism
888.624.6345, 207.624.7483
www.visitmaine.com

Maryland
Maryland Office of Tourism
866.639.3526
www.mdisfun.org

Massachusetts
Mass. Office of Travel & Tourism
800.227.6277, 617.973.8500
www.massvacation.com

Michigan
Travel Michigan
800.644.2489
www.michigan.org

Minnesota
Explore Minnesota Tourism
888.868.7476, 651.296.5029
www.exploreminnesota.com

Mississippi
Mississippi Div. of Tourism Development
866.733.6477, 601.359.3297
www.visitmississippi.org

Missouri
Missouri Division of Tourism
800.519.2100, 573.751.4133
www.visitmo.com

Montana
Travel Montana
800.847.4868, 406.841.2870
www.visitmt.com

Nebraska
Nebraska Travel & Tourism
877.632.7275, 402.471.3796
www.visitnebraska.org

Nevada
Nevada Commission on Tourism
800.638.2328, 775.687.4322
www.travelnevada.com

New Hampshire
New Hampshire Division of Travel & Tourism Development
800.386.4664, 603.271.2665
www.visitnh.gov

New Jersey
New Jersey Tourism Commission
800.847.4865, 609.777.0885
www.visitnj.org

New Mexico
New Mexico Tourism Department
800.545.2070, 505.827.7400
www.newmexico.org

New York
New York State Division of Tourism
800.225.5697, 518.474.4116
www.iloveny.com

North Carolina
North Carolina Division of Tourism, Film & Sports Development
800.847.4862, 919.733.4171
www.visitnc.com

North Dakota
North Dakota Tourism Division
800.435.5663, 701.328.2525
www.ndtourism.com

Ohio
Ohio Division of Travel & Tourism
800.282.5393, 614.466.8844
www.discoverohio.com

Oklahoma
Oklahoma Dept. of Tourism & Rec.
800.652.6552, 405.230.8400
www.travelok.com

Oregon
Oregon Tourism Commission
800.547.7842, 503.378.8850
www.traveloregon.com

Pennsylvania
Pennsylvania Tourism Office
800.847.4872, 717.787.5453
www.visitpa.com

Rhode Island
Rhode Island Tourism Division
800.556.2484
www.visitrhodeisland.com

South Carolina
S.C. Dept. of Parks, Rec. & Tourism
866.224.9339, 803.734.1700
www.discoversouthcarolina.com

South Dakota
South Dakota Department of Tourism
800.732.5682, 605.773.3301
www.travelsd.com

Tennessee
Tenn. Dept. of Tourist Development
800.462.8366, 615.741.2159
www.tnvacation.com

Texas
Texas Dept. of Econ. Dev., Tourism Div.
800.888.8839
www.traveltex.com

Utah
Utah Office of Tourism
800.882.4386, 801.538.1900
www.utah.com

Vermont
Vermont Dept. of Tourism & Marketing
800.837.6668, 802.828.3237
www.vermontvacation.com

Virginia
Virginia Tourism Corporation
800.847.4882, 804.545.5500
www.virginia.org

Washington
Washington State Tourism Division
800.544.1800
www.experiencewashington.com

West Virginia
West Virginia Division of Tourism
800.225.5982, 304.558.2200
www.wvtourism.com

Wisconsin
Wisconsin Department of Tourism
800.432.8747, 608.266.2161
www.travelwisconsin.com

Wyoming
Wyoming Travel & Tourism
800.225.5996, 307.777.7777
www.wyomingtourism.org

Puerto Rico
Puerto Rico Tourism Company
800.866.7827, 787.721.2400
www.gotopuertorico.com

CANADA

Alberta
Travel Alberta Canada
800.252.3782, 780.427.4321
www.travelalberta.com

British Columbia
Tourism British Columbia
800.435.5622, 250.356.6363
www.hellobc.com

Manitoba
Travel Manitoba
800.665.0040
www.travelmanitoba.com

New Brunswick
Tourism Communication Center
800.561.0123
www.tourismnewbrunswick.ca

Newfoundland & Labrador
Newfoundland & Labrador Tourism
800.563.6353, 709.729.2830
www.gov.nl.ca/tourism

Nova Scotia
Tourism Nova Scotia
800.565.0000, 902.425.5781
www.novascotia.com

Ontario
Ontario Tourism
800.668.2746, 905.282.1721
www.ontariotravel.net

Prince Edward Island
Tourism PEI
800.463.4734, 902.368.4444
www.gov.pe.ca/visitorsguide

Québec
Tourisme Québec
877.266.5687, 514.873.2015
www.bonjourquebec.com

Saskatchewan
Tourism Saskatchewan
877.237.2273, 306.787.9600
www.sasktourism.com

MEXICO

Mexico Ministry of Tourism
800.446.3942
www.visitmexico.com

BORDER CROSSING INFORMATION

TRAVEL ADVISORY

All travelers journeying to or from Canada, Mexico, Central and South America, and the Caribbean by air are now required to have a valid passport. Travelers to or from these countries by land or sea (including ferries) will soon be required to have a valid passport. Implementation of this requirement may occur on January 1, 2008. U.S. citizens traveling directly to or from Puerto Rico and the U.S. Virgin Islands are not required to have a passport. For more detailed information and updated schedules, please see http://travel.state.gov.

CANADA

New air travel regulations are in effect (see Travel Advisory). For land and sea travel, the old regulations remain in effect (temporarily), requiring U.S citizens to present either a passport or other proof of U.S. citizenship accompanied by photo identification. U.S. citize[n]s entering from a third country must hav[e] a valid passport. Naturalized citizens and alien permanent residents should carry the appropriate official documen[ta]tation. Individuals under the age of 18 traveling alone, with one parent, or wi[th] other adults must carry notarized pare[n]tal/legal guardian authorization.

U.S. driver's licenses are valid in Canada. Drivers should be prepared t[o] present proof of their vehicle's registra[-]tion, ownership, and insurance.

UNITED STATES (FROM CANADA)

New air travel regulations are in effe[ct] (see Travel Advisory). For land and se[a] travel, the old regulations remain in effect (temporarily), requiring Canadia[n] citizens to present either a passport o[r] other proof of citizenship accompanie[d] by photo identification. Visas are not required for customary tourist travel. Individuals under the age of 18 traveli[ng] alone, with one parent, or with other adults must carry notarized parental/ legal guardian authorization.

Canadian driver's licenses are vali[d] in the U.S. for one year. Drivers should be prepared to present proof of their vehicle's registration, ownership, and insurance.

MEXICO

New air travel regulations are in effec[t] (see Travel Advisory). For land and sea travel, the old regulations remain in effect (temporarily), requiring U.S. citizens to present either a passport or other proof of U.S. citizenship accompanied by photo identification. Passports are strongly recommended. Visas are not required for stays of up to 180 days. Naturalized citizens and alien permanent residents should carr[y] the appropriate official documentation. Individuals under the age of 18 traveling alone, with one parent, or with other adults must carry notarized parental/legal guardian authorization. All U.S. citizens visiting for up to 180 days must also procure a tourist card, obtainable from Mexican consulates, tourism offices, border crossing points and airlines serving Mexico. However, tourist cards are not needed for visits shorter than 72 hours to areas within t[he] Border Zone (extending approximately 25 km into Mexico)

U.S. driver's licenses are valid in Mexico. Visitors who wish to drive beyond the Baja California Peninsula [or] the Border Zone must obtain a tempo[-]rary import permit for their vehicles. T[o] acquire a permit, one must submit evi[-]dence of citizenship and of the vehicl[e] title and registration, as well as a vali[d] driver's license. A processing fee mus[t] be paid. Permits are available at any Mexican Army Bank (Banjercito) loca[ted] at border crossings or selected Mexi[can] consulates. Mexican law also require[s] the posting of a refundable bond, via credit card or cash, at the Banjercito [to] guarantee the departure of the vehicl[e]. Do not deal with any individual opera[t]ing outside of official channels.

All visitors driving in Mexico shou[ld] be aware that U.S. auto insurance policies are not valid and that buying short-term tourist insurance is mand[a]tory. Many U.S. insurance companies sell Mexican auto insurance. Americ[an] Automobile Association (for members only) and Sanborn's Mexico Insuranc[e] (800.638.9423) are popular companies with offices at most U.S. border crossings.

GRADE 5

Parrot Fire Kris Northern

"Rather than zoom into the fractal you can zoom into the edge of it and continually find the same pattern repeating itself much like the shoreline of a lake viewed from a plane."– **Kris Northern**

Number Puzzles and Multiple Towers

Multiplication and Division 1

UNIT 1

Investigations

IN NUMBER, DATA, AND SPACE®

PEARSON
Scott Foresman
scottforesman.com

Editorial offices: Glenview, Illinois • Parsippany, New Jersey • New York, New York
Sales offices: Boston, Massachusetts • Duluth, Georgia
Glenview, Illinois • Coppell, Texas • Sacramento, California • Mesa, Arizona

TERC

The Investigations curriculum was developed by TERC, Cambridge, MA.

NSF

This material is based on work supported by the National Science Foundation ("NSF") under Grant No. ESI-0095450. Any opinions, findings, and conclusions or recommendations expressed in this material are those of the author(s) and do not necessarily reflect the views of the National Science Foundation.

ISBN: 0-328-23762-0

ISBN: 978-0-328-23762-3

4 5 6 7 8 9 10-V003-15 14 13 12 11 10 09 08 07

CC:N1

TERC

Co-Principal Investigators

Susan Jo Russell

Karen Economopoulos

Authors

Lucy Wittenberg
Director Grades 3–5

Karen Economopoulos
Director Grades K–2

Virginia Bastable
(SummerMath for Teachers, Mt. Holyoke College)

Katie Hickey Bloomfield

Keith Cochran

Darrell Earnest

Arusha Hollister

Nancy Horowitz

Erin Leidl

Megan Murray

Young Oh

Beth W. Perry

Susan Jo Russell

Deborah Schifter
(Education Development Center)

Kathy Sillman

Administrative Staff

Amy Taber
Project Manager

Beth Bergeron

Lorraine Brooks

Emi Fujiwara

Contributing Authors

Denise Baumann

Jennifer DiBrienza

Hollee Freeman

Paula Hooper

Jan Mokros

Stephen Monk
(University of Washington)

Mary Beth O'Connor

Judy Storeygard

Cornelia Tierney

Elizabeth Van Cleef

Carol Wright

Technology

Jim Hammerman

Classroom Field Work

Amy Appell

Rachel E. Davis

Traci Higgins

Julia Thompson

Note: Unless otherwise noted, all contributors listed above were staff of the Education Research Collaborative at TERC during their work on the curriculum. Other affiliations during the time of development are listed.

Collaborating Teachers

This group of dedicated teachers carried out extensive field testing in their classrooms, met regularly to discuss issues of teaching and learning mathematics, provided feedback to staff, welcomed staff into their classrooms to document students' work, and contributed both suggestions and written material that has been incorporated into the curriculum.

Bethany Altchek
Linda Amaral
Kimberly Beauregard
Barbara Bernard
Nancy Buell
Rose Christiansen
Chris Colbath-Hess
Lisette Colon
Kim Cook
Frances Cooper
Kathleen Drew
Rebeka Eston Salemi
Thomas Fisher
Michael Flynn
Holly Ghazey
Susan Gillis
Danielle Harrington
Elaine Herzog
Francine Hiller
Kirsten Lee Howard
Liliana Klass
Leslie Kramer
Melissa Lee Andrichak
Kelley Lee Sadowski
Jennifer Levitan
Mary Lou LoVecchio
Kristen McEnaney
Maura McGrail
Kathe Millett
Florence Molyneaux
Amy Monkiewicz
Elizabeth Monopoli
Carol Murray
Robyn Musser
Christine Norrman
Deborah O'Brien
Timothy O'Connor
Anne Marie O'Reilly
Mark Paige
Margaret Riddle
Karen Schweitzer
Elisabeth Seyferth
Susan Smith
Debra Sorvillo
Shoshanah Starr
Janice Szymaszek
Karen Tobin
JoAnn Trauschke
Ana Vaisenstein
Yvonne Watson
Michelle Woods
Mary Wright

Advisors

Deborah Lowenberg Ball,
University of Michigan

Hyman Bass, Professor of Mathematics and Mathematics Education
University of Michigan

Mary Canner, Principal, Natick Public Schools

Thomas Carpenter, Professor of Curriculum and Instruction,
University of Wisconsin-Madison

Janis Freckmann, Elementary Mathematics Coordinator,
Milwaukee Public Schools

Lynne Godfrey, Mathematics Coach,
Cambridge Public Schools

Ginger Hanlon, Instructional Specialist in Mathematics,
New York City Public Schools

DeAnn Huinker, Director, Center for Mathematics and
Science Education Research, University of Wisconsin-Milwaukee

James Kaput, Professor of Mathematics, University of
Massachusetts-Dartmouth

Kate Kline, Associate Professor, Department of Mathematics
and Statistics, Western Michigan University

Jim Lewis, Professor of Mathematics,
University of Nebraska-Lincoln

William McCallum, Professior of Mathematics,
University of Arizona

Harriet Pollatsek, Professor of Mathematics,
Mount Holyoke College

Debra Shein-Gerson, Elementary Mathematics Specialist,
Weston Public Schools

Gary Shevell, Assistant Principal,
New York City Public Schools

Liz Sweeney, Elementary Math Department,
Boston Public Schools

Lucy West, Consultant, Metamorphosis:
Teaching Learning Communities, Inc.

This revision of the curriculum was built on the work of the many authors who contributed to the first edition (published between 1994 and 1998). We acknowledge the critical contributions of these authors in developing the content and pedagogy of *Investigations*:

Authors

Joan Akers

Michael T. Battista

Douglas H. Clements

Karen Economopoulos

Marlene Kliman

Jan Mokros

Megan Murray

Ricardo Nemirovsky

Andee Rubin

Susan Jo Russell

Cornelia Tierney

Contributing Authors

Mary Berle-Carman

Rebecca B. Corwin

Rebeka Eston

Claryce Evans

Anne Goodrow

Cliff Konold

Chris Mainhart

Sue McMillen

Jerrie Moffet

Tracy Noble

Kim O'Neil

Mark Ogonowski

Julie Sarama

Amy Shulman Weinberg

Margie Singer

Virginia Woolley

Tracey Wright

Contents

UNIT 1

Number Puzzles and Multiple Towers

Investigations

C U R R I C U L U M

Overview of Program Components

FOR TEACHERS

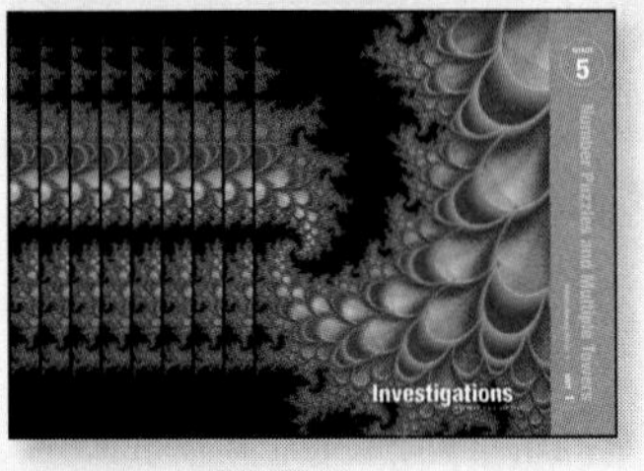

The **Curriculum Units** are the teaching guides. (See far right.)

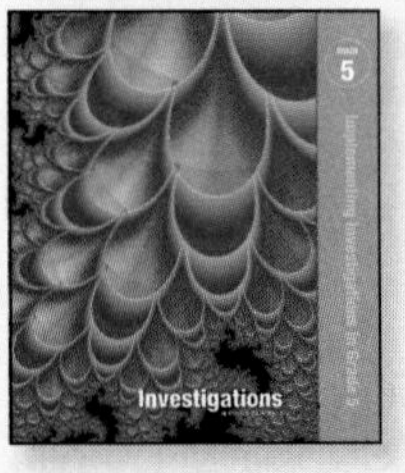

Implementing Investigations in Grade 5 offers suggestions for implementing the curriculum. It also contains a comprehensive index.

The **Resources Binder** contains all the Resource Masters and Transparencies that support instruction. (Also available on CD.) The binder also includes a student software CD.

FOR STUDENTS

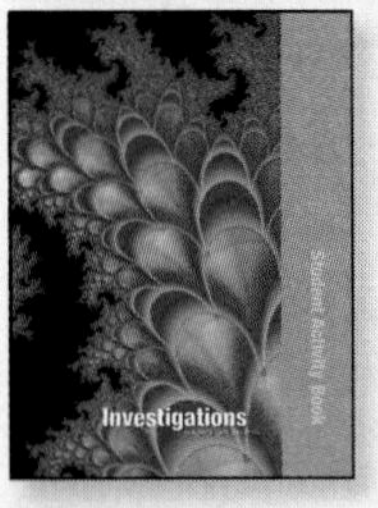

The **Student Activity Book** contains the consumable student pages (Recording Sheets, Homework, Practice, and so on).

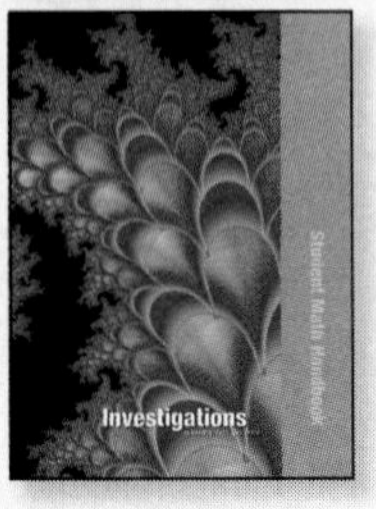

The **Student Math Handbook** contains Math Words and Ideas pages and Games directions.

The *Investigations* Curriculum

Investigations in Number, Data, and Space® is a K–5 mathematics curriculum designed to engage students in making sense of mathematical ideas. Six major goals guided the development of the *Investigations in Number, Data, and Space*® curriculum. The curriculum is designed to:

- Support students to make sense of mathematics and learn that they can be mathematical thinkers
- Focus on computational fluency with whole numbers as a major goal of the elementary grades
- Provide substantive work in important areas of mathematics—rational numbers, geometry, measurement, data, and early algebra—and connections among them
- Emphasize reasoning about mathematical ideas
- Communicate mathematics content and pedagogy to teachers
- Engage the range of learners in understanding mathematics

Underlying these goals are three guiding principles that are touchstones for the *Investigations* team as we approach both students and teachers as agents of their own learning:

1. *Students have mathematical ideas.* Students come to school with ideas about numbers, shapes, measurements, patterns, and data. If given the opportunity to learn in an environment that stresses making sense of mathematics, students build on the ideas they already have and learn about new mathematics they have never encountered. Students learn that they are capable of having mathematical ideas, applying what they know to new situations, and thinking and reasoning about unfamiliar problems.

2. *Teachers are engaged in ongoing learning* about mathematics content, pedagogy, and student learning. The curriculum provides material for professional development, to be used by teachers individually or in groups, that supports teachers' continued learning as they use the curriculum over several years. The *Investigations* curriculum materials are designed as much to be a dialogue with teachers as to be a core of content for students.

3. *Teachers collaborate with the students and curriculum materials* to create the curriculum as enacted in the classroom. The only way for a good curriculum to be used well is for teachers to be active participants in implementing it. Teachers use the curriculum to maintain a clear, focused, and coherent agenda for mathematics teaching. At the same time, they observe and listen carefully to students, try to understand how they are thinking, and make teaching decisions based on these observations.

Investigations is based on experience from research and practice, including field testing that involved documentation of thousands of hours in classrooms, observations of students, input from teachers, and analysis of student work. As a result, the curriculum addresses the learning needs of real students in a wide range of classrooms and communities. The investigations are carefully designed to invite all students into mathematics—girls and boys; members of diverse cultural, ethnic, and language groups; and students with a wide variety of strengths, needs, and interests.

Based on this extensive classroom testing, the curriculum takes seriously the time students need to develop a strong conceptual foundation and skills based on that foundation. Each curriculum unit focuses on an area of content in depth, providing time for students to develop and practice ideas across a variety of activities and contexts that build on each other. Daily guidelines for time spent on class sessions, Classroom Routines (K–3), and Ten-Minute Math (3–5) reflect the commitment to devoting adequate time to mathematics in each school day.

About This Curriculum Unit

This **Curriculum Unit** is one of nine teaching guides in Grade 5. The first unit in Grade 5 is *Number Puzzles and Multiple Towers.*

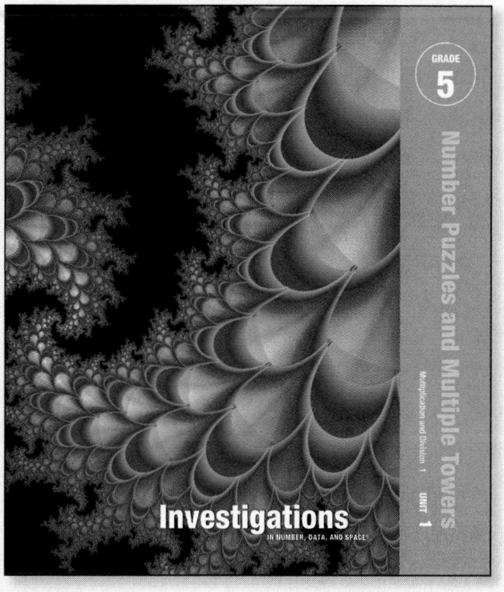

- The **Introduction and Overview** section organizes and presents the instructional materials, provides background information, and highlights important features specific to this unit.
- Each Curriculum Unit contains several **Investigations.** Each Investigation focuses on a set of related mathematical ideas.
- Investigations are divided into one-hour **Sessions,** or lessons.
- Sessions have a combination of these parts: **Activity, Discussion, Math Workshop, Assessment Activity,** and **Session Follow-Up.**
- Each session also has one or more **Ten-Minute Math** activities that are done outside of math time.
- At the back of the book is a collection of **Teacher Notes** and **Dialogue Boxes** that provide professional development related to the unit.
- Also included at the back of the book are the **Student Math Handbook** pages for this unit.
- The **Index** provides a way to look up important words or terms.

Overview

OF THIS UNIT

Investigation	Session	Day
INVESTIGATION 1 **Finding Factors and Prime Factors** Students use arrays and number puzzles to learn about factors, multiples, and other properties of numbers.	**1.1** Building and Using Arrays	1
	1.2 Identifying Properties of Numbers	2
	1.3 What Numbers Have Which Properties?	3
	1.4 Multiplying with More Than Two Numbers	4
	1.5 Assessment: Number Puzzles and Finding Factors	5
	1.6 Number Puzzles and Finding Factors, *continued*	6
	1.7 Prime Factorization	7
INVESTIGATION 2 **Multiplication Strategies** Students develop a number of strategies for solving 2-digit by 2-digit multiplication problems, including breaking numbers apart, solving an equivalent problem, and solving related problems.	**2.1** Naming Multiplication Strategies	8
	2.2 Comparing Representations	9
	2.3 Which Product Is Greater?	10
	2.4 Multiplication Cluster Problems	11
	2.5 Multiplication Cluster Problems, *continued*	12
	2.6 How Do I Start?	13
	2.7 Assessment: What Is the Answer?	14
INVESTIGATION 3 **Division Strategies** Students develop various strategies for solving division problems with 2-digit divisors and for interpreting the results.	**3.1** Solving a Division Problem	15
	3.2 Multiple Towers	16
	3.3 Solving More Division Problems	17
	3.4 Multiplication and Division Relationships on the Multiple Tower	18
	3.5 Division Cluster Problems	19
	3.6 Practicing Division Strategies	20
	3.7 Practicing Division Strategies, *continued*	21
	3.8 End-of-Unit Assessment	22

Each *Investigations* session has some combination of these five parts: **Activity, Discussion, Math Workshop, Assessment Activity,** and **Session Follow-Up.** These session parts are indicated in the chart below. Each session also has one or more **Ten-Minute Math** activities that are done outside of math time.

Activity	Discussion	Math Workshop	Assessment Activity	Session Follow-Up
●●●				●
●	●●			●
●●●				●
●				●
●	●	●		●
	●	●		●
●	●			●
●	●			●
●●	●			●
●	●●			●
●		●		●
	●	●		●
●	●			●
●			●	●
●	●			●
●●●				●
●	●			●
●	●			●
●	●			●
●	●	●		●
	●	●		●
			●	●

Ten-Minute Math

Quick Images	*Number Puzzles*
●	
●	
●	
●	
●	
●	
●	
●	
	●
	●
	●
	●
	●
●	
●	
●	
●	
	●
	●
	●
	●
	●

Mathematics

I N T H I S U N I T

Number Puzzles and Multiple Towers, which focuses on the operations of multiplication and division, is the first Grade 5 unit in the number and operations strand of *Investigations.* These units develop ideas about the meaning of operations with whole numbers, the development of computational fluency, the structure of place value and the base-ten number system, and generalizations about numbers and operations.

LOOKING BACK

In Grades 3 and 4, students learned that multiplication is used to combine a number of equal-sized groups. They also learned that division is used when the number of something is known, and either the number of equal-sized groups or the size of those groups is to be determined. Students have had experience representing multiplication with rectangular arrays and other representations and have related multiplication and division expressions to story-problem contexts. Students have discussed and used the relationship between multiplication and division to solve problems. In Grade 4, students worked on breaking numbers apart by 10s and 1s in order to more easily multiply large numbers, and they will continue that work in this unit.

The work in this unit assumes that students are fluent in the basic combinations for addition, subtraction, and multiplication. (The basic multiplication combinations will be reviewed in this unit for any students still needing to learn them.) It is expected that students are able to add and subtract fluently. They should understand the meaning of the operations of multiplication and division and be able to use some computational strategies to solve multiplication problems. They should be able to find the solution for a division problem, but may not yet have a fluent strategy.

This unit focuses on 4 Mathematical Emphases:

1 Whole-Number Operations **Reasoning about numbers and their factors**

Math Focus Points

- Determining whether one number is a factor or multiple of another
- Identifying prime, square, even, and odd numbers
- Using known multiplication combinations to find equivalent multiplication combinations (e.g., $18 = 3 \times 6 = 3 \times (2 \times 3)$)
- Using known multiplication combinations to find multiplication combinations for numbers related by place value (e.g., $3 \times 6 = 18$; $3 \times 6 \times 10 = 180$)
- Finding all the ways to multiply whole numbers for a given product
- Finding all the factors of a number
- Using properties (even, odd, prime, square) and relationships (factor, multiple) of numbers to solve problems
- Determining the prime factorization of a number

Understanding the relationship between a number and its factors, and being able to recognize patterns in the multiples of a given number, are critical elements in developing useful mental strategies. The first Investigation in this unit focuses on finding all of the ways to multiply whole numbers to equal a given product. Fifth graders are intrigued by finding all of the ways a number can be broken into more than two factors by multiplication ($18 = 2 \times 3 \times 3$) and by finding the longest factor string, or the *prime factorization of,* a number.

The purpose of this work for fifth graders is to increase their knowledge of and flexibility with multiplication relationships and to learn more about how one multiplication combination can generate other equivalent combinations.

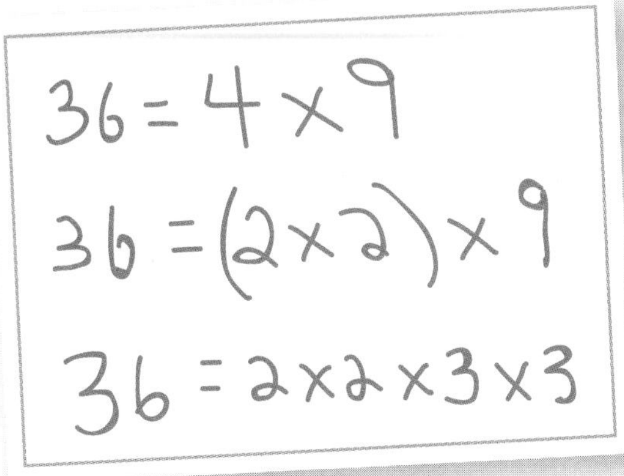
36 = 4 x 9
36 = (2 x 2) x 9
36 = 2 x 2 x 3 x 3

Sample Student Work

Such work with multiplication relationships supports students' mental computation strategies with whole numbers. Later, factorization will become important for their work with fractions. Through this work, fifth graders also begin to understand ideas that underlie the Fundamental Theorem of Arithmetic. See the **Teacher Note:** Finding Prime Factors, page 154, for more information on prime factorization and this important mathematical idea.

Students continue their work on factors and multiples in the third Investigation, as they build "multiple towers" (sequences of the multiples of numbers) that further their understanding of the relationship between multiplication and division.

2 Computational Fluency **Solving multiplication problems with 2-digit numbers**

Math Focus Points

- Solving 2-digit by 2-digit multiplication problems
- Describing and comparing strategies used to solve multiplication problems
- Breaking up multiplication problems efficiently
- Multiplying fluently by multiples of 10
- Estimating the product of two numbers
- Comparing multiplication problems to determine which product is greater

When students solve multiplication problems, they most often use strategies that involve breaking numbers apart to create problems that are manageable and that make use of familiar number relationships. They break one or both factors apart, multiply each part of one factor by each part of the other factor, and then combine all of the partial products.

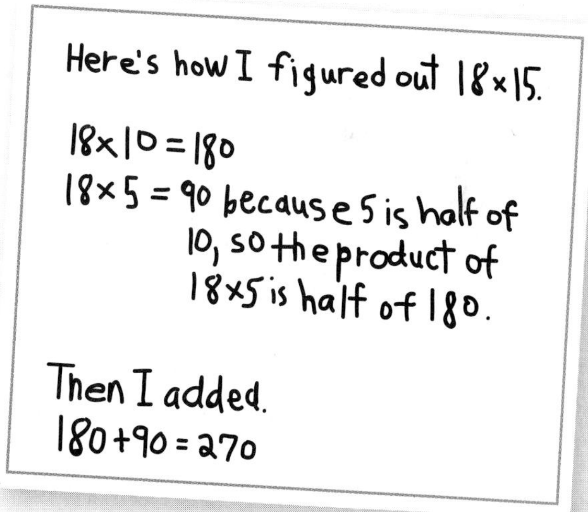
Here's how I figured out 18×15.
18×10 = 180
18×5 = 90 because 5 is half of 10, so the product of 18×5 is half of 180.
Then I added.
180+90 = 270

Sample Student Work

Students should come into Grade 5 with good strategies for multiplying 2-digit numbers. This first unit of the year gives you the opportunity to assess students' understanding of multiplication and the fluency of their problem-solving strategies.

For most students, this unit provides review and practice of the work they did in Grade 4, solidifying their understanding of the computational strategies they are using and, through the use of representations and story contexts, connecting these strategies to the meaning of multiplication. By the end of this unit, students should be able to solve multiplication problems accurately and efficiently.

Throughout the unit, ask students to explain the way they break apart problems into subproblems by using representations, such as arrays, and by talking through how their strategies relate to story contexts. This work helps all students develop mental images of multiplication that are critical to solving multidigit problems. See the **Teacher Note:** Visualizing Arrays, page 159, and the **Teacher Note:** Developing Computation Strategies That Make Sense, page 171, for more about the use of arrays and story contexts.

Underlying the strategy of breaking apart problems is the distributive property of multiplication.

Although it is not important for fifth graders to identify the distributive property by name, it is important that teachers understand how students' strategies relate to this property. Teachers also need to understand how the property enables students to keep track of which numbers must be multiplied when they break numbers apart in a multidigit multiplication problem. The distributive property is a foundation for students' computational fluency in whole-number multiplication and division and for later work in algebra. Read more about how students are using the distributive property in their work in **Algebra Connections in this Unit,** page 16.

Most students solve multidigit multiplication problems by breaking up one or both factors, but some students use other methods, such as doubling and halving. See the **Teacher Note:** Multiplication Strategies, page 161, for an overview of student strategies.

3 Whole-Number Operations **Understanding and using the relationship between multiplication and division to solve division problems**

Math Focus Points

- Solving division problems with 2-digit divisors
- Using knowledge of multiples of 10 to solve division problems
- Using and interpreting notation that represents division and relating division and multiplication notations (e.g., $170 \div 15 = \_\_\_$ and $\_\_\_ \times 15 = 170$)
- Describing and comparing strategies used to solve division problems
- Comparing division problems to determine which quotient is greater
- Solving a division problem by breaking the dividend into parts

Students continue to learn ways to solve division problems efficiently. This unit focuses on the relationship between multiplication and division. Students build multiple towers—sequences of multiples of a number—and use these towers to solve multiplication and division problems.

The work focuses on two key ideas. One is that division problems can be solved by relating them to missing-factor problems (e.g., $462 \div 21 = \_\_\_$ and $\_\_\_ \times 21 = 462$). Many students learn to solve division problems efficiently by building up groups of the divisor, using multiplication.

Students might say:

"To figure out 462 ÷ 21, I thought about multiplying by 21. I know that ten 21s make 210. 210 and another 210 are 420—that's twenty 21s. There's 42 left to get to 462, and that's two more 21s. So that's twenty-two 21s. The answer is 22."

The other key idea focuses on multiplying by multiples of 10 to solve problems more efficiently. A student could solve $462 \div 21$ even more efficiently by recognizing that $20 \times 21 = 420$. By thinking about multiples of 10, students are better prepared to decompose division problems to make the problems easier to solve.

The inverse relationship between multiplication and division is one of the building blocks of later work in algebra. See **Algebra Connections in This Unit,** page 16.

As in their work on multiplication, students develop representations and story contexts for division that help them think through and keep track of the parts of a problem. Here, too, the distributive property is central.

4 Whole Number Operations **Representing the meaning of multiplication and division**

Math Focus Points

- Writing multiplication equations that describe dot arrangements
- Using arrays to model multiplication
- Representing a multiplication or division problem with a picture or diagram
- Creating a story problem represented by a multiplication or division expression
- Making sense of remainders in terms of problem contexts

Students continue to draw arrays and pictures to model and solve problems and to clarify and communicate their thinking. Students also create story problems to represent the total number of groups and the numbers in each group. As students solve multiplication problems with 2-digit and 3-digit numbers, they often use strategies that involve breaking apart one or both numbers. It is important that students find effective ways to keep track of what parts of the problem they have solved and what is left to solve. This is done effectively by sketching an array, drawing groups, or creating a context.

This Unit also focuses on

- Using clear and concise notation
- Identifying and learning multiplication combinations ("facts") not yet known fluently

Ten-Minute Math activities focus on

- Organizing and analyzing visual images
- Developing language and concepts needed to communicate spatial relationships
- Writing equations to describe dot patterns
- Identifying prime, square, even, and odd numbers
- Determining if one number is a factor or multiple of another

LOOKING FORWARD

Students continue practicing multiplication and division throughout the year in homework, practice pages, and in solving problems in later units. In *How Many People? How Many Teams?* students continue studying multiplication with an emphasis on larger numbers. In that unit, they also study the U.S. algorithm for multiplication and become more fluent with division strategies. Students continue to investigate the properties of whole numbers and operations. This supports their current work on computational fluency and understanding the base-ten number system. It also builds knowledge they can apply in later years as they work with integers, rational numbers, and variables.

Assessment

IN THIS UNIT

ONGOING ASSESSMENT: Observing Students at Work

The following sessions provide **Ongoing Assessment: Observing Students at Work** opportunities:

- **Session 1.1, pp. 31 and 35**
- **Session 1.2, p. 39**
- **Session 1.3, pp. 43 and 45**
- **Session 1.4, p. 50**
- **Session 1.5, p. 55**
- **Session 1.6, p. 58**
- **Session 1.7, p. 64**
- **Session 2.1, p. 74**
- **Session 2.2, pp. 79 and 83**
- **Session 2.3, p. 89**
- **Session 2.4, pp. 94 and 96**
- **Session 2.6, p. 102**
- **Session 2.7, pp. 106 and 107**
- **Session 3.1, p. 117**
- **Session 3.2, p. 124**
- **Session 3.3, p. 128**
- **Session 3.4, p. 133**
- **Session 3.5, p. 139**
- **Session 3.6, p. 144**
- **Session 3.8, p. 152**

WRITING OPPORTUNITIES

The following sessions have **writing** opportunities for students to explain their mathematical thinking:

- **Session 1.4, p. 49**
  Student Activity Book, p. 12
- **Session 1.6, pp. 58 and 59**
  Student Activity Book, pp. 19–20
- **Session 2.3, p. 89**
  Student Activity Book, p. 31
- **Session 3.6, p. 146**
  Student Activity Book, pp. 69–70

PORTFOLIO OPPORTUNITIES

The following sessions have work appropriate for a **portfolio**:

- **Session 2.2, p. 83**
  Student Activity Book, p. 27
- **Session 2.7, p. 107**
  M51, Assessment: What Is the Answer?
- **Session 3.1, p. 119**
  Student Activity Book, p. 48
- **Session 3.4, p. 132**
  Student Activity Book, p. 55
- **Session 3.8, p. 152**
  M54–M55, End-of-Unit Assessment

Assessing the Benchmarks

Observing students as they engage in conversation about their ideas is a primary means to assess their mathematical understanding. Consider all of your students' work, not just the written assessments. See the chart below for suggestions about key activities to observe.

Checklist Available

Benchmarks in This Unit	Key Activities to Observe	Assessment
1. Find the factors of a number.	**Session 1.4:** Finding Multiplication Combinations for 18 and 180 **Session 1.7:** Factors of Larger Numbers	**Sessions 1.5–1.6 Assessment Activity:** Number Puzzles **Session 3.8 Unit Assessment:** Problem 2
2. Solve multiplication problems efficiently.	**Session 2.1:** Solving 35 × 28 **Session 2.6:** Starter Problems	**Session 2.7 Assessment Activity:** What Is the Answer? **Session 3.8 Unit Assessment:** Problem 1
3. Solve division problems with 1-digit and 2-digit divisors.	**Session 3.1:** Solving a Division Problem **Session 3.7:** Practicing Division Strategies	**Session 3.8 Unit Assessment:** Problem 3

Relating the Mathematical Emphases to the Benchmarks

Mathematical Emphases	Benchmarks
Whole Number Operations Reasoning about numbers and their factors	1
Computational Fluency Solving multiplication problems with 2-digit numbers	2
Whole Number Operations Understanding and using the relationship between multiplication and division to solve division problems	2, 3
Whole Number Operations Representing the meaning of multiplication and division	2, 3

Benito: My method would work with other numbers too. I just have to make sure I have all of the people and they have all of the paper. It would work with all numbers.

In the vignette, Benito uses a story context to keep track of all the parts of the multiplication problem when he splits up both factors. The story context not only supports his thinking about this problem, but also illustrates how the distributive property can be applied twice. Consider (12×26). First split apart the 12 into $10 + 2$: $(12 \times 26) = (10 \times 26) + (2 \times 26)$. Then split apart the 26 into $20 + 6$. $(12 \times 26) = (10 \times 20) + (10 \times 6) + (2 \times 20) + (2 \times 6)$. Nora's diagram is a visual image that matches both the story context and the arithmetic expressions that add to the final product. Both the diagram and the story context serve as an expression of the distributive property.

Understanding Division and Multiplication as Inverse Operations

Just as younger children often use an addition strategy to solve problems adults might consider subtraction, so will your students use multiplication to solve problems that you might consider division.

Consider the following vignette:

Alicia, Lourdes, and Renaldo worked on a story problem.

There are 168 fifth graders. For field day we need to have teams of 14. How many teams can we make?

Alicia: I know that 14×10 is 140. Two 14s is 28 and so I'm up to 168. That means $10 + 2$, or 12 teams.

Lourdes: I used multiple towers for 14. 14, 28, 42, 56, 70—I can double that to get 140. Then 154, 168. So it's 5, 10 . . . then another 2; the answer is 12.

Renaldo: I took cubes 14 at a time and lined them up until I had 168. I made 12 stacks of cubes.

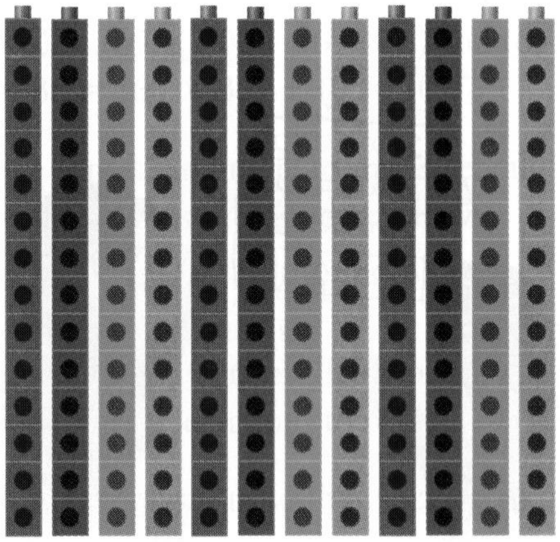

In this vignette, Alicia, Lourdes, and Renaldo used what they know about multiplication to solve a division problem. Alicia used the distributive property of multiplication, Lourdes used multiple towers, and Renaldo made groups of equal size. The fact that these students can interpret their multiplication strategies to answer a division problem correctly is based on the *inverse relationship* between multiplication and division. In this vignette, since $12 \times 14 = 168$, then $168 \div 14 = 12$. This can be stated in more general terms.

If $a \times b = c$, then $c \div b = a$, for $b \neq 0$.

Encourage students to articulate their reasoning by asking questions such as the following:

"Why is it that a strategy using multiplication solves this problem?"

"How is the multiple tower strategy related to the original problem?"

"How does the doubling method relate to the original problem?"

"How does the cube arrangement illustrate multiplication?"

"How does the cube arrangement illustrate division?"

Student responses to such questions will help them articulate and clarify their own thinking about the relationship between multiplication and division, and will form the basis for future work examining these operations.

Investigations students are encouraged to verbalize the generalizations they see about numbers and operations, and to explain and justify them using materials and tools, such as cubes or diagrams. For most adults, notation such as the use of variables, operations, and equal signs is the chief identifying feature of algebra. The notation, however, expresses rules about how operations work that students can reason out for themselves. This reasoning—about how numbers can be put together and taken apart under different operations—not the notation, is the work of elementary students in algebra.

Note: In the text of the sessions, you will find Algebra Notes in the sidebar that identify where these early algebra discussions are likely to arise. Some of the **Teacher Notes** and **Dialogue Boxes** further elaborate the ideas and illustrate students' conversations about them.

Ten-Minute Math

IN THIS UNIT

Ten-Minute Math offers practice and review of key concepts for this grade level. These daily activities, to be done in ten minutes outside of math class, are introduced in a unit and repeated throughout the grade. Specific directions for the day's activity are provided in each session. For the full description and variations of each classroom activity, see *Implementing Investigations in Grade 5.*

Activity	Introduced	Full Description of Activity and Its Variations
Quick Images	Unit 1, Session 1.1 (this unit)	*Implementing Investigations in Grade 5*
Number Puzzles	Unit 1, Session 2.2 (this unit)	*Implementing Investigations in Grade 5*

Quick Images

Students visualize and analyze images of shape patterns. After briefly viewing an image, students determine the number of shapes in a pattern and write multiplication and/or division equations to represent how they organized their count.

Math Focus Points

- Organizing and analyzing visual images
- Writing multiplication and division equations to represent the total number of shapes in a pattern

Number Puzzles

Students use reasoning and knowledge of properties and relationships of numbers to find a given number or numbers that fit three given clues. Each clue gives one characteristic of the number.

Math Focus Points

- Identifying prime, square, even, and odd numbers
- Determining if one number is a factor or multiple of another

Practice and Review

IN THIS UNIT

Practice and review play a critical role in the *Investigations* program. The following components and features are available to provide regular reinforcement of key mathematical concepts and procedures.

Books	Features	In This Unit . . .
Curriculum Unit	**Ten-Minute Math** offers practice and review of key concepts for this grade level. These daily activities, to be done in ten minutes outside of math class, are introduced in a unit and repeated throughout the grade. Specific directions for the day's activity are provided in each session. For the full description and variations of each classroom activity, see *Implementing Investigations in Grade 5.*	• **All sessions**
Student Activity Book	**Daily Practice** pages in the *Student Activity Book* provide one of three types of written practice: **reinforcement** of the content of the unit, **ongoing review,** or **enrichment** opportunities. Some Daily Practice pages will also have Ongoing Review items with multiple-choice problems similar to those on standardized tests.	• **All sessions**
	Homework pages in the *Student Activity Book* are an extension of the work done in class. At times they help students prepare for upcoming activities.	• **Session 1.1** • **Session 1.3** • **Session 1.4** • **Session 1.5** • **Session 1.6** • **Session 1.7** • **Session 2.1** • **Session 2.2** • **Session 2.3** • **Session 2.4** • **Session 3.1** • **Session 3.2** • **Session 3.4** • **Session 3.5** • **Session 3.6**
Student Math Handbook	**Math Words and Ideas** in the *Student Math Handbook* are pages that summarize key words and ideas. Most Words and Ideas pages have at least one exercise.	• **Student Math Handbook, pp. 14–39**
	Games pages are found in a section of the *Student Math Handbook.*	• **Student Math Handbook, pp. G6, G10**

Differentiation

IN THIS UNIT

Supporting the Range of Learners

Sessions	1.1	1.2	1.3	1.4	1.7	2.1	2.2	2.3	2.4	2.6	2.7	3.1	3.2	3.3	3.4	3.5	3.6	3.8
Intervention	•	•		•	•	•	•	•	•	•	•		•	•	•	•	•	•
Extension		•	•	•	•				•	•	•					•	•	
ELL	•					•		•				•						

Intervention

Suggestions are made to support and engage students who are having difficulty with a particular idea, activity, or problem.

Extension

Suggestions are made to support and engage students who finish early or may be ready for additional challenge.

English Language Learners (ELL)

Like their English-speaking classmates, English Language Learners will be called upon to develop story contexts for multiplication and division problems and to develop and refine strategies for solving these problems. These tasks present challenges for some English Language Learners, who may have an understanding of the mathematical concepts, but lack the language to explain their reasoning. You can help by using visuals to promote understanding of contexts and by highlighting or reviewing key vocabulary and language structures.

To participate fully in discussions about multiplication and division strategies, English Language Learners may need assistance forming questions in English. You can model simple questions using various question words such as *what, why,* and *how.*

Students must use sequential terminology as they explain the various steps used to solve complex problems. English Language Learners may need help expressing a sequence with words such as *first, next, then,* and *finally.* In activities that require students to compare their reasoning with that of their classmates, English Language Learners may also need to review expressions such as *same, similar, different, more than,* and *less than.*

To help English Language Learners with the Ten-Minute Math activity, *Quick Images: Seeing Numbers,* present words that describe how figures are arranged. You can use manipulatives or real-life objects to illustrate words such as *cluster, clump, row, column, group, section, line, portion, part, top,* and *bottom.*

Working with the Range of Learners: Classroom Cases is a set of episodes written by teachers that focuses on meeting the needs of the range of learners in the classroom. In the first section, *Setting up the Mathematical Community,* teachers write about how they create a supportive and productive learning environment in their classrooms. In the next section, *Accommodations for Learning,* teachers focus on specific modifications they make to meet the needs of some of their learners. In the last section, *Language and Representation,* teachers share how they help students use representations and develop language to investigate and express mathematical ideas. The questions at the end of each case provide a starting point for your own reflection or for discussion with colleagues. See *Implementing Investigations in Grade 5* for this set of episodes.

INVESTIGATION 1

Mathematical Emphases

Whole-Number Operations Reasoning about numbers and their factors

Math Focus Points

- Determining whether one number is a factor or multiple of another
- Identifying prime, square, even and odd numbers
- Using known multiplication combinations to find equivalent multiplication combinations (e.g., $18 = 3 \times 6 = 3 \times (2 \times 3)$)
- Using known multiplication combinations to find multiplication combinations for numbers related by place value (e.g., $3 \times 6 = 18$; $3 \times 6 \times 10 = 180$)
- Finding all the ways to multiply whole numbers for a given product
- Finding all the factors of a number
- Using properties (even, odd, prime, square) and relationships (factor, multiple) of numbers to solve problems
- Determining the prime factorization of a number

Whole Number Operations Representing the meaning of multiplication and division

Math Focus Points

- Writing multiplication equations that describe dot arrangements
- Using arrays to model multiplication

This Investigation also focuses on

- Identifying and learning multiplication combinations ("facts") not yet known fluently

Finding Factors and Prime Factors

Session	Student Activity Book	Student Math Handbook	Professional Development: Read Ahead of Time
SESSION 1.1 p. 28 **Building and Using Arrays** Students use multiplication equations to describe arrays of figures. They build arrays and identify factors and multiples of numbers.	1–4	16, 17	• **Mathematics in This Unit,** p. 10 • **Teacher Note:** Using Mathematical Vocabulary, p. 153 • **Part 4: Ten-Minute Math** in *Implementing Investigations in Grade 5,* Quick Images: Seeing Numbers
SESSION 1.2 p. 36 **Identifying Properties of Numbers** Students identify and discuss prime and square numbers. They use rectangular arrays to find numbers with given factors.	1–2, 5–7	21–22	
SESSION 1.3 p. 41 **What Numbers Have Which Properties?** Students use their knowledge of the meaning of *even, odd, prime, square, factor,* and *multiple* to identify numbers with particular combinations of these characteristics.	8–11	18, 19, 21–22	• **Dialogue Box:** Solving a Number Puzzle, p. 178
SESSION 1.4 p. 48 **Multiplying with More Than Two Numbers** Students use combinations they know to generate as many ways as possible of multiplying two, three, or more whole numbers to make the products 18 and 180.	12–14	23–24	• **Teacher Note:** Finding Prime Factors, p. 154 • **Dialogue Box:** Multiplying With More Than Two Numbers, p. 181 • **Part 5: Technology in Investigations: Calculators and Computers** in *Implementing Investigations in Grade 5:* Using Calculators with the Curriculum

Ten-Minute Math See page 20 for an overview.

Quick Images: Seeing Numbers

- **T1–T7, *Quick Images: Seeing Numbers***

Materials to Gather	Materials to Prepare
• **T1, *Quick Images: Seeing Numbers*** • **T13, Centimeter Grid Paper** • **Blank transparency** (1 per class) • **Color tiles** (70 per pair)	• **Chart paper** Make a chart titled "Math Vocabulary." • **M15, Centimeter Grid Paper** Make copies. (2 per student plus extras) • **M13 –M14, Family Letter** Make copies. (1 per student)
• **Chart: "Math Vocabulary"** (from Session 1.1) • **Color tiles** (as needed)	• **M15, Centimeter Grid Paper** Make copies. (as needed)
• **T16, 300 Chart** • **T14, Multiplication Combinations 1** • **T15, Multiplication Combinations Recording Sheet** • **Paper clips or envelopes** (about 15 per pair) • **Color tiles** (as needed) • **Calculators** (optional)	• **M16, Sample Number Puzzle** Make copies. Note that the page has two identical copies of the puzzle. Cut out the clues. Paper clip each set of four clues or place in envelopes. (1 set per group of 4) • **M31, 300 Chart** Make copies. (1 per student plus extras) • **M17–M22, *Number Puzzles* 1–12** Make copies. Cut out the clues. Paper clip each set or place in envelopes. (1 set per pair; Puzzles 8–12 optional in this session) • **M23–M26, Family Letter** Make copies. (1 per student)
• **Calculators** (optional)	• **M30, Multiplication Combinations Recording Sheet** Make copies. (as needed) • **M27–M28, Family Letter** Make copies. (1 per student)

Overhead Transparency

Finding Factors and Prime Factors, *continued*

Session	Student Activity Book	Student Math Handbook	Professional Development: Read Ahead of Time	
SESSION 1.5 p. 52				
Assessment: Number Puzzles and Finding Factors In the first Math Workshop of the year, students practice identifying numbers with common characteristics and continue working on ways to multiply two, three, or more numbers to make a given product. The Math Workshop includes an observed assessment.	8, 12, 15–17	25–29	• **Part 2: How to Use Investigations** in *Implementing Investigations in Grade 5*	
SESSION 1.6 p. 57				
Number Puzzles and Finding Factors, *continued* Students continue the Math Workshop and observed assessment from Session 1.5. The discussion focuses on finding all the factors for a given number.	12, 15, 19–22	18		
SESSION 1.7 p. 61				
Prime Factorization Students discuss prime factorization of numbers and find factors of larger numbers.	12, 15, 19–20, 23–24	23–24		

Materials to Gather	Materials to Prepare
• **M17–M22, Number Puzzles 1–12** (from Session 1.3) • **Color tiles** (as needed) • **Calculators** (optional)	• **M31, 300 Chart** Make copies. (as needed) • **M32, Number Puzzles Recording Sheet** Make copies. (as needed) • **M33, Number Puzzles 13–14** Make copies. Cut out the clues. Paper clip each set or place in envelopes. (1 set per student) • **M34, Assessment Checklist: Number Puzzles** ☑ Make several copies per class.
• **M17–M22, M33, Number Puzzles 1–14** (from Sessions 1.3 and 1.5) • **M34, Assessment Checklist: Number Puzzles** ☑ (several per class; from Session 1.5) • **Color tiles** (as needed) • **Calculators** (optional)	• **M30, Multiplication Combination Recording Sheet** Make copies. (as needed; from Session 1.4) • **M31, 300 Chart** Make copies. (as needed) • **M32, Number Puzzle Recording Sheet** Make copies. (as needed; from Session 1.5)
• **Calculators** (optional)	• **M30, Multiplication Combinations Recording Sheet** Make copies. (as needed; from Session 1.4)

☑ Checklist Available

Building and Using Arrays

Math Focus Points

- Writing multiplication equations that describe dot arrangements
- Using arrays to model multiplication
- Determining whether one number is a factor or multiple of another

Vocabulary

multiplication
division
factor
product
array
dimensions (of an array)
unmarked array
multiplication combination
multiple

Today's Plan		Materials
ACTIVITY **1** *Quick Images: Seeing Numbers*	15 MIN, CLASS	• T1* • Chart paper*
ACTIVITY **2** Building Arrays	20 MIN, CLASS, PAIRS	• M15 • Blank transparency; color tiles
ACTIVITY **3** Number Puzzles: 1 Clue	25 MIN, CLASS, PAIRS	• *Student Activity Book*, pp. 1–2 • M15 (as needed)* • Color tiles (as needed)
SESSION FOLLOW-UP **4** Daily Practice and Homework		• *Student Activity Book*, pp. 3–4 • M13–M14*, Family Letter • *Student Math Handbook*, pp. 16, 17

*See *Materials to Prepare*, p. 25.

Ten-Minute Math

Note: The Ten-Minute Math activity for this unit, *Quick Images: Seeing Numbers*, is introduced in this session. See the full write-up for this activity in *Implementing Investigations in Grade 5*.

1 ACTIVITY
Quick Images: Seeing Numbers

15 MIN CLASS

Tell the class that this first math unit, *Number Puzzles and Multiple Towers,* focuses on multiplication and division. Explain to students that they can refer to their *Student Math Handbook* for descriptions of the terms used throughout *Investigations.*

In this first activity, *Quick Images: Seeing Numbers,* which will continue throughout this unit and other units as a Ten-Minute Math activity, students create multiplication equations to represent different arrangements of figures. 1 2

Display Image 1 of *Quick Images: Seeing Numbers* (T1), on the overhead.

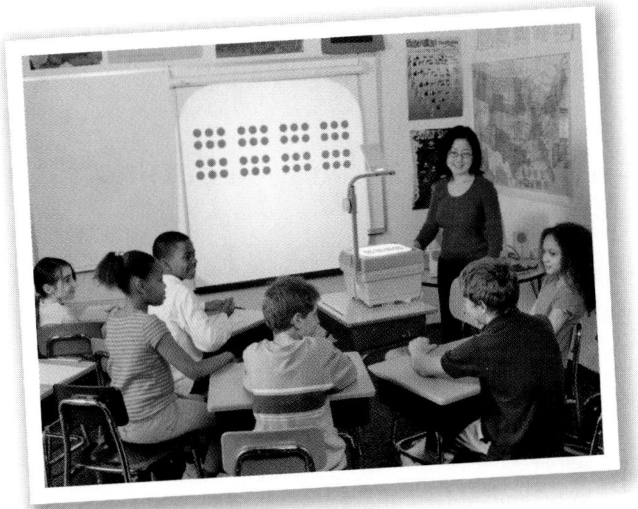

In Quick Images: Seeing Numbers, *students write multiplication equations that describe arrangements of dots and other figures.*

I am going to briefly display a picture of some dots. They are arranged in a way to help you determine how many there are. I will give you a few seconds to look at the picture and see how the dots are arranged.

Uncover Figure 1 for 5 seconds, and then cover it again.

How were the dots arranged? How many dots were there in all? You may sketch your mental picture of the arrangement or talk with your neighbor about what you saw. Then I will show you the picture again so that you can check.

Uncover Image 1 for another 5 seconds, and then cover it again. Students can add to or change their visual image. When most students have finished, uncover the picture again and leave it uncovered.

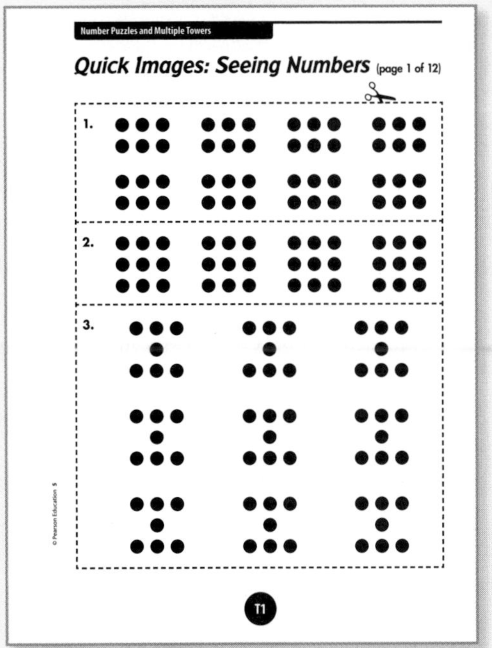

▲ Transparencies, T1

Professional Development

1 **Part 4: Ten-Minute Math** in *Implementing Investigations in Grade 5:* Quick Images: Seeing Numbers

Differentiation

2 **English Language Learners** English Language Learners may have trouble explaining the reasoning they used to create the equations. Before introducing this activity, preview terms such as *cluster, clump, row, column, group, section, line, portion, part, top,* and *bottom* that describe different arrangements of shapes. Have students demonstrate their understanding using pictures or small objects. You can also help English Language Learners create an illustrated vocabulary poster to use as a reference throughout the unit.

Teaching Notes

❸ **Writing Multiplication Combinations** In Session 1.4, students will be writing multiplication combinations with more than 2 factors, so it is important to introduce the idea here.

❹ **Using Vocabulary** Begin a vocabulary list on the "Math Vocabulary" chart you prepared. Post it on the wall and ask students to contribute their explanations of what each term means.

Call on several students to tell you the total number of dots and to explain how they figured it out. As students are explaining their thinking, record their statements as equations.

Students might say:

"I saw them as groups of 6, and there were 8 groups, so 8 × 6 = 48."

"I saw two groups of 6 in a clump and that's 12, and there are 4 groups of 12, so it's 48."

"I saw the top group as 4 groups of 6 and that is 24, and since the bottom group is the same, I just doubled the answer."

We can write Stuart's method like this. (Write 8 × 6 = 48 on the board or overhead.)

We can write Olivia's method in two ways. We can use two equations, like this. [Write 2 × 6 = 12 and 4 × 12 = 48.] Or, we can write one equation, like this. [Write (2 × 6) × 4 = 48.]❸

And this is how we can show Deon's method. [Write (4 × 6) × 2 = 48.]

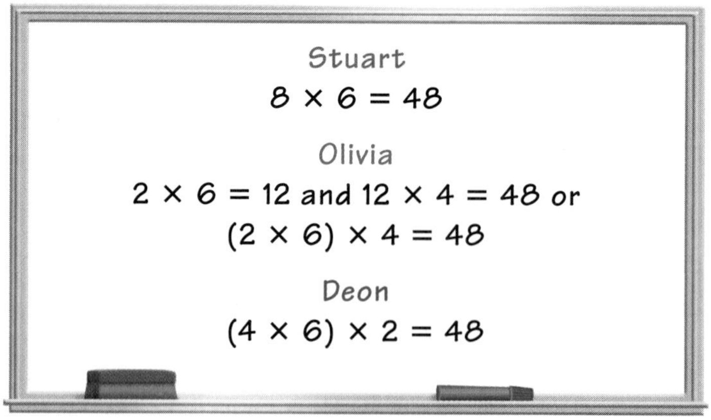

Have students look at the group of dots again. Ask them to think about what factors they see in the arrangement and to write equations that represent their thinking on their paper. If students do not remember what factors are, remind them that factors are numbers you multiply to get a product.❹

After a minute or so, call on a few students to give their equations. If no student has yet to mention an equation with more than two factors, ask whether anyone can do so.

It is important that the equations be seen in the drawings, so as students share their equations, ask them to explain where the numbers are in the arrangement of the dots.

Olivia, you said $(2 \times 6) \times 4 = 48$. Where are the two groups of 6? Where is the group of 4?

Follow the same process for Images 2 and 3 on the transparency. Ask students to write equations, including some that use more than two factors, and then discuss students' equations as a whole class.

ONGOING ASSESSMENT: Observing Students at Work

Students view arrays of dots or other figures and write multiplication equations to represent them.

- **Can students find the total number of figures in the arrangement by multiplying?**
- **Can students write equations to represent the way they see the figures organized?**

ACTIVITY 2

Building Arrays

20 MIN CLASS PAIRS

Students used arrays in Grades 3 and 4 to represent multiplication. This representation will continue to be useful in Grade 5 to help students visualize complex multiplication problems. In this activity, students review the use of arrays to represent multiplication. They will use arrays in upcoming activities to describe number relationships (factors and multiples) and number properties (odd, even, prime, square, and so on). If you have new students who have not used arrays before, this activity, in which they use them with small numbers, is a good introduction.

We are going to build arrays to help us think about numbers. If I wanted to build an array with 24 tiles, what are two factors I could use?

Call on one student for a suggestion, and build the array on the overhead. For example, if the student says 3×8, build a 3×8 array (3 rows with 8 tiles in each row). If students also suggest an 8×3 array (8 rows with 3 tiles in each row), acknowledge that these two arrays are different

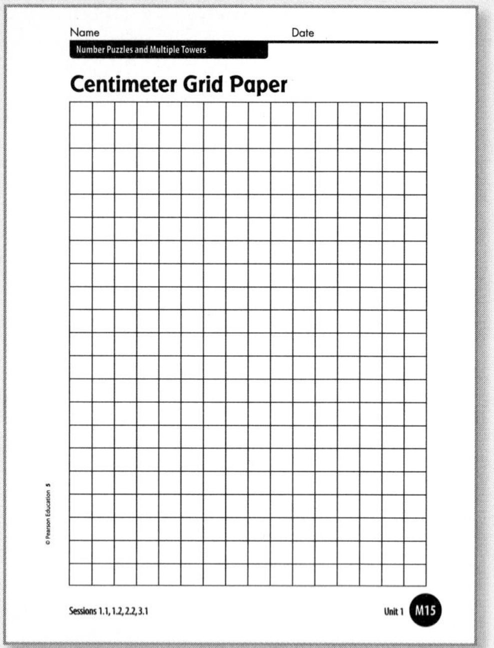
Name Date
Number Puzzles and Multiple Towers
Centimeter Grid Paper
© Pearson Education 5
Sessions 1.1, 1.2, 2.2, 3.1 Unit 1 M15

▲ Resource Masters, M15; T13

Teaching Notes

❺ **Transparencies** You can use the transparency of Centimeter Grid Paper (T13) to draw the array with grid lines and a blank transparency for the unmarked array.

❻ **Unmarked Arrays** Students will be encouraged to use unmarked arrays in Investigation 2, so modeling their use here is important.

❼ **Early Finishers** If some students finish quickly, have them work on another number (28, 36, 42, and so on) while other students finish.

in some ways, but that for the activities in this unit, two arrays that are congruent (exactly the same size and shape) will be considered the same.

On the board or overhead, draw two copies of this same array, one marked with grid lines to show the rows and columns, the other without grid lines.❺ Label and discuss the dimensions of each rectangular array.

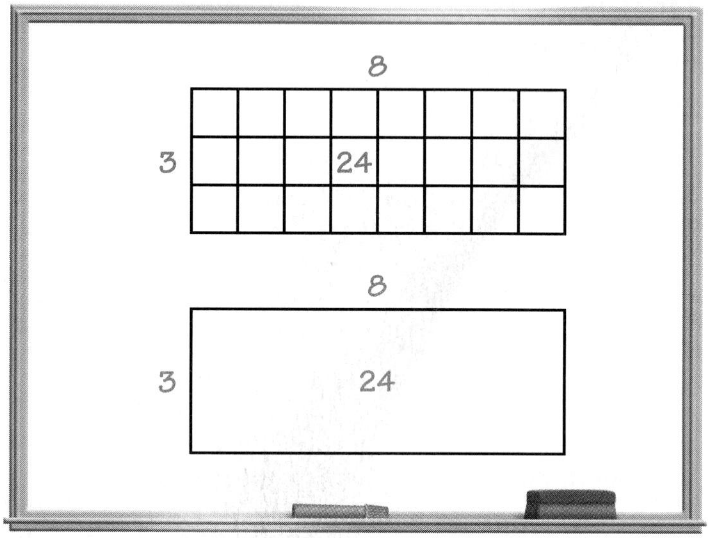

Sometimes we can't use grid paper to draw arrays because the numbers in the arrays are too large. So we use these "blank" rectangles, called unmarked arrays.❻

Have each student work with a partner to find other arrays that have 24 tiles. For each array they build, they draw the array on blank paper and label the dimensions.❼

After most students have finished working, or after about 10 minutes, call the class back together and have a brief discussion about the factors of 24. Ask students whether they have found all possible arrays and how they know that they have found them all.

When you label the dimensions of an array, you are naming multiplication combinations for the number. What are some other multiplication combinations you found for rectangles made with 24 tiles?

After you have listed all of the student combinations, remind students that these combinations include all of the factors of 24. Ask a volunteer to name all of the factors of 24 in order from smallest to largest while you list them on the board.

Students identify multiplication combinations for 24 and factors of 24.

Have a brief discussion about factors and multiples.

We have just listed the factors of 24. What are some multiples of 24? Sometimes people confuse factors and multiples. What are ways we can remember which are factors and which are multiples?

As you list students' responses, emphasize the meanings of and differences between factors and multiples.⑧

Multiples

- Multiples of a number are what you say when you count by that number.
- Multiples of 18: 18, 36, 54, 72 . . .
- Multiples are times a number, and you keep going up.

Factors

- Factors of a number are the sides of rectangles with that number of tiles.
- Factors of 18: 1, 2, 3, 6, 9, 18
- When you times the factors of a number, they equal that number.
- You find factors on the calculator by doing divide by.
- A factor is a number that fits into another number evenly.

Professional Development

⑧ **Teacher Note:** Using Mathematical Vocabulary, p. 153

Name Date

Number Puzzles and Multiple Towers

Number Puzzles: 1 Clue (page 1 of 2)

For each number puzzle, follow these steps.

a. Find two numbers that fit each clue.
b. Draw rectangles, and label the dimensions to show that your numbers fit the clue.
c. List other numbers that also fit the clue.

1. This number of tiles will make a rectangle that is 2 tiles wide.

Number: ______ Rectangle: Number: ______ Rectangle:

What other numbers fit this clue? ______

2. This number of tiles will make a rectangle that is 5 tiles wide.

Number: ______ Rectangle: Number: ______ Rectangle:

What other numbers fit this clue? ______

3. This number of tiles will make only one rectangle.

Number: ______ Rectangle: Number: ______ Rectangle:

What other numbers fit this clue? ______

© Pearson Education 5

Sessions 1.1, 1.2 Unit 1 1

▲ Student Activity Book, p. 1

Name Date

Number Puzzles and Multiple Towers

Number Puzzles: 1 Clue (page 2 of 2)

4. This number of tiles will make a square.

Number: ______ Rectangle: Number: ______ Rectangle:

What other numbers fit this clue? ______

5. There are some numbers that can be made into only one rectangle (Problem 3). Find all of these numbers up to 50.

6. There are some numbers that can make a square (Problem 4). Find all of these numbers up to 100.

© Pearson Education 5

2 Unit 1 Sessions 1.1, 1.2

▲ Student Activity Book, p. 2

ACTIVITY

3 Number Puzzles: 1 Clue

25 MIN CLASS PAIRS

Introduce the Number Puzzle activity by writing the following clue on the board or overhead:

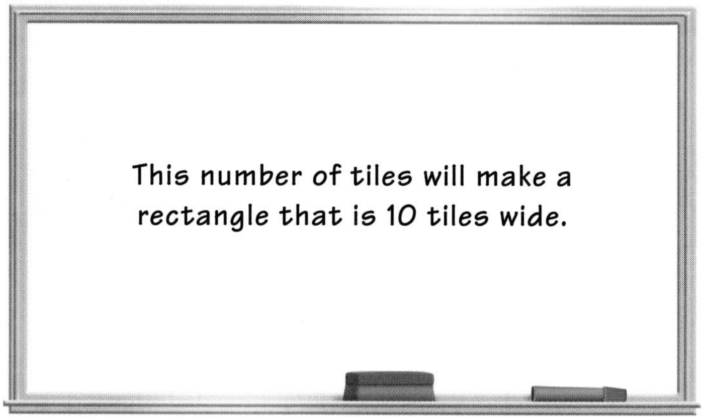

Allow pairs of students a minute to think of a number that works. Then they should draw the rectangle and label the dimensions.

Ask for answers. List the total number of tiles, draw the array that students describe, and label the dimensions.⑨

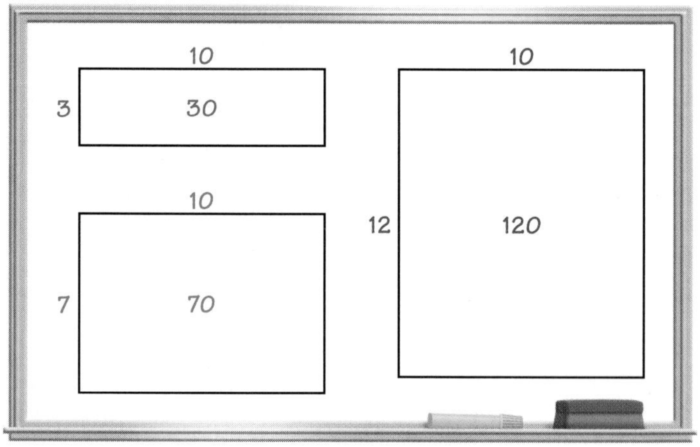

Ask students how many numbers there are that are multiples of 10. Students should realize that any multiple of 10 fits the clue and that there is an infinite number of possibilities.

Direct students' attention to *Student Activity Book* pages 1 and 2. Students determine numbers on the basis of the clue provided and draw arrays to represent the numbers.⑩

Teaching Note

⑨ **Larger Numbers** For some of these problems (finding primes, squares, and so on), all students will benefit from using the tiles or grid paper to help them determine the answer. However, also encourage students to start thinking about bigger numbers that fit the clue. For these numbers, building the actual rectangle may not be a good use of time if students can visualize and reason about the numbers.

Algebra Note

⑩ **Models** An important part of algebraic thinking is to be able to represent or to make models that illustrate the structure of an operation or a class of numbers. In this case, as students use tiles to represent prime and square numbers, they are also building mental models associated with each type of number. For instance, prime numbers are numbers greater than 1 that can be represented only by rectangles having 1 for one of their dimensions, and square numbers can be represented by rectangles with dimensions that are the same. When students are asked to make or justify generalizations about prime or square numbers, they will call on these models.

Students work in pairs, but each student completes an individual sheet. Have available Centimeter Grid Paper (M15) and square tiles. Students will have time to continue working on this sheet in Session 1.2.

ONGOING ASSESSMENT: Observing Students at Work

Students identify and examine properties of numbers.

- **Can students reason about whether a number fits a particular clue?**

As you move around the room observing students, ask questions such as the following:

- How do you know that the number fits the clue? Do you think there are more than two numbers that have rectangles that fit the clue?

DIFFERENTIATION: Supporting the Range of Learners

Intervention If students need help remembering what particular math terms mean, remind them that they can refer to the class vocabulary chart and to the *Student Math Handbook*.

4 SESSION FOLLOW-UP

Daily Practice and Homework

Daily Practice: For reinforcement of this unit's content, have students complete *Student Activity Book* page 3.

Homework: On *Student Activity Book* page 4, students write equations that represent different ways of visualizing dot arrangements.

Student Math Handbook: Students and families may use *Student Math Handbook* pages 16, 17 for reference and review. See pages 190–197 in the back of this unit.

Family Letter: Send home copies of Family Letter (M13–M14).

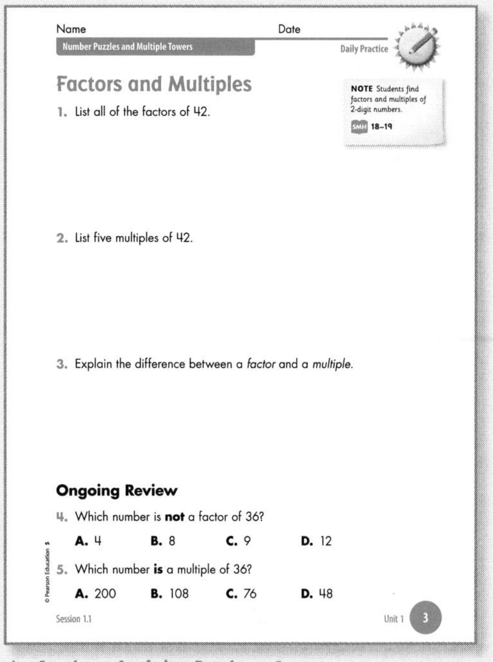

Name Date

Number Puzzles and Multiple Towers — Daily Practice

Factors and Multiples

NOTE Students find factors and multiples of 2-digit numbers. SMH 18–19

1. List all of the factors of 42.

2. List five multiples of 42.

3. Explain the difference between a *factor* and a *multiple*.

Ongoing Review

4. Which number is **not** a factor of 36?
A. 4 **B.** 8 **C.** 9 **D.** 12

5. Which number **is** a multiple of 36?
A. 200 **B.** 108 **C.** 76 **D.** 48

© Pearson Education 5

Session 1.1 Unit 1 3

▲ Student Activity Book, p. 3

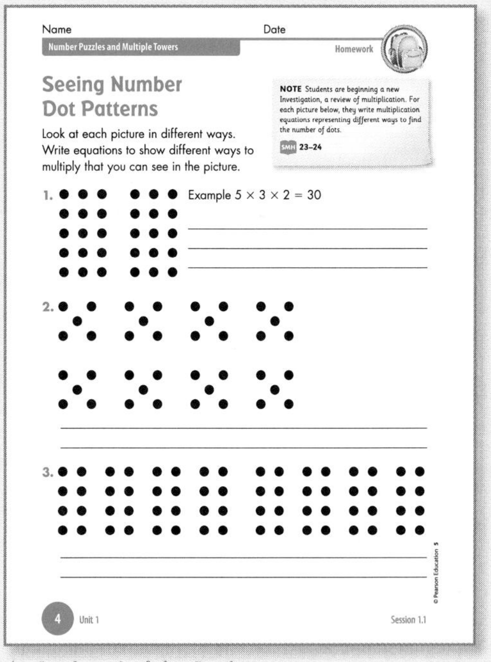

Name Date

Number Puzzles and Multiple Towers — Homework

Seeing Number Dot Patterns

Look at each picture in different ways. Write equations to show different ways to multiply that you can see in the picture.

NOTE Students are beginning a new Investigation, a review of multiplication. For each picture below, they write multiplication equations representing different ways to find the number of dots. SMH 23–24

1. Example $5 \times 3 \times 2 = 30$

2.

3.

© Pearson Education 5

4 Unit 1 Session 1.1

▲ Student Activity Book, p. 4

Identifying Properties of Numbers

Math Focus Points

- Identifying prime, square, even and odd numbers
- Determining whether one number is a factor or multiple of another
- Using arrays to model multiplication

Vocabulary

even number
odd number
prime number
composite number
square number

Today's Plan		Materials
DISCUSSION 1 Primes and Squares	15 MIN CLASS	• *Student Activity Book,* pp. 1–2 (from Session 1.1) • Chart: "Math Vocabulary" (from Session 1.1)
ACTIVITY 2 Number Puzzles: 2 Clues	35 MIN PAIRS	• *Student Activity Book,* pp. 5–6 • M15 (as needed)* • Color tiles (as needed)
DISCUSSION 3 How Many Answers?	10 MIN CLASS PAIRS	• *Student Activity Book,* pp. 5–6
SESSION FOLLOW-UP 4 Daily Practice		• *Student Activity Book,* p. 7 • *Student Math Handbook,* pp. 21–22

*See *Materials to Prepare,* p. 25.

Ten-Minute Math

Quick Images: Seeing Numbers Show Images 4–6 (one at a time) on *Quick Images: Seeing Numbers* (T2). For each pattern, ask students to write several different *multiplication* equations, including combinations with three or more numbers, to find the total number of shapes. For the first two viewings, give students 3 seconds to look at the pattern; the third time, leave the image displayed. Have two or three students explain how they saw the images (including any revisions they made) and their equations, showing how their numbers match the patterns. For example, Image 4 is an arrangement showing 8 groups of 7. Students might say $8 \times 7 = 56$, or $4 \times 7 \times 2 = 56$.

DISCUSSION
1 Primes and Squares

15 MIN CLASS

Math Focus Points for Discussion

- Identifying prime numbers and square numbers

This discussion focuses on Problems 3–6 on *Student Activity Book* pages 1–2. If necessary, give students an additional 5–10 minutes of work time before beginning the discussion.❶

What are some examples of numbers that have only one array? What do these arrays look like?

As students suggest numbers, quickly sketch (or have them sketch) the array. Make certain that the class agrees that only one rectangle can be made for these numbers.

Students use arrays to illustrate prime numbers and square numbers.

These numbers have a special name. Does anyone know what it is?

Collect a few examples, and establish that these are prime numbers. Ask students to say in their own words what a prime number is. Students might say that a prime number has two factors, 1 and itself. Write their descriptions on the class vocabulary chart started in Session 1.1. Also establish that composite numbers are those numbers with more than two factors.❷

What are some examples of numbers that make a square? What do these arrays look like?

As students suggest numbers, sketch (or have them sketch) the array. Make certain that the class agrees that these numbers would make a square.

Teaching Note

❶ **Even and Odd Numbers** It is assumed that students in Grade 5 can recognize and define even and odd numbers. If you think your students need to review these definitions, spend a few minutes doing so, using Problem 1 on *Student Activity Book* page 1. Add students' explanations of even and odd to the vocabulary list you started in Session 1.1. Students can also refer to their *Student Math Handbook* page 7.

Math Note

❷ **Neither Prime nor Composite** The number 1 is considered a special number. It is neither prime nor composite.

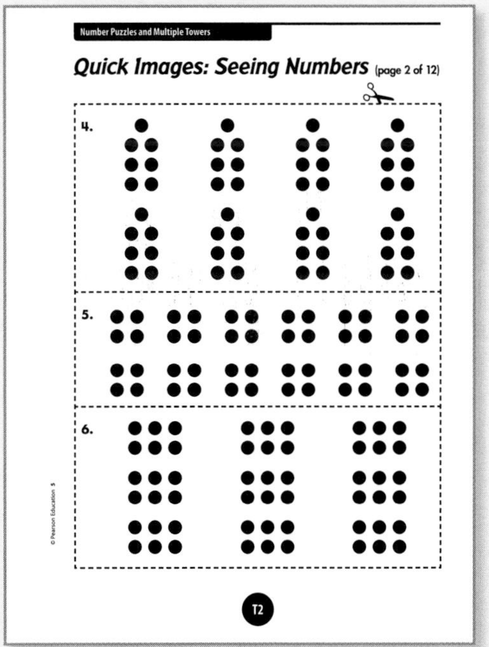

▲ Transparencies, T2

These numbers also have a special name. What is it?

Establish that these are square numbers. Ask students to say in their own words what a square number is. Students might say that a number that can be arranged in a square using tiles or that a number multiplied by itself is a square number. Write their descriptions on the class vocabulary chart.

ACTIVITY

35 MIN PAIRS

2 Number Puzzles: 2 Clues

Tell students that today they will work on number puzzles that have 2 clues, and that the answer must fit *both* clues. Write these 2 clues on the board or overhead.

> This number of tiles will make a rectangle that is 2 tiles wide.
>
> This number of tiles will make a rectangle that is 3 tiles wide.

Allow students a minute to think about this with a partner. Ask them to try modeling some numbers with tiles. Then have them draw the rectangles and label the dimensions. Ask for answers and have students demonstrate how that number of tiles can be arranged in a rectangle with one dimension of either 2 or 3.

Students might not understand at first that they have to draw two different rectangles for the same number—one that is 2 tiles wide and one that is 3 tiles wide. Six (a 2 × 3 array) will be an obvious answer, but encourage them to think of other numbers that work. Students may discover that any multiple of 6 (or the even multiples of 3) fits this clue.

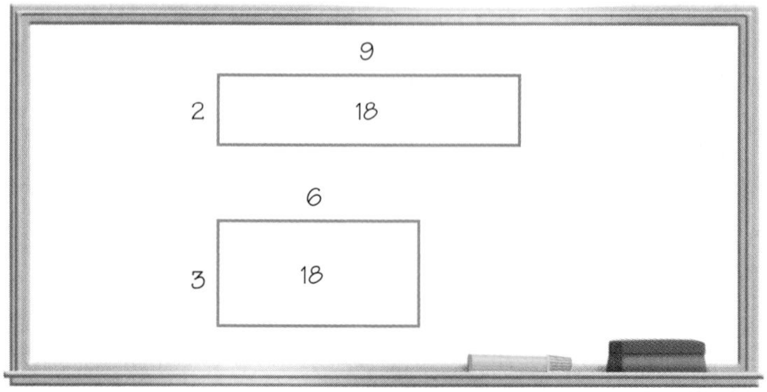

Students work in pairs to complete *Student Activity Book* pages 5 and 6. All students will benefit from using square tiles and/or Centimeter Grid Paper (M15). Tell students that the upcoming discussion will focus on Problem 2 on page 6.

ONGOING ASSESSMENT: Observing Students at Work

Students find numbers that share common factors.

- **Can students reason about whether a number fits a particular clue?**

As you circulate, check for understanding.

- How do you know that the number fits the clue (or both clues)? Do you think there are other numbers that have rectangles that fit?

DIFFERENTIATION: Supporting the Range of Learners

Intervention Students who are still working on the idea of arrays should build or draw the arrays and focus mostly on *Student Activity Book* pages 1 and 2.

Extension Students who finish early can extend Problems 5 and 6 on *Student Activity Book* pages 1 and 2 by finding prime numbers greater than 50 and square numbers greater than 100.

3 DISCUSSION

How Many Answers?

10 MIN CLASS PAIRS

Math Focus Points for Discussion

- Determining whether one number is a factor or a multiple of another

Ask students what numbers they found for Problem 2 (numbers that can be represented by a rectangle 3 tiles wide and a rectangle 4 tiles wide), and write answers on the board. Ask students whether these numbers are factors or multiples of 3 and 4. Also ask the class whether all the answers given fit both clues.

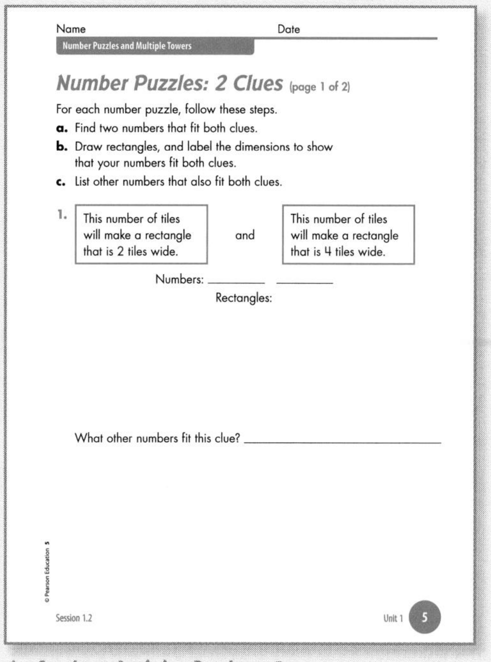
Name Date

Number Puzzles and Multiple Towers

Number Puzzles: 2 Clues (page 1 of 2)

For each number puzzle, follow these steps.

a. Find two numbers that fit both clues.

b. Draw rectangles, and label the dimensions to show that your numbers fit both clues.

c. List other numbers that also fit both clues.

1. This number of tiles will make a rectangle that is 2 tiles wide. and This number of tiles will make a rectangle that is 4 tiles wide.

Numbers: ______ ______

Rectangles:

What other numbers fit this clue? ______

© Pearson Education 5

Session 1.2 Unit 1 5

▲ Student Activity Book, p. 5

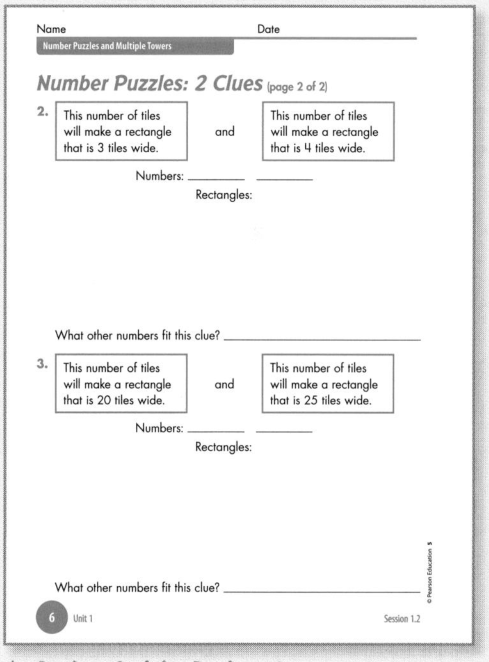
Name Date

Number Puzzles and Multiple Towers

Number Puzzles: 2 Clues (page 2 of 2)

2. This number of tiles will make a rectangle that is 3 tiles wide. and This number of tiles will make a rectangle that is 4 tiles wide.

Numbers: ______ ______

Rectangles:

What other numbers fit this clue? ______

3. This number of tiles will make a rectangle that is 20 tiles wide. and This number of tiles will make a rectangle that is 25 tiles wide.

Numbers: ______ ______

Rectangles:

What other numbers fit this clue? ______

© Pearson Education 5

6 Unit 1 Session 1.2

▲ Student Activity Book, p. 6

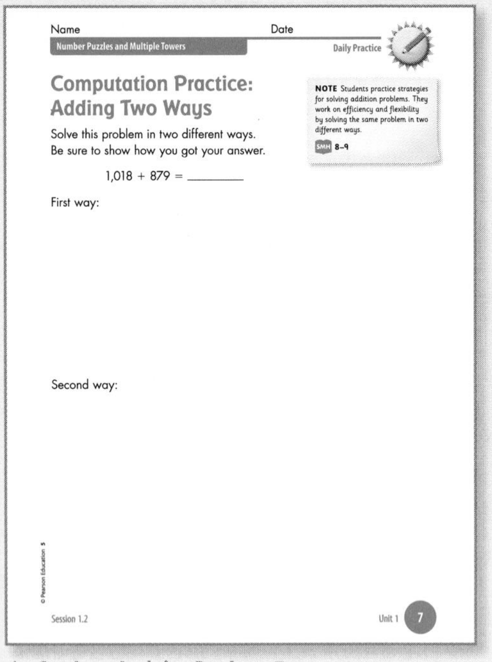
Name Date

Number Puzzles and Multiple Towers

Daily Practice

Computation Practice: Adding Two Ways

Solve this problem in two different ways.
Be sure to show how you got your answer.

1,018 + 879 = ________

NOTE Students practice strategies for solving addition problems. They work on efficiency and flexibility by solving the same problem in two different ways.

SMH 8–9

First way:

Second way:

© Pearson Education 5

Session 1.2

Unit 1 7

▲ Student Activity Book, p. 7

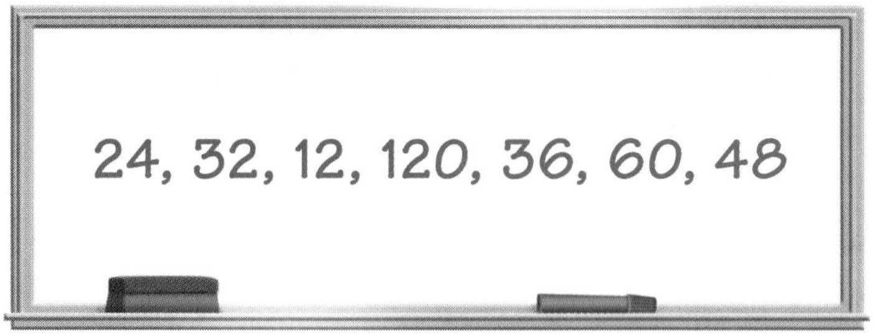

Does everyone agree that all these numbers fit both clues? Talk to a neighbor and decide. Who can explain which numbers work and which ones don't?

Students might say:

"32 doesn't work because nothing times 3 equals 32."

"60 is OK, because 3 × 20 = 60, and 4 × 15 = 60."

Ask the students whether they agree with the explanations offered by their classmates. Adjust the list to show only numbers that fit both of these clues.

Look at all of the numbers. What do they have in common? What statement can you make about these numbers that is also true about any number that would fit both of these clues? Talk to a neighbor.

Hana and Yumiko say that all these numbers are multiples of 12 and that any solution to this number puzzle will be a multiple of 12. Do you agree? Why or why not?

4 SESSION FOLLOW-UP
Daily Practice

Daily Practice: For ongoing review, have students complete *Student Activity Book* page 7.

Student Math Handbook: Students and families may use *Student Math Handbook* pages 21–22 for reference and review. See pages 190–197 in the back of this unit.

SESSION 1.3

What Numbers Have Which Properties?

Math Focus Points

- Identifying prime, square, even and odd numbers
- Determining whether one number is a factor or multiple of another
- Identifying and learning multiplication combinations ("facts") not yet known fluently

Today's Plan		Materials
1 ACTIVITY **Introducing Number Puzzles: 4 Clues**	20 MIN GROUPS CLASS	• M16*; M31*; T16 • Paper clips or envelopes*; Color tiles (as needed); Calculators (optional)
2 ACTIVITY **Solving Number Puzzles: 4 Clues**	30 MIN PAIRS	• *Student Activity Book,* p. 8 • M17–M22*; M31* • Paper clips or envelopes*; Color tiles (as needed); Calculators (optional)
3 ACTIVITY **Explaining Homework**	10 MIN CLASS	• *Student Activity Book,* pp. 9–10 • T14; T15
4 SESSION FOLLOW-UP **Daily Practice and Homework**		• *Student Activity Book,* pp. 9–11 • *Student Math Handbook,* pp. 18, 19, 21–22 • M23–M26, Family Letter

*See *Materials to Prepare,* p. 25.

Ten-Minute Math

Quick Images: Seeing Numbers Show Images 7–9 (one at a time) on *Quick Images: Seeing Numbers* (T3). For each pattern, ask students to write several different *multiplication* equations, including combinations with three or more numbers, to find the total number of shapes. Have two or three students explain how they saw the images (including any revisions they made) and their equations, showing how their numbers match the patterns. For example, Image 7 is an arrangement showing 8 groups of 8. Students might say $8 \times 8 = 64$, or $16 \times 4 = 64$, or $8 \times 2 \times 4 = 64$, or $8 \times 2 \times 2 \times 2 = 64$.

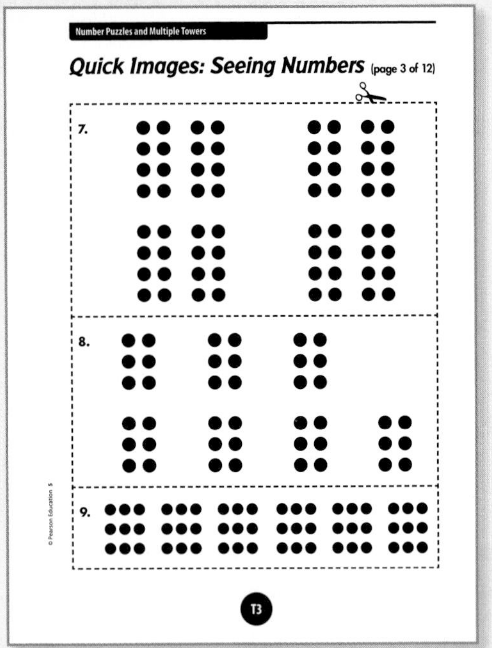
Number Puzzles and Multiple Towers

Quick Images: Seeing Numbers (page 3 of 12)

7.

8.

9.

© Pearson Education 5

T3

▲ Transparencies, T3

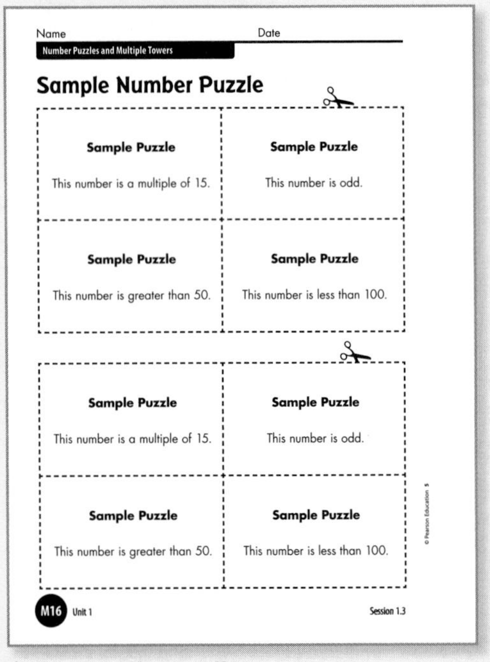
Name Date

Number Puzzles and Multiple Towers

Sample Number Puzzle

Sample Puzzle This number is a multiple of 15.	**Sample Puzzle** This number is odd.
Sample Puzzle This number is greater than 50.	**Sample Puzzle** This number is less than 100.
Sample Puzzle This number is a multiple of 15.	**Sample Puzzle** This number is odd.
Sample Puzzle This number is greater than 50.	**Sample Puzzle** This number is less than 100.

© Pearson Education 5

M16 Unit 1 Session 1.3

▲ Resource Masters, M16

ACTIVITY

20 MIN GROUPS CLASS

1 Introducing Number Puzzles: 4 Clues

In this session, as well as in the Math Workshops in Sessions 1.5 and 1.6, students work on solving number puzzles with 4 clues. A variation of this activity is introduced as a Ten-Minute Math activity in Investigation 2 of this unit, and in other units throughout the year.

Try to have 4 students in each group, so that each student is given exactly 1 clue.

Distribute 1 set of Sample Number Puzzle (M16) to each group. Have 300 charts (M31), color tiles, and calculators available.

Each person in the group gets 1 clue. The group shares clues by reading them to one another. As a group, they find a number that fits all the clues.[1]

If students are unfamiliar with the 300 chart, explain that some people find it useful in solving number puzzles. Without explaining its use further, give students a chance to find their own ways to use the chart. There are different ways to solve the sample puzzle, and various strategies that involve the 300 chart.[2]

Students use their knowledge of factors, multiples, and other number properties to find numbers that fit a set of four clues.

After groups have worked on the puzzle for about 10 minutes, call students together for a brief discussion.

After you read all of the clues, how did you start? What clue or clues did you use first to solve this puzzle? What clue did you use next? How did that clue help you eliminate some numbers?

Many students will say that they listed the multiples of 15. Ask whether they stopped listing multiples of 15 when they got to 100. If a group started with odd numbers, what was their experience? Did they also notice that they needed to consider only numbers between 50 and 100? If no one used the 300 chart, use T16 to model a possible way of using it to help solve the problem.

It is important to establish that students need to consider all of the clues to choose a good starting place. Help students think about the importance of the combination of clues.

How would your answer change if the clue "My number is odd" were not there?

Students should realize that there would be more possibilities. (60 and 90 are the even multiples of 15 between 50 and 100.)

How would the answer change if one of the other clues were missing?

Collect some responses, making sure that students are clear about what each clue means and how it affects the answer. They might mention that if the clue "less than 100" were missing, there would be an infinite number of answers, whereas if the clue "greater than 50" were missing, there would be only two more answers.

Collect the sample clues before you distribute clues for the next activity.

ONGOING ASSESSMENT: Observing Students at Work

Students use reasoning and knowledge of properties of numbers to find a given number.

- **Do students understand that the number(s) they are looking for must fit all the clues?**
- **Do students use each clue to help them eliminate certain numbers?**
- **Do students understand the meaning of the terms?** Do they know *even, odd, factor, multiple, square,* and *prime*?
- **As students work together, are they listening to others and making progress in solving the puzzle?**

Teaching Note

1 Positive Group Interaction As students work on this puzzle, circulate to look for the following signs of positive group interaction:

- Everyone is participating.
- Students are listening to one another.
- No one is dominating the clues in a way that does not allow other students to fully participate.
- Students are discussing any questions group members have.

Professional Development

2 Dialogue Box: Solving a Number Puzzle, p. 178

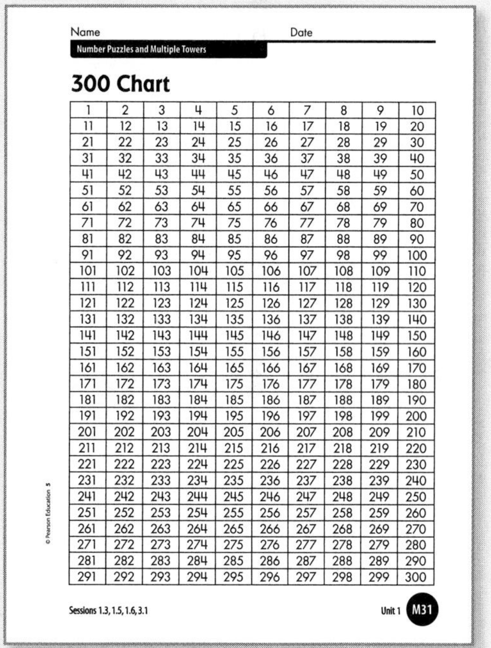

Name Date

Number Puzzles and Multiple Towers

300 Chart

1	2	3	4	5	6	7	8	9	10
11	12	13	14	15	16	17	18	19	20
21	22	23	24	25	26	27	28	29	30
31	32	33	34	35	36	37	38	39	40
41	42	43	44	45	46	47	48	49	50
51	52	53	54	55	56	57	58	59	60
61	62	63	64	65	66	67	68	69	70
71	72	73	74	75	76	77	78	79	80
81	82	83	84	85	86	87	88	89	90
91	92	93	94	95	96	97	98	99	100
101	102	103	104	105	106	107	108	109	110
111	112	113	114	115	116	117	118	119	120
121	122	123	124	125	126	127	128	129	130
131	132	133	134	135	136	137	138	139	140
141	142	143	144	145	146	147	148	149	150
151	152	153	154	155	156	157	158	159	160
161	162	163	164	165	166	167	168	169	170
171	172	173	174	175	176	177	178	179	180
181	182	183	184	185	186	187	188	189	190
191	192	193	194	195	196	197	198	199	200
201	202	203	204	205	206	207	208	209	210
211	212	213	214	215	216	217	218	219	220
221	222	223	224	225	226	227	228	229	230
231	232	233	234	235	236	237	238	239	240
241	242	243	244	245	246	247	248	249	250
251	252	253	254	255	256	257	258	259	260
261	262	263	264	265	266	267	268	269	270
271	272	273	274	275	276	277	278	279	280
281	282	283	284	285	286	287	288	289	290
291	292	293	294	295	296	297	298	299	300

Sessions 1.3, 1.5, 1.6, 3.1 Unit 1 M31

▲ Resource Masters, M31; T16

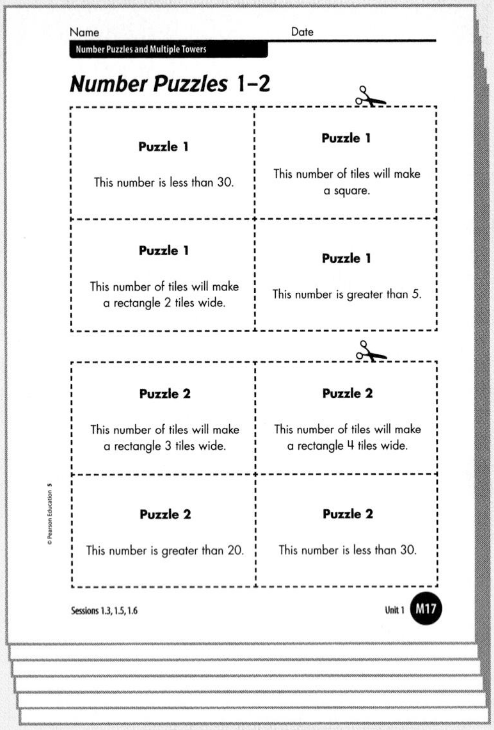
Name Date

Number Puzzles and Multiple Towers

Number Puzzles 1–2

Puzzle 1 This number is less than 30.	**Puzzle 1** This number of tiles will make a square.
Puzzle 1 This number of tiles will make a rectangle 2 tiles wide.	**Puzzle 1** This number is greater than 5.

Puzzle 2 This number of tiles will make a rectangle 3 tiles wide.	**Puzzle 2** This number of tiles will make a rectangle 4 tiles wide.
Puzzle 2 This number is greater than 20.	**Puzzle 2** This number is less than 30.

© Pearson Education 5

Sessions 1.3, 1.5, 1.6 Unit 1 M17

▲ Resource Masters, M17–M22

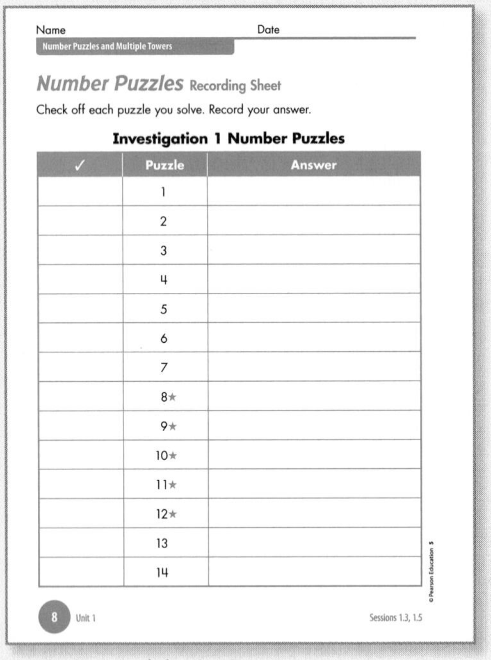
Name Date

Number Puzzles and Multiple Towers

Number Puzzles Recording Sheet

Check off each puzzle you solve. Record your answer.

Investigation 1 Number Puzzles

✓	Puzzle	Answer
	1	
	2	
	3	
	4	
	5	
	6	
	7	
	8*	
	9*	
	10*	
	11*	
	12*	
	13	
	14	

© Pearson Education 5

8 Unit 1 Sessions 1.3, 1.5

▲ Student Activity Book, p. 8; Resource Masters, M32

ACTIVITY

2 Solving Number Puzzles: 4 Clues

30 MIN PAIRS

Explain to students how to complete the activity and where the puzzles will be located during the next several days. Give each pair a set of Number Puzzles 1–7 (M17–M20).[3]

Students solve the problem and write the answer(s) on their recording sheet, *Student Activity Book* page 8.[4] They put the clues back in the envelope, return it to the puzzle pool, and get a different set of clues. Decide whether you will tell students ahead of time that some puzzles have more than one answer (or may be impossible) or whether you will discuss those issues as they naturally surface. Remind students to use math tools, such as 300 charts, color tiles, and calculators. Point out that the *Student Handbook* is also a helpful resource.

Let students know that they will work on only a few puzzles today, but will have the chance to work on these and other number puzzles later in the unit.

Circulate as students work.[5] Students may need help in using the tools and in using specific clues to eliminate numbers. For example, many students use clues that describe the range of numbers—less than 100, greater than 50—first. Talk with students about choosing clues that eliminate many numbers, and help students figure out ways of keeping track of what numbers they can either eliminate or include as they consider each clue. Some students may find it helpful to use the 300 chart to keep track. For example, they can cross out 50 and all the numbers less than 50, if a clue is "greater than 50." Or they can circle numbers they should consider, such as all the numbers that are factors of 42. Students might use the calculator to skip count by 60s in order to identify all the multiples of 60.

Following are the answers to number puzzles 1–12:

Puzzle 1: 16	Puzzle 7: 36
Puzzle 2: 24	Puzzle 8★: 6, 14
Puzzle 3: 20, 40	Puzzle 9★: impossible
Puzzle 4: 1, 3	Puzzle 10★: 350
Puzzle 5: 150	Puzzle 11★: 99
Puzzle 6: 120, 240	Puzzle 12★: 154

ONGOING ASSESSMENT: Observing Students at Work

Students use their knowledge of prime, square, odd, even, factors, and multiples to identify a number with certain characteristics.

Continue to observe and assess your students as suggested on page 39. As you circulate, ask questions such as the following:

- Which clue do you think will help eliminate many numbers? Does your puzzle only have one answer, or could there be more than one?

DIFFERENTIATION: Supporting the Range of Learners

Extension Students who finish the challenge puzzles, Puzzles 8–12, can make up their own puzzles—some that have one answer and some that have a few answers, and perhaps one that is impossible.

ACTIVITY 3

10 MIN CLASS

Explaining Homework

In Grade 4, students spent a significant amount of time organizing, learning, and practicing the multiplication combinations up to 12 × 12. Students are expected to enter Grade 5 knowing all the multiplication combinations through 12 × 12 fluently. It is important that they know these combinations as they continue to work on large multiplication problems later in this unit. However, it is likely that there will be a few of these combinations that students need to continue to practice. In today's homework, students begin to identify the combinations they do not yet know fluently. Students are expected to spend time at home or outside math class practicing these combinations until they know them fluently.

As we all know, some of these multiplication pairs are easier to remember than others.

Use an example of your own, such as the following:

For me, 7 × 7 has always been easy to remember because it is a square number, but 7 × 9 has always made me stop for an instant before I can say 63.

Each of you is going to spend some time finding out which combinations you still need to practice.

Teaching Notes

3 **Challenge Puzzles** Number Puzzles 8–12 (shown as 8★–12★ on M20–M22) are challenge puzzles. Their use today is optional. These challenging puzzles are more difficult (clues have bigger numbers, use a bigger variety of types of clues, and so on), and one of the puzzles is impossible. Put the starred puzzles in a separate place for students to work on as a challenge if they finish the other puzzles.

4 **Puzzles 13 and 14** The recording sheet includes Puzzles 13 and 14. These will be used in an observed assessment activity in Session 1.5 and 1.6.

5 **Listening and Explaining** Make sure that students are explaining their thinking and listening to the thinking of their partners. You may want to ask a specific student for an explanation to help his or her partner listen better. Listen for students using terms correctly in context (such as *factor, multiple, array, dimension, square,* and *prime*).

Name Date

Number Puzzles and Multiple Towers — Homework

Multiplication Combinations 1

Multiply each number in the first column of the table with the number at the top. For example, the answer for the first blank space in Table A is 14, which is 2 × 7. Circle any combinations you do not know immediately, and record them on *Student Activity Book* page 10.

NOTE Fifth-grade students are expected to know their multiplication combinations (facts). This page helps students determine whether they remember their combinations and identify any combinations they still need to practice.

SMH 25–29

Table A		Table B		Table C		Table D	
× 7		× 8		× 6		× 9	
2	14	2		10		5	
6		9		4		2	
8		4		2		12	
3		11		8		4	
10		8		3		10	
11		10		6		7	
7		6		9		3	
12		5		5		6	
4		3		12		11	
9		12		11		8	
5		7		7		9	

© Pearson Education 5

Session 1.3 Unit 1 9

▲ **Student Activity Book, p. 9; Resource Masters, M29; T14**

Name Date

Number Puzzles and Multiple Towers — Homework

Multiplication Combinations
Recording Sheet

Example: 7 × 9 = 63 and 9 × 7 = 63

Clue: 7 × 10 = 70 70 − 7 = 63

NOTE Fifth-grade students are expected to know their multiplication combinations (facts). On this sheet, students write any combination they still need to practice and a clue to help them learn it. An example is shown on the left.

SMH 25–29

___ × ___ = ___ and ___ × ___ = ___

Clue: ___

___ × ___ = ___ and ___ × ___ = ___

Clue: ___

___ × ___ = ___ and ___ × ___ = ___

Clue: ___

___ × ___ = ___ and ___ × ___ = ___

Clue: ___

___ × ___ = ___ and ___ × ___ = ___

Clue: ___

___ × ___ = ___ and ___ × ___ = ___

Clue: ___

© Pearson Education 5

10 Unit 1 Session 1.3

▲ **Student Activity Book, p. 10; Resource Masters, M30; T15**

Place the transparency of Multiplication Combinations 1 (T14), on the overhead or copy Table A on the board. (This is the same as *Student Activity Book* page 9.)

Tell students that they can use these tables to test themselves on the 6s, 7s, 8s, and 9s combinations because these are the hardest for many people. In Table A, they are to multiply the number in each line by 7 (the number at the top).

Quickly look through the numbers in the multiply-by-7 table, and tell me one that you know the answer to without having to calculate.

Fill in the answers on the transparency as students respond. Probably only a few problems will be left.

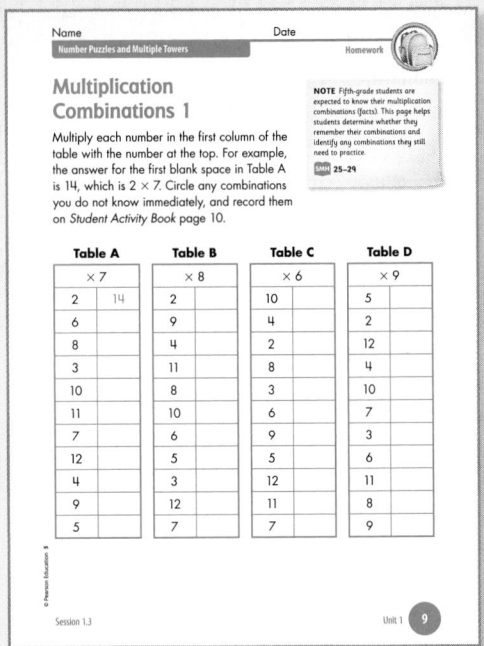

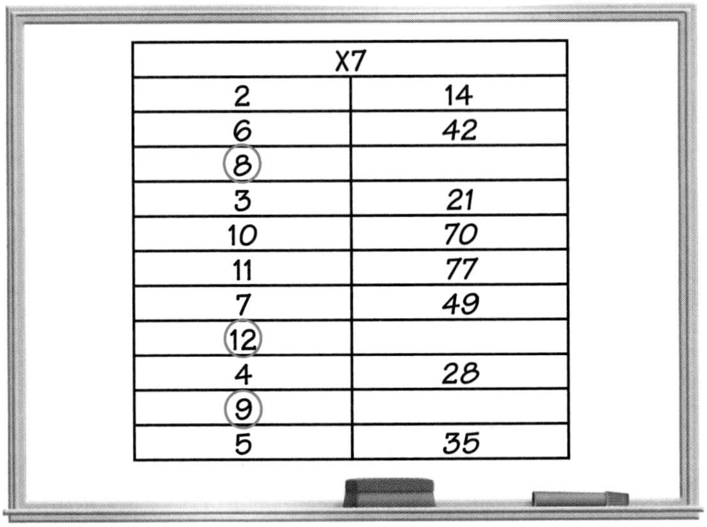

X7	
2	14
6	42
(8)	
3	21
10	70
11	77
7	49
(12)	
4	28
(9)	
5	35

When you work on this yourself, go down the list and answer the combinations you know quickly. Then circle the ones you don't know.

Place the transparency of Multiplication Combinations Recording Sheet (T15) on the overhead to demonstrate how to complete this sheet. (This is the same as *Student Activity Book* page 10.)

On the transparency, fill in the problems you circled on T15, such as 7 × 8 or 8 × 7, 9 × 7 or 7 × 9, 12 × 7 or 7 × 12.

Is there a way to use something you already know to figure these out?

Students might say:

"For 7 × 12, I can do 5 × 12 and 2 × 12 and add."

"For 7 × 9, I know 7 × 10 is 70, and I subtract one 7 to get 63."

Following the example at the top of the transparency, write in one of the clues students suggest for each of the combinations you listed.

Summarize for students the tasks they need to do for homework on *Student Activity Book* pages 9 and 10.⑥

- Begin filling in the tables on *Student Activity Book* page 9.
- Circle the ones that seem hard.
- Write these on the recording sheet, *Student Activity Book* page 10.
- Fill in a clue to help you quickly remember the product.

Have students begin these tasks in class so that you can help students get started. Remind students that they need to be honest in identifying multiplication combinations that are troublesome for them so that you can support them in this ongoing practice.

4 SESSION FOLLOW-UP
Daily Practice and Homework

Daily Practice: For reinforcement of this unit's content, have students complete *Student Activity Book* page 11.

Homework: Students complete *Student Activity Book* pages 9 and 10 as explained on pages 46–47.

Student Math Handbook: Students and families may use *Student Math Handbook* pages 18, 19, 21–22 for reference and review. See pages 190–197 in the back of this unit.

Family Letter: Send home copies of two Family Letters (M23–M24 and M25–M26).

Teaching Note

⑥ **Learning Multiplication Combinations** The homework sheets in this Investigation help your students and you identify which multiplication combinations they do not know fluently. Students use clues on *Student Activity Book* page 10 to help them practice. You may have other favorite practice methods or activities that you want to suggest for particular students. Help students keep track of which combinations they need to work on at home, and enlist parents or other family members in helping with practice. For many students, it is useful to put these "facts" into related sets, as they did in Grade 4. These groupings provide order and structure so that learning 144 combinations becomes less overwhelming.

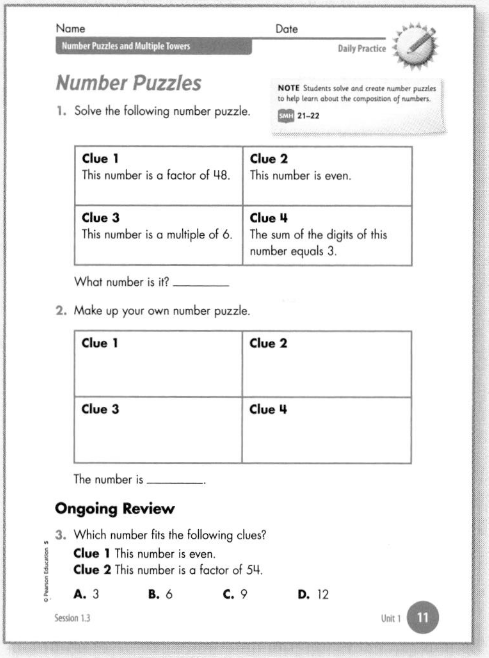

Name Date

Number Puzzles and Multiple Towers — Daily Practice

Number Puzzles

NOTE Students solve and create number puzzles to help learn about the composition of numbers.

SMH 21–22

1. Solve the following number puzzle.

Clue 1 This number is a factor of 48.	**Clue 2** This number is even.
Clue 3 This number is a multiple of 6.	**Clue 4** The sum of the digits of this number equals 3.

What number is it? ______

2. Make up your own number puzzle.

Clue 1	**Clue 2**
Clue 3	**Clue 4**

The number is ______.

Ongoing Review

3. Which number fits the following clues?
Clue 1 This number is even.
Clue 2 This number is a factor of 54.

A. 3 **B.** 6 **C.** 9 **D.** 12

© Pearson Education 5

Session 1.3 Unit 1 11

▲ Student Activity Book, p. 11

Multiplying with More Than Two Numbers

Math Focus Points

- Using known multiplication combinations to find equivalent multiplication combinations (e.g., $18 = 3 \times 6 = 3 \times (2 \times 3)$)
- Using known multiplication combinations to find multiplication combinations for numbers related by place value (e.g., $3 \times 6 = 18$, $3 \times 6 \times 10 = 180$)
- Finding all the ways to multiply whole numbers for a given product

Today's Plan	Materials
1 ACTIVITY **Finding Multiplication Combinations for 18 and 180** 60 MIN · CLASS · INDIVIDUALS	• *Student Activity Book,* p. 12 • Calculators (optional)
2 SESSION FOLLOW-UP **Daily Practice and Homework**	• *Student Activity Book,* pp. 13–14 • *Student Math Handbook,* pp. 23–24 • M27–M28*, Family Letter; M30*

*See *Materials to Prepare,* p. 25.

Ten-Minute Math

Quick Images: Seeing Numbers Show Images 10–12 (one at a time) on *Quick Images: Seeing Numbers* (T4). For each pattern, ask students to write several different *multiplication* equations, including combinations with three or more numbers, to find the total number of shapes. Have two or three students explain how they saw the images (including any revisions they made) and their equations, showing how their numbers match the patterns.

1 ACTIVITY

60 MIN | CLASS | INDIVIDUALS

Finding Multiplication Combinations for 18 and 180

In the next three sessions, students will be finding prime factorization of numbers. This work builds on the ideas that students have been using to solve number puzzles, particularly factors and multiples.①

You've been finding numbers with different characteristics—prime numbers, square numbers, and numbers that have certain factors. In the next few sessions, you'll be solving more number puzzles, and we're going to do some more work on finding factors. To start out, what whole numbers can you multiply to make 18?

Write ____ × ____ = 18 on the overhead or the board.

Let students know that they are looking for *different* combinations of factors, regardless of the order of the numbers, and that 2 × 9 and 9 × 2 will be considered the same number combination for this activity. If students suggest pairs that include decimals or fractions (e.g., 1.8 × 10, or $\frac{1}{2}$ × 36), explain that although those combinations do make a product of 18, when mathematicians talk about "factors of a number," they are referring to whole numbers only.

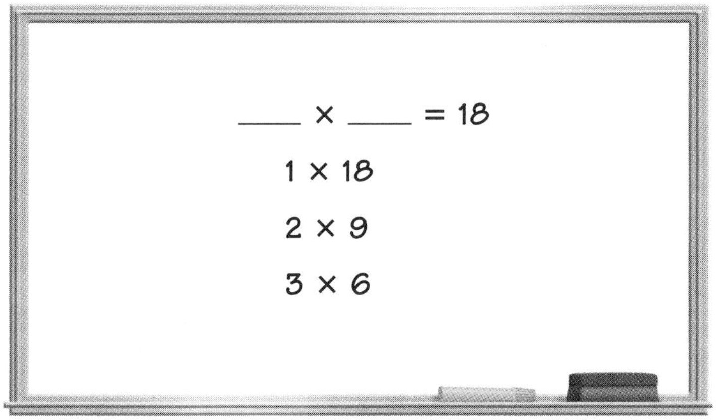

Ask students to find a way to multiply 3 numbers to make 18. Allow time for most students to find one combination with three factors and see if they can find another combination. Then ask students to talk together about how they determined the combination (2 × 3 × 3 in any order). Briefly discuss this as a group.

Students might have started with 2 × 9 and split the 9 into 3 × 3. Others might have started with 3 × 6 and split the 6 into 2 × 3.

Professional Development

① Teacher Note: Finding Prime Factors, p. 154

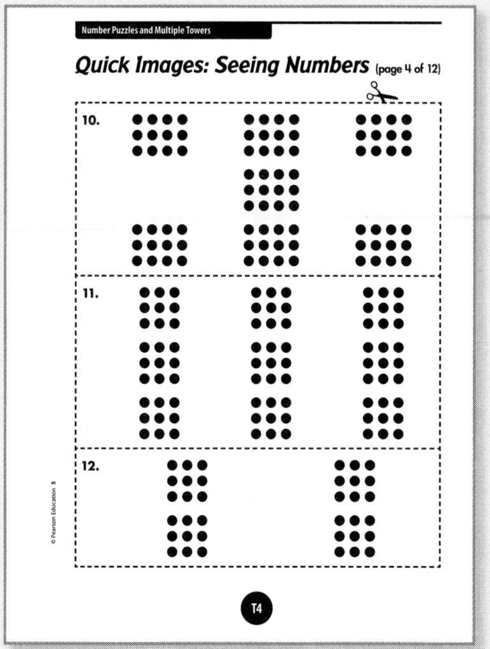

▲ Transparencies, T4

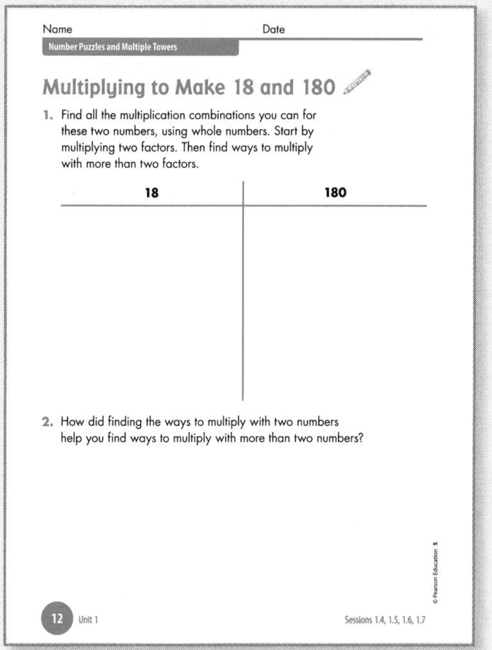

▲ Student Activity Book, p. 12 WRITING

Math Notes

❷ **Using 1 as a Factor** Some students may want to use 1 as one of the factors—1 × 2 × 9. If no one raises this issue, ask whether 1 should be considered. Students may see that since multiplying by 1 does not change the value of any multiplication combination, using 1 would allow them to generate an infinite number of multiplication combinations for any number, even prime numbers (e.g., 1 × 7, 1 × 1 × 7, 1 × 1 × 1 × 7, and so on). Therefore, 1 will not be used as part of the multiplication combinations students will be creating in the next few sessions except in the combination of 1 times the number itself (e.g., 1 × 18).

❺ **How Many Combinations?** Do not expect that all students will find all of the possible combinations. Focus students first on finding all of the combinations with two factors and then generate others in the following ways:

- Use one multiplication combination to create others for the same number. For example, students might break up the 20 in 9 × 20 to create 9 × 2 × 10.
- Use the combinations for 18 to create combinations for 180. For example, use 3 × 6 = 18 to create 3 × 6 × 10 = 180.

Technology Note

❸ **Using Calculators** At this point, students should be familiar with how to use calculators. As students find factors of large numbers, it is fine if they choose to use calculators. But observe students to make sure that the tool is being used in a sensible and meaningful way. For example, if students use the calculator to divide 180 by random numbers, looking for those that divide into it evenly, suggest that they consider what numbers they know that are factors, such as 10 or 2. Then students can use this knowledge to try other related factors such as 4, 5, or 20.

Read **Part 5: Technology in *Investigations:* Calculators and Computers** in *Implementing Investigations in Grade 5:* Using Calculators with the Curriculum

Are there any more combinations with three whole numbers that you can multiply that will equal 18? Is there any way to multiply more than three numbers to equal 18?❷

Students work on *Student Activity Book* page 12. They may continue to work on this sheet in the next session. Students write down the four combinations the class just found for 18 (1 × 18, 2 × 9, 3 × 6, 2 × 3 × 3) underneath 18. Under 180, students list all the combinations with two factors for 180 and then work on finding combinations with more than two factors.❸

When most students have five or six combinations with two factors for 180 and a few combinations with more than two factors, stop briefly to ask students to share their strategies for how they found the combinations with two factors.

I'd like to know how you figured out some combinations with 180 as the product. What is a combination with two factors that works, and how did you think of it?

What combinations did you make with more than two factors, and how did you figure them out?❹

Write students' examples of combinations for 180 with two, three, and more factors on the board or overhead, and allow students to copy down solutions they do not yet have.

Check with a partner to make sure that you have all of the ways to multiply two factors to get 180. Then compare and share combinations you have found with three or more factors. Organize your work in some way to help you find more combinations.❺

ONGOING ASSESSMENT: Observing Students at Work

Students find and organize multiplication combinations for 18 and 180.

- **How is the student using one solution to find others?**
- **Does the student use strategies for generating more solutions?** Does the student use doubling and halving, testing factors in order, or splitting numbers in one combination to make combinations with more factors?
- **To find combinations for 180, does the student know and use multiplying by 10 and multiples of 10?**

As you circulate, question students about their work.

- Did you find all of the combinations with two factors? How do you know?
- [Point to a combination.] Can any of these numbers be broken into smaller factors? Which ones? How do the combinations with two factors help you find combinations with more than two factors? How do the combinations with more than two factors help you find combinations with two factors?
- How do you know that each of your multiplication combinations has a product of 180?

DIFFERENTIATION: Supporting the Range of Learners

Intervention Allow students who are having difficulty finding more than one or two combinations to work with a smaller number or a number with which they may be more familiar, such as 100.

Extension Students who quickly find many combinations with the product 180 should try to find *all* of the possibilities with two, three, four, or more factors. Suggest that they organize their combinations in some way to see whether they can generate any more combinations, or whether they think they have all possible combinations.

2 SESSION FOLLOW-UP

Daily Practice and Homework

Daily Practice: For ongoing review, have students complete *Student Activity Book* page 13.

Homework: On *Student Activity Book* page 14, students practice their multiplication combinations. Make available extra copies of Multiplication Combinations Recording Sheet (M30).

Student Math Handbook: Students and families may use *Student Math Handbook* pages 23–24 for reference and review. See pages 190–197 in the back of this unit.

Family Letter: Send home copies of Family Letter (M27–M28).

Professional Development

4 **Dialogue Box:** Multiplying with More Than Two Numbers, p. 181

Name Date

Number Puzzles and Multiple Towers — Daily Practice

Factors of 2-Digit Numbers

NOTE Students find factors of 2-digit numbers. SMH 18

1. Find all of the factors of 36.

2. Find all of the factors of 72.

3. How are the factors of 36 related to the factors of 72?

Ongoing Review

4. Which number is **not** a factor of 124?

A. 31 **B.** 12 **C.** 4 **D.** 2

© Pearson Education 5

Session 1.4 Unit 1 13

▲ Student Activity Book, p. 13

Name Date

Number Puzzles and Multiple Towers — Homework

Multiplication Combinations 2

NOTE Fifth-grade students are expected to know their multiplication combinations (facts). This page helps students determine whether they remember their combinations and identify any combinations they still need to practice. SMH 25–29

Multiply each number in the first column of the table with the number at the top. For example, find the product of 3 × 7 for the first blank space in Table A.

In Tables C and D, write a number at the top for a group of combinations you need to practice.

Circle any combinations you do not know immediately, and record them on *Student Activity Book* page 10 or on M30 that your teacher will give you.

Table A		Table B		Table C		Table D	
× 7		× 8		× ___		× ___	
3		2		10		5	
6		7		4		2	
9		3		2		12	
2		5		8		4	
4		6		3		10	
10		10		6		7	
8		8		9		3	
12		11		5		6	
5		12		12		11	
7		4		11		8	
11		9		7		9	

© Pearson Education 5

14 Unit 1 Session 1.4

▲ Student Activity Book, p. 14

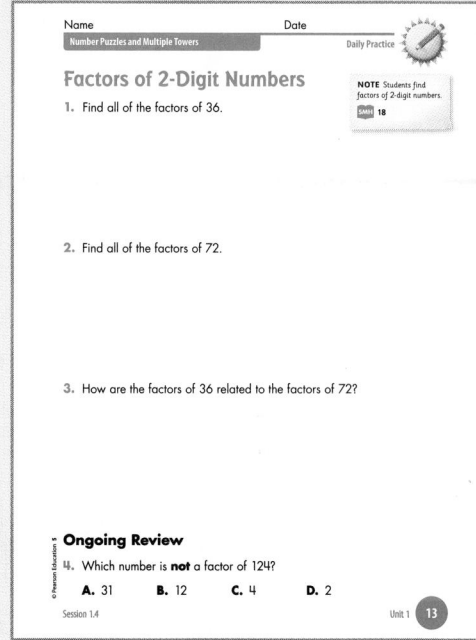

Assessment: Number Puzzles and Finding Factors

Math Focus Points

- Using known multiplication combinations to find multiplication combinations for numbers related by place value (e.g., $3 \times 6 = 18$, $3 \times 6 \times 10 = 180$)
- Finding all the factors of a number
- Finding all the ways to multiply whole numbers for a given product
- Using properties (even, odd, prime, square) and relationships (factor, multiple) of numbers to solve problems

Today's Plan		Materials
1 ACTIVITY **Introducing Math Workshop**	10 MIN CLASS	
2 MATH WORKSHOP ***Number Puzzles* and Finding Factors** 2A Assessment: *Number Puzzles* 2B Finding Factors	40 MIN	2A • Materials from Session 1.3, Activity 2 • M32*; M33*; M34* ✓ 2B • *Student Activity Book*, pp. 12 (from Session 1.4); 15 • Calculators (optional)
3 DISCUSSION **Solving a *Number Puzzle***	10 MIN CLASS	
4 SESSION FOLLOW-UP **Daily Practice and Homework**		• *Student Activity Book*, pp. 16–17 • M30 (as needed; from Session 1.4)* • *Student Math Handbook*, pp. 25–29

*See *Materials to Prepare*, p. 27.

Ten-Minute Math

Quick Images: Seeing Numbers **Show Images 13–15 on *Quick Images: Seeing Numbers* (T5) and follow the procedure described on page 48.**

ACTIVITY 1
Introducing Math Workshop

10 MIN CLASS

Although students will be familiar with the structure of Math Workshop from previous grades, it is important to spend some time in this session laying the foundation for what it means to have a productive Math Workshop. You may frequently need to remind students to work cooperatively with a partner and to stay focused on the mathematics of each activity. Bring up the following important matters as you introduce the first Math Workshop of the year: 1

- Where will students find the materials for the activities?
- How do you expect students to clean up and return materials at the end of each session?
- How will students be keeping track of the work that they do?
- How will students determine when to move to a new activity?

Math Workshop materials are placed in a convenient location for students.

After this first experience with Math Workshop for the year, plan to spend some time outside math time debriefing students about how it went. Some questions to ask might include the following:

What helped you work independently? What made it hard?

What helped you work cooperatively with your partner? What were any difficulties that you had to work out together?

How did you manage your time so that you accomplished the work that you needed to do?

Professional Development

1 ***Part 2: How to Use Investigations** in Implementing Investigations in Grade 5:* Using the Curriculum Units

Name Date

Number Puzzles and Multiple Towers

Multiplication Combinations 1

Multiply each number in the first column of the table by the number at the top. Write the product in the second column. For example, the product of the first row in Table A is 14, which is 2×7.

Circle any combinations you do not know immediately, and record them on *Student Activity Book* page 10.

Table A		Table B		Table C		Table D	
×7		×8		×6		×9	
2	14	2		10		5	
6		9		4		2	
8		4		2		12	
3		11		8		4	
10		8		3		10	
11		10		6		7	
7		6		9		3	
12		5		5		6	
4		3		12		11	
9		12		11		8	
5		7		7		9	

© Pearson Education 5

Session 1.3 Unit 1 M29

▲ Resource Masters, M29

Number Puzzles and Multiple Towers

Quick Images: Seeing Numbers (page 5 of 12)

13.
14.
15.

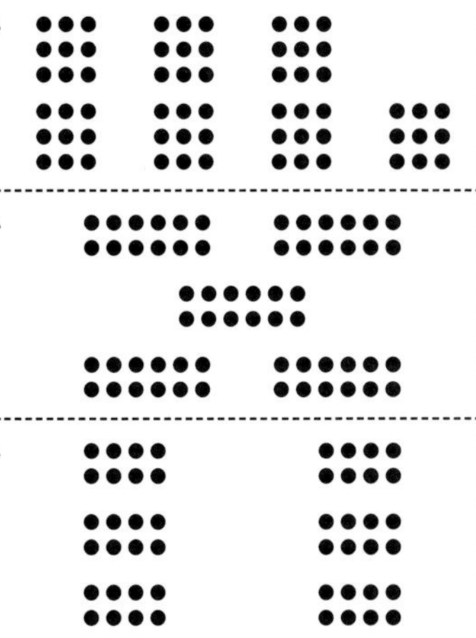

© Pearson Education 5

T5

▲ Transparencies, T5

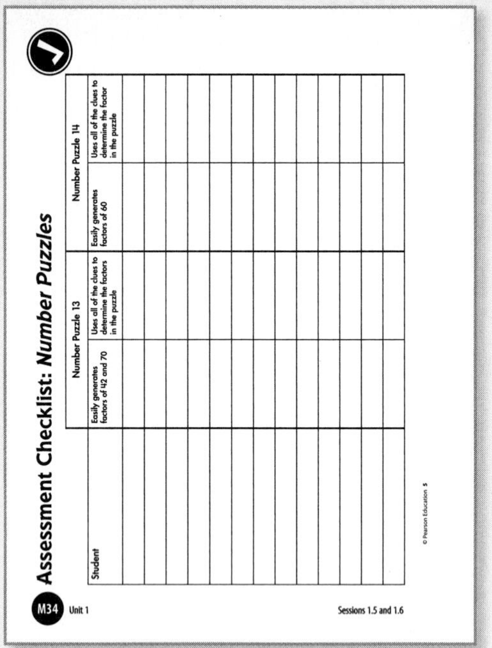
Assessment Checklist: *Number Puzzles*

Student	Number Puzzle 13: Easily generates factors of 42 and 70	Number Puzzle 13: Uses all of the clues to determine the factors in the puzzle	Number Puzzle 14: Easily generates factors of 60	Number Puzzle 14: Uses all of the clues to determine the factor in the puzzle

M34 Unit 1 Sessions 1.5 and 1.6

© Pearson Education 5

▲ Resource Masters, M34 ☑

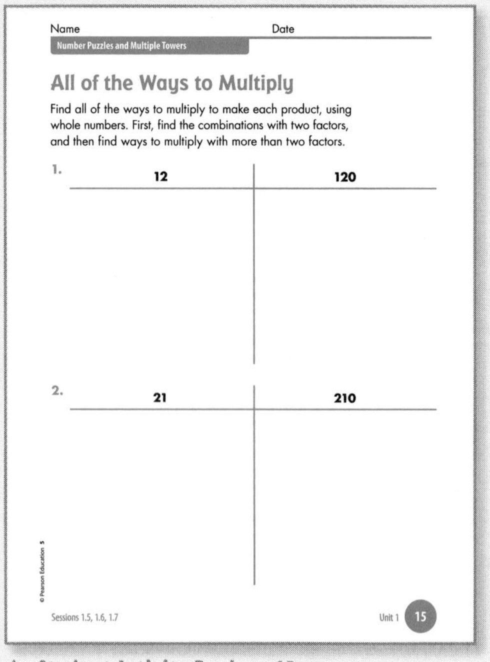
Name Date

Number Puzzles and Multiple Towers

All of the Ways to Multiply

Find all of the ways to multiply to make each product, using whole numbers. First, find the combinations with two factors, and then find ways to multiply with more than two factors.

1. 12 | 120

2. 21 | 210

© Pearson Education 5

Sessions 1.5, 1.6, 1.7 Unit 1 15

▲ Student Activity Book, p. 15

MATH WORKSHOP

2 Number Puzzles and Finding Factors

40 MIN

Students' work focuses on solving number puzzles and on finding factors and the prime factorization of given numbers. Students should spend equal time on these activities today. Tell them that they will be discussing the solution to Puzzle 9 at the end of the session today, so everyone should do it. Students will continue with both activities in the next session.

This Math Workshop as well as the one in Session 1.6 include an observed assessment that assesses Benchmark 1: Find the factors of a number.

2A Assessment: *Number Puzzles*

INDIVIDUALS PAIRS

Make available *Number Puzzles* 1–12 (M17–M22), as well as *Number Puzzles* 13–14 (M33). Students complete the *Number Puzzles* they have not finished already, including the challenge puzzles, Puzzles 8–12.

For the assessment, have students solve *Number Puzzles* 13 and 14. You will need to observe each student individually while his or her partner works on the other puzzles. Pay attention to how students work on the clues involving factors. Can they easily identify the factors of the numbers in the clues? Use Assessment Checklist: *Number Puzzles* (M34) to record your observations as students work.

See Session 1.3, pages 42–45, for complete details about *Number Puzzles,* including the answers to Puzzles 1–12. Use *Number Puzzles* Recording Sheet (M32) for additional copies of the recording sheet on *Student Activity Book* page 8.

Following are the answers to *Number Puzzles* 13 and 14:

Puzzle 13: 2, 7   Puzzle 14: 12

2B Finding Factors

PAIRS

Students complete their work on *Student Activity Book* page 12 from Session 1.4 and also work on *Student Activity Book* page 15. Students may use calculators for this activity. Decide whether you need to have a brief class discussion on how to use calculators to find factors.

Students will benefit from organizing their lists in some way. If they do not figure this out on their own, suggest that they first list ways with two factors, then ways with three factors, and so on. Students should also think about multiplying by 10s and multiples of 10 to help them find factors (e.g., $6 \times 30 = 180$ and $60 \times 3 = 180$).

ONGOING ASSESSMENT: Observing Students at Work

Students find as many multiplication combinations as they can, using two or more factors, for given products. As students find multiplication combinations for 18 and 180, 12 and 120, and 21 and 210, continue to observe and assess your students as suggested on pages 50–51.

3 DISCUSSION

Solving a *Number Puzzle*

10 MIN CLASS

Math Focus Points for Discussion

- Using properties (even, odd, prime, square) and relationships (factor, multiple) of numbers to solve problems

Ask students for the answer to *Number Puzzle* 9. Students should have determined that the puzzle is impossible to solve. If someone thinks that there is a solution, ask that student to explain his or her thinking to the group and have the class discuss it. Ask why no number can fit all the clues.

Students might say:

"Square numbers can't be prime numbers because they all have more than 2 factors, except for the square number 1. Like 9. It has three factors: 1, 3, and 9."

"The only even prime number is 2, and 2 isn't a square number."

Ask pairs of students to change one of the clues so that this puzzle will have a possible solution. After several minutes, ask pairs to give their new clue. Have the class solve each new puzzle.

Name Date

Number Puzzles and Multiple Towers — Daily Practice

Factors of 3-Digit Numbers

NOTE Students find factors of 3-digit numbers.

SMH 18

1. Find all of the factors of 100.

2. Find all of the factors of 200.

3. Did you use the factors of 100 to find the factors of 200? If so, how?

Ongoing Review

4. Which number is **not** a factor of 150?

A. 3 **B.** 15 **C.** 75 **D.** 125

© Pearson Education 5

16 Unit 1 Session 1.5

▲ Student Activity Book, p. 16

Name Date

Number Puzzles and Multiple Towers — Homework

Multiplication Combinations 3

NOTE Fifth-grade students are expected to know their multiplication combinations (facts). This page helps students determine whether they remember their combinations and identify any combinations they still need to practice.

SMH 25–29

Multiply each number in the first column of the table with the number at the top. For example, find the product of 9 × 6 for the first blank space in Table A.

In Tables C and D, write a number at the top for a group of combinations you need to practice.

Circle any combinations you do not know immediately, and record them on *Student Activity Book* page 10 or on M30 which your teacher will give you.

Table A: × 6		Table B: × 8		Table C: × ___		Table D: × ___	
9		4		10		12	
4		5		5		9	
7		3		12		8	
5		2		7		7	
8		6		4		2	
3		10		2		11	
12		7		8		3	
2		12		6		4	
11		8		3		10	
6		4		12		5	
10		9		9		6	

© Pearson Education 5

Session 1.5 Unit 1 17

▲ Student Activity Book, p. 17

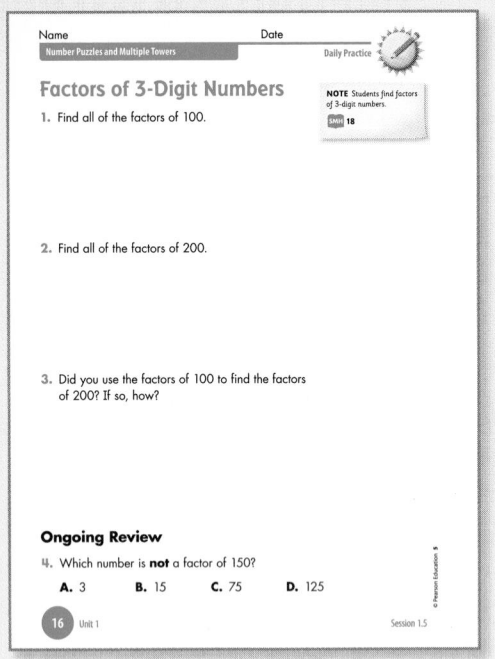

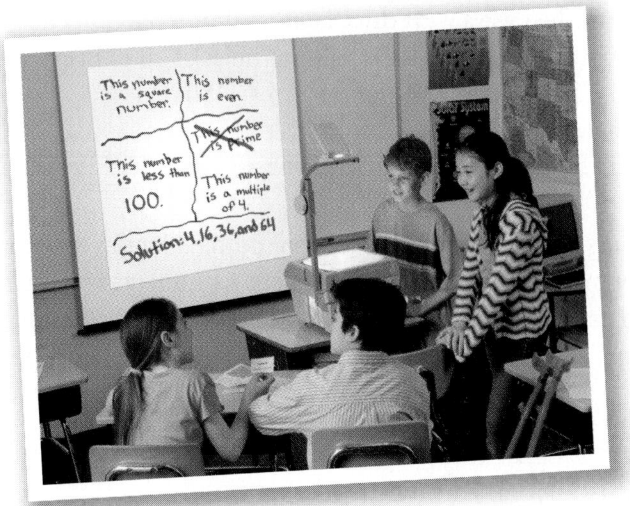

Students change one of the clues in Number Puzzle 9 so that an impossible puzzle becomes one that the class can solve.

4 SESSION FOLLOW-UP

Daily Practice and Homework

Daily Practice: For reinforcement of this unit's content, have students complete *Student Activity Book* page 16.

Homework: On *Student Activity Book* page 17, students practice their multiplication combinations. Make available extra copies of Multiplication Combinations Recording Sheet (M30).

Student Math Handbook: Students and families may use *Student Math Handbook* pages 25–29 for reference and review. See pages 190–197 in the back of this unit.

Number Puzzles and Finding Factors, *continued*

Math Focus Points

- Finding all the factors of a number
- Finding all the ways to multiply whole numbers for a given product

Today's Plan		Materials
1 MATH WORKSHOP **Number Puzzles and Finding Factors, *continued*** 1A Assessment: *Number Puzzles* 1B Finding Factors	45 MIN	1A • Materials from Session 1.5 • *Student Activity Book,* pp. 19–20 1B • Materials from Session 1.5
2 DISCUSSION **Finding All the Factors**	15 MIN PAIRS CLASS	• *Student Activity Book,* p. 19
3 SESSION FOLLOW-UP **Daily Practice and Homework**		• *Student Activity Book,* pp. 21–22 • M24, (as needed; from Session 1.4)* • *Student Math Handbook,* p. 18

*See *Materials to Prepare,* p. 27.

Ten-Minute Math

Quick Images: Seeing Numbers Show Images 16–18 (one at a time) on *Quick Images: Seeing Numbers* (T6). For each pattern, ask students to write several different *multiplication* equations, including combinations with three or more numbers, to find the total number of shapes. Have two or three students explain how they saw the images (including any revisions they made) and their equations, showing how their numbers match the patterns.

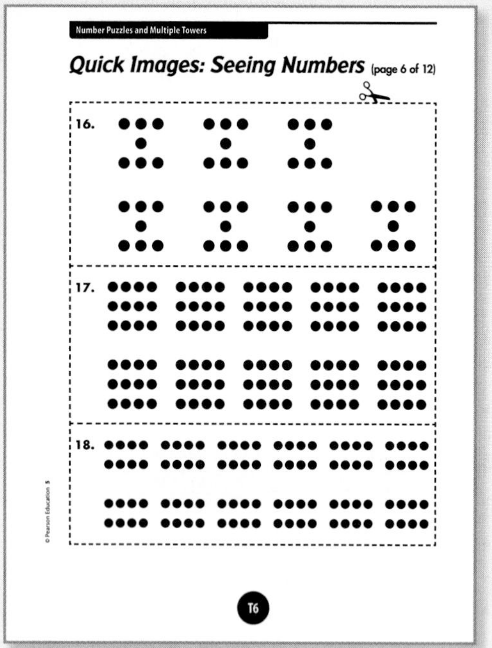
Number Puzzles and Multiple Towers

Quick Images: Seeing Numbers (page 6 of 12)

16.

17.

18.

© Pearson Education 5

T6

▲ Transparencies, T6

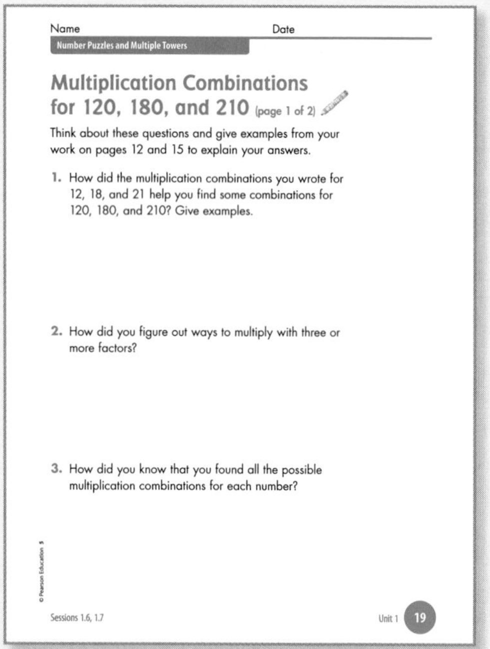
Name Date

Number Puzzles and Multiple Towers

Multiplication Combinations for 120, 180, and 210 (page 1 of 2)

Think about these questions and give examples from your work on pages 12 and 15 to explain your answers.

1. How did the multiplication combinations you wrote for 12, 18, and 21 help you find some combinations for 120, 180, and 210? Give examples.

2. How did you figure out ways to multiply with three or more factors?

3. How did you know that you found all the possible multiplication combinations for each number?

© Pearson Education 5

Sessions 1.6, 1.7 Unit 1 19

▲ Student Activity Book, p. 19 WRITING

MATH WORKSHOP

1 Number Puzzles and Finding Factors, *continued*

45 MIN

Students continue the work they began in the Math Workshop in the previous session. They should spend more time in this session on the Finding Factors activity. Use this time to complete the observed assessment.

Tell students that the discussion at the end of the session focuses on Problem 3 on *Student Activity Book* page 19.

1A Assessment: *Number Puzzles*

INDIVIDUALS PAIRS

Students complete their work with *Number Puzzles* 1–14 (M17–M22, M33). As they do so, continue to assess students on Puzzles 13–14.

For complete details about this activity, see Session 1.3, pages 42–45, and Session 1.5, page 54.

1B Finding Factors

PAIRS

Students complete *Student Activity Book* pages 12 and 15, as well as *Student Activity Book* pages 19 and 20. The discussion at the end of this session focuses on how students know that they have found all of the multiplication combinations with two factors for a particular product (Problem 3 on *Student Activity Book* page 19). Finding the prime factorization (Problems 4 and 5 on *Student Activity Book* page 20) will be discussed at the beginning of Session 1.7.

For details about this activity, see Session 1.4, pages 49–51, and Session 1.5, pages 54–55.

ONGOING ASSESSMENT: Observing Students at Work

Students find as many multiplication expressions using two or more factors as they can for given products. Continue to observe and assess your students as suggested on pages 50–51. Help students focus on how they know that they have found all of the expressions with two factors for given products.

DISCUSSION

2 Finding All the Factors

15 MIN PAIRS CLASS

Math Focus Points for Discussion

- Finding all the factors of a number

We've been working on finding factors of numbers, and today you were asked to think about how you know that you have found all of the multiplication combinations with two factors for a given product. With a partner, find all of the multiplication combinations with two factors for 36. After you've found them all, be ready to explain how you know that there are no more.

Give students time to complete the task, and then ask a pair to share the factors they found. Ask whether anyone found any other factors.

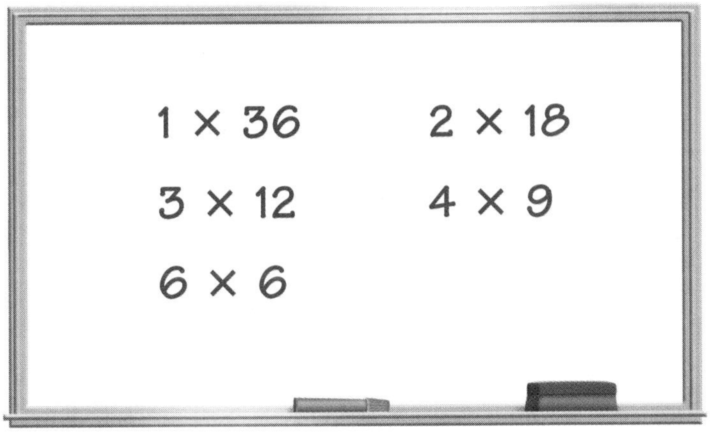

Ask students to explain how they know that they have found all of the multiplication combinations with two factors. Some students may argue that the only way to be sure is to try every possible whole number through 36. Point out that as they work with larger numbers, trying every number is difficult and tedious. Ask students whether there are other ways to know whether they have found all of the combinations. Encourage explanations based on ideas about mathematical relationships.

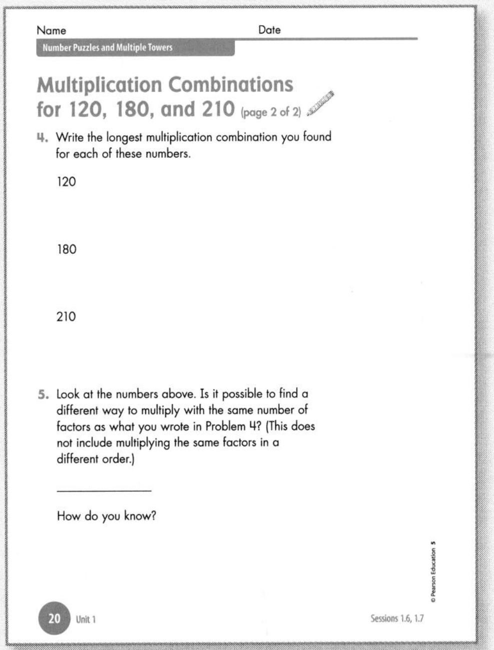
Name Date

Number Puzzles and Multiple Towers

Multiplication Combinations for 120, 180, and 210 (page 2 of 2)

4. Write the longest multiplication combination you found for each of these numbers.

120

180

210

5. Look at the numbers above. Is it possible to find a different way to multiply with the same number of factors as what you wrote in Problem 4? (This does not include multiplying the same factors in a different order.)

______

How do you know?

© Pearson Education 5

20 Unit 1 Sessions 1.6, 1.7

▲ Student Activity Book, p. 20 WRITING

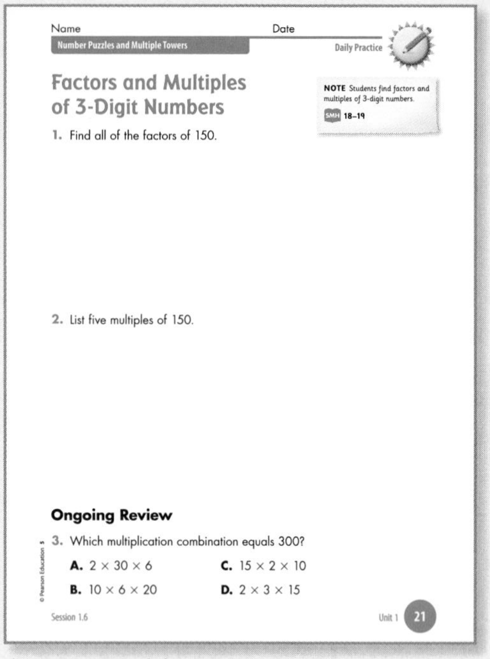
Name Date

Number Puzzles and Multiple Towers

Daily Practice

Factors and Multiples of 3-Digit Numbers

NOTE Students find factors and multiples of 3-digit numbers.

SMH 18–19

1. Find all of the factors of 150.

2. List five multiples of 150.

Ongoing Review

3. Which multiplication combination equals 300?

A. 2 × 30 × 6 **C.** 15 × 2 × 10

B. 10 × 6 × 20 **D.** 2 × 3 × 15

© Pearson Education 5

Session 1.6 Unit 1 21

▲ Student Activity Book, p. 21

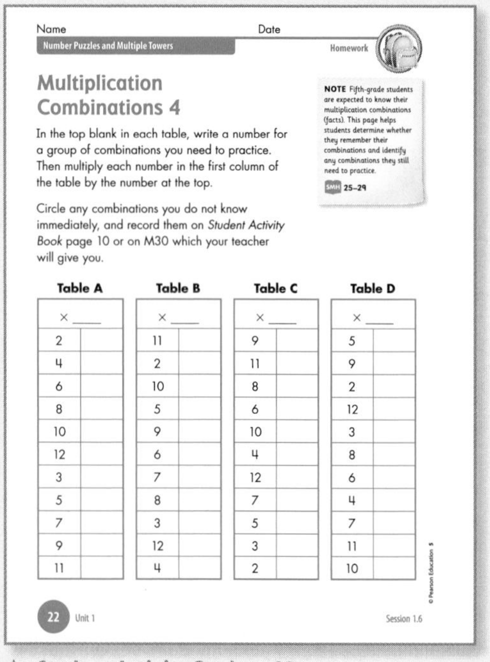
Name Date

Number Puzzles and Multiple Towers

Homework

Multiplication Combinations 4

NOTE Fifth-grade students are expected to know their multiplication combinations (facts). This page helps students determine whether they remember their combinations and identify any combinations they still need to practice.

SMH 25–29

In the top blank in each table, write a number for a group of combinations you need to practice. Then multiply each number in the first column of the table by the number at the top.

Circle any combinations you do not know immediately, and record them on *Student Activity Book* page 10 or on M30 which your teacher will give you.

Table A		Table B		Table C		Table D	
× ___		× ___		× ___		× ___	
2		11		9		5	
4		2		11		9	
6		10		8		2	
8		5		6		12	
10		9		10		3	
12		6		4		8	
3		7		12		6	
5		8		7		4	
7		3		5		7	
9		12		3		11	
11		4		2		10	

© Pearson Education 5

22 Unit 1 Session 1.6

▲ Student Activity Book, p. 22

Students might say:

"I know that 1, 2, 3, and 4 work. 5 doesn't work, and when you get to 6, it's 6 × 6, so there aren't any more because if you had something bigger than 6, you'd have to multiply it by something smaller than 6, and we already have all of those."

"5, 7, and 8 don't work, and then you're at 9, and you already have 4 × 9, so you know that there aren't any more."

3 SESSION FOLLOW-UP
Daily Practice and Homework

Daily Practice: For reinforcement of this unit's content, have students complete *Student Activity Book* page 21.

Homework: On *Student Activity Book* page 22, students practice their multiplication combinations. Make available extra copies of Multiplication Combinations Recording Sheet (M30).

Student Math Handbook: Students and families may use *Student Math Handbook* page 18 for reference and review. See pages 190–197 in the back of this unit.

Prime Factorization

Math Focus Points

- Determining the prime factorization of a number
- Using known multiplication combinations to find equivalent multiplication combinations (e.g., $18 = 3 \times 6 = 3 \times (2 \times 3)$)
- Using known multiplication combinations to find multiplication combinations for numbers related by place value (e.g., $3 \times 6 = 18$; $3 \times 6 \times 10 = 180$)

Vocabulary

prime factorization

Today's Plan		Materials
1 DISCUSSION **The Longest Multiplication Combination**	20 MIN CLASS	• *Student Activity Book,* p. 20 (from Session 1.6) • Calculators (optional)
2 ACTIVITY **Factors of Larger Numbers**	40 MIN INDIVIDUALS PAIRS	• *Student Activity Book,* pp. 12, 15, 19–20 (from Session 1.6) • Calculators (optional)
3 SESSION FOLLOW-UP **Daily Practice and Homework**		• *Student Activity Book,* pp. 23–24 • M30 (from Session 1.4)* • *Student Math Handbook,* pp. 23–24

*See *Materials to Prepare,* p. 27.

Ten-Minute Math

Quick Images: Seeing Numbers Show Images 19–21 (one at a time) on *Quick Images: Seeing Numbers* (T7). For each pattern, ask students to write several different *multiplication* equations, including combinations with three or more numbers, to find the total number of shapes. Have two or three students explain how they saw the images (including any revisions they made) and their equations, showing how their numbers match the patterns.

DISCUSSION

1 The Longest Multiplication Combination

20 MIN CLASS

Math Focus Points for Discussion

- Determining the prime factorization of a number

Ask students to give their answers to Problem 4 on *Student Activity Book* page 20. Write these on the board or overhead.

Students discuss the longest multiplication combinations they found for 120, 180, and 210.

If students suggest the same numbers in a different order (e.g., for 120, $2 \times 2 \times 2 \times 3 \times 5$ and $2 \times 3 \times 2 \times 5 \times 2$), list them, and then remind students that combinations of the same numbers in a different order are not considered different combinations. Ask students what they notice about the lists of longest combinations.❶

If students have not identified the "longest combination" for any of the numbers, focus first on the longest combination they have found so far. For example, if they have $2 \times 3 \times 4 \times 5$ for 120, ask whether any of the factors could be split into smaller numbers.

When students are sure that they have found the longest possible combination, ask questions such as the following:

How do you know that this is the only set of five whole numbers that has a product of 180? Why can't we make a combination for 180 with five factors that are different from these? Do you think it's possible to make an even longer combination of numbers? Why or why not?

Math Note

❶ **Longest Combinations** These are the longest multiplication combination of whole numbers, or prime factorization, for each of the numbers on *Student Activity Book* page 20.
120: $2 \times 2 \times 2 \times 3 \times 5$
180: $2 \times 2 \times 3 \times 3 \times 5$
210: $2 \times 3 \times 5 \times 7$

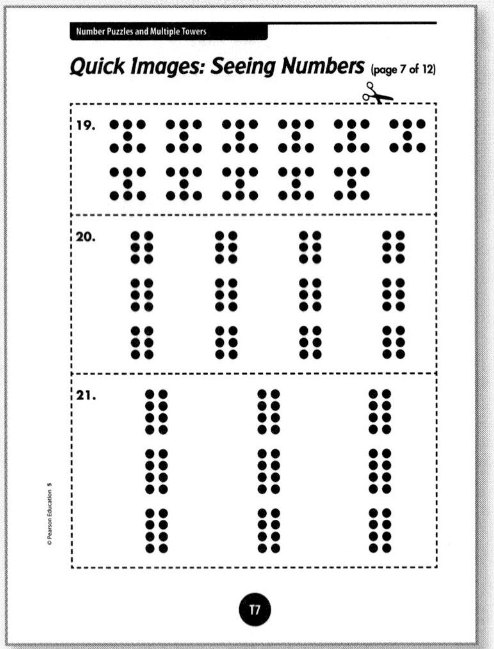
Number Puzzles and Multiple Towers

Quick Images: Seeing Numbers (page 7 of 12)

19.

20.

21.

© Pearson Education 5

T7

▲ Transparencies, T7

Although students at this grade are just beginning to develop an understanding that every whole number has a unique prime factorization, a very important characteristic of our number system, they can make the following observations about the "longest combination":

- The "longest combination" of whole numbers that they can multiply to make a certain product involves only prime numbers.
- No matter what multiplication combination they start out with for that product and no matter how they break it into smaller factors, they always end up with the same combination of prime numbers.❷

2 ACTIVITY
Factors of Larger Numbers

40 MIN INDIVIDUALS PAIRS

Have students work alone or with a partner to find multiplication combinations with two, three, or more factors for one of the following numbers:

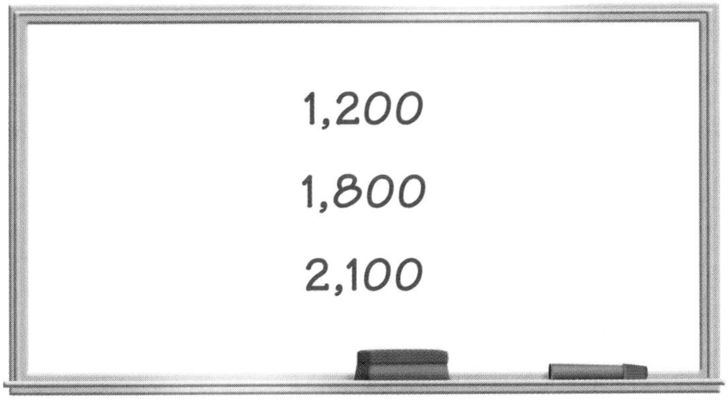

Students should use the work they have already done on the smaller numbers (*Student Activity Book* pages 12, 15, 19–20) to help them find expressions for these larger numbers. Challenge students to find the prime factorization for these larger numbers.❸

Professional Development

❷ **Teacher Note:** Finding Prime Factors, p. 154

Math Note

❸ **Generating Prime Factorization** Students have determined that the prime factorization of 18 is $2 \times 3 \times 3$. Because $180 = 18 \times 10$, the prime factorization for 180 can be found by multiplying the prime factorization for 18 by the prime factors of 10, as shown below.

$180 = 18 \times 10$
$180 = (2 \times 3 \times 3) \times 10$
$180 = (2 \times 3 \times 3) \times (2 \times 5)$

Some students might determine that the prime factorization of 1,800 includes another 2×5.

$1{,}800 = 180 \times 10$
$1{,}800 = (2 \times 2 \times 3 \times 3 \times 5) \times 10$
$1{,}800 = (2 \times 2 \times 3 \times 3 \times 5) \times (2 \times 5)$

Challenge these students to use the prime factorization to help them find all the multiplication combinations for 1,800.

Name Date

Number Puzzles and Multiple Towers — Daily Practice

Multiplying to Make 60 and 90

NOTE Students find multiplication combinations with two factors and with more than two factors for 60 and for 90.

SMH 23–24, 25–29

Find as many ways as you can to multiply whole numbers to make each product.

1. Multiplying to make 60

Ways to multiply with two factors:	Ways to multiply with more than two factors:

2. Multiplying to make 90

Ways to multiply with two factors:	Ways to multiply with more than two factors:

Ongoing Review

3. Which multiplication combination equals 150?

A. 10 × 5 × 10 **C.** 25 × 2 × 3

B. 75 × 2 × 10 **D.** 10 × 5 × 5

Session 1.7 Unit 1 23

▲ Student Activity Book, p. 23

Name Date

Number Puzzles and Multiple Towers — Homework

Multiplication Combinations 5

NOTE Fifth-grade students are expected to know their multiplication combinations (facts). This homework helps students determine whether they remember their combinations and identify any combinations they still need to practice.

SMH 25–29

In the top blank in each table, write a number for a group of combinations you need to practice. Then multiply each number in the first column of the table by the number at the top.

Circle any combinations you do not know immediately, and record them on *Student Activity Book* page 10 or on M30 which your teacher will give you.

Table A		Table B		Table C		Table D	
× ___		× ___		× ___		× ___	
4		6		5		3	
8		10		7		5	
6		2		8		2	
2		11		3		6	
9		4		9		9	
12		12		4		10	
3		9		11		4	
5		8		6		11	
7		5		12		12	
11		7		10		8	
10		3		2		7	

24 Unit 1 Session 1.7

▲ Student Activity Book, p. 24

ONGOING ASSESSMENT: Observing Students at Work

Students find multiplication combinations with two or more factors for given products. As students are working on the numbers 1,200, 1,800, and 2,100, continue to observe and assess your students as suggested on pages 50–51. For these larger numbers, take note of the following:

- **Do students use the relationship between a number and its multiple (e.g., use multiplication combinations for 180 to generate combinations for 1,800)?**

DIFFERENTIATION: Supporting the Range of Learners

Intervention If students are still working on finding the prime factors, have them use numbers that are less than 100, but are also related in some way, such as 24, 36, and 48.

Extension Students who easily find the prime factors of 1,200, 1,800 and 2,100 can choose numbers that they think might have a longer list of prime factors or that might have many factors.

3 SESSION FOLLOW-UP
Daily Practice and Homework

Daily Practice: For ongoing review, have students complete *Student Activity Book* page 23.

Homework: On *Student Activity Book* page 24, students practice their multiplication combinations. Make available extra copies of Multiplication Combinations Recording Sheet (M30).

Student Math Handbook: Students and families may use *Student Math Handbook* pages 23–24 for reference and review. See pages 190–197 in the back of this unit.

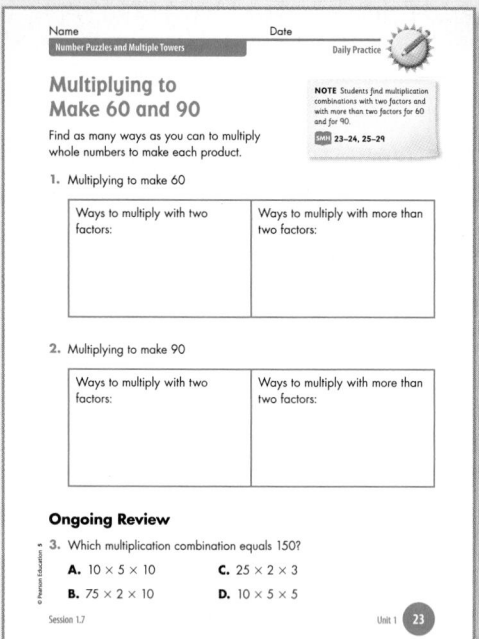

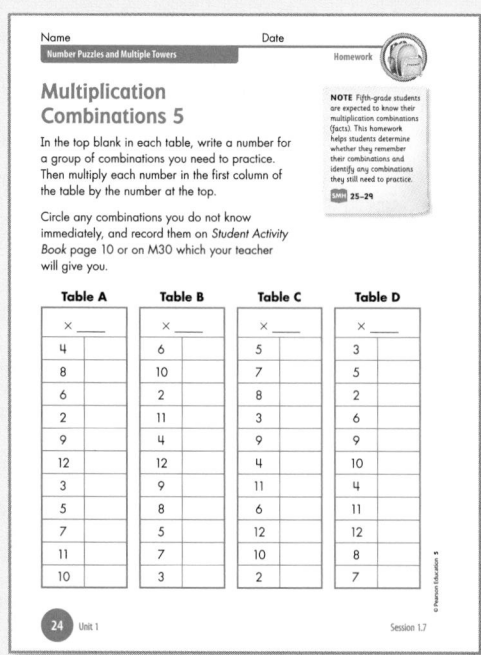

INVESTIGATION 2

Mathematical Emphases

Computational Fluency Multiplication problems with 2-digit numbers

Math Focus Points

- Solving 2-digit by 2-digit multiplication problems
- Describing and comparing strategies used to solve multiplication problems
- Breaking up multiplication problems efficiently
- Multiplying fluently by multiples of 10
- Comparing multiplication problems to determine which product is greater
- Estimating the product of two numbers

Whole Number Operations Representing the meaning of multiplication and division

Math Focus Points

- Creating a story problem represented by a multiplication or division expression
- Representing a multiplication or division problem with a picture or diagram
- Using arrays to model multiplication

This Investigation also focuses on

- Using clear and concise notation

Multiplication Strategies

	Student Activity Book	Student Math Handbook	Professional Development: Read Ahead of Time
SESSION 2.1 p. 70			
Naming Multiplication Strategies Students represent and solve multiplication problems and compare strategies.	25–26	30–32	• **Teacher Note:** Representing Multiplication with Arrays, p. 156; Visualizing Arrays, p. 159; Multiplication Strategies, p. 161; Division Notation, p. 163 • **Dialogue Box:** Naming Multiplication Strategies, p. 183
SESSION 2.2 p. 77			
Comparing Representations Students compare representations used for solving multiplication problems. They use arrays to represent different ways of breaking up a problem.	27–29	17	• **Part 4: Ten-Minute Math:** in *Implementing Investigations in Grade 5: Number Puzzles*
SESSION 2.3 p. 85			
Which Product Is Greater? Students estimate and compare products involving multiples of 10.	31–33	G11	• **Dialogue Box:** Playing *Multiplication Compare,* p. 185
SESSION 2.4 p. 91			
Multiplication Cluster Problems Students solve multiplication problems by breaking up each one or by using related problems.	35–38	35; G11	• **Teacher Note:** About Cluster Problems, p. 165

Ten-Minute Math See page 20 for an overview.

Quick Images: Seeing Numbers

- T8–T9, *Quick Images: Seeing Numbers*

Number Puzzles

- M35–M43, *Number Puzzle* Clue Cards Make one copy. Cut the cards apart and put into three piles, Set A, Set B, and Set C.

Materials to Gather	Materials to Prepare
• **12″ x 18″ construction paper** (1 per student) • **Markers** (1 set per student) • **Color tiles** (as needed) • **Connecting cubes** (as needed)	• **M15, Centimeter Grid Paper** Make copies. (as needed) • **M31, 300 Chart** Make copies. (as needed)
• **Posters of 35 × 28** (from Session 2.1) • **Color tiles** (as needed) • **Calculators** (optional)	• **M15, Centimeter Grid Paper** Make copies. (as needed) • **M35–M43, Number Puzzle Clue Cards** Make copies. Cut the cards apart and put into three piles, Set A, Set B, and Set C. (1 set per class)
	• **M44–M45, Compare Cards** Make copies and cut apart. (1 deck per pair) • **M46, *Multiplication Compare*** Make copies. (as needed)
• **M44–M45, Compare Cards** (1 deck per pair; from Session 2.3) • **M46, *Multiplication Compare*** (as needed; from Session 2.3)	• **M47, *Multiplication Compare* Recording Sheet** Make copies. (as needed) • **M48–M50, Digit Cards** Make copies and cut apart. (1 deck per pair; optional)

Overhead Transparency

Multiplication Strategies, *continued*

	Student Activity Book	Student Math Handbook	Professional Development: Read Ahead of Time
SESSION 2.5 p. 97			
Multiplication Cluster Problems, *continued* Students continue to solve multiplication problems by breaking up each one or by using related problems.	35–36, 39–40	35; G11	
SESSION 2.6 p. 100			
How Do I Start? Students solve multiplication problems by using a given first step for solving each one.	41–43	30–32	
SESSION 2.7 p. 105			
Assessment: What Is the Answer? Students practice and refine strategies for multiplication. They are assessed on multiplication strategies.	44–46	30–32	• **Teacher Note:** Assessment: What Is the Answer?, p. 166

Materials to Gather	Materials to Prepare
• **M44–M45, Compare Cards** (1 deck per pair; from Session 2.3) • **M46, *Multiplication Compare*** (as needed; from Session 2.3) • **M48–M50, Digit Cards** (1 deck per pair; from Session 2.4; optional)	• **M47, *Multiplication Compare* Recording Sheet** Make copies. (as needed; from Session 2.4)
• **Blank transparency** (optional)	
	• **M51, Assessment: What Is the Answer?** Make copies. (1 per student)

SESSION 2.1

Naming Multiplication Strategies

Math Focus Points

- Solving 2-digit by 2-digit multiplication problems
- Describing and comparing strategies used to solve multiplication problems
- Creating a story problem represented by a multiplication or division expression

Vocabulary

representation

Today's Plan	Materials
1 ACTIVITY **Solving 35 × 28** — 30 MIN, INDIVIDUALS, PAIRS	• M15 (as needed)*; M31 (as needed)* • 12″ x 18″ construction paper; markers; color tiles (as needed); connecting cubes (as needed)
2 DISCUSSION **Naming Strategies** — 30 MIN, CLASS	• Posters of 35 × 28 (from Activity 1)
3 SESSION FOLLOW-UP **Daily Practice and Homework**	• *Student Activity Book,* pp. 25–26 • *Student Math Handbook,* pp. 30–32

*See *Materials to Prepare,* p. 67.

Ten-Minute Math

Quick Images: Seeing Numbers Show Images 22–24 (one at a time) on *Quick Images: Seeing Numbers* (T8). Ask students to find the total number of shapes and write at least one *multiplication* and one *division* equation for each pattern. Have two or three students explain how they saw the images (including any revisions they made) and their equations, showing how their numbers match the patterns. For example, Image 22 is an arrangement showing 10 groups of 6. Student might say $10 \times 6 = 60$, or $5 \times 6 \times 2 = 60$, or $60 \div 6 = 10$.

1 ACTIVITY
Solving 35 × 28

30 MIN INDIVIDUALS PAIRS

It is expected that students enter Grade 5 with at least one efficient strategy for multiplying large numbers. This unit continues to work on improving fluency and focuses on using and comparing several efficient strategies.❶

Today you're going to make a poster showing your solution to a multiplication problem. First we're going to remind ourselves about ways to keep track of all of the parts in a multiplication problem, especially when the numbers get large. Let's start with a smaller problem: 18 × 14. Solve this problem, and then compare solutions with a neighbor.

Give students a few minutes to work.

What's the first thing you did to solve 18 × 14? For now, I only want your first step.

Record a few beginning steps from students. For example, these might include 10 × 10, 18 × 10, or 10 × 14.

Ask students to suggest a simple story context in which you might need to solve 18 × 14.❷ Select one to talk through the problem. Using the first steps that students suggested, briefly review how the story context can help you think through what parts of a problem you have done and what parts you have not done.

For the discussion below, 18 baskets with 14 apples in each of them is used as the context.

Cecilia, you suggested starting with 18 × 10. So if we're thinking about 18 baskets with 14 apples in each one, what part of the problem have we solved if we start with 18 × 10? *(18 baskets with 10 apples in each)* What part of the problem do we still have to do? *(4 more apples for each basket)*

Let's try Walter's idea, starting with 10 × 14. What part of the problem have we taken care of? *(10 baskets with 14 apples in each)* What part do we still need to solve? *(8 more baskets with 14 apples in each)*

Review with students how to draw a representation of their work. Ask students to give you ideas for sketching each part of the problem as it is solved in terms of baskets and apples (or whatever context you use). If needed, follow up with questions such as the following:

Teaching Notes

❶ **Assessing Multiplication Skills** This activity, in which students solve and represent a multiplication problem, provides an opportunity to assess what students understand about multiplication and what strategies they are comfortable using.

❷ **Story Contexts** By using story contexts and drawing pictures or other representations, students connect their strategies and notation with the meaning of multiplication. Choose contexts that are familiar to your students and that they can visualize easily. For example, baskets of apples may be familiar to students in some communities, and rows of apples in bins outside stores may be more familiar in other communities. Keep contexts simple so that students are not distracted by an overly elaborate story. When you first introduce a context, talk it through and, if feasible, sketch it so that students can really "see" it. Ask students to come up with their own contexts that make sense to them.

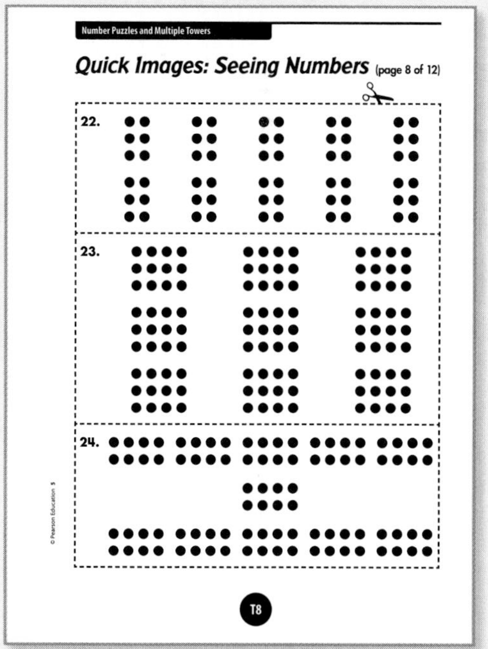

▲ Transparencies, T8

Professional Development

❸ **Teacher Note:** Representing Multiplication with Arrays, p. 156

❹ **Teacher Note:** Visualizing Arrays, p. 159

So if we're trying to figure out 18 baskets with 14 apples in each one, how does 18 × 10 help? What does the 18 mean? What does the 10 mean? How could I sketch what we have so far?

For example, a student might represent a first step of 18 × 10 as shown below.

Now ask students what other ways they have of keeping track of the parts of the problem. Sketch these students' ideas as well. If no one brings up the work they did with arrays in Grade 4, ask for a volunteer to use an array and explain how to show a way to break up the problem. A students might suggest a drawing like the following:❸ ❹

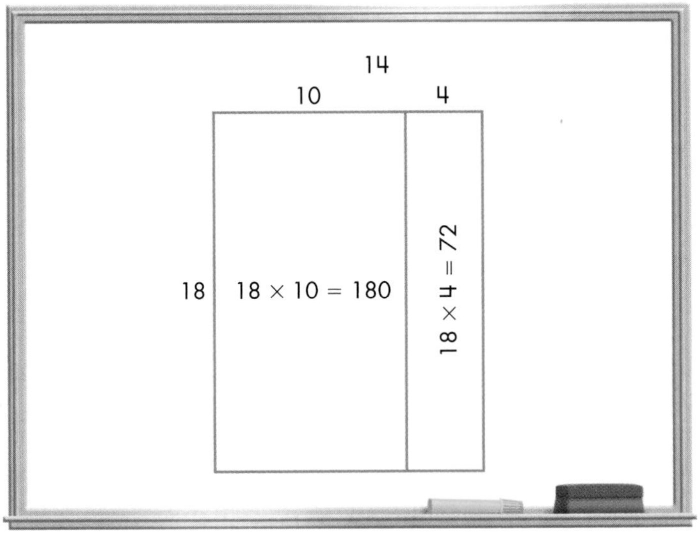

Write 35 × 28 on the board. Explain to students that they will make a poster that includes the following three parts.

- Students create a story problem for 35 × 28.
- Students solve the problem and show their solution clearly.5
- Students draw a representation for the problem by using cubes, an array, or groups.

Allow students to use whatever math tools they find useful, such as tiles, cubes, grid paper, or 300 charts. It is not necessary for students to use any of these tools, but if they do, they should show how they used them on their poster.

Let students know that using clear and concise notation is an expectation this year.6 7 8 After students have made their posters, ask them to share their solution and representation with a partner.

As students are working, circulate and identify two or three students who are using different strategies. In particular, look for examples of breaking numbers apart, changing one number to create an easier problem, and creating an equivalent problem. Sample solutions that illustrate these strategies are shown below.

Breaking the numbers apart

Charles	Yumiko
35 × 28	35 × 28
35 × 20 = 700	30 × 20 = 600
30 × 8 = 240	30 × 8 = 240
5 × 8 = + 40	5 × 20 = 100
980	5 × 8 = + 40
	980

Changing one number to make an easier problem

Renaldo

35 × 28

35 × 30 = 1,050 I multiplied 35 × 10 and added 350 three times.

35 × 2 = 70 It's not 30, it's only 28, so I had to find two 35s.

1,050 − 70 = 980

Technology Note

5 **Using Calculators** Students should not be using calculators as they multiply 2-digit and 3-digit numbers. The work in this Investigation supports students to develop efficient and accurate strategies (as well as solidify their fluency with single-digit multiplication combinations). It is important for students to use a combination of mental strategies with pencil and paper work.

Math Note

6 **Clear and Concise Notation** Clear and concise notation shows the steps of a student's solution so that they can be easily followed, and the relation of the steps to the solution is clear. Every detail need not be shown, especially parts of the problem the student does mentally, as long as the overall solution method is clear. For example, if the student is solving 35 × 28 by thinking of it as (35 × 30) − (35 × 2), the student might calculate 35 × 2 mentally and simply show the subtraction of 70.

Professional Development

7 **Teacher Note:** Multiplication Strategies, p. 161

8 **Teacher Note:** Division Notation, p. 163

Creating an equivalent problem

Tamira

$35 \times 28 = 70 \times 14$ I doubled 35, and took half of 28.

$70 \times 10 = 700$

$70 \times 4 = 280$

$700 + 280 = 980$

Note that there are a variety of ways to use each of these strategies, which students may see as being different. Ask the students you select to be prepared to explain their solutions to the rest of the class.

ONGOING ASSESSMENT: Observing Students at Work

Students solve a 2-digit by 2-digit multiplication problem.

- **What strategies are students using?** Are they keeping track of all parts of the problem? Are they using clear and concise notation?
- **Can students write a story problem for 35 × 28?**
- **Do students use the story context to help them keep track of their solutions?**
- **Are students able to follow the strategy their partner is using?**

As students are working, ask questions such as the following:

- What part of the problem have you solved? What remains to be solved? How does your story problem help you determine the answer? How does your representation help you figure out what you still need to solve? [Point to either the story context and/or the representation.] What does this number represent? Where is the 35 in your solution? The 28?

DIFFERENTIATION: Supporting the Range of Learners

Intervention If some students are still developing their understanding of multiplication and fluency with multiplication, you can change the problem to 35 × 8 or 23 × 5 to allow students to work with smaller numbers.

ELL English Language Learners may need assistance developing their story contexts. Have English Language Learners start by drawing a visual that represents a related multiplication expression, such as 18 × 14 or 35 × 20. Say, "I see you've drawn a picture of [18] boxes with [14] balls in each one of them. What story can you tell about that?"

DISCUSSION

2 Naming Strategies

30 MIN CLASS

Math Focus Points for Discussion

- Describing and comparing strategies used to solve multiplication problems

Several students are going to explain their solutions. As we look at each poster, we're going to try to name the strategy after seeing the first step, and we'll decide who else solved the problem by using the same strategy.

If you're explaining a strategy, your job is to help us follow your steps and to use your story or picture to explain how you are keeping track of the parts. If you're listening to someone else's strategy, your job is to listen carefully, try to understand the strategy being explained, compare it with your own strategy, and ask questions when you need a clearer explanation.

As students explain strategies, you can introduce phrases that help students clarify and describe their approaches.

I see you first *broke the numbers apart* by place.

Alex is *doubling* one number and *halving* the other one to create an *equivalent problem*—an easier problem with the same answer.

Ask questions that prompt students to use their stories, pictures, arrays, or other representations to justify their solutions. Suppose that a student started with 35 × 20.

If we think about Georgia's story context, how much of the problem has she done? What does the 35 mean? What does the 20 mean? Where is that in her representation? What does she still have to figure out?

As you discuss each strategy, ask students to agree on a short mathematical phrase that names it, such as those given on pages 73–74. Explain that having a name or phrase attached to each strategy will make it easier to refer to the strategies. 9 10

As you discuss a particular strategy, ask whether other students used that same strategy or a different one. 11

Display students' posters. Group those showing the same strategy together. Keep the posters up for the next session. 12

Professional Development

9 **Dialogue Box:** Naming Multiplication Strategies, p. 183

Differentiation

10 **English Language Learners** Help English Language Learners understand that the class will be developing its own name for multiplication strategies. You can make these terms more accessible by recording them on a chart, illustrated with examples.

Math Notes

11 **Same or Different Strategies** Help students decide whether strategies are the same or different. For example, if one student starts with 35 × 20 and another starts with 30 × 20, decide as a class whether students want to classify these strategies as the same (the first step involves breaking the numbers apart) or different (breaking both numbers apart or breaking one number apart).

12 **Other Strategies** Some strategies, such as changing one number to make an easier problem, or creating an equivalent problem, may not be brought up by your students. These strategies will be included in the cluster problems and starter problems later in this Investigation.

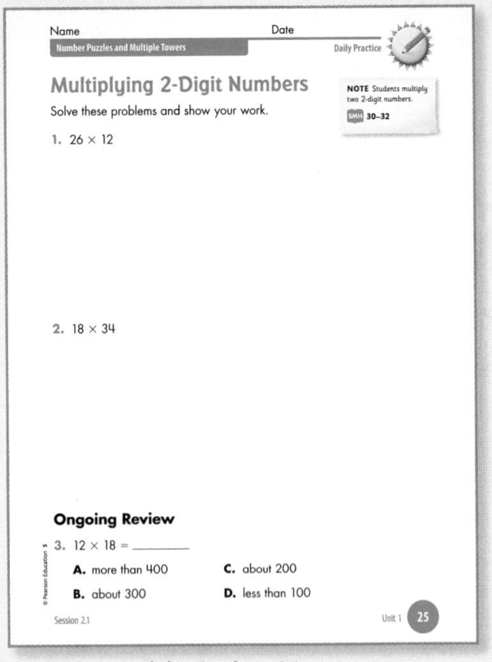
Name Date
Number Puzzles and Multiple Towers
Daily Practice

Multiplying 2-Digit Numbers

NOTE Students multiply two 2-digit numbers.
SMH 30–32

Solve these problems and show your work.

1. 26 × 12

2. 18 × 34

Ongoing Review

3. 12 × 18 = ______

A. more than 400
B. about 300
C. about 200
D. less than 100

© Pearson Education 5

Session 2.1 Unit 1 25

▲ Student Activity Book, p. 25

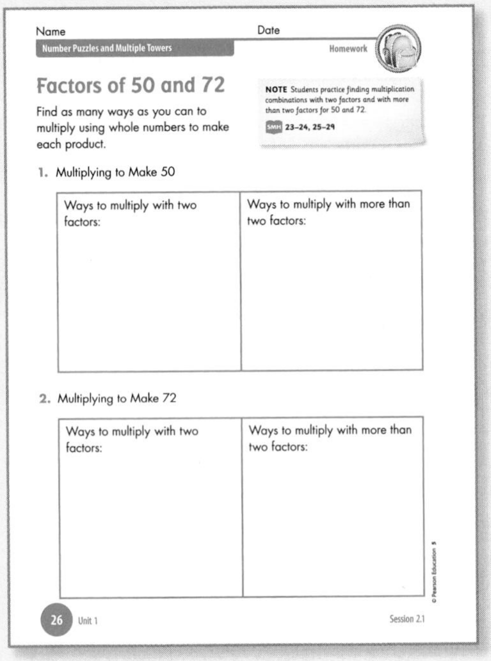
Name Date
Number Puzzles and Multiple Towers
Homework

Factors of 50 and 72

NOTE Students practice finding multiplication combinations with two factors and with more than two factors for 50 and 72.
SMH 23–24, 25–29

Find as many ways as you can to multiply using whole numbers to make each product.

1. Multiplying to Make 50

Ways to multiply with two factors:	Ways to multiply with more than two factors:

2. Multiplying to Make 72

Ways to multiply with two factors:	Ways to multiply with more than two factors:

© Pearson Education 5

26 Unit 1 Session 2.1

▲ Student Activity Book, p. 26

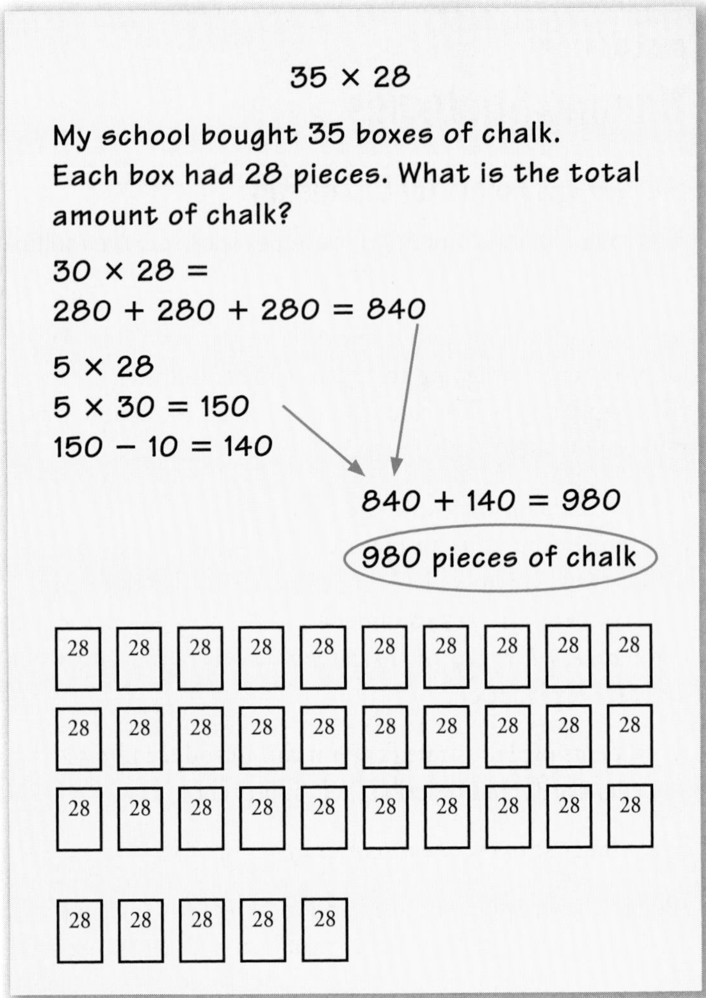
35 × 28
My school bought 35 boxes of chalk.
Each box had 28 pieces. What is the total amount of chalk?
30 × 28 =
280 + 280 + 280 = 840
5 × 28
5 × 30 = 150
150 − 10 = 140
840 + 140 = 980
980 pieces of chalk
28 28 28 28 28 28 28 28 28 28
28 28 28 28 28 28 28 28 28 28
28 28 28 28 28 28 28 28 28 28
28 28 28 28 28

SESSION FOLLOW-UP

3 Daily Practice and Homework

Daily Practice: For reinforcement of this unit's content, have students complete *Student Activity Book* page 25.

Homework: On *Student Activity Book* page 26, students find multiplication combinations for 50 and 72 having two factors and more than two factors.

Student Math Handbook: Students and families may use *Student Math Handbook* pages 30–32 for reference and review. See pages 190–197 in the back of this unit.

Comparing Representations

Math Focus Points

- Representing a multiplication or division problem with a picture or diagram
- Using arrays to model multiplication
- Solving 2-digit by 2-digit multiplication problems

Vocabulary

distributive property

Today's Plan		Materials
1 ACTIVITY **More *Number Puzzles***	15 MIN CLASS PAIRS	• M15 (as needed); M35–M43* • Color tiles (as needed); Calculators (optional)
2 DISCUSSION **Using Arrays to Represent Solutions**	20 MIN CLASS	• Posters of 35 × 28 (from Session 2.1)
3 ACTIVITY **Solving Multiplication Problems**	25 MIN CLASS PAIRS	• *Student Activity Book,* p. 27
4 SESSION FOLLOW-UP **Daily Practice and Homework**		• *Student Activity Book,* pp. 28–29 • *Student Math Handbook,* p. 17

*See *Materials to Prepare,* p. 67.

Ten-Minute Math

Note: The Ten-Minute Math activity for this unit, *Number Puzzles,* is introduced in this session. Plan to do today's Ten-Minute Math sometime after math class.

Number Puzzles **Draw one card from each set (A, B, C) of *Number Puzzle* Clue Cards (M35–M43) and read them aloud. Students work with a partner and find the number(s) that fit the set of three clues. If the solution is "inconclusive," students modify the clues. Have students share their strategies for solving the puzzle.**

Professional Development

① **Part 4: Ten-Minute Math** in *Implementing Investigations in Grade 5:* Number Puzzles

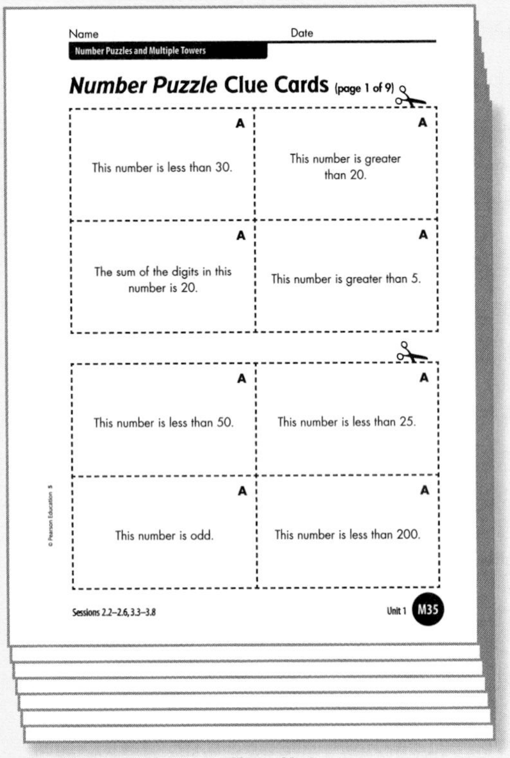

Name Date

Number Puzzles and Multiple Towers

***Number Puzzle* Clue Cards** (page 1 of 9)

A This number is less than 30.	A This number is greater than 20.
A The sum of the digits in this number is 20.	A This number is greater than 5.
A This number is less than 50.	A This number is less than 25.
A This number is odd.	A This number is less than 200.

Sessions 2.2–2.6, 3.3–3.8 Unit 1 M35

▲ Resource Masters, M35–M43

ACTIVITY

15 MIN CLASS PAIRS

1 More Number Puzzles

This activity introduces *Number Puzzles,* which will continue throughout this unit and other units as a Ten-Minute Math activity.① The activity builds on the work students did in Investigation 1.

To introduce students to the activity, start with the following three *Number Puzzle* Clue Cards (M35–M43):

Card A: This number is less than 50.

Card B: This number is not a factor of 100.

Card C: This number of tiles will make only one rectangle.

(Answers include all the prime numbers less than 50, except 2 and 5: 3, 7, 11, 13, 17, 19, 23, 29, 31, 37, 41, 43, 47.)

Read (or have volunteers read) each individual clue aloud, and write a brief description of each on the board or overhead.

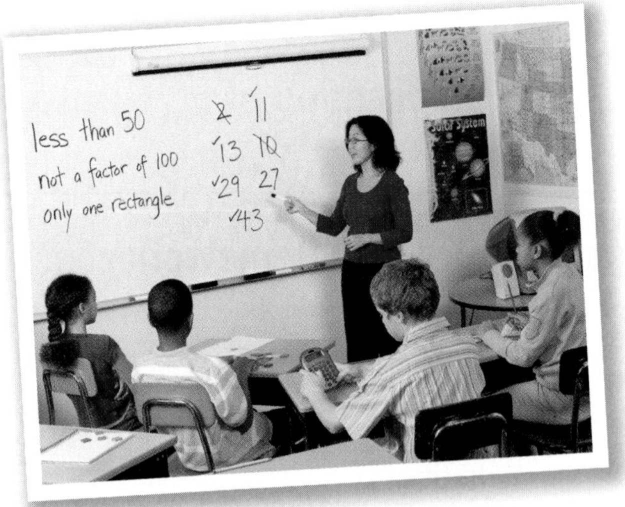

Students determine numbers that fit the descriptions given on Number Puzzle *Clue Cards.*

Ask students to interpret each clue and, if necessary, to restate it in their own words. Students may choose to rephrase Card B to say that it does not go "evenly" into 100, and to say for Card C that "it's a prime number."

Give students between 3 and 5 minutes to work with a partner on finding numbers they think will fit all of the clues. They may choose to use math tools such as tiles, grid paper, or calculators.

Call the class back together, and ask for answers. Make a list of students' suggestions next to the clues.

How did you go about finding the solution? Which clue did you start with? Why?

As students share solutions, invite the rest of the class to confirm or challenge the proposed answers. Challengers must explain their reasoning.

Students might say:

"2 doesn't work. The second clue says that the number is not a factor of 100, so even though 2 is a prime number, it doesn't fit all three clues."

"I don't think 27 is right. 27 isn't a factor of 100, but it makes more than one rectangle."

If time remains, do another puzzle. Draw one card from Set A, one from Set B, and one from Set C. Because you are drawing cards randomly, it is possible that some sets of clues will have no solution or an infinite number of solutions.

Let students know that you will be doing more *Number Puzzles* as Ten-Minute Math activities throughout the year.

ONGOING ASSESSMENT: Observing Students at Work

Students use reasoning and knowledge of properties and relationships of numbers to find a given number.

- **Do students understand that the number(s) they are looking for must fit all the clues?**
- **Do students use each clue to help them eliminate certain numbers?**
- **Do students understand the meaning of the terms?** Do they know *even, odd, factor, multiple, square,* and *prime?*

DISCUSSION

20 MIN CLASS

2 Using Arrays to Represent Solutions

Math Focus Points for Discussion

- Using arrays to model multiplication

Spend the first 10 minutes having students look at the representations they used to solve 35 × 28 on the posters they made in Session 2.1.

Which of these representations are the same? Which are different? Are there any representations that seem particularly clear to you? Why? Are there ways we can use "shortcuts" on our representations?

For example, instead of drawing 35 circles, you can draw 4 or 5 circles, use an ellipsis, and then draw a few more. Instead of drawing dots or objects inside the circles, you can write numerals.

How do these representations help us keep track of our work? Where is the 35 in this problem? Where is the 28? How does this representation show the product?

If any students have used arrays, spend some time discussing how an array can represent different ways to break up a multiplication problem. If no student has used an array, bring up this representation yourself. In either case, first draw a rectangle on the board or overhead and ask students how to label the dimensions for this problem.

Let's look at two different strategies and how they can be represented on an array.

Talk through the two strategies illustrated below. For each, start with the first step (35 × 20 in the first example). Ask students what part of the array represents 35 × 20. Draw a segment to show that section of the array, and record the multiplication combination, as shown below.

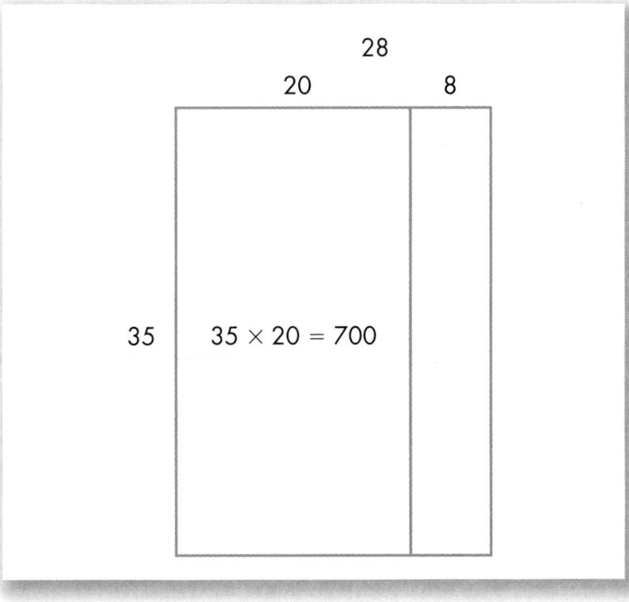

Before going on to the other steps of the problem, introduce a story context that students can relate to the array, such as the following:

Let's think about teams. It's Field Day and all of the teams need to line up in the field to make a huge rectangle. There are 35 teams, with 28 students in each team. The problem is to figure out how many students there are on all of the teams. So if we start with 35 × 20, who can explain what 35 × 20 means in terms of the teams? What part of the problem have we solved? What else do we need to do? Where do you see the part we've solved in the array? What is this other part of the array?

Record each part of each solution, and ask students how to determine the product.

Pages 82–83 show two strategies to talk through with a sequence of diagrams showing how the array is gradually filled in.

Breaking one number apart by place value

35×28

$35 \times 20 = 700$

(35×8)

$30 \times 8 = 240$

$5 \times 8 = 40$

980

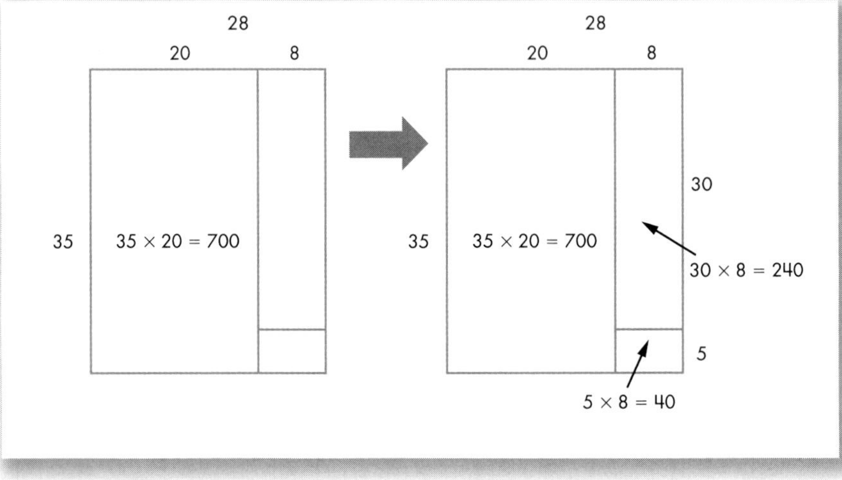

Breaking both numbers apart by place value

35×28

$30 \times 20 = 600$

$30 \times 8 = 240$

$5 \times 20 = 100$

$5 \times 8 = 40$

980

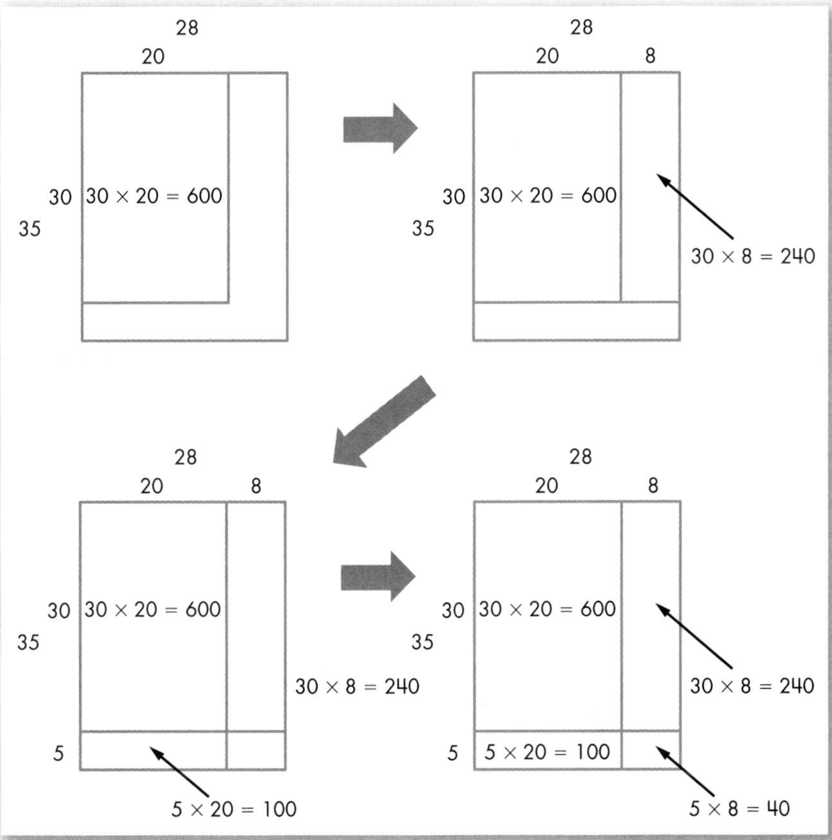

Take a few minutes and talk to your neighbor about what you see. What is the same and different about how these two solutions are represented on arrays?❷

After a few minutes, bring the class back together and ask what they noticed.

ACTIVITY 3

Solving Multiplication Problems

25 MIN | CLASS | PAIRS

Students complete three problems out of the five on *Student Activity Book* page 27. They pick one problem and use a representation to show their solution. Encourage students to try using an array.

ONGOING ASSESSMENT: Observing Students at Work

Students solve and represent multiplication problems.

- **Which problems do students select?** Are they choosing problems that provide them with an appropriate challenge?

Algebra Note

❷ **Distributive Property** Although students' explanations about breaking apart the factors in a multiplication problem and recombining them to get the product will be based on story contexts and array diagrams, the mathematical principle they are illustrating is expressed in formal terms as the distributive property. See **Algebra Connections in This Unit** on page 16 for more information.

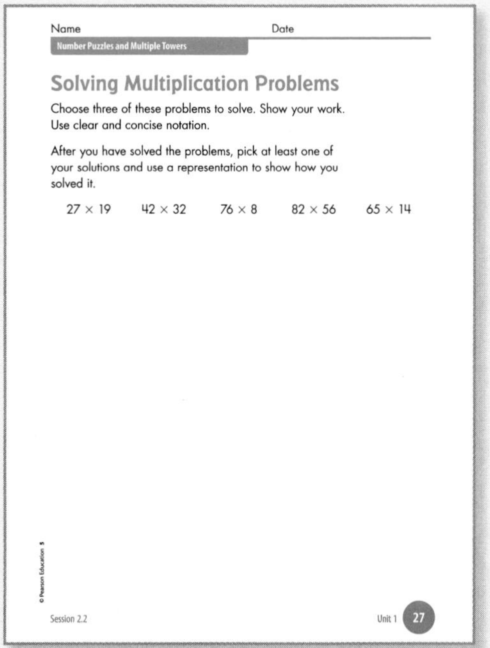

Name Date

Number Puzzles and Multiple Towers

Solving Multiplication Problems

Choose three of these problems to solve. Show your work. Use clear and concise notation.

After you have solved the problems, pick at least one of your solutions and use a representation to show how you solved it.

27 × 19   42 × 32   76 × 8   82 × 56   65 × 14

© Pearson Education 5

Session 2.2   Unit 1 27

▲ Student Activity Book, p. 27 PORTFOLIO

Name Date

Number Puzzles and Multiple Towers — Daily Practice

Multiplying Two Ways

NOTE Students multiply a 2-digit number in two different ways.

SMH 30–32

1. Solve this problem in two different ways. Show each solution clearly.

$26 \times 19 =$ ______

First way:

Second way:

Ongoing Review

2. $6 \times 3 \times 10 \times 2 =$ ______

A. 120 **B.** 300 **C.** 360 **D.** 630

28 Unit 1 Session 2.2

▲ Student Activity Book, p. 28

Name Date

Number Puzzles and Multiple Towers — Homework

Multiplication Practice

Solve the problems below. Show your solution clearly.

NOTE In class, students are solving multiplication problems. They are using story contexts and representations to help them solve the problems and explain their solutions. As they solve the problems below, ask them what part of the problem they have solved and what part they still need to solve.

SMH 30–32

1. 24×15

2. 49×9

3. 36×25

Session 2.2 Unit 1 29

▲ Student Activity Book, p. 29

- **Do all students have at least 1 strategy for solving a multiplication problem?** How efficiently are they using that strategy?
- **Are students using story contexts and representations to help them solve the problem and keep track of their work?**
- **Can students use a story context or a representation to show the meaning of the multiplication problem?**

While the class is working, circulate and talk to individual students about the problems.

- What do you already know about these numbers that might help you? Is there a way to shorten the number of steps you used to solve the problem? Does your answer seem reasonable?
- Can you put these numbers into a story context? How many groups of what?
- How does your representation show your solution? Where are the numbers in your representation? What do those numbers mean?

DIFFERENTIATION: Supporting the Range of Learners

This work time is an opportunity to work with small groups of students.

Intervention Some students should continue working with smaller numbers. As you talk with students who need support to keep track of the parts of a multiplication problem, help them visualize the problem by using a story context or a picture.

4 SESSION FOLLOW-UP
Daily Practice and Homework

Daily Practice: For ongoing review, have students complete *Student Activity Book* page 28.

Homework: Students solve three multiplication problems on *Student Activity Book* page 29.

Student Math Handbook: Students and families may use *Student Math Handbook* page 17 for reference and review. See pages 190–197 in the back of this unit.

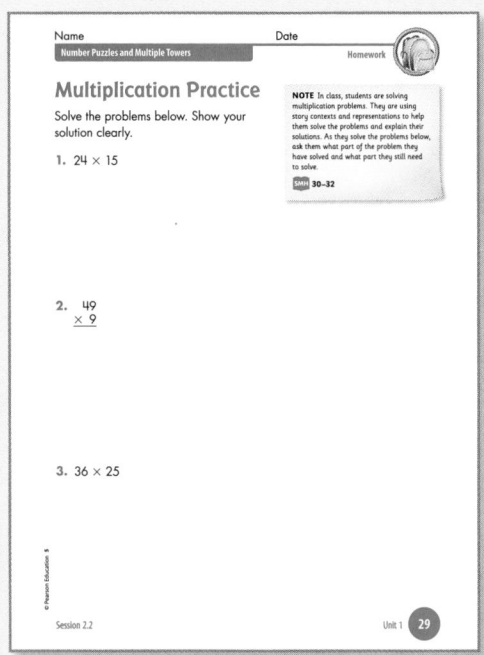

Which Product Is Greater?

Math Focus Points

- Multiplying fluently by multiples of 10
- Comparing multiplication problems to determine which product is greater
- Estimating the product of two numbers

Vocabulary

associative property
less than
greater than

Today's Plan		Materials
1 DISCUSSION **Multiplying by 10s**	25 MIN CLASS PAIRS	
2 ACTIVITY ***Multiplication Compare***	25 MIN PAIRS	• *Student Activity Book,* p. 31 • M44–M45*; M46 (as needed)*
3 DISCUSSION **Making Estimates**	10 MIN CLASS PAIRS	
4 SESSION FOLLOW-UP **Daily Practice and Homework**		• *Student Activity Book,* pp. 32–33 • *Student Math Handbook,* p. G11

*See *Materials to Prepare,* p. 67.

Ten-Minute Math

Number Puzzles Draw one card from each set (A, B, C) of *Number Puzzle* Clue Cards (M35–M43) and read them aloud. Students work with a partner and find the number(s) that fit the set of three clues. If the solution is "inconclusive," students modify the clues. Have students share their strategies for solving the puzzle.

25 MIN   CLASS   PAIRS

DISCUSSION

Multiplying by 10s

Math Focus Points for Discussion

◆ Multiplying fluently by multiples of 10

Students often describe solving a problem such as 6×40 as "multiply 6 times 4 and then add a zero." In this discussion, students consider what "add a zero" really means, and what the relationship is between expressions such as 6×4, 6×40, and 6×400.

Write the following problems on the board or overhead, and ask students to solve them:

$$6 \times 4 = \underline{}$$

$$6 \times 40 = \underline{}$$

$$6 \times 400 = \underline{}$$

$$60 \times 400 = \underline{}$$

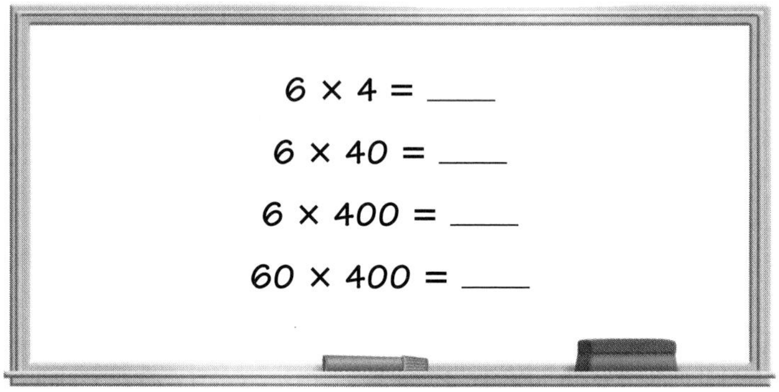

Solving these problems and sharing solutions should take about 10 minutes. Spend most of the discussion on why the pattern of zeros occurs.

Ask students to share with a partner how they solved each problem and what they noticed about the answers. After a few minutes, ask students to explain how they solved these four problems.

Students might say:

"I just figured out each problem."

"$6 \times 4 = 24$, and 6×40 is 10 times more than 6×4. So 6×40 is 240."

"I just added a zero each time."

Insert the answers to the problems you wrote on the board. Then focus the discussion on the idea of "adding a zero."

When people look at equations like these, I've often heard them say that there's a pattern here, and that the pattern is that we just "add zeros." But what happens if we *really* add 24 + 0? It doesn't change, does it? So I want to know what you mean when you say "add a zero" and why it works. Why is it that in these problems you can multiply the first digits of the numbers and then "add zeros" to get the answer?

Let's look at 6 × 400. Work with your partner to find a way to explain why you can solve it by first thinking of it as 6 × 4. If you have time, think about the other problems, too.

When we come back together, you will share the ideas you have about this and any picture or diagram or story context that helps explain your ideas. Some of you might also show something with the numbers themselves that helps you explain what is happening.

Give students 10 minutes or so to discuss this question with a partner, and then call the class back together. As students are working, identify explanations students are using that would be helpful for the whole class to consider. Ask students to share these explanations, and ask them to point out how they are thinking about groups of 10 and groups of 100. Acknowledge other students who have similar explanations each time a different explanation is shared.

Conclude the discussion by asking students to explain the pattern or rule in their own words.

So when one factor is a multiple of 100, such as 6 × 400, will it always work to think of the solution as 6 × 4 and then put two zeros on the number? Who thinks they can explain whether this method will always work and why?❶

When students understand why the "add zeros" rule of thumb works, they can use this as a shortcut for solving a number of multiplication problems. However, if students use it without understanding the underlying knowledge about why this method works, it is likely that they will misapply it as problems get more difficult.

Algebra Note

❶ **Associative Property** Students reason about the relationship between multiplication expressions such as 6 × 4 and 6 × 40 by using what they know about numbers. This relationship is also an illustration of the associative property of multiplication, $a \times (b \times c) = (a \times b) \times c$. In this special case, $c = 10$ or $c = 100$. For example, $6 \times 40 = 6 \times (4 \times 10) = (6 \times 4) \times 10 = 24 \times 10$. So, multiplying 6 × 40 results in a product that is the same as 24 × 10; that is, 10 times greater than the product of 6 × 4.

Professional Development

❷ **Dialogue Box:** Playing *Multiplication Compare,* p. 185

Differentiation

❸ **English Language Learners** The term *greater* may be confusing to English Language Learners, who may be familiar with the conversational use of the word *great.* You can use a number line to illustrate the mathematical meaning of the term.

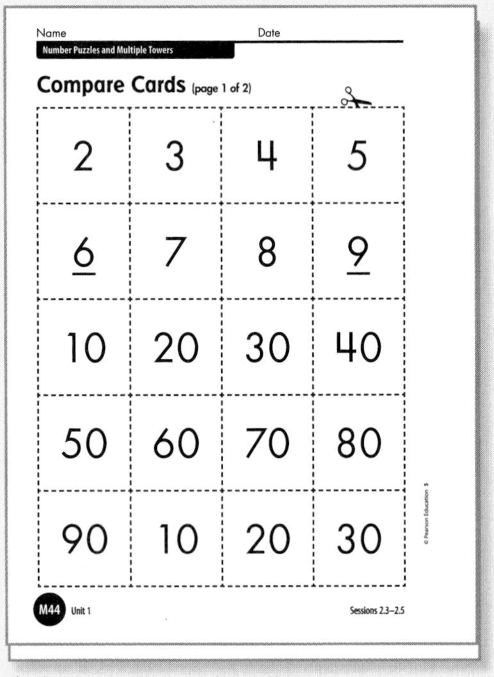
Name Date

Number Puzzles and Multiple Towers

Compare Cards (page 1 of 2)

2	3	4	5
6	7	8	9
10	20	30	40
50	60	70	80
90	10	20	30

© Pearson Education 3

M44 Unit 1 Sessions 2.3–2.5

▲ Resource Masters, M44–M45

ACTIVITY

2 Multiplication Compare

25 MIN PAIRS

Students continue to solve problems that involve multiples of 10 in a game in which they determine which product is greater.❷ ❸

We are going to play a game called *Multiplication Compare* in which you will compare two multiplication problems. Let's play a couple of rounds together before you play in pairs.

Introduce the game to students by playing two rounds with the class. Pick two pairs of Compare Cards (M44–M45), one pair for you and the other for the class. For example, the teacher draws 40 and 60, and the class draws 100 and 30. Use these numbers to record two multiplication problems on the board as they would appear on the recording sheet, *Student Activity Book* page 31 (40 × 60 ☐ 100 × 30).

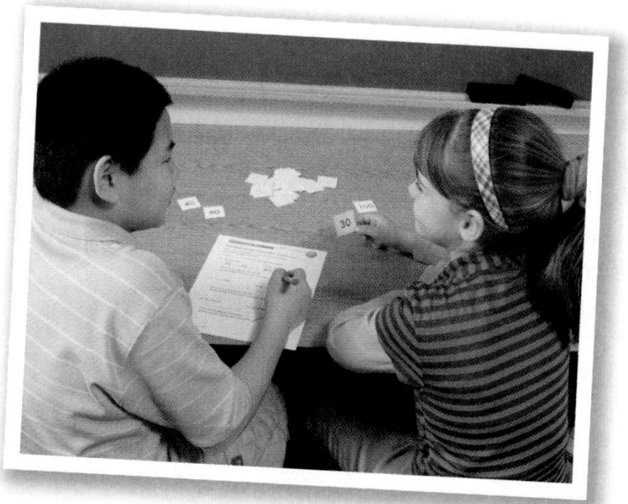

In Multiplication Compare, *students compare two multiplication problems.*

Which of these problems has a greater product? How do you know? Are there other ways we could think about these two problems to decide which is greater? So we could say that 40 × 60 is less than 100 × 30.

If necessary, review the meaning of the less than sign, <, and the greater than sign, >, and fill in the box between the problems you wrote on the board.

Play one more round with the class in the same way. Explain that in each round, the player with the larger product takes all the cards for that round.

For the remainder of the activity, each student plays *Multiplication Compare* with a partner. Whoever has collected more cards by the end of the activity is the winner. Students may find a summary of the rules on page G11 of the *Student Math Handbook*. (You may use M46 to provide extra copies.)

Have students play the first few rounds without using the recording sheet. Then, when you think they are ready, have students complete the *Multiplication Compare* Recording Sheet, *Student Activity Book* page 31.

ONGOING ASSESSMENT: Observing Students at Work

Students compare multiplication problems involving multiples of 10 and decide which product is greater.

- **Can students easily multiply numbers that are multiples of 10 without breaking the problem into smaller parts?**

DIFFERENTIATION: Supporting the Range of Learners

Intervention Some students may need more study of the patterns of multiplying by 10s before they can solve these easily. You may want to give them some related sets of problems such as the following:

10×5 $\quad$ 10×50 $\quad$ 20×50 $\quad$ 30×50

As students work on these problems, point out how the number grows with each multiple of 10. Students will continue to practice this as they solve large multiplication problems.

3 DISCUSSION

Making Estimates

10 MIN CLASS PAIRS

Math Focus Points for Discussion

- Comparing multiplication problems to determine which product is greater

When you were playing *Multiplication Compare,* I noticed that some of you did not find the exact solution to each problem in order to decide which product was greater.

Write "30 × 70 □ 32 × 50" on the board.

Name Date

Number Puzzles and Multiple Towers

Multiplication Compare

You need

- Compare Cards (1 deck per pair)

Play with a partner.

1. Divide the deck of cards evenly so that both players have the same number of cards. Place the cards facedown in a stack in front of you.
2. Each player turns over the top two cards in his or her stack.
3. Determine which player has the greater product. Discuss how you know which product is greater.
4. The person with the greater product takes all of the cards that have been turned over and places them at the bottom of his or her stack.
5. If the products are equal, players turn over two new cards. The person with the greater product takes all of the cards.
6. Play for a given amount of time or until one player has all of the cards. The player with more cards wins.

Variation

You need

- Digit Cards (1 deck per pair)

Play the same game using Digit Cards with the "0" cards removed. Each player draws 4 cards. Using the cards in the order they were picked, each player forms two 2-digit factors. (For example, if you picked 7, 2, 3, 8, your multiplication problem would be 72 × 38.)

M46 Unit 1 Sessions 2.3, 2.4, 2.5

▲ Resource Masters, M46

Name Date

Number Puzzles and Multiple Towers

Multiplication Compare Recording Sheet

After you have played a few rounds of *Multiplication Compare,* complete this sheet.

Place a <, >, or = in the box between the problems.

1. Your problem: Partner's problem:

 _____ × _____ □ _____ × _____

 How did you decide whose product is greater? Explain your reasoning.

2. Your problem: Partner's problem:

 _____ × _____ □ _____ × _____

 How did you decide whose product is greater? Explain your reasoning.

3. Your problem: Partner's problem:

 _____ × _____ □ _____ × _____

 How did you decide whose product is greater? Explain your reasoning.

Session 2.3 Unit 1 31

▲ Student Activity Book, p. 31
Resource Masters, M47 WRITING

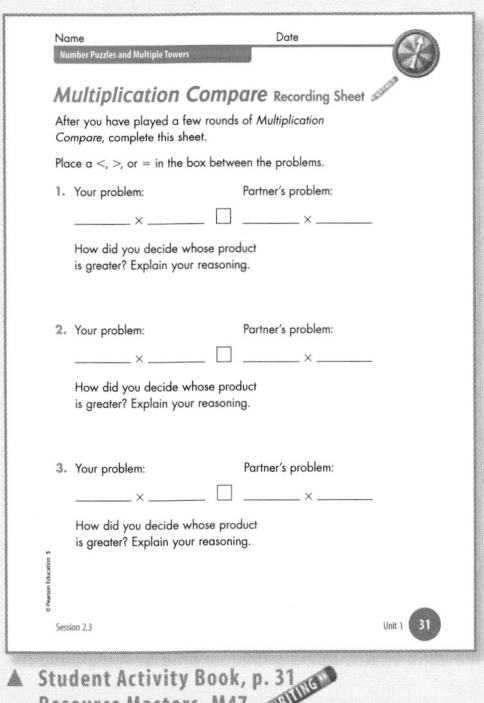

Math Note

4 Analyzing the Problems Some students may solve each problem to see which product is greater, but many students will be able to analyze the problems to determine which is greater. That is, 30 × 70 is greater because 20 additional groups of 30 is far more than 50 additional groups of 2.

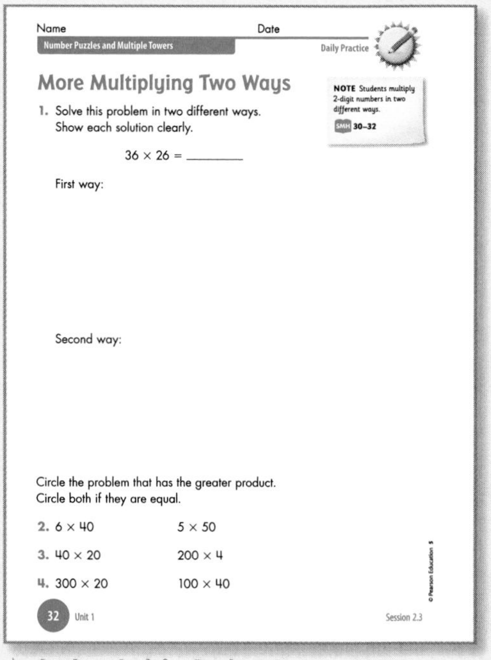

Name Date

Number Puzzles and Multiple Towers — Daily Practice

More Multiplying Two Ways

NOTE Students multiply 2-digit numbers in two different ways.

SMH 30–32

1. Solve this problem in two different ways. Show each solution clearly.

   36 × 26 = ______

   First way:

   Second way:

Circle the problem that has the greater product. Circle both if they are equal.

2. 6 × 40 5 × 50
3. 40 × 20 200 × 4
4. 300 × 20 100 × 40

32 Unit 1 Session 2.3

© Pearson Education 5

▲ Student Activity Book, p. 32

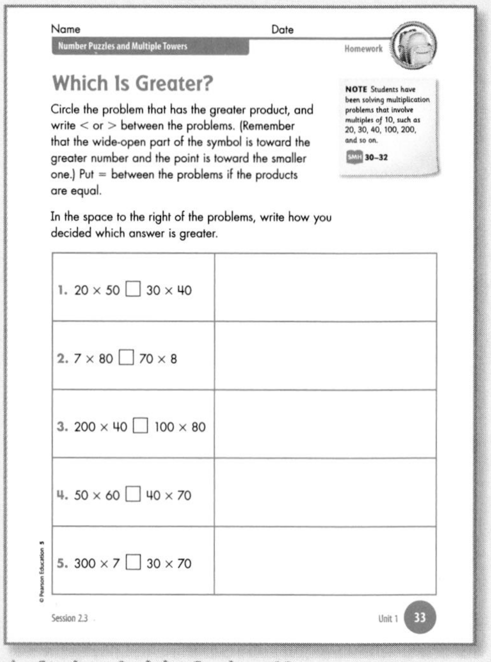

Name Date

Number Puzzles and Multiple Towers — Homework

Which Is Greater?

NOTE Students have been solving multiplication problems that involve multiples of 10, such as 20, 30, 40, 100, 200, and so on.

SMH 30–32

Circle the problem that has the greater product, and write < or > between the problems. (Remember that the wide-open part of the symbol is toward the greater number and the point is toward the smaller one.) Put = between the problems if the products are equal.

In the space to the right of the problems, write how you decided which answer is greater.

1. 20 × 50 ☐ 30 × 40	
2. 7 × 80 ☐ 70 × 8	
3. 200 × 40 ☐ 100 × 80	
4. 50 × 60 ☐ 40 × 70	
5. 300 × 7 ☐ 30 × 70	

© Pearson Education 5

Session 2.3 Unit 1 33

▲ Student Activity Book, p. 33

Imagine that these are two hands in *Multiplication Compare.* Which one has the greater product and how do you know?

Allow students a couple of minutes to work. Then collect a few examples of how students determined which is greater. 4 Ask students to explain their thinking by using a story context, array, or drawings.

Students might say:

"I thought about 70 teams with 30 kids and 50 teams with 32 kids. 70 teams is a lot more than 50. And the 2 extra kids on the 50 teams will only add up to 100 more."

"I drew an array. You can see that 30 x 20 is extra for the 30 × 70 array, but there's only 2 × 50 extra for the other one. So 30 × 70 is greater."

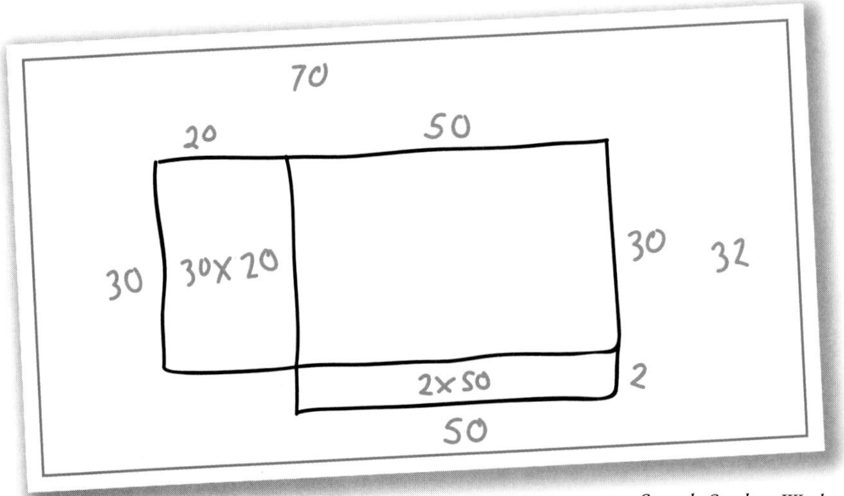

Sample Student Work

SESSION FOLLOW-UP

4 Daily Practice and Homework

Daily Practice: For reinforcement of this unit's content, have students complete *Student Activity Book* page 32.

Homework: Students compare multiplication problems and decide which is greater on *Student Activity Book* page 33.

Student Math Handbook: Students and families may use *Student Math Handbook* page G11 for reference and review. See pages 190–197 in the back of this unit.

SESSION 2.4

Multiplication Cluster Problems

Math Focus Points

- Solving 2-digit by 2-digit multiplication problems
- Breaking up multiplication problems efficiently
- Multiplying fluently by multiples of 10

Today's Plan		Materials
1 ACTIVITY **Introducing Multiplication Cluster Problems**	15 MIN CLASS PAIRS	
2 MATH WORKSHOP **Multiplication Practice** 2A Multiplication Cluster Problems 2B *Multiplication Compare* 2C Problems About Teams	45 MIN	2A • *Student Activity Book,* p. 35 2B • Materials from Session 2.3, Activity 2 • M47* (as needed); M48–M50 (optional)* 2C • *Student Activity Book,* p. 36
3 SESSION FOLLOW-UP **Daily Practice and Homework**		• *Student Activity Book*, pp. 37–38 • *Student Math Handbook,* pp. 35; G11

*See *Materials to Prepare,* p. 67.

Ten-Minute Math

Number Puzzles Draw one card from each set (A, B, C) of *Number Puzzle* Clue Cards (M35–M43) and read them aloud. Students work with a partner and find the number(s) that fit the set of three clues. If the solution is "inconclusive," students modify the clues. Have students share their strategies for solving the puzzle.

Professional Development

(1) **Teacher Note:** About Cluster Problems, p. 165

Teaching Notes

(2) **Multiplication Skills** The problems in the cluster often are ones that students can solve mentally. Many involve multiplying by multiples of 10 or solving half of the problem and doubling that product (e.g., $2 \times 21 = 42$, so $4 \times 21 = 84$).

(3) **Contexts** The context of teams is suggested. Use any context familiar to your students that helps them visualize equal groups.

(4) **Relating the Problems** Remind students to think about the story of teams (or any other context that makes sense) to help them figure out how each of the cluster problems may help solve the final problem. Students may also sketch unmarked arrays.

ACTIVITY 1

15 MIN CLASS PAIRS

Introducing Multiplication Cluster Problems

In this session, students are introduced to cluster problems.(1) Cluster problems provide different ways to break up a final problem or use related problems to solve it. These are two skills that are critical for work on more difficult multiplication later in the year.(2)

When students are solving multiplication problems, it is often useful to create a context for the problem. This helps students visualize the action of the problem and identify what part of the problem they have solved.

Write "$42 \times 19 =$ ____" on the board.

Who can think of a story problem for 42×19 that is about teams? Let's use Martin's story. For Field Day there will be 42 teams with 19 students on each team.(3) We need to figure out how many students will be playing so that the school can make a T-shirt for each player. How can we solve this?

Zachary says that he knows that 42×10 is 420. So what does that tell us so far? What part of the problem have we solved already? *(10 students on each of the 42 teams)* What do we still need to figure out? *(9 more students on each of the teams)*

Write the following set of problems on the board or overhead and ask students to solve them mentally:

$42 \times 19 =$ ____

$42 \times 10 =$ ____ $42 \times 20 =$ ____

$42 \times 5 =$ ____

$42 \times 2 =$ ____

Then focus students' attention on the original problem you wrote, $42 \times 19 =$.

Solve this problem. See whether some of the problems you just did mentally might help you find the product of 42×19.(4) If you have time, solve the problem a second way.

Give students a minute or two to work, and then have them share their solutions with a partner.

Let's look at the problems you had to solve first. What are the answers?

Record the answers.

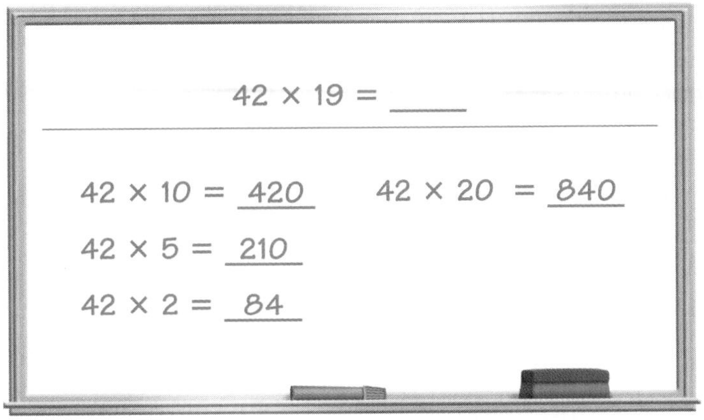

Look at the first column of problems. How do these problems help you solve 42 × 19? *(42 × 10 represents 42 teams with 10 on a team, 42 × 5 represents 42 teams with 5 on a team, and doubling 42 × 2 represents 42 teams with 4 on a team. Add the products to get the number of students on 42 teams with 10 + 5 + 4 = 19 on each team.)*

Refer back to the team story (or other context) as needed, as students consider what part of the problem has been solved and what parts remain to be solved.

How can we use 42 × 20 to help us solve 42 × 19? *(There is one group of 42 too many, so 42 needs to be subtracted.)*

You've been working with multiplication cluster problems. They are a set of problems that can help you solve the final problem. When you work on cluster problems today, you'll see problems written like these—two columns of problems that you need to solve mentally and then a final problem to solve. Use what you know to solve the cluster problems. Then see how they can help you solve the final problem.

Name Date

Number Puzzles and Multiple Towers

Multiplication Cluster Problems

1. Solve these problems.
   10 × 12 = ______ 20 × 12 = ______
   20 × 10 = ______ 8 × 10 = ______
   28 × 2 = ______
   Now solve 28 × 12 = ______.
   How did you solve it?

2. Solve these problems.
   35 × 10 = ______ 10 × 25 = ______
   35 × 20 = ______ 20 × 25 = ______
   30 × 25 = ______
   Now solve 35 × 25 = ______.
   How did you solve it?

3. Solve these problems.
   10 × 21 = ______ 20 × 20 = ______
   20 × 21 = ______ 7 × 20 = ______
   5 × 21 = ______
   Now solve 27 × 21 = ______.
   How did you solve it?

4. Solve these problems.
   100 × 7 = ______ 15 × 7 = ______
   40 × 7 = ______ 150 × 7 = ______
   Now solve 146 × 7 = ______.
   How did you solve it?

Sessions 2.4, 2.5 Unit 1 35

▲ Student Activity Book, p. 35

MATH WORKSHOP

2 Multiplication Practice

45 MIN

The Math Workshops today and tomorrow allow time for students to practice and refine their multiplication strategies. They also provide an opportunity for you to meet with students in a wide range of groupings—those needing extra support, and those needing more of a challenge. Students who need practice multiplying by multiples of 10 should spend more time playing *Multiplication Compare.* Otherwise, students should spend about equal amounts of time on each activity.

Explain to students that there will be a class discussion at the beginning of tomorrow's session that will focus on one of the cluster problems, Problem 3 on *Student Activity Book* page 35. Students should be sure to complete that problem during class today.

2A Multiplication Cluster Problems

INDIVIDUALS

On *Student Activity Book* page 35, students first solve a set of problems that they should be able to solve mentally. They use these to solve a more difficult, related problem. Students will have an opportunity to complete page 35 tomorrow, as well as work on additional cluster problems.

ONGOING ASSESSMENT: Observing Students at Work

Students solve related multiplication problems.

- **Are students able to mentally solve the cluster problems?** If not, how are they solving those problems?
- **Do students see and use the relationships in one problem in the cluster to solve others?** For example, 10 × 18 = 180. So, 20 × 18 would be double that, and 5 × 18 would be half.
- **Do students use the problems they have already solved to solve the final problem?** Do they recognize whether the problems in the cluster are the only problems they need to solve the final problem? How do they figure out what they still need to solve?

DIFFERENTIATION: Supporting the Range of Learners

Intervention If students are not able to use the relationships between the cluster of problems and the final problem, point to one of the problems in the cluster and ask questions such as following:

If this were your first step to solve the final problem, what else would you still need to do? Do any of the other problems in this set help you figure that out?

Extension Students who would like a challenge may create their own cluster problems for a multiplication problem that is difficult for them, or they may draw the array representation for the product.

2B *Multiplication Compare*

PAIRS

Students play *Multiplication Compare.* See Session 2.3, pages 88–89 for complete details. Extra copies of *Multiplication Compare* Recording Sheet (M47) can be made available.

DIFFERENTIATION: Supporting the Range of Learners

Intervention Adjust the cards students are using to help them benefit most from this activity. Students who need more practice should use only the cards that have 1 or 2 digits.

Extension Students who can easily multiply by multiples of 10 and 100 should begin playing the variation of the game in which they use the Digit Cards (M48–M50) to create two 2-digit numbers and determine which product is greater. See *Student Math Handbook* page G11 for more details.

2C Problems Involving Teams

INDIVIDUALS

Students solve problems in the context of teams on *Student Activity Book* page 36. You may choose to spend some time first reviewing the multiplication strategies you identified in Session 2.1, pages 73–74, by having students explain their solution to 1 of the problems, either in a small group or as a whole class. Work with students who have good strategies so that they can carry them out efficiently and notate them clearly. 5

Professional Development

5 **Teacher Note:** Multiplication Strategies, p. 183

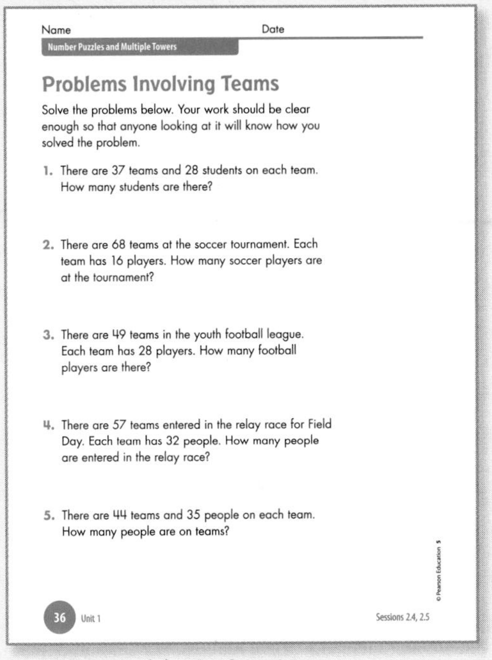
Name Date

Number Puzzles and Multiple Towers

Problems Involving Teams

Solve the problems below. Your work should be clear enough so that anyone looking at it will know how you solved the problem.

1. There are 37 teams and 28 students on each team. How many students are there?
2. There are 68 teams at the soccer tournament. Each team has 16 players. How many soccer players are at the tournament?
3. There are 49 teams in the youth football league. Each team has 28 players. How many football players are there?
4. There are 57 teams entered in the relay race for Field Day. Each team has 32 people. How many people are entered in the relay race?
5. There are 44 teams and 35 people on each team. How many people are on teams?

36 Unit 1 Sessions 2.4, 2.5

▲ Student Activity Book, p. 36

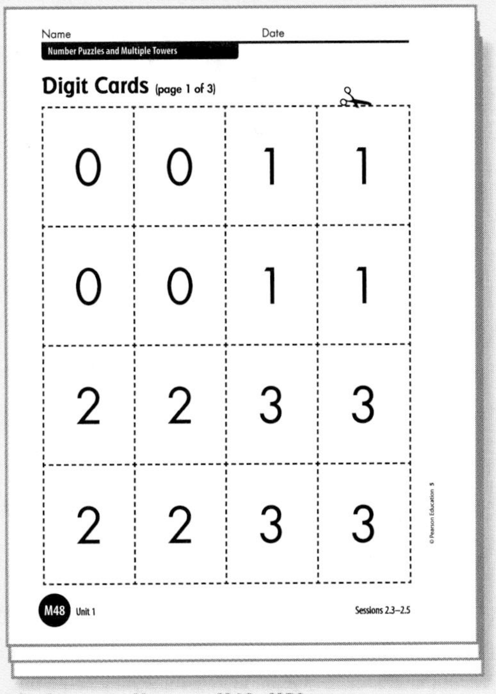
Name Date

Number Puzzles and Multiple Towers

Digit Cards (page 1 of 3)

0	0	1	1
0	0	1	1
2	2	3	3
2	2	3	3

M48 Unit 1 Sessions 2.3–2.5

▲ Resource Masters, M48–M50

Name Date

Number Puzzles and Multiple Towers — Daily Practice

More Multiplying 2-Digit Numbers

NOTE Students multiply 2-digit numbers. SMH 30–32

Solve these problems. Show each solution clearly.

1. 48 × 34

2. 28 × 21

Ongoing Review

3. Is the product of 32 × 28
   A. more than 1,000?
   B. between 500 and 1,000?
   C. between 100 and 500?
   D. less than 100?

© Pearson Education 5

Session 2.4 Unit 1 37

▲ Student Activity Book, p. 37

Name Date

Number Puzzles and Multiple Towers — Homework

Many Ways to Multiply

NOTE Students find ways to multiply to make each product. SMH 33–34

Find all of the ways to multiply to make each product. First, find the ways with two factors, and then find ways to multiply with more than two factors.

1. 144

2. 300

© Pearson Education 5

38 Unit 1 Session 2.4

▲ Student Activity Book, p. 38

ONGOING ASSESSMENT: Observing Students at Work

Students solve 2-digit by 2-digit multiplication problems in a story context.

- **Do all students have at least one strategy for solving a multiplication problem?** How efficiently are they using that strategy?
- **Are students using clear and concise notation?**

As students are solving problems, circulate and ask questions such as the following:

- What do you already know about these numbers that might help you? What parts of the problem can you solve easily? Is there a way to shorten the number of steps you used to solve the problem? Does your answer seem reasonable? What part of the problem have you solved? What do you still need to solve?

3 SESSION FOLLOW-UP

Daily Practice and Homework

Daily Practice: For reinforcement of this unit's content, have students complete *Student Activity Book* page 37.

Homework: Students find multiplication combinations for 144 and 300 on *Student Activity Book* page 38.

Student Math Handbook: Students and families may use *Student Math Handbook* pages 35 and G11 for reference and review. See pages 190–197 in the back of this unit.

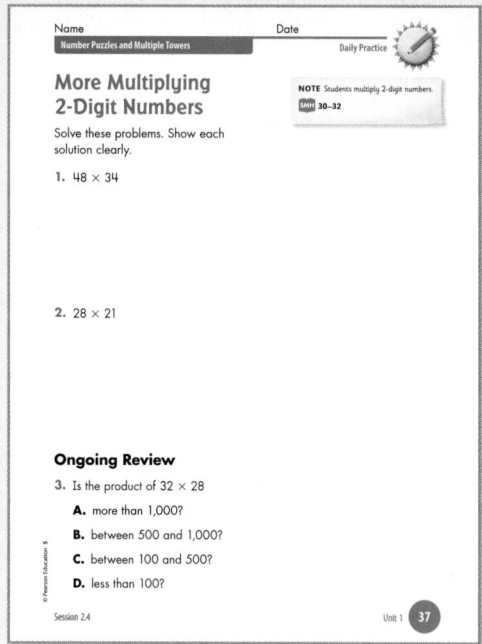

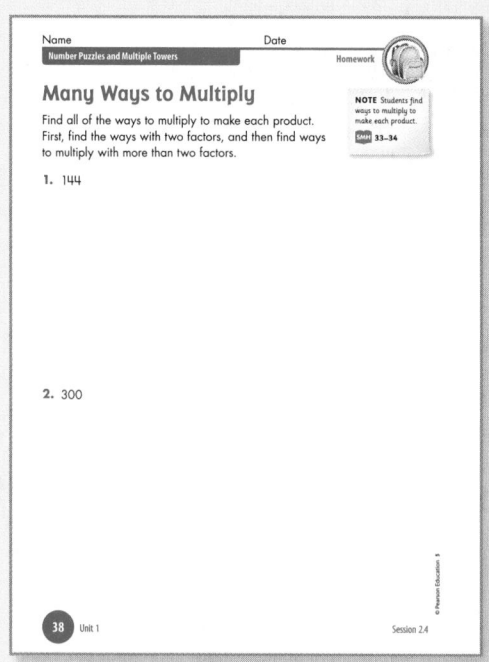

Multiplication Cluster Problems, *continued*

Math Focus Points

- Solving 2-digit by 2-digit multiplication problems
- Breaking up multiplication problems efficiently
- Multiplying fluently by multiples of 10
- Using clear and concise notation

Today's Plan		Materials
1 DISCUSSION **Multiplication Clusters**	15 MIN CLASS	• *Student Activity Book*, p. 35 (from Session 2.4)
2 MATH WORKSHOP **More Multiplication Practice** 2A Multiplication Cluster Problems 2B *Multiplication Compare* 2C Problems Involving Teams	45 MIN	2A • *Student Activity Book*, p. 39 2B • Materials from Session 2.3 2C • *Student Activity Book*, p. 36 (from Session 2.4)
3 SESSION FOLLOW-UP **Daily Practice**		• *Student Activity Book*, p. 40 • *Student Math Handbook*, pp. 35; G11

*See *Materials to Prepare*, p. 69.

Ten-Minute Math

Number Puzzles Draw one card from each set (A, B, C) of *Number Puzzle* Clue Cards (M35–M43) and read them aloud. Students work with a partner and find the number(s) that fit the set of three clues. If the solution is "inconclusive," students modify the clues. Have students share their strategies for solving the puzzle.

Teaching Note

➊ **Relating the Problems** Remind students to think about context to help them figure out how each of the cluster problems may help them solve the final problem.

DISCUSSION

1 Multiplication Clusters

15 MIN CLASS

Math Focus Points for Discussion

- Breaking up multiplication problems efficiently

Write Problem 3 from *Student Activity Book* page 35 on the board. Ask students to give you the answers to the problems in the cluster and record them.

10 × 21 = 210	20 × 20 = 400
20 × 21 = 420	7 × 20 = 140
5 × 21 = 105	

27 × 21 = 567

Ask students questions to explain how the problems in the cluster helped solve the final problem.➊ Ask questions such as the following about the problems in the first column:

How did you solve 5 × 21? Did 10 × 21 help you? *(5 × 21 is half of 10 × 21.)* How did you use 5 × 21 to solve the final problem, 27 × 21? *(20 groups of 21 is 420, and 5 more groups of 21 is 105. Find just 2 more groups of 21 to solve the problem.)*

Ask about the problems in the second column as well.

Who used the second column of problems to help you solve 27 × 21? What part of the problem have you solved if you solve 20 × 20 and 7 × 20? *(27 groups of 20)* What else do you need to multiply to finish the problem? *(27 × 1)*

2 MATH WORKSHOP
More Multiplication Practice

45 MIN

A new page of multiplication cluster problems is added to yesterday's Math Workshop activities. See Session 2.4, page 94, for more details.

2A Multiplication Cluster Problems

INDIVIDUALS

On *Student Activity Book* page 39, students continue their work on cluster problems. Some students might need more time to complete the cluster problems on *Student Activity Book* page 35. See Session 2.4, pages 92–95, for more details.

2B Multiplication Compare

PAIRS

Students play *Multiplication Compare.* See Session 2.3, pages 88–89 and Session 2.4, page 95, for complete details.

2C Problems Involving Teams

INDIVIDUALS

Students continue to solve problems in the context of teams on *Student Activity Book* page 36. See Session 2.4, pages 95–96, for more details.

3 SESSION FOLLOW-UP
Daily Practice

Daily Practice: For ongoing review, have students complete *Student Activity Book* page 40.

Student Math Handbook: Students and families may use *Student Math Handbook* pages 35 and G11 for reference and review. See pages 190–197 in the back of this unit.

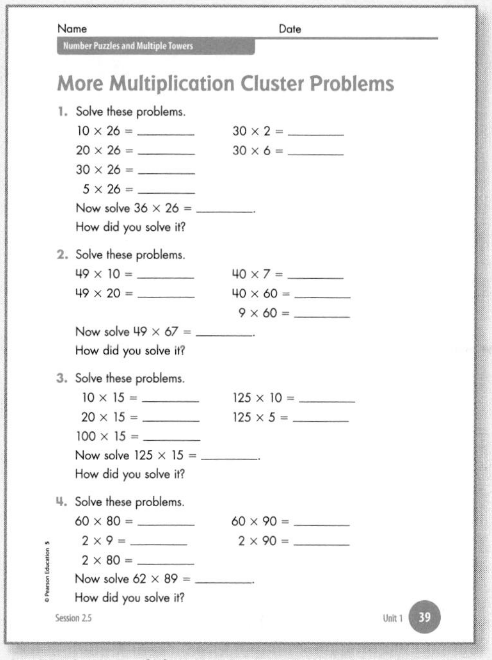

Name Date

Number Puzzles and Multiple Towers

More Multiplication Cluster Problems

1. Solve these problems.

10 × 26 = ____ 30 × 2 = ____
20 × 26 = ____ 30 × 6 = ____
30 × 26 = ____
5 × 26 = ____
Now solve 36 × 26 = ____.
How did you solve it?

2. Solve these problems.

49 × 10 = ____ 40 × 7 = ____
49 × 20 = ____ 40 × 60 = ____
9 × 60 = ____
Now solve 49 × 67 = ____.
How did you solve it?

3. Solve these problems.

10 × 15 = ____ 125 × 10 = ____
20 × 15 = ____ 125 × 5 = ____
100 × 15 = ____
Now solve 125 × 15 = ____.
How did you solve it?

4. Solve these problems.

60 × 80 = ____ 60 × 90 = ____
2 × 9 = ____ 2 × 90 = ____
2 × 80 = ____
Now solve 62 × 89 = ____.
How did you solve it?

© Pearson Education 5

Session 2.5 Unit 1 39

▲ Student Activity Book, p. 39

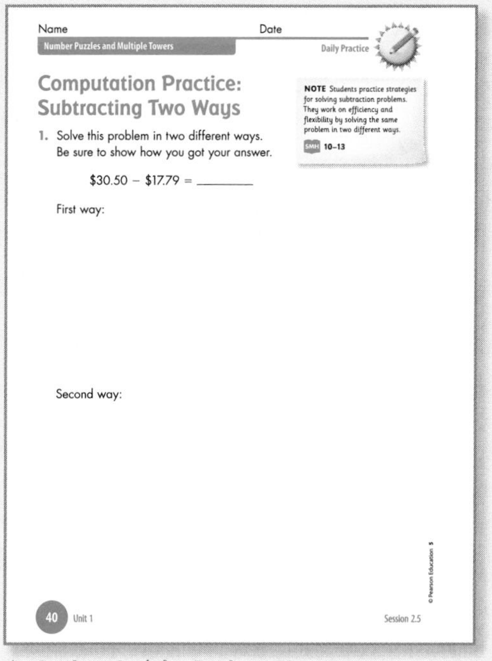

Name Date

Number Puzzles and Multiple Towers

Daily Practice

Computation Practice: Subtracting Two Ways

NOTE Students practice strategies for solving subtraction problems. They work on efficiency and flexibility by solving the same problem in two different ways.

SMH 10–13

1. Solve this problem in two different ways. Be sure to show how you got your answer.

$30.50 − $17.79 = ____

First way:

Second way:

© Pearson Education 5

40 Unit 1 Session 2.5

▲ Student Activity Book, p. 40

How Do I Start?

Math Focus Points

- Solving 2-digit by 2-digit multiplication problems
- Breaking up multiplication problems efficiently
- Using clear and concise notation

Today's Plan		Materials
1 ACTIVITY **Starter Problems**	45 MIN, CLASS, INDIVIDUALS	• *Student Activity Book,* pp. 41–42
2 DISCUSSION **Multiplication Strategies**	15 MIN, PAIRS, CLASS	• *Student Activity Book,* pp. 41–42 • Blank transparency (optional)
3 SESSION FOLLOW-UP **Daily Practice**		• *Student Activity Book,* p. 43 • *Student Math Handbook,* pp. 30–32

Ten-Minute Math

Number Puzzles **Play the variation, "Draw Three, Trade One." Draw one card from each set (A, B, C) of *Number Puzzle* Clue Cards (M35–M43) and read them aloud. If the solution is "inconclusive," students decide which of the three cards should be removed, and a new card (from the same set) is selected. Students work with a partner and find the number(s) that fit the set of three clues. Have students share their strategies for solving the puzzle.**

1 ACTIVITY
Starter Problems

45 MIN CLASS INDIVIDUALS

Students work on problems in which they are given a particular way to start. Finishing a problem that has been started helps students identify which parts of the problem have been solved and what else needs to be solved. For some students, these problems will help them become more efficient (e.g., starting with 12×30 instead of 12×10 to solve 12×36). In other cases, students work on flexibility when they finish a problem by using a strategy that they would not typically use.

We have been solving many multiplication problems by using several different strategies. Today we are going to focus on first steps that lead to different ways to solve the problem.

Choose a problem that is of medium difficulty for your class and write it on the board. 54×48 will be used as an example here.

Ask students to solve this problem and record their steps. Allow students a few minutes to work.

Instead of sharing how you solved the whole problem, share the first step you took. What did you do first to solve this problem?

Collect and record one example, such as $54 \times 10 = 540$.

Who else started with 54×10? What did you do next?

Allow someone to finish solving the problem. Record it on the board for everyone to see, modeling clear and concise notation.

(First step) $54 \times 10 = 540$

$54 \times 20 = 1{,}080$

$54 \times 40 = 2{,}160$

$54 \times 8 = 432$

$2{,}160 + 432 = 2{,}592$

Who started a different way?

Teaching Note

➊ **Advanced Strategies** It is not expected that students use this as a strategy, but some may notice that it is possible and sometimes quite useful.

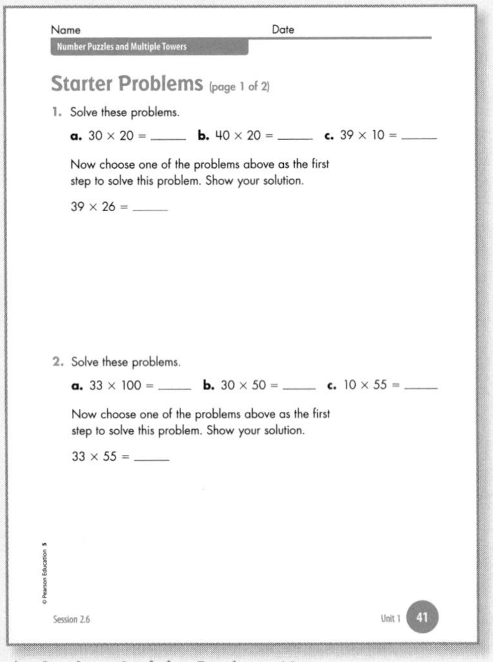
Name Date
Number Puzzles and Multiple Towers

Starter Problems (page 1 of 2)

1. Solve these problems.

   a. 30 × 20 = ____ **b.** 40 × 20 = ____ **c.** 39 × 10 = ____

   Now choose one of the problems above as the first step to solve this problem. Show your solution.

   39 × 26 = ____

2. Solve these problems.

   a. 33 × 100 = ____ **b.** 30 × 50 = ____ **c.** 10 × 55 = ____

   Now choose one of the problems above as the first step to solve this problem. Show your solution.

   33 × 55 = ____

© Pearson Education 5

Session 2.6 Unit 1 41

▲ **Student Activity Book, p. 41**

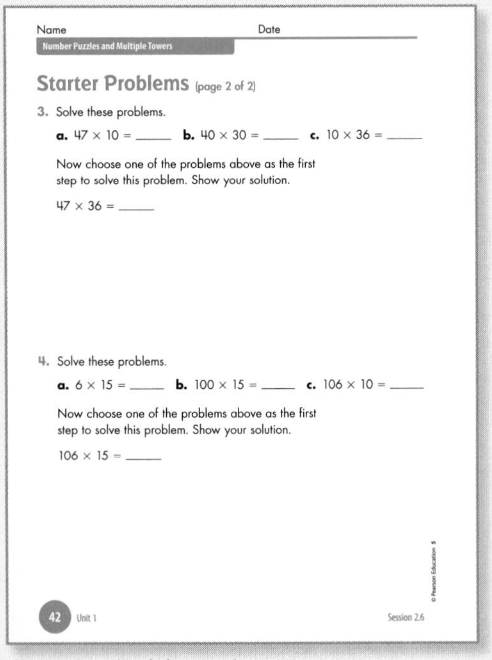
Name Date
Number Puzzles and Multiple Towers

Starter Problems (page 2 of 2)

3. Solve these problems.

   a. 47 × 10 = ____ **b.** 40 × 30 = ____ **c.** 10 × 36 = ____

   Now choose one of the problems above as the first step to solve this problem. Show your solution.

   47 × 36 = ____

4. Solve these problems.

   a. 6 × 15 = ____ **b.** 100 × 15 = ____ **c.** 106 × 10 = ____

   Now choose one of the problems above as the first step to solve this problem. Show your solution.

   106 × 15 = ____

© Pearson Education 5

42 Unit 1 Session 2.6

▲ **Student Activity Book, p. 42**

Collect one or two more ways to start. Again, for each way, ask a student to follow through the whole solution. Other possible solutions are shown below.

$$\begin{array}{rrcr} \text{(First step)} & 50 \times 40 & = & 2{,}000 \\ & 50 \times 8 & = & 400 \\ & 4 \times 40 & = & 160 \\ & 4 \times 8 & = & \underline{32} \\ & & & 2{,}592 \end{array}$$

$$\begin{array}{rll} \text{(First step)} & 54 \times 50 & = 27 \times 100 \\ & 27 \times 100 & = 2{,}700 \\ & 54 \times 2 & = 108 \end{array}$$

$$\begin{array}{r} 2{,}700 \\ \underline{-108} \\ 2{,}592 \end{array}$$

In the last method, the student first changed 54 × 48 to an easier problem, 54 × 50, and then multiplied one factor by 2 and divided the other factor by 2 to create an equivalent problem.➊

Students complete *Student Activity Book* pages 41–42. Each set of problems includes three possible “starter problems” (first steps). Students complete each of the three starter problems and choose one of them as the way to start their solution.

ONGOING ASSESSMENT: Observing Students at Work

Students study ways to begin multiplication problems as a way to develop efficiency and fluency.

- **Can students see the relationship between the starter problems and the final problem?**
- **Can students finish solving a multiplication problem on the basis of a certain first step?** That is, can students determine what parts of the problem remain to be solved after the first step?

As students are working, circulate and ask questions about their solutions.

- Which starter problem did you use? How does it help you solve the main problem? What part of the problem have you solved? What part remains to be solved?

Remind students that thinking about the problem in a story context can help them figure out what pieces are solved and what is remaining. Ask questions about their stories and representations.

- What story context can you use to help you figure this out? What representation can you use to help you keep track of your work?

DIFFERENTIATION: Supporting the Range of Learners

Intervention Starter problems can be a way to support students in learning multiplication strategies. If they have no way to begin a problem, a fairly simple starter may help. For example, if a student has difficulty solving 18×4, provide 18×2 or 10×4 to solve first. Then guide students through the solution.

- How can 18×2 (or 10×4) help you solve 18×4? What part of the problem have you solved?

You may want students to use cubes or an array to demonstrate.

Students who are able to start a problem may lose track of what parts they have solved and still need to solve. For these students, choose a familiar story context that they can visualize easily. Talk through the problem with them using that context.

Extension If students finish *Student Activity Book* pages 41–42 quickly, provide a less familiar first step such as doubling and halving.

2 DISCUSSION
Multiplication Strategies

15 MIN PAIRS CLASS

Math Focus Points for Discussion

- Solving 2-digit by 2-digit multiplication problems

Bring the class together when most students have finished *Student Activity Book* page 41. Choose one problem to work on as a class, even if everyone has not yet solved it. In addition to discussing the strategy, emphasize clear notation. Problem 1 is used as the example for this discussion.

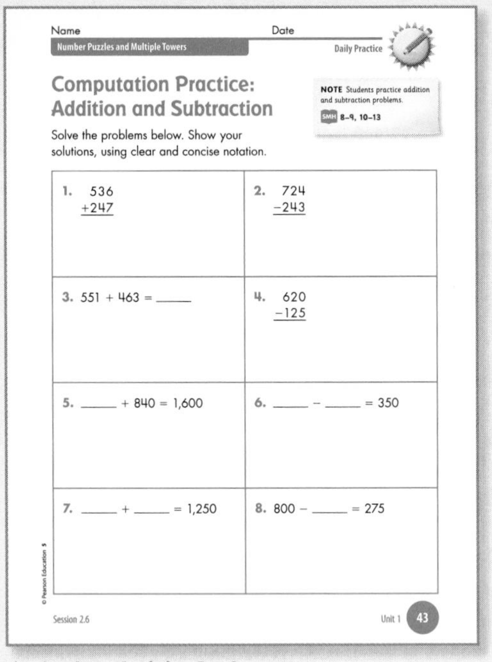

Name Date

Number Puzzles and Multiple Towers — Daily Practice

Computation Practice: Addition and Subtraction

NOTE Students practice addition and subtraction problems.

SMH 8–9, 10–13

Solve the problems below. Show your solutions, using clear and concise notation.

1. 536 + 247	2. 724 − 243
3. 551 + 463 = ____	4. 620 − 125
5. ____ + 840 = 1,600	6. ____ − ____ = 350
7. ____ + ____ = 1,250	8. 800 − ____ = 275

© Pearson Education 5

Session 2.6 Unit 1 43

▲ Student Activity Book, p. 43

Write the three starter problems on the board or on a blank transparency. For each of the starter problems, collect an example of what a student did as a second step.

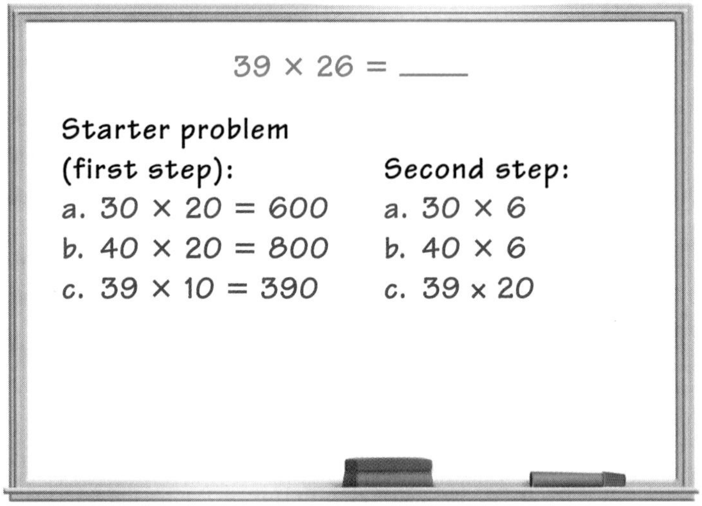

Ask the rest of the class to finish solving the problem, using one of those pairs of first and second steps. As students finish solving each problem, ask them to share their solutions with a partner and think about the following:

- Can they read their partner's work?
- Can they follow each other's strategy on paper?
- Are any steps unclear? Can they make their solutions clearer?

3 SESSION FOLLOW-UP
Daily Practice

Daily Practice: For ongoing review, have students complete *Student Activity Book* page 43.

Student Math Handbook: Students and families may use *Student Math Handbook* pages 30–32 for reference and review. See pages 190–197 in the back of this unit.

SESSION 2.7

Assessment: What Is the Answer?

Math Focus Points

- Solving 2-digit by 2-digit multiplication problems
- Breaking up multiplication problems efficiently

Today's Plan		Materials
ACTIVITY **1 More Starter Problems**	40 MIN INDIVIDUALS	• *Student Activity Book,* pp. 44–45
ASSESSMENT ACTIVITY **2 What Is the Answer?**	20 MIN INDIVIDUALS	• *Student Activity Book,* pp. 44–45 • M51*
SESSION FOLLOW-UP **3 Daily Practice**		• *Student Activity Book,* p. 46 • *Student Math Handbook,* pp. 30–32

*See *Materials to Prepare,* p. 69.

Ten-Minute Math

Quick Images: Seeing Numbers Show Images 25–27 (one at a time) on *Quick Images: Seeing Numbers* (T9). Ask students to find the total number of shapes and write at least one *multiplication* and one *division* equation for each pattern. Have two or three students explain how they saw the images (including any revisions they made) and their equations, showing how their numbers match the patterns. For example, Image 25 is an arrangement showing 6 groups of 12. Student might say $6 \times 12 = 72$, or $6 \times 2 \times 6 = 72$, or $72 \div 6 = 12$.

Name Date

Number Puzzles and Multiple Towers

More Starter Problems (page 1 of 2)

Finish solving the problems with the first step that is given. Then solve the same problem in your own way. Record both of your solutions, using clear and concise notation.

1. 68 × 75 = ______
   a. Start with 60 × 70.
   b. Solve 68 × 75 another way.
2. 98 × 36 = ______
   a. Start with 100 × 36.
   b. Solve 98 × 36 another way.

© Pearson Education 5

44 Unit 1 Session 2.7

▲ Student Activity Book, p. 44

Name Date

Number Puzzles and Multiple Towers

More Starter Problems (page 2 of 2)

3. 16 × 128 = ______
   a. Start with 4 × 128.
   b. Solve 16 × 128 another way.
4. 207 × 46 = ______
   a. Start with 207 × 10.
   b. Solve 207 × 46 another way.

© Pearson Education 5

Session 2.7 Unit 1 45

▲ Student Activity Book, p. 45

1 ACTIVITY

More Starter Problems

40 MIN INDIVIDUALS

In this activity, students solve a multiplication problem two ways. One way utilizes a given "starter problem" (first step), and the other way is of the student's own choosing.

ONGOING ASSESSMENT: Observing Students at Work

Students solve multiplication problems by using a given first step.

- **Can students see the relationship between the starter problems and the final problem?**
- **Can students finish solving a multiplication problem on the basis of a certain first step?** That is, can students determine what parts of the problem still remain to be solved after the first step?
- **Are students able to solve the problem in a different way?**

As students work, circulate and ask questions such as the following:

- How does the problem you were given to start help you solve the main problem? What part of the problem have you solved? What part remains to be solved?

Help guide students who are having a difficult time making sense of what they should do next.

- What story context can you use to help you figure this out? What representation can you use to help you keep track of your work?

DIFFERENTIATION: Supporting the Range of Learners

Intervention Students who are still working hard on developing their own efficient strategy for multiplication may not be ready to solve problems by using unfamiliar first steps. In that case, students can solve the problem in whatever way they can and then consider how the starter problem is the same as or different from their own first step.

Extension Encourage students who are efficient in using one strategy to try a different strategy.

2 ASSESSMENT ACTIVITY
What Is the Answer?

20 MIN INDIVIDUALS

Students complete Assessment: What is the Answer? (M51).❶ Explain to students that each should work on this multiplication assessment individually. When they are finished, they can continue working on *Student Activity Book* pages 44–45.

In this assessment, you have to figure out what is wrong with the strategy a fourth grader used to solve a multiplication problem. You will explain what is wrong and then solve the problem correctly.

ONGOING ASSESSMENT: Observing Students at Work

Students use knowledge of multiplication to solve a multiplication problem.

- **Can students use their knowledge of multiplication to evaluate the incorrect strategy in Problem 1?**
- **Can students solve a 2-digit multiplication problem efficiently and accurately?** Can they keep track of all the parts? Can they record their solution clearly?

3 SESSION FOLLOW-UP
Daily Practice

Daily Practice: For ongoing review, have students complete *Student Activity Book* page 46.

Student Math Handbook: Students and families may use *Student Math Handbook* pages 30–32 for reference and review. See pages 190–197 in the back of this unit.

Professional Development

❶ **Teacher Note:** Assessment: What is the Answer? p. 166

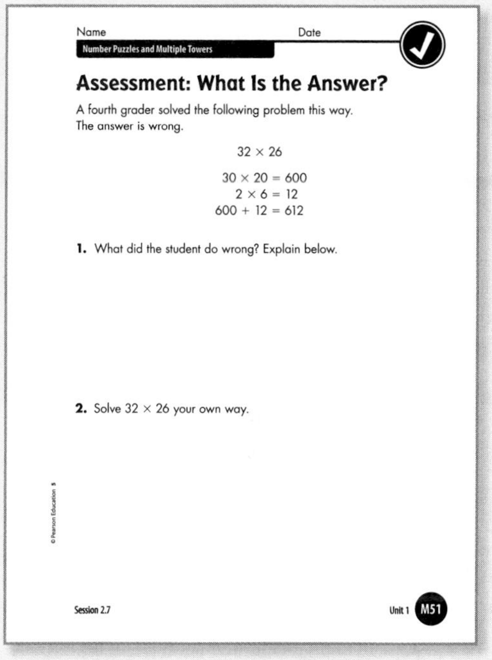
Name Date
Number Puzzles and Multiple Towers

Assessment: What Is the Answer?

A fourth grader solved the following problem this way. The answer is wrong.

32×26

$30 \times 20 = 600$
$2 \times 6 = 12$
$600 + 12 = 612$

1. What did the student do wrong? Explain below.

2. Solve 32×26 your own way.

© Pearson Education 5

Session 2.7 Unit 1 M51

▲ Resource Masters, M51 PORTFOLIO

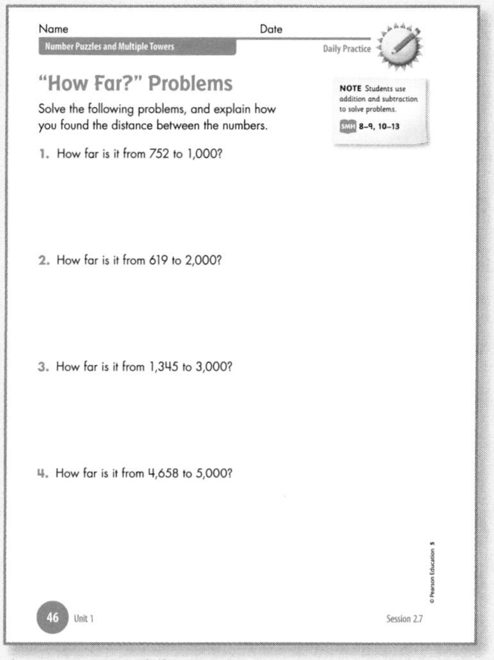
Name Date
Number Puzzles and Multiple Towers Daily Practice

"How Far?" Problems

NOTE Students use addition and subtraction to solve problems.
SMH 8–9, 10–13

Solve the following problems, and explain how you found the distance between the numbers.

1. How far is it from 752 to 1,000?
2. How far is it from 619 to 2,000?
3. How far is it from 1,345 to 3,000?
4. How far is it from 4,658 to 5,000?

© Pearson Education 5

46 Unit 1 Session 2.7

▲ Student Activity Book, p. 46

INVESTIGATION 3

Mathematical Emphases

Whole-Number Operations Reasoning about numbers and their factors

Math Focus Points

- Finding all the factors of a number

Computational Fluency Multiplication problems with 2-digit numbers

Math Focus Points

- Solving 2-digit by 2-digit multiplication problems

Whole-Number Operations Understanding and using the relationship between multiplication and division to solve division problems

Math Focus Points

- Describing and comparing strategies used to solve division problems
- Using knowledge of multiples of 10 to solve division problems
- Using and interpreting notation that represents division and relating division and multiplication notations (e.g., 170 ÷ 15 = _____ and 15 × _____ = 170)
- Solving division problems with 2-digit divisors
- Solving a division problem by breaking the dividend into parts
- Comparing division problems to determine which quotient is greater

Whole Number Operations Representing the meaning of multiplication and division

Math Focus Points

- Representing a multiplication or division problem with a picture or diagram
- Creating a story problem represented by a multiplication or division expression
- Making sense of remainders in terms of problem contexts

This Investigation also focuses on

- Using clear and concise notation

Division Strategies

	Student Activity Book	Student Math Handbook	Professional Development: Read Ahead of Time
SESSION 3.1 p. 114 **Solving a Division Problem** Students solve and represent a division problem. Then they share their solutions.	47–48	14, 37	• **Teacher Note:** The Relationship Between Multiplication and Division, p. 169 • **Teacher Note:** Division Strategies, p. 170 • **Dialogue Box:** Solving a Division Problem, p. 186
SESSION 3.2 p. 120 **Multiple Towers** Students list a sequence of multiples of a number and solve multiplication and division problems on the basis of the numbers in the list.	49–52	20	
SESSION 3.3 p. 126 **Solving More Division Problems** Using teams as a context, students solve division problems. They compare their solutions and strategies.	53–54	38–39	• **Teacher Note:** Developing Computation Strategies That Make Sense, p. 171
SESSION 3.4 p. 131 **Multiplication and Division Relationships on the Multiple Tower** Students solve multiplication and division problems for multiples greater than those given on a multiple tower. Then they discuss notation of solutions.	55–58	20	• **Teacher Note:** Division Notation, p. 163 • **Dialogue Box:** How Many 21s are in 1,344?, p. 188

Ten-Minute Math See page 20 for an overview.

Quick Images: Seeing Numbers

- **T10–T12, *Quick Images: Seeing Numbers***

Number Puzzles

- **M35–M43, Number Puzzle Clue Cards** Gather the cards from Investigation 2, and put them into three piles, Set A, Set B, and Set C.

Materials to Gather	Materials to Prepare
• **12″ x 18″ Construction paper** (1 per student) • **Markers** (1 set per student) • **Color tiles** (as needed) • **Connecting cubes** (as needed)	• **M15, Centimeter Grid Paper** Make copies. (2 per student plus extras) • **M31, 300 Chart** Make copies. (as needed)
• **Self-stick notes** (1 pad per class; optional)	• **Adding Machine Tape** Cut a 6-foot strip of adding machine tape (or butcher/chart paper cut and taped to make a strip). Post this on a classroom wall where students can see it from a distance.
• **Multiple tower for 21** (from Session 3.2) • **Calculators** (optional) • **Chart paper**	

Overhead Transparency

Division Strategies, *continued*

	Student Activity Book	Student Math Handbook	Professional Development: Read Ahead of Time
SESSION 3.5 p. 136			
Division Cluster Problems Students solve division cluster problems and discuss how to break up a division problem into manageable parts.	59–62	35	
SESSION 3.6 p. 141			
Practicing Division Strategies Students compare division problems that involve multiples of 10 and work on strategies for solving division problems.	59–60, 63–70	38–39; G6	
SESSION 3.7 p. 147			
Practicing Division Strategies, *continued* Students continue to work on strategies for solving division problems.	59–60, 63–66, 71	38–39	
SESSION 3.8 p. 151			
End-of-Unit Assessment Students are assessed on the work they have done in this unit on multiplication and division.	72	15	• **Teacher Note:** End-of-Unit Assessment, p. 174 • **Assessment in This Unit,** p. 14

Materials to Gather	Materials to Prepare
• **Calculators** (optional)	
• **M44–M45, Compare Cards** (1 deck per pair; from Session 2.3) • **M48–M50, Digit Cards** (1 deck per pair; from Session 2.4; optional)	• **M52, *Division Compare*** Make copies. (as needed) • **M53, *Division Compare* Recording Sheet** Make copies. (as needed)
• **M44–M45, Compare Cards** (1 deck per pair; from Session 2.3) • **M48–M50, Digit Cards** (1 deck per pair; from Session 2.4; optional) • **M52, *Division Compare*** (as needed; from Session 3.6)	• **M53, *Division Compare* Recording Sheet** Make copies. (as needed; from Session 3.6)
	• **M54–M55, End-of-Unit Assessment** Make copies. (1 per student)

Solving a Division Problem

Math Focus Points

- Representing a multiplication or division problem with a picture or diagram
- Creating a story problem represented by a multiplication or division expression
- Describing and comparing strategies used to solve division problems

Vocabulary

dividend
divisor
quotient

Today's Plan		Materials
ACTIVITY **1 Solving a Division Problem**	40 MIN · INDIVIDUALS · PAIRS	• M15 (as needed)*; M31 (as needed)* • 12″ x 18″ construction paper; markers; color tiles (as needed); calculators (optional); connecting cubes (as needed)
DISCUSSION **2 Naming Strategies**	20 MIN · CLASS	• Posters of 170 ÷ 15 (from Activity 1)
SESSION FOLLOW-UP **3 Daily Practice and Homework**		• *Student Activity Book,* pp. 47–48 • *Student Math Handbook,* pp. 14, 37

*See *Materials to Prepare,* p. 111.

Ten-Minute Math

Quick Images: Seeing Numbers **Show Images 28–30 (one at a time) on *Quick Images: Seeing Numbers* (T10). Ask students to find the total number of shapes and write at least one *multiplication* and one *division* equation for each pattern. Have two or three students explain how they saw the images (including any revisions they made) and their equations, showing how their numbers match the patterns.**

1 ACTIVITY
Solving a Division Problem

40 MIN INDIVIDUALS PAIRS

In this activity, students represent and solve a division problem. The purpose is to provide examples of different strategies that students might use. At this point, it is expected that students are still learning and refining strategies for solving division problems efficiently.

Write the following problems on the board:

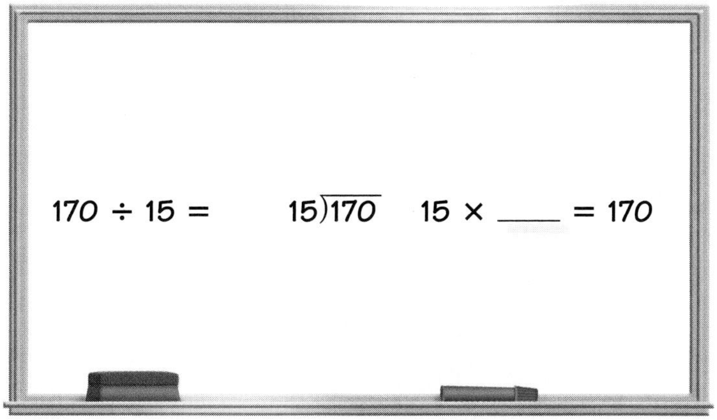

Ask students to share a story with a partner that fits each of these problems and to consider whether the same story can fit all three problems. For example, students might come up with stories such as this one: The principal ordered 170 computers for our school. There are 15 classrooms in our school. If each classroom receives the same number of computers, how many computers does a classroom receive?

Have a brief discussion, no more than 5 minutes, about how these different notations all mean "170 divided by 15." Remind students that in this case 170 is the dividend, the number being divided; 15 is the divisor, the number by which they are dividing; and the answer they are going to find is the quotient.

When students begin to solve $170 \div 15$ by thinking about 10×15, they are using what they know about multiplication to help them solve a division problem.❶ This is an example of the inverse relationship between multiplication and division.❷ ❸

In a story context, all of these notations mean that students are finding either how many groups of 15 are in 170, or if there are 15 groups that make up 170, how many are in each group.

Algebra Note

❶ **Relating Multiplication and Division** Students might complete the problem this way: $10 \times 15 = 150$. Adding another 15 makes 165. Therefore, $170 \div 15 = 11$ with a remainder of 5. The division equation, $165 \div 15 = 11$, and the multiplication equation, $15 \times 11 = 165$, both express the relationship between 11, 15, and 165. Refer to **Algebra Connections in This Unit**, page 16, for more information.

Math Note

❷ **Multiplication Combinations and Related Division Facts** This is a good opportunity to remind students of what they already know about how multiplication combinations are related to division. That is, if $7 \times 8 = 56$, then $56 \div 7 = 8$ and $56 \div 8 = 7$. Students worked on this relationship at the end of Grade 4 and will be assessed on division facts at the end of Grade 5. As students continue to practice their multiplication combinations, look for opportunities to identify related division facts.

Professional Development

❸ **Teacher Note:** The Relationship Between Multiplication and Division, p. 169

Teaching Notes

④ Getting Started If students seem unsure how to get started with their representation, ask them about their story context and whether they can use that. The representation does not need to be a literal picture of their story context. For example, if the story context is about teams, the representation does not need to show drawings of people, but could show a circle for each team and numbers written inside the circles to indicate the number of people on a team.

⑤ Stories and Remainders Story contexts can be particularly useful in helping students understand the meaning of remainders in division problems. Throughout this Investigation, some problems involve remainders. Encourage students to use, or create, a story context to help them determine the meaning of the quotient and the meaning of the remainder.

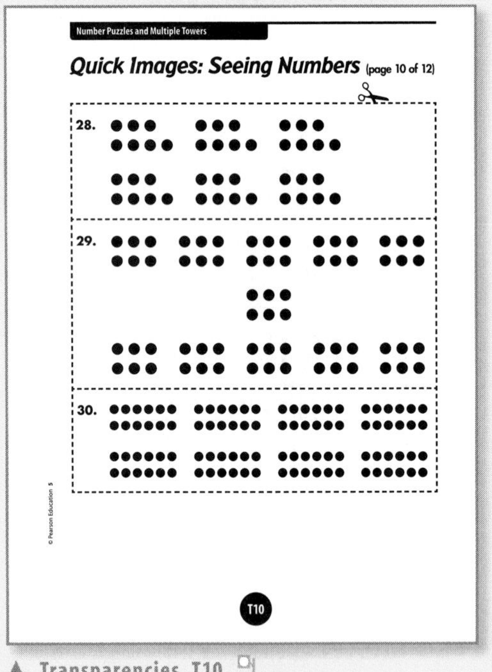

▲ Transparencies, T10

Remind students of the multiplication posters they made in Session 2.1. Now they are going to solve the division problem 170 ÷ 15 and make a similar poster showing their solution. Their poster should include the following three elements:

- A story problem for 170 ÷ 15
- A solution to the problem shown clearly
- A visual representation for the problem that uses, for example, cubes, an array, or groups④

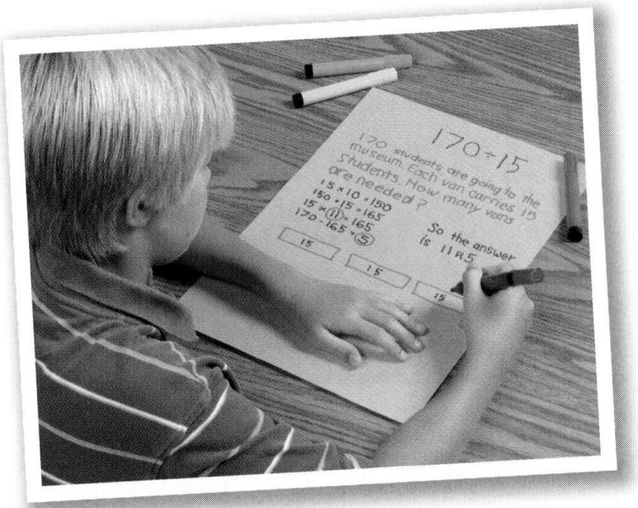

Students make posters showing a context, a solution, and a representation for 170 ÷ 15.

The quotient to 170 ÷ 15 involves a remainder. The convention of recording the remainder with a capital "R" is used in these materials: 170 ÷ 15 = 11 R5.

Talk with students as they work on their posters.

In your story context, what does the 11 mean? What does the 5 (or R5) mean? What do you have to do with the 5 to answer the problem you posed in your story context?⑤

Where is the 15 in your representation? Where is the 170? How does your representation show the answer? How do you know that it's correct?

When students finish their posters, they should find another person who is finished and compare posters.

Allow students to use whatever math tools they find useful. If they use math tools such as square tiles, interlocking cubes, grid paper, or 300 charts, they should show how they used them on their poster. It is not necessary for students to use any of these tools, but the tools should be available for those students who want to use them.[6]

Before the discussion, identify students who used a story context that involves the number in a group (e.g., how many on each team?) and others who used a story context that involves the number of groups (e.g., how many teams?).[7]

ONGOING ASSESSMENT: Observing Students at Work

Students solve a division problem with a 3-digit dividend and a 2-digit divisor.

- **How are students thinking about division?** Are they finding groups of 15? Or 15 groups? Are they thinking about an array?
- **Do students' representations show an understanding of division?** Do their partial solutions (e.g., $10 \times 15 = 150$) match their drawing?

2 DISCUSSION

Naming Strategies

20 MIN CLASS

Math Focus Points for Discussion

- Describing and comparing strategies used to solve division problems

Ask the students you previously identified to share their posters. For each student, ask questions such as the following:[8]

What's the answer to $170 \div 15$?

What story context did you use? In your story context, what does the 11 mean? What does the 5 mean?

Who has questions about Samantha's solution? Who else solved the problem in a similar way? Did you use the same representation or a different one? What does your representation show? How do the pictures match the numbers?

Professional Development

[6] **Dialogue Box:** Solving a Division Problem, p. 186

[7] **Teacher Note:** Division Strategies, p. 170

Differentiation

[8] **English Language Learners** You may need to rephrase some of the questions, using simpler vocabulary and language structures, so English Language Learners can understand them. For example, instead of "What story context did you use?" say, "Tell us about the story you used. What is your story about?" Instead of "What does the 11 mean?" ask, "What does the 11 tell us? Can you point to 11 of something on your poster?" You may also want to work with English Language Learners ahead of time to help them practice answering the questions.

You might choose to group together all the posters that use the same strategy and leave these posters up for the rest of the unit.

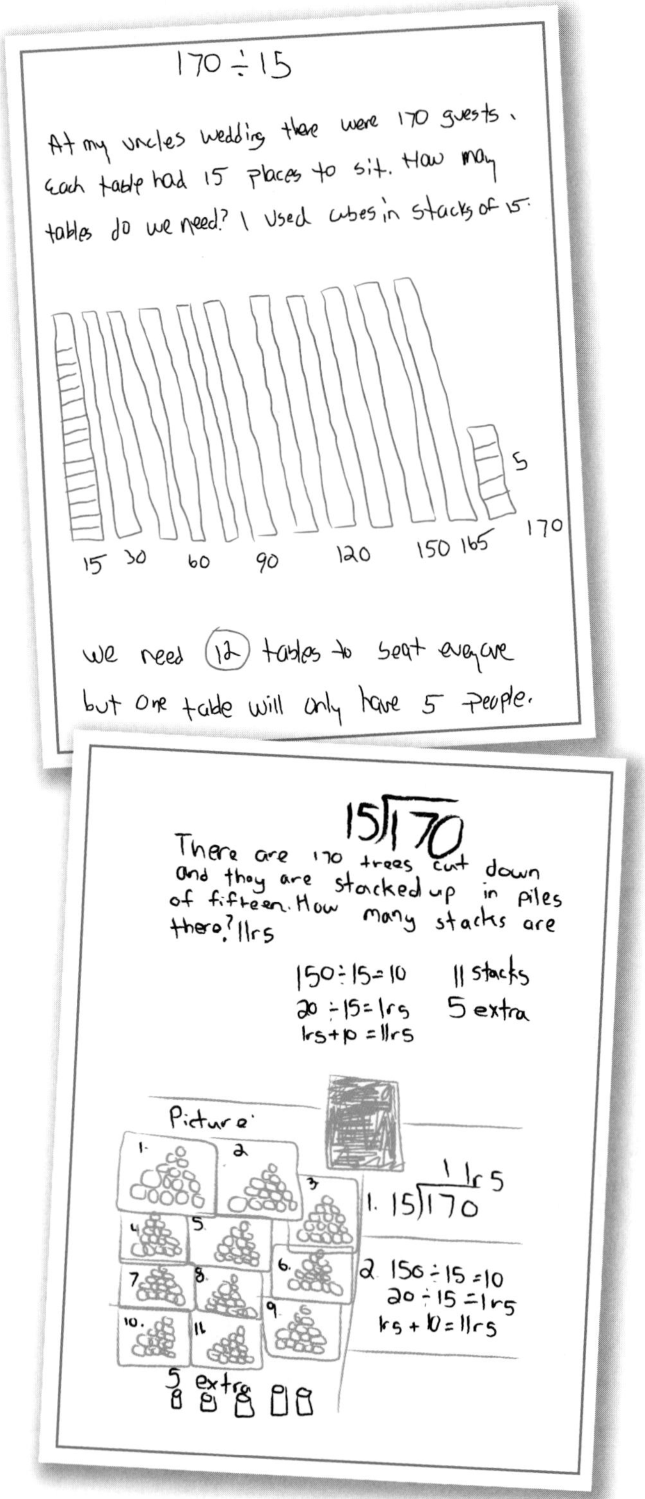

Sample Student Work

SESSION FOLLOW-UP

3 Daily Practice and Homework

Daily Practice: For reinforcement of this unit's content, have students complete *Student Activity Book* page 47.

Homework: Students solve a multiplication problem in two ways on *Student Activity Book* page 48.

Student Math Handbook: Students and families may use *Student Math Handbook* pages 14, 37 for reference and review. See pages 190–197 in the back of this unit.

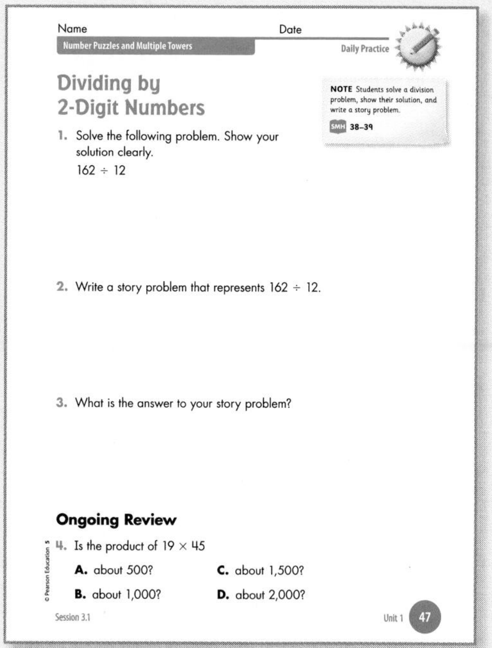

Name Date

Number Puzzles and Multiple Towers — Daily Practice

Dividing by 2-Digit Numbers

NOTE Students solve a division problem, show their solution, and write a story problem.

SMH 38–39

1. Solve the following problem. Show your solution clearly.
   162 ÷ 12

2. Write a story problem that represents 162 ÷ 12.

3. What is the answer to your story problem?

Ongoing Review

4. Is the product of 19 × 45
   A. about 500? **C.** about 1,500?
   B. about 1,000? **D.** about 2,000?

© Pearson Education 5

Session 3.1 Unit 1 47

▲ Student Activity Book, p. 47

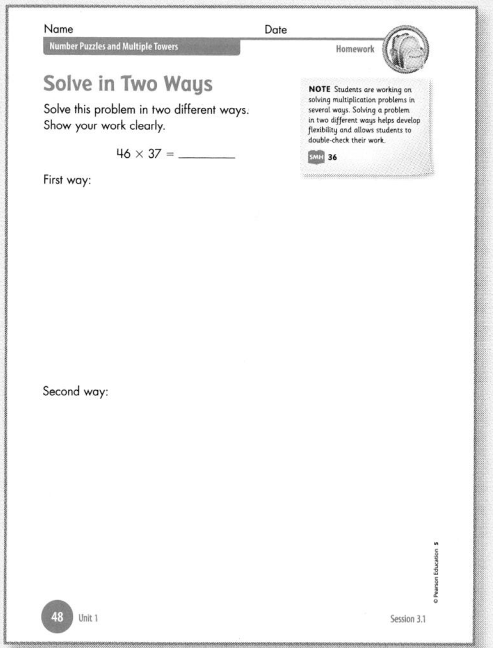

Name Date

Number Puzzles and Multiple Towers — Homework

Solve in Two Ways

Solve this problem in two different ways. Show your work clearly.

46 × 37 = ______

NOTE Students are working on solving multiplication problems in several ways. Solving a problem in two different ways helps develop flexibility and allows students to double-check their work.

SMH 36

First way:

Second way:

© Pearson Education 5

48 Unit 1 Session 3.1

▲ Student Activity Book, p. 48 PORTFOLIO

Multiple Towers

Math Focus Points

- Using knowledge of multiples of 10 to solve division problems
- Using and interpreting notation that represents division and relating division and multiplication notations (e.g., 170 ÷ 15 = ____ and ____ × 15 = 170)

Vocabulary

multiple

Today's Plan		Materials
ACTIVITY 1 Counting by 21s	10 MIN CLASS PAIRS	
ACTIVITY 2 Introducing Multiple Towers	20 MIN CLASS INDIVIDUALS	• Lists of multiples of 21 (from Activity 1); 6-foot strip of adding machine tape (or other strip of paper)*; Self-stick notes (1 pad per class; optional)
ACTIVITY 3 Using Multiple Towers	30 MIN INDIVIDUALS	• *Student Activity Book,* pp. 49–50
SESSION FOLLOW-UP 4 Daily Practice and Homework		• *Student Activity Book,* pp. 51–52 • *Student Math Handbook,* p. 20

*See *Materials to Prepare,* p. 111.

Ten-Minute Math

Quick Images: Seeing Numbers **Show Images 31–33 (one at a time) on *Quick Images: Seeing Numbers* (T11). Ask students to find the total number of shapes and write at least one *multiplication* and one *division* equation for each pattern. Have two or three students explain how they saw the images (including any revisions they made) and their equations, showing how their numbers match the patterns.**

ACTIVITY 1

10 MIN CLASS PAIRS

Counting by 21s

In this session, students create a "tower" of the multiples of 21. This brief activity helps students start thinking about these multiples.

When we skip count by a certain number, we are finding multiples of that number. What are the first few multiples of 21?

Multiples of 21

1 × 21 = 21

2 × 21 = 42

3 × 21 = 63

4 × 21 = 84

5 × 21 = 105

6 × 21 = 126

Ask students to work with a partner and list some more multiples of 21 in order. Students do not need to write the equations. Tell them that you will be interested in hearing how they find the next multiple.

After a few minutes, stop students and ask how they found the next multiples.

Students might say:

"I thought of 21 as 20 and 1. I found multiples of 20 and then added on a number 1 larger each time. So 20 is the first one, and add 1. Then 40, add 2, then 60, add 3."

"I noticed that each digit increased with each count, but by a different amount. The ones digit goes up by one with each number. The rest of the number goes up by 2, except when the ones digit goes from 9 to 0."

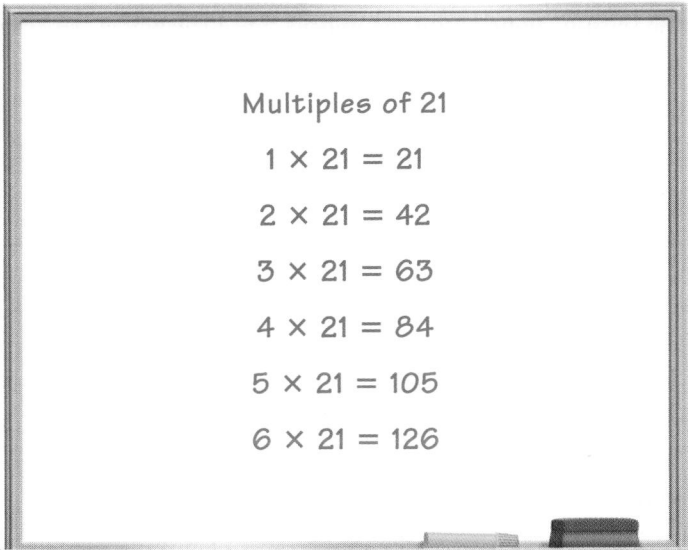

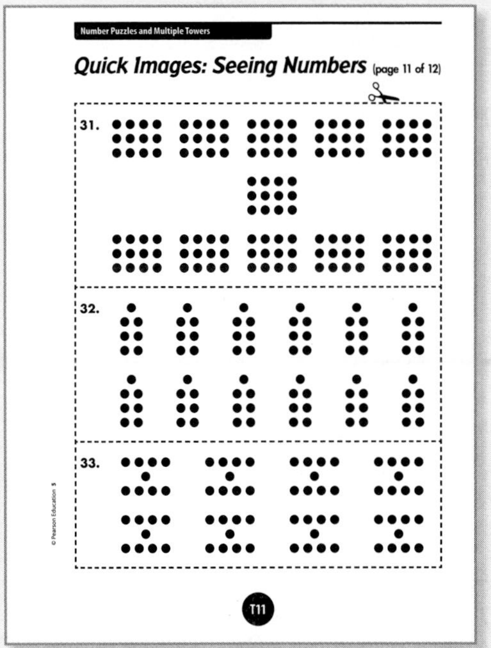

▲ Transparencies, T11

Teaching Note

① **Classroom Management** You might want to ask a student who has clear writing to write the multiples on self-stick notes and have another student stick them on the tower.

Give students a few more minutes to find more multiples of 21.

If students make a mistake early on in the sequence of multiples, they are likely to continue the mistake as they find higher multiples. Check with students, and encourage them to check with one another to make sure that their lists are accurate.

Ask students to think ahead about the answer to 10 × 21. Many students will know this product easily and they can use it to check that it is their 10th number. Students keep these lists to use later in the session, along with the multiple tower the whole class creates in the next activity.

ACTIVITY 2

Introducing Multiple Towers

20 MIN CLASS INDIVIDUALS

In this activity, students examine the relationship between multiplication and division. Students focus their thinking on the 10th, 20th, 30th, (and so on) multiples of a number. They build on their understanding of multiplying by multiples of 10 from Investigation 2. Knowing how to find these multiples is useful when breaking apart multidigit multiplication and division problems.

Show students the 6-foot paper strip you have posted.

We're going to count again by 21s, but now when you say a number, I will write it on this strip of paper, which we're going to call a multiple tower. Later we'll use this list of multiples to help solve multiplication and division problems.

Collect multiples of 21 in order from various students. Starting at the bottom of the strip, write the first few multiples of 21.

Write so that the numbers can be seen from a distance, spacing them evenly, one above the other.①

Stop at about 210 or a little higher.

How many 21s are in 210? Can you tell me how you know without counting?

Write these problems on the board, and fill in the answers when students agree.

$\_\_\_\_ \times 21 = 210 \qquad 210 \div 21 = \_\_\_\_$

After 210, what's the next multiple of 21 that ends in 0? (420) How do you know? How many 21s are in that number? How can we write a division problem with that answer? A multiplication problem?

Write these problems and their answers on the board.❷

$$\begin{array}{r} 20 \\ 21\overline{)420} \end{array} \qquad 20 \times 21 = 420$$

Tell students that you want to circle or highlight all of the numbers that are 21 multiplied by a multiple of 10. Circle or highlight 210 on the multiple tower, and ask students to remind you to do the same when you get to 420. If you are using self-stick notes, you can use a different color for the 10th, 20th, and 30th multiples, and so on.

As students count higher by 21s, continue to write the numbers on the tower. Stop at 315.

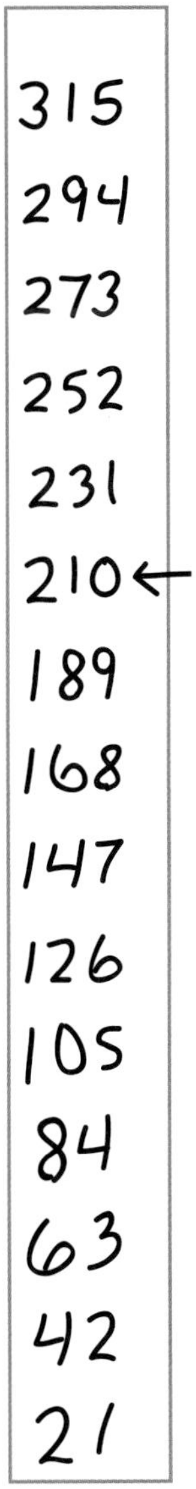

How many 21s are in 315? How do you know? Discuss it with your neighbor, and then let's hear your answers.

Stand next to the tower.

If we continue to list multiples of 21 until the tower is as tall as I am, what number do you think we will end with? Talk to a neighbor and make an estimate.

Give students a minute to discuss this. Then collect a few answers and write them on the board.

Is each of these estimates a multiple of 21? How do you know?

Ask students to continue counting by 21s while you write the multiples. Have them remind you to highlight the multiples of 10.

Ask students to stop you when they think the tower is at your height. Using the last number listed, discuss how many multiples of 21 are on the tower. In the next example, the top number on the tower is 1,029.

Math Note

❷ **Division Notation** Throughout this Investigation, sometimes use this form of division notation, $\begin{array}{r} 20 \\ 21\overline{)420} \end{array}$, and sometimes use $420 \div 21 = 20$.

Name Date

Number Puzzles and Multiple Towers

Problems about Multiples of 21 (page 1 of 2)

Use the multiple tower for 21 or your list of multiples of 21 to help you with these problems. Be sure to use your answers to the earlier problems to help you with the later problems.

1. 10 × 21 = ______
2. 105 ÷ 21 = ______
3. 315 ÷ ______ = 21
4. ______ × 21 = 420
5. 5 × 21 = ______
6. 210 ÷ 21 = ______
7. 15 × 21 = ______

© Pearson Education 5

Session 3.2 Unit 1 49

▲ Student Activity Book, p. 49

Name Date

Number Puzzles and Multiple Towers

Problems about Multiples of 21 (page 2 of 2)

8. ______ × 21 = 630
9. 945 ÷ ______ = 21
10. 441 ÷ 21 = ______
11. Write and solve two division problems using multiples of 21.

© Pearson Education 5

50 Unit 1 Session 3.2

▲ Student Activity Book, p. 50

We counted 21s and stopped at 1,029. Without counting from the bottom, figure out how many numbers are in the tower; that is, how many 21s are in 1,029. How does knowing how many 21s are in 420 or 630 or 840 help you figure this out?

Ask students to share strategies. As students explain how they used the 10th multiple, 20th multiple, 30th multiple, and so on, point out that these are called landmark multiples.

On the board, write these equations for the final number in your tower.

$$49 \times 21 = 1{,}029 \qquad 1{,}029 \div 21 = 49$$

Keep the tower posted in the classroom for the remainder of the Investigation.

1,029
1,008
987
966
945
924
903
882
861
840
819
798
777
756
735
714
693
672
651
630
609
588
567
546
525
504
483
462
441
420
399
378
357
336
315
294
273
252
231
210
189
168
147
126
105
84
63
42
21

ACTIVITY 3

Using Multiple Towers

30 MIN INDIVIDUALS

On *Student Activity Book* pages 49–50, students answer multiplication and division problems, highlighting the relationship between multiplication and division. They use the multiple tower for 21 and what they know about these multiples, particularly the landmark multiples.

As students are working, circulate and observe whether they are using the multiple tower to answer the questions. Ask them to show you on the tower how they got their answers and whether they notice any relationships between problems.

ONGOING ASSESSMENT: Observing Students at Work

Students solve multiplication and division problems, using knowledge of the multiples of 21.

- **Can students solve the problem by using relationships on the multiple tower?**
- **Do they use the solution to one problem to help them solve others?** Do students see the relationship between the division equation and a missing factor problem, such as 126 ÷ 21 = ______ and ______ × 21 = 126?

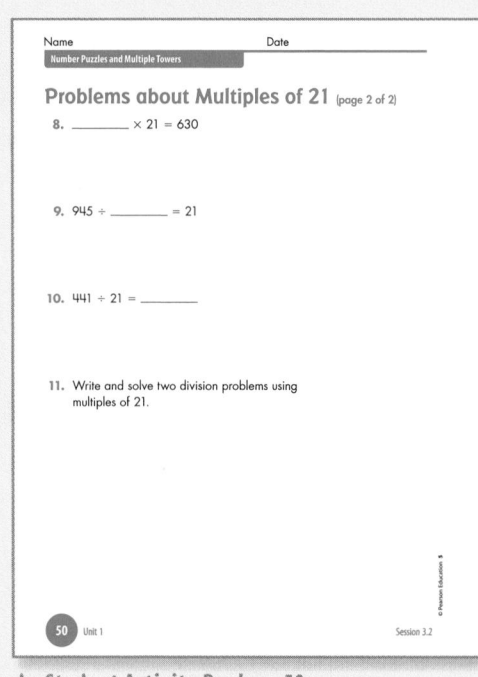

DIFFERENTIATION: Supporting the Range of Learners

Intervention You may choose to meet in a small group with students who find this task difficult, and work on the questions together. These students may also find it helpful to make a list of the multiples, using multiplication equations like the following:

$$3 \times 21 = 63$$
$$2 \times 21 = 42$$
$$1 \times 21 = 21$$

Students who are struggling to see the relationships that exist on the multiple tower, particularly the landmark multiples, may benefit from making their own multiple tower with a smaller number.

4 SESSION FOLLOW-UP

Daily Practice and Homework

Daily Practice: For ongoing review, have students complete *Student Activity Book* page 51.

Homework: Students write a story problem that represents 315 ÷ 21 and then solve the problem on *Student Activity Book* page 52.

Student Math Handbook: Students and families may use *Student Math Handbook* page 20 for reference and review. See pages 190–197 in the back of this unit.

Name Date

Number Puzzles and Multiple Towers — Daily Practice

Story Problems: Reading a Long Book

NOTE Students practice solving addition and subtraction problems in a story problem context.

SMH 8–9, 10–13

1. Noemi borrowed a new book from the library. At 1,000 pages, it is the longest book she has ever tried to read! The first day, she read 115 pages. How many more pages does she have to read to reach the end?

2. During the next week, Noemi read 388 pages. How many pages has she read altogether?

3. At the end of 2 weeks, Noemi had read 816 pages. How many pages does she have left to finish the book?

© Pearson Education 5

Session 3.2 Unit 1 51

▲ Student Activity Book, p. 51

Name Date

Number Puzzles and Multiple Towers — Homework

Solving 315 ÷ 21

NOTE Students have been solving division problems. Students should think about what multiplication combinations they know that can help them solve this problem.

SMH 14

1. Write a story problem for 315 ÷ 21.

2. Solve 315 ÷ 21.

© Pearson Education 5

52 Unit 1 Session 3.2

▲ Student Activity Book, p. 52

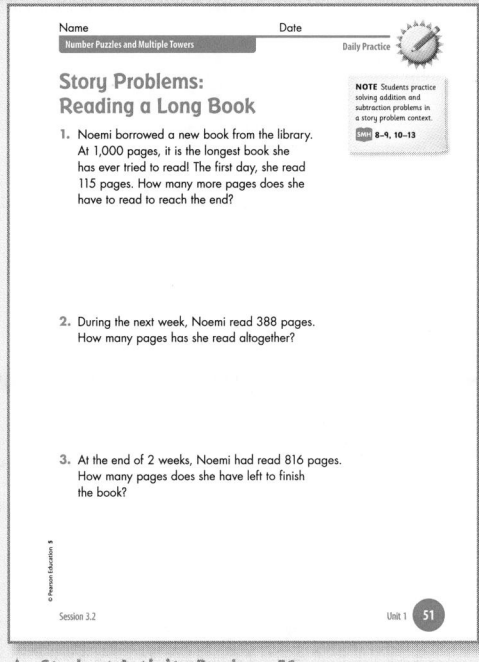

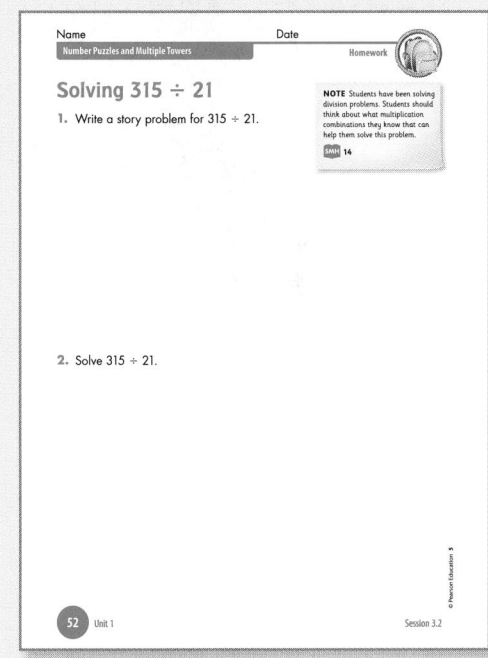

Solving More Division Problems

Math Focus Points

- Describing and comparing strategies used to solve division problems
- Using knowledge of multiples of 10 to solve division problems
- Solving division problems with 2-digit divisors
- Making sense of remainders in terms of problem contexts

Today's Plan		Materials
ACTIVITY 1 Division Problems	40 MIN · CLASS · INDIVIDUALS	• *Student Activity Book,* p. 53
DISCUSSION 2 Division Strategies	20 MIN · CLASS · PAIRS	• *Student Activity Book,* p. 53
SESSION FOLLOW-UP 3 Daily Practice		• *Student Activity Book,* p. 54 • *Student Math Handbook,* pp. 38–39

Ten-Minute Math

Quick Images: Seeing Numbers **Show Images 34–36 (one at a time) on *Quick Images: Seeing Numbers* (T12). Ask students to find the total number of shapes and write at least one *multiplication* and one *division* equation for each pattern. Have two or three students explain how they saw the images (including any revisions they made) and their equations, showing how their numbers match the patterns.**

1 ACTIVITY

Division Problems

40 MIN CLASS INDIVIDUALS

Write the following problems on the board or overhead:

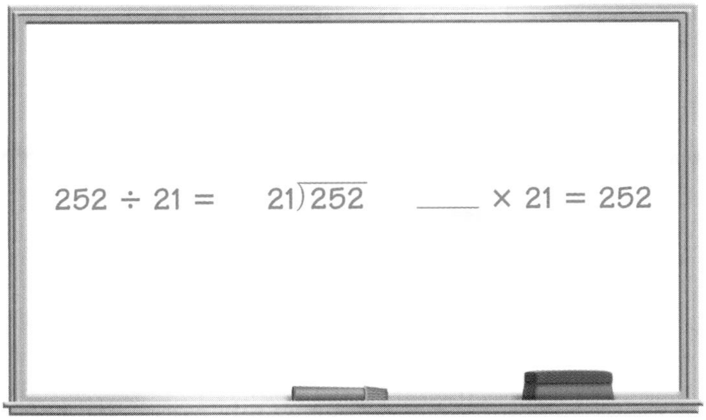

Ask students to share a story with a partner that will fit each of these problems. Also ask if the same story can fit all three problems.

Let's use Shandra's story. There are 252 fourth and fifth graders who need to be placed in 21 equal-sized teams for Field Day. How many people will be on each team?❶

Review with the students what the numbers in the problem mean.

What does the number 252 represent? *(the number of fourth and fifth graders)* What does the number 21 represent? *(the number of teams)* What will our answer represent? *(the number of people on a team)*

Discuss how the story problem fits all three problems, so the different notations you wrote all mean 252 ÷ 21.

Ask students to think about how they would start the problem.

Walter pointed out that he knows that 210 is 10 × 21, and this is close to 252.

When you solve division problems, you can think about a story context like Shandra's, and decide what each number in the story means, and what question you are trying to answer. Remember that a good starting place is the multiples of 10, 20, 30, and so on.

Tell students that as they start working on *Student Activity Book* page 53, they should think about what strategy they are using to solve division problems. How do they start? What do they do next? When they finish, they should find a partner and share their solutions. Remind students that the posters of division strategies they created in Session 3.1 may be

Teaching Note

❶ **Team Stories** Teams are used as an example because it is a context that students know and understand and that may help them solve the problem. If your class comes up with a different context that works for your students, use that instead.

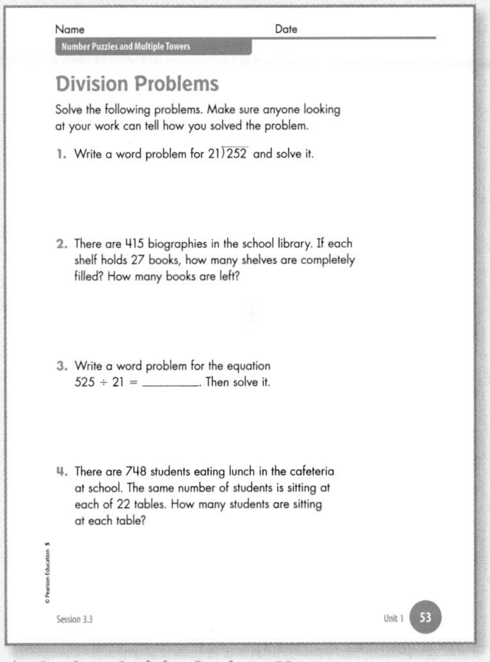

Name Date

Number Puzzles and Multiple Towers

Division Problems

Solve the following problems. Make sure anyone looking at your work can tell how you solved the problem.

1. Write a word problem for 21)252 and solve it.

2. There are 415 biographies in the school library. If each shelf holds 27 books, how many shelves are completely filled? How many books are left?

3. Write a word problem for the equation 525 ÷ 21 = ______. Then solve it.

4. There are 748 students eating lunch in the cafeteria at school. The same number of students is sitting at each of 22 tables. How many students are sitting at each table?

© Pearson Education 5

Session 3.3 Unit 1 53

▲ Student Activity Book, p. 53

Teaching Note

❷ **Understanding Remainders** Continue to use story contexts to facilitate understanding of the meaning of remainders in division problems. Remind students to use, or create, the story context to help them determine what needs to be done with the remainder.

Professional Development

❸ **Teacher Note:** Division Strategies, p. 170

❹ **Teacher Note:** Developing Computation Strategies that Make Sense, p. 171

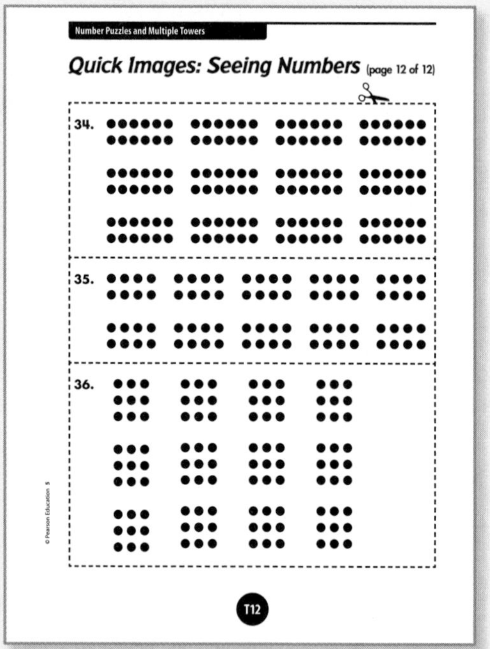

▲ Transparencies, T12

helpful if they are not sure how to get started in solving a division problem. If necessary, remind students of the convention of recording the remainder with a capital "R," for example, 170 ÷ 15 = 11 R5.❷

As students work on the division problems, circulate and notice what division strategies and notation students are using.❸ ❹ The discussion at the end of the session will focus on Problem 1. Find several students who have solved the problem in different ways, and ask them to be prepared to explain their thinking.

ONGOING ASSESSMENT: Observing Students at Work

Students solve division problems.

- **Are students able to solve division problems accurately?** What strategies are students using?
- **Are students using relationships between numbers to solve the problem? (e.g., 210 ÷ 21 = 10, 420 ÷ 21 = 20, 630 ÷ 21 = 30, and so on)**
- **What size numbers are they using?** Groups of 10? Groups of multiples of 10?
- **Are students able to keep track of their work?** Do they know what part of the problem they have solved and what they still have to solve? What their answer is?

As students are working, circulate around the room and ask questions such as the following:

- What do the numbers mean? Where is your answer? How do you know that it's correct? Did you answer the question to the word problem?

DIFFERENTIATION: Supporting the Range of Learners

Intervention If students seem unsure how to start, ask them whether any of the work with the multiple towers and finding every 10th multiple can help.

Students who are still developing strategies for division might need to work with smaller numbers and use tools. Use problems such as 105 ÷ 7, 96 ÷ 6, and 72 ÷ 3. The questions you ask students can remain the same.

DISCUSSION

2 Division Strategies

20 MIN CLASS PAIRS

Math Focus Points for Discussion

- Describing and comparing strategies used to solve division problems

Write Problem 1 from *Student Activity Book* page 53 on the board.

Call on students you identified to explain their strategy for doing the computation. Either you or the student should record the strategy on the board. Ask questions as each student explains a strategy.

How did you decide where to start?

How did you decide what to do next?

How did you know what your answer was?

Does anyone have questions about this strategy?

Who else solved the problem in this way?

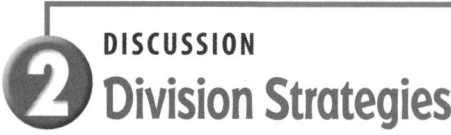

Students record and explain the strategies they used to solve 252 ÷ 21.

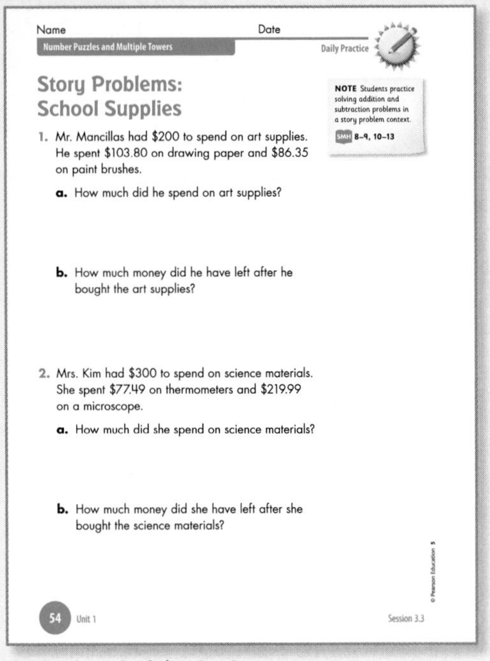

Name Date

Number Puzzles and Multiple Towers Daily Practice

Story Problems: School Supplies

NOTE Students practice solving addition and subtraction problems in a story problem context.

SMH 8–9, 10–13

1. Mr. Mancillas had $200 to spend on art supplies. He spent $103.80 on drawing paper and $86.35 on paint brushes.
   a. How much did he spend on art supplies?
   b. How much money did he have left after he bought the art supplies?
2. Mrs. Kim had $300 to spend on science materials. She spent $77.49 on thermometers and $219.99 on a microscope.
   a. How much did she spend on science materials?
   b. How much money did she have left after she bought the science materials?

© Pearson Education 5

54 Unit 1 Session 3.3

▲ Student Activity Book, p. 54

After several strategies are on the board, have students discuss with a partner how the strategies are similar or different.

Students might say:

"Janet multiplied 10 × 21 = 210, and Felix divided, 210 ÷ 21 = 10."

"They used different numbers of steps. Janet showed the subtraction, 252 − 210 = 42, and Alicia didn't."

"Avery knew that 21 × 2 was 42, so he just added 210 and 42. Mitch added 21 at a time: 210 + 21 = 231, then 231 + 21 = 252."

SESSION FOLLOW-UP

3 Daily Practice

Daily Practice: For ongoing review, have students complete *Student Activity Book* page 54.

Student Math Handbook: Students and families may use *Student Math Handbook* pages 38–39 for reference and review. See pages 190–197 in the back of this unit.

Multiplication and Division Relationships on the Multiple Tower

Math Focus Points

- Using and interpreting notation that represents division and relating division and multiplication notations (e.g., 170 ÷ 15 = ____ and ____ × 15 = 170)
- Solving division problems with 2-digit divisors
- Using clear and concise notation

Today's Plan		Materials
ACTIVITY **1 Numbers Off the Tower**	45 MIN CLASS INDIVIDUALS	• *Student Activity Book,* p. 55 • Multiple tower for 21 (from Session 3.2) • Calculators (optional)
DISCUSSION **2 Division Notation**	15 MIN CLASS	• *Student Activity Book,* p. 55 • Chart paper
SESSION FOLLOW-UP **3 Daily Practice and Homework**		• *Student Activity Book,* pp. 56–58 • *Student Math Handbook,* p. 20

Ten-Minute Math

Number Puzzles **Play the variation, "Draw Three, Trade One." Draw one card from each set (A, B, C) of *Number Puzzle* Clue Cards (M35–M43) and read them aloud. If the solution is "inconclusive," students decide which of the three cards should be removed, and a new card (from the same set) is selected. Students work with a partner and find the number(s) that fit the set of three clues. Have students share their strategies for solving the puzzle.**

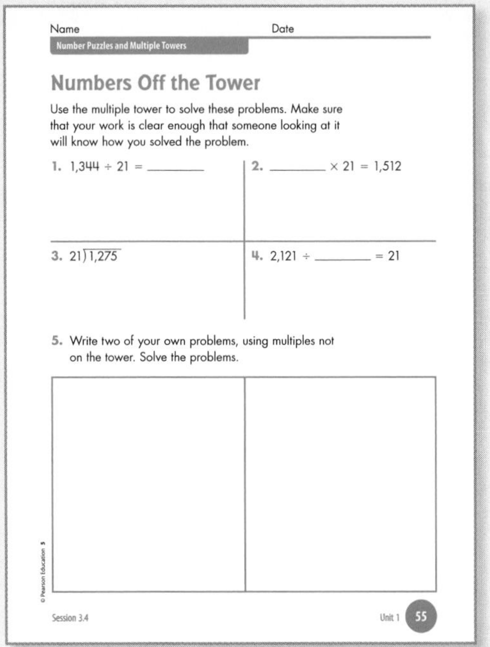

Name Date

Number Puzzles and Multiple Towers

Numbers Off the Tower

Use the multiple tower to solve these problems. Make sure that your work is clear enough that someone looking at it will know how you solved the problem.

1. 1,344 ÷ 21 = ______
2. ______ × 21 = 1,512
3. 21)1,275
4. 2,121 ÷ ______ = 21
5. Write two of your own problems, using multiples not on the tower. Solve the problems.

© Pearson Education 5

Session 3.4 Unit 1 55

▲ Student Activity Book, p. 55 PORTFOLIO

ACTIVITY

45 MIN CLASS INDIVIDUALS

1 Numbers Off the Tower

Students have been solving problems using the multiple tower they made in Session 3.2. In this activity, students use the numbers on the multiple tower for 21 to solve problems about multiples of 21 that are *not* on the tower. Work focuses on solving division problems by breaking the dividend into parts.

Remind students about the multiple tower for 21, and ask a few questions reviewing previous work.

How many 21s are in 315? In 945?

Direct students' attention to Problem 1 on *Student Activity Book* page 55, 1,344 ÷ 21 = ____.

Our tower doesn't go all the way to 1,344. Can you use the numbers on our tower to solve this problem? If you want, you may work with a partner. When you've finished with Problem 1, don't go on to the rest of the problems yet.

Even though 1,344 is not on the multiple tower for 21, students develop strategies for using the tower to solve 1,344 ÷ 21.

Give students 5 minutes or so to solve this problem. As you circulate, identify several students who have broken the dividend, 1,344, into smaller parts to help them solve the problem. Ask these students to be ready to explain their thinking when the class comes back together.

To solve this problem, some students might start with the last number on the multiple tower, and then add 21 or multiples of 21 until they reach 1,344. Other students may multiply 21 by a multiple of 10 to get close to 1,344.①

As students explain how they solved the problem, record their solutions. Ask students to make their thinking clear to the class by pointing out where parts of their solutions appear on the tower, and ask students whether they have questions.

Students complete the remainder of *Student Activity Book* page 55. Tell students that as they work, they should focus on how to write the steps of their work so that anyone looking at it will know what they did.

Tell students that as a class they will discuss Problem 3 at the end of the session. As students are working, identify for the upcoming discussion several students who have clear notation in their solutions.②

To prepare for the discussion, stop students about 5 minutes before the discussion. Tell students to share their solution to Problem 3 with a partner. Their task is to see whether they can follow the strategy their partner used. They should think about whether the strategy is clear and whether they can understand each step.

ONGOING ASSESSMENT: Observing Students at Work

Students solve division problems by breaking the problem into manageable parts.

- **How are students solving the problems?** Are they using the relationships on the tower such as the landmark multiples? Are they using the last (top) number on the tower and adding multiples of 21?
- **Do they know when they have solved the problem?**
- **How do students determine multiples that are off the tower for their own problems in Problem 5?** Do they use a calculator and multiply by 21? Do they add numbers together that are on the chart?

DIFFERENTIATION: Supporting the Range of Learners

Intervention Some students will need to continue practicing with multiples of 21 that are on the tower. Give these students a list of numbers on the tower that are the 10th, 20th, or 30th multiples (or one multiple above or below, such as 231, 441, 651 or 189, 399, 609), and have them write multiplication and/or division sentences, using those numbers.

Professional Development

① **Dialogue Box:** How Many 21s are in 1,344?, p. 188

② **Teacher Note:** Division Notation, p. 163

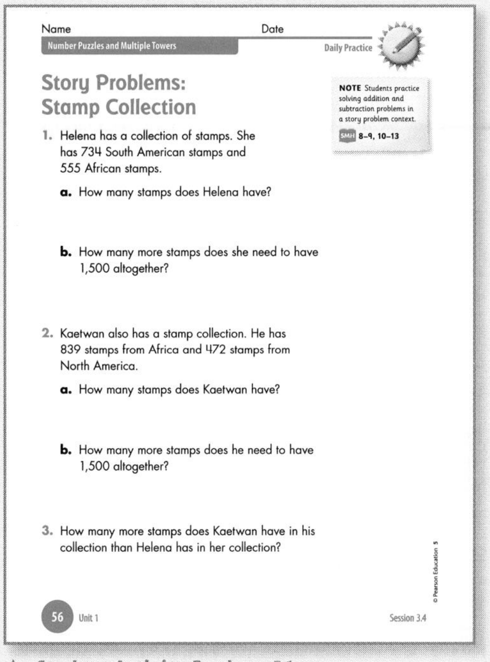
Name Date
Number Puzzles and Multiple Towers Daily Practice

Story Problems: Stamp Collection

NOTE Students practice solving addition and subtraction problems in a story problem context.
SMH 8–9, 10–13

1. Helena has a collection of stamps. She has 734 South American stamps and 555 African stamps.
   a. How many stamps does Helena have?
   b. How many more stamps does she need to have 1,500 altogether?
2. Kaetwan also has a stamp collection. He has 839 stamps from Africa and 472 stamps from North America.
   a. How many stamps does Kaetwan have?
   b. How many more stamps does he need to have 1,500 altogether?
3. How many more stamps does Kaetwan have in his collection than Helena has in her collection?

© Pearson Education 5

56 Unit 1 Session 3.4

▲ Student Activity Book, p. 56

Math Note

❸ **Showing the Steps** Clear and concise notation does not necessarily show every step that is needed for a solution. Some steps are carried out mentally and are not written down. For example, a student might show that $10 \times 15 = 150$ and then mentally subtract 150 from 170, without writing down $170 - 150 = 20$. During this session and throughout the year, when you talk with students about notation, help them think through which steps they need to write down to make their solution clear and to keep track of their steps, and which steps can be carried out mentally.

DISCUSSION

15 MIN CLASS

2 Division Notation

Math Focus Points for Discussion

- Using clear notation

Let students know that they are going to share strategies for solving Problem 3 on *Student Activity Book* page 55 and that they should focus particular attention to how the strategy is notated.

As you look at the solutions, decide whether all the steps are clear. Can you understand what each person did?

Have the students you identified earlier write their solutions to Problem 3 on the board, explaining how they solved the problem.

As each student explains a solution, ask other students to ask clarifying questions. Discuss whether the steps are clear.❸ You may also choose to have another student rephrase the solution being discussed.

Write the solutions on chart paper, and leave the chart posted in the classroom for students to use during the rest of the unit. A sample chart is given below. If necessary, remind students of the notation used for remainders: $1{,}275 \div 21 = 60$ R15.

___ × 21 = 1,275 6 × 21 = 126 60 × 21 = 1,260 1,275 − 1,260 = 15 1,275 ÷ 21 = 60 R15	1,275 ÷ 21 = 630 ÷ 21 = 30 1,275 − 630 = 645 630 ÷ 21 = 30 645 − 630 = 15 1,260 ÷ 21 = 60 1,275 ÷ 21 = 60 R15
21)1,275 10 × 21 = 210 20 × 21 = 420 30 × 21 = 630 60 × 21 = 1,260 60 R15 21)1,275	21)1,275 − 420 20 × 21 = 420 855 − 630 30 × 21 = 630 225 − 210 10 × 21 = 210 15 20 + 30 + 10 = 60 1,275 ÷ 21 = 60 R15

After the solutions have been posted, focus on the notation that students used to show their solutions. Ask questions such as the following:

What makes the notation of the solution clear? How does that help us understand how someone solved the problem?

Students might say:

"The order of the steps makes sense. They're organized so that we can see how the work was done."

"I can see that there are not too many steps."

"It is easy to see where the answer is."

Tell students that as they continue working in this Investigation, they should continue to focus on how they notate their answers. Clear notation helps them keep track of what part of the problem they have solved so far. It also helps other people follow and understand the strategy.

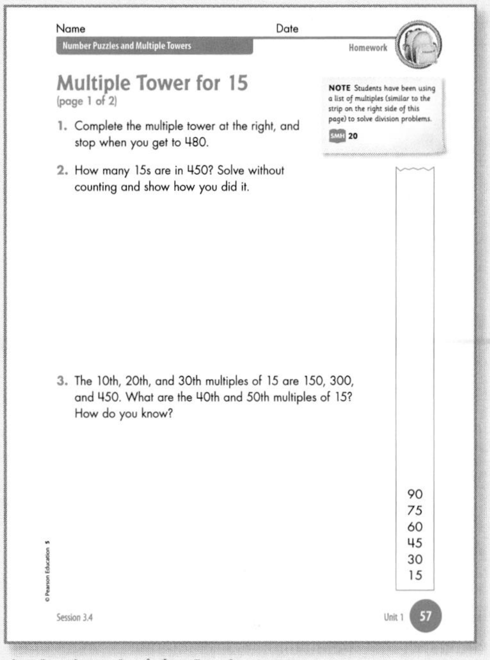

Name Date

Number Puzzles and Multiple Towers Homework

Multiple Tower for 15
(page 1 of 2)

NOTE Students have been using a list of multiples (similar to the strip on the right side of this page) to solve division problems.

SMH 20

1. Complete the multiple tower at the right, and stop when you get to 480.

2. How many 15s are in 450? Solve without counting and show how you did it.

3. The 10th, 20th, and 30th multiples of 15 are 150, 300, and 450. What are the 40th and 50th multiples of 15? How do you know?

90
75
60
45
30
15

© Pearson Education 5

Session 3.4 Unit 1 57

▲ Student Activity Book, p. 57

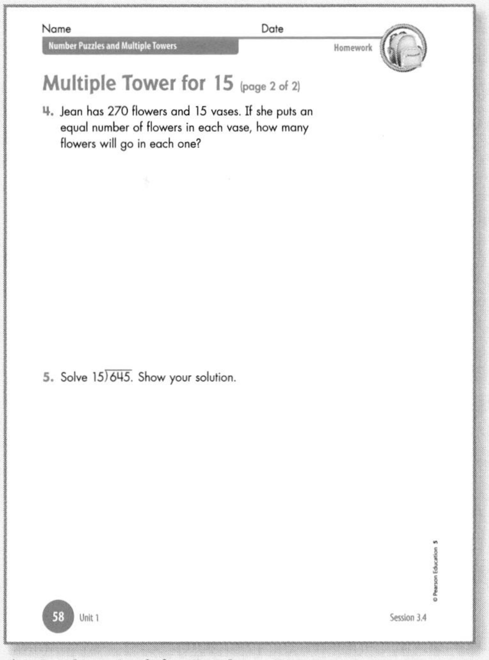

Name Date

Number Puzzles and Multiple Towers Homework

Multiple Tower for 15 (page 2 of 2)

4. Jean has 270 flowers and 15 vases. If she puts an equal number of flowers in each vase, how many flowers will go in each one?

5. Solve $15\overline{)645}$. Show your solution.

58 Unit 1 Session 3.4

© Pearson Education 5

▲ Student Activity Book, p. 58

SESSION FOLLOW-UP
3 Daily Practice and Homework

Daily Practice: For ongoing review, have students complete *Student Activity Book* page 56.

Homework: Students create a multiple tower for 15 on *Student Activity Book* pages 57–58 and use those numbers to solve division problems.

Student Math Handbook: Students and families may use *Student Math Handbook* page 20 for reference and review. See pages 190–197 in the back of this unit.

SESSION 3.5

Division Cluster Problems

Math Focus Points

- Solving division problems with 2-digit divisors
- Using and interpreting notation that represents division and relating division and multiplication notations (e.g., $170 \div 15 =$ ____ and ____ $\times 15 = 170$)
- Solving a division problem by breaking the dividend into parts

Today's Plan		Materials
1 ACTIVITY **Division Cluster Problems**	50 MIN, CLASS, INDIVIDUALS	• *Student Activity Book,* pp. 59–60 • Calculators (optional)
2 DISCUSSION **Breaking Up Numbers**	10 MIN, CLASS	
3 SESSION FOLLOW-UP **Daily Practice and Homework**		• *Student Activity Book,* pp. 61–62 • *Student Math Handbook,* p. 35

Ten-Minute Math

Number Puzzles **Play the variation "Random Clues." Combine all the clues from the three sets (A, B, C) of *Number Puzzle* Clue Cards (M35–M43). Draw the first three cards from the deck. This variation has greater potential for resulting in "inconclusive" solutions, providing students more opportunities to analyze, discuss, and modify clues. Students work with a partner and find the number(s) that fit the set of three clues. If the solution is "inconclusive," students modify the clues. Have students share their strategies for solving the puzzle.**

ACTIVITY

1 Division Cluster Problems

50 MIN CLASS INDIVIDUALS

Tell students that today they will be solving division cluster problems, and that just like the multiplication clusters, they solve the first problems mentally, and use those problems to solve the final problem. Because students are still developing and refining strategies for division, these clusters, unlike the multiplication clusters, have only one path to the solution.

Continue to use contexts for the problems. This helps students visualize the action of the problem and use division or multiplication relationships they know to solve the problem. The context of teams used below is a suggestion. Use any context that your students used earlier or that is easy for them to visualize.

Write the following problems on the board or overhead:

$10 \times 18 =$

$5 \times 18 =$

$290 \div 18 =$

Our final problem is 290 ÷ 18. Who can think of a story problem for 290 ÷ 18 that involves teams? Let's use Renaldo's story that 290 boys and girls signed up for baseball and there are 18 teams. If each team has the same number of players, how many children will be on each team?

Ask students for the products of 10 × 18 *(180)* and 5 × 18 *(90),* and write those products in the equations on the board. Then ask students to think about how these problems help solve the final problem.

Let's think about these smaller problems in terms of Renaldo's story. If there are 18 teams, how many children are on teams so far?

If students seem unsure of this, make a quick sketch like the one on page 138. Ask students what numbers go in each circle and what the numbers represent. *(A "10" for 10 children would go in each circle and then a "5" for another 5 children. 270 children are on 18 teams.)* ❶

Math Note

❶ **Meanings of Division** A multiplication expression, such as 10 × 18, is often interpreted in these materials and by students as meaning "10 groups of 18." However, in this case, Renaldo's context calls for students to think about 18 groups, not groups of 18. Renaldo is correctly using one of the meanings of division to create his story for 290 ÷ 18: if 290 is split into 18 groups, how many are in each group? Using 10 × 18 to help solve this problem requires an interpretation of 10 × 18 as "10 in each of 18 groups." Many students may implicitly be using the commutative property of multiplication here, knowing that 10 × 18 = 18 × 10, and that 18 × 10 means for them "18 groups of 10." Other students may just directly interpret 10 × 18 as "10 in each of 18 groups" for this problem because that interpretation matches the problem context.

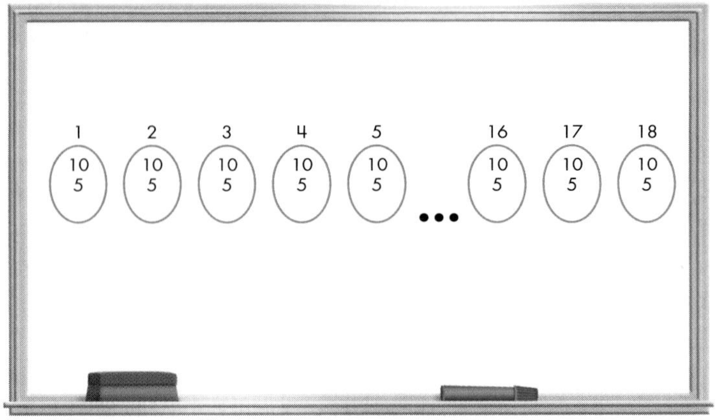

How many children are left to be placed on teams? *(20 children)* How did you figure that out? What else do you need to do to solve the problem 290 ÷ 18 = ____?

Again, if students seem unsure, continue with the sketch showing that one more child can go on each team.

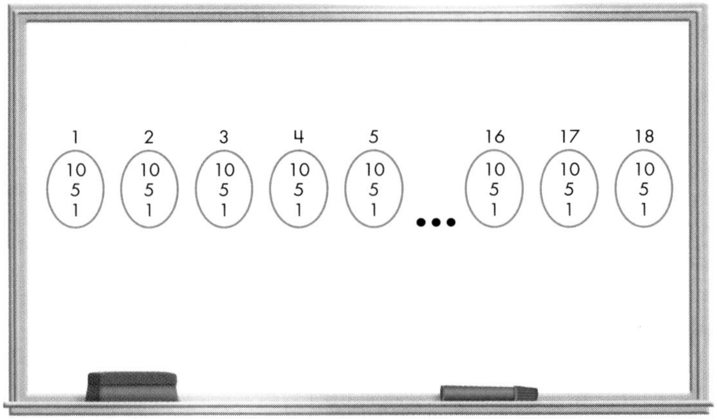

How many children are on each team? Are there any children left? *(16 on each team, with 2 children left)*

What's the answer to 290 ÷ 18? What's the answer to Renaldo's story problem?

Discuss the difference between the answer to the problem in mathematical notation and the answer to the story problem. In written notation, students can write 290 ÷ 18 = 16 R2, indicating that when 290 is divided into 18 equal groups, there are 18 groups with 16 in each group plus 2 ones. Another way of thinking of this is that 290 = $(18 \times 16) + 2$. However, when answering a problem in a context, it is often not sufficient to give the answer in terms of equal groups plus a remainder. The remainder must be considered in context. In this case, the answer 16 R2 suggests that there are 16 children on each team, with

2 children not playing. However, it is more likely that all children are placed on a team; in that case, one solution would be to have 16 teams of 16 and 2 teams of 17.

On *Student Activity Book* pages 59 and 60, students work on cluster problems like the one they just solved. They use the cluster problems they have solved mentally to help them solve the final problem. Continue to encourage students to use a story context and representation to help them keep track of what part of the problem they have solved.

The class will discuss Problem 4 later in the session, so tell students that they should be sure to complete it. They will have time during the Math Workshop in the next 2 sessions to complete all the problems.

ONGOING ASSESSMENT: Observing Students at Work

Students solve division problems by using related problems.

- **Can students use the set of cluster problems to find the answer to the final problem?**
- **Do students make use of a story context and/or representations to help them keep track of what part of the problem they have solved?**

DIFFERENTIATION: Supporting the Range of Learners

Intervention If students are not able to use the relationships between the cluster problems and the final problem, help them focus first on one of the smaller problems.

- If this were the first step to solve the final problem, what else would you still need to do? Do any of the other problems here help you? What is a story context you can use?

Talk through a story context with students, and ask them to draw a representation of what is happening in the story.

Extension Some students may have their own efficient methods for solving the problem and may not find the given "path" to the solution useful. For example, they may always multiply to solve a division problem: 20 × 15 = 300, not 300 ÷ 15 = 20. However, they can benefit from thinking through a different method in order to expand their understanding of division and to enhance their flexibility in solving problems. These students, as well as students who finish early, should be encouraged to make up their own cluster problems.

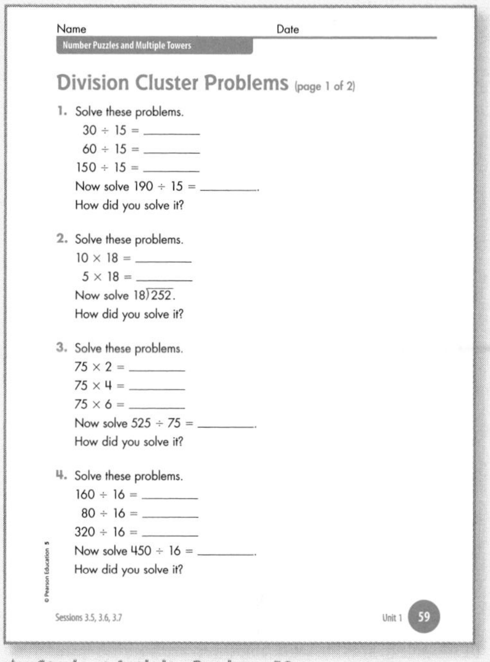

Name Date

Number Puzzles and Multiple Towers

Division Cluster Problems (page 1 of 2)

1. Solve these problems.
   30 ÷ 15 = ______
   60 ÷ 15 = ______
   150 ÷ 15 = ______
   Now solve 190 ÷ 15 = ______.
   How did you solve it?

2. Solve these problems.
   10 × 18 = ______
   5 × 18 = ______
   Now solve $18\overline{)252}$.
   How did you solve it?

3. Solve these problems.
   75 × 2 = ______
   75 × 4 = ______
   75 × 6 = ______
   Now solve 525 ÷ 75 = ______.
   How did you solve it?

4. Solve these problems.
   160 ÷ 16 = ______
   80 ÷ 16 = ______
   320 ÷ 16 = ______
   Now solve 450 ÷ 16 = ______.
   How did you solve it?

© Pearson Education 5

Sessions 3.5, 3.6, 3.7 Unit 1 59

▲ Student Activity Book, p. 59

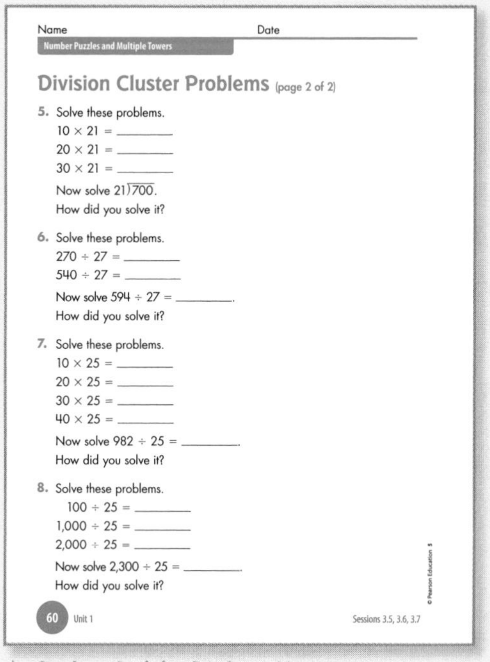

Name Date

Number Puzzles and Multiple Towers

Division Cluster Problems (page 2 of 2)

5. Solve these problems.
   10 × 21 = ______
   20 × 21 = ______
   30 × 21 = ______
   Now solve $21\overline{)700}$.
   How did you solve it?

6. Solve these problems.
   270 ÷ 27 = ______
   540 ÷ 27 = ______
   Now solve 594 ÷ 27 = ______.
   How did you solve it?

7. Solve these problems.
   10 × 25 = ______
   20 × 25 = ______
   30 × 25 = ______
   40 × 25 = ______
   Now solve 982 ÷ 25 = ______.
   How did you solve it?

8. Solve these problems.
   100 ÷ 25 = ______
   1,000 ÷ 25 = ______
   2,000 ÷ 25 = ______
   Now solve 2,300 ÷ 25 = ______.
   How did you solve it?

© Pearson Education 5

60 Unit 1 Sessions 3.5, 3.6, 3.7

▲ Student Activity Book, p. 60

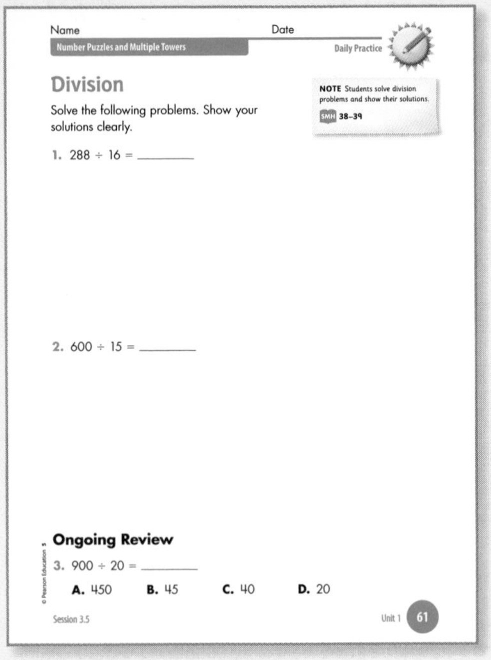

Name Date

Number Puzzles and Multiple Towers — Daily Practice

Division

Solve the following problems. Show your solutions clearly.

NOTE Students solve division problems and show their solutions.

SMH 38–39

1. 288 ÷ 16 = ______

2. 600 ÷ 15 = ______

Ongoing Review

3. 900 ÷ 20 = ______

A. 450 B. 45 C. 40 D. 20

© Pearson Education 5

Session 3.5 Unit 1 61

▲ Student Activity Book, p. 61

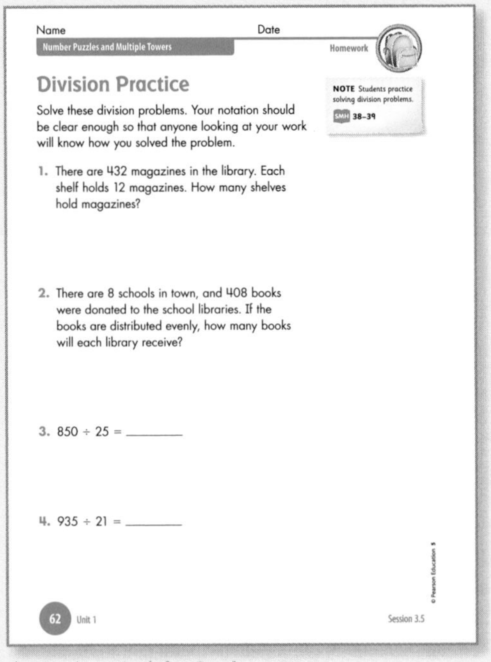

Name Date

Number Puzzles and Multiple Towers — Homework

Division Practice

Solve these division problems. Your notation should be clear enough so that anyone looking at your work will know how you solved the problem.

NOTE Students practice solving division problems.

SMH 38–39

1. There are 432 magazines in the library. Each shelf holds 12 magazines. How many shelves hold magazines?

2. There are 8 schools in town, and 408 books were donated to the school libraries. If the books are distributed evenly, how many books will each library receive?

3. 850 ÷ 25 = ______

4. 935 ÷ 21 = ______

© Pearson Education 5

62 Unit 1 Session 3.5

▲ Student Activity Book, p. 62

2 DISCUSSION

Breaking Up Numbers

10 MIN CLASS

Math Focus Points for Discussion

- Solving division problems with 2-digit divisors
- Solving a division problem by breaking the dividend into parts

On the board, write Problem 4 from *Student Activity Book* page 59. Ask students to give you the answers, and record them on the board.

Ask students how the cluster problems helped them solve the final problem. Call on a few students to explain their thinking.

Students might say:

"To figure out 450 ÷ 16, first I added 320 and 80 and got 400. So that's twenty-five 16s so far, and I have another 50 to divide. I divided 50 ÷ 16, which is 3 with 2 left, so the answer is 28 R2."

"I used 160 ÷ 16 = 10 to help me solve 80 ÷ 16. Since 80 is half of 160, the quotient is half of 10, or 5. Since 320 is double 160, the quotient of 320 ÷ 16 is 10 doubled, or 20."

3 SESSION FOLLOW-UP

Daily Practice and Homework

Daily Practice: For reinforcement of this unit's content, have students complete *Student Activity Book* page 61.

Homework: Students solve division problems with story contexts on *Student Activity Book* page 62.

Student Math Handbook: Students and families may use *Student Math Handbook* page 35 for reference and review. See pages 190–197 in the back of this unit.

SESSION 3.6

Practicing Division Strategies

Math Focus Points

- Solving division problems with 2-digit divisors
- Comparing division problems to determine which quotient is greater

Today's Plan		Materials
1 ACTIVITY **Introducing *Division Compare***	10 MIN · CLASS · PAIRS	• M44–M45 (from Session 2.3)*; M52 (as needed)*
2 MATH WORKSHOP **Practicing Division Strategies** 2A Division Cluster Problems 2B *Division Compare* 2C Solving Division Problems	40 MIN	2A • *Student Activity Book,* pp. 59–60 (from Session 3.5) 2B • *Student Activity Book,* pp. 63–64 • *Student Math Handbook,* p. G6 • M44–45; M48–M50 (optional); M52; M53 2C • *Student Activity Book,* pp. 65–66
3 DISCUSSION **Reasoning about *Division Compare***	10 MIN · CLASS	• *Student Activity Book,* p. 64
4 SESSION FOLLOW-UP **Daily Practice and Homework**		• *Student Activity Book,* pp. 67–69 • *Student Math Handbook,* pp. 38–39; G6

*See *Materials to Prepare,* p. 113.

Ten-Minute Math

Number Puzzles Play the variation "Random Clues." Combine all the clues from the three sets (A, B, C) of *Number Puzzle* Clue Cards (M35–M43). Draw the first three cards from the deck. This variation has greater potential for resulting in "inconclusive," solutions, providing students more opportunities to analyze, discuss, and modify clues. Students work with a partner and find the number(s) that fit the set of three clues. If the solution is "inconclusive," students modify the clues. Have students share their strategies for solving the puzzle.

Math Note

➊ **Which Number Is the Dividend?** Watch for students' misconceptions that the larger number in a division problem must always be the dividend. Although they are not solving problems like this now, they should know that a division problem such as 10 ÷ 24 has meaning. For example, they know that 10 pizzas can be shared by 24 fifth graders.

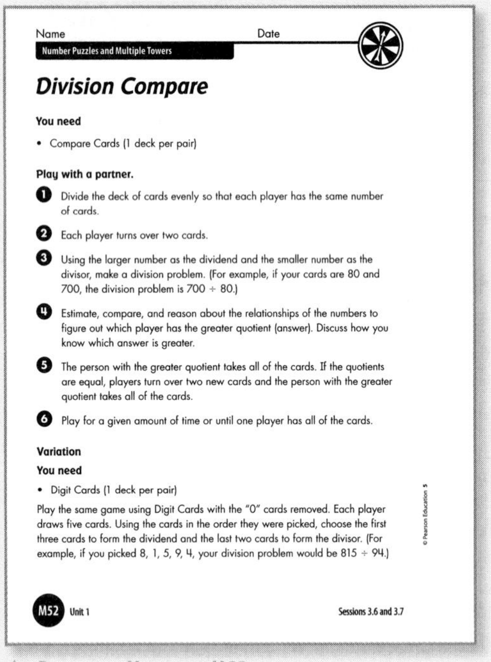

Name Date

Number Puzzles and Multiple Towers

Division Compare

You need

- Compare Cards (1 deck per pair)

Play with a partner.

1. Divide the deck of cards evenly so that each player has the same number of cards.
2. Each player turns over two cards.
3. Using the larger number as the dividend and the smaller number as the divisor, make a division problem. (For example, if your cards are 80 and 700, the division problem is 700 ÷ 80.)
4. Estimate, compare, and reason about the relationships of the numbers to figure out which player has the greater quotient (answer). Discuss how you know which answer is greater.
5. The person with the greater quotient takes all of the cards. If the quotients are equal, players turn over two new cards and the person with the greater quotient takes all of the cards.
6. Play for a given amount of time or until one player has all of the cards.

Variation

You need

- Digit Cards (1 deck per pair)

Play the same game using Digit Cards with the "0" cards removed. Each player draws five cards. Using the cards in the order they were picked, choose the first three cards to form the dividend and the last two cards to form the divisor. (For example, if you picked 8, 1, 5, 9, 4, your division problem would be 815 ÷ 94.)

© Pearson Education 5

M52 Unit 1 Sessions 3.6 and 3.7

▲ Resource Masters, M52

ACTIVITY

1 Introducing *Division Compare*

10 MIN CLASS PAIRS

Inform students that they are going to learn a new game, *Division Compare,* which is similar to the game *Multiplication Compare* that they played in the previous Investigation.

When you play *Division Compare,* you pick two cards, just as you did in *Multiplication Compare.* The larger number will be the dividend, the number you're dividing, and the smaller number will be the divisor, the number you're dividing by.➊

Turn over two Compare Cards (M44–M45) to be your hand, and two cards to be the class's hand. For example, the teacher draws 400 and 70, and the class draws 10 and 80. Ask students what the division problems would be for each hand, and write these on the board.

In Division Compare, *students compare two division problems.*

Students turn to a neighbor and discuss which problem has the greater quotient. Let them know that they do not have to find the exact answer to compare the quotients. Give students a minute or two to do this, and then ask for someone to explain their thinking.

Students might say:

"80 ÷ 10 is 8; that's easy, because I know 8 × 10 = 80. 400 ÷ 70 is between 5 and 6, so 80 ÷ 10 must be greater."

"I agree. 80 ÷ 10 is 8. So I looked at the other problem. I know that 8 × 70 is 560. 400 is less than 560, so the answer to 400 ÷ 70 has to be less than 8."

It may be necessary to play another round with the whole class, both to help students understand what division problems to make, and to provide more examples of how to think through to the answer. Discuss with students how to use *Student Activity Book* page 63 to record each round. Explain that the player with the larger quotient takes all the cards for that round.

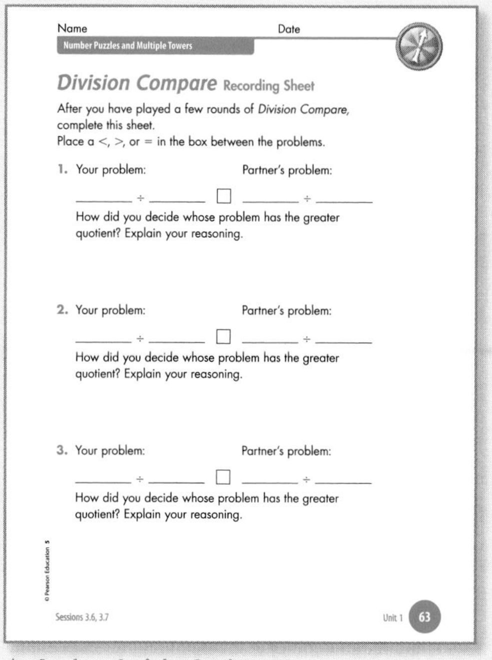

Name Date

Number Puzzles and Multiple Towers

Division Compare Recording Sheet

After you have played a few rounds of *Division Compare*, complete this sheet.

Place a <, >, or = in the box between the problems.

1. Your problem: Partner's problem:

_____ ÷ _____ ☐ _____ ÷ _____

How did you decide whose problem has the greater quotient? Explain your reasoning.

2. Your problem: Partner's problem:

_____ ÷ _____ ☐ _____ ÷ _____

How did you decide whose problem has the greater quotient? Explain your reasoning.

3. Your problem: Partner's problem:

_____ ÷ _____ ☐ _____ ÷ _____

How did you decide whose problem has the greater quotient? Explain your reasoning.

© Pearson Education 5

Sessions 3.6, 3.7 Unit 1 63

▲ Student Activity Book, p. 63; Resource Masters, M53; Transparencies, T24

MATH WORKSHOP

40 MIN

2 Practicing Division Strategies

Students begin a Math Workshop that continues in the next session. Each of the activities provides division practice for students.

Tell students that the discussion at the end of today's session will focus on the *Division Compare* problems on *Student Activity Book* page 64. They should spend most of their time today playing *Division Compare* and completing that page as well as the recording sheet, page 63. They will have additional time in the next session to complete the division problems on *Student Activity Book* pages 65 and 66.

2A Division Cluster Problems

INDIVIDUALS

Students continue their work on division cluster problems on *Student Activity Book* pages 59–60. See Session 3.5, pages 137–139, for complete details.

2B *Division Compare*

PAIRS

Students play *Division Compare* with a partner. Tell students to play a few rounds to learn the rules of the game as given on page G6 of the *Student Math Handbook*. (You may use M52 to provide extra copies.) Then students should complete *Student Activity Book* pages 63–64. See pages 142–143 for more details.

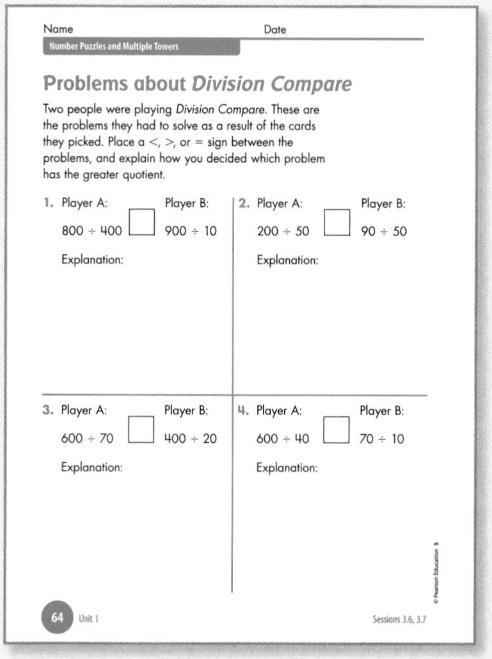

Name Date

Number Puzzles and Multiple Towers

Problems about *Division Compare*

Two people were playing *Division Compare*. These are the problems they had to solve as a result of the cards they picked. Place a <, >, or = sign between the problems, and explain how you decided which problem has the greater quotient.

1. Player A: 800 ÷ 400 ☐ Player B: 900 ÷ 10
Explanation:

2. Player A: 200 ÷ 50 ☐ Player B: 90 ÷ 50
Explanation:

3. Player A: 600 ÷ 70 ☐ Player B: 400 ÷ 20
Explanation:

4. Player A: 600 ÷ 40 ☐ Player B: 70 ÷ 10
Explanation:

© Pearson Education 5

64 Unit 1 Sessions 3.6, 3.7

▲ Student Activity Book, p. 64

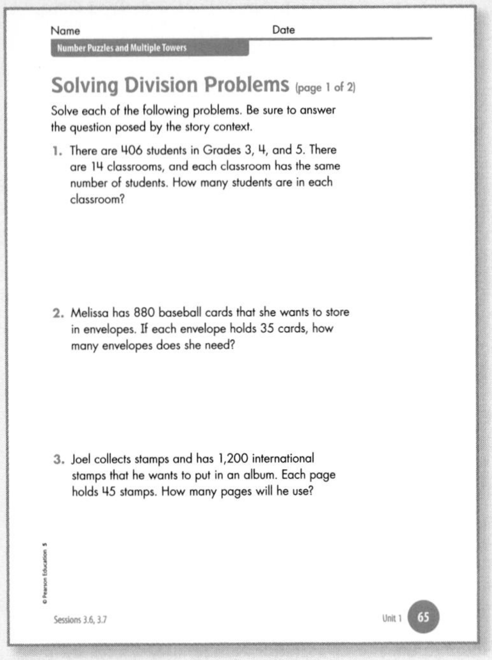

Name Date

Number Puzzles and Multiple Towers

Solving Division Problems (page 1 of 2)

Solve each of the following problems. Be sure to answer the question posed by the story context.

1. There are 406 students in Grades 3, 4, and 5. There are 14 classrooms, and each classroom has the same number of students. How many students are in each classroom?

2. Melissa has 880 baseball cards that she wants to store in envelopes. If each envelope holds 35 cards, how many envelopes does she need?

3. Joel collects stamps and has 1,200 international stamps that he wants to put in an album. Each page holds 45 stamps. How many pages will he use?

© Pearson Education 5

Sessions 3.6, 3.7 Unit 1 65

▲ Student Activity Book, p. 65

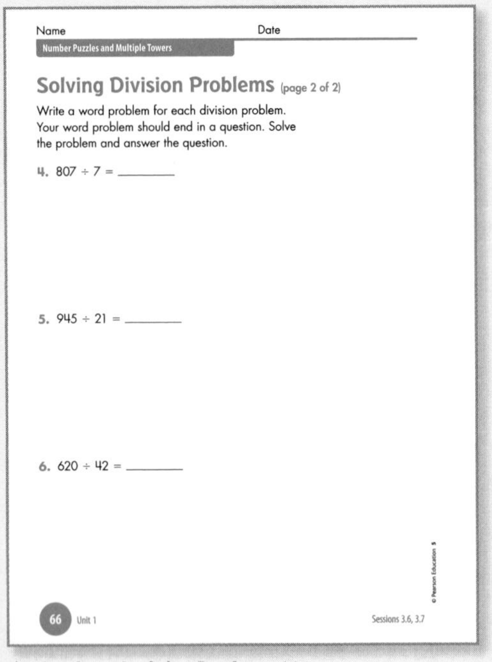

Name Date

Number Puzzles and Multiple Towers

Solving Division Problems (page 2 of 2)

Write a word problem for each division problem. Your word problem should end in a question. Solve the problem and answer the question.

4. $807 \div 7 =$ ______

5. $945 \div 21 =$ ______

6. $620 \div 42 =$ ______

© Pearson Education 5

66 Unit 1 Sessions 3.6, 3.7

▲ Student Activity Book, p. 66

ONGOING ASSESSMENT: Observing Students at Work

In Activity 2B, students compare quotients of two division problems to decide which is greater.

- **How do students decide which quotient is greater?** Do they solve both problems and compare the answers? Do they use estimates based on multiplication and division relationships they know?❷

DIFFERENTIATION: Supporting the Range of Learners

Intervention Students who are having difficulty estimating quotients to problems such as $900 \div 60$ can use fewer Compare Cards so that they are solving only problems with a 2-digit dividend and a 1-digit divisor. Remove the 3-digit Compare Cards. Sort the remaining cards into 2-digit cards and 1-digit cards, and have each player draw one card from each pile.

Extension Students who can easily compare quotients with multiples of 10 should play the variation of the game, which involves using Digit Cards (M48–M50). If students play this variation, they need to remove the "0" cards before playing.

2C Solving Division Problems

INDIVIDUALS

Students solve division word problems on *Student Activity Book* pages 65–66. For this session, students should spend the least amount of time on this activity.

ONGOING ASSESSMENT: Observing Students at Work

Students solve division word problems.

- **What strategies are students using to solve division problems?** Are they using these strategies efficiently (e.g., using large chunks of the problem)?
- **Are students able to notate their solutions so that they can keep track of their work, know what their answer is, and communicate their reasoning to others?**
- **Are students able to make sense of the story context and use the remainder to answer the question posed by the story context?**
- **Are students able to write a word problem for a division expression?**

3 DISCUSSION

Reasoning about *Division Compare*

10 MIN CLASS

Math Focus Points for Discussion

- Comparing division problems to determine which quotient is greater

As time permits, go over each of the 4 problems on *Student Activity Book* page 64. Ask students to explain how they decided which player would win.

Some sample strategies and explanations are given below.

- Solving the division problems and directly comparing quotients

Problem 1

Player A: Player B:

$800 \div 400 \boxed{<} 900 \div 10$

$800 \div 400 = 2$ and $900 \div 10 = 9$; $2 < 9$

Problem 4

Player A: Player B:

$600 \div 40 = \boxed{>} 70 \div 10$

$600 \div 40 = 15$ and $70 \div 10 = 7$; $15 \boxed{>} 7$

- Reasoning about the relationships between dividend and divisor

Problem 2

Player A: Player B:

$200 \div 50 \boxed{>} 90 \div 50$

The divisor in both problems is 50, Player A would win because 200 is greater than 90. If you think of groups of 50, there are more groups of 50 in 200 than there are in 90.

Problem 3

Player A: Player B:

$600 \div 70 \boxed{<} 400 \div 20$

The answer for $600 \div 70$ is between 8 and 9 (because 8×70 is 560 and 9×70 is 630), and $400 \div 20$ is 20, so Player B would win.

Math Note

2 **Using Multiplication and Division Relationships** Suppose that students are comparing $700 \div 70$ and $800 \div 90$. They might say that they know that $700 \div 70$ is 10 and that 90×9 is 810. Because 810 is larger than 800, $800 \div 90$ will be less than 9. So $700 \div 70$ will have a larger quotient.

4 SESSION FOLLOW-UP

Daily Practice and Homework

Daily Practice: For ongoing review, have students complete *Student Activity Book* page 67.

Homework: Students solve multiplication and division problems on *Student Activity Book* pages 69–70.

Student Math Handbook: Students and families may use *Student Math Handbook* pages 38–39 and G6 for reference and review. See pages 190–197 in the back of this unit.

Name Date

Number Puzzles and Multiple Towers — Homework

Practicing Multiplication and Division (page 1 of 2)

NOTE Students continue to practice solving multiplication and division problems. SMH 14

Solve the following problems. Make sure that anyone looking at your work can tell how you solved the problem.

1. There are 64 teams at the basketball tournament. Each team has 12 players. How many players are at the basketball tournament?

2. Write a word problem for $35\overline{)490}$ and solve it.

3. Michael has 275 baseball cards that he wants to store in envelopes. If he puts 25 cards in each envelope, how many envelopes does he need?

© Pearson Education 5

Session 3.6 Unit 1 69

▲ Student Activity Book, p. 69 WRITING

Name Date

Number Puzzles and Multiple Towers — Homework

Practicing Multiplication and Division (page 2 of 2)

4. There are 118 rows in the auditorium. If 29 people sit in each row, how many people are in the auditorium?

5. Mrs. Garcia teaches fifth grade. She has 720 pattern blocks and 24 plastic containers. If she wants to divide the pattern blocks evenly, how many will she put in each container?

6. Write a word problem for 13×42 and show your solution.

© Pearson Education 5

70 Unit 1 Session 3.6

▲ Student Activity Book, p. 70 WRITING

Name Date

Number Puzzles and Multiple Towers — Daily Practice

Story Problems: Selling Fruit

NOTE Students practice solving addition and subtraction problems in a story problem context. SMH 8–9, 10–13

1. On Monday, a grocery store received a shipment of 1,000 apples. The apples were quite tasty, and the store sold 346 of them that day. How many apples were left to sell?

2. On Wednesday, the store received a shipment of 1,200 oranges. The store sold 263 oranges that day. How many oranges were left to sell?

3. On Saturday, the store received a shipment of 2,000 mangos. The store sold 415 mangos on Saturday and 680 mangos on Sunday.
   a. How many mangos did the store sell on the weekend (Saturday and Sunday)?
   b. How many mangos were left to sell?

© Pearson Education 5

Session 3.6 Unit 1 67

▲ Student Activity Book, p. 67

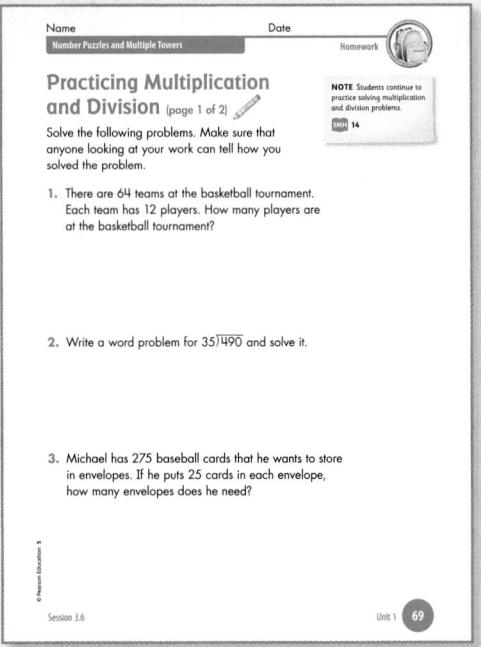

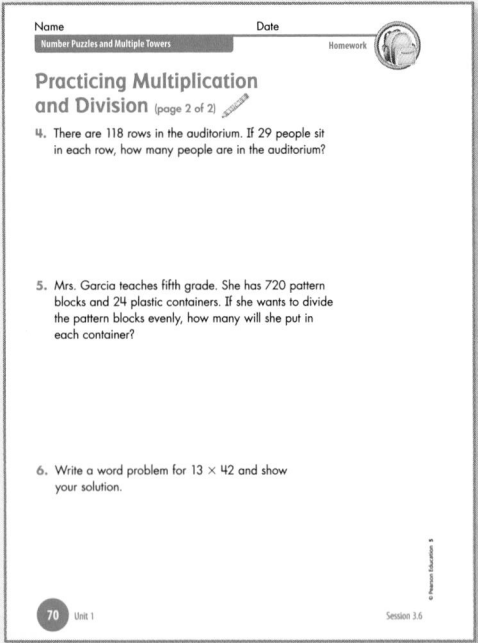

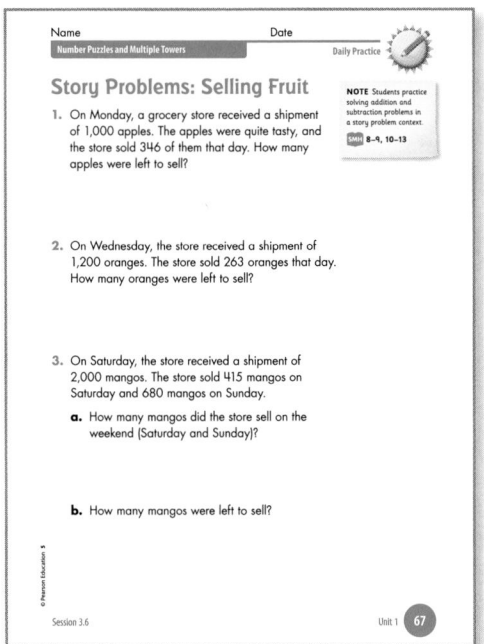

Practicing Division Strategies, *continued*

Math Focus Points

- Solving division problems with 2-digit divisors
- Making sense of remainders in terms of problem contexts
- Describing and comparing strategies used to solve division problems

Today's Plan		Materials
1 MATH WORKSHOP **Practicing Division Strategies** 1A Division Cluster Problems 1B *Division Compare* 1C Solving Division Problems	45 MIN	1A • *Student Activity Book,* pp. 59–60 (from Session 3.5) 1B • Materials from Session 3.6, Activity 2B 1C • *Student Activity Book,* pp. 65–66 (from Session 3.6)
2 DISCUSSION **Division Strategies**	15 MIN CLASS	• *Student Activity Book,* p. 65 (from Session 3.6)
3 SESSION FOLLOW-UP **Daily Practice**		• *Student Activity Book,* p. 71 • *Student Math Handbook,* pp. 38–39

*See *Materials to Prepare,* p. 113.

Ten-Minute Math

Number Puzzles Play the variation "Random Clues." Combine all the clues from the three sets (A, B, C) of *Number Puzzle* Clue Cards (M35–M43). Draw the first three cards from the deck. This variation has greater potential for resulting in "inconclusive," solutions, providing students more opportunities to analyze, discuss, and modify clues. Students work with a partner and find the number(s) that fit the set of three clues. If the solution is "inconclusive," students modify the clues. Have students share their strategies for solving the puzzle.

Teaching Note

① **Expectations** Remember that students will do a significant amount of division work in *How Many People? How Many Teams?* The expectation in this unit is that students are increasing their fluency with solving division problems.

1 MATH WORKSHOP

45 MIN

Practicing Division Strategies, *continued*

Students complete the Math Workshop that began in the last session. Each of the activities provides division practice for students.①

Tell students that the discussion at the end of today's session will focus on Problem 2 on *Student Activity Book* page 65. Most students should spend most of their time completing that page and the rest of the division problems on *Student Activity Book* page 66. However, students who are still struggling with how to get started solving a division problem should concentrate on finishing the division cluster problems on *Student Activity Book* pages 59–60. Students who complete all of the division problems and division cluster problems can play more rounds of *Division Compare.*

While students are working, identify three or four students who will share their solution to Problem 2 on page 65 at the end of this session. Choose examples that address the needs of your students, such as the following:

- If students are having problems notating their solutions, pick examples that will help students think about the characteristics of clear notation.
- If students are using only the 10th multiple to solve the problem, pick an example of a solution that uses a larger multiple.
- If most students are using the same strategy, pick an example of a solution that uses a different strategy.

1A Division Cluster Problems

INDIVIDUALS

Students continue their work on division cluster problems on *Student Activity Book* pages 59–60. See Session 3.5, pages 137–139 for complete details.

1B *Division Compare*

PAIRS

Students play *Division Compare* with a partner when they have completed the other two activities. See Session 3.6, pages 142–143 for complete details.

1C Solving Division Problems

INDIVIDUALS

Students solve division word problems on *Student Activity Book* pages 65–66. For this session, students should spend most of their time on this activity. This is a good time to talk to individual students as they are working.

What do you already know about these numbers that might help you? What smaller parts of the problem can you solve easily?

What part of the problem have you solved? What do you still need to solve? Would drawing a representation help you keep track?

What's the answer to the division problem? If there's a remainder, how do you use the remainder to answer the question posed in the story context?

If someone else looked at the steps of your solution, would they be able to follow your thinking? If not, how can you make it clearer?

See Session 3.6, page 144 for more information.

2 DISCUSSION
Division Strategies

15 MIN CLASS

Math Focus Points for Discussion

- Describing and comparing strategies used to solve division problems

Write Problem 2 from *Student Activity Book* page 65 on the board.

Melissa has 880 baseball cards she wants to store in envelopes. If each envelope holds 35 cards, how many envelopes does she need?

Have the students you identified earlier come to the board and explain their solutions to the class. Tell class members that their job is to listen carefully, try to understand the strategy being explained, compare it with their own solution, and ask questions when they need a clearer explanation.

As each student shares, establish that the answer to the division problem $880 \div 35$ is 25 R5, but that the answer in the story context is 26 envelopes.

Because you chose these students for specific reasons (e.g., clear notation, using a strategy efficiently, using a different strategy, and so on), highlight these reasons for the class by asking questions such as the following:

Look at how [Rachel] solved this problem. Is her notation clear? What makes it easy to see how she solved each step of the problem?

How did [Tyler] use multiples of 10 to solve this problem? What relationships did he use?

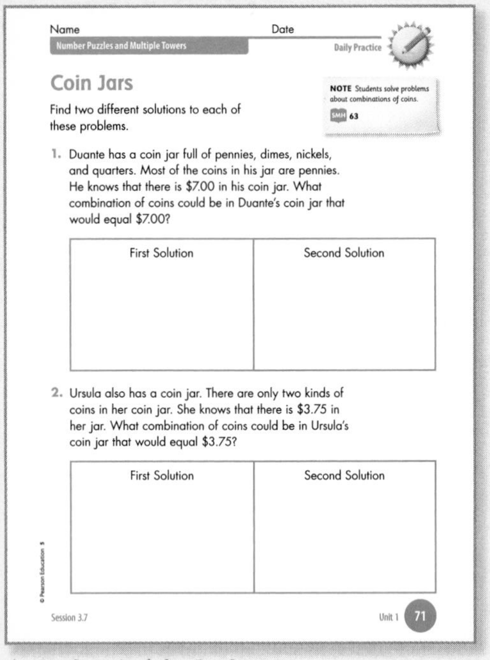

Name Date

Number Puzzles and Multiple Towers

Daily Practice

Coin Jars

Find two different solutions to each of these problems.

NOTE Students solve problems about combinations of coins.

SMH 63

1. Duante has a coin jar full of pennies, dimes, nickels, and quarters. Most of the coins in his jar are pennies. He knows that there is $7.00 in his coin jar. What combination of coins could be in Duante's coin jar that would equal $7.00?

First Solution	Second Solution

2. Ursula also has a coin jar. There are only two kinds of coins in her coin jar. She knows that there is $3.75 in her jar. What combination of coins could be in Ursula's coin jar that would equal $3.75?

First Solution	Second Solution

© Pearson Education 5

Session 3.7 Unit 1 71

▲ Student Activity Book, p. 71

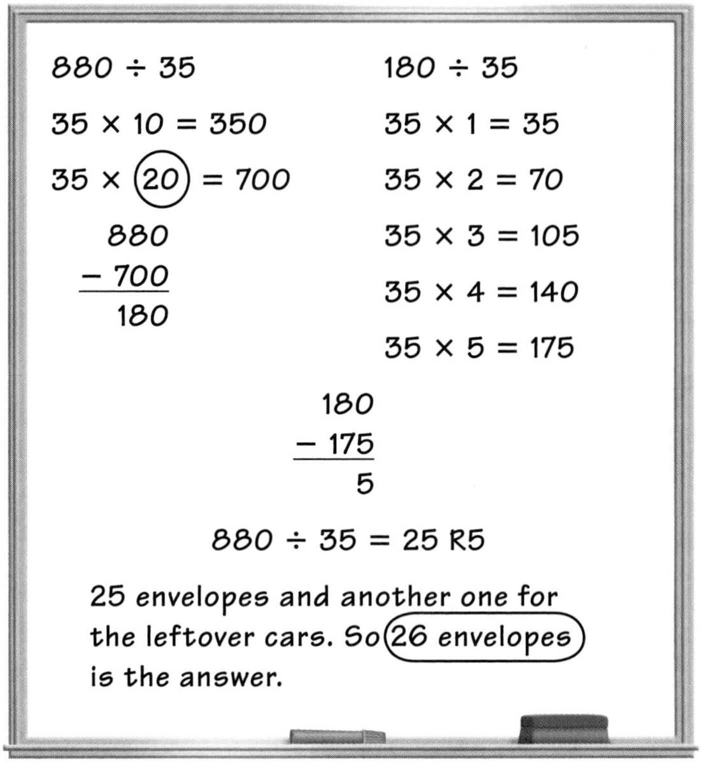

Have a brief discussion with students about multiplication and division. This is a good chance for students to reflect on what they learned during this unit and what they need to practice.

What multiplication strategies do you most often use? Do you feel confident about solving a problem such as 47 × 63? What helps you solve this kind of problem?

What about division? Do you feel confident solving a problem such as 126 ÷ 14? What helps you solve that kind of problem?

What is still difficult for you in multiplication and division? What work do you need to do to become more efficient at solving multiplication problems? What about division problems?

3 SESSION FOLLOW-UP
Daily Practice

Daily Practice: For ongoing review, have students complete *Student Activity Book* page 71.

Student Math Handbook: Students and families may use *Student Math Handbook* pages 38–39 for reference and review. See pages 190–197 in the back of this unit.

SESSION 3.8

End-of-Unit Assessment

Math Focus Points

- Solving 2-digit by 2-digit multiplication problems
- Finding all the factors of a number
- Solving a division problem by breaking the dividend into parts

Today's Plan	Materials
1 ASSESSMENT ACTIVITY **End-of-Unit Assessment** 60 MIN INDIVIDUALS	• M54–M55*
2 SESSION FOLLOW-UP **Daily Practice**	• *Student Activity Book,* p. 72 • *Student Math Handbook,* p. 15

*See *Materials to Prepare,* p. 113.

Ten-Minute Math

Number Puzzles Draw one card from each set (A, B, C) of *Number Puzzle* Clue Cards (M35–M43) and read them aloud. Students work with a partner and find the number(s) that fit the set of three clues. If the solution is "inconclusive," students modify the clues. Have students share their strategies for solving the puzzle.

Professional Development

1 **Teacher Note:** End-of-Unit Assessment, p. 174

2 **Assessment in This Unit,** p. 14

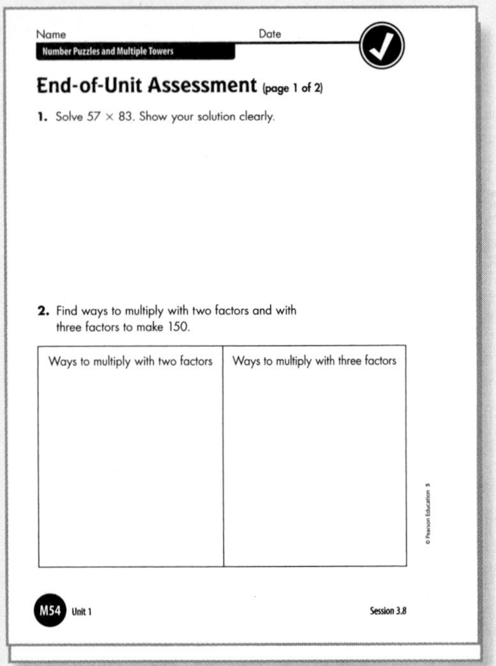
Name Date
Number Puzzles and Multiple Towers

End-of-Unit Assessment (page 1 of 2)

1. Solve 57 × 83. Show your solution clearly.

2. Find ways to multiply with two factors and with three factors to make 150.

Ways to multiply with two factors	Ways to multiply with three factors

© Pearson Education 5

M54 Unit 1 Session 3.8

▲ Resource Masters, M54–M55 PORTFOLIO

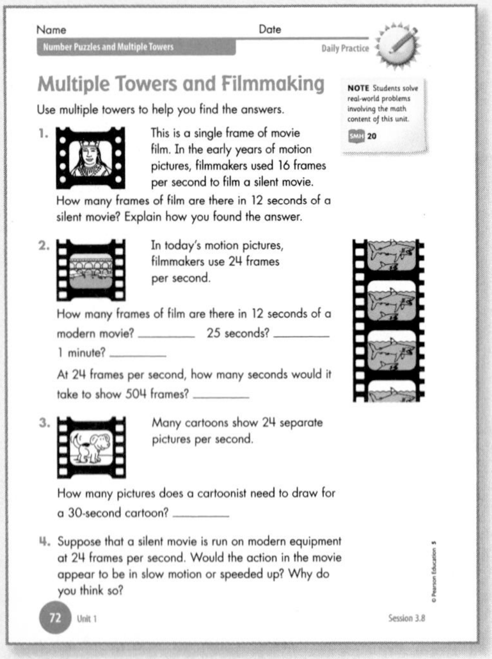
Name Date
Number Puzzles and Multiple Towers Daily Practice

Multiple Towers and Filmmaking

NOTE Students solve real-world problems involving the math content of this unit.

SMH 20

Use multiple towers to help you find the answers.

1. This is a single frame of movie film. In the early years of motion pictures, filmmakers used 16 frames per second to film a silent movie.
How many frames of film are there in 12 seconds of a silent movie? Explain how you found the answer.

2. In today's motion pictures, filmmakers use 24 frames per second.
How many frames of film are there in 12 seconds of a modern movie? ______ 25 seconds? ______
1 minute? ______
At 24 frames per second, how many seconds would it take to show 504 frames? ______

3. Many cartoons show 24 separate pictures per second.
How many pictures does a cartoonist need to draw for a 30-second cartoon? ______

4. Suppose that a silent movie is run on modern equipment at 24 frames per second. Would the action in the movie appear to be in slow motion or speeded up? Why do you think so?

© Pearson Education 5

72 Unit 1 Session 3.8

▲ Student Activity Book, p. 72

1 ASSESSMENT ACTIVITY

End-of-Unit Assessment

60 MIN INDIVIDUALS

Students work individually on three problems to assess the benchmarks for this unit. You may find it helpful to familiarize yourself with this assessment and view samples of student work. 1 2

Problem 1 addresses Benchmark 2: Solve multiplication problems efficiently. Problem 2 addresses Benchmark 1: Find the factors of a number. Problem 3 addresses Benchmark 3: Solve division problems with 1-digit and 2-digit divisors.

ONGOING ASSESSMENT: Observing Students at Work

Students use their knowledge of multiplication and division to solve three problems on the end-of-unit assessment.

- **What strategies are students using?** Are they keeping track of all parts of the problem? Are they using clear and concise notation?
- **Can students find two or more numbers that, when multiplied together, equal 150?**

DIFFERENTIATION: Supporting the Range of Learners

Intervention If students are having a difficult time solving the first problem, help them use a context to think through the problem. If the numbers are too large, choose smaller ones and see whether students can find a way to solve the problem.

If a student can solve a problem but cannot record the strategy used, ask the student to explain the method to you, write it down for the student, and make a note to work with this student on recording clearly.

2 SESSION FOLLOW-UP

Daily Practice

Daily Practice: For enrichment, have students complete *Student Activity Book* page 72. This page provides real-world problems involving the math content of this unit.

Student Math Handbook: Students and families may use *Student Math Handbook* page 15 for reference and review. See pages 190–197 in the back of this unit.

Professional Development

UNIT 1

Number Puzzles and Multiple Towers

Teacher Notes

In Part 6 of *Implementing Investigations in Grade 5,* you will find a set of Teacher Notes that addresses topics and issues applicable to the curriculum as a whole rather than to specific curriculum units. They include the following:

Computational Fluency and Place Value

Computation Algorithms and Methods

Representations and Contexts for Mathematical Work

Foundations of Algebra in the Elementary Grades

Discussing Mathematical Ideas

Racial and Linguistic Diversity in the Classroom: Raising Questions About What Equity in the Math Classroom Means Today

Dialogue Boxes

Teacher Note

Using Mathematical Vocabulary

Students learn mathematical words the same way they learn other vocabulary—by hearing them used correctly and frequently. Use mathematical vocabulary as often as you can. Connect mathematical terms to more familiar words that students know and to the mathematical ideas and images with which students are working.

Consider the following classroom scene:

During class discussions of student work, the teacher introduces and reinforces mathematical vocabulary.

Cecilia and Talisha made a rectangle from 24 tiles to solve a number puzzle. The rectangle is 6 tiles across and 4 tiles down. Its *dimensions* are 4 by 6.

Talisha used the clue that it is a multiple of 6. Let's list the multiples of 6.

I hear some people saying 2 and 3 and some other people saying 6, 12, 18, 24, 30. Which are *factors* of 6, and which are *multiples* of 6?

Students' use of correct terms does not always match their understanding of a mathematical idea. For example, although a student may understand what a factor is, he or she may mistakenly call it a multiple. In such a case, remind students of the correct term and connect it to what they know.

Do you mean that 24 is a *factor* of 6, that it fits into 6 evenly? Or do you mean that it's a *multiple* of 6, that you can count by 6 and land exactly on 24?

Also refer students to the Math Words and Ideas section of the *Student Math Handbook* when they are trying to remember a term.

Throughout Grade 5, mathematical vocabulary is reviewed from previous years (e.g., *factor* and *multiple*), and new terms are introduced. It is important that students express their ideas clearly and accurately, using whatever words are comfortable for them. However, as students gain familiarity with mathematical terms, expect them to use these terms appropriately.

Teacher Note

Finding Prime Factors

One way to categorize numbers is to examine their factors. From work in previous grades, students are familiar with *square numbers,* numbers that equal a whole number multiplied by itself. They are also familiar with *prime numbers,* numbers that have only two factors, 1 and the number itself. They know that some numbers, such as 36, have many factors, and others, such as 21, have few factors. They have had experience modeling factors and products by making rectangular arrays with tiles or on grid paper.

Most of students' past work has focused on finding pairs of factors for a given product. For example, ways to make 36 with two whole-number factors are: 1×36, 2×18, 3×12, 4×9, and 6×6. In this unit, students start with such familiar pairs of factors, and then build on these to find all the ways to multiply whole numbers to produce a certain product, using three, four, five, or more factors. For 36, the complete list is given below.

1×36	$2 \times 2 \times 9$	$2 \times 2 \times 3 \times 3$
2×18	$2 \times 3 \times 6$	
3×12	$3 \times 3 \times 4$	
4×9		
6×6		

When students work on *Quick Images* patterns, they find different multiplication combinations that represent the same array of figures, depending on how they picture the arrangements. For example, Problem 3 (shown in the next column) on *Student Activity Book* page 4, which shows 64 dots, can be described in a number of ways.

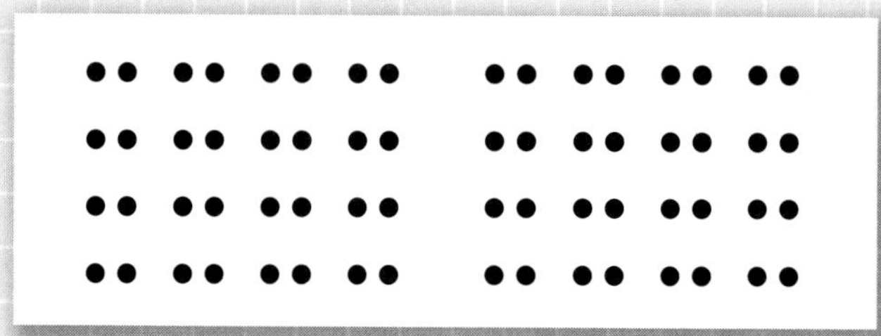

$(4 \times 2) \times 4 \times 2$
(4 groups of 2) $\times$ 4 rows $\times$ 2 sections

$8 \times 4 \times 2$
8 in each double column $\times$ 4 columns $\times$ 2 sections

16×4
16 in each row all the way across $\times$ 4 rows

8×8
8 in each double column $\times$ 8 double columns

Through visualizing representations such as this one and building on the pairs of factors they know, students generate as many multiplication combinations as they can (using whole numbers greater than 1) for a given product. See **Dialogue Box:** Multiplying with More Than Two Numbers, page 181.

Many fifth graders are fascinated with questions about factors. How many different ways can you find to multiply whole numbers to make 180? What is the longest combination you can find? Are there any other combinations that use the same number of factors? As students investigate these questions, they find that they can generate new combinations by "breaking up" ones they already know: if $2 \times 90 = 180$, then $2 \times (2 \times 45) = 180$. By working on such problems, students are calling on the associative property, a property they will examine formally in future years. It can be expressed in algebraic notation as $(a \times b) \times c = a \times (b \times c)$.

As they use one multiplication combination to generate others, students notice that at some point, they can no longer split up the numbers in their combinations into other factors. For example, while working with $2 \times 2 \times 3 \times 3 \times 5 = 180$, some students find that they cannot generate any multiplication combinations with more than five factors because 2, 3, and 5 are prime numbers. Prime factors cannot be split up into other factors (except by using 1). Some students also notice that, discounting different orders of the same numbers, there is only one "longest" multiplication combination for any number.

These students are beginning to notice an important property of whole numbers—that each whole number greater than 1 can be factored into a product of prime numbers in only one way. In other words, there is one and only one multiplication combination consisting of prime numbers for each whole number greater than 1. This principle, the Fundamental Theorem of Arithmetic, expresses the idea that every whole number can be uniquely named. The prime numbers in the combination for a given whole number are its prime factors, and the combination itself is called the *prime factorization* for that number. For example, the prime factorization of 30 is $2 \times 3 \times 5$, and the prime factors of 30 are 2, 3, and 5. The prime factorization of 90 is $2 \times 3 \times 3 \times 5$, and its prime factors are also 2, 3, and 5. (The prime factor 3 is used twice in the prime factorization of 90.)

In this unit, students are challenged to find the "longest way to multiply," using whole numbers. Fifth graders can find the longest combination for given products, but do not yet necessarily understand that there is one such unique combination for each whole number. It is not necessary to stress that some factors are prime numbers or to emphasize the term *prime factorization* for fifth-grade students.

Goals for the work on factors in this unit include the following:

1. Engagement and interest in noticing properties of number
2. Reasoning about number relationships (e.g., using a known multiplication combination, such as 3×60, to find equivalent combinations, such as $3 \times (20 \times 3)$)
3. Increasing flexibility in mental computation

This work also develops building blocks for future work with prime factors in computation with fractions and in algebra.

Teacher Note

Representing Multiplication with Arrays

Representing mathematical relationships is a key element for developing mathematical understanding. For multiplication, the rectangular array is a powerful mathematical representation because it highlights important relationships and provides a tool for solving problems. Arrays can be extended when students encounter new multiplication situations, such as multiplying with fractions or with algebraic expressions.

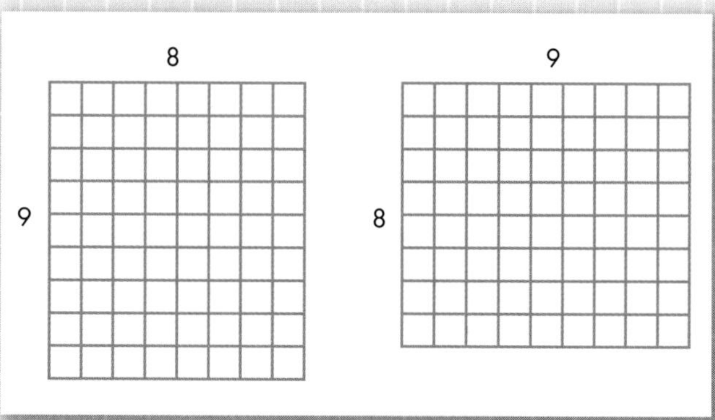

Why Arrays for Multiplication?

As students came to understand the operation of multiplication in Grades 3 and 4, they gradually moved away from thinking of multiplication only as repeated addition. They learned that multiplication has particular properties that distinguish it from addition. Although a number line or hundred chart can be used to show how multiplication can be viewed as adding equal groups, neither of these tools provides easy access to other important aspects of multiplication, such as commutativity or distributivity. The rectangular array provides a window into these properties, which are central to students' work in learning the multiplication combinations and in solving multidigit multiplication and division problems.

The rectangular array makes it clearer why the product of 9×8, for example, is equal to the product of 8×9. As shown in the diagram, the array with 9 rows of 8 in each row can be rotated to produce an array with 8 rows of 9 in each row. Both arrays have the same number of squares. A column in one array becomes a row in the other and vice versa.

Arrays are particularly useful for solving or visualizing how to solve multidigit multiplication problems. After students have worked with rectangular arrays for single-digit multiplication combinations and thoroughly understand how an array represents the factors and product, they can use arrays to solve harder problems. See **Teacher Note:** Visualizing Arrays, page 159, for more about helping students understand how arrays represent multiplication.

Consider the array for 28×25. It can be broken up in many ways, 3 of which are shown below.

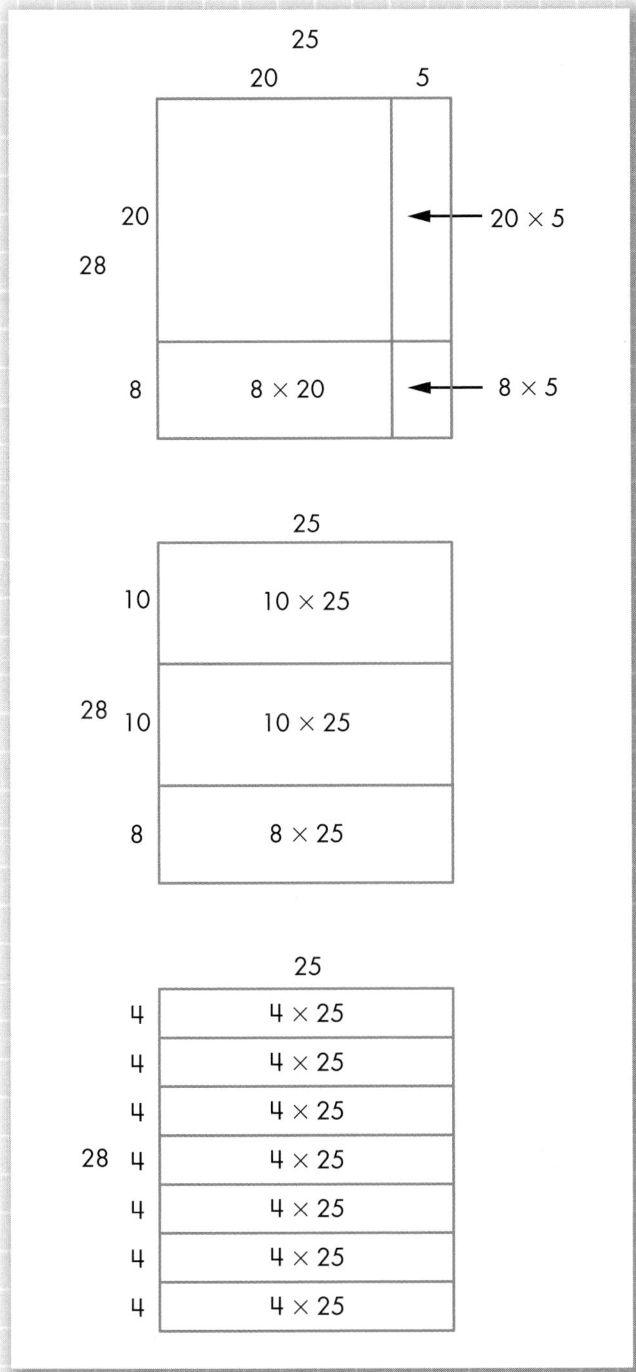

These representations show how multiplication expressions can be broken into a combination of simpler expressions using the distributive property, and how the sum of the products of those simpler expressions equals the product of the original problem.

Arrays also support students' learning about the relationship between multiplication and division. In division, we know the total amount (the area of the array) and one factor (one side of the array), and we are trying to find the other factor.

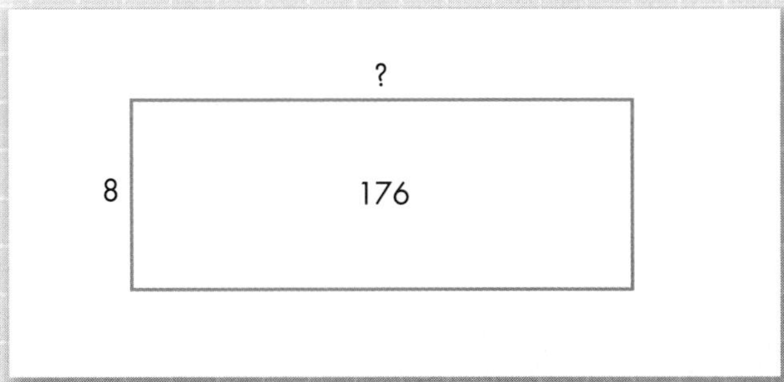

Students can think of "slicing off" pieces of the rectangle as they gradually figure out the other factor.

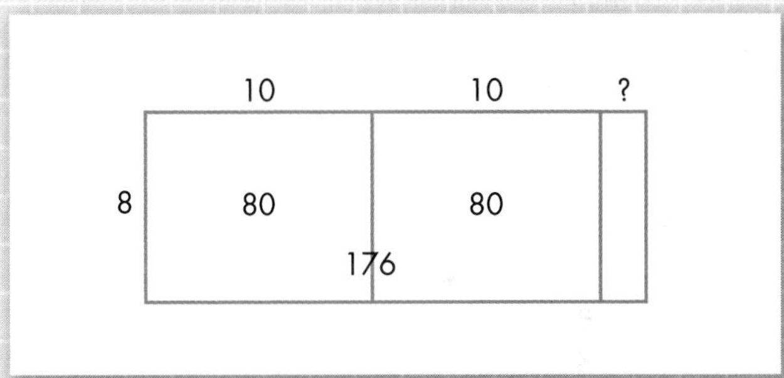

Finally, the use of the rectangular array can be extended in later grades as students work with multiplication of fractions and, later yet, multiplication of algebraic expressions.

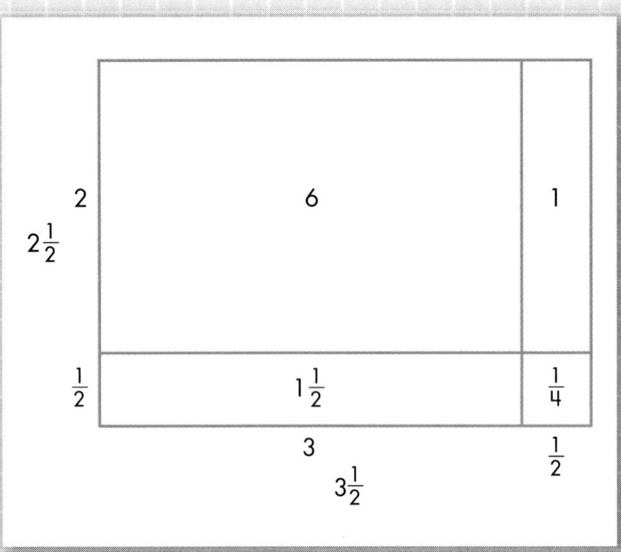

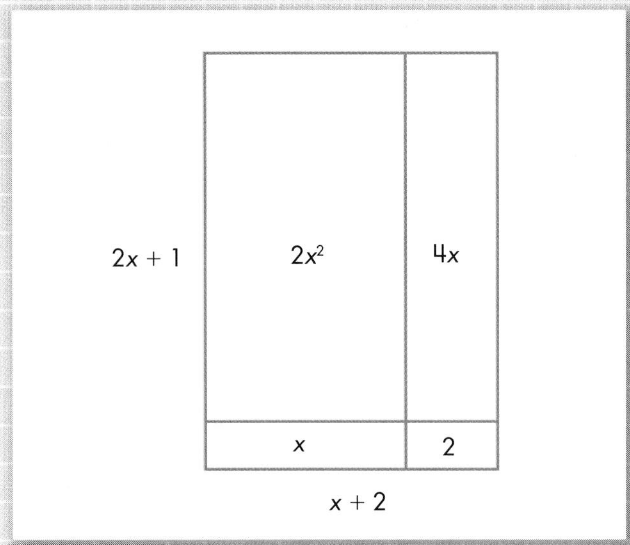

Labeling Arrays

When using multiplication notation to describe arrays, the *Investigations* curriculum uses the convention of designating the number of rows first and the number in each row second (e.g., 3 × 2 indicates 3 rows with 2 in each row). This convention is consistent with using 3 × 2 to indicate 3 groups of 2 in other multiplication situations (e.g., 3 pots with 2 flowers in each pot). However, at this age level, it is not necessary for students to follow this system rigidly; trying to remember which number stands for the number of rows and which for the number in each row can be unnecessarily distracting for students. When students suggest a multiplication expression for an array, what is important is that they understand what the numbers mean. Note that in other cultures, conventions about interpreting multiplication expressions differ. In some countries, the convention for interpreting 3 × 2 is not "3 groups of 2," but "3 taken 2 times."

Teacher Note

Visualizing Arrays

In order for students to use rectangular arrays to visualize how to break up a multiplication problem, they must be able to see how the lengths of the sides of the rectangle represent the factors in a multiplication problem and how the area represents the product. Evidence from research and practice indicate that really seeing this relationship takes time and experience. As adults, we are so familiar with the relationship between the area of a rectangle and the lengths of its sides that we may not realize that this relationship does not necessarily seem obvious to students.

If some of your students have never used rectangular arrays to represent multiplication, borrow some Array Cards and instructions for some of the Array Card games from a Grade 4 teacher. Although these games involve only single-digit multiplication combinations, by playing these games, students can learn to visualize how an array represents a multiplication expression. As an alternative, students can use tiles to build arrays in order to gain this experience.

In Grades 3 and 4, students encountered arrays as drawings, showing all of the individual units in each array.

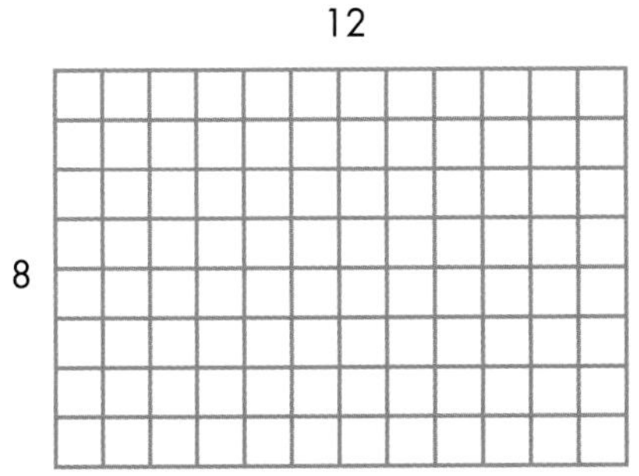

Seeing all of the units helps students learn why the product of the lengths of two sides of the rectangle equals the area of the rectangle. However, to be most useful, a good mathematical representation must become part of students' repertoire of mental images—a representation they can visualize in order to help them think through a problem.

As students think through problems with larger numbers, it is cumbersome to draw and count the individual units in an array. Rather, students use the idea of an array to help them visualize how to break up the numbers in multiplication or division problems and to keep track of which parts have been multiplied by which other parts.

In Grade 4, students worked on visualizing arrays to represent multidigit multiplication problems by making the transition from marked to unmarked arrays.

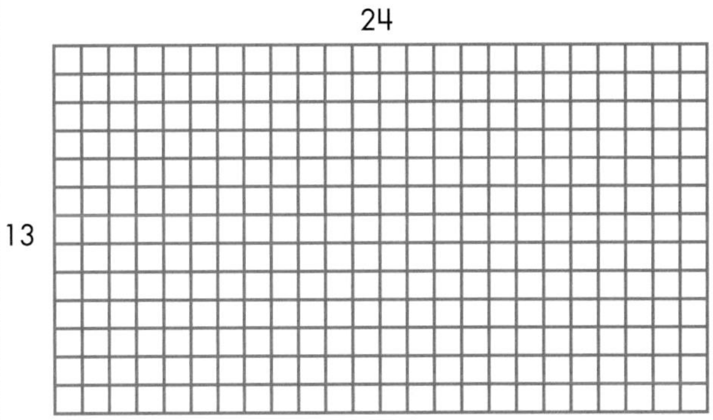

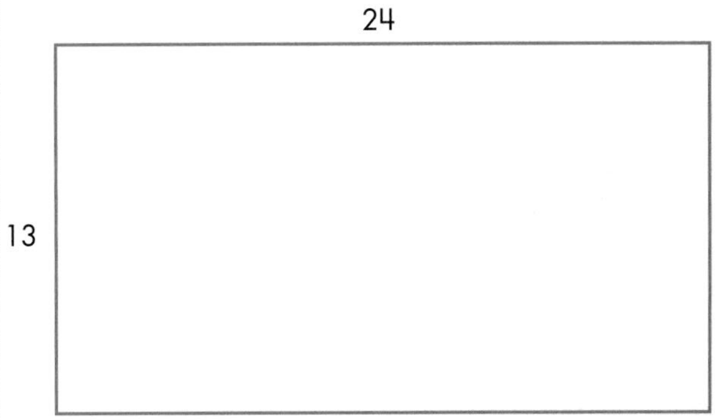

As a transition between marked and unmarked arrays, some teachers use tick marks along two sides of the rectangle.

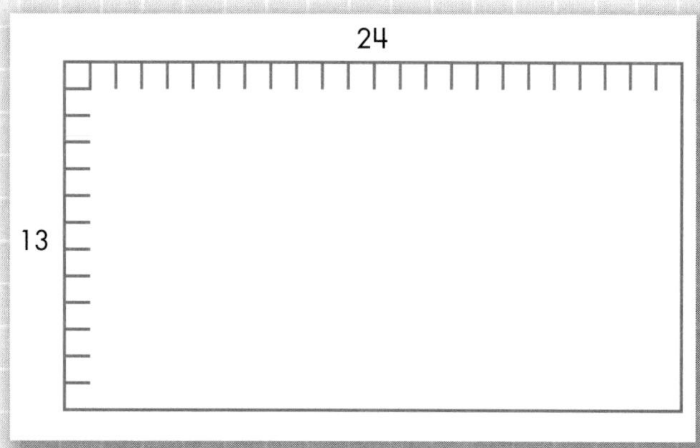

In Grades 4 and 5, students use and create unmarked arrays to help them visualize how to break apart multiplication problems.

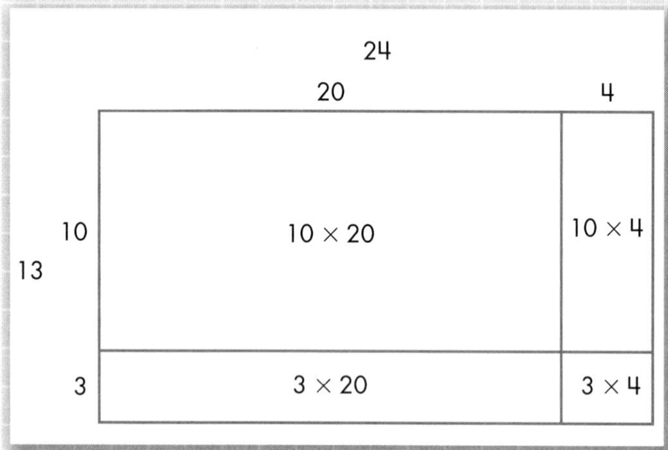

At first, unmarked arrays are presented with sides drawn to scale, as in the array for 13 × 24 above. Because students are learning to use and visualize arrays, drawing the diagram to scale helps some students visualize how the array represents multiplication. Eventually, when students are confident about breaking apart arrays into partial products, you and your students may sketch the arrays without trying to be accurate about the scale.

Teacher Note

Multiplication Strategies

Judging from the first step students take, their strategies for multiplication fall into the following three basic categories:

1. Breaking the numbers apart by addition
2. Changing one number to make an easier problem
3. Creating an equivalent problem

In the first strategy, students break the given numbers apart and create a series of partial products. In the second and third strategies, students change the problem in some way to create a problem that is easier to solve. Students often use a combination of these approaches when solving a single problem and may see their own variations or combinations as different strategies.

In order to use the strategies described in this **Teacher Note,** students need to understand the meaning of multiplication and have a good mental model of what is happening in the problem. They need to look at the problem as a whole, think about the relationships of the numbers in the problem, and choose an approach they can carry out easily and accurately.

Using the distributive property is essential in solving multiplication problems. When using this property, students can break the numbers in any multiplication problem into parts and then multiply each part of one number by each part of the other number(s). For example, if the problem is 48 × 42, they might think of 48 as 40 + 8, and/or think of 42 as 40 + 2. This breaking apart is usually done by place, but not always. For example, for the problem 27 × 18, the student might think of 27 as 25 + 2. It is not necessary for students to use the term distributive property, or the notation (40 + 8) × (40 + 2). What is important is that they realize that the numbers can be broken apart by addition, that they know how to keep track of multiplying the parts, and that they add all the partial products to find the final product.

By the end of Grade 5, students should be fluent with one or more of these strategies. They should be able to work with the largest or most reasonable parts of the number, using a small number of steps in both solving problems and recording their solutions. They may use one strategy most of the time, but during the year they should encounter and understand all of these strategies.

Here are examples of the three strategies.

1. Breaking the numbers apart by addition

Many students choose to break the numbers apart by place and find all the partial products. Here are two ways a student might record this approach:

$$\begin{array}{l} 48 \times 42 \\ \\ 40 \times 40 = 1{,}600 \\ 40 \times 2 = 80 \\ 8 \times 40 = 320 \\ 8 \times 2 = \underline{16} \\ 2{,}016 \end{array} \qquad \begin{array}{rl} 48 & \\ \underline{\times\ 42} & \\ 1{,}600 & 40 \times 40 \\ 320 & 40 \times 8 \\ 80 & 2 \times 40 \\ \underline{16} & 2 \times 8 \\ 2{,}016 & \end{array}$$

Note that because of the commutative and associative properties of multiplication and addition, numbers can be multiplied or added in any order.

For a simpler problem, students might break up only one number. For example, to solve 22 × 13, a student might break up only the 13, thinking of the problem as (22 × 10) + (22 × 3), since both of these partial products are solved easily.

2. Changing one number to make an easier problem

In this sample solution, a student changes 48 × 42 to 50 × 42, solves 50 × 42, and then compensates for the initial change.

$$48 \times 42$$

$$50 \times 42 = 2{,}100$$

$$2{,}100 - 84 = 2{,}016$$

Changing 48×42 to 50×42 results in a problem that is easier to solve. The new numbers can be either broken apart by addition, yielding $(50 \times 40) + (50 \times 2)$, or thought of as $\frac{1}{2}$ of 100×42. Then the student has to decide how to adjust the answer to 50×42. Because 50×42 is two more groups of 42 than 48×42, subtracting two groups of 42, or 84, gives the final answer.

It should be noted that although this strategy will always work, it does not always make the problem easier to solve. Also, whereas it is possible to solve a multiplication problem by changing *both* numbers to make an easier problem (e.g., changing 48×42 to 50×40), it is difficult to figure out how to adjust that answer in order to solve the original problem.

3. Creating an equivalent problem

Here are two ways students might create an equivalent problem:

$48 \times 42 = 96 \times 21$ $\quad$ $48 \times 42 = 16 \times 126$

Students often call this strategy "doubling and halving" because that is how they most often create equivalent problems in multiplication. In these examples, however, you can see that the first student "doubled and halved," and the second student actually "tripled and took one-third." Using the same strategy, the students also could have changed the problem to 24×84 or 144×14.

As with changing one number to create an easier problem, it makes sense to use this strategy only if it does indeed result in a problem that is easier to solve. Sometimes a series of steps of doubling and halving can result in a much easier problem, such as $48 \times 42 = 24 \times 84 = 12 \times 168$.

When a student's first step is to create an equivalent problem, the next steps often include a combination of the other strategies. After a student has changed the problem to 96×21, it can be solved by breaking the numbers apart: $96 \times 21 = (96 \times 20) + (96 \times 1)$.

One way you might want to think about how this strategy works is that one of the numbers is first broken apart by multiplication and then the associative property is applied. Two examples are shown below.

$$48 \times 42 = 48 \times (2 \times 21) = (48 \times 2) \times 21 = 96 \times 21$$

$$48 \times 42 = (16 \times 3) \times 42 = 16 \times (3 \times 42) = 16 \times 126$$

Division Notation

When solving division problems, students often use words to explain their thinking, as this student does for $374 \div 12$.

"I start with 12 × 10 which equals 120. Then I triple 120 which is 360. So that's thirty 12s in 360. I have 14 left. There is one 12 in 14 with 2 left over. Then I add 30 + 1 to get thirty-one 12s and I still have the 2 left over, so my answer is 31 remainder 2."

Although this thinking is easy to follow, writing solutions in this way is not efficient in the long run. As students explain their strategies for solving a problem, model how to write the steps of their problem-solving process with equations. Recording in this way helps students keep track of their process, facilitates comparison of strategies, and provides examples of ways in which students can record strategies themselves. The convention for recording the remainder with a capital "R" is used in these materials: $374 \div 12 = 31$ R2.

In order to help the class follow a student's strategy, you may need to record all of the steps in the student's procedure. For the student's explanation given above, for example, you might record the following:

$$10 \times 12 = 120$$
$$10 \times 12 = 120$$
$$10 \times 12 = 120$$

$$30 \times 12 = 360$$

$$374 - 360 = 14$$

$$\begin{array}{r} 30 \times 12 = 360 \\ 1 \times 12 = 12 \\ \hline 31 \times 12 = 372 \end{array}$$

$$374 \div 12 = 31 \text{ R2}$$

You need to decide the importance of asking the student to record every step in different situations. Writing down every step can be cumbersome, but may be necessary when strategies are being shared with other students. You might choose to supply some of the missing steps after hearing students' oral explanations.

At other times, when you are helping the class or individual students summarize their solution strategy, encourage them to use more compact notation that shows the essence of their approach. Here are two more concise notations for this solution:

$$374 \div 12$$

$$\begin{array}{ll} 120 & 10 \times 12 \\ 360 & 30 \times 12 \\ 372 & 31 \times 12 \end{array}$$

$$374 \div 12 = 31 \text{ R2}$$

$$374 \div 12$$

$$\begin{array}{r} 30 \times 12 = 360 \\ 1 \times 12 = 12 \\ \hline 31 \times 12 = 372 \end{array}$$

$$374 \div 12 = 31 \text{ R2}$$

In these summaries, some of the steps that were done mentally, such as $374 - 372$, are not recorded, but we can easily follow the student's thinking. Remind students to make clear what the answer to the problem is, because the answer may be embedded in the computation steps.

Here are several other examples of clear notations that are often used by students:

374 ÷ 12
120 ÷ 12 = 10
120 ÷ 12 = 10
120 ÷ 12 = 10
360 ÷ 12 = 30

374 ÷ 12 = 31 R2

374 ÷ 12
10 × 12 = 120
20 × 12 = 240
30 × 12 = 360
~~40 × 12 = 480~~
31 × 12 = 372

374 ÷ 12 = 31 R2

374 ÷ 12

120	10
240	20
360	30
12	1
372	31

374 ÷ 12 = 31 R2

Using a table helps some students organize their work.

374 ÷ 12

	× 12
10	120
10	120
20	240
10	120
30	360
1	12
31	372
374 ÷ 12 = 31 R2	

Consider if, when, and how to introduce these notations (or others your students devise) to your students. Students need not learn many different notations. What is expected is that students learn to clearly and unambiguously notate their solutions so that they can keep track of the work and know when they have completed the problem, and so that anyone looking at the work can understand their reasoning.

Teacher Note

About Cluster Problems

Cluster problems are a set of problems that encourage students to think about what they know in order to solve harder problems. The cluster problems in this unit are designed to help students become familiar with different strategies for multiplying and dividing. Through these clusters, students learn how they can pull apart problems into more manageable components. They learn to use multiplication combinations they know as well as the relationship between multiplication and division. Cluster problems also employ understanding multiplying and dividing by multiples of 10 and 100, a critical skill in solving multidigit problems.

Different clusters suggest different strategies. For example, here is a cluster of problems that includes two different ways to think about solving 24×31:

24×10	24×3
24×20	24×30
24×30	

The problems in the column on the left are designed to encourage students to consider how their knowledge of 24×10 can help them solve 24×31: $24 \times 31 = (24 \times 10) + (24 \times 10) + (24 \times 10) + (24 \times 1)$. The problems on the right encourage students to think about how 24×3 is related to 24×30.

Students solve all the problems in the cluster and then think about what strategy makes the most sense and is the most efficient for solving the final problem.

Here is an example of a division cluster for $767 \div 36$:

10×36	$360 \div 36$
20×36	$720 \div 36$
	$47 \div 36$

The problems in the left column provide a way of solving the problem by considering the relationship between division and multiplication: $20 \times 36 = 720$, so $720 \div 36 = 20$. Now what remains to be solved is $47 \div 36$. The problems in the right column suggest breaking the dividend into parts that are more easily divisible by 36.

As students solve multiplication and division problems in this unit, observe their strategies and ways of thinking about multiplication and division. Encourage them to use what they already know to create their own clusters and to find solutions. Ask them to explain what parts of the problem they have solved and what parts are still remaining.

Students build on what they already understand well, such as the basic multiplication combinations, multiplying by multiples of 10, and other multiplication or division problems they can solve easily. As they do this, they deepen their understanding of multiplication and division and become increasingly efficient and flexible in solving problems.

Assessment: What Is the Answer?

Students have been solving multiplication problems, working toward a clear understanding of how to break apart a problem, solve each subproblem, and combine the products for an accurate solution. These notes include a discussion of each of the two assessment problems and provide information on interpreting student work. The examples of student work provide a range of typical responses.

Problem 1

Benchmark addressed:

Benchmark 2: Solve multiplication problems efficiently.

In order to meet the benchmark, students' work should show that they can:

- Identify what pieces of a solution are missing from a multiplication problem.

Name Date

Number Puzzles and Multiple Towers

Assessment: What Is the Answer?

A fourth grader solved the following problem this way.
The answer is wrong.

32 × 26

30 × 20 = 600
2 × 6 = 12
600 + 12 = 612

1. What did the student do wrong? Explain below.

▲ Resource Masters, M51

To solve this problem, students must be able to interpret the meaning of 32 × 26 in a way that allows them to reason about how to break up the two factors by place and solve all the necessary subproblems. They should recognize that multiplying only the two numbers in the tens place and then the two numbers in the ones place and adding those two products does not result in 32 groups of 26. The fourth grader's solution is missing two other subproblems: (2 × 20) and (30 × 6).

Some students may mention that the answer of 612 is too low. One possible way to make a quick estimate for 32 × 26 is: 32 × 20 = (32 × 10) + (32 × 10) = 320 + 320 = 640, which is already greater than the fourth grader's answer. Another 6 groups of 32 is about 200 more, so the correct product is closer to 800 than to 600.

Meeting the Benchmark

Both Alicia and Martin met the benchmark—they were able to explain what pieces of the solution are missing.

Alicia shows how to break the problem correctly into four partial products. She knows that (30 × 6) and (20 × 2) were missing from the incorrect solution.

30x6=180
20x2= 40
30x20=600
2x6= 12

180
40
600
+ 12
832

He forgot to do 20x2 and 30x6 and if he didnt he would of got the rite anwer,

Alicia's Work

Martin uses an array to show which parts of the problem are solved and which were omitted. He goes on to multiply each part of the problem, using equations in the same way as Alicia does.

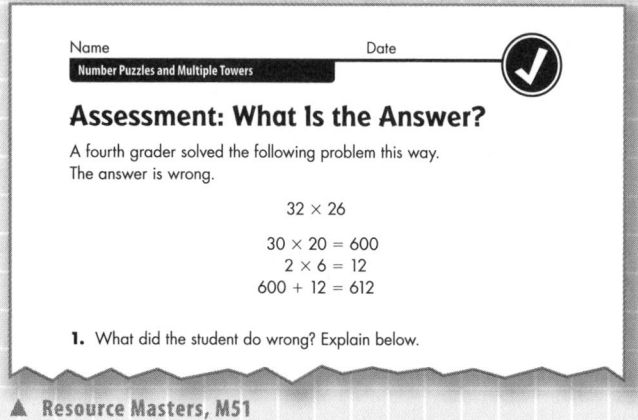

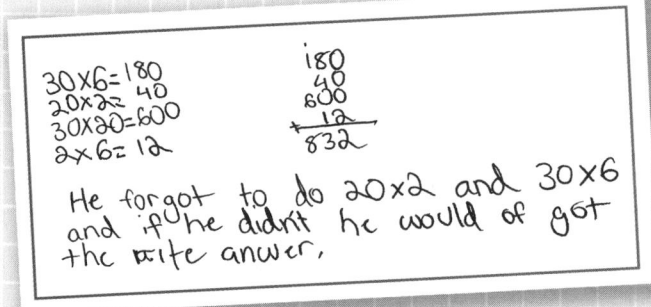

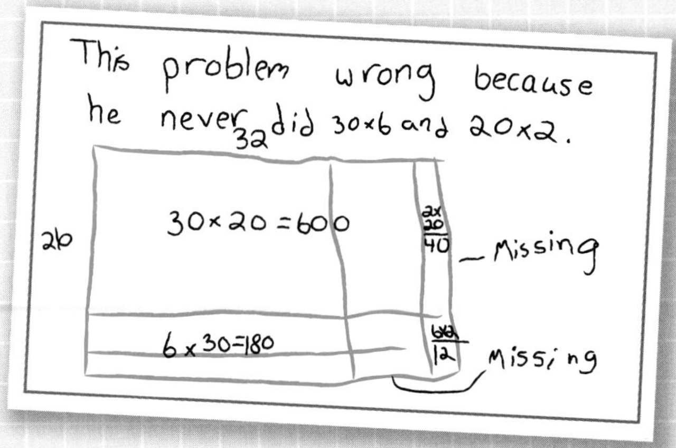

Martin's Work

Partially Meeting the Benchmark

Some students know that the answer of 612 is too small, but are not clear about what pieces of the solution are missing. For example, Hana wrote, "It would be more than 612 because he didn't do 2 × 20." However, she does not identify that 6 × 30 is also missing from the solution.

Other students explain that all the pieces are not finished, but they do not identify which ones.

Not Meeting the Benchmark

These students cannot identify what is wrong with the solution. These students are not able to interpret 32 × 26 in a way that allows them to keep track of all parts of the problem.

Continue to help these students visualize the meaning of multiplication by using a story context, a picture, or an array. They may also need to work with smaller numbers.

Problem 2

Benchmark addressed:

Benchmark 2: Solve multiplication problems efficiently.

In order to meet the benchmark, students' work should show that they can:

- Break the problem into easily manageable parts;
- Solve all of the subproblems, keeping track of what has been solved and what remains to be solved;
- Add the products of the subproblems back together accurately.

Alternatively, students' work should show that they solved the problem in another efficient way, such as creating an equivalent problem.

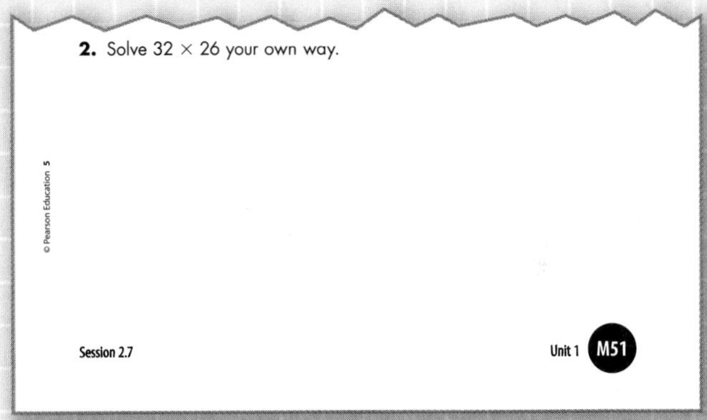
2. Solve 32 × 26 your own way.

© Pearson Education 5

Session 2.7 Unit 1 M51

▲ **Resource Masters, M51**

Meeting the Benchmark

Some students break this problem into four partial products as discussed above: (30 × 20) + (20 × 2) + (6 × 30) + (2 × 6). Others break the problem into different but equally efficient pieces.

Deon breaks the problem into (10 × 26) + (10 × 26) + (10 × 26) + (2 × 26).

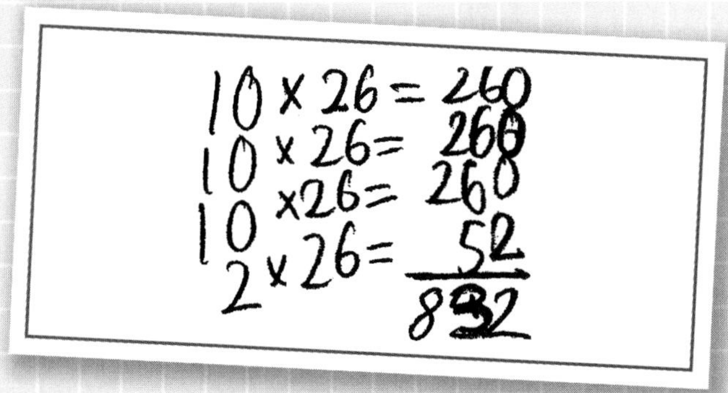

Deon's Work

Samantha breaks up the 26 and multiplies each part by 32. She thinks of the problem as (32 × 10) + (32 × 10) + (32 × 6).

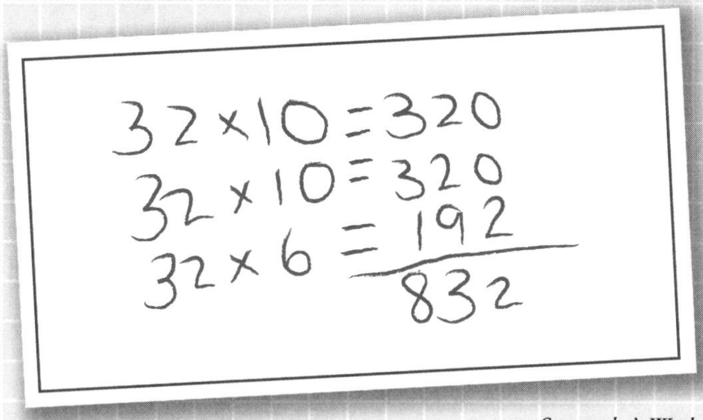

Samantha's Work

Partially Meeting the Benchmark

These students break apart the problem efficiently but make minor errors in multiplication or addition.

Not Meeting the Benchmark

Some students get the correct answer but do not break the problem apart efficiently.

Rachel cannot solve each partial product without breaking each into smaller pieces. She breaks apart 30 × 20 into (10 × 20) + (10 × 20) + (10 × 20) and 30 × 6 into (30 × 3) + (30 × 3). By Grade 5, all students should be able to multiply 30 × 20 or 30 × 6 without breaking it apart further.

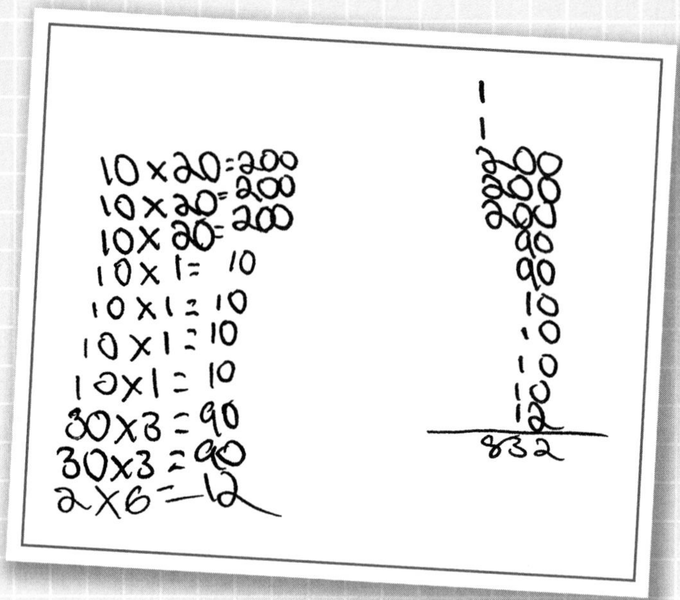

Rachel's Work

Students who use addition to solve the problem also do not meet the benchmark, even if they solve the problem correctly. Although adding 32 twenty-six times indicates a correct interpretation of the multiplication expression, this method is inefficient and prone to computational errors.

Students who have no correct method for this problem also do not meet the benchmark.

Alex shows one step of the problem but is not able to complete it.

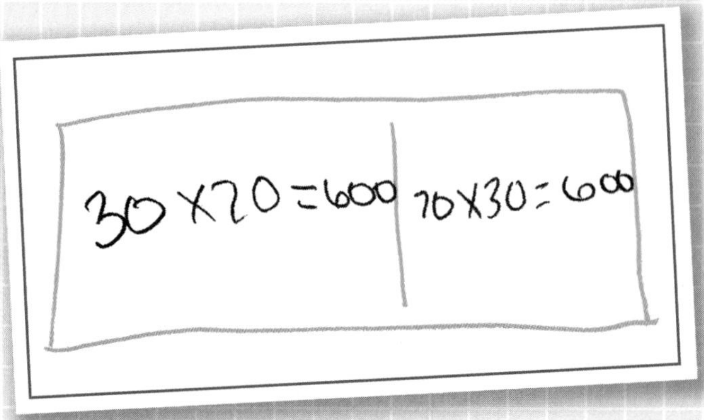

Alex's Work

The Relationship Between Multiplication and Division

Multiplication and division are related operations. For example, here is a set of linked multiplication and division equations:

$$8 \times 3 = 24 \qquad 3 \times 8 = 24$$

$$24 \div 8 = 3 \qquad 24 \div 3 = 8$$

The multiplication equations show the multiplication of two factors to equal a product. The division equations show that the product divided by one of the factors is equal to the other factor.

Some problem situations your students encounter in this unit can be described as both multiplication and division. Consider the following story problem:

I have a supply of 336 treats for my dog. If I give her 14 treats every week, how many weeks will the supply last?

The quantities in this problem are 336 treats, 14 treats per week, and a number of weeks to be determined. This problem can be written in standard notation as either division or multiplication.

$$336 \div 14 = \underline{\quad\quad} \text{ or } 14 \times \underline{\quad\quad} = 336$$

After the answer to the problem has been found, both division and multiplication equations can be written to show the relationship of the three quantities.

336 treats divided into groups of 14 treats per week result in 24 weeks.	Groups of 14 treats per week for 24 weeks result in 336 treats.
$336 \div 14 = 24$	$14 \times 24 = 336$

When students solve a problem like this one, they might write either a division equation or a multiplication equation to express the answer and its relationship to the quantities in the problem. Both notations represent the problem, depending on whether the student is thinking of the problem as division or as a multiplication problem with a missing factor. They should be able to read and interpret both of these notations, explaining what each number in the equation represents and relating the equation to the original problem.

Teacher Note

Division Strategies

In order to solve division problems efficiently, students need to understand the meaning of division and have a good mental model of the problem. They need to look at the problem as a whole, think about the relationships of the numbers in the problem, and choose an approach they can carry out easily and accurately.

Most students think about solving division problems in one of two basic ways, both of which are emphasized in this unit.

1. Using groups of the divisor

Students think of the division problem as a question about the number of groups. For example, if the problem is $159 \div 13 = \_\_\_\_$, the student thinks, "How many 13s are in 159?" Then the student uses multiplication to create groups of the divisor, keeping track of what part of the problem remains to be solved. The student might first think about 10 groups of 13, or 10×13. Then the student would subtract the product of 10×13 from 159 mentally and recognize that 2 more groups of 13 can "fit" into what is left over, 29.

Help students use story contexts or representations.

Alex is thinking about teams. We have 159 students and want to make teams of 13. Janet's first step is to multiply 10 and 13. How can you explain her first step by using this story? What would be the next step?

Story contexts can be particularly useful in helping students understand the meaning of remainders in division.

What does a remainder of 3 mean?

Use some story contexts in which the divisor represents the number in a group, as in the example above. At other times, use stories such as the one below, in which the divisor represents the number of groups.

Here's a different story about 159 divided by 13. I have 159 oranges that I want to put into 13 bags. I want to know how many oranges I can put in each bag if I want to have the same number in each bag. Now how can you explain Janet's first step of 10 times 13 using this story?

2. Breaking the dividend into parts

In this strategy, students think about the dividend first, and how it can be broken up into numbers that are easier to divide. For example, for the problem $159 \div 13 = \_\_\_\_$, the student thinks, "How can I break 159 apart to make easier problems to solve?" The student might think of the problem as $159 \div 13 = (130 \div 13) + (29 \div 13)$.

The actual computations are closely related to those used in the first strategy. The difference is that students thinking about groups of the divisor usually multiply, and students using the second strategy think of the dividend first. Both of these methods depend on the distributive property, just as many of the multiplication methods do. For more information about the distributive property, see **Algebra Connections in This Unit,** page 16.

There are at least two other strategies students use, although they are not emphasized in this unit.

3. Making an equivalent problem

A division problem can be changed into an equivalent problem by dividing both numbers by the same number, thus maintaining the ratio between them. In the following problem, both numbers were divided by 7:

$$1{,}400 \div 35 = 200 \div 5 = 40$$

4. Solving an easier, related problem and then compensating

Students solve a related problem for which they already know the answer and then adjust the result. For example, when solving $159 \div 13$, one student "just knew" that $13 \times 13 = 169$ and reasoned that subtracting 13 from 169 is 156, the closest multiple of 13 to 169. $12 \times 13 = 156$, 3 less than 159, so $159 \div 3 = 12$ R3.

Developing Computation Strategies That Make Sense

In this unit and throughout Grade 5, students solve computation problems by building on numbers and relationships they know. By the end of Grade 5, students should have computation strategies that they can justify and carry out easily for addition, subtraction, multiplication, and division of whole numbers. Most students will have one algorithm or procedure for a particular operation that they use often, but they should also be familiar with, understand, and be able to explain more than one procedure for each operation.

Watch and listen carefully as your students solve computation problems. Find out what makes sense to them. If a student's approach is unfamiliar to you, do a problem or two yourself, using the student's approach. Common procedures for solving multiplication and division problems are described in **Teacher Note:** Multiplication Strategies, page 161, and **Teacher Note:** Division Strategies, page 170. Some of these strategies may be different from the procedures that are most comfortable for you. Do not assume that what seems easy and efficient to you is necessarily the best or most efficient approach for your students.

If students have procedures that they can apply easily and accurately, that do not bog them down in laborious calculations, and that they know how to apply to a variety of problems, then they have the tools they need to solve virtually any problem. If, on the other hand, students lose track of their own procedures, make many errors, or end up with calculations that are very difficult, they need to become more proficient at decomposing their problems into manageable parts. They do this in order to develop strategies that both make sense to them and are easy to carry out.

Here are some guidelines for helping students refine their strategies:

Provide time to work with different strategies. When students first meet a new strategy, they need opportunities to try it with a range of problems before they begin to see how efficient it is in different situations. Through their work in Grade 5, some students will develop a repertoire of strategies, others only one. What is important is that students understand why their strategies work and be able to use them with confidence.

Encourage students to think about ways to start a problem. Students need support in order to work with unfamiliar strategies that can eventually give them tools to become more fluent problem solvers. In this unit, activities such as Multiplication Clusters, Starter Problems, and Division Clusters are designed to help students consider different building blocks for solving multiplication and division problems. You can use these ideas throughout Grade 5 to help students approach a problem. For example, suggest a first step and ask students to talk in pairs about what else they would need to know in order to find the answer.

Terrence wanted to solve 374 ÷ 12. He started by doing 20 × 12. Does that first step help? What else would he need to find? What would be *your* first step?

When a student seems stuck, ask the student to think about known number relationships that can help.

I noticed that a few of you are having a hard time thinking of a first step to solve 374 ÷ 12. What numbers do you know that you can divide by 12? Let's brainstorm a few.

I hear 24, 36, 48, 120, 1,200, 360, 480. Which of these can help with this problem?

Emphasize representations and story contexts. Frequently ask students to use a diagram or a story to help them think through or explain how they are breaking up a problem. The goal is for students to learn how to use representations and story contexts as part of their own mental repertoire for visualizing problems. If students are working on keeping track of all the parts of a multiplication problem, help them use stories, pictures, and arrays to visualize the parts.

If students are solving 64 × 27 and first multiply 60 × 20, you might pose the following question:

The other day, we were imagining a grocer at Ted's Fruits and Vegetables filling bags with oranges. If the grocer needs to fill 64 bags with 27 oranges in each bag, and we start with 60 × 20, what have we done so far? *(filled 60 bags with 20 oranges each)* **What would you do next? OK, Mercedes suggests 60 × 7. So now what have we done?** *(put 7 more oranges into those 60 bags)* **What have we done so far?** *(put 27 oranges into 60 bags)*

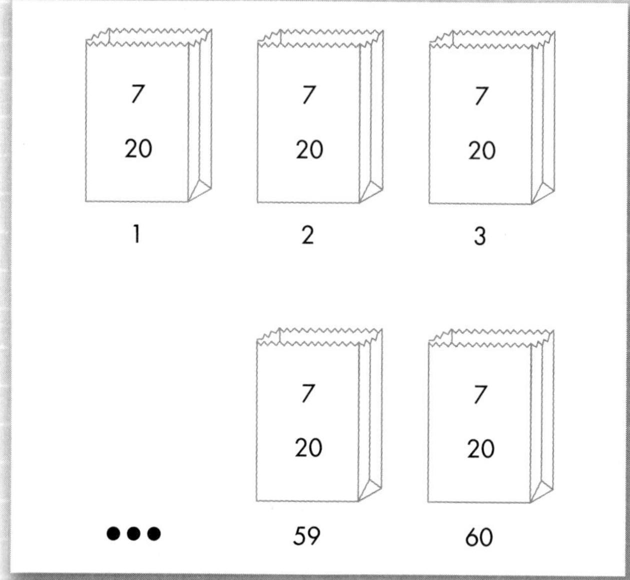

What is left to do? *(put 27 oranges into each of 4 more bags)*

Also ask students to use an array to show the same partial products, and ask questions that relate the story to the array.

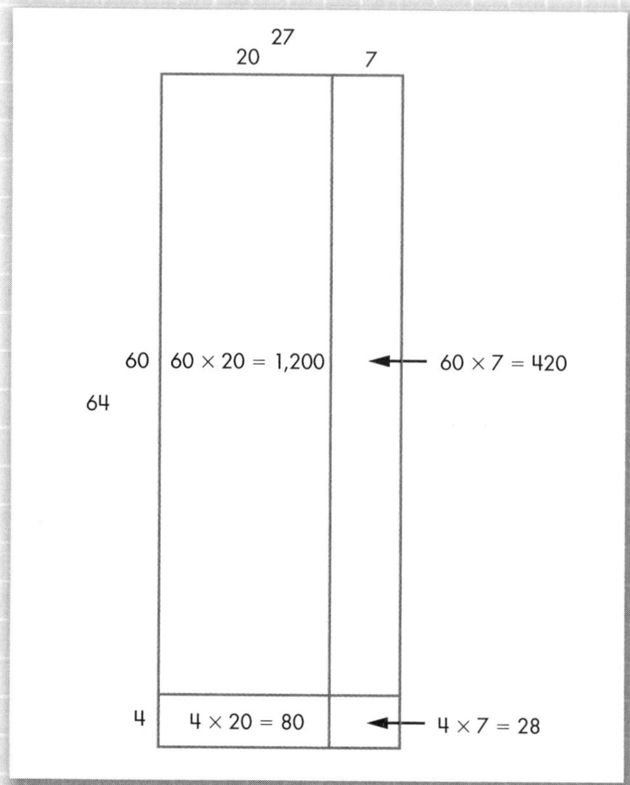

So where does the 60 × 20 from our story show up in the array?

Ask students to come up with their own contexts. Even if the student who is explaining a solution is easily keeping track of all the parts of the problem, the other class members may benefit more from the student's explanation if you discuss how to represent the solution with a story and a representation. Refer students to the Math Words and Ideas section of the *Student Math Handbook* for examples of representations and stories.

Work with students to develop clear and concise notation. Clear and concise notation helps students keep track of the steps in solving problems, allows them to check their work, and allows others to understand their solution process. By Grade 5, students need to develop ways to notate their computation strategies that are easy for them and others to follow. One of the most important ways you can help students develop good notation habits is by modeling clear and concise notation as you record students' strategies during class discussions. You will also need to identify individual students with whom you need to work more closely. See **Teacher Note:** Division Notation, page 163, for more about notation.

Help students think about why and when certain strategies are easy to use. Students need to understand that there is no one best strategy for every person and every problem. As students reflect on why and when strategies are easy to use and share their thinking with others, they come to see that their choice of strategy depends on the numbers in the problem, the kinds of computations with which a person is most comfortable, and the number relationships a person knows very well.

Many students will solve most multiplication problems by breaking up the numbers by place and creating all the necessary partial products. However, for some problems, other methods are easier to apply. Consider 52×36, for example. One student might first think about an easier problem, 50×36, then change this into the equivalent problem 100×18, find the products of 100×18 and 2×36, and, finally, add those products. Another student might double 52 and halve 36 and then double and halve again to create the simpler problem, 208×9.

Teacher Note

End-of-Unit Assessment

These notes include a discussion of each of the three assessment problems and provide information on interpreting student work.

Problem 1

Benchmark addressed:

Benchmark 2: Solve multiplication problems efficiently.

In order to meet the benchmark, students' work should show that they can:

- Break the problem into easily manageable parts;
- Solve all the subproblems, keeping track of what has been solved and what remains to be solved;
- Add the products of the subproblems back together accurately.

Alternatively, students' work should show that they solved the problem in another efficient way, such as creating an equivalent problem.

In Grade 5, the focus of the work on multiplication is to help students use strategies efficiently and accurately and to notate their work clearly.

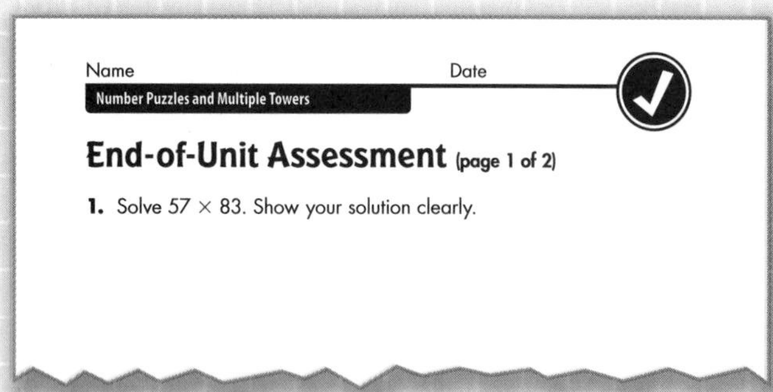
Name Date

Number Puzzles and Multiple Towers

End-of-Unit Assessment (page 1 of 2)

1. Solve 57×83. Show your solution clearly.

▲ Resource Masters, M54

Meeting the Benchmark

Yumiko breaks each number into tens and ones and then multiplies each part of one number by each part of the other. She thinks of the problem as $(50 \times 3) + (7 \times 3) + (50 \times 80) + (7 \times 80)$. She multiplies accurately, keeps track of all the parts of the problem, and adds them back together.

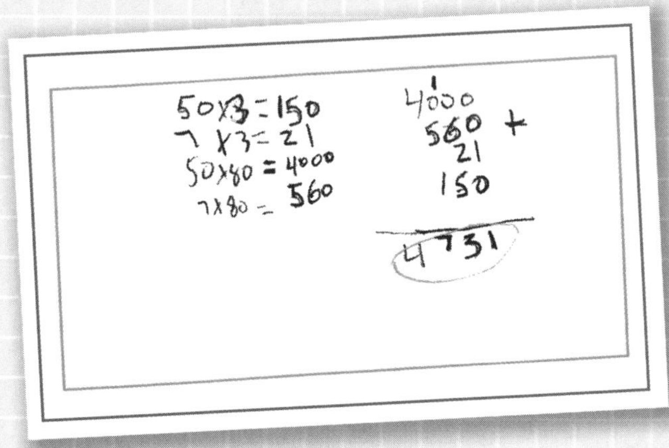

Yumiko's Work

Georgia also breaks the problem up efficiently. She shows how each partial product relates to the product of 57×83 on an array.

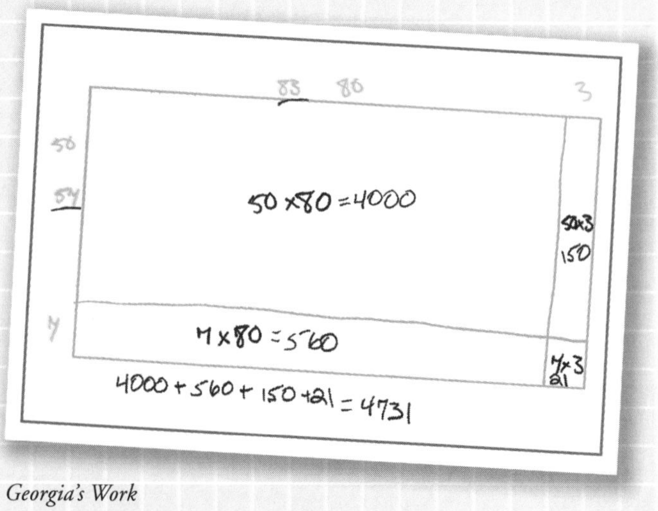

Georgia's Work

Partially Meeting the Benchmark

Some fifth graders understand the meaning of multiplication, can break multidigit multiplication problems correctly into subproblems, solve those problems, and add the partial products correctly. However, they are not yet fluent in multiplying by multiples of 10 or do not know all the multiplication combinations through 12 × 12.

Renaldo shows a clear understanding of how to break the numbers apart and keep track of the parts. He thinks of the problem as (50 × 83) + (7 × 83).

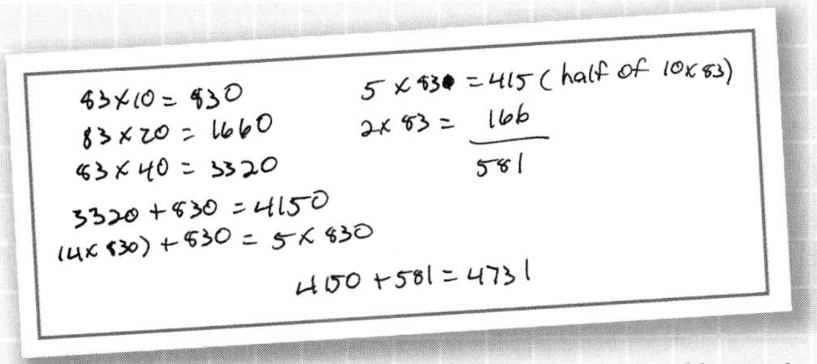

Renaldo's Work

However, Renaldo does not multiply 50 × 83 directly, but figures out this product by starting with 10 × 83. Similarly, he does not multiply 7 × 80. Renaldo's work shows a great deal of understanding about multiplication. For example, he knows that the product of 5 × 83 is half of the product of 10 × 83. Also, he is accurate in his computation. However, he needs to work on breaking problems apart in more efficient ways. Renaldo is well on his way to developing fluent multiplication strategies. It will be important for his teacher to work with him on multiplying by multiples of 10 and to check how well he knows his multiplication combinations.

Here are other ways students show that they partially meet the benchmark:

- Some students may break apart the problem in a reasonable way, but make errors in multiplication or addition.
- Some students can successfully break the problem into smaller pieces, but either lose track of the pieces or add them back together incorrectly.
- Some students understand how to solve the problem correctly, but do not use clear notation and thus lose track of their own steps.

Students will continue to work on clear and concise notation in *How Many People? How Many Teams?*

Not Meeting the Benchmark

Some students may not yet have a solid understanding of multiplication as involving equal groups. Here are some examples:

- Some students might change 57 × 83 to 57 × 100. However, they may then wonder how much to subtract from 5,700 and may not be clear that the problem was increased by seventeen 57s, not just by 17.
- Some students may not understand how to multiply by multiples of 10 as, for example, a student who incorrectly computes the product of 80 × 50 as 400.
- Some students may attempt to add instead of multiply (e.g., adding ten 80s or ten 83s and then adding those sums to build up to fifty 80s or fifty 83s). Although adding 83 fifty-seven times is a correct way to solve the problem, Grade 5 students should be using multiplication and a more efficient strategy.

By the end of Grade 5, all students should have at least one strategy for solving multiplication problems efficiently, so it is important to work individually with students who do not have these skills and understandings.

Problem 2

Benchmark addressed:

Benchmark 1: Find the factors of a number.

In order to meet the benchmark, students' work should show that they can:

- Find the factors of a number;
- Use reasoning to find some combinations with more than 2 factors.

2. Find ways to multiply with two factors and with three factors to make 150.

Ways to multiply with two factors	Ways to multiply with three factors

▲ Resource Masters, M54

Meeting the Benchmark

These students find all of the pairs of factors for 150: 1 × 150, 2 × 75, 3 × 50, 5 × 30, 6 × 25, and 10 × 15. They also find three or four multiplication combinations with three factors to make 150. (There are four possible combinations, not counting the same factors in a different order: 2 × 3 × 25, 2 × 5 × 15, 3 × 5 × 10, and 5 × 5 × 6. Note that some students might also generate additional expressions using 1, such as 1 × 10 × 15. These are not incorrect, but it is important that they generate several that do not use 1.) They are able to use factors they know to create multiplication expressions. They carefully look at each expression to see whether they can use it to generate other possibilities. They might reason as follows:

Hana: I knew that 50 fits into 150 *(50 × 3)*. 50 can be broken up into 10 × 5, so 10 × 5 × 3 works.

Terrence: I looked at all the factors of 150 *(1, 2, 3, 5, 6, 10, 15, 25, 30, 75, 150)* and broke some up into smaller pieces. I used 75 × 2 to build 3 × 25 × 2 and 3 × 5 × 5 × 2.

Partially Meeting the Benchmark

These students find only a few expressions with two or three factors for 150 and are generally less systematic in using what they know to generate new expressions. They may take a random approach, as illustrated by Mercedes' reasoning.

Mercedes: I tried different numbers and kept multiplying until I found some that equaled 150.

Not Meeting the Benchmark

Some students may find only the most obvious factors, such as 150 × 1 and 75 × 2. They may find at most one longer expression. It may help to ask these students whether they can use 75 × 2 to find any other factors. What are some factors of 75? How many times do they fit into 150?

Problem 3

Benchmark addressed:

Benchmark 3: Solve division problems with 1-digit and 2-digit divisors.

In order to meet the benchmark, students' work should show that they can:

- Solve the problem correctly by using division or multiplication (that is, building up to the dividend by using groups of the divisor or splitting the dividend into parts);
- Identify the remainder: 12 bags will be filled up, with 3 oranges left.

Name Date

Number Puzzles and Multiple Towers

End-of-Unit Assessment (page 2 of 2)

3. There are 159 oranges to be packed into bags. Each bag holds 13 oranges. How many bags will be filled?

Show how you figured this out.

▲ Resource Masters, M55

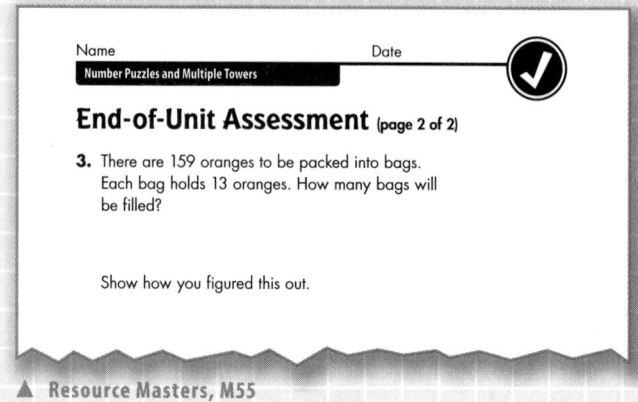

Meeting the Benchmark

Avery knew that 13 + 13 = 26, so two 13s equal 26. He added 26s together until he got close to 159.

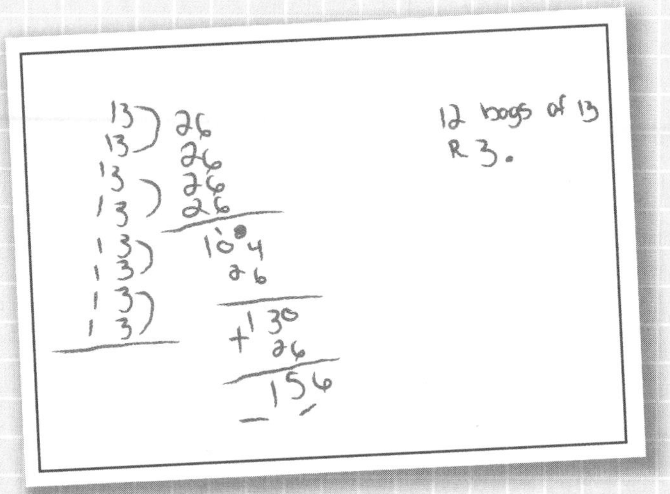

Avery's Work

Walter built up to 159 with groups of 13 by using multiplication. He knew that 10 groups of 13 was equal to 130. He then counted by 13s from 130 to 156 (143, 156). He knew that he had counted 12 groups of 13 (12 bags) and had 3 oranges left over.

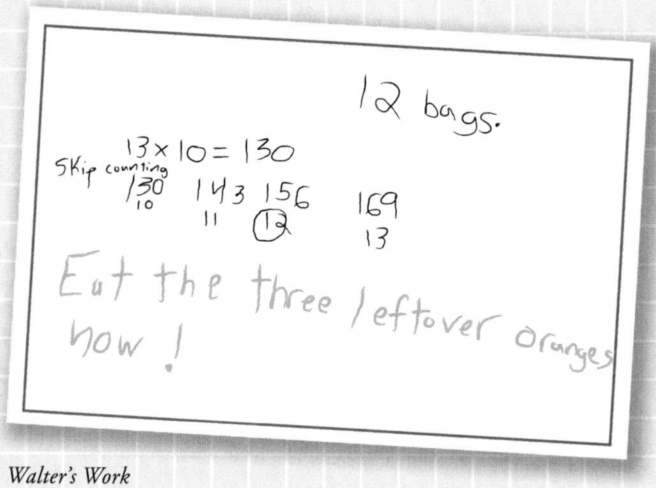

Walter's Work

Margaret's strategy is similar to Walter's but notated differently. You can see her group of ten 13s and two more on the side of her problem.

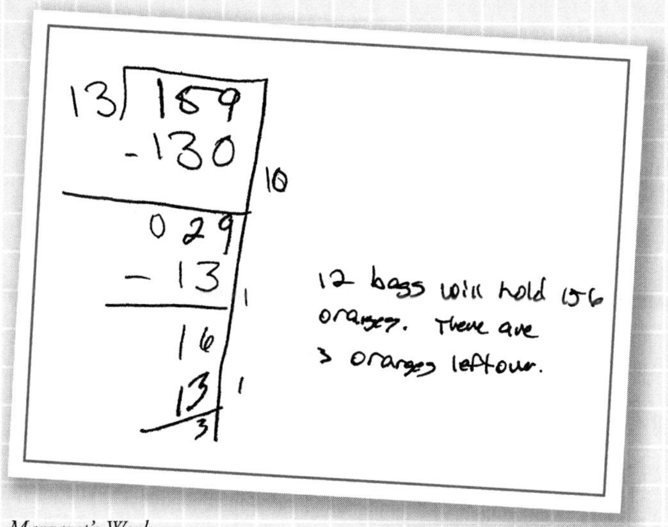

Margaret's Work

Partially Meeting the Benchmark

Some students may have a solid understanding of what division means but only partially meet the benchmark. Here are some examples:

- Some students may have difficulty keeping track of the parts they have multiplied or divided, which results in one (or more) part(s) being left out of the solution.
- Some students may have kept track of all the pieces but added incorrectly.
- Some students have carried out correct computation but are unable to identify the quotient and remainder in their calculations.

Not Meeting the Benchmark

A few students may still rely on using materials and counting to solve a division problem. For example, a student might understand the problem and solve it correctly by dealing out 159 cubes into 13 groups. In this case, the student needs to work on thinking about a group of ten groups of 13s.

Students will spend a significant amount of time on division strategies in *How Many People? How Many Teams?*.

Dialogue Box

Solving a Number Puzzle

Students are solving the Sample Number Puzzle (M16) in the first activity of Session 1.3, *Introducing Number Puzzles: 4 Clues.* After each group works through a puzzle, a few groups will share the ways in which they determined which numbers fit all the clues. As students share their strategies with the class, some use the transparency of the 300 Chart (T16), and others show their work on the board.

Teacher: Let's review the clues for this sample puzzle.

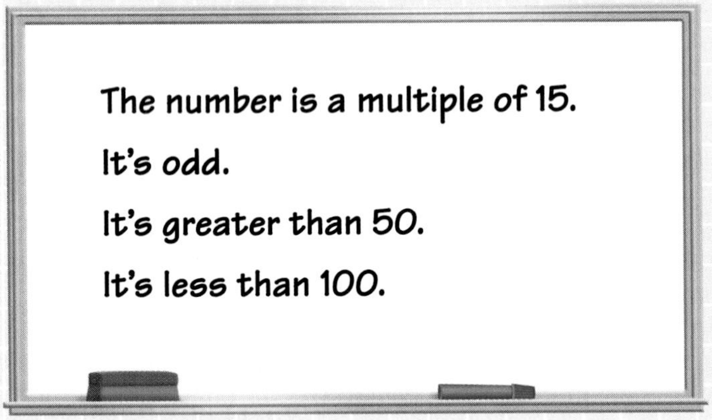
The number is a multiple of 15.
It's odd.
It's greater than 50.
It's less than 100.

1	2	3	4	5	6	7	8	9	10
11	12	13	14	15	16	17	18	19	20
21	22	23	24	25	26	27	28	29	30
31	32	33	34	35	36	37	38	39	40
41	42	43	44	45	46	47	48	49	50
51	52	53	54	55	56	57	58	59	60
61	62	63	64	65	66	67	68	69	70
71	72	73	74	75	76	77	78	79	80
81	82	83	84	85	86	87	88	89	90
91	92	93	94	95	96	97	98	99	100
101	102	103	104	105	106	107	108	109	110
111	112	113	114	115	116	117	118	119	120
121	122	123	124	125	126	127	128	129	130
131	132	133	134	135	136	137	138	139	140
141	142	143	144	145	146	147	148	149	150
151	152	153	154	155	156	157	158	159	150
161	162	163	164	165	166	167	168	169	160
171	172	173	174	175	176	177	178	179	170
181	182	183	184	185	186	187	188	189	180
191	192	193	194	195	196	197	198	199	190
201	202	203	204	205	206	207	208	209	200
211	212	213	214	215	216	217	218	219	210
221	222	223	224	225	226	227	228	229	220
231	232	233	234	235	236	237	238	239	230
241	242	243	244	245	246	247	248	249	250
251	252	253	254	255	256	257	258	259	260
261	262	263	264	265	266	267	268	269	270
271	272	273	274	275	276	277	278	279	280
281	282	283	284	285	286	287	288	289	290
291	292	293	294	295	296	297	298	299	300

Teacher: Let's hear how all the groups solved the puzzle.

Tyler [from Group A]: We used the 300 chart. We circled the multiples of 15 that are less than 100 and greater than 50. [circles 60, 75, and 90 on the transparency] Then it tells you to only use the odd ones, so we went back and crossed out 60 and 90. [crosses these out] So it's 75.

Martin [from Group B]: We used the chart too, but we crossed out the numbers under 50 and over 100 [crosses them out on the transparency]. Then we crossed out all the even numbers [crosses these out]. We were left with the multiples of 15 clue. Between 50 and 100, there's 60, 75, and 90. The only one of those that wasn't crossed out was 75 [draws a box around 75].

1	2	3	4	5	6	7	8	9	10
11	12	13	14	15	16	17	18	19	20
21	22	23	24	25	26	27	28	29	30
31	32	33	34	35	36	37	38	39	40
41	42	43	44	45	46	47	48	49	50
51	52	53	54	55	56	57	58	59	60
61	62	63	64	65	66	67	68	69	70
71	72	73	74	75	76	77	78	79	80
81	82	83	84	85	86	87	88	89	90
91	92	93	94	95	96	97	98	99	100
101	102	103	104	105	106	107	108	109	110
111	112	113	114	115	116	117	118	119	120
121	122	123	124	125	126	127	128	129	130
131	132	133	134	135	136	137	138	139	140
141	142	143	144	145	146	147	148	149	150
151	152	153	154	155	156	157	158	159	150
161	162	163	164	165	166	167	168	169	160
171	172	173	174	175	176	177	178	179	170
181	182	183	184	185	186	187	188	189	180
191	192	193	194	195	196	197	198	199	190
201	202	203	204	205	206	207	208	209	200
211	212	213	214	215	216	217	218	219	210
221	222	223	224	225	226	227	228	229	220
231	232	233	234	235	236	237	238	239	230
241	242	243	244	245	246	247	248	249	250
251	252	253	254	255	256	257	258	259	260
261	262	263	264	265	266	267	268	269	270
271	272	273	274	275	276	277	278	279	280
281	282	283	284	285	286	287	288	289	290
291	292	293	294	295	296	297	298	299	300

Olivia [from Group C]: We wrote down the multiples of 15 less than 100. That's 15 [writes as she says them], 30, 45, 60, 75, 90, and, that's all we can go. Then it said more than 50 [crosses out the numbers less than 50], and it can't be 60 or 90 because that's even [crosses out 60 and 90].

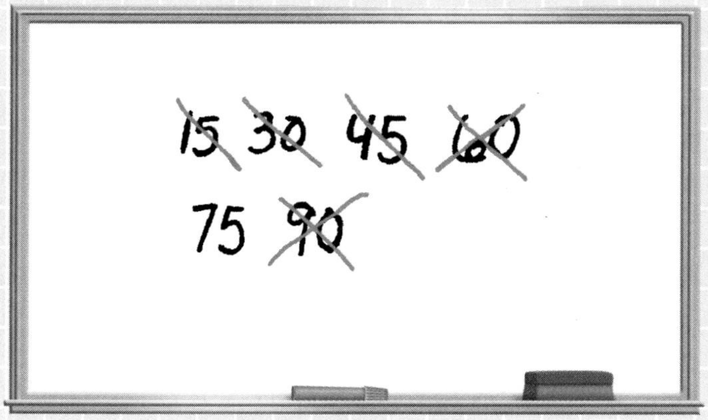

Janet [from Group D]: We wrote down the multiples of 15 under 100. Then we crossed out the even numbers [writes 15, 30, 45, 60, 75, and 90, then crosses out 30, 60, and 90]. Then our paper was a mess, so we wrote the numbers that were left [writes 15, 45, and 75 underneath]. That was 15, 45, and 75. We crossed out the ones less than 50 [crosses out 15 and 45], so we had 75.

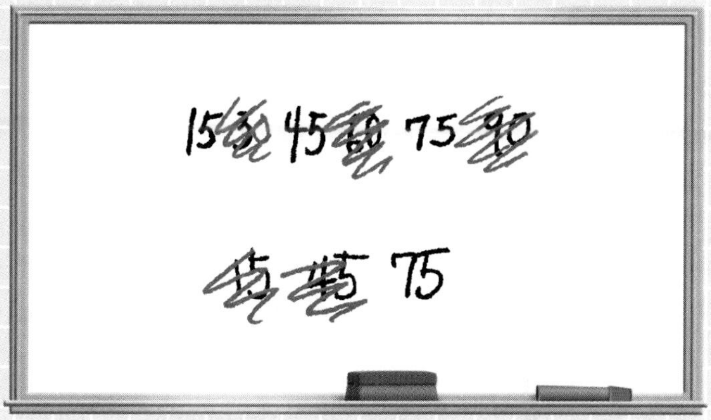

Stuart [from Group E]: We also wrote down the multiples of 15 under 100. But we crossed out the ones under 50 first [crosses out 15, 30, and 45]. Then we crossed out the even ones [crosses out 60 and 90], so we got 75 left.

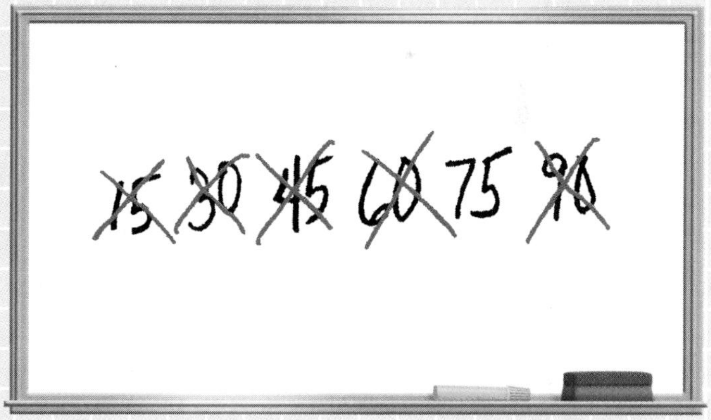

Teacher: What a lot of different ways to solve the same problem! What do you notice about how different groups found the answer?

Hana: Two different groups used the 300 charts.

Felix: A lot of the groups started with multiples of 15.

Janet: One of the groups listed the numbers they found partway through.

Benito: Some of the groups did sort of the same thing in different orders.

The teacher briefly records different strategies and leaves this list posted, as students solve more puzzles in pairs in the second activity of the session.

Examples of strategies:

- 300 Chart
- List multiples of 15
- Circle multiples of 15 < 100 and > 50
- Cross out all odds
- Cross out numbers > 100

Teacher: Did you need everyone's clue to solve the puzzle? Do you think you would have all come up with 75 for your answer if you had only 3 of the clues?

As students work, the teacher reminds students to refer to this chart for ideas about how to eliminate certain numbers and limit the range of possibilities. The teacher also asks students to consider whether there is only one answer or whether there could be more than one number that fits the set of clues.

Dialogue Box

Multiplying With More Than Two Numbers

In Session 1.4, students are working on finding ways to multiply with whole numbers to make the products 18 and 180. In the middle of the session, the teacher stops the class for a few minutes to share how they are generating different ways to multiply more than two whole numbers to make 180. She starts by asking students about their work on 18.

Teacher: At the beginning of class, we found all the ways of multiplying two whole numbers to make 18. (Points to this list on the board: 1 × 18, 2 × 9, 3 × 6). How did you find three numbers that you can multiply together to equal 18?

Lourdes: 3 × 3 × 2 will make 18.

Teacher: How do you know?

Lourdes: I multiplied it out and also used 9 × 2. I know that 9 = 3 × 3 so 3 × 3 × 2 will equal 18.

Teacher: [to the class] Does 3 × 3 × 2 = 18?

Class: Yes!

Stuart: There are more that will work, too! 2 × 3 × 3—oh wait, those are the same factors, just in a different order. I got them from starting with 3 × 6.

Teacher: How about 180? What numbers can you multiply together to get 180?

Tamira: All you need to do is add a 0 to one of the numbers in the ones we have for 18.

Stuart: What do you mean, "add a 0"?

Tamira: You multiply by 10. So, instead of 3 × 3 × 2, you could have 30 × 3 × 2. 30 × 3 = 90 and 90 × 2 = 180!

Mercedes: I got a whole list of them. I kept making each number smaller and smaller. [The teacher records Mercedes' ways to make 180 on the board as she says them.] I had 2 × 90, so I broke the 90 into 2 × 45, to get 2 × 2 × 45. Then I broke the 45 into 9 × 5, so I got 2 × 2 × 9 × 5. Then I broke the 9 into 3 × 3, and I got 2 × 2 × 3 × 3 × 5.

Teacher: Why did you stop there?

Mercedes: I couldn't break any of the numbers down any more.

Teacher: Why not? Why can't Mercedes break 2 × 2 × 3 × 3 × 5 into more numbers?

Margaret: Because, like 2, you can't split it. You can only say 2 × 1, and we decided not to use 1s because they'd just make it go on forever.

Olivia: Those are called square numbers, I think.

Janet: No, square numbers are like 3 × 3 = 9 or 2 × 2 = 4.

Rachel: They're prime numbers. They're the ones where we could only make one rectangle.

Teacher: That's right. These are prime numbers because the only factors they have are 1 and themselves. Keep thinking about how you can break up one way of multiplying to get other ways as you work with 180. Also keep thinking about what happens when you end up with prime numbers, like Mercedes did. Does that happen for other numbers? We'll be talking more about these ideas in the next few sessions.

When Mercedes noticed that her multiplication combination consisted of only prime numbers (2 × 2 × 3 × 3 × 5) and that she could not break down any of the factors further by using multiplication of whole numbers, Mercedes had found the *prime factorization* for 180. Mercedes and her classmates may not yet realize that 2 × 2 × 3 × 3 × 5 is not only the longest multiplication combination for 180 that is possible using only whole numbers greater than 1, but it is also the only way to make a product of 180 by multiplying five whole numbers greater than 1. For mathematicians, the fact that every whole number has a unique prime factorization is a central law of arithmetic. See **Teacher Note:** Finding Prime Factors, page 154.

For fifth graders, this work provides an opportunity to develop building blocks for understanding the properties of numbers in a context that is intriguing and engaging. At the same time, students continue to improve their flexibility in solving multiplication problems by using one multiplication combination to find other equivalent combinations. In Sessions 1.6 and 1.7, after students have done more work on finding ways to multiply whole numbers to create different products, they continue this discussion.

Dialogue Box

Naming Multiplication Strategies

A Grade 5 class is working on multiplication at the beginning of the school year. Students are solving problems and focusing on the strategies they are using. As they share strategies for multiplying 35 × 28, they comment on similarities and differences in their methods. At this point in the discussion, there are several strategies on the board, each one labeled with a student's name. Students are organizing and naming their strategies. (See the sample board at the end of this **Dialogue Box.**)

Teacher: We have quite a few solutions on the board right now. I can see that in some ways they are the same and in some ways they are different. Let's look at Cecilia's strategy. She kept the 35 whole and broke the 28 into 20 and 8. Can anyone describe in general words what Cecilia did? I mean, don't use numbers, just words. What did she do to the numbers?

Stuart: She pulled apart one number and left the other one alone.

Teacher: Then what did she do?

Stuart: She multiplied the two parts.

Teacher: Wouldn't it be nice if we had a name for this strategy?

Cecilia: How about breaking numbers?

[General nods from the class]

Teacher: OK, so Breaking Numbers Apart is one group of strategies (writes this on the board). Are there any other strategies on the board that fit into the Breaking Numbers Apart strategy?

Zachary: I broke up all the pieces—I broke the 35 into 30 and 5 and the 28 into 20 and 8.

Teacher: OK, Let's add that one to Breaking Numbers Apart. Charles, does yours fit into this strategy? I see you started with 10 × 28.

Charles looks unsure and does not answer.

Teacher: Can anyone tell me why Charles might have started with 10 × 28? It certainly worked and he got the right answer!

Walter: I think he did it because 28 × 10 is easy. It's 280.

Teacher: But where did the 10 come from? I don't see any 10 in the problem.

Walter: He broke 35 into 10 + 10 + 10 + 5.

Teacher: That sure does look like Breaking Numbers Apart. It is interesting that there are several ways to break these numbers up into smaller problems that you can solve easily. The important part is keeping track of all the parts so that nothing is left out at the end!

Teacher: Do all these other strategies fit into Breaking Numbers Apart?

Cecilia: Nora's strategy is weird. Her numbers are 70 and 14.

Teacher: Hmmm. Did she break apart numbers or do something else?

Nora: I did what I remember doing last year. I like to multiply by 10s. So I changed the 35 to 70. I doubled it. And then I knew that I had to cut the other number in half. So 28 became 14. Then I first did 70 × 10, which is 700 and then 70 × 4, which is 280.

Cecilia: Wait! She broke the numbers apart also! She broke the 14 apart.

Teacher: Good observation, Cecilia. I wonder whether others of you noticed that as well. Let's make another category, though, because Cecilia really started in a different way. She broke the numbers apart to finish off her problem, but basically she changed her problem into a whole new one and it worked!

> I am going to call this category Doubling and Halving. We are going to talk more about that strategy in the middle of the year when we do more multiplication work.

As the teacher asks students to decide what is the same and what is different about these strategies, the teacher is listening for an understanding of the general approach that is taken by each student. Does the student notice that in several of the posted methods, students broke the numbers apart and created subproblems, although the particular ways they broke up the numbers might be different? Students may have a difficult time identifying the mathematics of their strategy, so the teacher focuses them on how they started their work—their first step—as one way to classify their methods. Naming strategies and keeping them posted provide students with some language to describe their multiplication methods. They also raise students' awareness of approaches to consider when they are solving problems.

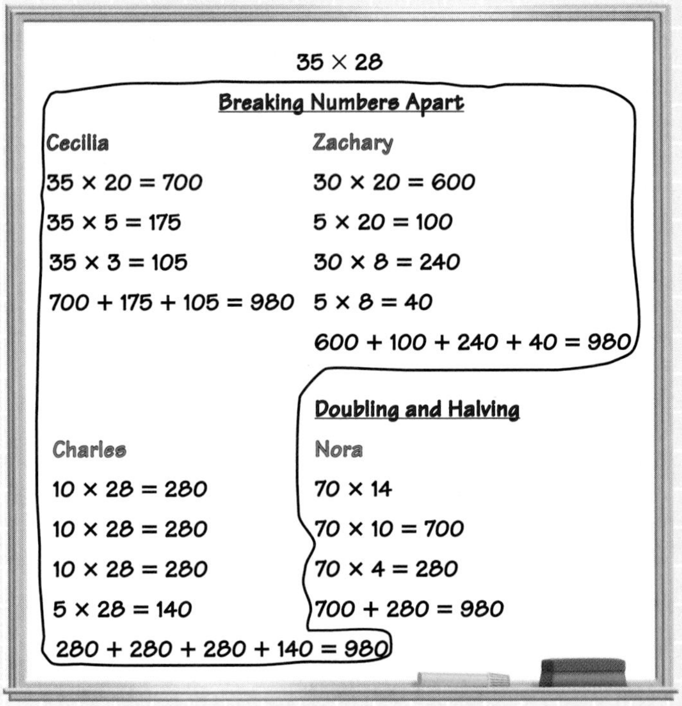

Playing *Multiplication Compare*

After learning how to play *Multiplication Compare* in Session 2.3, some students are playing the game as part of the Math Workshop in Sessions 2.4 and 2.5. The teacher moves around the classroom, stopping to check in on each pair and asking them to articulate their reasoning.

The teacher is watching Zachary and Benito play. Zachary draws 20 and 700, and Benito draws 70 and 700.

$$20 \times 700 \qquad 70 \times 700$$

Benito: I win!

Teacher: You said that really fast—how do you know you won?

Benito: We both have 700, so I just looked at the 20 and 70. I have 70, so my product will be bigger.

The teacher is now watching Olivia and Avery play. Olivia draws a 20 and 50, and Avery draws 30 and 40. After studying the cards for a few moments, Avery says he wins, and Olivia agrees.

$$20 \times 50 \qquad 30 \times 40$$

Teacher: How do you know that 30 × 40 would be greater?

Olivia: I couldn't tell by just looking at the numbers, so I had to think some more. I know that 30 × 40 would be 1,200 because 3 × 4 equals 12, and 20 × 50 would be 1,000.

Teacher: So you used a basic combination, 3 × 4, to help you solve 30 × 40. How did you know 20 × 50?

Avery: I just know that one because we worked on factors of 1,000 last year, but I can tell you another way, too—ten 50s is 500, so just double that.

Rachel and Alex are playing. Rachel draws 70 and 8 and Alex draws 80 and 7.

$$70 \times 8 \qquad 80 \times 7$$

Alex: Both of these are the same because of the zeros and the eight times sevens. That is why the answers are the same. We have to pick new cards.

Teacher: Is there any other way you can explain why they are the same?

Rachel: It's like if you were thinking about times 10. That if you think about the 70 as 7 × 10 then it's 7 × 10 × 8. And Alex's would be the same as 8 × 10 × 7. So you're multiplying the same numbers.

The teacher is watching Margaret and Tyler play. Margaret draws 4 and 200, and Tyler draws 20 and 40.

$$4 \times 200 \qquad 20 \times 40$$

Margaret: 4 × 200 is the same as 4 × 2, which is 8, and then I put the two zeros in the end, and it's 800. I do the same with 20 × 40. 2 × 4 = 8, and then I put the two zeros at the end, and it's 800. So I'd put an equal sign. No one wins, and we have to draw more cards.

Teacher: Hmmm. Rachel and Alex had one like this, and I want to ask you the same question I asked them. Why can you put those two zeros on the end for both of these? One has 100s, and the other one doesn't have any 100s. Why is it they both come out as 800?

As the teacher questions each pair of students, he makes note of which strategies some students use easily and which combinations are more difficult than others for them.

The teacher notices that quite a few students talk about "adding on zeros" or "counting the zeros." He wonders whether students are making sense of why this rule works. Can they reason why the product of 20 × 40 will be in the 100s, or are they arbitrarily "adding on zeros"? He decides that he will focus on this idea when he brings the students together for a discussion, posing questions such as these: Will 20 × 40 have an answer less than 100, in the 100s, or in the 1,000s? How do you know? What about 2 × 400? What about 200 × 40?

Dialogue Box

Solving a Division Problem

In Session 3.1, students are solving the division problem 170 ÷ 15, and they are using a representation to show their solution. The assignment is a useful assessment for the teacher to determine in what ways students are thinking about division. The teacher is circulating around the classroom, asking students to explain their work.

Tamira has written the following story problem. She is sorting through boxes of connecting cubes to find 15 yellow cubes and 170 red ones.

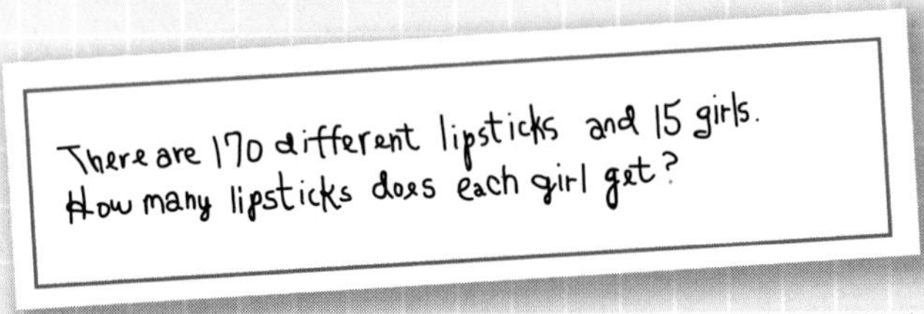
There are 170 different lipsticks and 15 girls.
How many lipsticks does each girl get?

Tamira's Work

Teacher: What do the yellow cubes represent? What about the reds?

Tamira: The yellows are the girls, and the red are all the lipsticks.

Tamira spreads out the yellow cubes and begins to sort out the reds.

Teacher: Will you give all of the lipsticks out one at a time?

Tamira: No. I know that they will each get at least 10 so I can start there.

Teacher: How do you know they get 10?

Tamira: I know that 15 × 10 = 150.

Tamira makes 15 cube towers with 11 cubes in each tower and arranges them on her desk.

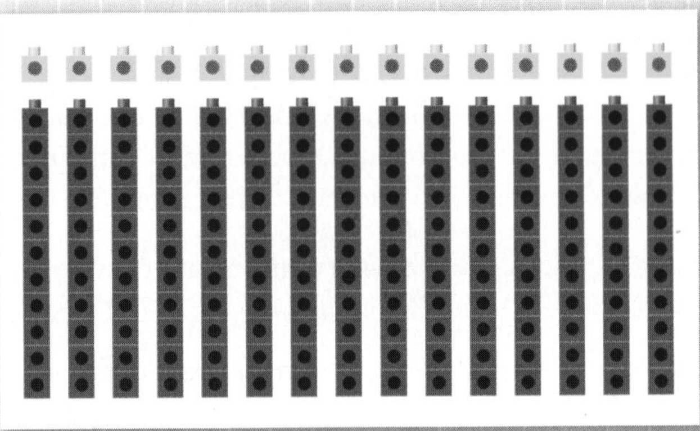

Teacher: That representation is very clear. I can see that the yellow cubes represent girls, and the red are the lipsticks.

Tamira: I had leftovers though. They could each get 11, but there were still 5 more that I couldn't give out because there wasn't enough for everyone to have 12.

Teacher: So what is the answer to your story problem?

Tamira: Each girl gets 11 lipsticks, and there are 5 left.

The teacher moves on to the next student. Felix is not using any math tools. He wrote the following story problem and equations.

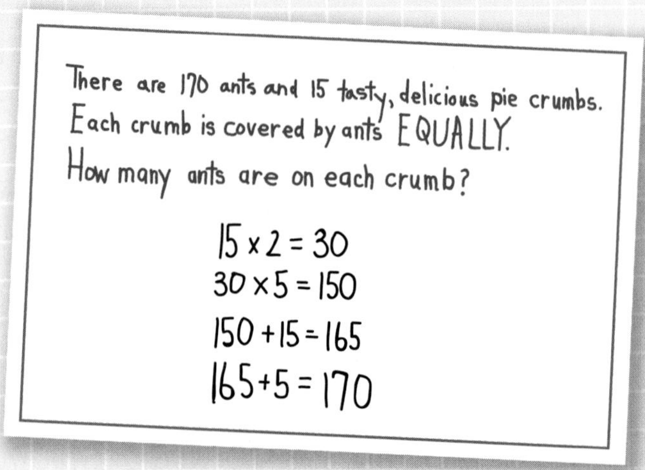
There are 170 ants and 15 tasty, delicious pie crumbs.
Each crumb is covered by ants EQUALLY.
How many ants are on each crumb?

15 x 2 = 30
30 x 5 = 150
150 + 15 = 165
165 + 5 = 170

Felix's Work

As the teacher watches, Felix begins to draw 15 crumbs, each with 11 ants munching on it.

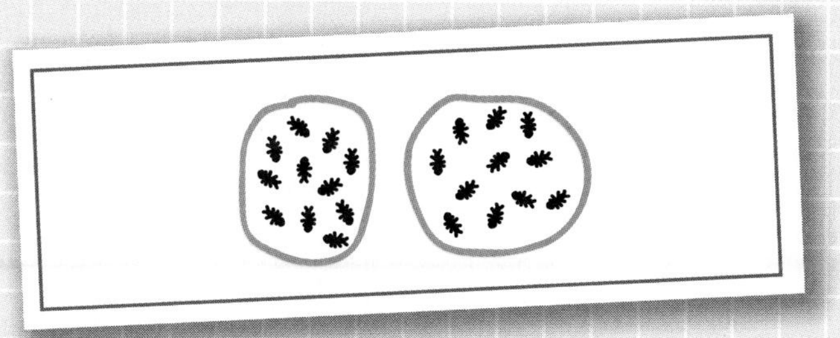

Felix's Work

Teacher: You don't have to draw 11 ants on each crumb. You could just draw the crumb, and write the number "11" on it.

Felix: That would be a lot easier!

Felix's Work

Teacher: Where is your answer within all these equations?

Felix: 150 + 15 got me to 165 and I added 5. So there's ten 15s and one more 15, so it's 11 ants on each crumb. And there's 5 left. Those ants don't get any crumbs!

Teacher: I'm not sure I understand where you got 150.

Felix: I started with 15 × 2 and that was 30. Then I timesed that by 5, so that would be ten 15s, and that equals 150. Then I figured out 15 × 10 was also 150.

As the teacher is working with students, she is doing an informal assessment of what students understand about division and how they solve division problems. The teacher is making certain that Tamira knows what her cubes represent, and also asks her about the remainder of 5. As she asks Felix about his work, she has several other points in mind. In Grade 5, students do not need to draw a literal representation, so she suggests that Felix use numbers. The teacher also knows that when solving division problems, students often lose track of their work and what the division question is, so she asks him where his answer is.

All of these ideas are explored further in the remainder of this Investigation. Also, because this is the first session about division, the teacher is laying the foundation for some important ideas that students will continue to study.

How Many 21s Are In 1,344?

The students in one class have built a multiple tower for 21. The highest number on their tower is 1,029. In Session 3.4, they are considering division problems in which the dividend is not on the multiple tower. Students have solved 1,344 ÷ 21, the first problem on *Student Activity Book* page 55, and are discussing their strategies. The teacher has identified students who used the multiple tower in different ways to share their solutions.

Mercedes started with 1,029, the last number on the class tower.

Mercedes: We already figured out that 1,029 was 49 × 21, so it's easy to know that 50 × 21 = 1,050. I need to get to 1,344, so I added 10 more 21s and got 1,260. I added by 21 and got 1,281; another 21 is 1,302. Then I'm 42 away from 1,344, so that's two more 21s. So let's see, that's 50, 60, 61, 62, 64.

Teacher: What do you mean, 50, 60, 61, 62, 64? Where did those numbers come from?

Mercedes: I was just keeping track of how many 21s I had. First I had 50, then I add 10 more 21s, like that.

The teacher records Mercedes' strategy and labels it A. (See the chart at the end of this **Dialogue Box.**)

Samantha used numbers that are already on the tower and broke the dividend into two pieces, 840 and 504.

Samantha: From the multiple tower, I know that 840 ÷ 21 = 40. I did 1,344 − 840 and it's 504. I looked at the multiple tower and found 504. It's 4 more 21s from 420, so that makes it 24. I added 40 and 24; it equals 64.

The teacher records Samantha's strategy and labels it B.

Renaldo used the multiple tower and a relationship about multiples of 10 to help him find the answer.

Renaldo: I looked at the bottom of the tower, but instead of thinking 21, 42, 63, I thought of them as multiples of 10, so I thought 210, 420, 630, 840, 1,050, 1,260. If 126 is the 6th multiple, then 1,260 is the 60th multiple. Then I wanted to see how many further I had to go. 40 more than 1,260 is 1,300, and 44 more than that is 1,344, so I knew I needed to go 84 more. That's 4 more 21s, so my answer is 64.

Teacher: Can someone else explain how Renaldo used the numbers at the bottom of the tower to figure out that 1,260 is the 60th multiple of 21?

The teacher records Renaldo's strategy and labels it C.

The teacher makes deliberate choices about asking Mercedes, Samantha, and Renaldo to explain their strategies. Samantha simply broke the dividend into two pieces, 840 and 504, and the teacher wants students to see this as a possible strategy for solving division problems. Mercedes and Renaldo both finished in a similar way (going up 84 more from 1,260), but Mercedes added up by 21s to get to 1,344 and then added her partial answers. The teacher understands why Mercedes said, "50, 60, 61, 62, 64" and wants Mercedes to explain her thinking more fully to the class. The teacher also thinks Renaldo noticed an important relationship: that because the 6th multiple of 21 is 126, the 60th multiple is 1,260. So she asks another student to explain Renaldo's idea, to give time for other students to think this through.

(A)

50 × 21 = 1,050

10 × 21 = 210	1,050 + 210 = 1,260
1 × 21 = 21	1,260 + 21 = 1,281
1 × 21 = 21	1,281 + 21 = 1,302
2 × 21 = 42	1,302 + 42 = 1,344

64 × 21 = 1,344

(B)

840 ÷ 21 = 40

504 ÷ 21 = 24	1,344 − 840 = 504

1,344 ÷ 21 = 164

(C)

60 × 21 = 1,260	6 × 21 = 126
4 × 21 = 84	1,260 + 84 = 1,344

64 × 21 = 1,344

Student Math Handbook

The *Student Math Handbook* pages related to this unit are pictured on the following pages. This book is designed to be used flexibly: as a resource for students doing classwork, as a book students can take home for reference while doing homework and playing math games with their families, and as a reference for families to better understand the work their children are doing in class.

When students take the *Student Math Handbook* home, they and their families can discuss these pages together to reinforce or enhance students' understanding of the mathematical concepts and games in this unit.

Math Words and Ideas

Multiplication and Division

Math Words
- **multiplication**
- **division**

× Use multiplication when you want to combine groups that are the same size.

Number of groups	Size of group	Number in all the groups	
28 teams	18 players on each team	*unknown*	There are 28 youth soccer teams in our town, and there are 18 players on each team. How many players are there on all of the teams? $28 \times 18 =$ **504** Answer: There are **504** players in all.

÷ Use division when you want to separate a quantity into equal-sized groups.

Number of groups	Size of group	Number in all the groups	
28 teams	*unknown*	504 players	There are 28 soccer teams in our town and 504 players altogether on all the teams. Each team has the same number of players. How many players are there on each team? $504 \div 28 =$ **18** Answer: Each team has **18** players.

Number of groups	Size of group	Number in all the groups	
unknown	18 players on each team	504 players	There are 504 soccer players in our town, and there are 18 players on each team. How many teams are there? $504 \div 18 =$ **28** Answer: There are **28** teams.

SMH 14 fourteen

◀ Math Words and Ideas, p. 14

Math Words and Ideas

Mathematical Symbols and Notation

Math Words
- **factor**
- **product**
- **dividend**
- **divisor**
- **quotient**
- **equal to (=)**
- **greater than (>)**
- **less than (<)**

Multiplication

$28 \times 18 = 504$ — 28 and 18: **factors**; 504: **product**

$\begin{array}{r} 18 \\ \times\ 28 \\ \hline 504 \end{array}$ — 18 and 28: **factors**; 504: **product**

Division

$120 \div 15 = 8$ — 120: **dividend**; 15: **divisor**; 8: **quotient**

$15\overline{)120}$ with 8 above — 15: **divisor**; 8: **quotient**; 120: **dividend**

Comparing values

equal to	$2 \times 40 = 2 \times 5 \times 8$	$80 = 80$
greater than	$10 \times 10 > 11 \times 9$	$100 > 99$
less than	$4 \times 30 < 3 \times 50$	$120 < 150$

What are the factors in $14 \times 20 = 280$?
What is the quotient in $144 \div 16 = 9$?

fifteen SMH 15

◀ Math Words and Ideas, p. 15

Math Words and Ideas

Arrays

Math Words
- **array**
- **dimensions**

Arrays can be used to represent multiplication.

This is one of the rectangular arrays you can make with 24 tiles.

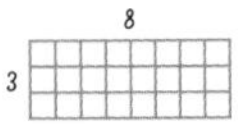

The dimensions of the array are 3 × 8 (or 8 × 3, depending on how you are looking at the array).

This array shows that
- 3 and 8 are two of the factors of 24.
- 24 is a multiple of 8.
- 24 is a multiple of 3.

This array shows a way to solve 8 × 12.

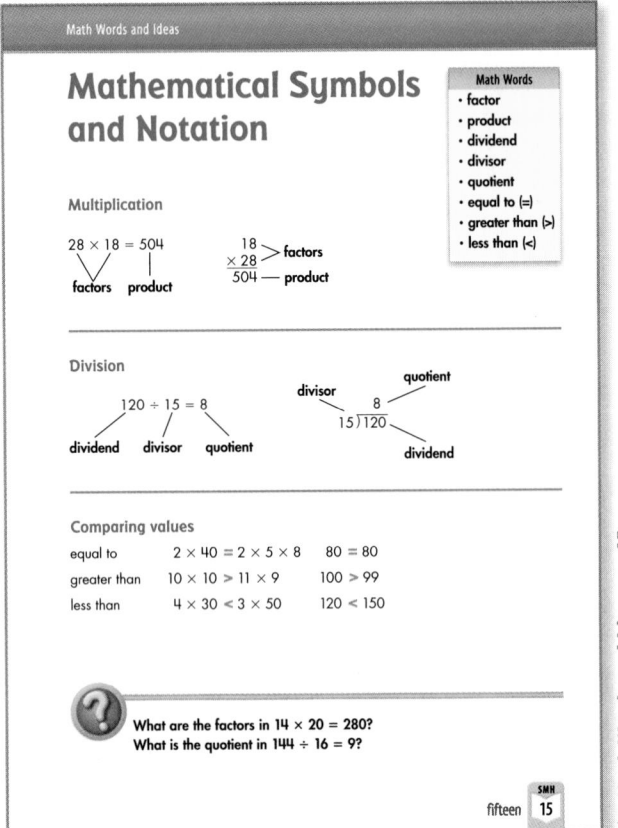

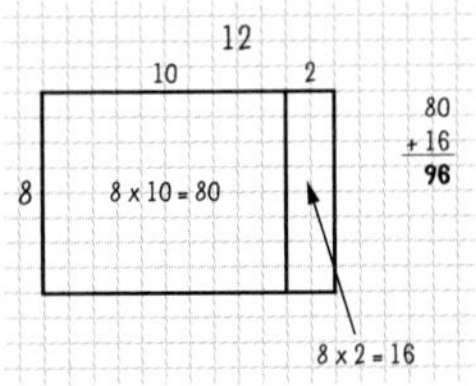

$8 \times 12 = (8 \times 10) + (8 \times 2) = 80 + 16 =$ **96**

Draw an array with dimensions 5 by 9.

SMH 16 sixteen

◀ Math Words and Ideas, p. 16

Math Words and Ideas

Unmarked Arrays

For larger numbers, arrays without grid lines can be easier to use than arrays with grid lines.

Look at how unmarked arrays are used to show different ways to solve the problem 9 × 12.

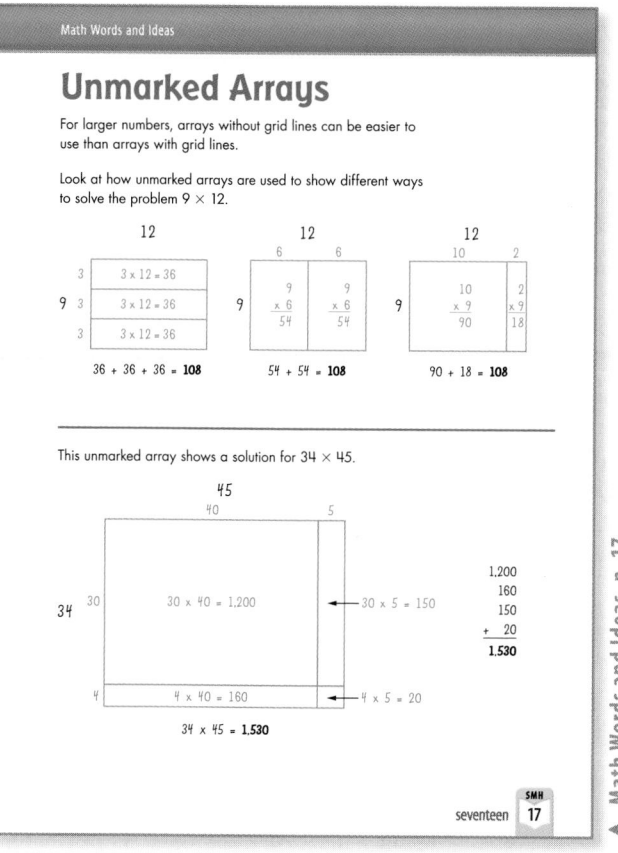

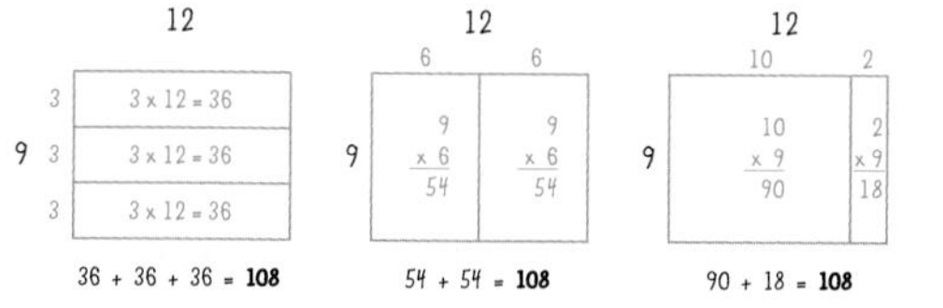

This unmarked array shows a solution for 34 × 45.

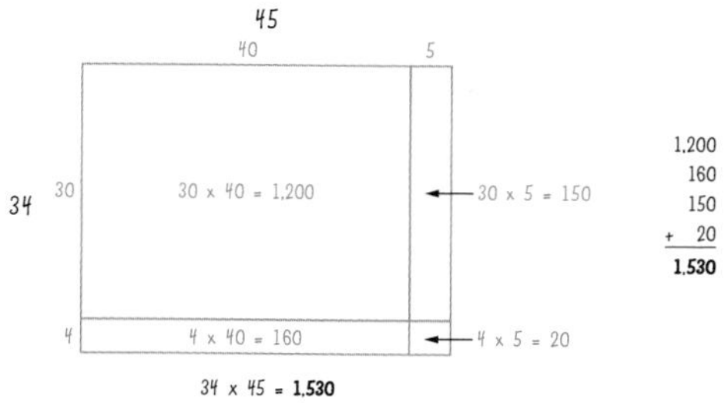

Math Words and Ideas, p. 17

Math Words and Ideas

Factors

Math Words
- **factor**

These are all the possible whole-number rectangular arrays for the number 36, using whole numbers.

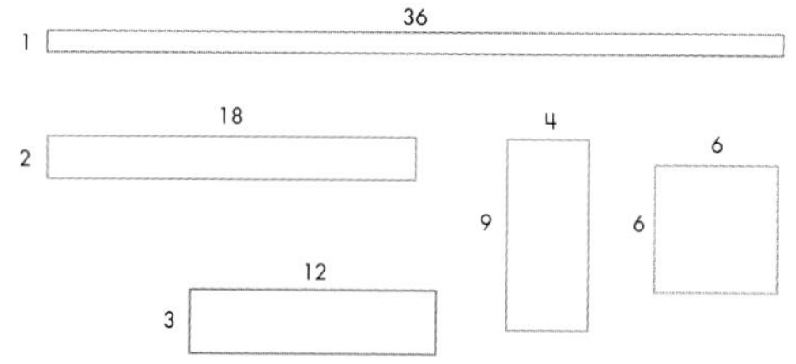

Each dimension of these rectangles is a factor of 36.

Listed in order, the factors of 36 are

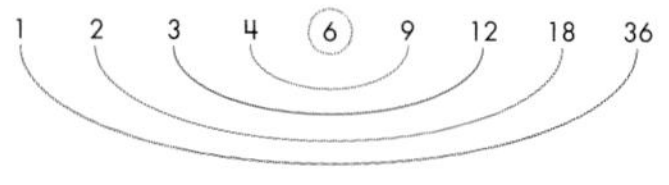

Pairs of factors can be multiplied to get a product of 36.

1 × 36 = 36	2 × 18 = 36	3 × 12 = 36
4 × 9 = 36	6 × 6 = 36	9 × 4 = 36
12 × 3 = 36	18 × 2 = 36	36 × 1 = 36

Use the factors of 36 to find the factors of 72.
Use the factors of 36 to find the factors of 360.

Math Words and Ideas, p. 18

Math Words and Ideas

Multiples

Math Words
- **multiple**

This 300 chart shows skip counting by 15. The shaded numbers are multiples of 15. A multiple of 15 is a number that can be divided evenly into groups of 15.

1	2	3	4	5	6	7	8	9	10
11	12	13	14	15	16	17	18	19	20
21	22	23	24	25	26	27	28	29	30
31	32	33	34	35	36	37	38	39	40
41	42	43	44	45	46	47	48	49	50
51	52	53	54	55	56	57	58	59	60
61	62	63	64	65	66	67	68	69	70
71	72	73	74	75	76	77	78	79	80
81	82	83	84	85	86	87	88	89	90
91	92	93	94	95	96	97	98	99	100
101	102	103	104	105	106	107	108	109	110
111	112	113	114	115	116	117	118	119	120
121	122	123	124	125	126	127	128	129	130
131	132	133	134	135	136	137	138	139	140
141	142	143	144	145	146	147	148	149	150
151	152	153	154	155	156	157	158	159	160
161	162	163	164	165	166	167	168	169	170
171	172	173	174	175	176	177	178	179	180
181	182	183	184	185	186	187	188	189	190
191	192	193	194	195	196	197	198	199	200
201	202	203	204	205	206	207	208	209	210
211	212	213	214	215	216	217	218	219	220
221	222	223	224	225	226	227	228	229	230
231	232	233	234	235	236	237	238	239	240
241	242	243	244	245	246	247	248	249	250
251	252	253	254	255	256	257	258	259	260
261	262	263	264	265	266	267	268	269	270
271	272	273	274	275	276	277	278	279	280
281	282	283	284	285	286	287	288	289	290
291	292	293	294	295	296	297	298	299	300

20 × **15** = 300

300 ÷ **15** = 20

Use the shaded numbers on the 300 chart to write other multiplication and division equations about the multiples of 15.

? × 15 = ? ? ÷ 15 = ?

Math Words and Ideas, p. 19

Math Words and Ideas

Multiple Towers

When you skip count by a certain number, you are finding multiples of that number.

Nora's class made a multiple tower for the number 16. They recorded the multiples of 16 on a paper strip, starting at the bottom.

They circled every 10th multiple of 16 and used them as landmark multiples to solve the following problems.

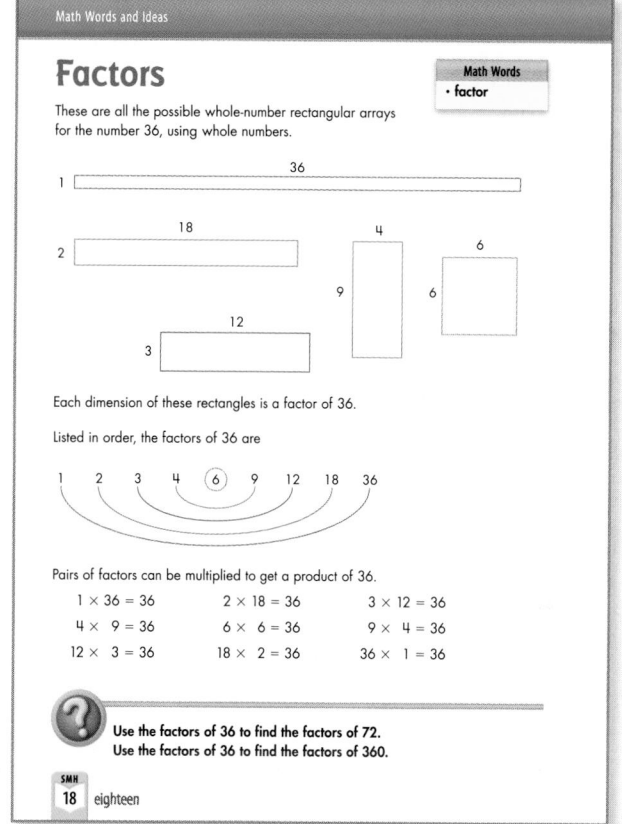

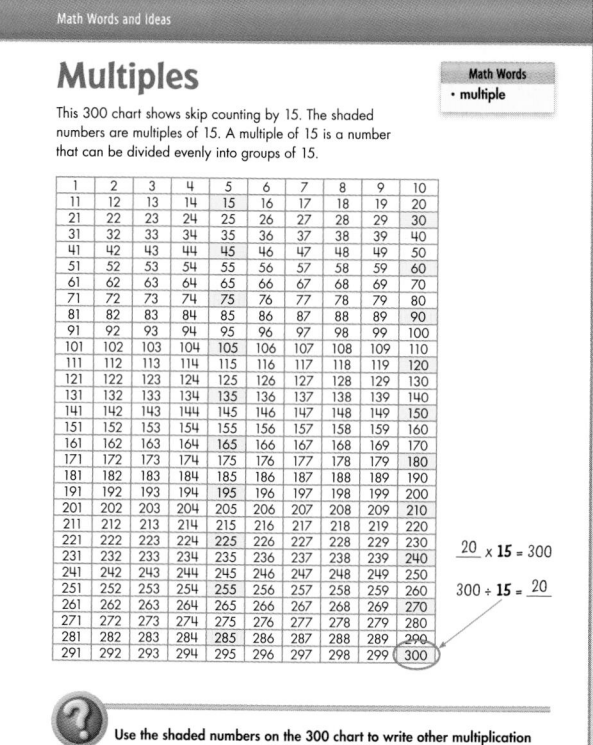

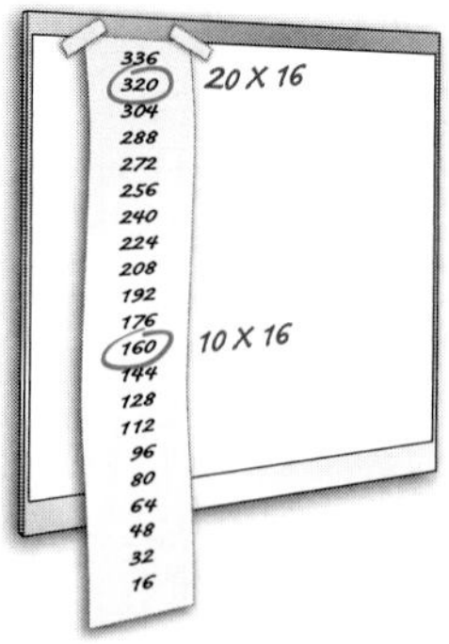

21 × 16 = 336

Nora's solution

We know that 20 × 16 = 320. 336 is next on the tower after 320, so it is one more 16.

30 × 16 = 480

Georgia's solution

30 × 16 would be the next landmark multiple on our tower. Since 3 × 16 = 48, then 30 × 16 = 48 × 10.

208 ÷ 16 = 13

Renaldo's solution

Ten 16s land on 160. Three more 16s will go to 208.

How would you use this multiple tower to solve this problem?
18 × 16 = ____

Math Words and Ideas, p. 20

Math Words and Ideas

Properties of Numbers

(page 1 of 2)

Math Words
- prime number
- composite number
- square number

When a number is represented as an array, you can recognize some of the special properties of that number.

Prime numbers have exactly two factors: 1 and the number itself.

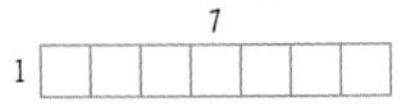

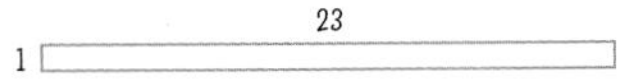

7 and 23 are examples of prime numbers.

Numbers that have more than two factors are called composite numbers.

The number 1 has only one factor. It is neither a prime number nor a composite number.

A square number is the result when a number is multiplied by itself.

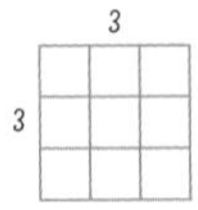

$9 = 3 \times 3$

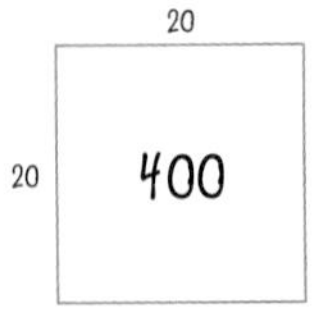

9 and 400 are examples of square numbers.

Math Words and Ideas

Properties of Numbers

(page 2 of 2)

Math Words
- even number
- odd number

An even number is composed of groups of 2. One of the factors of an even number is 2.

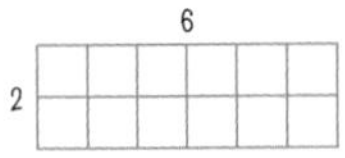

$12 = 2 \times 6$

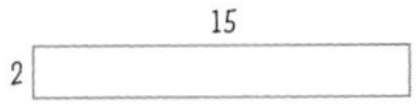

$30 = 2 \times 15$

12 and 30 are examples of even numbers.

An odd number is composed of groups of 2 plus 1. An odd number does not have 2 as a factor.

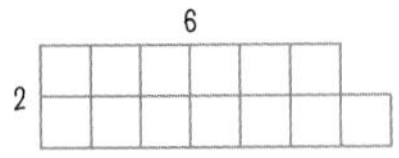

$13 = (2 \times 6) + 1$

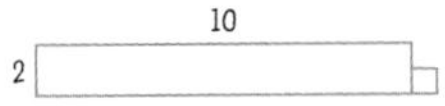

$21 = (2 \times 10) + 1$

13 and 21 are examples of odd numbers.

Find all the prime numbers up to 50.
Find all the square numbers up to 100.

Math Words and Ideas

Multiplying More than Two Numbers (page 1 of 2)

There are 36 dots in this arrangement.

You can visualize the total number of dots in many ways.

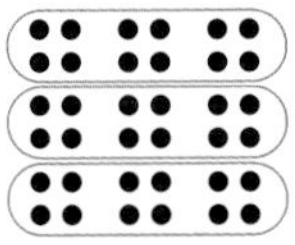

3 x 12
3 groups of 12

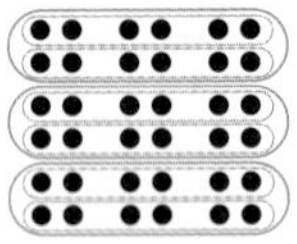

3 x (2 x 6)
3 groups of 12
Each group of 12 is made up of 2 groups of 6.

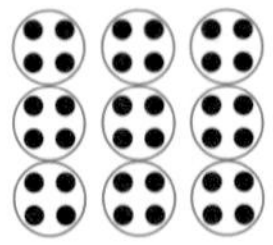

9 x 4
9 groups of 4

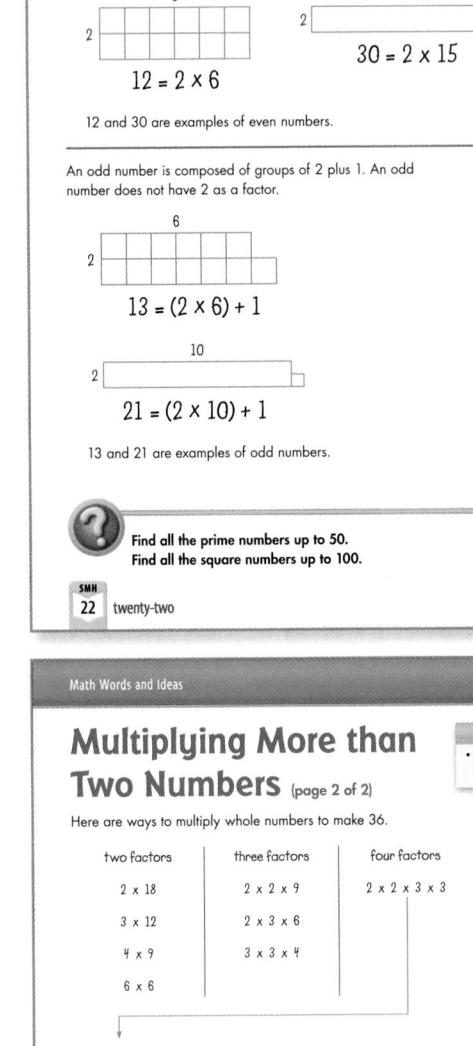

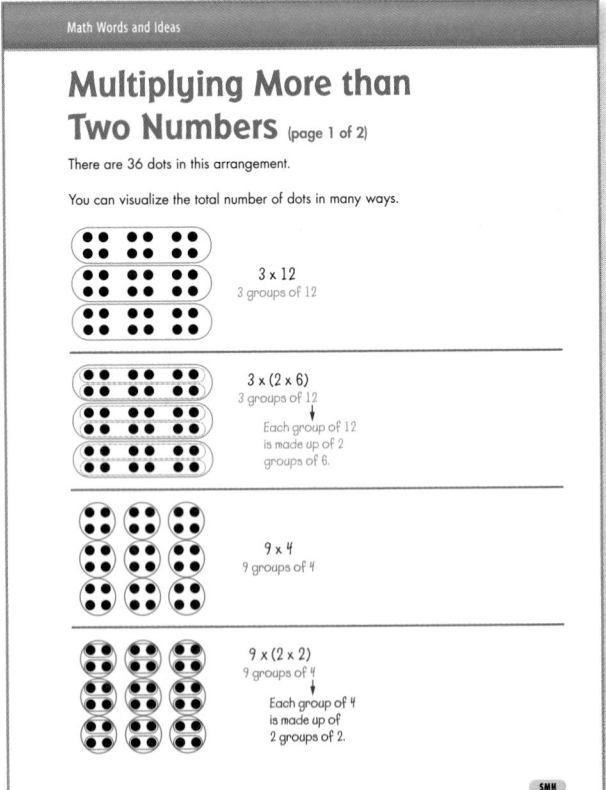

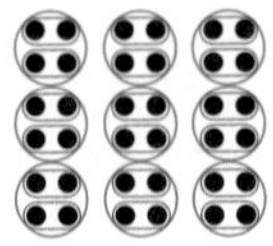

9 x (2 x 2)
9 groups of 4
Each group of 4 is made up of 2 groups of 2.

Math Words and Ideas

Multiplying More than Two Numbers (page 2 of 2)

Math Words
- prime factorization

Here are ways to multiply whole numbers to make 36.

two factors	three factors	four factors
2 x 18	2 x 2 x 9	2 x 2 x 3 x 3
3 x 12	2 x 3 x 6	
4 x 9	3 x 3 x 4	
6 x 6		

$2 \times 2 \times 3 \times 3$ is the longest multiplication expression with a product of 36 using only whole numbers greater than 1.

(2) x (2) x (3) x (3)

Notice that these factors are prime numbers.

$2 \times 2 \times 3 \times 3$ is the prime factorization of 36.

Find the prime factorization of 120.

Multiplication Combinations

(page 1 of 5)

One of your goals in math class this year is to review and practice all the multiplication combinations up to 12 × 12.

1 x 1	1 x 2	1 x 3	1 x 4	1 x 5	1 x 6	1 x 7	1 x 8	1 x 9	1 x 10	1 x 11	1 x 12
2 x 1	2 x 2	2 x 3	2 x 4	2 x 5	2 x 6	2 x 7	2 x 8	2 x 9	2 x 10	2 x 11	2 x 12
3 x 1	3 x 2	3 x 3	3 x 4	3 x 5	3 x 6	3 x 7	3 x 8	3 x 9	3 x 10	3 x 11	3 x 12
4 x 1	4 x 2	4 x 3	4 x 4	4 x 5	4 x 6	4 x 7	4 x 8	4 x 9	4 x 10	4 x 11	4 x 12
5 x 1	5 x 2	5 x 3	5 x 4	5 x 5	5 x 6	5 x 7	5 x 8	5 x 9	5 x 10	5 x 11	5 x 12
6 x 1	6 x 2	6 x 3	6 x 4	6 x 5	6 x 6	6 x 7	6 x 8	6 x 9	6 x 10	6 x 11	6 x 12
7 x 1	7 x 2	7 x 3	7 x 4	7 x 5	7 x 6	7 x 7	7 x 8	7 x 9	7 x 10	7 x 11	7 x 12
8 x 1	8 x 2	8 x 3	8 x 4	8 x 5	8 x 6	8 x 7	8 x 8	8 x 9	8 x 10	8 x 11	8 x 12
9 x 1	9 x 2	9 x 3	9 x 4	9 x 5	9 x 6	9 x 7	9 x 8	9 x 9	9 x 10	9 x 11	9 x 12
10 x 1	10 x 2	10 x 3	10 x 4	10 x 5	10 x 6	10 x 7	10 x 8	10 x 9	10 x 10	10 x 11	10 x 12
11 x 1	11 x 2	11 x 3	11 x 4	11 x 5	11 x 6	11 x 7	11 x 8	11 x 9	11 x 10	11 x 11	11 x 12
12 x 1	12 x 2	12 x 3	12 x 4	12 x 5	12 x 6	12 x 7	12 x 8	12 x 9	12 x 10	12 x 11	12 x 12

There are 144 multiplication combinations on this chart. You may think that remembering all of them is a challenge, but you should not worry. On the next few pages you will find some suggestions for learning many of them.

Multiplication Combinations

(page 2 of 5)

Learning Two Combinations at a Time

To help you review multiplication combinations, think about two combinations at a time, such as 8 × 3 and 3 × 8.

These two problems look different but have the same answer.

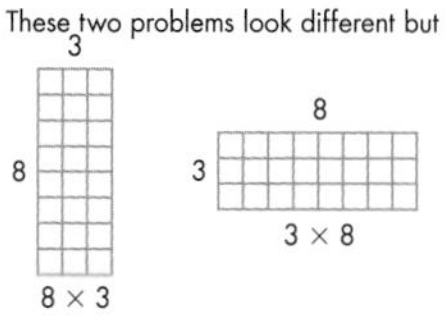

When you know that 8 × 3 = 24, you also know that 3 × 8 = 24.

You've learned two multiplication combinations!

By "turning around" combinations and learning them two at a time, the chart of multiplication combinations is reduced from 144 to 78 combinations to learn.

1 x 1	1 x 2	1 x 3	1 x 4	1 x 5	1 x 6	1 x 7	1 x 8	1 x 9	1 x 10	1 x 11	1 x 12
2 x 1 1 x 2	2 x 2	2 x 3	2 x 4	2 x 5	2 x 6	2 x 7	2 x 8	2 x 9	2 x 10	2 x 11	2 x 12
3 x 1 1 x 3	3 x 2 2 x 3	3 x 3	3 x 4	3 x 5	3 x 6	3 x 7	3 x 8	3 x 9	3 x 10	3 x 11	3 x 12
4 x 1 1 x 4	4 x 2 2 x 4	4 x 3 3 x 4	4 x 4	4 x 5	4 x 6	4 x 7	4 x 8	4 x 9	4 x 10	4 x 11	4 x 12
5 x 1 1 x 5	5 x 2 2 x 5	5 x 3 3 x 5	5 x 4 4 x 5	5 x 5	5 x 6	5 x 7	5 x 8	5 x 9	5 x 10	5 x 11	5 x 12
6 x 1 1 x 6	6 x 2 2 x 6	6 x 3 3 x 6	6 x 4 4 x 6	6 x 5 5 x 6	6 x 6	6 x 7	6 x 8	6 x 9	6 x 10	6 x 11	6 x 12
7 x 1 1 x 7	7 x 2 2 x 7	7 x 3 3 x 7	7 x 4 4 x 7	7 x 5 5 x 7	7 x 6 6 x 7	7 x 7	7 x 8	7 x 9	7 x 10	7 x 11	7 x 12
8 x 1 1 x 8	8 x 2 2 x 8	8 x 3 3 x 8	8 x 4 4 x 8	8 x 5 5 x 8	8 x 6 6 x 8	8 x 7 7 x 8	8 x 8	8 x 9	8 x 10	8 x 11	8 x 12
9 x 1 1 x 9	9 x 2 2 x 9	9 x 3 3 x 9	9 x 4 4 x 9	9 x 5 5 x 9	9 x 6 6 x 9	9 x 7 7 x 9	9 x 8 8 x 9	9 x 9	9 x 10	9 x 11	9 x 12
10 x 1 1 x 10	10 x 2 2 x 10	10 x 3 3 x 10	10 x 4 4 x 10	10 x 5 5 x 10	10 x 6 6 x 10	10 x 7 7 x 10	10 x 8 8 x 10	10 x 9 9 x 10	10 x 10	10 x 11	10 x 12
11 x 1 1 x 11	11 x 2 2 x 11	11 x 3 3 x 11	11 x 4 4 x 11	11 x 5 5 x 11	11 x 6 6 x 11	11 x 7 7 x 11	11 x 8 8 x 11	11 x 9 9 x 11	11 x 10 10 x 11	11 x 11	11 x 12
12 x 1 1 x 12	12 x 2 2 x 12	12 x 3 3 x 12	12 x 4 4 x 12	12 x 5 5 x 12	12 x 6 6 x 12	12 x 7 7 x 12	12 x 8 8 x 12	12 x 9 9 x 12	12 x 10 10 x 12	12 x 11 11 x 12	12 x 12

Multiplication Combinations

(page 3 of 5)

Another helpful way to learn multiplication combinations is to think about one category at a time. Here are some categories you may have seen before.

Learning the ×1 Combinations

You may be thinking about only one group.

1 group of 9 equals 9.

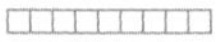

→ 1 x 9 = 9

You may also be thinking about several groups of 1.

6 groups of 1 equal 6.

→ 6 x 1 = 6

Learning the ×2 Combinations

Multiplying by 2 is the same as doubling a number.

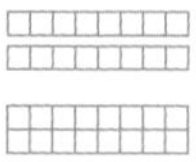

→ 8 + 8 = 16

→ 2 x 8 = 16

Learning the ×10 and ×5 Combinations

You can learn these combinations by skip counting by 10s and 5s.

10, 20, 30, 40, 50, 60 → 6 x 10 = 60

5, 10, 15, 20, 25, 30 → 6 x 5 = 30

Another way to find a ×5 combination is to remember that it is half of a ×10 combination.

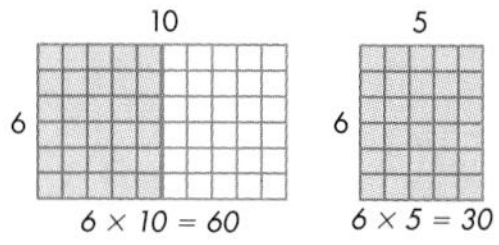

6 x 5 (or 30) is half of 6 x 10 (or 60).

Multiplication Combinations

(page 4 of 5)

Here are some more categories to help you learn the multiplication combinations.

Learning the ×11 Combinations

Many students learn these combinations by noticing the double-digit pattern they create.

11	11	11	11	11
x 3	x 4	x 5	x 6	x 7
33	44	55	66	77

Learning the ×12 Combinations

Many students multiply by 12 by breaking the 12 into 10 and 2.

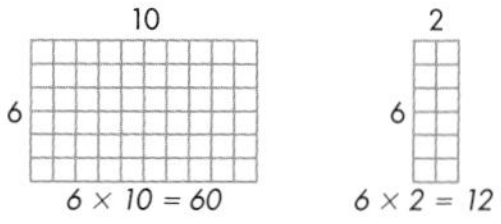

6 x 12 = (6 x 10) + (6 x 2)
6 x 12 = 60 + 12
6 x 12 = 72

Learning the Square Numbers

Many students remember the square number combinations by building the squares with tiles or drawing them on grid paper.

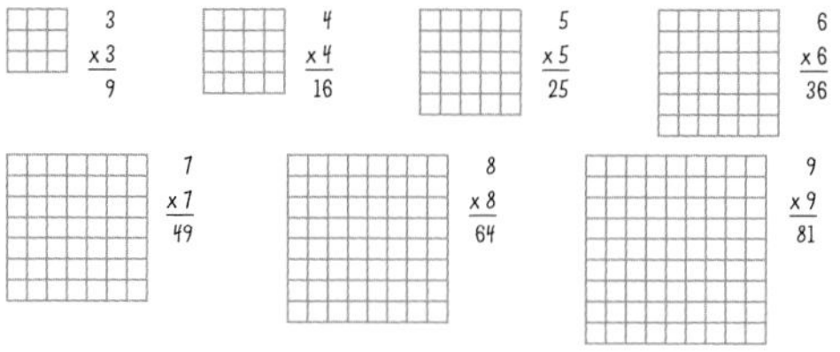

Multiplication Combinations

(page 5 of 5)

After you have used all these categories to practice the multiplication combinations, you have only a few more to learn.

1 x 1	1 x 2	1 x 3	1 x 4	1 x 5	1 x 6	1 x 7	1 x 8	1 x 9	1 x 10	1 x 11	1 x 12
2 x 1	2 x 2	2 x 3	2 x 4	2 x 5	2 x 6	2 x 7	2 x 8	2 x 9	2 x 10	2 x 11	2 x 12
3 x 1	3 x 2	3 x 3	3 x 4	3 x 5	3 x 6	3 x 7	3 x 8	3 x 9	3 x 10	3 x 11	3 x 12
4 x 1	4 x 2	4 x 3 3 x 4	4 x 4	4 x 5	4 x 6	4 x 7	4 x 8	4 x 9	4 x 10	4 x 11	4 x 12
5 x 1	5 x 2	5 x 3	5 x 4	5 x 5	5 x 6	5 x 7	5 x 8	5 x 9	5 x 10	5 x 11	5 x 12
6 x 1	6 x 2	6 x 3 3 x 6	6 x 4 4 x 6	6 x 5	6 x 6	6 x 7	6 x 8	6 x 9	6 x 10	6 x 11	6 x 12
7 x 1	7 x 2	7 x 3 3 x 7	7 x 4 4 x 7	7 x 5	7 x 6 6 x 7	7 x 7	7 x 8	7 x 9	7 x 10	7 x 11	7 x 12
8 x 1	8 x 2	8 x 3 3 x 8	8 x 4 4 x 8	8 x 5	8 x 6 6 x 8	8 x 7 7 x 8	8 x 8	8 x 9	8 x 10	8 x 11	8 x 12
9 x 1	9 x 2	9 x 3 3 x 9	9 x 4 4 x 9	9 x 5	9 x 6 6 x 9	9 x 7 7 x 9	9 x 8 8 x 9	9 x 9	9 x 10	9 x 11	9 x 12
10 x 1	10 x 2	10 x 3	10 x 4	10 x 5	10 x 6	10 x 7	10 x 8	10 x 9	10 x 10	10 x 11	10 x 12
11 x 1	11 x 2	11 x 3	11 x 4	11 x 5	11 x 6	11 x 7	11 x 8	11 x 9	11 x 10	11 x 11	11 x 12
12 x 1	12 x 2	12 x 3	12 x 4	12 x 5	12 x 6	12 x 7	12 x 8	12 x 9	12 x 10	12 x 11	12 x 12

As you practice all of the multiplication combinations, there will be some that you "just know" and others that you are "working on" learning. To practice the combinations that are difficult for you to remember, think of a combination that you know as a clue to help you. Here are some suggestions.

$9 \times 8 = 72$ $8 \times 9 = 72$	Clue:	10 x 8 = 80	80 − 8 = 72	
$6 \times 7 = 42$ $7 \times 6 = 42$	Clue:	6 x 5 = 30	6 x 2 = 12	30 + 12 = 42
$4 \times 8 = 32$ $8 \times 4 = 32$	Clue:	2 x 8 = 16	16 + 16 = 32	

Math Words and Ideas, p. 29

Multiplication Strategies (page 1 of 3)

In Grade 5, you are learning how to solve multiplication problems efficiently.

There are 38 rows in an auditorium, and 26 chairs in each row. How many people can sit in the auditorium?

Breaking the Numbers Apart

Georgia solved the problem 38×26 by breaking apart both factors.

Georgia's solution

First I'll figure out how many people are in the first 30 rows.

30 x 20 = 600 — *That's the first 30 rows, with 20 people in each row.*

30 x 6 = 180 — *That's 6 more people in each of those 30 rows, so now I've filled up 30 rows.*

There are 8 more rows to fill.

8 x 20 = 160 — *That's 20 people in those last 8 rows.*

8 x 6 = 48 — *I've filled up the last 8 rows with 6 more people in each row.*

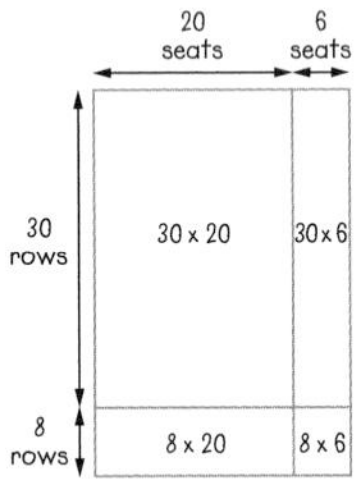

Now I add together all the parts I figured out to get the answer.

600 + 180 + 160 + 48 = **988**

988 people can sit in the auditorium.

Solve 14 × 24 by using this first step: $14 \times 20 =$ __?__

Math Words and Ideas, p. 30

Multiplication Strategies (page 2 of 3)

There are 38 rows in the auditorium, and 26 chairs in each row. How many people can sit in the auditorium?

Changing One Number to Make an Easier Problem

Benson solved the auditorium problem, 38×26, by changing the 38 to 40 to make an easier problem.

Benson's solution

I'll pretend that there are 40 rows in the auditorium instead of 38.

40 x 26 = 1,040 — *I knew that 10 × 26 = 260. I doubled that to get 520, and doubled that to get 1,040.*

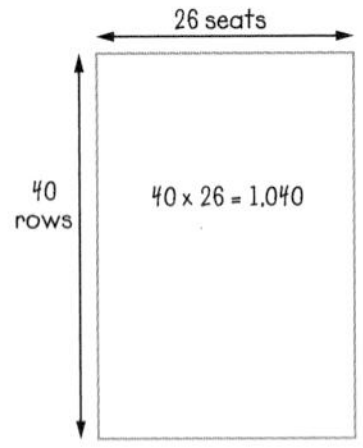

So, if there were 40 rows, 1,040 people could sit in the auditorium. But there are really only 38 rows, so I have 2 extra rows of 26 chairs. I need to subtract those.

2 x 26 = 52 — *I need to subtract 52. I'll do that in two parts.*

1,040 − 40 = 1,000 — *First I'll subtract 40.*

1,000 − 12 = 988 — *Then I'll subtract 12.*

So, **988** people can sit in the auditorium.

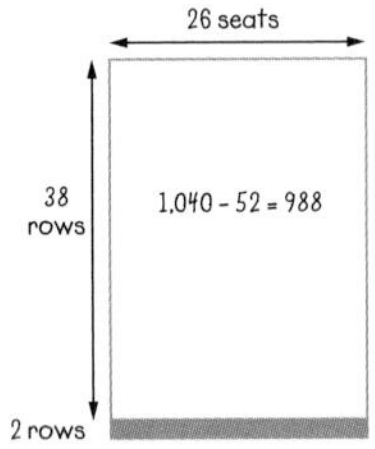

Solve 19 × 14 by using this first step: $20 \times 14 =$ __?__

Math Words and Ideas, p. 31

Multiplication Strategies (page 3 of 3)

A classroom measures 36 feet by 45 feet. How many 1-foot-square tiles will cover the floor?

Creating an Equivalent Problem

Nora's solution

I can double 45 and take half of 36 and pretend to change the shape of the classroom.

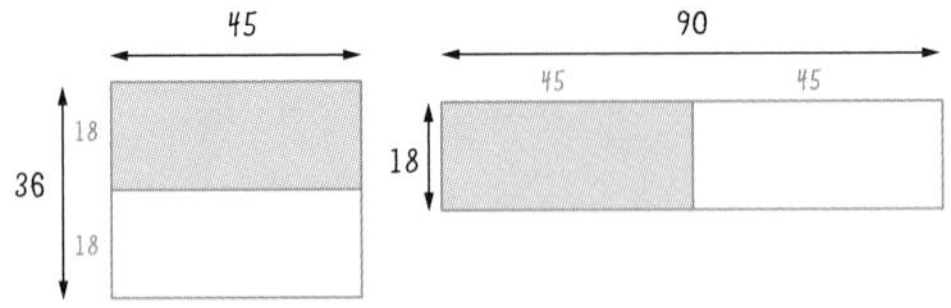

A 36-foot by 45-foot classroom needs the same amount of floor tiles as a 18-foot by 90-foot classroom.

For me, 18 × 90 is an easier problem to solve.

10 x 90 = 900

8 x 90 = 720

18 x 90 = **1,620**

1,620 tiles will cover the floor.

Solve: $35 \times 22 =$ __?__ $\times 11$

Math Words and Ideas, p. 32

Equivalent Expressions in Multiplication (page 1 of 2)

A large box holds twice as many muffins as a small box.

large box   small box

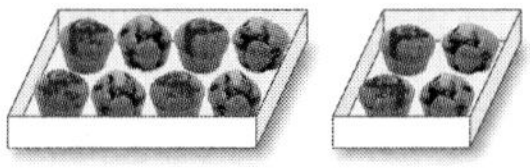

A customer ordered 7 large boxes of muffins at the bakery. The baker only had small boxes. How many small boxes of muffins should the customer buy to get the same number of muffins?

The small boxes are half the size of the large boxes. The customer should buy twice as many small boxes.

double

$7 \times 8 = 14 \times 4$

half

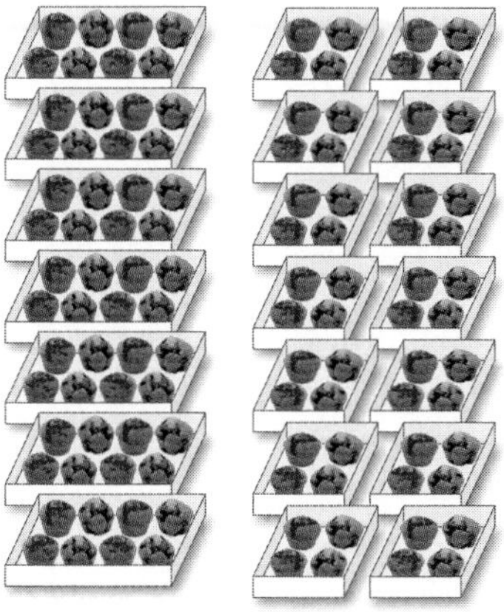

7 boxes with 8 muffins in each box

14 boxes with 4 muffins in each box

thirty-three **SMH 33**

◀ Math Words and Ideas, p. 33

Equivalent Expressions in Multiplication (page 2 of 2)

The fifth grade is going on a field trip. The teachers planned to take 4 buses. Instead they need to take vans. How many vans do they need?

A bus holds three times as many students as a van.

21 students   7 students 7 students 7 students

21 students   7 students 7 students 7 students

21 students   7 students 7 students 7 students

21 students   7 students 7 students 7 students

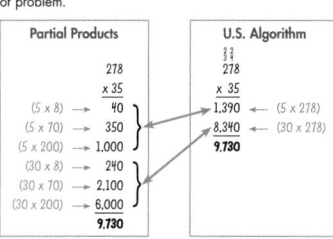

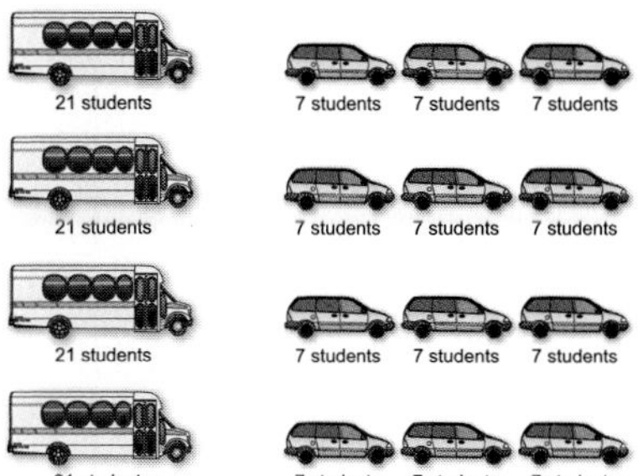

four buses with 21 students in each bus

twelve vans with 7 students in each van

The vans hold one third as many students as the buses do. The teachers need three times as many vans.

triple

$4 \times 21 = 12 \times 7$

third

Create an equivalent problem:
$4 \times 12 = \underline{\hspace{1cm}} \times \underline{\hspace{1cm}}$

SMH 34 thirty-four

◀ Math Words and Ideas, p. 34

Multiplication and Division Cluster Problems

Cluster problems help you use what you know about easier problems to solve harder problems.

1. Solve the problems in each cluster.

2. Use one or more of the problems in the cluster to solve the final problem, along with other problems if you need them.

Solve these cluster problems:	How did you solve the final problem?
$24 \times 10 = \underline{240}$ $24 \times 3 = \underline{72}$ $24 \times 20 = \underline{480}$ $24 \times 30 = \underline{720}$ Now solve this problem: $24 \times 31 = \underline{744}$	I figured out that 24×30 would be 720 because $24 \times 10 = 240$, and $240 + 240 + 240 = 720$. I need one more group of 24. That's $720 + 24 = 744$. So, $24 \times 31 = 744$.

Solve these cluster problems:	How did you solve the final problem?
$10 \times 12 = \underline{120}$ $5 \times 12 = \underline{60}$ Now solve this problem: $192 \div 12 = \underline{16}$	I thought of $192 \div 12$ as $\underline{\hspace{0.5cm}} \times 12 = 192$. $10 \times 12 = 120$ and $5 \times 12 = 60$, so $15 \times 12 = 120 + 60 = 180$. I need one more 12 to get to 192. $16 \times 12 = 192$ So, $192 \div 12 = 16$.

Solve these cluster problems:	How did you solve the final problem?
$54 \div 6 = \underline{9}$ $540 \div 6 = \underline{90}$ Now solve this problem: $6 \overline{)546}$ $\overset{91}{}$	After I knew $540 \div 6 = 90$, then I knew I needed one more group of 6 because $546 = 540 + 6$. So, $546 \div 6 = 91$.

thirty-five **SMH 35**

◀ Math Words and Ideas, p. 35

Comparing Multiplication Algorithms

Math Words
• algorithm

Some fifth graders compared these two algorithms.

An algorithm is a step-by-step procedure to solve a certain kind of problem.

Partial Products		U.S. Algorithm
		$\overset{2\ 2}{\underset{}{}}$
	278	278
	× 35	× 35
(5 × 8) →	40	1,390 ← (5 × 278)
(5 × 70) →	350	8,340 ← (30 × 278)
(5 × 200) →	1,000	**9,730**
(30 × 8) →	240	
(30 × 70) →	2,100	
(30 × 200) →	6,000	
	9,730	

Here are some of the things the students noticed.

• Both solutions involve breaking apart numbers.

• The first three numbers in the partial products algorithm are combined in the first number in the solution using the U.S. algorithm.

• The algorithms are mostly the same, but the U.S. algorithm notation combines steps $(40 + 350 + 1,000 = 1,390)$.

• The little numbers in the U.S. algorithm stand for tens and hundreds. The 4 and the 2 above the 7 are really 40 and 20.

SMH 36 thirty-six

◀ Math Words and Ideas, p. 36

Math Words and Ideas

Remainders: What Do You Do with the Extras?

Math Words
- remainder

When you are asked to solve division problems in context, it is important to consider the remainder to correctly answer the question asked. Here are some different story problem contexts for the division problem 186 ÷ 12 = 15 R6.

186 people are taking a trip. One van holds 12 people. How many vans do they need?

15 vans will hold 15 × 12 or 180 people, but the other 6 people still need a ride. They need 1 more van.

Answer: **They need 16 vans.**

There are 186 pencils and 12 students. A teacher wants to give the same number of pencils to each student. How many pencils will each student get?

It does not make sense to give students half a pencil, so the teacher can keep the remaining 6 pencils.

Answer: **Each student will get 15 pencils.**

Twelve friends earned $186 by washing cars. They want to share the money equally. How much money should each person get?

Dollars can be split up into smaller amounts. Each person can get $15. The remaining $6 can be divided evenly so that every person gets another 50¢.

Answer: **Each person gets $15.50.**

Twelve people are going to share 186 crackers evenly. How many crackers does each person get?

Each person gets 15 crackers. Then the last 6 crackers can be split in half. Each person gets another half cracker.

Answer: **Each person gets $15\frac{1}{2}$ crackers.**

Write and solve a story problem for 153 ÷ 13.

Math Words and Ideas, p. 37

Math Words and Ideas

Division Strategies (page 1 of 2)

In Grade 5, you are learning how to solve division problems efficiently.

Here is an example of a division problem.

Janet has 1,780 marbles. She wants to put them into bags, each of which holds 32 marbles. How many full bags of marbles will she have?

Samantha solved this problem by multiplying groups of 32 to reach 1,780.

Samantha's solution

30 x 32 =	960	There are 960 marbles in 30 bags of 32.
20 x 32 =	640	There are 640 marbles in 20 bags of 32.
5 x 32 =	160	There are 160 marbles in 5 bags of 32.
55	1,760	There are 1,760 marbles in 55 bags of 32.

1,760 is as close as I can get to 1,780 with groups of 32.

1,780 ÷ 32 = 55 R20

Janet can fill 55 bags, and she will have 20 extra marbles.

Talisha solved this problem by subtracting groups of 32 from 1,780.

Talisha's solution

```
32)1,780
   -640    20 bags
   1,140
   -640    20 bags
    500                 → 55 bags
   -320    10 bags
    180
   -160     5 bags
     20                 → 20 extra marbles
```


Math Words and Ideas, p. 38

Math Words and Ideas

Division Strategies (page 2 of 2)

Here is another division example.

54)2,500

Hana solved this problem by subtracting groups of 54 from 2,500.

Hana's solution

```
54)2,500
  - 1,080  (20)
    1,420
  - 1,080  (20)
      340
    - 216  (4)
      124
    - 108  (2)
       16  46 R16
```

Walter solved this problem by multiplying groups of 54 to reach 2,500.

Walter's solution

```
10 x 54 = 540
20 x 54 = 1,080
(40) x 54 = 2,160 → 2,160
(4)  x 54 = 216   → 216
(1)  x 54 = 54    → 54
(1)  x 54 = 54    → 54
                    2,484
```

2,500 ÷ 54 = **46 R16**

How would you solve this problem? 54)2,500

Math Words and Ideas, p. 39

Games

Division Compare

You need

- Compare Cards (1 deck per pair)

Play with a partner.

1. Divide the deck of cards evenly so that each player has the same number of cards.
2. Each player turns over two cards.
3. Using the larger number as the dividend and the smaller number as the divisor, make a division problem. (For example, if your cards are 80 and 700, the division problem is 700 ÷ 80.)
4. Estimate, compare, and reason about the relationships of the numbers to figure out which player has the greater quotient (answer). Discuss how you know which answer is greater.
5. The person with the greater quotient takes all of the cards. If the quotients are equal, players turn over two new cards and the person with the greater quotient takes all of the cards.
6. Play for a given amount of time or until one player has all of the cards.

Variation

You need

- Digit Cards (1 deck per pair)

Play the same game using Digit Cards with the "0" cards removed. Each player draws five cards. Using the cards in the order they were picked, choose the first three cards to form the dividend and the last two cards to form the divisor. (For example, if you picked 8, 1, 5, 9, 4, your division problem would be 815 ÷ 94.)

Games, G6

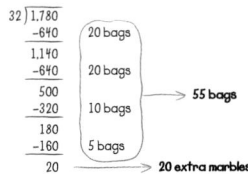

Games

Multiplication Compare

You need

- Compare Cards (1 deck per pair)

Play with a partner.

1. Divide the deck of cards evenly so that both players have the same number of cards. Place the cards facedown in a stack in front of you.
2. Each player turns over the top two cards in his or her stack.
3. Determine which player has the greater product. Discuss how you know which product is greater.
4. The person with the greater product takes all of the cards that have been turned over and places them at the bottom of his or her stack.
5. If the products are equal, players turn over two new cards. The person with the greater product takes all of the cards.
6. Play for a given amount of time or until one player has all of the cards. The player with more cards wins.

Variation

You need

- Digit Cards (1 deck per pair)

Play the same game using Digit Cards with the "0" cards removed. Each player draws 4 cards. Using the cards in the order they were picked, each player forms two 2-digit factors. (For example, if you picked 7, 2, 3, 8, your multiplication problem would be 72×38.)

◄ Games, G11

Index

IN THIS UNIT

O

P

Q

R

S

T

U

V

W